Donald E. Kirk

BORN: Baltimore, Maryland

FORMAL EDUCATION: Worcester Polytechnic Institute (BSEE), Naval Postgraduate School (MSEE), University of Illinois at Urbana-Champaign (Ph.D EE)

FAMILY: Wife (Judy), three daughters (Kara, Valerie, Dana), golden retriever (Daisy)

PROFESSIONAL EXPERIENCE: San Jose State University (8 years), Naval Postgraduate School (25 years), MIT Lincoln Laboratory (1 year), Quantum Corporation, Sangamo Electric, Grumman Aircraft

PROFESSIONAL CONCENTRATIONS: Digital signal processing, linear systems, digital systems design, optimal control and estimation

PROFESSIONAL HONORS: Fellow of the IEEE for contributions to signals and systems education, 1992; ASEE-Dow Outstanding Young Faculty Member, 1969

OTHER BOOKS: *First Principles of Discrete Systems and Digital Signal Processing,* with R. D. Strum, Addison-Wesley, 1988; *Optimal Control Theory: An Introduction,* Prentice-Hall, 1970

PERSONAL PASTIMES: Reading, walking, golf, cross-country skiing

E-MAIL: dkirk@isc.sjsu.edu

LIST OF NOTATION

$\delta(t), \delta(n)$	Unit impulse, unit sample	$u(t), u(n)$	Unit step, continuous/discrete
LTI	Linear, time-invariant system	$x(t), x(n)$	System inputs
$y(t), y(n)$	System outputs	$h(t), h(n)$	Unit impulse and unit sample response

DE — System differential/difference equation

$$\sum_{k=0}^{N} a_k \frac{d^k y(t)}{dt^k} = \sum_{k=0}^{L} b_k \frac{d^k x(t)}{dt^k} \quad \text{and}$$
$$\sum_{k=0}^{N} a_k y(n-k) = \sum_{k=0}^{L} b_k x(n-k)$$

$\mathbf{v}(t), \mathbf{v}(n)$ — State-variable vector, continuous/discrete

Matrix DE — State or matrix differential/difference equation

$$\dot{\mathbf{v}}(t) = \mathbf{A}\mathbf{v}(t) + \mathbf{B}\mathbf{x}(t) \quad \text{and}$$
$$\mathbf{v}(n+1) = \mathbf{A}\mathbf{v}(n) + \mathbf{B}\mathbf{x}(n)$$

$\mathbf{y}(t), \mathbf{y}(n)$ — System output equation, continuous/discrete

$$\mathbf{y}(t) = \mathbf{C}\mathbf{v}(t) + \mathbf{D}\mathbf{x}(t) \quad \text{and}$$
$$\mathbf{y}(n) = \mathbf{C}\mathbf{v}(n) + \mathbf{D}\mathbf{x}(n)$$

$F(s), f(t)$ — Bilateral Laplace transform and inverse

$$\mathcal{L}[f(t)] = F(s) = \int_{-\infty}^{\infty} f(t)e^{-st}dt \quad \text{and}$$
$$\mathcal{L}^{-1}[F(s)] = f(t) = \frac{1}{2\pi j}\int_{c-j\infty}^{c+j\infty} F(s)e^{st}ds$$

$F(z), f(t)$ — Bilateral z transform and inverse

$$\mathcal{Z}[f(n)] = F(z) = \sum_{n=-\infty}^{\infty} f(n)z^{-n} \quad \text{and}$$
$$\mathcal{Z}^{-1}[F(z)] = f(n) = \frac{1}{2\pi j}\oint_C F(z)z^{n-1}dz$$

$X(\omega), x(t)$	Continuous-Time Fourier Transform (CTFT) and inverse

$$\mathcal{F}[x(t)] = X(\omega) = \int_{-\infty}^{\infty} x(t)e^{-j\omega t}dt \quad \text{and}$$

$$\mathcal{F}^{-1}[X(\omega)] = x(t) = \frac{1}{2\pi}\int_{-\infty}^{\infty} X(\omega)e^{j\omega t}d\omega$$

$F_k, f(t)$	Fourier coefficients and series

$$F_k = \frac{1}{T_0}\int_{t_0}^{t_0+T_0} f(t)e^{-j2\pi k f_0 t}dt \quad \text{and}$$

$$f(t) = \sum_{k=-\infty}^{\infty} F_k e^{j2\pi k f_0 t}$$

$X(e^{j\theta}), x(n)$	Discrete Time Fourier Transform (DTFT) and inverse

$$\mathcal{D}[x(n)] = X(e^{j\theta}) = \sum_{-\infty}^{\infty} x(n)e^{-jn\theta} \quad \text{and}$$

$$\mathcal{D}^{-1}[X(e^{j\theta})] = x(n) = \frac{1}{2\pi}\int_{\theta_0}^{\theta_0+2\pi} X(e^{j\theta})e^{jn\theta}d\theta$$

$X(k), x(n)$	Discrete Fourier Transform (DFT)

$$\text{DFT}[x(n)] = X(k) = \sum_{n=0}^{N-1} x(n)W_N^{nk}, \qquad k = 0, 1, \ldots, N-1 \quad \text{and}$$

$$\text{IDFT}[X(k)] = x(n) = \frac{1}{N}\sum_{k=0}^{N-1} X(k)W_N^{-nk}, \qquad n = 0, 1, \ldots, N-1$$

$$W_N = e^{-j2\pi/N}$$

$H(s), H(z)$	System TF (SISO), $H(s) = Y(s)/X(s)$; $H(z) = Y(z)/X(z)$, ICs = 0
$\mathbf{H}(s), \mathbf{H}(z)$	System TF (MIMO), $\mathbf{Y}(s) = \mathbf{H}(s)\mathbf{X}(s)$; $\mathbf{Y}(z) = \mathbf{H}(z)\mathbf{X}(z)$, ICs = 0
$H(j\omega) = H(\omega)$, $H(e^{j\theta})$	System (SISO) frequency response, $H(\omega) = \mathcal{F}[h(t)]$ $H(e^{j\theta}) = \mathcal{D}[h(n)]$
$y_{ss}(t)$, $y_{ss}(n)$	Sinusoidal steady-state:
input:	$x(t) = A\cos(\omega t + \alpha)$
output:	$y_{ss}(t) = A\lvert H(j\omega)\rvert\cos(\omega t + \alpha + \angle H(j\omega)), H(j\omega) = \int_{-\infty}^{\infty} h(t)e^{-j\omega t}dt = H(s)\rvert_{s=j\omega}$
input:	$x(n) = A\cos(n\theta + \alpha)$
output:	$y_{ss}(n) = A\lvert H(e^{j\theta})\rvert\cos(n\theta + \alpha + \angle H(e^{j\theta})), H(e^{j\theta}) = \sum_{n=-\infty}^{\infty} h(n)e^{-jn\theta} = H(z)\rvert_{z=e^{j\theta}}$

$y(t) = h(t) * x(t)$, System (SISO) output via linear convolution

$$y(t) = \int_{-\infty}^{\infty} h(\tau)x(t-\tau)d\tau = \mathcal{L}^{-1}[Y(s) = H(s)X(s)] = \mathcal{F}^{1}[H(\omega)X(\omega)]$$

$y(n) = h(n) * x(n)$

$$y(n) = \sum_{-\infty}^{\infty} h(m)x(n-m) = \mathcal{Z}^{-1}[Y(z) = H(z)X(z)]$$

$x_3(n) = x_1(n) \circledast x_2(n)$ Circular convolution

$$x_3(n) = \sum_{m=0}^{N-1} x_1(m)x_2 = (n \ominus m) = \mathcal{D}^{-1}[X_1(k)X_2(k)]$$

$R_{xy}(\tau)$ Cross correlation—finite energy signals/sequences

$$R_{xy}(\tau) = \int_{-\infty}^{\infty} x(t)y(t+\tau)\,dt = \mathcal{F}^{-1}[X(-\omega)Y(\omega) = X^*(\omega)Y(\omega)]$$

$$R_{xy}(p) = \sum_{m=-\infty}^{\infty} x(m)y(p+m), \qquad -\infty \leq p \leq \infty$$

$\tilde{R}_{xy}(p)$ Circular cross correlation

$$\tilde{R}_{xy}(p) = \mathcal{D}^{-1}[X^*(k)Y(k)]$$

$\mathbf{v}(t), \mathbf{v}(n)$ Solution of state equation

$$\mathbf{v}(t) = \mathbf{e}^{\mathbf{A}t}\mathbf{v}(0) + \int_0^t \mathbf{e}^{\mathbf{A}(t-\tau)}\mathbf{B}\mathbf{x}(\tau)\,d\tau = \mathcal{L}^{-1}[\mathbf{\Phi}(s)\mathbf{v}(0)] + \mathcal{L}^{-1}[\mathbf{\Phi}(s)\mathbf{B}\mathbf{X}(s)]$$

$$\mathbf{v}(n) = \mathbf{A}^n\mathbf{v}(0) + \sum_{m=0}^{n-1} \mathbf{A}^{n-m-1}\mathbf{B}\mathbf{x}(m) = \mathcal{Z}^{-1}[\mathbf{\Phi}(z)\mathbf{v}(0)] + \mathcal{Z}^{-1}[z^{-1}\mathbf{\Phi}(z)\mathbf{B}\mathbf{X}(z)]$$

$\boldsymbol{\phi}(t), \boldsymbol{\phi}(n)$ State transition matrix

$$\boldsymbol{\phi}(t) = \mathbf{e}^{\mathbf{A}t} = \mathcal{L}^{-1}[(s\mathbf{I} - \mathbf{A})^{-1} = \mathbf{\Phi}(s)],$$

$$\boldsymbol{\phi}(n) = \mathbf{A}^n = \mathcal{Z}^{-1}[z(z\mathbf{I} - \mathbf{A})^{-1} = \mathbf{\Phi}(z)]$$

MGR Mason's Gain Rule

$$H(\sigma) = \frac{\sum_{k=1}^{M} P_k(\sigma)\Delta_k(\sigma)}{\Delta(\sigma)}$$

A BC Note

The BookWare Companion Series consists of a set of self-contained, correlated volumes covering fundamental topics in engineering and applied mathematics. Each volume in the series contains previews, essential topics, illustrative examples with step-by-step solutions, exploration examples, summaries of key ideas, equations and software functions used, procedures for assessing computer results, and bound-in software scripts for running electronic examples on both the PC and Macintosh platforms. Written for the student, BCs are designed to be used with reliable, readily available commercial software.

Students learn in a number of ways and in a variety of settings. They learn through lectures, in informal study groups, or alone at their desks or in front of a computer terminal. Wherever the location, students learn most efficiently by solving problems, with frequent feedback from an instructor, following a worked-out problem as a model. Worked-out problems have a number of positive aspects. They can capture the essence of a key concept—often better than paragraphs of explanation. They provide methods for acquiring new knowledge and for evaluating its use. They provide a taste of real-life issues and demonstrate techniques for solving real problems. Most importantly, they encourage active participation in learning.

We created the BookWare Companion Series because we saw an unfulfilled need for computer-based learning tools that address the computational aspects of problem solving across the curriculum. The BC series concept was also shaped by other forces: A general agreement among instructors that students learn best when they are actively involved in their own learning, and the realization that textbooks have not kept up with or matched student learning needs. Educators and publishers are just beginning to understand that the amount of material crammed into most textbooks cannot be absorbed, let alone the knowledge to be mastered in four years of undergraduate study. Rather than attempting to teach students all the latest knowledge, colleges and universities are now striving to teach them to reason: to understand the relationships and connections between new information and existing knowledge; and to cultivate problem-solving skills, intuition, and critical thinking. The BookWare Companion Series was developed in response to this changing mission.

Specifically, the BookWare Companion Series was designed for educators who wish to integrate their curriculum with computer-based learning tools, and for students who find their current textbooks overwhelming. The former will find in the BookWare Companion Series the means by which to use powerful software tools to support their course activities, without having to customize the applications themselves. The latter will find relevant problems and examples quickly and easily and have instant electronic access to them.

We hope that the BC series will become a clearing house for the exchange of reliable teaching ideas and a baseline series for incorporating learning advances from emerging technologies. For example, we intend to reuse the kernel of each BC volume and add electronic scripts from other software programs as desired by customers. We are pursuing the addition of AI/Expert System technology to provide an intelligent tutoring capability for future iterations of BC volumes. We also anticipate a paperless environment in which BC content can flow freely over high-speed networks to support remote learning activities. In order for these and other goals to be realized, educators, students, software developers, network administrators and publishers will need to communicate freely and actively with each other. We encourage you to participate in these exciting developments and become involved in the BC Series today. If you have an idea for improving the effectiveness of the BC concept, an example problem, a demonstration using software or multimedia, or an opportunity to explore, contact us (an insert card for your feedback is attached in the back of this volume).

Thank you one and all for your continuing support.

The PWS Electrical Engineering Team

Bill_Barter@PWS.Com — Acquisitions Editor
Ken_Morton@PWS.Com — Assistant Editor
Nathan_Wilbur@PWS.Com — Marketing Manager
Pam_Rockwell@PWS.Com — Production Editor
Lai_Wong@PWS.Com — Editorial Assistant

The PWS

BookWare Companion Series™

Contemporary Linear Systems

Using MATLAB® 4.0

Robert D. Strum
Naval Postgraduate School, emeritus

Donald E. Kirk
San Jose State University

PWS Publishing Company

I(T)P An International Thomson Publishing Company

Boston · Albany · Bonn · Cincinnati · Detroit · London · Madrid · Melbourne
Mexico City · New York · Paris · San Francisco · Singapore · Tokyo · Toronto
Washington

PWS Publishing Company

PWS Publishing Company is a division of Wadsworth, Inc.

Library of Congress Cataloging-in-Publication Data

Strum, Robert D.
Contemporary linear systems using MATLAB 4.0 / Robert D. Strum, Donald E. Kirk.
p. cm.—(The PWS BookWare companion series)
Includes bibliographical references (p. –) and index.
ISBN 0-534-94710-7
1. Linear systems—Data processing. 2. MATLAB. I. Kirk, Donald E., 1937- . II. Title. III. Series.
QA402.S858 1995
003′.74′02855369—dc20 95-14865
CIP

Printed in the United States of America
96 97 98 99—10 9 8 7 6 5 4 3 2

Series Co-originators: Tom Robbins and Robert D. Strum
Acquisitions Editor: Bill Barter
Assistant Editor: Ken Morton
Editorial Assistant: Lai Wong
Production Editor: Pamela Rockwell
Marketing Manager: Nathan Wilbur
Manufacturing Coordinator: Wendy Kilborn
Interior Design: Eve Lehmann
Cover Design: Stuart Paterson, Image House, Inc.
Cover Printer: Henry N. Sawyer, Inc.
Text Printer and Binder: R. R. Donnelly & Sons/ Crawfordsville

About the Cover: The BookWare Companion Series cover was created on a Macintosh Quadra 700, using Aldus FreeHand and Quark XPress. The surface plot on the cover, provided courtesy of The MathWorks, Inc., Natick, MA, was created with MATLAB® and was inserted on the cover mockup with a HP ScanJet IIP Scanner. It represents a surface created by assigning the values of different functions to specific matrix elements.

Printed on recycled paper

Brief Contents

Normal paths are shown as heavy arrows, but other arrangements are possible

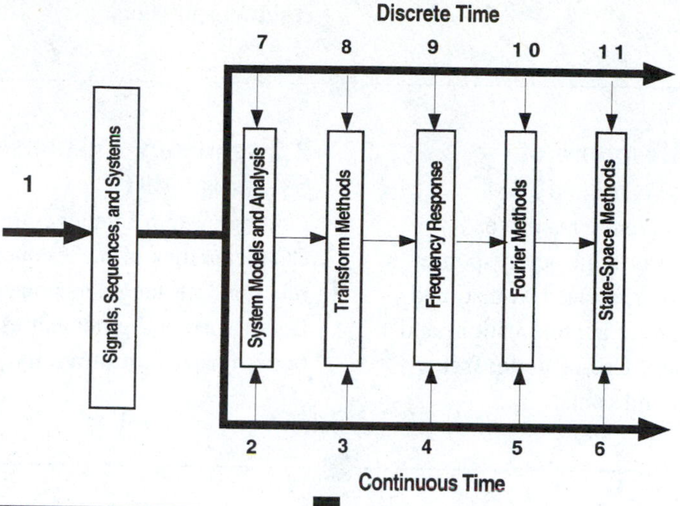

1 Signals and Sequences 1
Reasons for studying linear systems theory; Mathematical description and MATLAB generation of important signals and sequences; A road map of domains, models, and operations; Use of **displot** and **plot** for graphing signals and sequences.

2 Continuous Systems 37
Properties and characteristics; Applications including convolution, unit impulse response, sinusoidal steady-state, and state-space models; System simulation using **conv, impulse, kslsim, lsim, rlocus, roots**, and **step.**

7 Discrete Systems 363
Properties and characteristics; Applications including convolution, unit sample response, sinusoidal steady-state, and state-space models; System simulation using **conv, dlsim, dstep, filter, ksdlsim, rlocus,** and **roots.**

3 Laplace Transforms and Applications 99
Bilateral and unilateral pairs and properties; Applications including solution of DEs, convolution, transfer functions, sinusoidal

8 *z* Transforms and Applications 419
Bilateral and unilateral pairs and properties; Applications including solution of DEs, convolution, transfer functions, sinusoidal

BRIEF CONTENTS

Contents

9 Frequency Response of Discrete Systems

Retrospective: Chapters 7, 8, and 9

10 Discrete Fourier Transforms

CONTENTS

Preface

Computers have been an indispensable engineering tool for more than three decades. They are used extensively for analysis and design and often as system components. Recently, inexpensive and easy-to-use software packages have become widely available, making it possible to focus on the problem being solved rather than on the programming necessary to obtain a solution. This book is built around one such software package, MATLAB®, which is in widespread use in both academia and industry.

Our goal is to provide an effective and efficient environment for students to learn the theory and problem-solving skills for linear systems. The material is designed to provide appropriate background to proceed into areas such as communications, control systems, digital filter design and signal processing, and analog filter design.

To accomplish this we have used a "computer-biased" approach in which computer solutions and theory are viewed as mutually reinforcing rather than as an either-or proposition. We adhere to the axiom that one learns by doing rather than by listening, and, consequently, more than 200 problems are provided for the reader to solve. We also believe that students need feedback on their work, preferably sooner rather than later. To that end, answers for all problems and approximately 250 MATLAB scripts (programs) are contained on the Contemporary Linear Systems (CLS) disk that comes with the book. We also have provided the briefest of introductions to the Symbolic Math Toolbox. For a complete presentation of the use of this toolbox, we recommend appropriate publications of The MathWorks, Inc.

The prerequisites are a basic background in calculus, experience with the formulation of differential equations, ability to manipulate complex numbers, and some familiarity with matrix operations. The background that comes with the completion of a first course in electrical engineering is also helpful. It is also assumed that the reader has used a computer. While expertise in programming is not required, familiarity with a programming language such as Basic, Pascal, Fortran, or C is helpful. In addition, the capability to perform some fundamental computer operations, such as creating, editing, printing, and managing files, is needed.

BOOK ORGANIZATION

We begin with an introductory chapter (Chapter 1) that presents the basics of how signals and sequences are described, both analytically and graphically. Some fundamental concepts, nomenclature, and notation for linear continuous-time and discrete-time systems are also discussed. Chapter 2 introduces several important aspects of continuous systems, including differential equations, unit impulse response, convolution, and state-space models. Laplace transforms are presented in Chapter 3 and are used to solve differential equations, perform convolution, and draw signal flowgraphs or block diagrams. Chapter 4 describes frequency response methods for continuous-time systems. Fourier series and Fourier transform methods for continuous systems are presented in Chapter 5, and these techniques are applied to determine the output of linear systems when periodic or nonperiodic signals are applied as inputs. State-variable methods for continuous systems are discussed in Chapter 6.

Chapter 7 begins the coverage of discrete-time systems with a discussion parallel to that of Chapter 2 for continuous systems. Included are difference equations, unit sample response, and state-space models, as well as the convolution operation. In Chapter 8 we enter the z transform domain and learn how to use z transforms to solve difference equations, carry out convolution, and create a signal flowgraph or block diagram model. Frequency response of discrete systems is the topic of Chapter 9 where we also see a simple approach for designing digital filters. Chapter 10 presents Discrete Fourier Transforms (DFTs) and introduces the Fast Fourier Transform (FFT) algorithm for finding DFTs. State-space techniques for discrete systems are introduced in Chapter 11 where we see that solutions can be obtained in either the time or z transform domains.

Although the principal coverage of state-space methods is given in Chapters 6 and 11, we have presented enough of the basics in Chapters 2 and 7 to enable the reader to use state-equation-oriented MATLAB functions.

USING THE BOOK

We are advocates of studying discrete systems first and have previously stated the case for doing this.[1] In fact, in an earlier draft Chapters 7–11 on discrete systems were Chapters 2–6. So why the change? We decided that in spite of our preference, the typical electrical engineering curriculum is strongly

[1] R.D. Strum and D.E. Kirk, "Linear Systems: Be Discrete—Then Continuous," *IEEE Transactions on Education,* vol. 32, no. 3, pp. 335–342, August 1989.

biased in its early phases toward continuous-time systems with differential and integral calculus, differential equations, and analog circuit theory. Even fearless authors like us shrink from the task of trying to re-orient the entire curriculum, though there are indications that it is beginning to happen. So, persuaded by comments from several reviewers, we decided to cover continuous-time systems first. We did, however, attempt to minimize cross references (leaving these to the instructor) so that it is quite natural to study the material in the order: Chapters 1–11; or Chapter 1, Chapters 7–11, Chapters 2–6. Another arrangement, which is based upon parallel coverage of continuous and discrete system concepts, would start with Chapter 1 (introductory material), followed by the chapter pairs (in either order) of 2 and 7 (time-domain treatment of continuous and discrete systems), 3 and 8 (Laplace and z transform applications), 4 and 9 (frequency response), 5 and 10 (Continuous and Discrete Fourier Transforms), and 6 and 11 (state-space topics).

CHAPTER ORGANIZATION

Each chapter is organized in the following format.

Preview: An overview, motivation, and an historical perspective.

Basic Concepts: A development of the important concepts and relationships, including *Illustrative Examples* to clarify the concepts and computational details.

Solved Examples and MATLAB Applications: Comprehensive illustrations of problem-solving techniques and additional concepts.

Reinforcement and Exploration Problems: Practice in applying the basic concepts (reinforcement) or guidance for the reader to probe new issues or make extensions (exploration).

Definitions, Techniques, and Connections: A summary of the important relationships of the chapter.

MATLAB Functions Used: A list of the computer functions that were used in the chapter.

Annotated Bibliography: Sources of additional information together with a brief statement of where the relevant topics are covered.

Answers: Results for all problems.

In addition, after Chapters 4, 6, 9, and 11, *Retrospectives* are provided which attempt to integrate the material by pausing and reconsidering the interrelationships of the material that has gone before.

Note: It should be emphasized that the *Solved Examples and MATLAB Applications* sections are important reading. Frequently, a point covered lightly, or not at all, in the *Basic Concepts* section of a chapter is amplified or introduced in a solved example. It is recommended that the reader attempt to work these solved examples as if solutions were not available, using the solutions to check or to get back on track when necessary.

CLS DISK

The Contemporary Linear Systems (CLS) disk is an important adjunct to the text and is provided in both Macintosh and DOS formats. The Macintosh version contains a folder of California Functions and 11 other folders (one for each chapter) that have the scripts for answers to problems, solved examples, chapter figures plotted with MATLAB, and plots used as problem statements. The DOS version has a directory containing the California Functions and a second directory with the script files for all 11 chapters. On both versions, the California Functions augment the capabilities of MATLAB. A few of these functions (e.g., **dzpresp, ksxcorr, sampfcn**) emulate MATLAB functions not available in the *Student Edition of MATLAB, Version 4*. For both the Macintosh and DOS versions, the script files for the chapters begin with the letters A (Answers), E (Examples), F (Figures), and P (Problems) followed by a one- or two-digit number 1 through 11 corresponding to the chapter. The final characters indicate the number of the appropriate answer (A5_2), example (E7_4), figure (F4_2), or problem (P8_20). The underbar "_" character is used where a period "." appears in a text caption because MATLAB doesn't like periods in file names. All but a handful of the scripts run with the *Student Edition of MATLAB;* the exceptions were included because we felt that they offered something special.

The CLS disk requires that either the *Student Edition of MATLAB, Version 4*[2] or the Professional version 4.X or later is installed on your system. Experience has shown that it is relatively easy to install MATLAB if the installation directions provided by the documentation are followed. Installation instructions for the CLS disk files are given in Appendix A along with a discussion of how to get started with MATLAB. The CLS installation instructions are also contained in the README text file on the disk.

OPEN MANUSCRIPT

It is our goal that this book be an "open" manuscript. In that spirit, the user sees everything, and we hope that faculty members and students pass on their suggestions for new and interesting problems and examples. These recommendations will receive careful consideration for inclusion in future versions of the book. And, of course, anything we use will be specifically acknowledged. The open-manuscript approach will also be fostered by publishing updated versions more frequently than traditional revised editions. The original *Contemporary Linear Systems Using MATLAB,* for example, was published in 1994.

[2]*The Student Edition of MATLAB, Version 4* Prentice-Hall, Inc., Englewood Cliffs, NJ, 1995.

ACKNOWLEDGMENTS

No effort to develop a textbook can be successful without the help of many people and we were extremely fortunate in this regard. Many classes of students ploughed through the various drafts of the manuscript pointing out errors and suggesting improvements. Special thanks go to Dwight Alexander of the Naval Postgraduate School and Alan Gale, Dan Naar, and Yu Peng of San Jose State University for their help. Our colleagues also provided considerable support, ranging from periodic inquiries and encouragement to specific recommendations for improvement. Especially helpful were comments from Raymond Bernstein, Michael Shields, Charles Therrien, and Murali Tummala of the Naval Postgraduate School. We also received the comments of many reviewers and they were generally encouraging and always constructively critical. We have incorporated many of their suggestions. The long list includes

Ashok Ambardar, Michigan Technological University
Roy Barnett, California State University/Los Angeles
Guy O. Beale, George Mason University
Christopher DeMarco, University of Wisconsin/Madison
David Farden, North Dakota State University
Keat-Choon Goh, University of Southern California
Martin Hagan, Oklahoma State University
Banu Onaral, Drexel University
Ahmed Tewfik, University of Minnesota/Twin Cities
Nadipuram Prasad, New Mexico State University
Chit-Sang Tsang, California State University/Long Beach
Guanghan Xu, University of Texas/Austin

We especially wish to acknowledge the assistance of Thomas E. Bullock, University of Florida; Steven Chin, Catholic University; Julio Gonzalez, Colorado State University; Jo W. Howze, Texas A&M University; Robert Kubicheck, University of Wyoming; Donald Scott, University of Massachusetts/Amherst; Henrik Sorensen, University of Pennsylvania; H. Joel Trussell, North Carolina State University; and Stephen Yurkovich, Ohio State University.

We thank our friend Gina M. Tarantino of New York City for creating the drawings that appear on the first page of most chapters. Tarantino's studies in Florence may have influenced her dashing Leonardo of Pisa of Chapter 7. The people whose likenesses appear helped advance knowledge to its current state. Of course, there are many others who have also contributed to the development of systems theory.

Tom Robbins, the initiator and former editor of the BookWare Companion Series, deserves a great deal of credit (and no blame) for our finished product. This was our second book with Tom, and we found him to be creative, innovative, considerate, and fun to work with. The support given to Tom and us by Edward F. Murphy, President of PWS, J.P. Lenney, Editor-in-Chief, and Peter Hoenigsberg, Executive Vice President, International Thomson Publishing, is very much appreciated. We also wish to thank Craig Borghesani, David R. Dietz, Cynthia Harris, Ken Morton, Mary Thomas, Warren Wake, and Lai Wong for their help. Pam Rockwell did a superb job in all aspects of supervising production and was always most considerate even of our sometimes inconsiderate requests for changes. We include Eve Lehmann in our nods of sincere appreciation, for she took leave from the project for something much more important than the birth of a book, the birth of a child. In short, we found that the support from PWS was very professional and all have earned our respect and gratitude.

Finally, we wish to acknowledge and thank our spouses, Judith Sand Kirk and Ione Thornton Strum, for their support and sacrifices during this venture. We also appreciate the continuing encouragement of Kara, Valerie, and Dana Kirk and Arthur Strum, Sarah Strum, and Judd Rose.

Carmel, California

R. D. S.
D. E. K.

Signals and Sequences

PREVIEW

Engineers, scientists, and mathematicians all use linear systems theory because it is the foundation for building many of the things we use in our daily lives. The theory of linear systems provides powerful tools for analysis and design, and many communications, control, and signal-processing systems can be approximated by linear models. By applying these tools to suitable mathematical models, we can design and develop better systems and also shorten the production cycle. Computer simulation plays a central role in applying linear systems theory and there are now available powerful and easy-to-use software packages. One of these, MATLAB, is used extensively in this book. Computers

FIGURE 1.1 *C.E. Shannon*

CHAPTER 1

are also extensively used as elements of systems. This means that we need to consider both systems whose signals change only at discrete-time instants, called discrete-time systems, and also those whose signals vary continuously with time, called continuous-time systems. To shorten the terminology a bit, we usually refer to such systems as simply discrete and continuous.

In this chapter we begin our study of linear systems theory by first considering the ways in which signals and sequences are described. Continuous systems operate on analog signals, whereas discrete systems operate on sequences. After considering the basic "building blocks" of signals and sequences, we turn our attention to their use in describing signals and sequences that occur in real-life systems. There are at least three good reasons for doing this. First, it is convenient to be able to describe signals and sequences analytically using compact notation. Second, analytical descriptions of signals and sequences are very useful when using the transform methods of Chapters 3, 5, 8, and 10 to analyze and design linear systems. Finally, analytical descriptions make system simulation much easier than if we had to use a tabulation of values as a description of a signal or a sequence.

Many systems contain both analog and discrete, or digital, elements. Thus, there is a need to be able to convert analog signals to sequences and vice versa. These conversions are accomplished by Analog-to-Digital Converters (ADCs) and Digital-to-Analog Converters (DACs), respectively. Following a brief discussion of ADCs and DACs, we'll take a glimpse at where the rest of the book is going.

BASIC CONCEPTS

SIGNALS, SEQUENCES, AND SYSTEMS

Continuous systems operate on and generate signals that may vary over the entire time interval rather than just at discrete instants—that is, analog signals. Figure 1.2 shows a typical analog input signal $x(t)$, a continuous system and an output analog signal $y(t)$. The signals occurring in nature are generally analog. As designers wanting to process and use such signals, we have a choice of whether to process these signals using continuous (analog) systems or discrete (digital) systems. If we choose a discrete system, the continuous signals must be converted to a discrete format. This is most often accomplished by simply sampling the analog signals at equally spaced time intervals to obtain

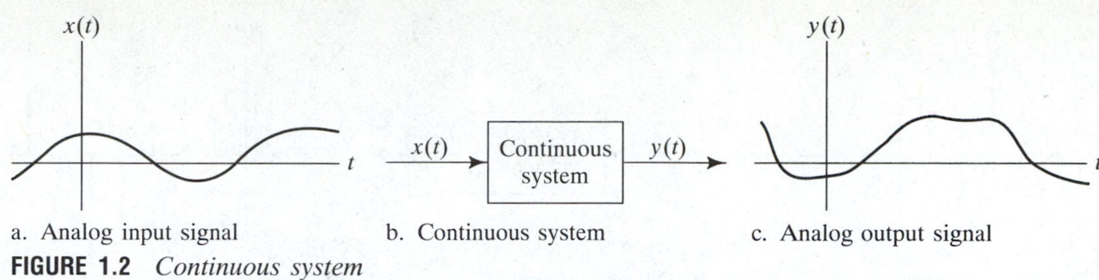

a. Analog input signal b. Continuous system c. Analog output signal

FIGURE 1.2 *Continuous system*

sequences, as illustrated in Figure 1.3. Collectively, these sample values make up the sequence $r(n)$.

Discrete systems generally receive inputs at equally spaced (uniform) time intervals. The inputs are simply numbers, but the order in which they are received makes a difference, so we represent the numbers as sequences. For example, $x(n) = \{8.0, -3.2, 2.1, 6.9\}$ represents a sequence with the value 8.0 occurring first, followed by -3.2, and so forth, as shown in Figure 1.4a. In addition to knowing the sequence values, we also need to know a time reference for its beginning, and we'll generally use (arbitrarily) $n = 0$ as the start of the sequence. Using the sequence $x(n)$ in Figure 1.4a as the input to a discrete system, such as a savings account (represented in Figure 1.4b) results in a different sequence, such as the output $y(n)$ shown in Figure 1.4c.

The abscissas of the graphs of $x(n)$ and $y(n)$ portray time. The sequences are shown only at a discrete set of values, because a system operates on or generates signals only at these discrete instants. Since the sequence values are often obtained by sampling a continuous-time, or analog, waveform at uniform intervals, we refer to the integer values of n as the sampling times and the amplitudes $8.0, -3.2, \ldots,$ as the sample values. We'll see later that the time separating the samples is an important parameter in the sampling process.

Now let's take a more detailed look at the representation of signals for continuous and discrete systems. We begin with continuous systems and see how to describe analog signals.

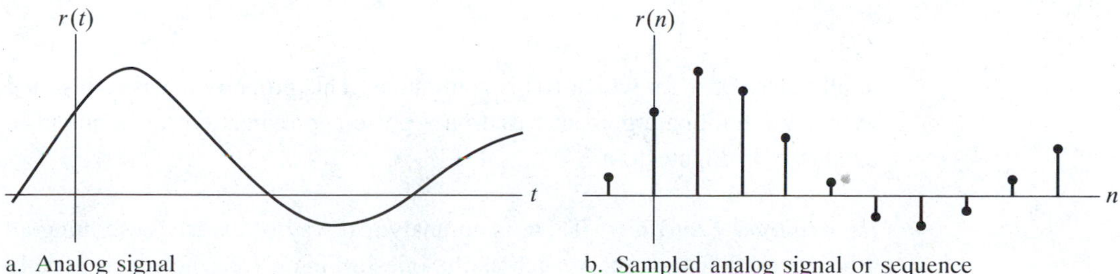

a. Analog signal b. Sampled analog signal or sequence

FIGURE 1.3 *Analog signal and sequence resulting from the sampling process*

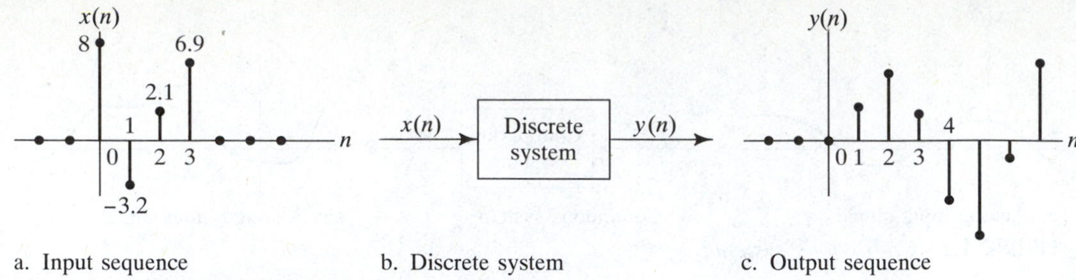

a. Input sequence b. Discrete system c. Output sequence

FIGURE 1.4 *Discrete system*

CONTINUOUS-TIME SIGNALS

(a) Unit Impulse The unit impulse is an important signal in the study of continuous-time systems. An arbitrary analog signal can be approximated by impulses, and knowing the response of a system to an impulse input enables us to find the response to all other inputs. Denoted by $\delta(\sigma)$, the unit impulse is not really a function in the normal sense. It is generally defined as the limit of a function or through its properties. One commonly used definition is

unit impulse

$$\delta(\sigma) = \lim_{\Delta \to 0} f(\Delta, \sigma)$$

where $f(\Delta, \sigma)$ is a function such as the pulse shown in Figure 1.5a. As $\Delta \to 0$, the height of this pulse becomes infinitely large, its base approaches zero, and its area is always 1. Thus we think of an impulse as having zero duration, infinite height, and unit area. Multiplying an impulse by a constant A simply makes its area equal to A. Rather than representing impulses as pulses, we find it convenient to use the symbol shown in Figure 1.5b, where the numeral 1 next to the arrow indicates the impulse's area. Of course, we can have impulses with different areas occurring at various times, as shown in Figure 1.5c.

A useful property of impulses is the *sifting property,* which is

sifting property

$$\int_{-\infty}^{\infty} f(t)\delta(t - t_1)\, dt = f(t_1)$$

at all values of t_1 for which $f(t)$ is continuous. This property is easily justified by using a limiting argument based on a pulse approximation to the impulse, as shown in Figure 1.5a.

(b) Arbitrary Functions There is no analytical way of exactly describing an arbitrary analog waveshape such as the one in Figure 1.6a; however, equally spaced impulses can be used as an approximate representation, shown as $s(t)$

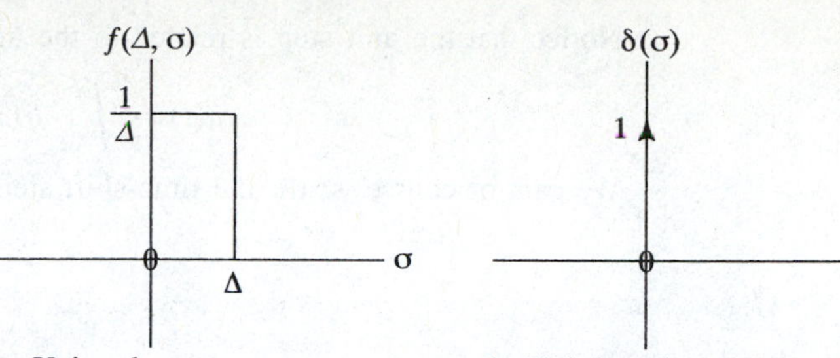

a. Unit pulse

b. Unit impulse

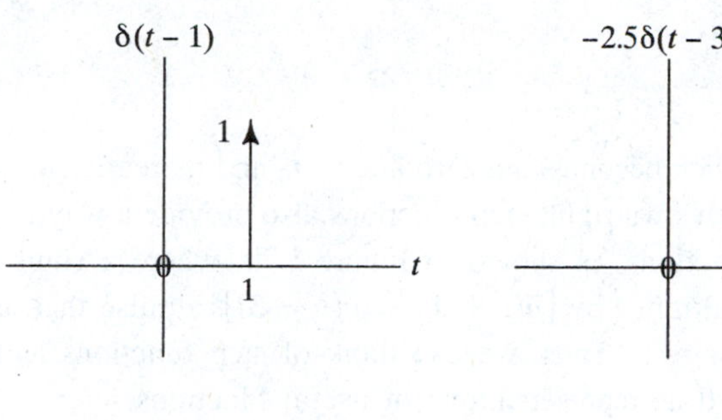

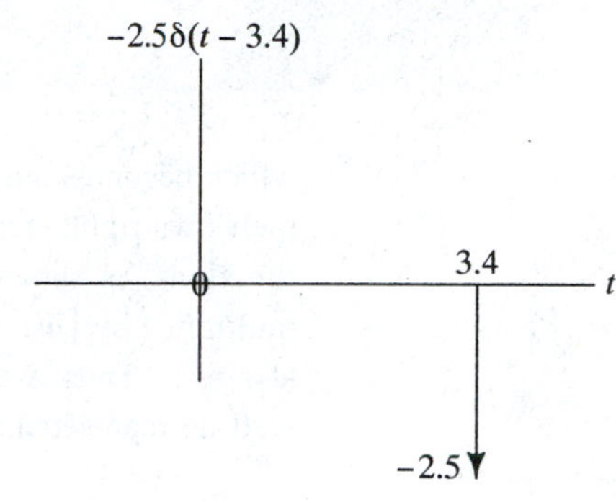

c. Shifted, scaled impulses

FIGURE 1.5 *Impulse functions*

in Figure 1.6b. The impulses have the areas associated with the function $f(t)$ in the intervals shown. As the number of impulses becomes infinite we have

$$f(t) = \int_{-\infty}^{\infty} f(\tau)\delta(t - \tau)\,d\tau\,.$$

(c) Unit Step Function Figure 1.7a pictures a unit step function that "turns on" at $t = 0$ and is defined by

$$u(t) = \begin{cases} 1, & t \geq 0 \\ 0, & t < 0. \end{cases}$$

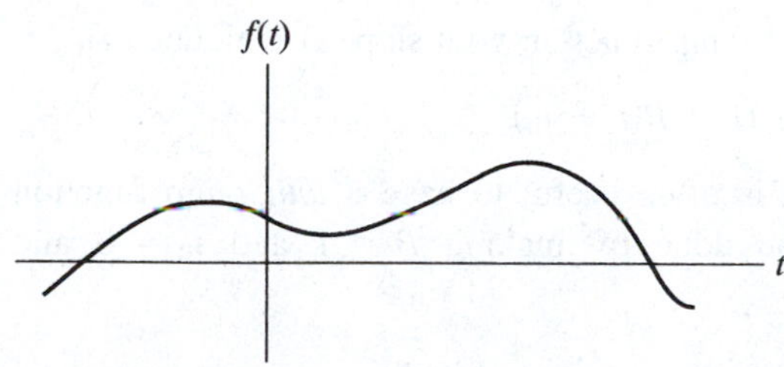

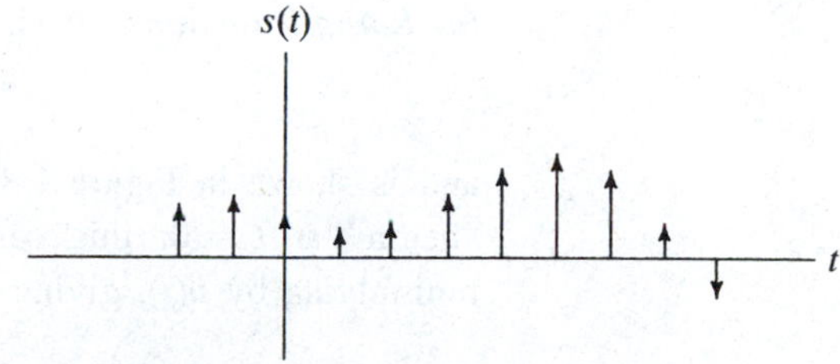

a. Arbitrary function

b. Approximation of sum of impulses

FIGURE 1.6 *Approximation of a function*

Notice that the unit step is related to the unit impulse by

$$u(t) = \int_{-\infty}^{t} \delta(\tau)\, d\tau\,.$$

We can, of course, scale and time-shift step functions to obtain

generic step function

$$Bu(t - t_0) = \begin{cases} B, & t \geq t_0 \\ 0, & t < t_0 \end{cases}$$

which becomes nonzero at $t = t_0$ and thereafter. In addition to being useful in their own right, step functions also provide a way to "turn on" and "off" other functions, as shown in Figure 1.7b, where a sinusoidal signal $g(t)$ has been multiplied by $[u(t + 1) - u(t - 3)]$, a pulse that turns on at $t = -1$ and off at $t = 3$. Thus, we can think of step functions as mathematical switches as well as representations of useful functions.

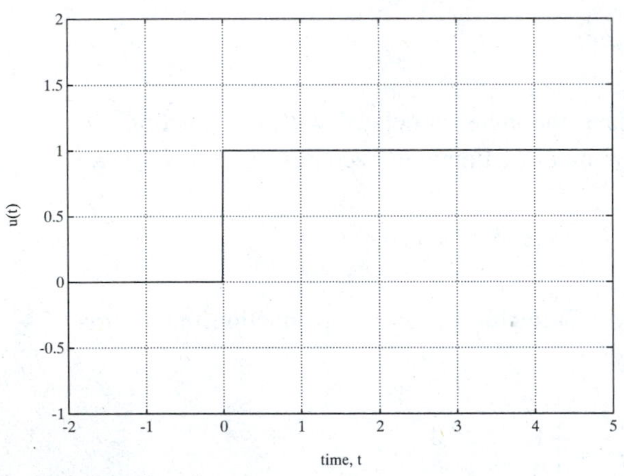

a. Unit step function

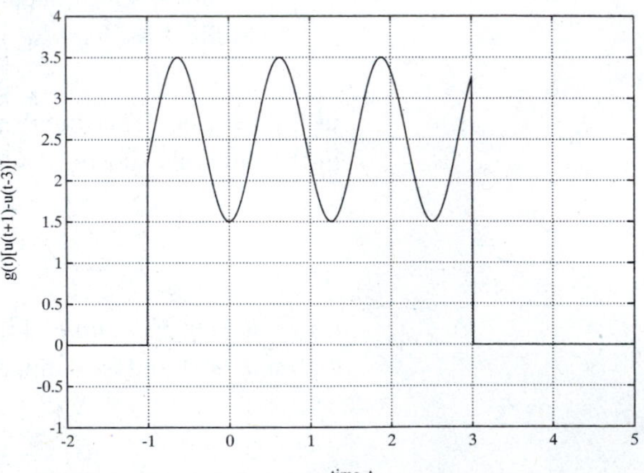

b. Signal $g(t)$ multiplied by a pulse function

FIGURE 1.7 *Step function and an application*

(d) Ramp Functions A shifted ramp function with slope B is defined as

$$g(t) = B(t - t_0)$$

and is shown in Figure 1.8. It is often useful to have a *unit* ramp function "begin" at $t = 0$; this can be done by making $B = 1$ and $t_0 = 0$ and multiplying by $u(t)$, giving

unit ramp function

$$r(t) = tu(t) = \begin{cases} 0, & t < 0 \\ t, & 0 \leq t. \end{cases}$$

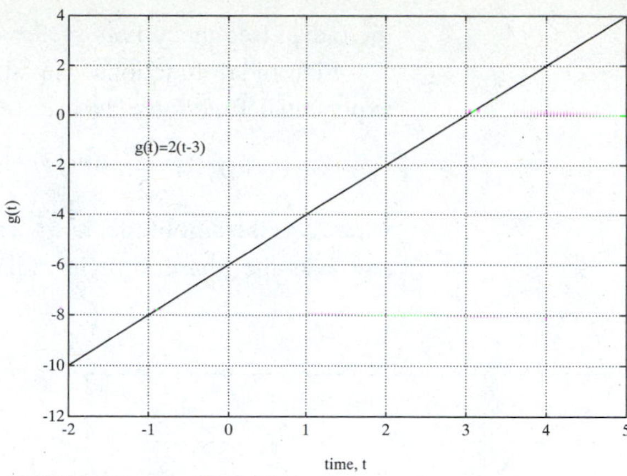

FIGURE 1.8 *Scaled, shifted ramp function*

(e) Real Exponential Functions Exponential signals are characterized by

exponential function

$$f(t) = Ae^{at}$$

where e is the Naperian constant 2.718 . . . and a and A are real constants. Two of the possibilities are illustrated in Figure 1.9.

(f) Sinusoidal Functions A sinusoidal function

$$f(t) = A\cos(2\pi ft + \alpha) = A\cos(\omega t + \alpha) = A\cos\left(\frac{2\pi t}{T_0} + \alpha\right)$$

is shown in Figure 1.10 with A (the amplitude) $= 2.0$, f (the frequency in hertz, or cycles per second) $= 0.25$ and α (the phase shift in radians) $= 1$. Recall that

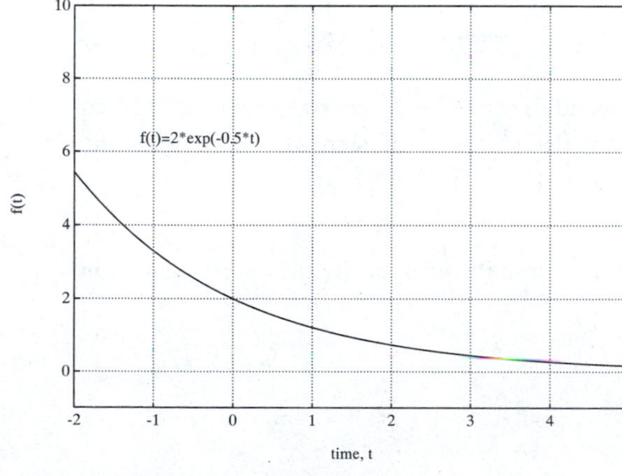

a. A decaying exponential function

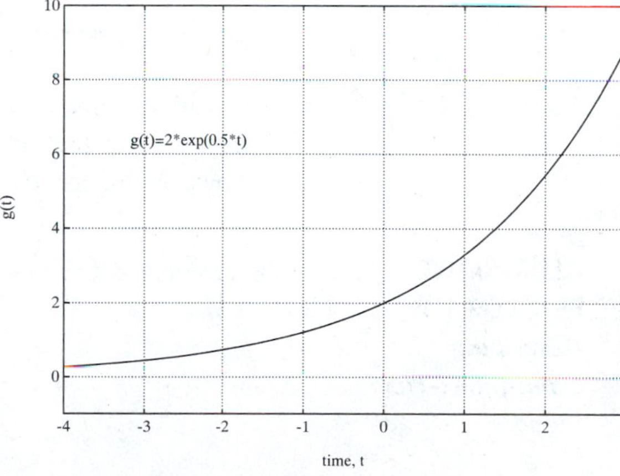

b. An increasing exponential function

FIGURE 1.9 *Exponential functions*

the radian frequency is $\omega = 2\pi \cdot f$ rad/s and that the period is $T_0 = 1/f$ s.[1]

Sinusoidal functions can also be formed by the sum of two complex exponential functions, as in

$$e(t) = \frac{B}{2}e^{j(\omega t+\rho)} + \frac{B}{2}e^{-j(\omega t+\rho)} = B\cos(\omega t + \rho)$$

where B is the amplitude, $\omega = 2\pi f$ is the angular frequency in radians/second, and ρ is the phase shift in radians.

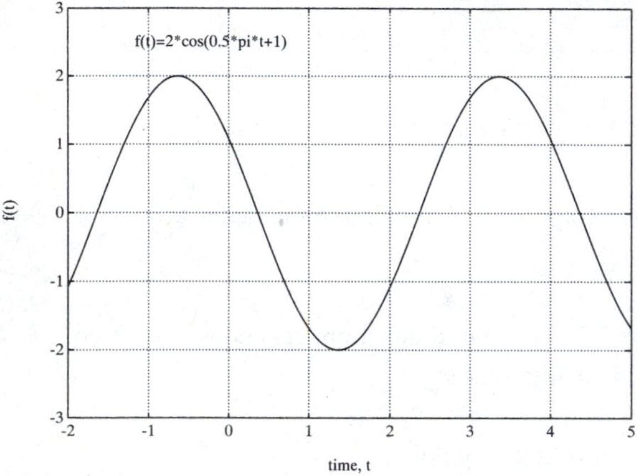

FIGURE 1.10 *A sinusoidal function*

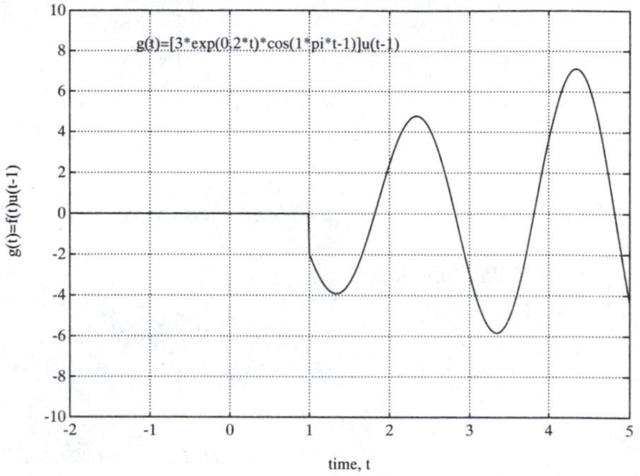

FIGURE 1.11 *Exponentially modulated sinusoidal function*

(g) Exponentially Modulated Sinusoidal Functions If a sinusoid is multiplied by a real exponential, we have an exponentially modulated sinusoid

$$f(t) = Ae^{at}\cos(2\pi ft + \alpha)$$

that also can arise as a sum of complex exponentials, as in

$$e(t) = \frac{B}{2}e^{at}e^{j(\omega t+\rho)} + \frac{B}{2}e^{at}e^{-j(\omega t+\rho)} = Be^{at}\cos(\omega t + \rho).$$

An exponentially modulated sinusoid, $f(t) = 3e^{0.2t}\cos(\pi t - 1)$, is "turned on" at $t = +1$ by multiplying by the shifted unit step $u(t-1)$, with the result shown in Figure 1.11.

ILLUSTRATIVE PROBLEM 1.1 ***Describing Continuous-Time Signals***

Use the functions defined previously to describe analytically the signals shown in Figure 1.12.

[1]For convenience on MATLAB plots, we often use pi to denote π, $\wedge$ to denote exponentiation, w for ω, phi for ϕ, and so forth.

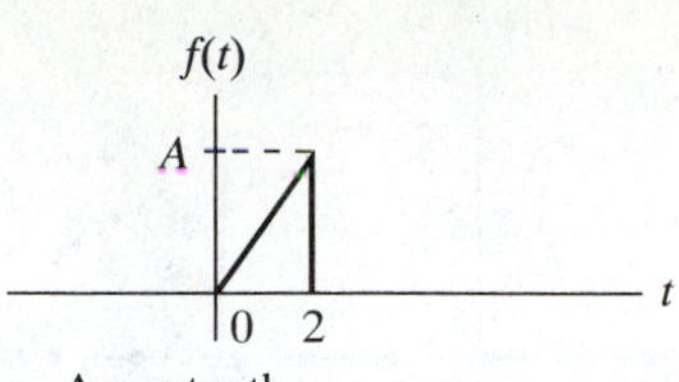

a. A sawtooth

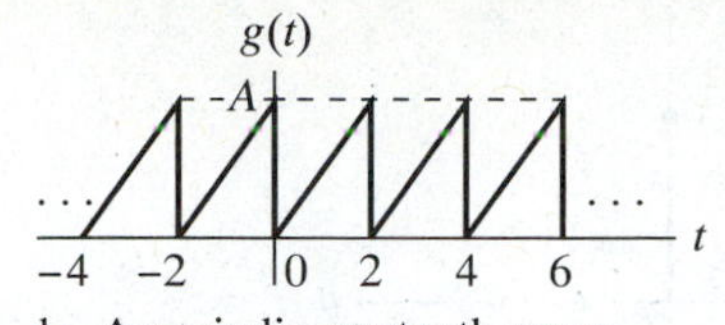

b. A periodic sawtooth wave

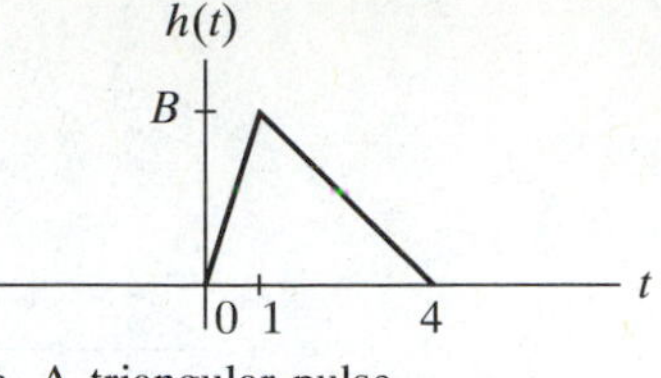

c. A triangular pulse

FIGURE 1.12 *Signals for Illustrative Problem 1.1*

Solution

a. The signal in the interval $0 \le t \le 2$ is described by $(A/2)t$, so we need only turn on this signal at $t = 0$ and turn it off again at $t = 2$. This gives

$$f(t) = \frac{A}{2}\,t[u(t) - u(t - 2)].$$

b. Here we need only replicate the pulse of part (a) an infinite number of times, with each replicate displaced by 2 time units from the previous one. The result is

$$g(t) = \sum_{m=-\infty}^{\infty} \frac{A}{2}(t - 2m)[u(t - 2m) - u(t - 2m - 2)],$$

where m is an integer.

c. We can represent the rising pulse edge by $Bt[u(t) - u(t - 1)]$ and the falling edge by $-[B/3][t - 4][u(t - 1) - u(t - 4)]$; adding these gives

$$h(t) = Bt[u(t) - u(t - 1)] - \frac{B}{3}[t - 4][u(t - 1) - u(t - 4)]$$

which can be simplified as

$$h(t) = Btu(t) - \frac{4B}{3}[t - 1]u(t - 1) + \frac{B}{3}[t - 4]u(t - 4).$$

SEQUENCES

(a) Unit Sample Sequence The unit sample sequence is the discrete-time version of the unit impulse in continuous-time situations. Thus it is possible to represent an arbitrary sequence as the weighted sum of unit sample sequences. In addition, if the response of a discrete system to a unit sample input is known, the system's response to an arbitrary input can be determined. The unit sample sequence $\delta(m)$ is defined as

unit sample sequence

$$\delta(m) = \begin{cases} 1, & m = 0 \\ 0, & m \neq 0 \end{cases}$$

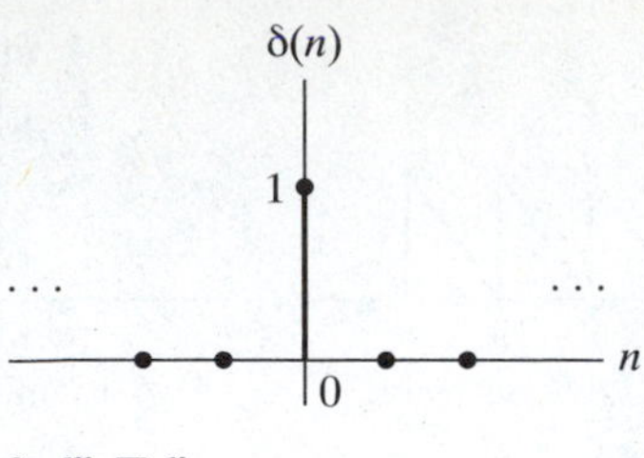

a. $m = n$

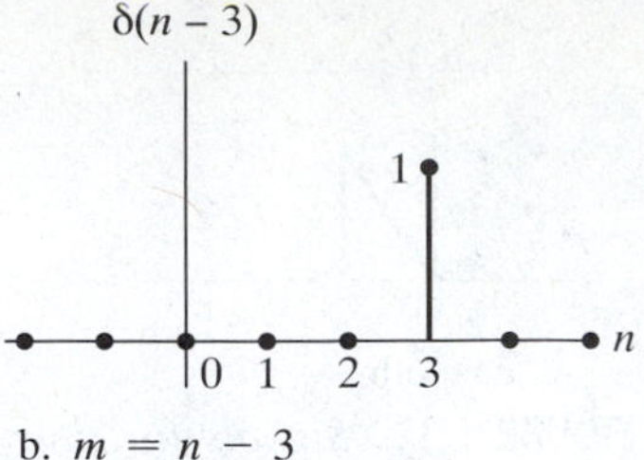

b. $m = n - 3$

FIGURE 1.13 *Plots of unit sample sequences*

where m is an integer. Figure 1.13 shows $\delta(m)$ for $m = n$ and $m = n - 3$, where $\delta(n - 3)$ is referred to as a shifted unit sample sequence. When $\delta(n)$ is used as the input to a discrete system, the resulting output is referred to as the unit sample response. $\delta(n)$ is also referred to as the unit impulse, in which case the response $y(n)$ of a system to the input $x(n) = \delta(n)$ is called the unit impulse response. We can represent a sample of arbitrary amplitude A by simply multiplying by A to obtain the sequence $A\delta(n)$ or $A\delta(n - n_0)$ as appropriate.

(b) Arbitrary Sequences An arbitrary sequence can always be described by a summation of weighted unit sample sequences; that is,

description of any sequence

$$f(n) = \sum_{m=-\infty}^{\infty} f(m)\delta(n - m).$$

For example, the system input $x(n)$ in Figure 1.4a on page 3 is described by

$$x(n) = 8.0\delta(n) - 3.2\delta(n - 1) + 2.1\delta(n - 2) + 6.9\delta(n - 3).$$

(c) Unit Step Sequence Figure 1.14a shows a unit step sequence defined as

unit step sequence

$$u(n) = \begin{cases} 1, & n \geq 0 \\ 0, & n < 0. \end{cases}$$

Notice that the unit step sequence $u(n)$ is related to the unit sample sequence $\delta(n)$ by

$$u(n) = \sum_{m=-\infty}^{n} \delta(m).$$

As with the unit sample sequence, we can shift and scale the step sequence to obtain

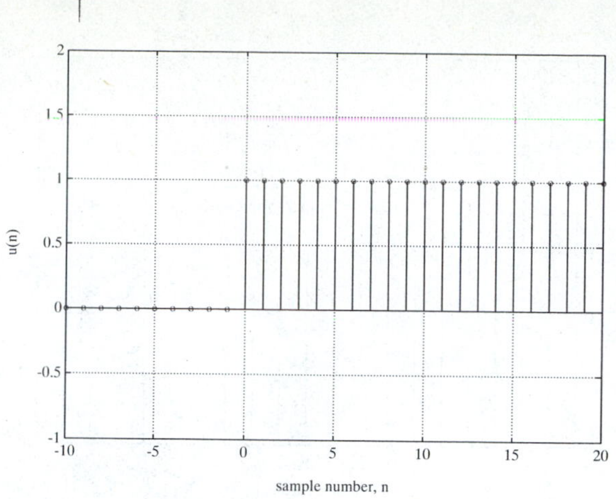

a. Unit step sequence

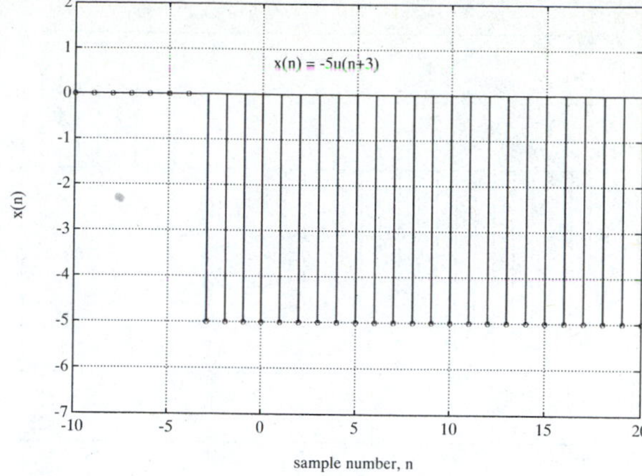

b. Shifted and scaled step sequence

FIGURE 1.14 *Step sequences*

generic step sequence

$$Bu(n - n_0) = \begin{cases} B, & n \geq n_0 \\ 0, & n < n_0 \end{cases}$$

as shown in Figure 1.14b for $B = -5$ and $n_0 = -3$. It is also possible to generalize the definition of the step sequence by considering the sequence given by $Cu(-n - n_0)$. This sequence will be zero for $-n - n_0 < 0$ or for $-n_0 < n$.

(d) Ramp Sequences A shifted ramp sequence with slope B and defined as

$$g(n) = B(n - n_0)$$

is shown in Figure 1.15. It is often useful to have a *unit* ramp sequence that begins at $n = 0$; this can be done by making $B = 1$ and $n_0 = 0$ and multiplying by $u(n)$.

(e) Real Exponential Sequences Figure 1.16 shows the exponential sequence $f(n) = A(a)^n$ with $A = 10$ and $a = 0.9$. Notice that in MATLAB $(0.9)^n$ is written as (0.9)∧n. As $n \to \infty$, this sequence approaches zero, whereas for $n \to -\infty$, the sequence approaches $+\infty$. We can also define a variant on this sequence by multiplying it (point by point) by the unit step sequence to form the composite sequence $p(n) = A(a)^n u(n)$, which is zero for $n < 0$.

(f) Sinusoidal Sequences A sinusoidal sequence may be described as

sinusoidal sequence

$$f(n) = A \cos\left[\frac{2\pi n}{N} + \alpha\right]$$

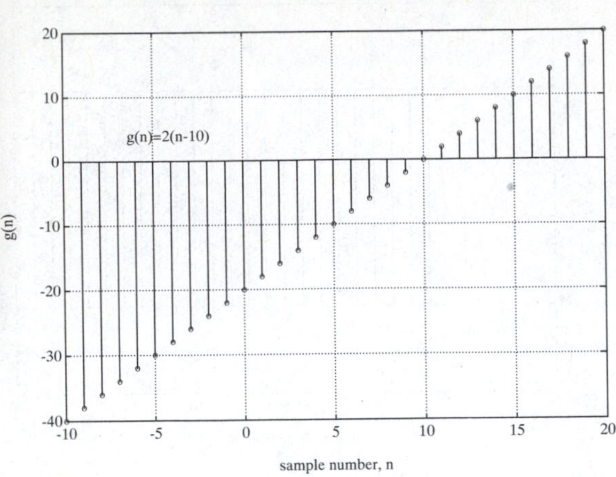

FIGURE 1.15 *Shifted ramp sequence*

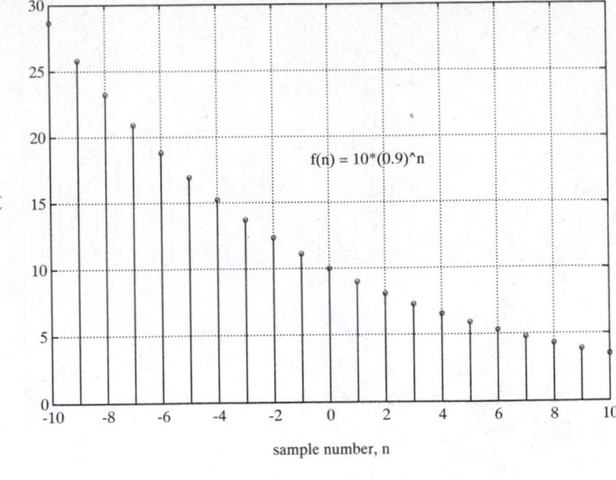

FIGURE 1.16 *Exponential sequence*

where A is a positive real number (the amplitude of the sinusoidal sequence), N is the period, and α is the phase in radians. A representative sinusoidal sequence is shown in Figure 1.17 for $A = 5$, $N = 16$, and $\alpha = \pi/4$.

(g) Exponentially Modulated Sinusoidal Sequences By multiplying an exponential sequence by a sinusoidal sequence, we obtain an exponentially modulated sequence described by

generic sinusoidal sequence

$$g(n) = A(a)^n \cos\left(\frac{2\pi n}{N} + \alpha\right)$$

as shown in Figure 1.18 with $N = 16$, $A = 10$, $a = 0.9$, and $\alpha = \pi/4$. The

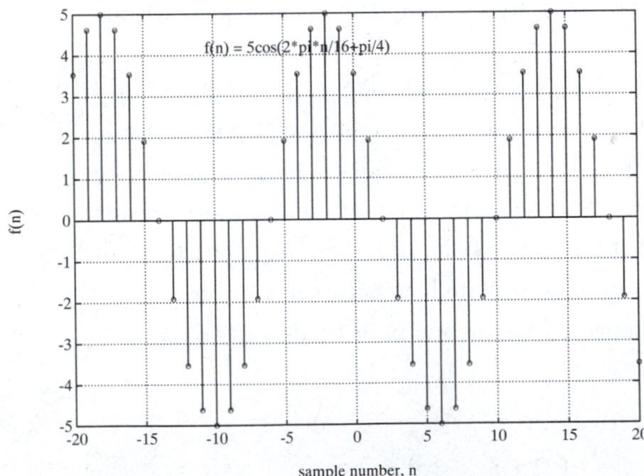

FIGURE 1.17 *Sinusoidal sequence*

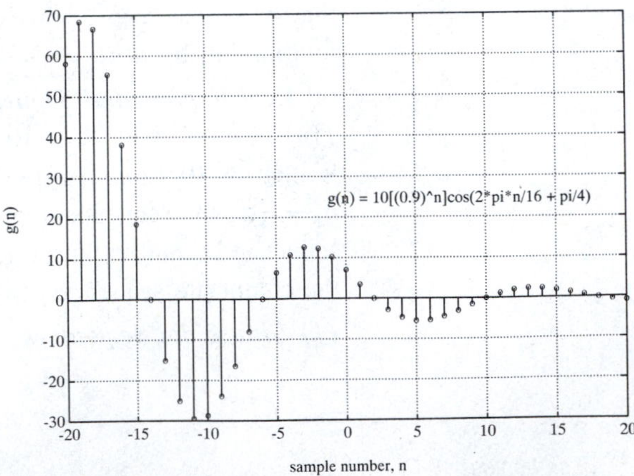

FIGURE 1.18 *Exponentially modulated sinusoidal sequence*

sequences we have discussed cover many of the important possibilities. In addition, it is possible to combine these sequences to describe other sequences as well.

ILLUSTRATIVE PROBLEM 1.2
Describing Sequences

Use the sequences defined previously to describe analytically the sequences shown in Figure 1.19.

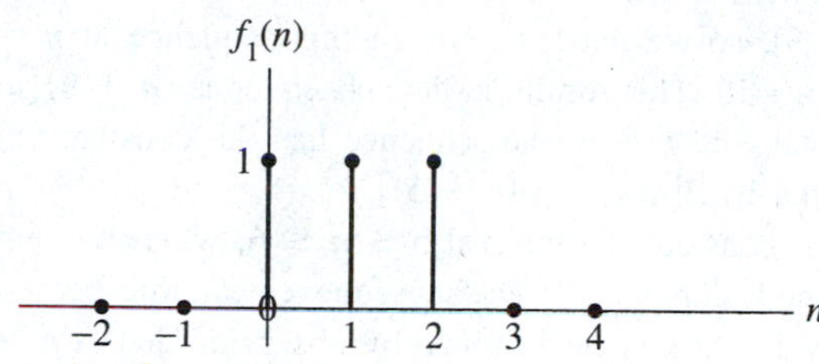

a. A pulse sequence

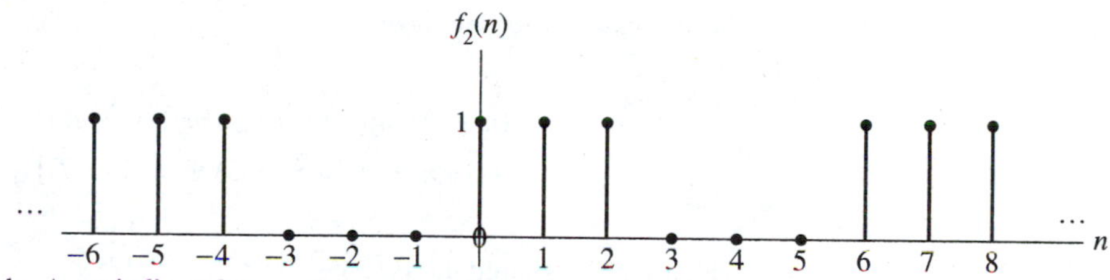

b. A periodic pulse sequence

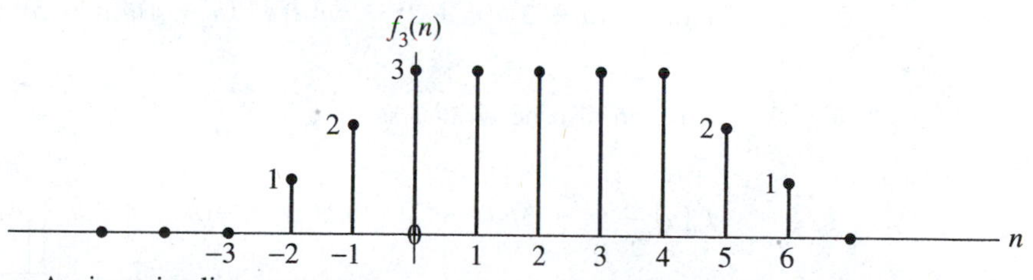

c. A piecewise linear sequence

FIGURE 1.19 *Sequences for Illustrative Problem 1.2*

Solution

a. The pulse sequence can be described by $f_1(n) = u(n) - u(n-3)$. The first step sequence turns on the pulse at $n = 0$ and the second step turns it off at $n = 3$.

CHAPTER 1

b. The periodic sequence $f_2(n)$ can be described by first determining the sequence for one period—for example, for $n = 0$ to $n = 5$ (the period is $N = 6$). This has already been done in part (a), so all we need to do is create replicas of $u(n) - u(n-3)$, each displaced by mN units, where m is a positive or negative integer. The result is

$$f_2(n) = \sum_{m=-\infty}^{\infty} f_1(n - 6m) \text{ where } f_1(n) = u(n) - u(n-3)\,.$$

c. We can use an approach similar to that of Illustrative Problem 1.1(c); that is, we determine the overall sequence by partitioning the time axis into segments during which the sequence is linear, or into the intervals $n \leq -3$, $-2 \leq n < 0$, $0 \leq n \leq 4$, $5 \leq n \leq 6$, and $7 \leq n$. In the interval $n \leq -3$, the sequence is identically zero. In the interval $-2 \leq n \leq 0$, the sequence values are given by $(n + 3)$, so we need to turn on this sequence at $n = -2$ and turn it off again at $n = 0$. This results in the subsequence $(n + 3)[u(n + 2) - u(n)]$. In the interval $0 \leq n \leq 4$, the sequence has the constant value 3 and so can be represented by $3[u(n) - u(n - 5)]$.

Next, consider the interval $5 \leq n \leq 6$, where the envelope[2] is a straight line having a slope of -1. The subsequence can thus be described as $(-n + k)$, and the value of k is easily found by observing that at $n = 5$, the subsequence has the value 2, which indicates that $k = 7$. So this piece is described by $(-n + 7)[u(n - 5) - u(n - 7)]$. For $7 \leq n$ the sequence is zero. Putting all the pieces together we have

$$\begin{aligned} f_3(n) = {} & (n + 3)[u(n + 2) - u(n)] + 3[u(n) - u(n - 5)] \\ & + (-n + 7)[u(n - 5) - u(n - 7)] \end{aligned}$$

which can be simplified as

$$f_3(n) = (n + 3)u(n + 2) - nu(n) - (n - 4)u(n - 5) + (n - 7)u(n - 7)\,.$$

This can also be written as

$$f_3(n) = (n + 3)u(n + 3) - nu(n) + -(n - 4)u(n - 4) + (n - 7)u(n - 7).$$

This latter form leads to another approach (explored in Problem 1.11) for writing the expressions describing piecewise linear sequences. ■

[2]The *envelope* is the curve joining the sequence values.

CONVERSION BETWEEN CONTINUOUS AND DISCRETE SIGNALS

In applications, systems often contain components, or subsystems, which are both discrete and continuous. A typical situation is illustrated in Figure 1.20, where a continuous-time, or analog, signal is to be processed using a digital processing algorithm implemented on a general-purpose digital computer or using a special-purpose Digital Signal Processing (DSP) chip set. It is also common for the sequence resulting from this digital processing to be converted back to an analog signal for the purpose of driving a display, a motor, a stereo system, or whatever. The conversion of an analog signal to a sequence is accomplished by a sample-and-hold circuit in conjunction with an Analog-to-Digital Converter (ADC), whereas conversion of a sequence to an analog signal requires a Digital-to-Analog Converter (DAC).

Let's consider the analog-to-digital conversion process in a bit more detail. The sample-and-hold element samples the analog input signal and holds the sample value constant long enough for the ADC to complete the conversion and transmit the resulting sequence to the following digital hardware. The ADC itself has basically two functions: quantizing the sampled analog signal into a discrete set of levels and coding the quantized representation into an acceptable format. The quantization operation is illustrated in Figure 1.21 (page 16), which shows the quantization characteristic, illustrates a typical input to the quantizer from the sample-and-hold element, and shows the quantizer's output. Notice that the quantization process introduces inaccuracies into the representation. This is unavoidable, but the negative effects can be reduced by using a converter with more precision—that is, a larger number of bits. Thus, it is important to consider the length of the word in bits produced by the ADC. Finally, the coding function of the ADC provides, for example, a 2's complement representation of the quantizer output.

The inverse process, digital-to-analog conversion, consists of using the numbers provided by the digital hardware to generate an analog output. This

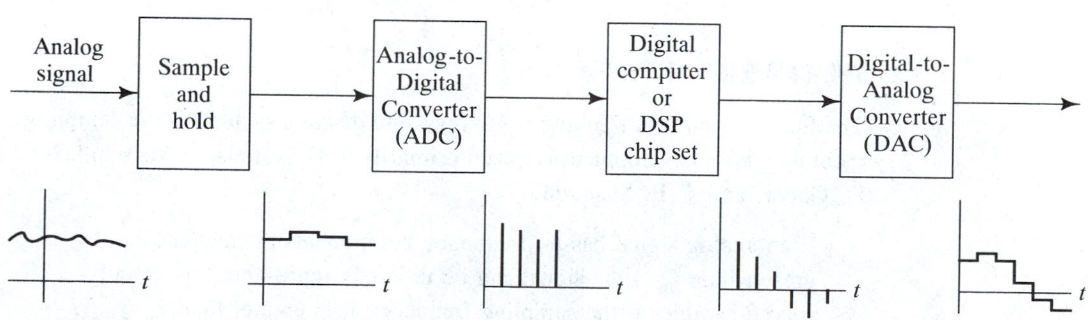

FIGURE 1.20 *Digital Signal Processing System*

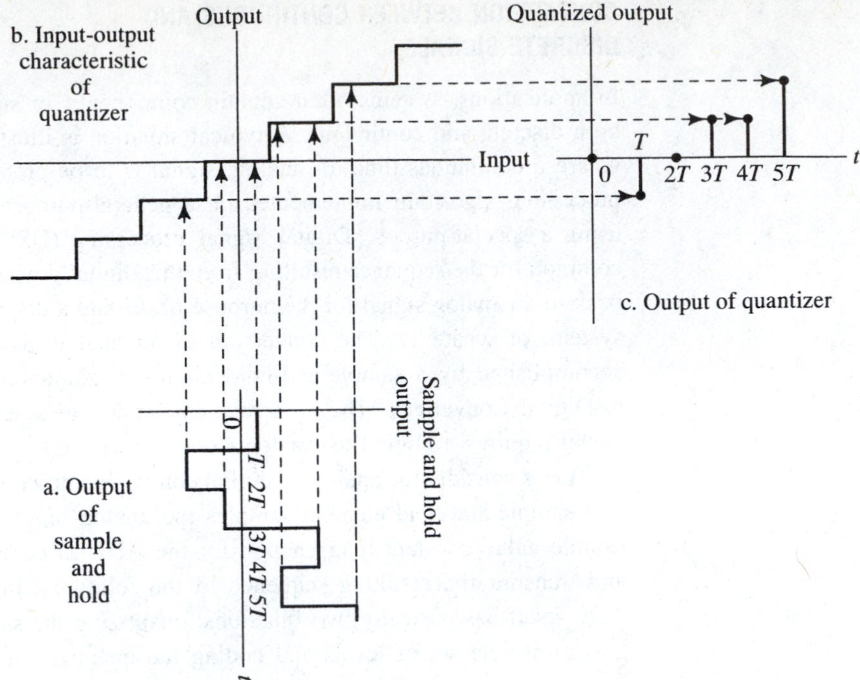

FIGURE 1.21 *The quantization process*

process typically involves scaling the digital values and converting them to a piecewise-constant output voltage, as shown in Figure 1.22.

For the most part in this book, we will assume that the quantization effects of analog-to-digital conversion can be ignored when dealing with sequences. This is equivalent to assuming that we have ADCs with an infinite number of bits, or that we are simply dealing with sampled analog signals. The effects of limited precision, however, have been extensively analyzed and the results can be used to determine the number of bits of precision required for coefficients and data.

THE SAMPLING THEOREM

The fundamental result linking continuous and discrete systems is the sampling theorem whose development is traced primarily to H. Nyquist, J. M. Whittaker, D. Gabor, and C. E. Shannon.

> If an analog signal has no frequency components at frequencies greater than $f_{\max}$, the signal can be uniquely represented by equally spaced samples if the sampling frequency f_s is greater than $2 \cdot f_{\max}$. Furthermore, the original analog signal can be recovered from the

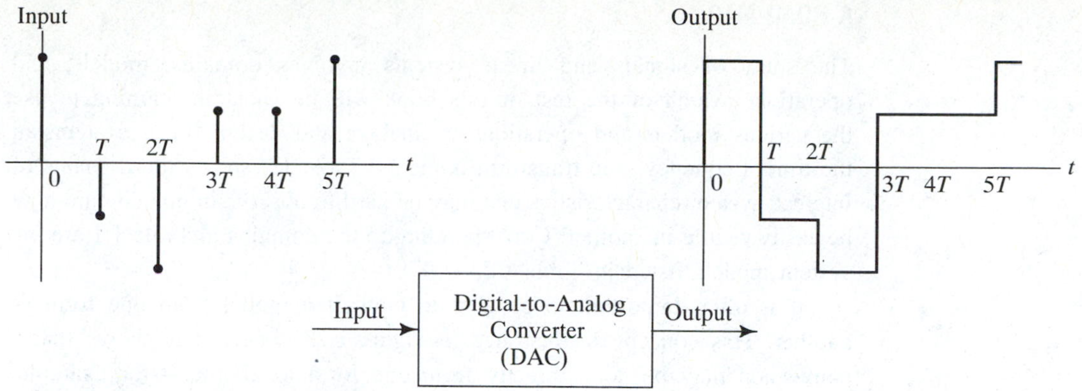

FIGURE 1.22 *Digital-to-analog conversion*

samples by passing them through an ideal lowpass filter having an appropriate bandwidth.

Comment: A detailed development of the sampling theorem requires a background in Fourier transform theory. See Chapter 5, "Continuous-Time Fourier Series and Transforms."

The minimum acceptable sampling frequency, $2 \cdot f_{\max}$, is known as the Nyquist frequency. Although real-life signals are not bandlimited by having no components at frequencies greater than $f_{\max}$, this hurdle can be overcome by prefiltering a signal to be sampled with a lowpass filter to greatly attenuate frequencies above a specified cutoff. This presampling filter is known as an *antialiasing* filter, since it prevents unwanted high frequencies from appearing in the sampled values.

It is instructive to consider what happens when a sinusoidal signal is sampled. If, for example, a sinusoidal input to the ideal sampler (ADC) of Figure 1.20 is sampled at an interval of T seconds or a sampling frequency $f_s = 1/T$ samples per second, the sampled output sequence is

$$x(nT) = x(t)|_{t=nT} = A\cos(\omega t + \alpha)|_{t=nT} = A\cos(\omega nT + \alpha)$$

which, by defining $\theta = \omega T$, can be written as the sinusoidal sequence

$$x(n) = A\cos(n\theta + \alpha)\,.$$

We define θ as the digital frequency of the discrete-time sinusoid, and θ, ω, T, and f_s have the following relationships: $\theta = \omega T = 2\pi f T = 2\pi f/f_s$. For $x(n)$ to be a periodic sinusoidal sequence, it is required that θ be a rational multiple of 2π. Clearly, if $\omega = 0$, this corresponds to $\theta = 0$. If $\omega = 2\pi f_s/2$, the highest frequency that can be represented by the specified sampling frequency, this corresponds to the digital frequency $\theta = \pi$. Thus, the useful range of digital frequencies for a discrete system is from 0 to π.

A ROAD MAP

The study of signals and linear systems involves domains, models, and operations. Much of the rest of this book will be spent in learning to use the various models and operations to analyze and design linear systems in the time, frequency, and transform domains. Table 1.1 shows the domains of interest; system characteristics that may be hard to observe in one domain may be easily visible in another. Corresponding to the domains in Table 1.1 are the system models listed in Table 1.2.

It is often important to be able to convert a model from one form to another. This concept is illustrated in Figure 1.23, where it is shown that a conversion may be done directly from one form to another—for example, from differential equations describing a continuous system to its frequency response—or indirectly, as shown in the conversion from differential (or difference) equations to frequency response via an intermediate transfer function. In Figure 1.23 we also see that the path to frequency response from other models can also be either direct or indirect. The model and domain used for a particular task depend on various factors, including what is known about the system; what information is desired; and what solution aids (for example, a computer with MATLAB) are available.

Along with the models and domains are various operations (given in Table 1.3) employed in the analysis and design of linear systems. Notice that several of these operations may be carried out in more than one domain—as the (very) old saying goes, "there's more than one way to. . . ."

Final Comment: Tables 1.1–1.3 and Figure 1.23 are offered in the spirit of telling you, the reader, where we're going. Depending of the state of your

TABLE 1.1 *Domains for continuous and discrete systems*

Continuous systems	*Discrete systems*
Time, t	Time, n
Frequency, f	Frequency, θ
Transform, s	Transform, z

TABLE 1.2 *Models for continuous and discrete systems*

Continuous systems	*Discrete systems*
Differential equations, DEs	Difference equations, DEs
Transfer functions, $H(s)$	Transfer functions, $H(z)$
Frequency response, $H(j\omega)$	Frequency response, $H(e^{j\theta})$
State differential equations	State difference equations
Unit impulse response, $h(t)$	Unit sample response, $h(n)$
Signal flowgraph or block diagram	Signal flowgraph or block diagram

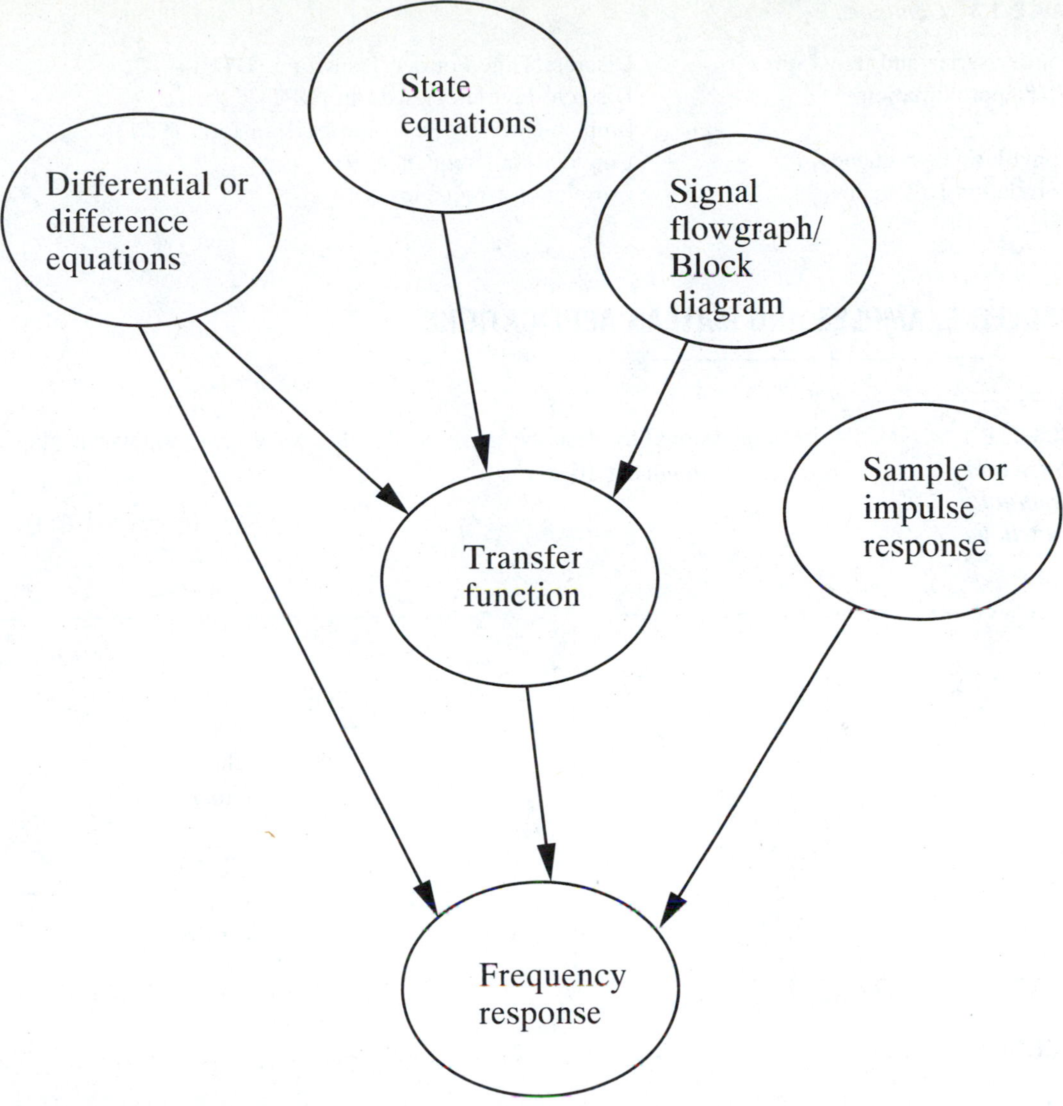

FIGURE 1.23 *Finding the frequency response from other system models*

current knowledge, the information in these tables and the figure may mean a lot or very little. We recommend that you periodically return to this section to gain perspective as you work your way through the following chapters.

TABLE 1.3 *Operations for linear systems*

Continuous systems	*Discrete systems*
Laplace (s) transform	z transform
Convolution integral (time)	Convolution sum (time)
Correlation integral (time)	Correlation sum (time)

Continues

TABLE 1.3 *Continued*

Fourier series and transform	Discrete Time Fourier Transform (DTFT)
DFT approximation	Discrete Fourier Transform (DFT) [implemented by Fast Fourier Transform (FFT)]
Convolution (frequency, f)	Convolution (frequency, θ)
Correlation (frequency, f)	Correlation (frequency, θ)

SOLVED EXAMPLES AND MATLAB APPLICATIONS

EXAMPLE 1.1 ***Pulse and Triangular Functions***

Write expressions to describe analytically the waveforms shown in Figures E1.1a through E1.1d.

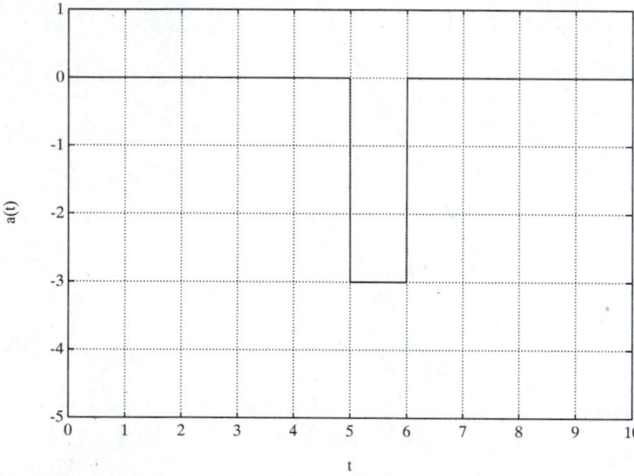

FIGURE E1.1a

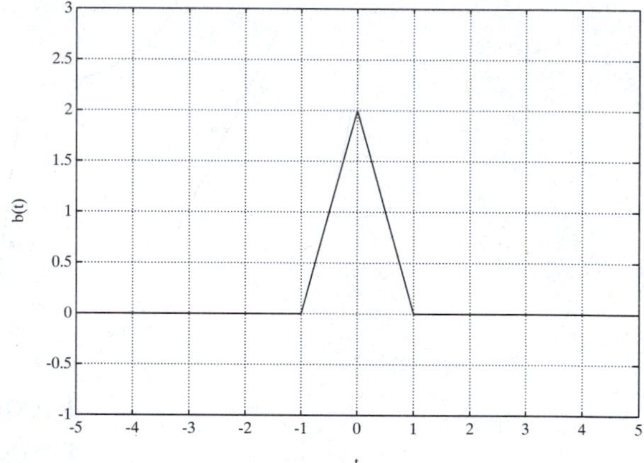

FIGURE E1.1b

Solution

a. A step that turns on at $t = 5$ and one that turns on at $t = 6$ are needed. The appropriate relationship is

$$a(t) = -3u(t - 5) + 3u(t - 6) = -3[u(t - 5) - u(t - 6)].$$

b. Here we need to turn on a ramp function having a slope of 2 at $t = -1$; we follow this by another ramp with a slope of -4 (-2 to "cancel" the positive-going ramp that started at $t = -1$, and -2 additional to make the overall waveform move downward) starting at $t = 0$ and, finally, by a ramp with a slope of $+2$ starting at $t = 1$ to cancel the net negative slope of -2 initiated at $t = 0$. The overall result is

$$b(t) = 2(t + 1)u(t + 1) - 4tu(t) + 2(t - 1)u(t - 1).$$

EXAMPLES

c. We could proceed as in part (b); however, it is easier to note that $c(t) = 2b(t - 4)$. Thus, by substituting $t - 4$ for t in $b(t)$ and scaling by a factor of 2, we obtain

$$\begin{aligned} c(t) &= 2b(t-4) \\ &= 2[2(t-4+1)u(t-4+1) - 4(t-4)u(t-4) \\ &\quad + 2(t-4-1)u(t-4-1)] \\ &= 2[2(t-3)u(t-3) - 4(t-4)u(t-4) + 2(t-5)u(t-5)]. \end{aligned}$$

d. Here again we could use the fact that $d(t)$ is a time- and magnitude-scaled time-shifted version of $b(t)$, but let's instead proceed from first principles, as in part (a). The result is

$$d(t) = 2.5[(t-4)u(t-4) - 2(t-6)u(t-6) + (t-8)u(t-8)].$$

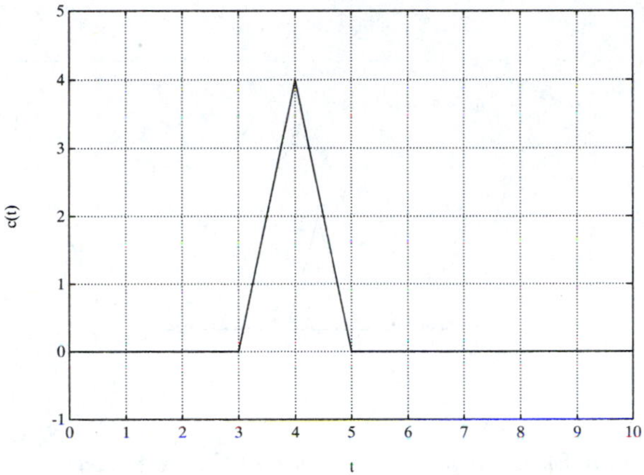

FIGURE E1.1c

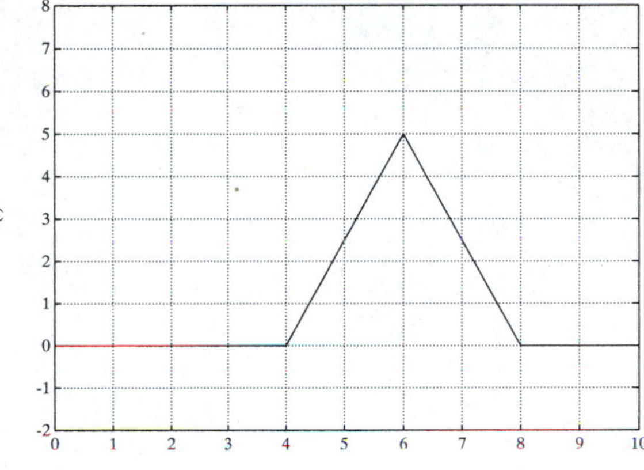

FIGURE E1.1d

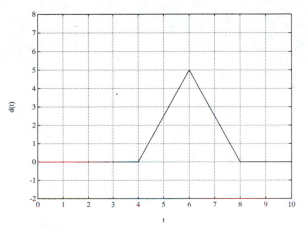

EXAMPLE 1.2
Sinusoidal Waveforms

a. The waveform shown in Figure E1.2a (page 22) has the form $f(t) = A\cos(\omega t + \phi)$. Find A, ω, and ϕ.

b. The waveform shown in Figure E1.2b has the form $g(t) = Ae^{at}\cos(\omega t + \phi)u(t)$. Find A, a, ω, and ϕ.

c. We want to compare the plot generated by the parameters identified in part (b) with the original plot. Write an appropriate MATLAB script and obtain the plot.

Solution

a. The original plot is shown in Figure E1.2c with some specific characteristics of interest added. In particular, we observe that the time separating corresponding points on the waveform is 1.0 time units. Thus the period of the wave is (assuming time is measured in seconds) $T = 1.0$ s and the

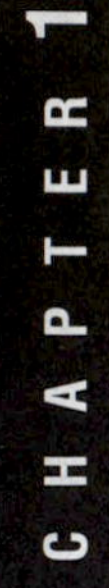

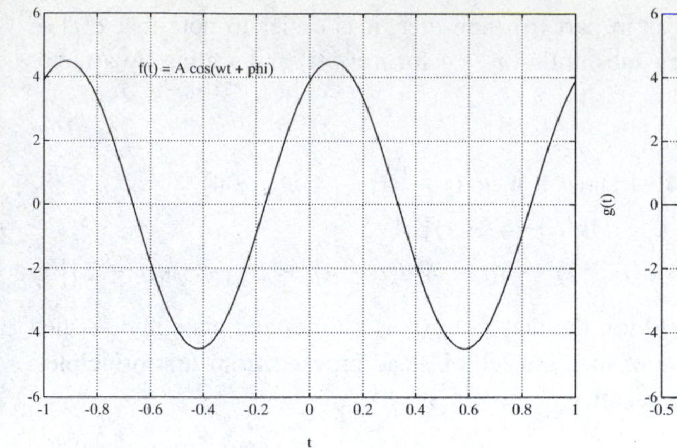

FIGURE E1.2a

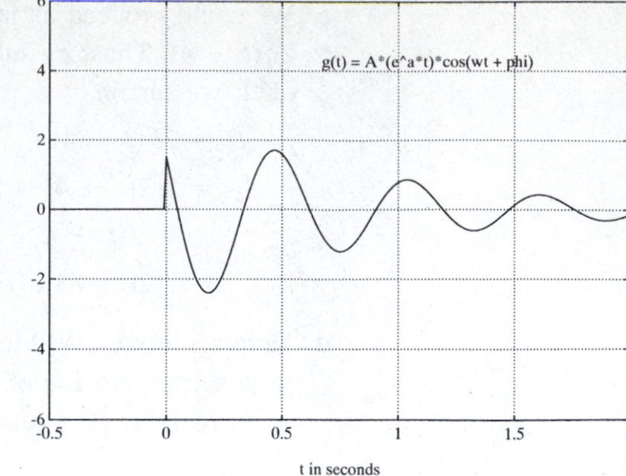

FIGURE E1.2b

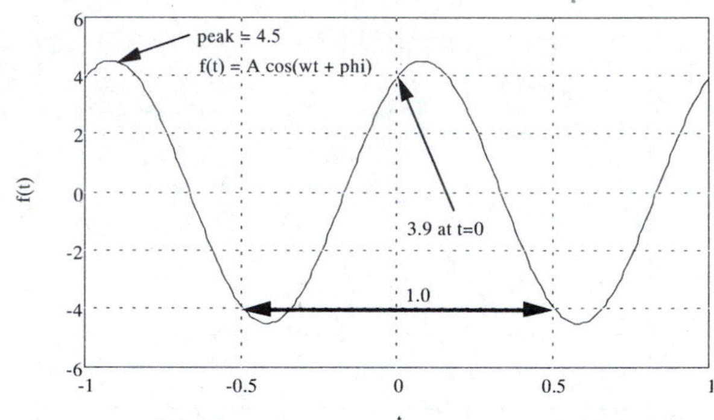

FIGURE E1.2c *Annotated version of Figure E1.2a*

frequency is $f = 1$ Hz. This means that $\omega = 2\pi f = 6.28$ rad/s. Since the peak value is 4.5, we know that $A = 4.5$. Finally, since $f(t)$ at $t = 0$ is 3.9, we have $3.9 = 4.5\cos(0 + \phi)$, or $\phi = \cos^{-1}(0.867) = \pm 0.522$ rad, which is approximately $\pm 30°$. Looking at the plot, we need to use $-30°$, giving $f(t) = 4.5\cos(6.28t - 30°)$.

b. A good starting point is first to find the angular frequency ω, which can be done by finding the period of the oscillation. Adjacent zero crossings are $T/2$ apart, where T is the period. From Figure E1.2b we estimate the value of T as 0.58 s, which gives $\omega = 2\pi/0.58 \approx 10.8$. Next, we observe that at any zero crossing t_i, it must be the case that $\cos(\omega t_i + \phi) = 0$. Having found ω and estimating the time of the third zero crossing as $t_1 = 0.62$ gives $\cos(6.7 + \phi) = 0$, or $\phi \approx 1.15$ rad, where we have selected the principal value of ϕ. Now let's find A by using the fact that at $t = 0$, $e^{a(0)} = 1$. Thus, $g(0) = A\cos(1.15)$, and from the plot $g(0) \approx 1.3$, so $A = 1.3/0.41 = 3.17$. Finally, we need to find a. We can do this by selecting any point on the curve, estimating the time and ordinate, and using the equation $g(t) = Ae^{at}\cos(\omega t + \phi)u(t)$. Using the first positive peak and denoting the time as t_p, we find that $t_p = 0.47$ s and $g(t_p) = 1.72$. To solve for a, we solve the equation $g(t_p) = Ae^{at_p}\cos(\omega t_p + \phi)$ for a, with the result

$$a = \left(\frac{1}{t_p}\right)\ln\left[\frac{g(t_p)}{A\cos(\omega t_p + \phi)}\right],$$

which gives $a \approx -1.3$. Thus the complete result is

$$g(t) = 3.17e^{-1.3t}\cos(10.8t + 1.15)u(t).$$

c. The MATLAB script and the resulting plot (Figure E1.2d, page 24) are given next.

Comment: For help in running scripts, see MATLAB: *An Overview* in Appendix A.

MATLAB Script

```
%E1_2 Sinusoidal waveforms

%E1_2d Plot of the results from the solution to E1.2b
ts=-0.5;                              % start time
tf=4;                                 % final time
dt=0.01;                              % time increment
t=ts:dt:tf;                           % time--start:increment:stop
tzro=0;                               % time step becomes 1
u=ustpfcnr(ts,tf,tzro,dt);            % A California function
gprime=3.17*exp(-1.3*t).*cos(10.8*t + 1.15).*u;

% NOTE the use of the .* operator. The terms 3.17*exp(-1.3*t),
% cos(10.8*t + 1.15), and u are all vectors.
% We want the components of these vectors to be multiplied by the
% corresponding components of the other
% vectors, hence the need to use .* rather than *

% The following statements plot the sequence and label the plot
plot(t,gprime);
axis([-.5,3,-3,2]);
title('Fig.E1.2d');
xlabel('t in seconds');
ylabel('gprime(t)');
grid;
```

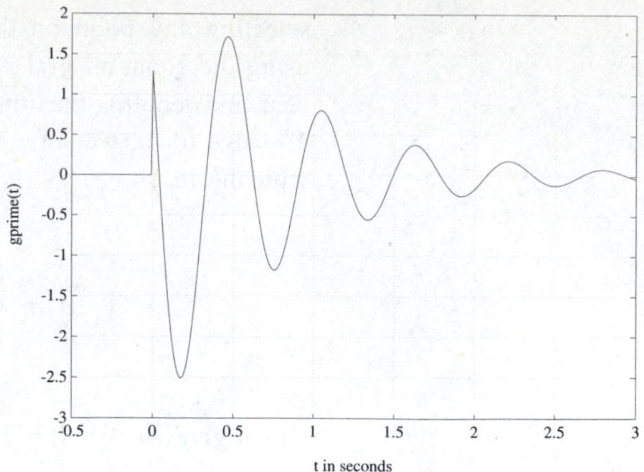

FIGURE E1.2d

A comparison of this plot with the one in Figure E1.2b shows them to be nearly identical. Indeed, if they are plotted on the same axes, only small differences are observed.

EXAMPLE 1.3
A Left-Sided Step Sequence

a. Describe analytically the sequence $f(n) = Cu(-n - n_0)$.

b. Write a MATLAB script to generate and plot this sequence for $C = 2$ and $n_0 = 3$.

Solution

a. From the definition of the unit step sequence, $u(-n - n_0)$ is 1 whenever $-n - n_0 \geq 0$ and it is 0 otherwise. Since $-n - n_0 \geq 0$ when $-n_0 \geq n$, we have

$$Cu(-n - n_0) = \begin{cases} C, & -n_0 \geq n \\ 0, & -n_0 < n. \end{cases}$$

b. The MATLAB script and the resulting plot (Figure E1.3) are as follows.

Comment: The graphics function **stem** will also produce discrete (or stem) plots such as Figure E1.3.

MATLAB Script

```
%E1_3 Left-sided step sequence
n=-10:1:10;                        % n=[-10 -9...9 10]
f=zeros(size(n));                  % f=[0(at n=-10) 0...0 0(at n=10)]
f(1:8)=2*ones(size(n(1:8)));       % Makes first 8 entries of f=2.0, i.e.,
                                   % f=[2 2 2 2 2 2 2 2(at n=-3) 0 0...0]
xlab='sample number, n';
ylab='f(n)';
ptitle='Fig.E1.3 Left-sided step sequence';
axis([-10, 10, -1,3]);
displot(n, f, xlab, ylab, ptitle);
text(2, 1.7, 'x(n)=2u(-n-3)');
grid;
```

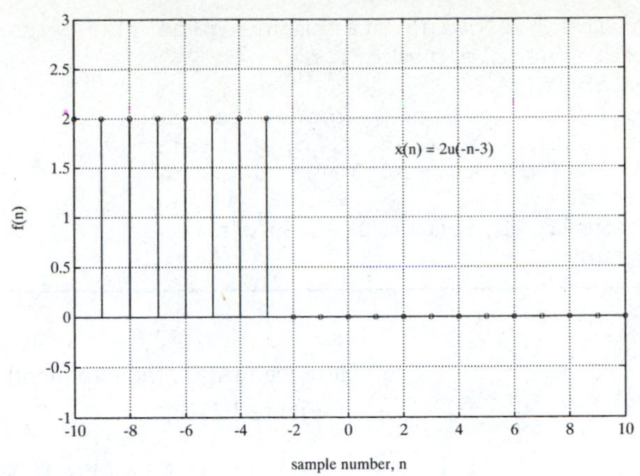

FIGURE E1.3 *Left-sided step sequence*

EXAMPLE 1.4
An Alternating Exponential Sequence

a. Write a MATLAB script to generate and plot the sequence $f(n) = A(a)^n$ with $A = 10$ and $a = -0.9$ for $-10 \le n \le 10$.

b. Write a MATLAB script to generate and plot the sequence $g(n) = A(a)^n u(n)$ with $A = 10$ and $a = -0.9$ for $-10 \le n \le 20$.

Solution

a. The MATLAB script and the resulting plot (Figure E1.4a) are as follows:

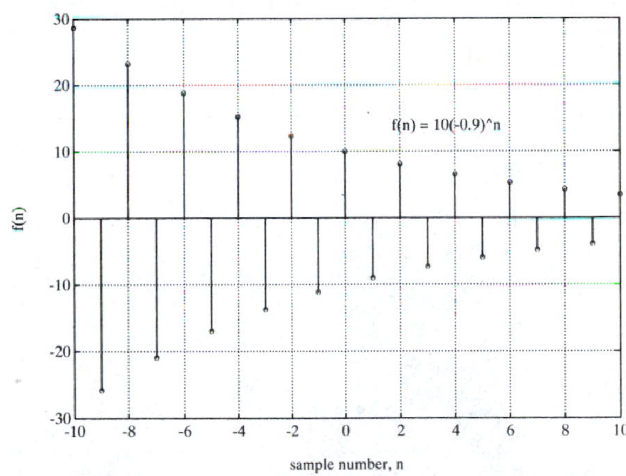

FIGURE E1.4a *Exponential sequence*

MATLAB Script

```
%E1_4 Alternating exponential sequences

%E1_4a
n=-10:1:10;                                    % n=[-10 -9 . . . 9 10]
f=10*(-0.9).^n; % f=[10*(-0.9)^(-10)...10*(-0.9)^10]
```

Continues

```
% The following statements plot the sequence and label the plot
xlab='sample number, n';
ylab='f(n)';
ptitle='Fig.E1.4a Exponential sequence';
displot(n, f, xlab, ylab, ptitle);
axis([-10, 10, -30,30]);
grid;
text(2, 12, 'f(n)=10(-0.9).^n');
pause;
```

b. Here we turn on the exponential at $n = 0$, with the result shown in the plot of Figure E1.4b.

MATLAB Script

```
%E1_4b Exponential sequence multiplied by a step sequence
n=-10:1:20;                     % n=[-10 -9...19 20]
f=10*(-0.9).^n;                 % f=[10*(-0.9)^(-10)...10*(-0.9)^10]
u=zeros(size(n));               % u=[0 0 ... 0]
u(11:31)=ones(size(n(11:31))); % u=[0...0(at n=-1) 1 1...1]
g=f.*u;  % Makes the first 10 elements of g equal to 0 and
% Makes the final 21 elements of g
% equal to 10(-0.9).^n, i.e., g=[0...0(n=-1) 10...10(-0.9)^20]
% Note the use of the .* operator--* won't work(try it and see)
% The following statements plot the sequence and label the plot
xlab='sample number, n';
ylab='g(n)';
ptitle='Fig.E1.4b Exponential sequence multiplied by step sequence';
displot(n, g, xlab, ylab, ptitle);
axis([-10, 20, -10,10]);
grid;
text(8, 7, 'g(n)=[10(-0.9).^n]u(n)');
```

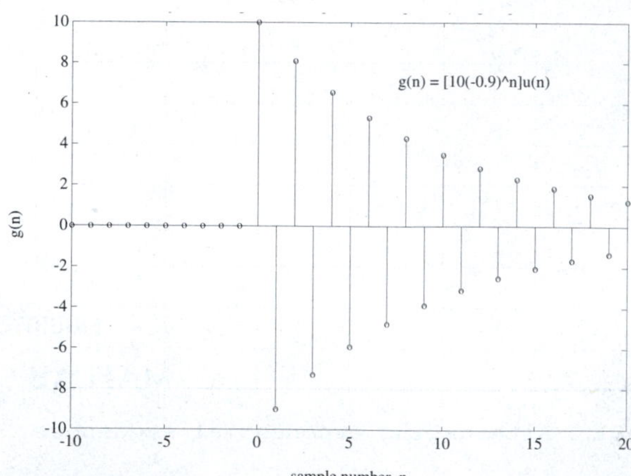

FIGURE E1.4b *Exponential sequence multiplied by step sequence*

EXAMPLE 1.5
Analytical Expressions for Pulse Sequences

a. Write an analytical expression for the pulse sequence $e(n)$ shown in Figure E1.5.

b. Write an analytical expression for the pulse sequence $f(n)$ that is 3 for $9 \le n \le 15$ and 0 otherwise;

c. Write an analytical expression for a sequence $g(n)$ that is 2.0 for $-3 \le n \le 3$ and for $12 \le n \le 18$ and 0 otherwise.

Solution

a. The pulse is to turn on at $n = -3$ and should turn off at $n = 4$; thus we have $e(n) = u(n + 3) - u(n - 4)$.

b. This pulse is to turn on at $n = 9$ and should turn off at $n = 16$, so the result is $f(n) = 3[u(n - 9) - u(n - 16)]$.

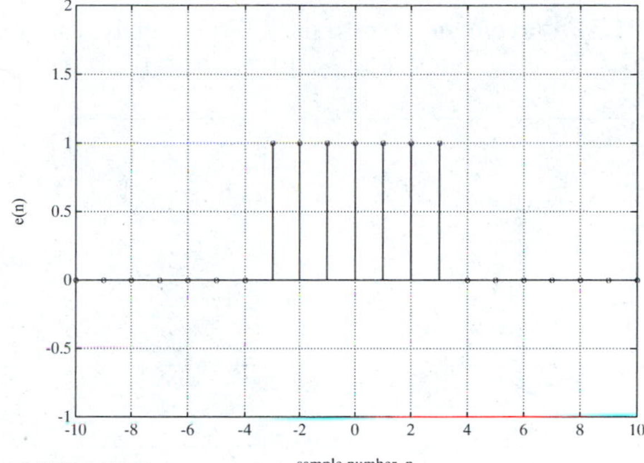

FIGURE E1.5

Notice also that $f(n)$ can be expressed as a scaled and shifted version of $e(n)$:

$$\begin{aligned} f(n) &= 3e(n - 12) = 3[u(n - 12 + 3) - u(n - 12 - 4)] \\ &= 3[u(n - 9) - u(n - 16)]. \end{aligned}$$

c. Here we can represent $g(n)$ as the sum of a scaled version of $e(n)$ and a scaled and shifted version of $e(n)$:

$$\begin{aligned} g(n) &= 2e(n) + 2e(n - 15) \\ &= 2[u(n + 3) - u(n - 4)] + 2[u(n - 12) - u(n - 19)]. \end{aligned}$$

Comment: The approach illustrated here of generating a sequence as a summation of shifted and scaled sequences can frequently be used to advantage.

REINFORCEMENT PROBLEMS

P1.1 Sketching step and pulse waveforms. Sketch the following waveforms.

a. $u(t - 4)$
b. $-2u(t - 1)$
c. $1.3u(t + 6)$
d. $u(t - 2) - u(t - 5)$

P1.2 Sketching waveforms. Sketch the following waveforms.

a. $-2u(t + 1) + 3u(t) - u(t - 2)$
b. $2.5t[u(t) - u(t - 2)]$
c. $-2.5t[u(t + 2) - u(t)]$
d. $-(t + 4)u(t + 4) + (t + 2)u(t + 2) + (t - 2)u(t - 2) - (t - 4)u(t - 4)$

P1.3 Waveform synthesis. Write analytical expressions for the waveforms shown in Figures P1.3a–P1.3d.

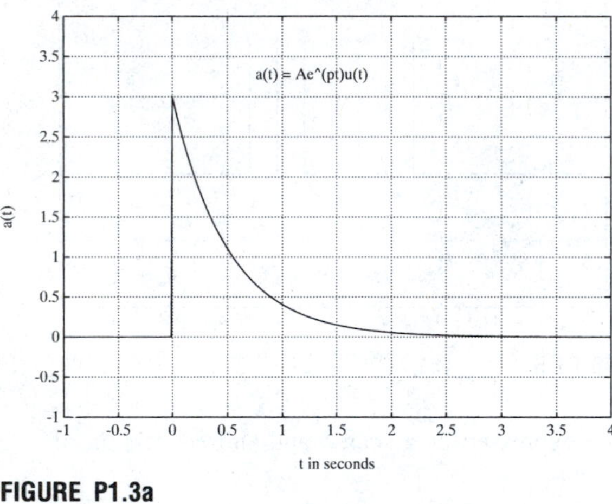

FIGURE P1.3a

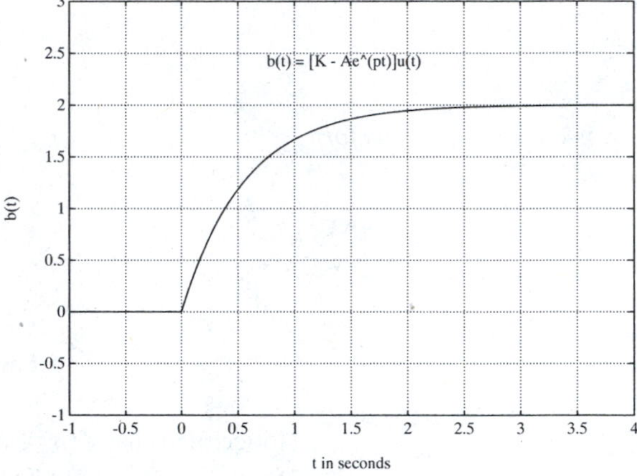

FIGURE P1.3b

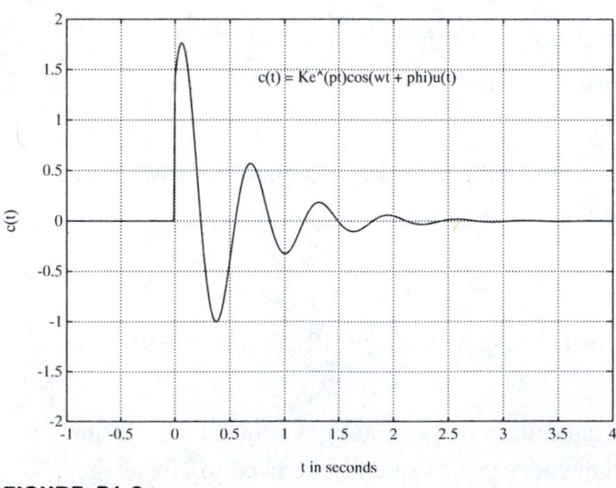

FIGURE P1.3c

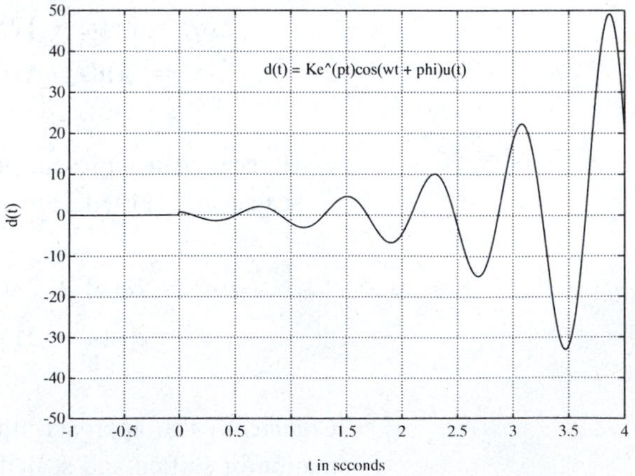

d(t) = Ke^(pt)cos(wt + phi)u(t)

FIGURE P1.3d

P1.4 Waveform synthesis. Write analytical expressions for the waveforms shown in Figure P1.4.

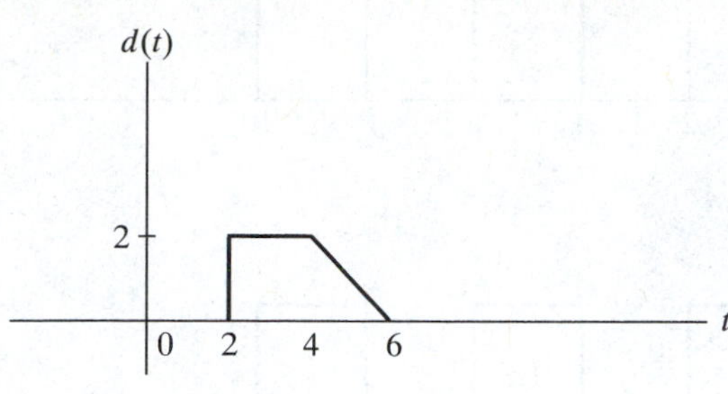

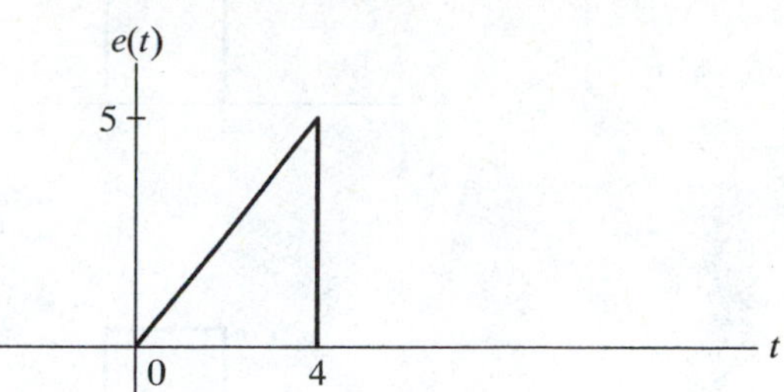

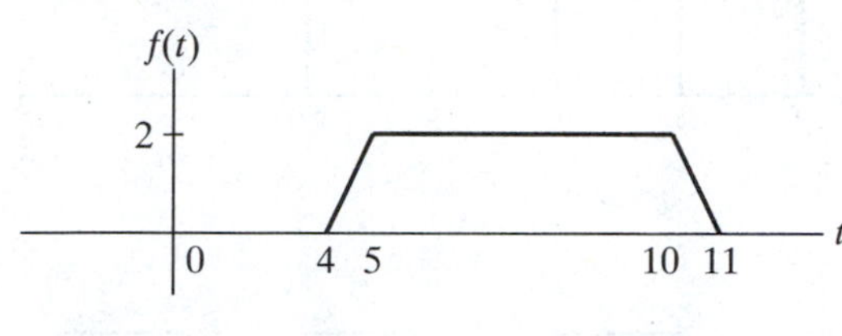

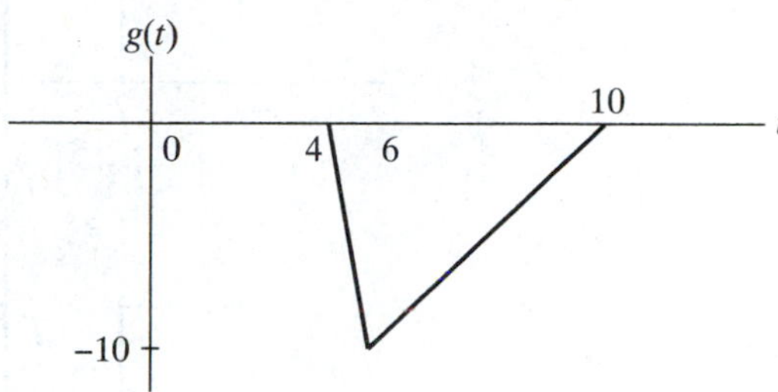

FIGURE P1.4 *Waveform synthesis*

P1.5 Describing periodic waveforms. Write analytical expressions to describe the periodic waveforms $d_p(t)$, $e_p(t)$, $f_p(t)$, and $g_p(t)$ shown in Figure P1.5 (page 30).

P1.6 MATLAB generation of waveforms. Write MATLAB scripts to generate and plot the waveforms of Problem P1.1.

P1.7 MATLAB generation of waveforms. Write MATLAB scripts to generate and plot the waveforms of Problem P1.2.

P1.8 MATLAB generation of waveforms. Write MATLAB scripts to generate and plot the following waveforms.

a. $d(t) = 3e^{-2t}u(t)$

b. $e(t) = 2[1 - e^{-1.8t}]u(t)$

c. $f(t) = 2e^{-1.8t}\cos(10t - \pi/4)u(t)$

d. $g(t) = e^{-1.01t}\cos(8t + \pi/5)u(t)$

P1.9 Sketching step and pulse sequences. Sketch the sequences

a. $3u(n - 4)$

b. $-2u(n - 1)$

c. $1.3u(n + 6)$

d. $u(n - 2) - u(n - 5)$

Compare your results with those of Problem P1.1.

P1.10 Sketching sequences. Sketch each sequence.

a. $4u(n - 3) - 2(n - 6)u(n - 6) + 2(n - 8)u(n - 8)$

b. $4[u(n - 3) - u(n - 6)] - 2(n - 6)[u(n - 6) - u(n - 8)]$

c. $2(n + 5)u(n + 5) - 3nu(n) + (n - 10)u(n - 10)$

d. $2(n + 5)[u(n + 5) - u(n - 10)] - 3n[u(n) - u(n - 10)]$

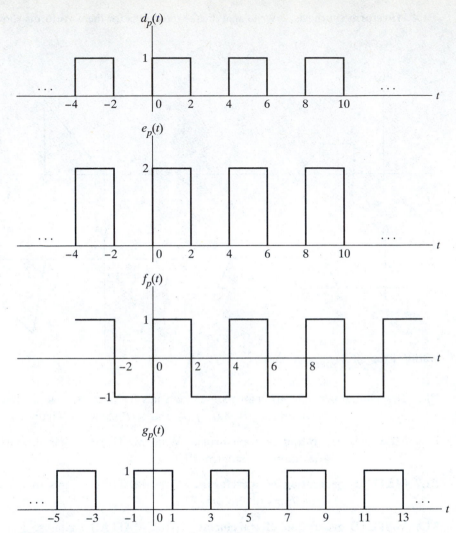

FIGURE P1.5 *Describing periodic waveforms*

P1.11 Sequence synthesis.

a. Describe the sequence shown in Figure P1.11 as the sum of subsequences consisting of (i) ramp sequences "turned on" at $n = 0, 2, 6$, and 8, respectively, or (ii) two triangular sequences and a rectangular pulse sequence.
b. Show that the two forms obtained in part (a) are equivalent.

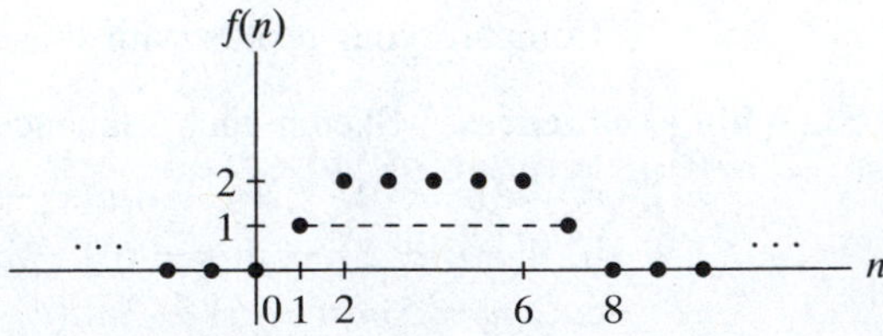

FIGURE P1.11 *Sequence synthesis*

P1.12 Sequence synthesis. Write analytical expressions for the sequences shown in Figure P1.12.

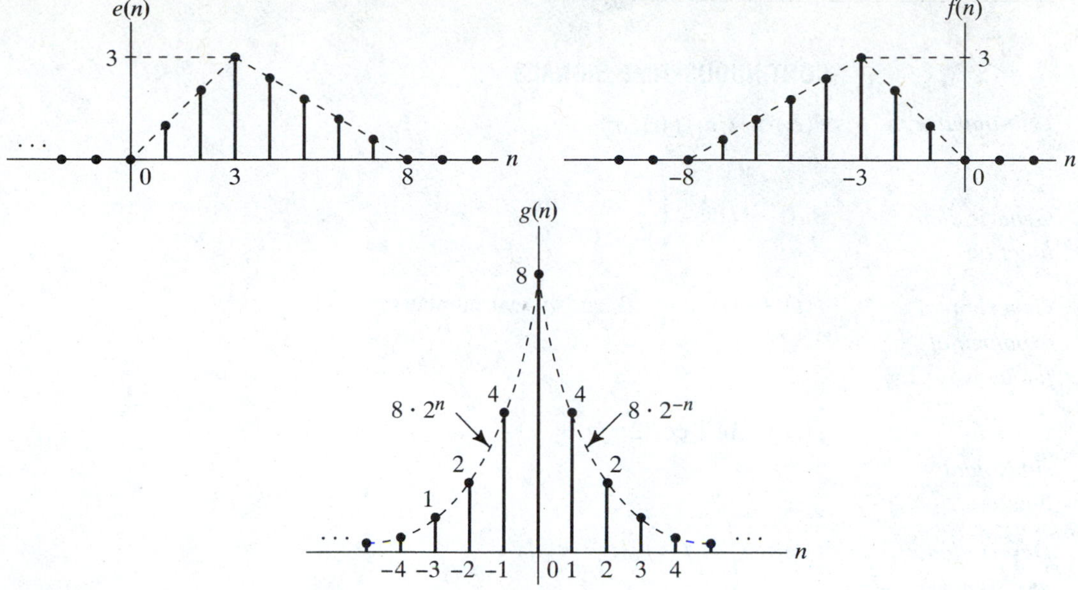

FIGURE P1.12 *More sequence synthesis*

P1.13 Describing periodic sequences. Write analytical expressions to describe the periodic sequences $f_p(n)$ and $g_p(n)$ shown in Figure P1.13.

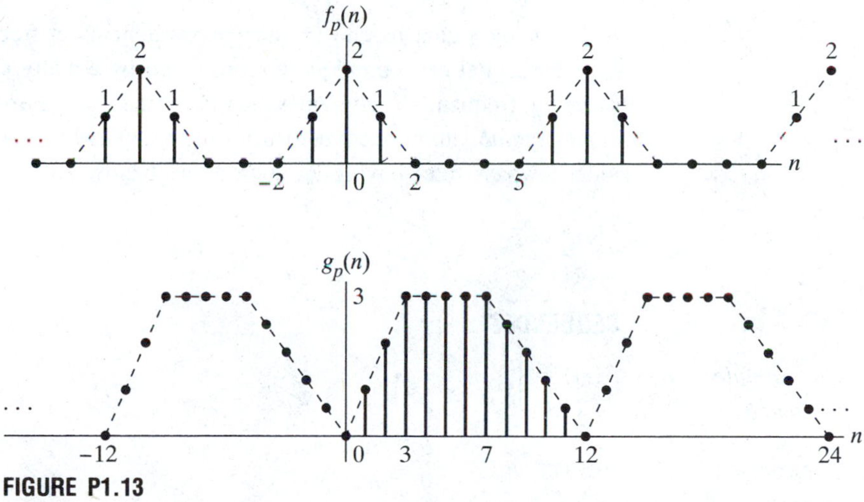

FIGURE P1.13

P1.14 MATLAB generation of sequences. Write MATLAB scripts to generate and plot the sequences of Problem P1.9.

P1.15 MATLAB generation of pulse sequences. Write MATLAB scripts to generate and plot the sequences of ProblemP1.10.

P1.16 MATLAB generation of pulse sequences. Write MATLAB scripts to generate and plot the sequences of Example 1.5.

CHAPTER 1

DEFINITIONS, TECHNIQUES, AND CONNECTIONS

CONTINUOUS-TIME SIGNALS

Unit impulse $\delta(\sigma) = \lim_{\Delta \to 0} f(\Delta, \sigma)$

Generic step function $Bu(t - t_0) = \begin{cases} B, & t \geq t_0 \\ 0, & t < t_0 \end{cases}$

Generic real exponential function $f(t) = Ae^{at}$, A and a real numbers

Generic sinusoidal function $f(t) = Ae^{at} \cos(2\pi ft + \alpha)$

Description of any signal $f(t) = \int_{-\infty}^{\infty} f(\tau)\delta(t - \tau)\, d\tau$

SAMPLING THEOREM

If an analog signal has no frequency components at frequencies greater than $f_{\max}$, the signal can be uniquely represented by equally spaced samples if the sampling frequency f_s is greater than two times $f_{\max}$. Furthermore, the original analog signal can be recovered from the samples by passing them through an ideal lowpass filter having an appropriate bandwidth.

SEQUENCES

Unit sample sequence $\delta(m) = \begin{cases} 1, & m = 0 \\ 0, & m \neq 0 \end{cases}$

Generic step sequence $Bu(n - n_0) = \begin{cases} B, & n \geq n_0 \\ 0, & n < n_0 \end{cases}$

Generic real exponential sequence $f(n) = A(a)^n$, A and a real numbers, n an integer

Generic sinusoidal sequence $g(n) = A(a)^n \cos\left(\dfrac{2\pi n}{N} + \alpha\right)$

Description of any sequence $$f(n) = \sum_{m=-\infty}^{\infty} f(m)\delta(n - m)$$

MATLAB FUNCTIONS USED

Function	*Purpose and Use*	*Toolbox*
displot	Plots discrete functions.	California Functions
plot	Provides continuous or discrete plots.	MATLAB

ANNOTATED BIBLIOGRAPHY

1. Gabel, Robert A., and Richard A. Roberts, *Signals and Linear Systems, 3rd ed.,* John Wiley & Sons, New York, 1987. *Chapter One includes a discussion of classification of systems, linearity, and properties and models of discrete-time systems.*

2. Oppenheim, Alan V., and Ronald W. Schafer, *Discrete-Time Signal Processing,* Prentice Hall, Inc., Englewood Cliffs, N.J., 1989. *This book is a new version of the landmark work* Digital Signal Processing *that was published by the same authors in 1975. Although* Discrete-Time Signal Processing *is used primarily in senior/graduate-level courses, the discussion on quantization illuminates some practical aspects of digital processing of signals that are of interest when studying introductory linear systems.*

3. Peled, Abraham, and Bede Liu, *Digital Signal Processing, Theory, Design, and Implementation,* John Wiley & Sons, New York, 1976. *Chapter One includes a discussion of the basic principles and characteristics of analog-to-digital (A/D) and digital-to-analog (D/A) conversions. Chapter Four is devoted to the hardware implementation of digital signal processors. The first two sections of this chapter review the basics of binary arithmetic and digital hardware.*

4. Soliman, Samir S., and Mandyam D. Srinath, *Continuous and Discrete Signals and Systems,* Prentice Hall, Inc., Englewood Cliffs, N.J., 1990. *Continuous-time signals are defined and described in Chapter One, with similar treatment accorded discrete-time signals in Chapter Six.*

ANSWERS

P1.1 See Figure A1.1.

P1.2 See Figure A1.2.

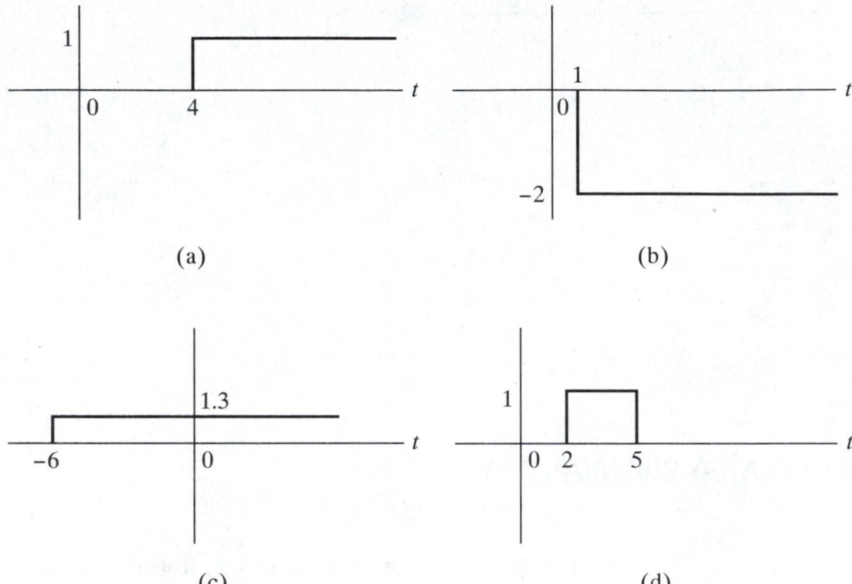

FIGURE A1.1

P1.3 $a(t) = 3e^{-2t}u(t)$; $b(t) = 2[1 - e^{-1.8t}]u(t)$;
$c(t) = 2e^{-1.8t}\cos(10t - \pi/4)u(t)$; $d(t) = e^{1.01t}\cos(8t + \pi/5)u(t)$

P1.4 $d(t) = 2u(t - 2) - (t - 4)u(t - 4) + (t - 6)u(t - 6)$;
$e(t) = 1.25t[u(t) - u(t - 4)]$; $f(t) = 2(t - 4)u(t - 4) - 2(t - 5)u(t - 5) - 2(t - 10)u(t - 10) + 2(t - 11)u(t - 11)$;
$g(t) = -5(t - 4)u(t - 4) + 7.5(t - 6)u(t - 6) - 2.5(t - 10)u(t - 10)$

P1.5

$$d_p(t) = \sum_{m=-\infty}^{\infty} [u(t - 4m) - u(t - 2 - 4m)];\ e_p(t) = 2d_p(t);$$

;

$$f_p(t) = e_p(t) - 1;\ g_p(t) = \sum_{m=-\infty}^{\infty} [u(t + 1 - 4m) - u(t - 1 - 4m)]$$

P1.6 See A1_6 on the CLS disk.

P1.7 See A1_7 on the CLS disk.

P1.8 See A1_8 on the CLS disk.

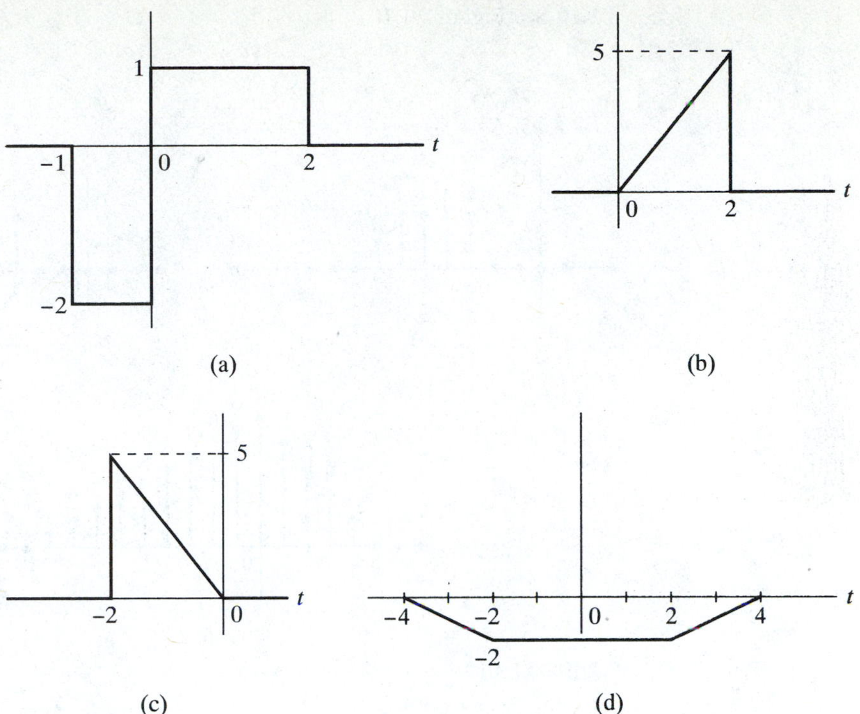

FIGURE A1.2

P1.9 See Figure A1.9.

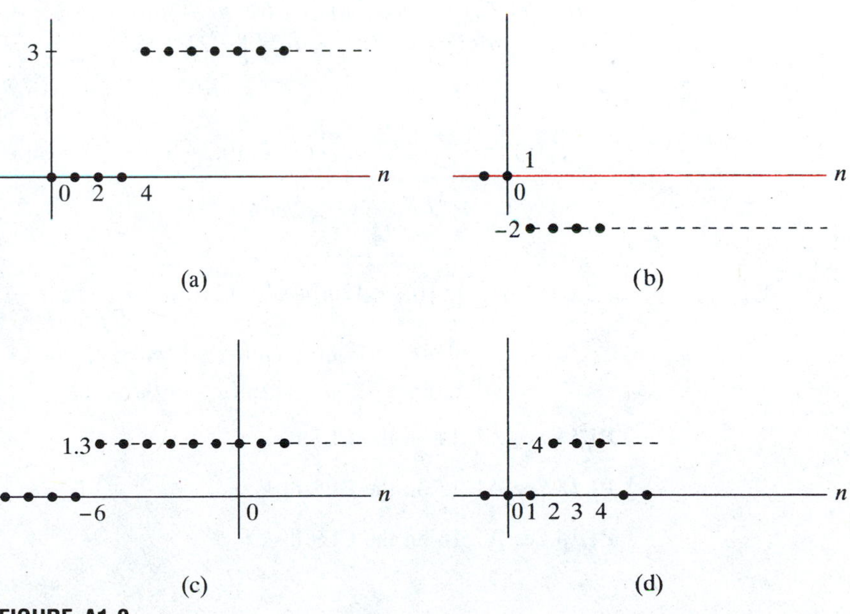

FIGURE A1.9

P1.10 See Figure A1.10.

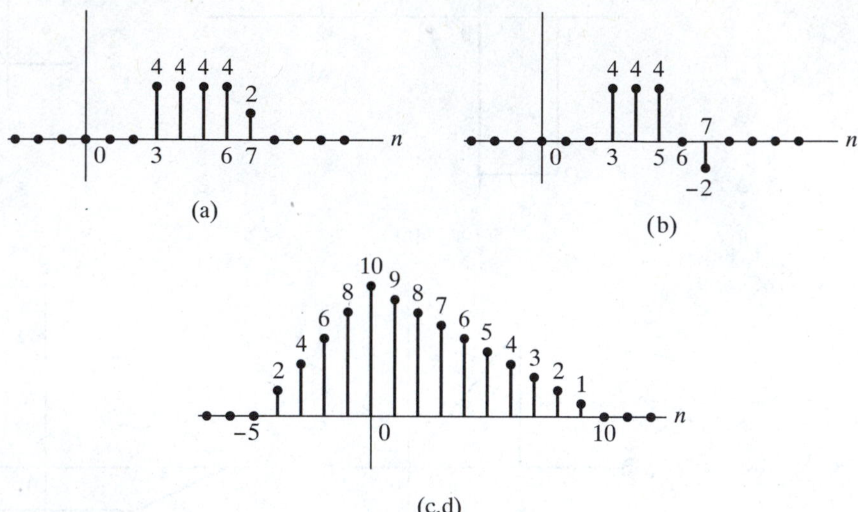

FIGURE A1.10

P1.11 a. (i) $f(n) = nu(n) - (n - 2)u(n - 2) - (n - 6)u(n - 6) + (n - 8)u(n - 8)$; (ii) $f(n) = n[u(n) - u(n - 2)] + 2[u(n - 2) - u(n - 6)] + (-n + 8)[u(n - 6) - u(n - 8)]$

P1.12 $e(n) = nu(n) - 1.6(n - 3)u(n - 3) + 0.6(n - 8)u(n - 8)$; $f(n) = e(-n) = -nu(-n) - 1.6(-n - 3)u(-n - 3) + 0.6(-n - 8)u(-n - 8)$; $g(n) = 8 \cdot 2^{-|n|}$

P1.13

$$f_p(n) = \sum_{m=-\infty}^{\infty} [2(n - 7m + 2)u(n - 7m + 2) - 4(n - 7m)u(n - 7m) + 2(n - 7m - 2)u(n - 7m - 2)]$$

$$g_p(n) = \sum_{m=-\infty}^{\infty} [(n - 12m)u(n - 12m) - (n - 12m - 3)u(n - 12m - 3) - 0.6(n - 12m - 7)u(n - 12m - 7) + 0.6(n - 12m - 12)u(n - 12m - 12)]$$

P1.14 See A1_14 on the CLS disk.

P1.15 See A1_15 on the CLS disk.

P1.16 See A1_16 on the CLS disk.

CHAPTER 2

Continuous Systems

PREVIEW

A continuous-time system receives an input signal $x(t)$ and generates an output signal $y(t)$, as illustrated in Figure 2.2, where the block labeled "continuous-time system" could, for example, represent a filter to eliminate unwanted signals in an instrumentation system or an autopilot used to keep an aircraft on a prescribed course. On a more familiar level, perhaps, are systems such as an automobile, the express elevator in an urban skyscraper, and the heating or cooling system in your living space. Systems are everywhere. According to one popular dictionary,[1]

FIGURE 2.1 *Leonhard Euler*

[1] *The Concise Oxford Dictionary, 1st ed.,* Oxford Press, 1911.

a system is defined as a "complex whole set of connected things or parts, organized body of material or immaterial things." We will generally live with this definition.

To aid in analyzing and designing systems, mathematical models are formulated. In this chapter we introduce several models for continuous systems and discuss techniques for system design and analysis by analytical methods. We begin by investigating the importance and application of system properties such as linearity, time invariance, causality, and stability. Systems having both the properties of linearity and time invariance are described as the important class of linear time-invariant (LTI) systems, which we shall concentrate on throughout this book.

Next, LTI systems are described by linear constant-coefficient differential equations (DEs), with examples to illustrate the general nature of these systems. The characteristic equation and its roots are determined and then used in the initial condition (IC) solution. Finding the forced solution, a difficult task when using time-domain procedures, is postponed until Chapter 3, "Laplace Transforms and Applications."

The unit impulse response, which plays a significant role in the analysis of linear systems, is defined and then applied in determining the response of a system to arbitrary inputs through the operation of convolution. Several different methods of performing convolution are presented, including taking advantage of the ubiquitous computer. Convolution is then used to develop the steady-state system response to the pervasive sinusoidal signal.

The chapter concludes with a discussion of the state-space or state-variable model for a linear system. Formal methods of solution of these state equations are postponed until Chapter 6, but solution using a MATLAB function is illustrated.

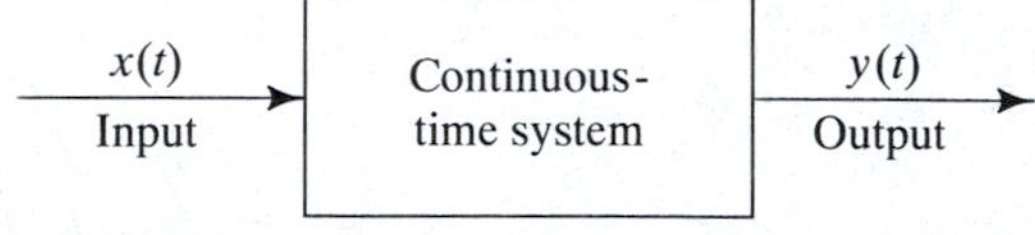

FIGURE 2.2 *System block diagram*

PROPERTIES

Let us now investigate some properties and characteristics of continuous systems.

Linearity One of the most important concepts in system theory is linearity. Many accepted forms of mathematical analysis are permitted with a linear system that allow for easier and more convenient methods of analysis and design. In addition, in many situations, practical engineering systems may be modeled as being linear. Linear systems possess the property of superposition, which leads to the following definition: *A system is linear if and only if it satisfies the principle of homogeneity and the principle of additivity.*

1. *Homogeneity.* In Figure 2.3a on page 40 we see that an input $x_1(t)$ applied to a linear system produces the output $y_1(t)$. When a new input $x(t) = C_1x_1(t)$ is applied to the linear system, the output is $y(t) = C_1y_1(t)$. That is, if the input is scaled by the complex constant C_1, the output will be scaled by the same constant. This is the homogeneity property.
2. *Additivity.* In Figure 2.3b $x_1(t)$ again produces $y_1(t)$ and $x_2(t)$ produces $y_2(t)$ in our same linear system. If a new input $x(t) = x_1(t) + x_2(t)$ is applied to the linear system, the output is $y(t) = y_1(t) + y_2(t)$. That is, the output due to the sum of two (or several) inputs is equal to the sum of the outputs produced by each separate input.
3. *Homogeneity and additivity.* Again with $x_1(t) \rightarrow y_1(t)$ and $x_2(t) \rightarrow y_2(t)$ as in Figure 2.3c, we make the input $x(t) = C_1x_1(t) + C_2x_2(t)$, combining the preceding items (1) and (2). In a linear system the output will be $y(t) = C_1y_1(t) + C_2y_2(t)$. Thus, linear continuous systems satisfy homogeneity and additivity. When both of these principles are satisfied, a system is said to satisfy the principle of superposition. A continuous system, therefore, is linear if and only if it satisfies the principle of superposition. A system that does not satisfy the principle of superposition is called *nonlinear*.
4. *Observations on linearity.* Linearity is a property of systems but it is also a property of mathematical operators, such as Laplace and Fourier transforms used to analyze systems. Indeed, linearity is necessary (but not in itself sufficient) to provide an important benefit of Laplace transforms in the transformation of linear differential equations to linear algebraic equations. Physically, linearity also has some very important consequences: (a) If we increase the amplitude

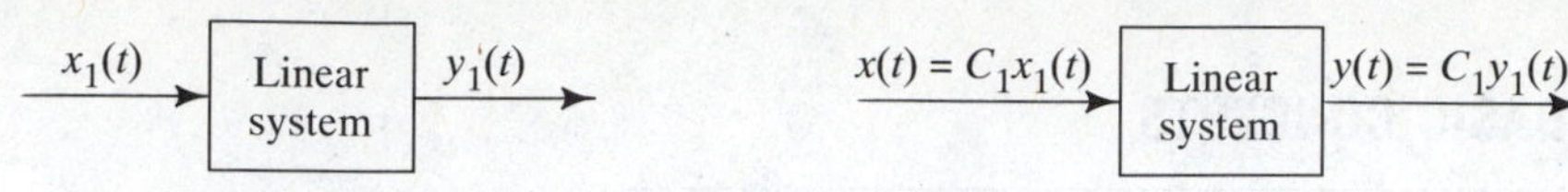

a. Principle of homogeneity

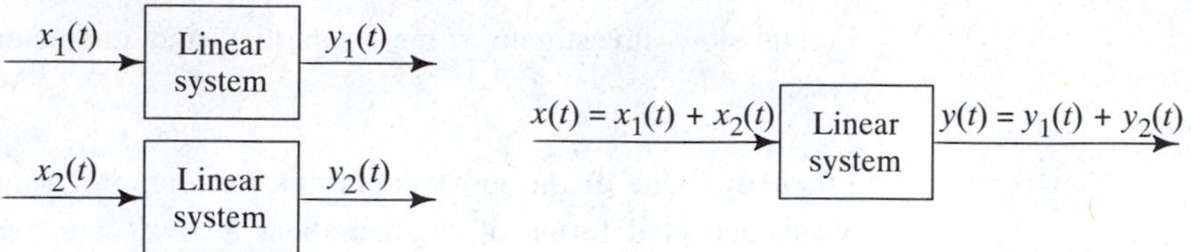

b. Principle of additivity

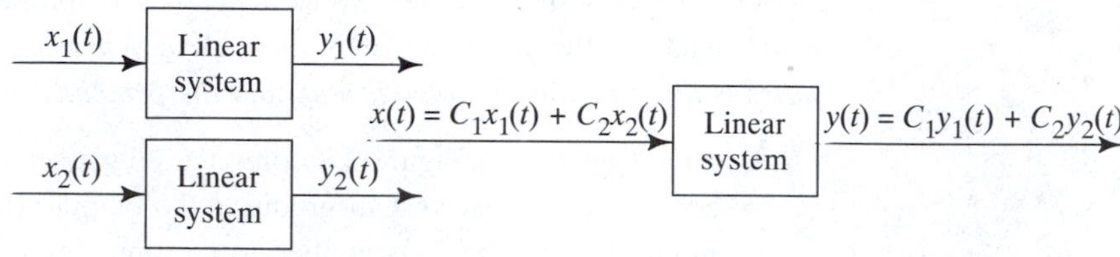

c. Principle of superposition

FIGURE 2.3 *Linearity*

of a linear system's input by a factor F, the output is also amplified by the same factor F but otherwise retains the same shape. (b) If we have the output responses $y_1(t)$, $y_2(t)$ of a system to two inputs $x_1(t)$, $x_2(t)$, linearity guarantees that exactly the same terms will appear in the output, perhaps in different amounts, if we apply an input $x(t) = C_1x_1(t) + C_2x_2(t)$. (c) Linearity also implies that if a system has several inputs or if it has one input that can be represented as a sum of signals, we can find the outputs due to each input separately and then simply add them.

Time Invariance Suppose an input $x_1(t)$ produces an output $y_1(t)$. Consider a second input $x_2(t)$ that is a time-shifted version of $x_1(t)$—that is,

shifted input

$$x_2(t) = x_1(t - t_0).$$

If the output $y_2(t)$ caused by $x_2(t)$ is a shifted version of $y_1(t)$—that is,

shifted output

$$y_2(t) = y_1(t - t_0)$$

for arbitrary $x_1(t)$ and t_0 and for all t—then the system is said to be a *time-invariant system.* Loosely speaking, the system parameters do not change with time. That is, the same input applied at different times will produce outputs that are identical in shape and size but shifted in time. A pictorial description of time invariance is given in Figure 2.4.

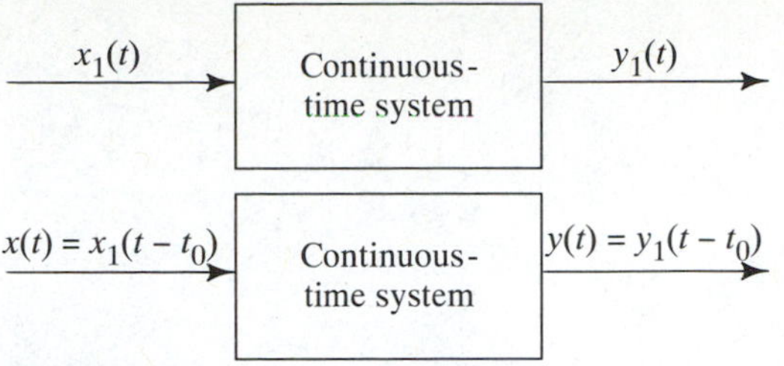

FIGURE 2.4 *Time invariance*

Linear Time-Invariant Systems When a system is both linear and time invariant, it is called a linear time-invariant (LTI) system and it is amenable to analysis using many techniques. In particular, Laplace and Fourier transforms are frequently used. LTI systems are typically described in the time domain by linear differential equations with constant coefficients. In transform domains (Laplace or Fourier) the equations are *linear* and *algebraic* in form. This means that we can use our (and the computer's) facility for doing algebra to study LTI systems.

Causality A causal (not casual!) system is a nonpredictive system in that the output does not precede, or anticipate, the input. See Figure 2.5 on page 42 for the output signals of causal, noncausal, and anticausal systems, respectively, where the input signal is the impulse $x(t) = A\,\delta(t)$. More is said about these three different kinds of systems in the next chapter.

Stability We normally do not design engineering systems that are inherently unstable, because we want (and expect), for example, a stable flight in a helicopter flying from one heliport to another. When several stable systems are connected together to form a larger single system, however, there is no guarantee that the overall system made up of the subsystems will be stable. In this book, we will define system stability from several points of view, with the first being in terms of input-output behavior. If the system input is bounded (the input magnitude does not approach $\pm\infty$) and if the system is *stable,* then the output must also be bounded.[2] This stability test is known as the bounded-input–bounded-output (BIBO) criterion. This is a difficult test to perform practically, for it implies that to be really certain that a system is stable, all possible bounded inputs must be applied if the output continues to meet the bounded test. It is quite difficult to have sufficient time to apply all possible inputs.

[2] It isn't always true that stable systems are more desirable than unstable ones. If the system of interest is our bank savings account, most of us would prefer it to be unstable.

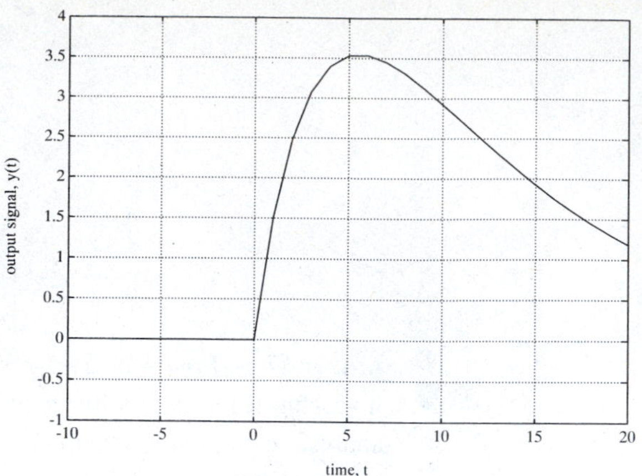

a. Causal system

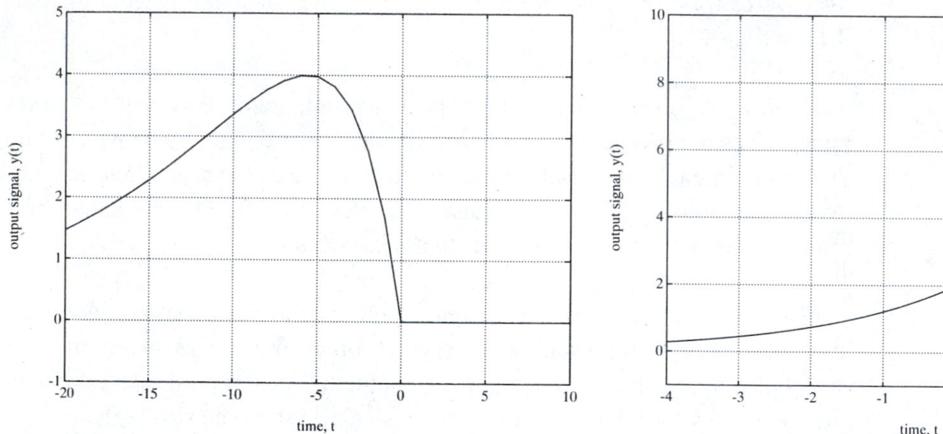

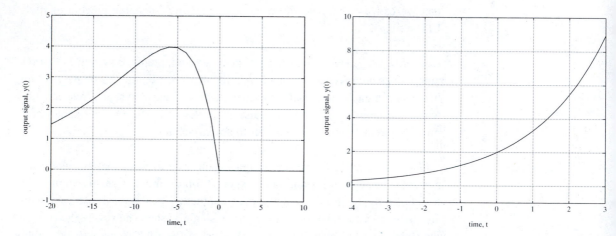

b. Anticausal system

c. Noncausal system

FIGURE 2.5 *Unit impulse responses of causal, anticausal, and noncausal systems*

ILLUSTRATIVE PROBLEM 2.1
System Properties

a. A continuous system is described by the input-output equation $y(t) = Kx(t) + A$, where K and A are real constants. (i) Investigate this system for linearity using the tests illustrated in Figure 2.3. (ii) Is this system time invariant or time varying?

b. A continuous-time integrator is mathematically defined by

$$y(t) = \int_0^t x(\tau)\,d\tau + y(0), \; t \geq 0$$

where $y(t)$ is the integrator output, $y(0)$ is the initial value of the integrator output, and $x(t)$ is the input. Is an integrator causal or noncausal?

c. The relationship $y(t) = t|x(t)|^2$ describes the input-output relationship of a continuous system. Is this a stable system?

Suggestion: A step function $x(t) = Au(t)$ is a bounded input that you might want to use.

Solution

a. (i) For the input $x_1(t)$, the output is $y_1(t) = Kx_1(t) + A$ and for the input $x_2(t)$ the output is $y_2(t) = Kx_2(t) + A$. For the input $x_3(t) = a_1x_1(t) + a_2x_2(t)$, the output is $y_3(t) = Kx_3(t) + A = K\{a_1x_1(t) + a_2x_2(t)\} + A$. But $a_1y_1(t) + a_2y_2(t) = a_1\{Kx_1(t) + A\} + a_2\{Kx_2(t) + A\}$ and $y_3(t) \neq a_1y_1(t) + a_2y_2(t)$. The system is nonlinear.

Comment: It would be linear if $A = 0$.

(ii) An arbitrary input $x_1(t)$ produces an output $y_1(t) = Kx_1(t) + A$, and a shifted version of this input, $x_2(t) = x_1(t - t_0)$, produces an output $Kx_1(t - t_0) + A$. But $y_1(t - t_0) = Kx_1(t - t_0) + A$ for all t_0, so the system is time invariant (or shift invariant).

b. The output of the integrator depends only on past input values and the initial value of the output, so the system is causal.

c. Using $x(t) = Au(t)$, the output signal is $y(t) = t|x(t)|^2 = tA^2$, a signal that is increasing without bound as $t \to \infty$. The system is unstable.

THE *N*th-ORDER DIFFERENTIAL EQUATION MODEL

One important way of modeling linear time-invariant (LTI) systems is by means of linear constant-coefficient differential equations. Here we consider a differential equation (DE) to be a relationship between a function of time and its derivatives. An example is a system such as the lowpass analog filter of Figure 2.6a (page 44). This filter is described by the differential equation

filter DE

$$\ddot{v}(t) + \frac{2}{RC_1}\dot{v}(t) + \frac{1}{R^2C_1C_2}v(t) = \frac{1}{R^2C_1C_2}e(t)$$

whereas the instrument servo of Figure 2.6b is governed by the three simultaneous equations

$$e(t) = k_p\{r(t) - \theta(t)\}, \qquad L\dot{i}(t) + Ri(t) = Ae(t),$$
$$\text{and} \quad \lambda(t) = ki(t) = J\ddot{\theta}(t) + B\dot{\theta}(t).$$

These three instrument servo equations can be reduced to the single differential equation

servo DE

$$\dddot{\theta}(t) + \left(\frac{RJ + BL}{LJ}\right)\ddot{\theta}(t) + \frac{RB}{LJ}\dot{\theta}(t) + \frac{kk_pA}{LJ}\theta(t) = \frac{kk_pA}{LJ}r(t).$$

In general, linear constant-coefficient *N*th-order DEs for Single-Input–Single-Output (SISO) systems are written as

$$a_0y(t) + a_1\frac{dy(t)}{dt} + \cdots + a_N\frac{d^Ny(t)}{dt^N} = b_0x(t) + b_1\frac{dx(t)}{dt} + \cdots + b_L\frac{d^Lx(t)}{dt^L}$$

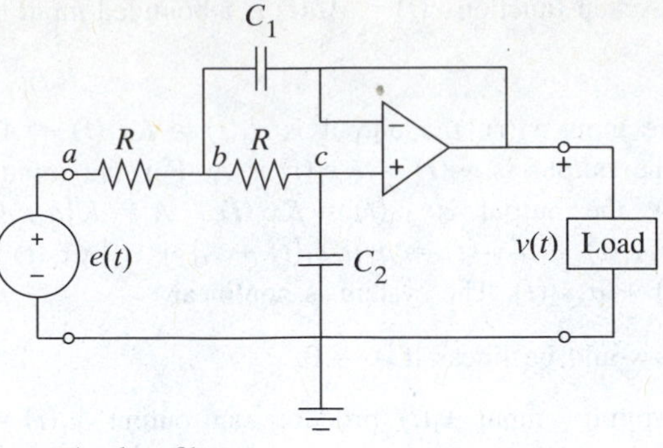

a. Analog filter

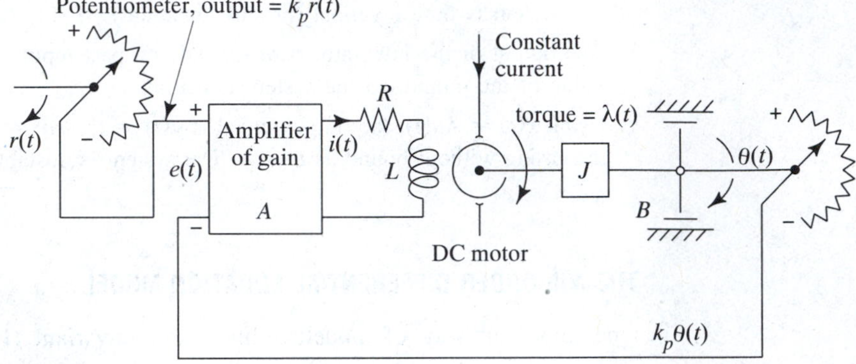

b. Instrument servo

FIGURE 2.6 *Systems for DE model*

where $y(t)$ and $x(t)$ represent the system output and input. A more concise representation using finite summations is

Nth-order DE

$$\sum_{k=0}^{N} a_k \frac{d^k y(t)}{dt^k} = \sum_{k=0}^{L} b_k \frac{d^k x(t)}{dt^k}.$$

We shall restrict our studies to the practical situation where the number of derivatives of the output is greater than or equal to the number of input derivatives, that is, $N \geq L$.

The *order* of a differential equation is the order of the highest derivative of the output function that appears in the equation. Thus, the lowpass analog filter is described by a second-order differential equation and the instrument servo, by a third-order DE.

Initial Condition Solution of a Differential Equation The *characteristic equation* (CE) of the system is found by substituting a trial solution $y(t) = Ce^{st}$ into the homogeneous differential equation

homogeneous DE

$$\sum_{k=0}^{N} a_k \frac{d^k y(t)}{dt^k} = 0$$

(obtained by setting to zero terms involving the input and its derivatives), with the result

$$a_0 s^0 (Ce^{st}) + a_1 s^1 (Ce^{st}) + \cdots + a_N s^N (Ce^{st}) = 0 .$$

Since Ce^{st} cannot be zero (this corresponds to the trivial and uninteresting solution $y(t) = 0$), it can be factored out. The remaining terms must satisfy the algebraic equation

$$a_0 s^0 + a_1 s^1 + \cdots + a_N s^N = 0$$

This result is known as the *characteristic equation* and can also be written

characteristic equation (CE)

$$\sum_{k=0}^{N} a_k s^k = 0 .$$

The N roots of this equation $s_1 = r_1, s_2 = r_2, \ldots, s_N = r_N$ are called the *characteristic roots*, and thus the characteristic equation can be written in factored form as

factored CE

$$\sum_{k=0}^{N} a_k s^k = a_N (s - r_1)(s - r_2) \cdots (s - r_N) = 0,$$

where $r_1, r_2, \ldots, r_N$ may be real or complex. If we are dealing with systems described by differential equations with real coefficients, complex roots must appear in conjugate pairs.

The solution of the homogeneous differential equation

homogeneous DE

$$\sum_{k=0}^{N} a_k \frac{d^k y(t)}{dt^k} = 0$$

for a given set of initial conditions $y(t)|_{t=0}, [dy(t)/dt]|_{t=0}, \ldots, [d^{N-1}y(t)/dt^{N-1}]|_{t=0}$ is called the *initial condition (IC) solution* and for simple (nonrepeating) roots is of the form

$$y_{\mathrm{IC}}(t) = C_1 e^{r_1 t} + C_2 e^{r_2 t} + \cdots + C_N e^{r_N t}$$

for an Nth-order system. Thus, for a system with no input(s), the output is

initial condition (IC) solution

$$y_{IC}(t) = \sum_{k=1}^{N} C_k e^{r_k t}$$

where the C_k, $k = 1, 2, \ldots, N$, are coefficients that must be determined in order to satisfy the given set of initial conditions and r_k, $k = 1, 2, \ldots, N$, are the characteristic roots. If the CE contains multiple roots indicated by the factor $(s - s_k)^q$, terms of the form

multiple roots

$$C_1 t^{q-1} e^{s_k t} + C_2 t^{q-2} e^{s_k t} + \cdots + C_q e^{s_k t}$$

will appear in the IC solution.

System stability is often defined in terms of this IC response, and a causal system is said to be stable if the IC response decays to zero as $t \to \infty$. This happens if and only if all the r_k's are *negative real,* or, if there are any complex roots, *their real parts must be negative.*

ILLUSTRATIVE PROBLEM 2.2 ***Characteristic Equation, the Initial Condition Response, and Stability***

The flexible shaft of Figure 2.7 is subjected to an applied torque $\lambda(t)$ with the shaft angle $\theta(t)$ described by the differential equation

$$\ddot{\theta}(t) + 3\dot{\theta}(t) + 2\theta(t) = (1/3)\lambda(t).$$

a. Find the system's characteristic equation.
b. What are the characteristic roots?
c. Is the system stable?
d. Give the algebraic form of the initial condition response.
e. Find the constants in the initial condition solution if $\theta(0) = 1$ and $[d\theta(t)/dt]|_{t=0} = 0$.

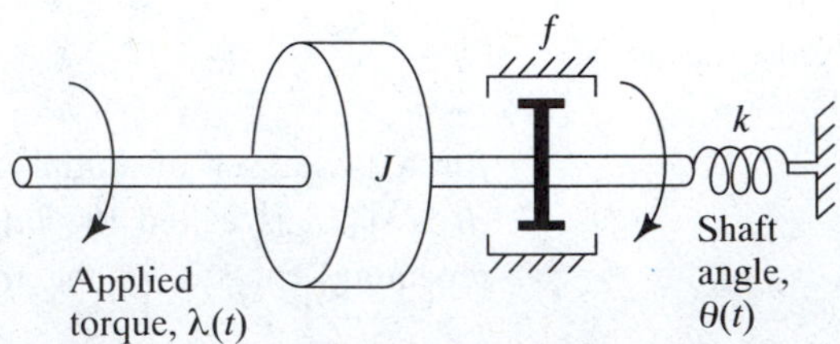

FIGURE 2.7 *Flexible shaft for Illustrative Problem 2.2*

Solution

a. The homogeneous differential equation is

$$\ddot{\theta}(t) + 3\dot{\theta}(t) + 2\theta(t) = 0$$

yielding the characteristic equation $s^2 + 3s + 2 = 0$.

b. Using the quadratic formula, $s_{1,2} = -1, -2$.

c. The system is stable because both roots are negative real.

d. $\theta_{IC}(t) = C_1 e^{-t} + C_2 e^{-2t}$

e. $\theta_{IC}(0) = 1 = C_1 + C_2$ and $[d\theta_{IC}(t)/dt]|_{t=0} = 0 = -C_1 - 2C_2$. Solving gives $C_1 = 2$ and $C_2 = -1$. Thus $\theta_{IC}(t) = 2e^{-t} - e^{-2t}$, $t \geq 0$.

Comment: We often need to find the roots of a characteristic equation of higher order than two. The function **roots** in the MATLAB Toolbox is available for this purpose. The equation's coefficients are entered as a vector, say **p**, and the roots are returned by the command **r = roots(p)**. If, for example, the CE is $5s^5 + 4s^4 + 3s^3 + 2s^2 + s = 0$, we write $\mathbf{p} = [5\,4\,3\,2\,1\,0]$ and **r = roots(p)** returns $r_1 = 0$, $r_{2,3} = 0.1378 \pm j0.6782$, and $r_{4,5} = -0.5378 \pm j0.3583$. Notice that the coefficient of s^0 was included in the vector **p**, which allowed us to find all five roots of the fifth-order equation.

Complete Solution of a Differential Equation The complete solution of a DE includes a component due to ICs and another caused by the forcing function, or input. We will postpone consideration of this topic until Chapter 3, "Laplace Transforms and Applications."

THE UNIT IMPULSE RESPONSE MODEL

Let us now consider another system model based on the response to a unit impulse input. The continuous-time unit impulse function was introduced in Chapter 1, but for convenience we will summarize its important properties here. The unit impulse function is defined implicitly by its sifting property:

sifting property

$$\int_{-\infty}^{\infty} \delta(t - t_0) f(t)\, dt = f(t_0)$$

where $f(t)$ is assumed to be continuous at $t = t_0$. Thus, the unit impulse function is defined in terms of its *area*. One representation of such an unusual function is as the limit of a very tall and very narrow pulse $p(t)$ having unit area, as shown in Figure 2.8a; that is,

impulse function

$$\delta(t - t_0) = \lim p(t) \text{ as } \Delta \to 0.$$

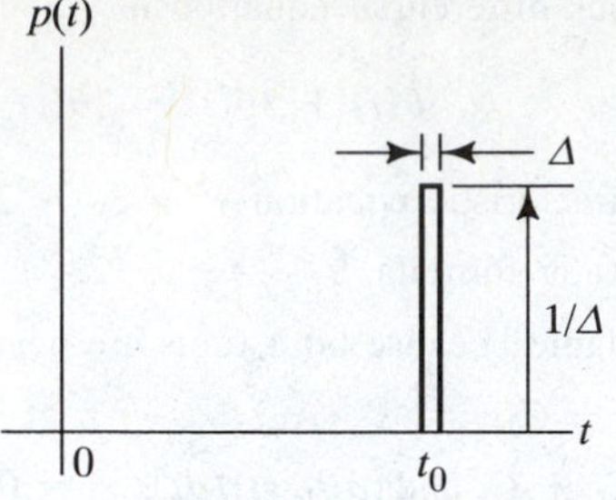

a. Approximation to impulse function

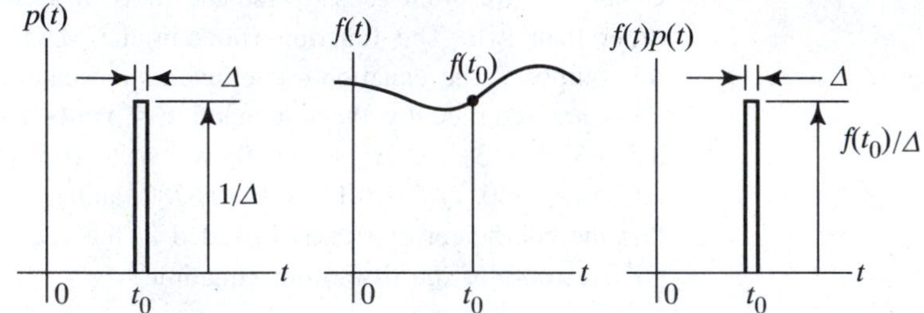

b. Approximation of the sifting property

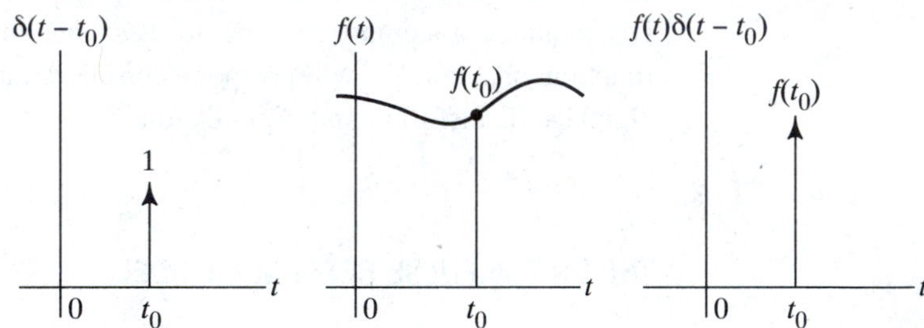

c. Impulse function representation and sifting property

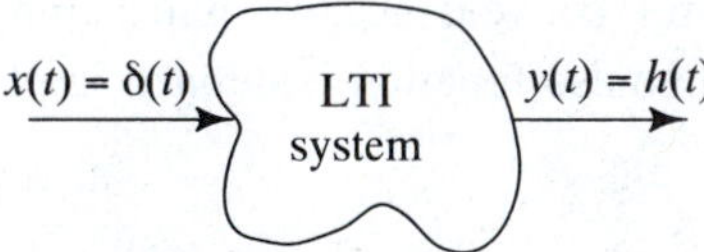

d. Unit impulse response

FIGURE 2.8 *Impulse function and response*

Using the sifting property leads to the product function shown in Figure 2.8b, and we see that the value of the integral is

$$\{f(t_0)/\Delta\} \cdot \Delta = f(t_0).$$

In shorthand form we generally represent a unit impulse as shown in Figure 2.8c, where the magnitude (1 in this case) is taken to mean the area of the impulse.

The response of an LTI system to an input of a unit impulse function, $x(t) = \delta(t)$, is called *the unit impulse response*. It is common practice to designate the output $y(t)$ as $h(t)$ when the input is a unit impulse $\delta(t)$. This is illustrated in Figure 2.8d, where we see that

$$\text{an input } x(t) = \delta(t) \xrightarrow{\text{produces}} \text{the output } y(t) = h(t).$$

When determining the unit impulse response $h(t)$ of an LTI system, it is necessary to make all initial conditions zero. That is, we must be certain that the response is due entirely to the input $x(t) = \delta(t)$ and not due to any energy that might be stored in the system. In simplistic terms, we must momentarily short-circuit all capacitors, momentarily open-circuit all inductors, drain all the reservoirs, stop all rotating masses, and so forth. We want the unit impulse response $h(t)$ to be a viable model of the system, which will occur if the system response is caused only by $x(t) = \delta(t)$. In general, it is not easy to find an analytic form for the unit impulse response $h(t)$ without the use of transforms. The simple system of Illustrative Problem 2.3 is an exception.

Earlier in this chapter, system stability was defined in terms of the system's input and output signals; that is, in a stable system a bounded input signal $x(t)$ produces a bounded output signal $y(t)$. Turning to the unit impulse response $h(t)$, we say that the system is stable if and only if its unit impulse response is *absolutely integrable:*

stability definition

$$\int_{-\infty}^{\infty} |h(\tau)|\, d\tau < \infty.$$

ILLUSTRATIVE PROBLEM 2.3
Finding an Impulse Response and Checking Stability

A continuous-time integrator is defined mathematically by

$$y(t) = \int_0^t x(\tau)\, d\tau + y(0)$$

where $t \geq 0$, $y(t)$ is the integrator output, $y(0)$ is the initial value of the integrator output, and $x(t)$ is the input.

a. Find the unit impulse response model $h(t)$ for this system.

b. Is the system stable?

Solution

a. First, the initial condition $y(0)$ is set to zero; then, with $x(t) = \delta(t)$, we have—by the sifting property of the impulse function—

$$y(t) = h(t) = \int_0^t \delta(\tau)\, d\tau = \begin{cases} 1, & \text{for } t \geq 0 \\ 0, & \text{for } t < 0, \end{cases}$$

which is actually the unit step function, $u(t)$.

b. Evaluation of

$$\int_{-\infty}^{\infty} |h(\tau)|\, d\tau \quad \text{gives} \quad \int_{-\infty}^{\infty} 1\, d\tau = \infty$$

and so the system is unstable. This agrees with the BIBO criterion because with a step input (a bounded input signal), for instance, the output is $y(t) = tu(t)$, an unbounded output signal.

Comment: Even though the definition indicates that integrators are unstable, they are widely used.

CONVOLUTION

If the unit impulse response $h(t)$ of a linear continuous system is known, the system output $y(t)$ can be found for any input $x(t)$. The solution can be obtained either in the time domain or by means of transforms. Here let us consider the time-domain solution, with the transform-domain solution treated in the next chapter.

A Continuum of Impulses In Figure 2.9a we have represented the function $x(t)$ between $t = -T$ and $t = +T$ by a sequence of $2n$ rectangular pulses that are spaced Δ units apart with a typical pulse, the kth one, designated as $p_k(t - k\Delta)$, as pictured in Figure 2.9b. Notice that these pulses do not, in general, have unit area, since their amplitudes are $x(k\Delta)$ and the widths are Δ, yielding an area of $x(k\Delta) \cdot \Delta$ units. A pulse of unit area is shown in Figure 2.9c, where we are using the notation $p_\Delta(t - k\Delta)$. Thus, to describe the pulse $p_k(t - k\Delta)$ of Figure 2.9b in terms of the unit pulse $p_\Delta(t - k\Delta)$, we write

$$p_k(t - k\Delta) = \frac{x(k\Delta)}{1/\Delta} p_\Delta(t - k\Delta) = x(k\Delta) \cdot p_\Delta(t - k\Delta) \cdot \Delta\,.$$

Now we sum the pulses and, to the extent that the staircase function of Figure 2.9a approximates $x(t)$, we have

approximation by a sum

$$x(t) \approx \sum_{k=-T/\Delta}^{T/\Delta} x(k\Delta) \cdot p_\Delta(t - k\Delta) \cdot \Delta\,.$$

Next we proceed to the limit, letting $\Delta \to 0$. We see that $k\Delta$ becomes the (continuous) running variable, which we shall call τ; Δ becomes $d\tau$; the pulse p_Δ becomes a unit impulse δ; and the summation becomes integration. Thus we now have

integral representation

$$x(t) = \int_{-T}^{T} x(\tau)\delta(t - \tau)\, d\tau\,.$$

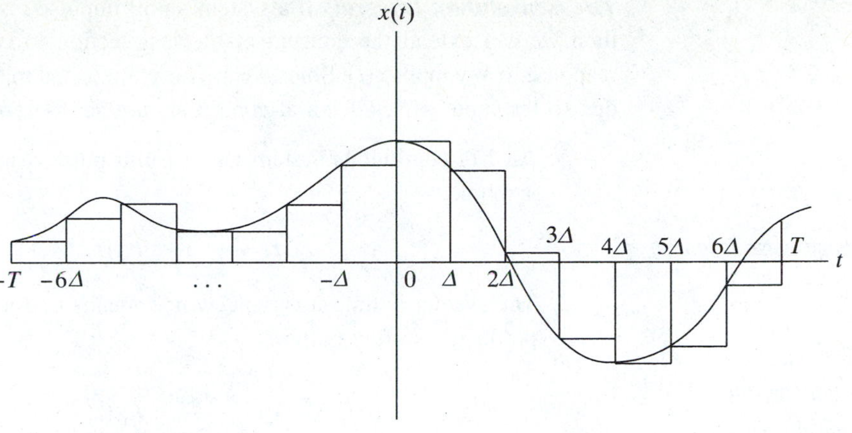

a. Approximation by pulses

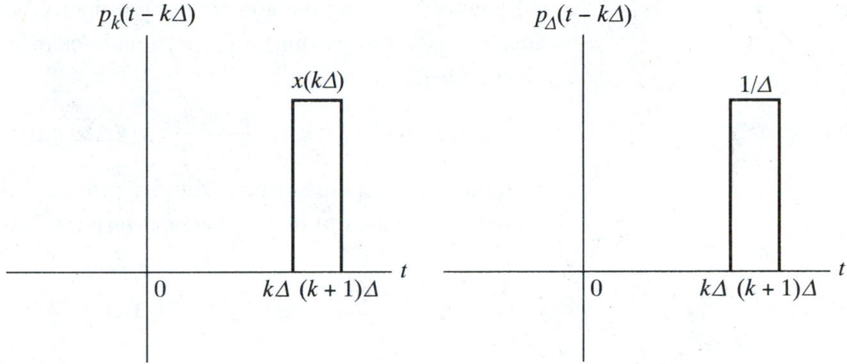

b. A typical pulse

c. A pulse of unit area

FIGURE 2.9 *Representation of any signal*

Finally, if we consider $x(t)$ as extending from $-\infty$ to $+\infty$, we can let $T \longrightarrow \infty$ to yield

sifting property of impulse

$$x(t) = \int_{-\infty}^{\infty} x(\tau)\delta(t - \tau)\,d\tau\,.$$

We have shown that a continuous-time function can be broken up into a continuum (an infinite sequence) of shifted impulse functions, and this leads to the sifting property of impulse functions.

The Convolution Integral If a system's unit impulse response $h(t)$ is known, then we can extend the concept of the last section to compute the system's response to any input $x(t)$. Since we are only interested in the system's response due to the input $x(t)$, all initial conditions are set to zero.

CHAPTER 2

1. An LTI continuous system has the unit pulse response $h_\Delta(t)$. Using symbols,

unit pulse response

$$p_\Delta(t) \xrightarrow{\text{produces}} h_\Delta(t)\,.$$

2. The system is time invariant, which means that a shifted input will produce a shifted output:

time or shift invariance

$$p_\Delta(t - k\Delta) \xrightarrow{\text{produces}} h_\Delta(t - k\Delta)\,.$$

3. Linearity consists of the two properties, (a) homogeneity and (b) additivity. Using the homogeneity property, a shifted pulse of area $x(k\Delta)\Delta$ will produce a proportionate change in the size of the output; that is,

homogeneity

$$x(k\Delta)\Delta p_\Delta(t - k\Delta) \xrightarrow{\text{produces}} x(k\Delta)\Delta h_\Delta(t - k\Delta)\,.$$

4. Summing the responses (the additivity property) due to the entire sequence of pulses to obtain the total response gives

additivity

$$y(t) \approx \sum_{k=-T/\Delta}^{T/\Delta} x(k\Delta)h_\Delta(t - k\Delta)\Delta, \quad -T \le t \le T$$

where the approximation symbol simply indicates that the input staircase function is an approximation to the actual input $x(t)$.

5. Finally, going to the limit as before, we have

$$y(t) = \int_{-T}^{T} x(\tau)h(t - \tau)\,d\tau$$

or, as $T \to \infty$,

convolution integral

$$y(t) = \int_{-\infty}^{\infty} x(\tau)h(t - \tau)\,d\tau$$

where $h(t)$ is the system's unit impulse response. The convolution integral is one of the most important results used in the study of the response of linear time-invariant systems.[3] This remarkable result

[3]Convolution also applies to time-varying systems, but its use in such systems will not be treated here.

states that if we know the unit impulse response $h(t)$ for a linear system, by using the convolution integral we can compute the system output for any known input $x(t)$. By change of variable, letting $\zeta = t - \tau$, we get

alternative form

$$y(t) = \int_{-\infty}^{\infty} x(t - \zeta)h(\zeta)\, d\zeta = \int_{-\infty}^{\infty} x(t - \tau)h(\tau)\, d\tau .$$

Convolution is a perfectly respectable mathematical operation that can be used with any functions, such as $c_1(t)$ and $c_2(t)$. Designating $c_3(t)$ as the convolution of $c_1(t)$ and $c_2(t)$, we have

$$c_3(t) = \int_{-\infty}^{\infty} c_1(\tau)c_2(t - \tau)\, d\tau = \int_{-\infty}^{\infty} c_2(\tau)c_1(t - \tau)\, d\tau .$$

Finally, convolution is often expressed operationally using a "star" as

operational notation

$$c_3(t) = c_1(t)*c_2(t) = c_2(t)*c_1(t) .$$

The convolution integral can be evaluated in three distinct ways. The following list is probably in order of increasing practicality.

1. *Analytical method.* If $h(t)$ and $x(t)$ are expressed analytically, it is usually possible to integrate. The integration may be difficult, however, and the transform methods of the next chapter are almost always simpler.
2. *Graphical method.* If $h(t)$ and $x(t)$ are available (known) in graphical form, then by "graphical integration" we can obtain an approximation to the convolution $y(t) = h(t)*x(t)$.
3. *Numerical convolution.* If $h(t)$ and $x(t)$ are approximated by numerical sequences, say $h(n)$ and $x(n)$ (most conveniently the values of $h(t)$ and $x(t)$ at equal intervals of time T), then an approximation to the convolution $y(t) = h(t)*x(t)$ may be found in the form of a third number sequence $y(n)$ by discrete convolution. We then multiply the sample $y(n)$ by the sampling interval T to obtain the approximate value of $y(t)$ at $t = nT$. This method is ideally suited for digital computer implementation.

Analytical Method of Evaluation Shown in Figure 2.10a is the block diagram representation of an LTI system having a known unit impulse response $h(t) = Ae^{\alpha t}u(t)$ and a given input signal $x(t) = Be^{\beta t}u(t)$ with $\alpha < 0$, $\beta < 0$, $A > 0$, $B > 0$, and $\alpha \neq \beta$; both signals are shown in Figure 2.10b. The two forms of the convolution integral that allow us to compute the system output are repeated for convenience.

$$y(t) = \int_{-\infty}^{\infty} x(\tau)h(t - \tau)\, d\tau \quad \text{or} \quad y(t) = \int_{-\infty}^{\infty} h(\tau)x(t - \tau)\, d\tau$$

CHAPTER 2

1. Using the integral on the right, we have

$$y(t) = \int_{-\infty}^{\infty} Ae^{\alpha\tau}u(\tau)Be^{\beta(t-\tau)}u(t - \tau)\,d\tau\,.$$

2. Adjust the limits of integration by sketching $Ae^{\alpha\tau}u(\tau)$ and $Be^{\beta(t-\tau)} \cdot u(t - \tau)$ for $t = 0$, as we have done in Figure 2.10c. From this we see that the product $Ae^{\alpha\tau}u(\tau) \cdot Be^{\beta(t-\tau)}u(t - \tau)$ is 0 for all values of t less than 0; consequently, the lower limit of the integral is $\tau = 0$. In Figure 2.10d the function $Be^{\beta(t-\tau)}u(t - \tau)$ is shown shifted to the right ($t > 0$), and we see that the product $Ae^{\alpha\tau}u(\tau) \cdot Be^{\beta(t-\tau)}u(t - \tau)$ is 0 for all values of τ greater than t; consequently, the upper limit of the integral is $\tau = t$. The convolution integral is now given by

$$y(t) = \int_{0}^{t} Ae^{\alpha\tau}Be^{\beta(t-\tau)}\,d\tau\,.$$

Notice that this is a general result for causal systems ($h(t) = 0$ for $t < 0$) that are subjected to inputs that satisfy $x(t) = 0$ for $t < 0$, a very common situation in the study of linear systems.

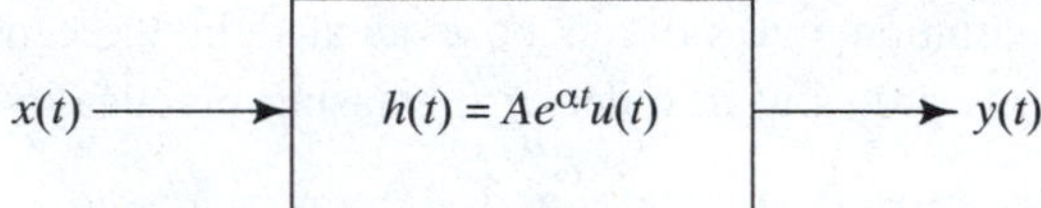

a. Block diagram for a system

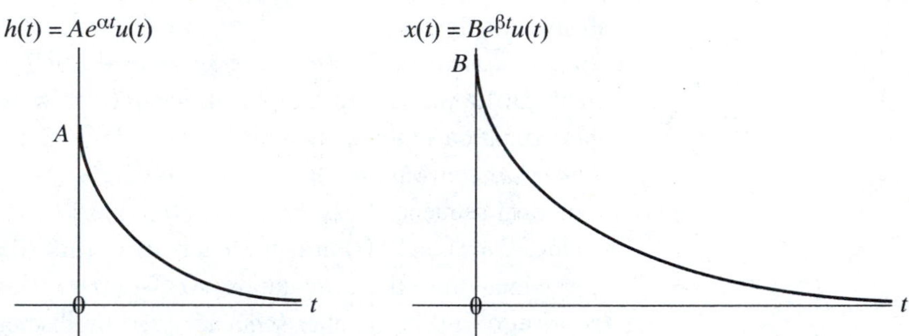

b. Unit impulse response and exponential input signal

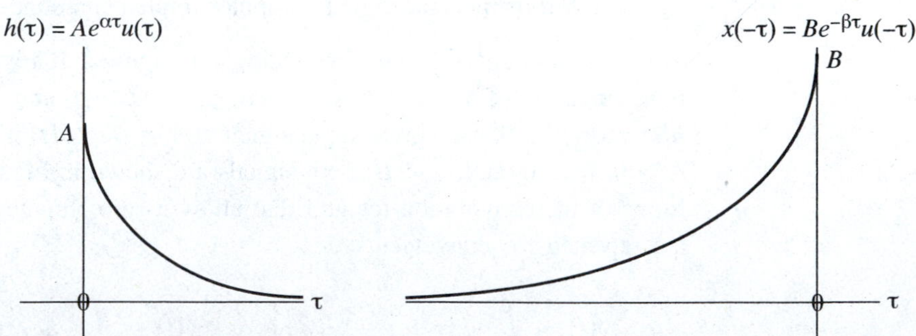

c. Unit impulse response and time-reversed input as functions of τ

FIGURE 2.10 *Analytical convolution*

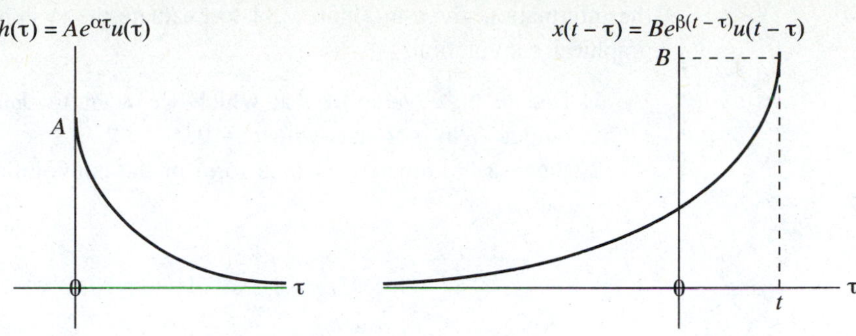

d. Impulse response and time-reversed and shifted input

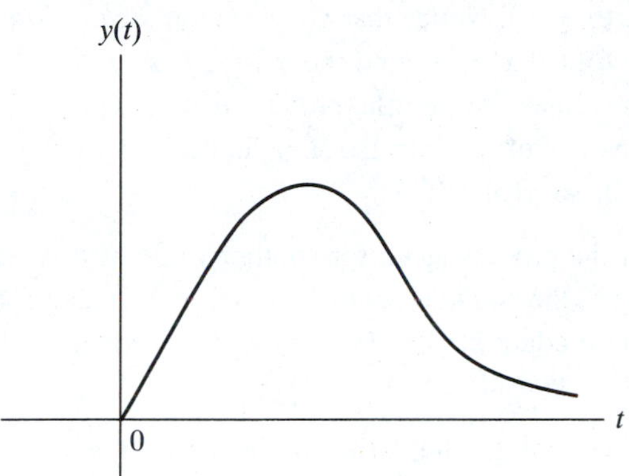

e. System output

FIGURE 2.10 *Continued*

3. Perform the integration. A and B are constants and the integration is with respect to τ, so we can take A, B, and $e^{\beta t}$ outside of the integral, giving

$$\begin{aligned} y(t) &= ABe^{\beta t}\int_0^t e^{\alpha\tau}e^{-\beta\tau}\,d\tau \\ &= \frac{ABe^{\beta t}}{\alpha-\beta}\left[e^{(\alpha-\beta)t} - e^{(\alpha-\beta)0}\right] \\ &= \frac{AB}{\alpha-\beta}\left[e^{\alpha t} - e^{\beta t}\right],\ t \geq 0, \end{aligned}$$

with the result sketched in Figure 2.10e for $\beta < \alpha < 0$.

Graphical Method of Evaluation For each value of time t for which we want the value of the system output $y(t)$, we need to compute the integral (area) of the product (integrand) $x(t) \cdot h(t - \tau)$ or $h(t) \cdot x(t - \tau)$. We use the information given in Figure 2.11 to evaluate the system output $y(t)$ using graphical convolution.

1. Decide on a value of t at which we want to determine the system output. Why not start with $t = 0$?
2. Since it is immaterial which form of the convolution integral is used, we'll use

$$y(t) = \int_{-\infty}^{\infty} h(\tau)x(t - \tau)\,d\tau$$

   because of the symmetry in $x(t)$. In Figure 2.12a we have plotted $h(\tau)$ and $x(0 - \tau)$. Notice that $x(-\tau)$ is a *folded version* of $x(\tau)$, where we have folded $x(\tau)$ around the $\tau = 0$ axis.
3. Next we need to plot the product $h(\tau) \cdot x(t - \tau)$, as in Figure 2.12b.
4. The value of $y(0)$ is the area under the $h(\tau)x(t - \tau)$ plot, in this case 0, so $y(0) = 0$.

We start the process again for another value of t. At this point it is worth noting that $y(t)$, the convolution of $h(t)$ and $x(t)$, is zero for all negative values of t, since the product $h(\tau) \cdot x(t - \tau)$ will be zero. Thus only positive values of t need to be considered.

1a. To keep things simple, we'll choose $y(1)$ as the next output point to be determined.
2a. In Figure 2.12c we have plotted $x(1 - \tau)$.
3a. See Figure 2.12d for the product $h(\tau) \cdot x(1 - \tau)$.
4a. The area under $h(\tau) \cdot x(1 - \tau)$, is easily determined as 2; consequently, $y(1) = 2$.

We can see that this same product, $h(\tau) \cdot x(1 - \tau)$, will result for all positive values of t from 1 to 4, giving $y(t) = 2$ for $1 \leq t \leq 4$.

1b. Let's refine things a bit and find the output for $t = 4.5$.
2b. In Figure 2.12e we have plotted $x(4.5 - \tau)$.
3b. In Figure 2.12f we have the product $h(\tau) \cdot x(4.5 - \tau)$.
4b. The area under $h(\tau) \cdot x(4.5 - \tau)$ is easily determined as 1; consequently, $y(4.5) = 1$.

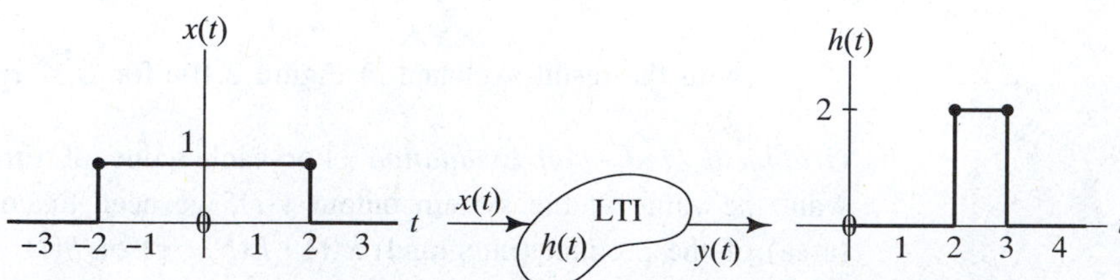

FIGURE 2.11 *System data*

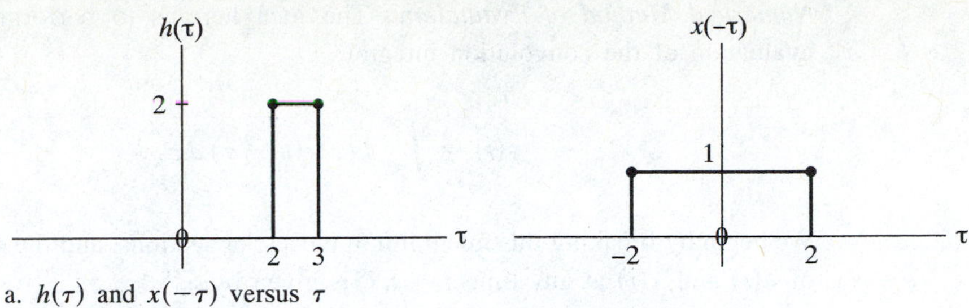

a. $h(\tau)$ and $x(-\tau)$ versus τ

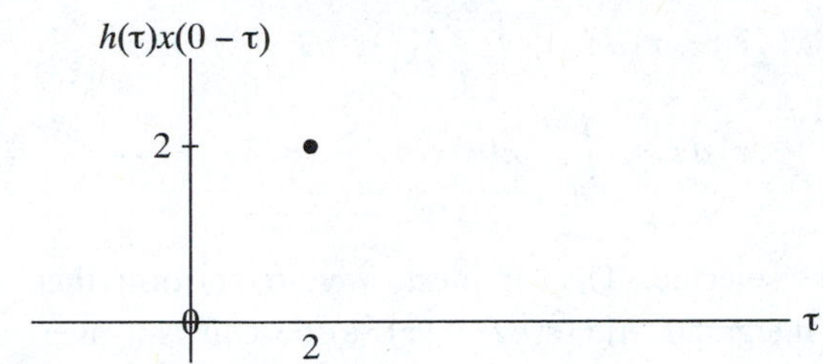

b. Product $h(\tau)x(0-\tau)$ versus τ

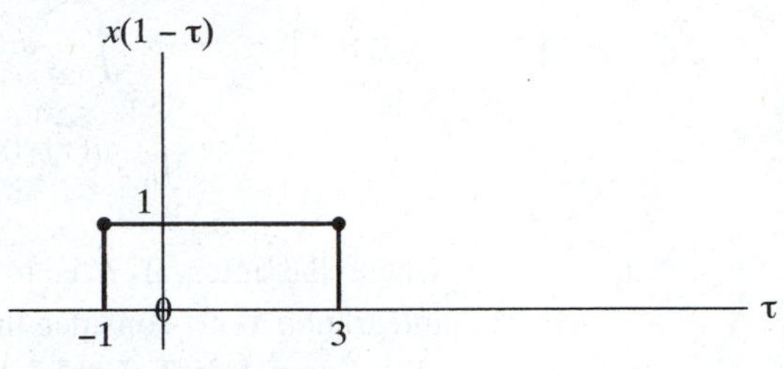

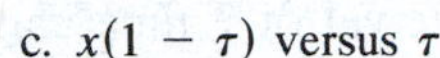

c. $x(1-\tau)$ versus τ

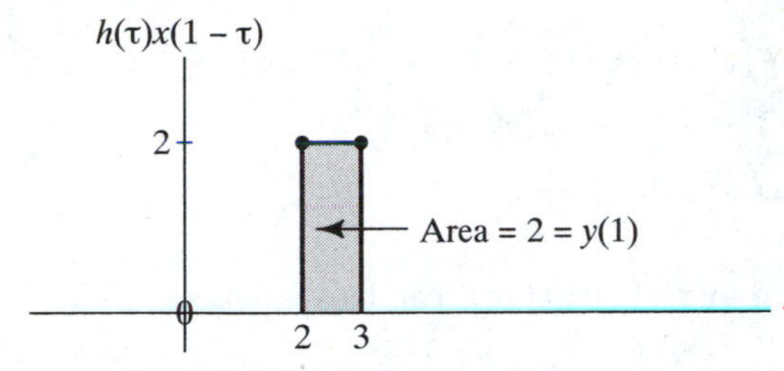

d. Product $h(\tau)x(1-\tau)$ versus τ

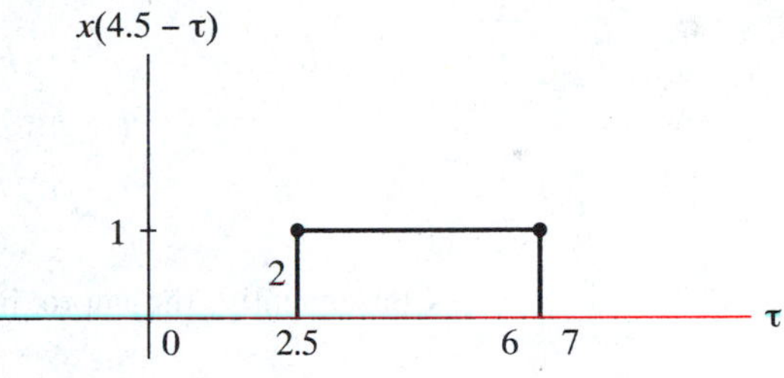

e. $x(4.5-\tau)$ versus τ

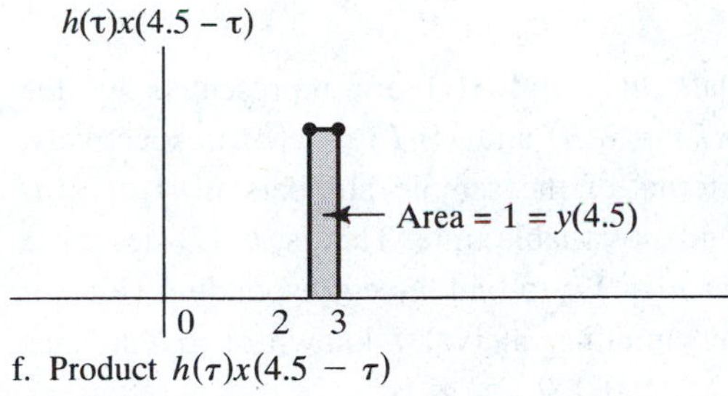

f. Product $h(\tau)x(4.5-\tau)$

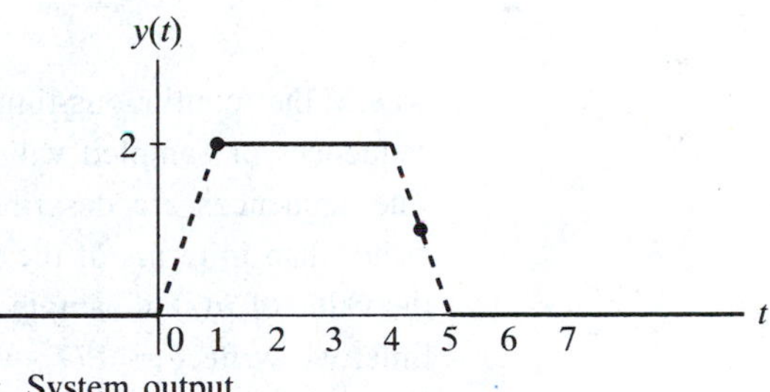

g. System output

FIGURE 2.12 *Steps in graphical convolution*

We also notice that for all $t \geq 5$, $h(\tau) \cdot x(t-\tau) = 0$, making $y(t) = 0$. In Figure 2.12g we have plotted the results obtained as well as the extrapolated results for $0 < t < 1$ and $4 < t < 4.5$.

Numerical Method of Evaluation The idea here is to perform numerical evaluation of the convolution integral

$$y(t) = \int_{-\infty}^{\infty} h(\tau)x(t - \tau)\,d\tau .$$

We begin by breaking this integral into pieces, or sections, and the convolution of $h(t)$ and $x(t)$ at any time $t = nT$ is given by

$$\begin{aligned} y(nT) = \cdots &+ \int_{-2T}^{-T} h(\tau)x(nT - \tau)\,d\tau + \int_{-T}^{0} h(\tau)x(nT - \tau)\,d\tau \\ &+ \int_{0}^{T} h(\tau)x(nT - \tau)\,d\tau + \int_{T}^{2T} h(\tau)x(nT - \tau)\,d\tau + \cdots \end{aligned}$$

where the interval T is to be selected. The simplest way to perform this integration is to consider the integrand, $h(\tau)x(nT - \tau)$, to be constant over each interval of T time units. Thus the integral is approximated by summing the areas of the rectangles formed by this procedure, and $y(nT)$ is given by

$$\begin{aligned} y(nT) \approx \cdots &+ T \cdot h(-2T)x(nT + 2T) + T \cdot h(-T)x(nT + T) \\ &+ T \cdot h(0)x(nT) + T \cdot h(T)x(nT - T) \\ &+ T \cdot h(2T)x(nT - 2T) + \cdots . \end{aligned}$$

Consequently, the approximation to $y(t)$ at $t = nT$ can be written as

$$y(nT) = T \sum_{m=-\infty}^{\infty} h(mT)x(nT - mT) = T \sum_{m=-\infty}^{\infty} h(m)x(n - m)$$

where the continuous-time signals $h(t)$ and $x(t)$ are represented by the sequences or sampled values $h(nT) = h(n)$ and $x(nT) = x(n)$, respectively. The sequences are described in terms of the sample numbers $h(n)$ or $x(n)$ rather than in terms of the independent variable time. That is, $h(17)$ describes the value of $h(t)$ at sample number $n = 17$; to find the corresponding value of time, we write $t = 17T$, with the sampling interval T known or given. Thus the sequence for $n = -\infty, \ldots, -2, -1, 0, 1, 2, \ldots, \infty$ is

$$\{h(-\infty) \quad \ldots \quad h(-2) \quad h(-1) \quad h(0) \quad h(1) \quad h(2) \quad \ldots \quad h(\infty)\}$$

where a similar description can be made for $x(n)$. To illustrate this numerical procedure we prepare sequences that adequately represent the unit impulse response $h(t)$ of Figure 2.13a and the cosine pulse input $x(t)$ of Figure 2.13b. If you are interested, the equations are $h(t) = 5(e^{-0.5t} - e^{-t})u(t)$ and $x(t) = \cos(0.5\pi t)\{u(t + 5) - u(t - 5)\}$.

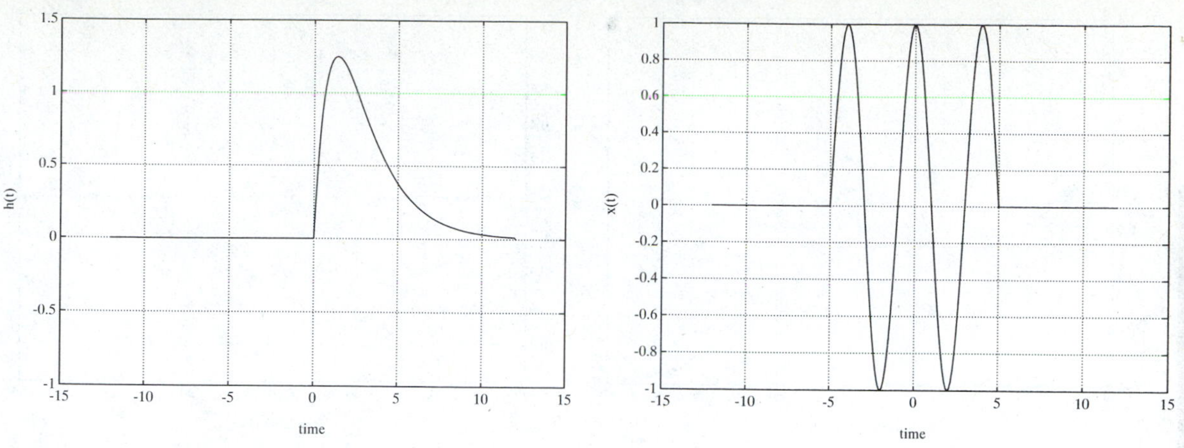

a. Unit impulse response
b. Input signal

FIGURE 2.13 *System impulse response $h(t)$ and input $x(t)$*

The sampled waveforms for $T = 0.2$ are shown in Figure 2.14a and b (page 60). With the sampled waveforms, the sequences $h(n)$ and $x(n)$, in hand, we need to evaluate the discrete convolution[4]

$$y(n) = \sum_{m=-\infty}^{\infty} x(m)h(n-m) = \sum_{m=-\infty}^{\infty} h(m)x(n-m).$$

The result of convolving the sequences of Figure 2.14a and b using the MATLAB function **conv** is given in Figure 2.14c.

The estimate of the continuous-time convolution at each sampling instant $t = nT$ is, consequently, given by

$$y(nT) = T \cdot \sum_{m=-\infty}^{\infty} x(m)h(n-m)$$

with a smooth curve that estimates the convolution of the continuous signals $h(t)$ and $x(t)$ given in Figure 2.14d. The accuracy of this method depends on the chosen sampling interval T and, with the computer, very good results can be obtained.

Comment: If you harbor any doubts about which of the three methods (analytical, graphical, or numerical) is the most useful or easiest, try the analytical or graphical procedure on this one.

[4]More information on the operation of discrete convolution is presented in Chapter 7.

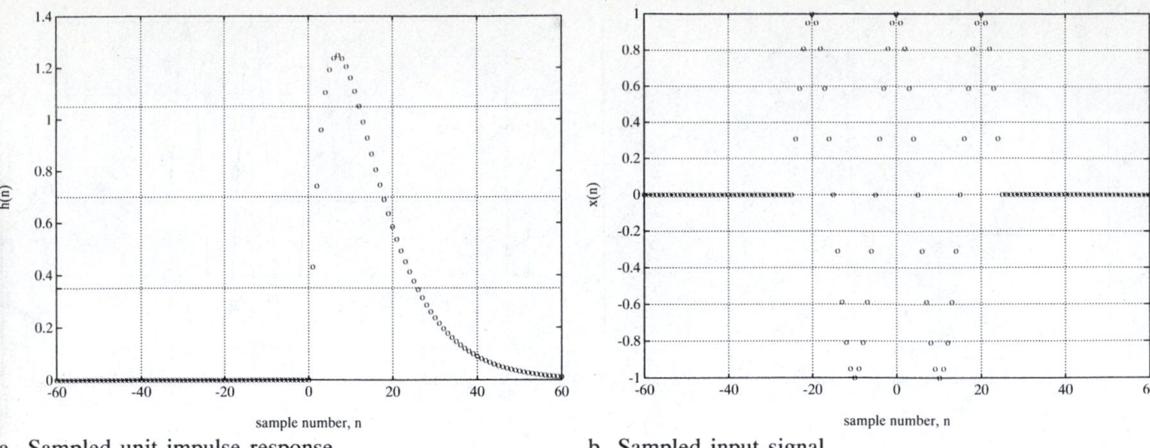

a. Sampled unit impulse response

b. Sampled input signal

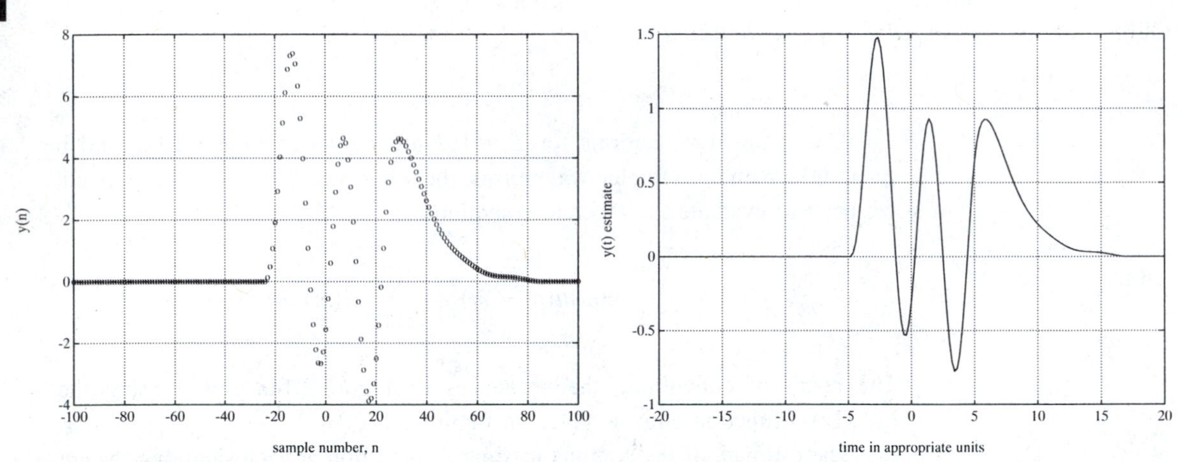

c. Discrete convolution

d. Result of numerical convolution

FIGURE 2.14 *Approximating the convolution integral by discrete convolution*

ILLUSTRATIVE PROBLEM 2.4 *Numerical Convolution*

A first-order circuit with the unit impulse response $h(t) = e^{-t}u(t)$ is subjected to a ramp voltage of $x(t) = tu(t)$. We want to find the output voltage at, say, $t = 1$ s by means of numerical convolution. For ease of hand computation, use a sampling interval of $T = 0.05$ s.

a. Find the data sequences $x(n)$ and $h(n)$ needed to find the approximate value of the output $y(t)$ at $t = 1.0$ s.

b. Show the computations required to estimate $y(t)$ at $t = 1.0$ s.

Solution

a. For the input $x(t) = tu(t)$, we have $x(n) = \{0, 0.05, 0.10, 0.15, \ldots, 0.95, 1.0\}$, and for the unit impulse response $h(t)$ we have $h(n) = e^{-0.05n}$ for $0 \le n \le 20$, giving

$$h(n) = \{1, e^{-0.05}, e^{-0.10}, e^{-0.15}, \ldots, e^{-0.95}, e^{-1.0}\}.$$

b. Using

$$y(nT) = T \cdot \sum_{m=-\infty}^{\infty} h(m)x(n - m)$$

for $n = 20$ gives

$$y(20 \cdot 0.05) = 0.05 \cdot [(1)(1.0) + (e^{-0.05})(0.95) + (e^{-0.10})(0.90) + \cdots + (e^{-1.0})(0)].$$

From a calculator or a computer, we find that $y(1.0) = (0.05)(7.86444) = 0.3932$. Analytically evaluating the convolution integral yields the expression

$$y(t) = \int_0^t e^{-\tau}(t - \tau)\,d\tau = [t - 1 + e^{-t}]u(t)$$

which when evaluated at $t = 1.0$ gives $y(1) = e^{-1} = 0.3679$.

Comment: We can use the MATLAB function **conv** to evaluate the convolution. The sampling interval can be made as small as necessary because we don't have to do the calculations by hand. Shown next is the script used for $T = 0.01$; the resulting plot for $0 \le t \le 1$ is given in Figure 2.15. The approximation for $y(t)$ at $t = 1.0$ s using this sampling interval is 0.3679.

MATLAB Script

```
%F2_15 Numerical convolution for Illustrative Problem 2.4
t=0:.01:1;
h=exp(-1*t); % generates h(n)
x=1*t; % generates x(n)
y=0.01*conv(h,x); % calculates y(n)=T·h(n)*x(n)
plot(t,y(1:101),'o');
axis([0,1,0,.4])
%...labeling statements
```

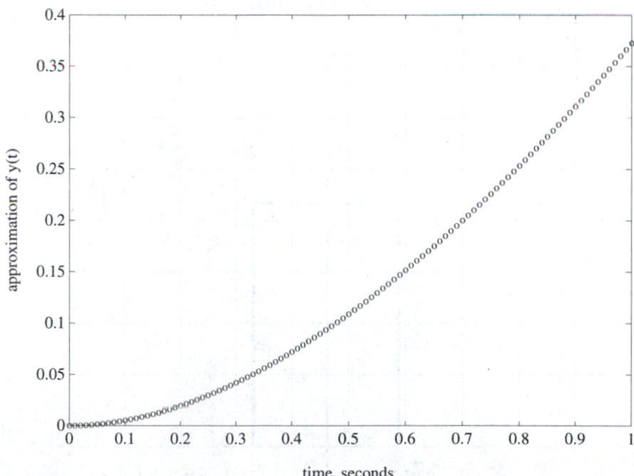

FIGURE 2.15 *Numerical convolution for Illustrative Problem 2.4*

CHAPTER 2

Convolution Involving Impulses The convolution $c_3(t)$ of the arbitrary signal $c_1(t)$ with the weighted (size or strength of D), shifted (located at $t = d$) impulse function $c_2(t) = D\delta(t - d)$ is given by the integral

$$c_3(t) = \int_{-\infty}^{\infty} c_1(\tau) D\delta(t - d - \tau)\, d\tau\,.$$

Using the sifting property of the impulse function, namely,

$$\int_{-\infty}^{\infty} \delta(\sigma - a) f_1(\sigma)\, d\sigma = f_1(a)$$

we have the convolution

$$c_3(t) = \int_{-\infty}^{\infty} D\delta(t - d - \tau) c_1(\tau)\, d\tau = Dc_1(t - d)\,.$$

Notice that this result is $c_1(t - d)$, the value of the signal at the time of occurrence of the impulse function scaled by the factor D.

ILLUSTRATIVE PROBLEM 2.5
Convolution with Impulse Functions

A single rectangular pulse described by $x(t) = A\{u(t + a/2) - u(t - a/2)\}$ is shown in Figure 2.16a. A periodic train of unit impulse functions described by the infinite summation

$$\delta_p(t) = \sum_{k=-\infty}^{\infty} \delta(t - kT_0)$$

is pictured in Figure 2.16b. Find the result of the convolution $y(t) = x(t) * \delta_p(t)$ for $T_0 >> a$.

Solution

Using the preceding results, the impulse at $t = 0$ when convolved with $p(t)$ will reproduce the pulse centered at $t = 0$. In a similar way the impulses at $t = \pm T_0$ will shift the pulse $x(t)$ to be centered at $\pm T_0$ as a result of convolution. The result of continuing this procedure is the pulse train shown in Figure 2.16c.

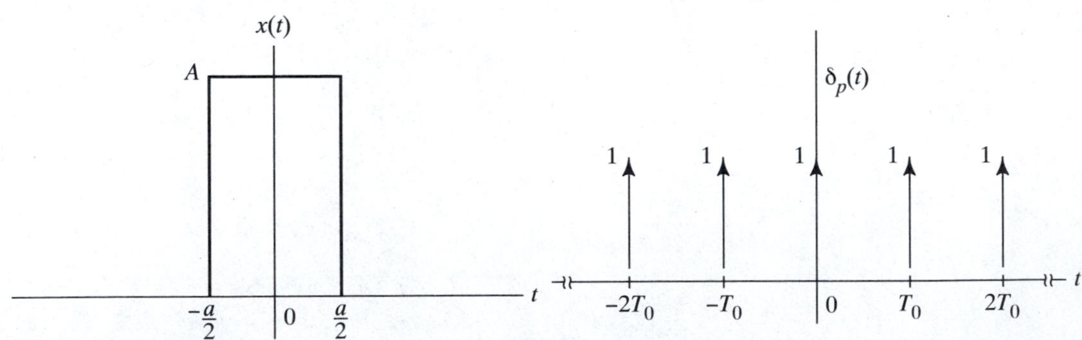

a. Rectangular pulse, $x(t)$

b. Impulse train, $\delta_p(t)$

FIGURE 2.16 *Convolution with impulse functions*

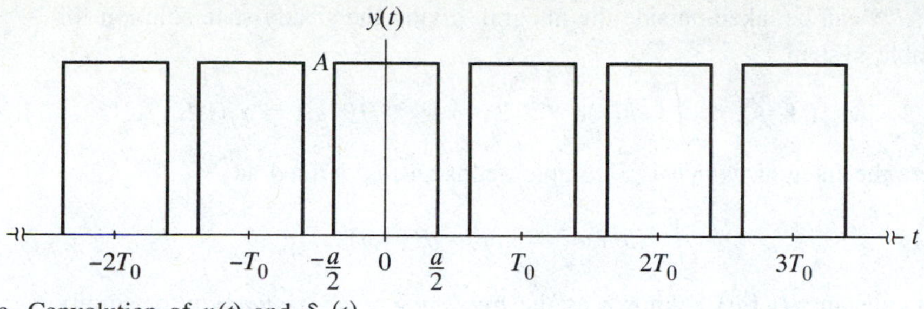

c. Convolution of $x(t)$ and $\delta_p(t)$

FIGURE 2.16 *Continued*

BASIC CONCEPTS

SINUSOIDAL STEADY-STATE RESPONSE

Let's use the convolution integral to develop an important characteristic of a stable LTI system: its steady-state response to a sinusoidal input signal. Using the Euler[5] relation, a sinusoidal signal such as $x(t) = 2\cos(\omega t)$ can be synthesized from two complex exponentials as $2\cos(\omega t) = e^{j\omega t} + e^{-j\omega t}$. We'll find the steady-state response to each exponential and then invoke linearity and add the results. Given the unit impulse response $h(t)$, the system output $y(t)$ is found from the convolution integral as

$$y(t) = \int_{-\infty}^{\infty} h(\tau)x(t-\tau)\,d\tau.$$

For the complex exponential input $x_1(t) = e^{j\omega t}$, $-\infty \le t \le \infty$, the steady-state output is

$$y_{1ss}(t) = \int_{-\infty}^{\infty} h(\tau)e^{j\omega(t-\tau)}\,d\tau, \ -\infty \le t \le \infty$$

[5]Leonhard Euler (1707–1783), a key figure in eighteenth-century mathematics, was the son of a Lutheran pastor who lived near Basel, Switzerland. Euler received a master's degree in mathematics at age 16 and when only 19 won a prize from the Académie des Sciences for a treatise on the most efficient arrangement of ship masts. He held appointments at the Academy of St. Petersburg and at the Royal Academy of Berlin, and for a while was a lieutenant in the Russian Navy. Euler, who became blind early in life, is accepted as being the most versatile and prolific writer in the history of mathematics. He wrote or dictated over 700 books and papers in his lifetime and was an inspiration to generations of younger mathematicians, including P. S. Laplace, whose advice was: "Read Euler, he is our master in all." From *The History of Mathematics,* David M. Burton, Allyn and Bacon, Inc., 1985.

and $e^{j\omega t}$ can be taken outside the integral, giving the steady-state solution for a stable system

steady-state output

$$y_{1ss}(t) = e^{j\omega t} \int_{-\infty}^{\infty} h(\tau)e^{-j\omega\tau}\,d\tau = e^{j\omega t}H(j\omega) = x_1(t)H(j\omega)$$

where the integral, which is a complex constant, is defined as

$$\int_{-\infty}^{\infty} h(\tau)e^{-j\omega\tau}\,d\tau = H(j\omega).$$

This constant $H(j\omega)$ is known as the *frequency response function* or simply the *frequency response* of the system. For the complex exponential input $x_2(t) = e^{-j\omega t}$, the steady-state output is found in the same manner as

steady-state output

$$y_{2ss}(t) = e^{-j\omega t} \int_{-\infty}^{\infty} h(\tau)e^{j\omega\tau}\,d\tau = e^{-j\omega t}H(-j\omega) = x_2(t)H(-j\omega).$$

Using linearity to sum the two responses gives

$$y_{ss}(t) = y_{1ss}(t) + y_{2ss}(t) = e^{j\omega t}H(j\omega) + e^{-j\omega t}H(-j\omega).$$

But $H(j\omega) = |H(j\omega)|e^{j\angle H(j\omega)}$ and $H(-j\omega) = |H(j\omega)|e^{-j\angle H(j\omega)}$ which gives

$$\begin{aligned} y_{ss}(t) &= e^{j\omega t}|H(j\omega)|e^{j\angle H(j\omega)} + e^{-j\omega t}|H(j\omega)|e^{-j\angle H(j\omega)} \\ &= |H(j\omega)|\{e^{j(\omega t+\angle H(j\omega))} + e^{-j(\omega t+\angle H(j\omega))}\} \\ &= 2|H(j\omega)|\cos(\omega t + \angle H(j\omega)), \quad -\infty \le t \le \infty. \end{aligned}$$

We recall that the system input was

$$x(t) = x_1(t) + x_2(t) = e^{j\omega t} + e^{-j\omega t} = 2\cos(\omega t), \quad -\infty \le t \le \infty.$$

Generalizing to a sinusoidal input amplitude of A and a phase of α, we see that when a sinusoidal signal

$$x(t) = A\cos(\omega t + \alpha), \quad -\infty \le t \le \infty,$$

is applied to a linear time-invariant system, the steady-state output signal is

$$y_{ss}(t) = A|H(j\omega)|\cos(\omega t + \angle H(j\omega) + \alpha), \quad -\infty \le t \le \infty.$$

This result makes the evaluation of a stable continuous system's steady-state response to a sinusoidal input signal very straightforward, and we refer to the result as the sinusoidal steady-state formula. In other words, it tells us that when a sinusoidal signal $x(t) = A\cos(\omega t + \alpha)$, $-\infty \le t \le \infty$, is applied to a *stable* linear time-invariant system, then to obtain the steady-state output $y_{ss}(t)$,

1. The input amplitude A is multiplied by the gain $|H(j\omega)|$ (magnitude of frequency response), and
2. The input phase α is shifted by the angle $\angle H(j\omega)$ (phase of frequency response) to yield the steady-state output $y_{ss}(t)$. This result also holds for a constant, or DC, input, since using $\omega = 0$ in the Euler relation yields $\cos(0t) = 0.5\{e^{j0t} + e^{-j0t}\} = 1$.

Comment: A bit later in this chapter and in Chapter 3 on Laplace transforms we'll see that the system frequency response $H(j\omega)$ can also be found without having to actually evaluate the integral

system frequency response

$$H(j\omega) = \int_{-\infty}^{\infty} h(\tau)e^{-j\omega\tau}\, d\tau .$$

ILLUSTRATIVE PROBLEM 2.6 ***Using the Sinusoidal Steady-State Formula***

A highpass analog filter that attenuates undesirable signal components at low frequencies and passes desirable high-frequency components has the unit impulse response $h(t) = \delta(t) - 10e^{-10t}u(t)$. Find the steady-state response of this filter to the input signal $x(t) = 5 + 5\cos(10t)$, $-\infty \le t \le \infty$.

Solution

First we find $H(j\omega)$ and then evaluate this frequency response at the two input frequencies of $\omega = 0$ and $\omega = 10$ rad/s. The system frequency response $H(j\omega)$ is

$$H(j\omega) = \int_{0}^{\infty} [\delta(\tau) - 10e^{-10\tau}]e^{-j\omega\tau}\, d\tau = 1 - \frac{10}{10 + j\omega} = \frac{j\omega}{10 + j\omega}.$$

For the DC term ($\omega = 0$), we find that $H(j0) = j0/(10 + j0) = 0$ and for $\omega = 10$, $H(j10) = j10/(10 + j10) = 0.707e^{j0.785}$. The filter output is $y_{SS}(t) = 3.53\cos(10t + 0.785)$, $-\infty \le t \le \infty$. The input and output signals are shown in Figure 2.17, where it is seen that this highpass filter does indeed eliminate the DC component. It also attenuates and shifts the phase of the sinusoid $5\cos(10t)$.

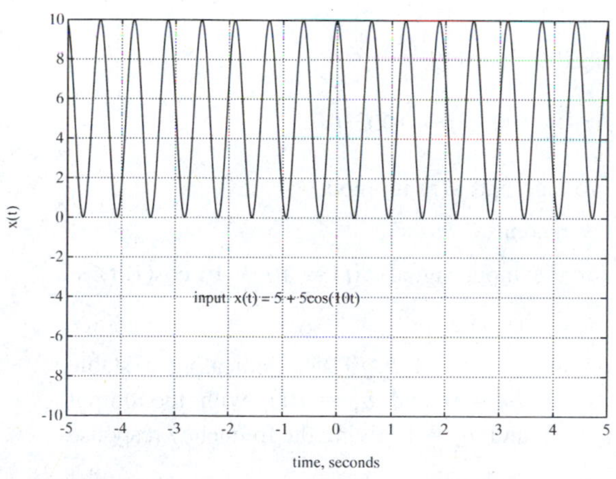

a. System input

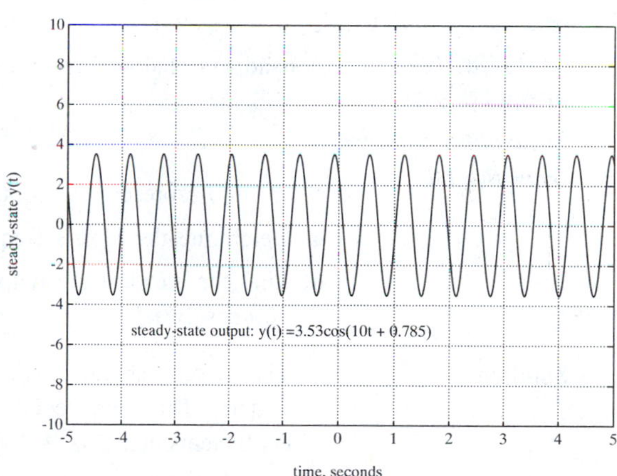

b. Steady-state output

FIGURE 2.17 *Input and steady-state output for a highpass filter*

Alternative Path to $H(j\omega)$ Frequently we would like to find a system's frequency response directly from its differential equation model,

$$\sum_{k=0}^{N} a_k \frac{d^k y(t)}{dt^k} = \sum_{k=0}^{L} b_k \frac{d^k x(t)}{dt^k}.$$

CHAPTER 2

For the complex exponential input $x(t) = e^{j\omega t}$, the kth derivative of the input is $(j\omega)^k e^{j\omega t}$. Knowing that the steady-state solution for this special input is $y(t) = e^{j\omega t}H(j\omega)$, it follows that the kth derivative of the output is $d^k y(t)/dt^k = (j\omega)^k e^{j\omega t} H(j\omega)$. Substituting the expressions for $x(t)$ and $y(t)$ into the general differential equation gives

$$\sum_{k=0}^{N} a_k H(j\omega)(j\omega)^k e^{j\omega t} = \sum_{k=0}^{L} b_k (j\omega)^k e^{j\omega t}$$

which can be solved for the frequency response $H(j\omega)$ as

system frequency response

$$H(j\omega) = \frac{\sum_{k=0}^{L} b_k (j\omega)^k}{\sum_{k=0}^{N} a_k (j\omega)^k}.$$

Thus, the frequency response can be found from the DE coefficients and the appropriate power of $(j\omega)$. Remember that the system must always be checked for stability before applying the frequency response to obtain the steady-state response.

ILLUSTRATIVE PROBLEM 2.7
Sinusoidal Analysis from the DE

A bandpass analog filter is described by the DE

$$\ddot{v}(t) + 2\dot{v}(t) + 100v(t) = 100\dot{x}(t)$$

where $v(t)$ represents the output voltage and $x(t)$ its input.

a. Determine the filter's frequency response.

b. Find the steady-state output for the input signal $x(t) = 10 + 10\cos(10t) + 100\cos(100t)$.

Solution

a. The system characteristic roots are $s_{1,2} = -1 \pm j9.950$, indicating a stable system. The input coefficients are $b_0 = 0$ and $b_1 = 100$, with the output coefficients being $a_0 = 100$, $a_1 = 2$, and $a_2 = 1$, giving the frequency response

$$H(j\omega) = \frac{100j\omega}{(j\omega)^2 + 2j\omega + 100}.$$

b. $H(j0) = 0, H(j10) = 50$, and $H(j100) = 0.01e^{-j1.55}$, making the steady-state output $y_{SS}(t) = 500\cos(10t) + 1\cos(100t - 1.55)$, where it is seen that the midfrequency of 10 rad/s is amplified quite nicely.

THE STATE-SPACE MODEL

A third description of linear continuous systems treated in this chapter is called the state-space, or first-order, model. Used extensively in computer simulation of systems, this model represents differential equations of arbitrary order as a set of first-order differential equations. A general Nth-order linear constant-coefficient differential equation is

$$\frac{d^N y(t)}{dt^N} + a_{N-1}\frac{d^{N-1}y(t)}{dt^{N-1}} + \cdots + a_1\frac{dy(t)}{dt} + a_0 y(t) = b_0 x(t)$$

where $y(t)$ is the output, $x(t)$ is the single (scalar) input, and there are N derivatives of the system output and no (zero) derivatives of the system input.[6] To represent this system by N first-order equations, we define N new variables $v_1(t), v_2(t), \ldots, v_N(t)$ as

$$v_1(t) = y(t),\ v_2(t) = \frac{dy(t)}{dt},\ v_3(t) = \frac{d^2y(t)}{dt^2}, \cdots,\ v_N(t) = \frac{d^{N-1}y(t)}{dt^{N-1}}.$$

Then we can establish the following set of first-order differential equations that are commonly called state equations.

$$\begin{aligned}
\frac{dv_1(t)}{dt} &= \frac{dy(t)}{dt} = v_2(t) \\
\frac{dv_2(t)}{dt} &= \frac{d^2y(t)}{dt^2} = v_3(t) \\
&\ \ \vdots \\
\frac{dv_{N-1}(t)}{dt} &= \frac{d^{N-1}y(t)}{dt^{N-1}} = v_N(t) \\
\frac{dv_N(t)}{dt} &= \frac{d^N y(t)}{dt^N} = -a_0 v_1(t) - a_1 v_2(t) - \cdots - a_{N-1} v_N(t) + b_0 x(t).
\end{aligned}$$

Putting all of this into matrix form, which is well suited for computation, we call $v_1(t), v_2(t), \ldots, v_N(t)$ the states or state variables and define the state vector as

state vector

$$\mathbf{v}(t) = [\,v_1(t) \quad v_2(t) \quad \cdots \quad v_N(t)\,]^T$$

where the superscript T stands for the matrix transpose operator. In some disciplines, notably control engineering, the symbols $x_1(t), x_2(t), \ldots, x_N(t)$ are used to denote the state variables. Digital signal processing specialists, however, generally use $x_1(t), \ldots,$ to denote the system inputs, and we have adopted this convention in the text. As a result, another symbol is needed for state variables, and we have selected $v_i(t), i = 1, \ldots, N$. This schizophrenia in notation also shows up in MATLAB, where the control notation is used for

[6] In Chapter 6, "State-Space Topics for Continuous Systems," we consider the general situation where the differential equation contains terms involving derivatives of the input and/or multiple inputs.

state equations, whereas $x(t)$ is frequently used as an input when using transfer functions or differential equations. Finally, the matrix state equation is

matrix state differential equation

$$\frac{d\mathbf{v}(t)}{dt} = \begin{bmatrix} 0 & 1 & 0 & \cdots & 0 \\ 0 & 0 & 1 & 0 & \cdots \\ \vdots & \vdots & \vdots & \vdots & \vdots \\ 0 & 0 & 0 & \cdots & 1 \\ -a_0 & -a_1 & \cdots & -a_{N-2} & -a_{N-1} \end{bmatrix} \mathbf{v}(t) + \begin{bmatrix} 0 \\ 0 \\ \vdots \\ 0 \\ b_0 \end{bmatrix} x(t)$$

$$= \mathbf{Av}(t) + \mathbf{B}x(t).$$

A common definition for the state of a system is as follows:

> *The state of a system is a minimum set of quantities $v_1(t), v_2(t), \ldots, v_N(t)$, which if known at $t = t_0$ are uniquely determined for $t \geq t_0$ by specifying the inputs to the system for $t \geq t_0$.*

The M *outputs* $\mathbf{y}(t)$ of a system are related to the states $\mathbf{v}(t)$ and a single (scalar) input $x(t)$ by the *output equation*

$$\mathbf{y}(t) = \mathbf{Cv}(t) + \mathbf{D}x(t)$$

where $\mathbf{C}$ is an M by N matrix and $\mathbf{D}$ is an M by 1 vector.

ILLUSTRATIVE PROBLEM 2.8
Putting Differential Equations in State-Space Form

A cart with an inverted pendulum is modeled physically by the drawing of Figure 2.18. For very small values of the angle $\theta(t)$, the describing differential equations are

$$\ddot{\theta}(t) = \theta(t) + x(t) \quad \text{and} \quad \ddot{p}(t) = \beta\theta(t) - x(t)$$

where $x(t)$ is the input force, $p(t)$ is the horizontal displacement, and β is a real constant.

a. Describe the system in state variable form with $v_1(t) = \theta(t)$ and $v_3(t) = p(t)$.

b. Find the output equation if the outputs are defined as $y_1(t) = \theta(t)$ and $y_2(t) = p(t)$.

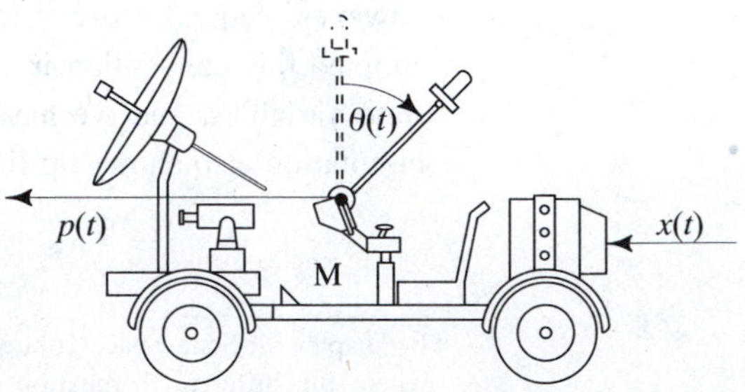

FIGURE 2.18 *Cart with inverted pendulum*

Solution **a.** With $v_1(t) = \theta(t)$, we have $dv_1(t)/dt = v_2(t)$ and $dv_2(t)/dt = d^2\theta(t)/dt^2 = v_1(t) + x(t)$. For $v_3(t) = p(t)$, the remaining state equations are $dv_3(t)/dt = v_4(t)$ and $dv_4(t)/dt = d^2p(t)/dt^2 = \beta v_1(t) - x(t)$. The state matrix differential equation is

$$\begin{bmatrix} \dot{v}_1(t) \\ \dot{v}_2(t) \\ \dot{v}_3(t) \\ \dot{v}_4(t) \end{bmatrix} = \begin{bmatrix} 0 & 1 & 0 & 0 \\ 1 & 0 & 0 & 0 \\ 0 & 0 & 0 & 1 \\ \beta & 0 & 0 & 0 \end{bmatrix} \begin{bmatrix} v_1(t) \\ v_2(t) \\ v_3(t) \\ v_4(t) \end{bmatrix} + \begin{bmatrix} 0 \\ 1 \\ 0 \\ -1 \end{bmatrix} x(t) .$$

b. For the defined outputs of $y_1(t) = v_1(t)$ and $y_2(t) = v_3(t)$, the matrix output equation is

$$\begin{bmatrix} y_1(t) \\ y_2(t) \end{bmatrix} = \begin{bmatrix} 1 & 0 & 0 & 0 \\ 0 & 0 & 1 & 0 \end{bmatrix} \begin{bmatrix} v_1(t) \\ v_2(t) \\ v_3(t) \\ v_4(t) \end{bmatrix} .$$

SYSTEM SIMULATION

To illustrate an important application of the state-space model, let's use the MATLAB function **lsim** (linear systems simulation) from the Signals and Systems Toolbox or the function **kslsim** from the California Functions file to plot (a) the initial condition response and (b) the unit step response of the system of Illustrative Problem 2.2 on page 46, where a flexible shaft in torsion was described by the second-order differential equation

$$\ddot{\theta}(t) + 3\dot{\theta}(t) + 2\theta(t) = 0.333\lambda(t) ,$$

where $\theta(t)$ represents the angular position, $d\theta(t)/dt = \omega(t)$ is the angular velocity, and $\lambda(t)$ is the driving torque. From the California Functions file we use the function

$$[\mathbf{y}, \mathbf{v}] = \mathbf{kslsim}(\mathbf{A}, \mathbf{B}, \mathbf{C}, \mathbf{D}, \mathbf{X}, \mathbf{t}, \mathbf{v0}) ,$$

where it is seen that if the matrices **A**, **B**, **C**, **D**, and **X** are supplied along with the initial condition vector **v0** and **t**, the time vector,[7] **kslsim** will return the time histories of the output **y** and of the states **v**. The matrix **X** must have as many columns as there are inputs, where each row of **X** corresponds to a new time point. Given next is a script that might be used along with some comments, thc input data as a check, and plots of the results in Figure 2.19a and b. Notice that this script returns the input data as well as the size of **X**,

[7] Since we want to determine the response for a range of values of t, a vector **t** is defined consisting of the values $[0, T, 2T, \ldots]$, where T is the interval between the computed values of $\mathbf{v}(t)$.

an indicator to detect a common source of errors. We define $v_1(t) = \theta(t)$ and $v_2(t) = \omega(t) = dv_1(t)/dt$, $y(t) = \theta(t) = v_1(t)$, and $\lambda(t) = x(t)$. Thus

$$\mathbf{A} = \begin{bmatrix} 0 & 1 \\ -2 & -3 \end{bmatrix}, \ \mathbf{B} = \begin{bmatrix} 0 \\ 1/3 \end{bmatrix}, \ \mathbf{C} = [1 \quad 0], \quad \text{and} \quad \mathbf{D} = d = 0.$$

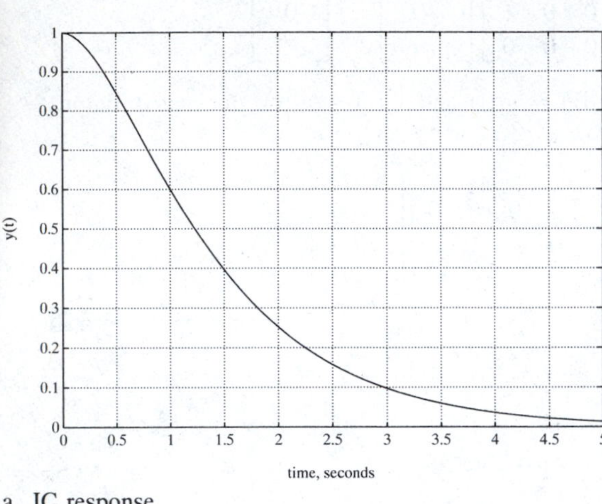

a. IC response

b. Unit step response

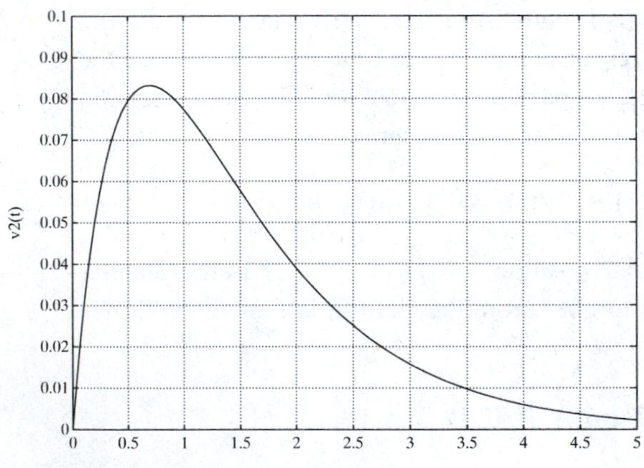

c. Angular velocity

FIGURE 2.19 *System responses using* **lsim** *or* **kslsim**

MATLAB Script

```
%F2_19 System response using lsim or kslsim

%F2_19a IC response
A=[0,1;-2,-3]                          % elements of A matrix
B=[0;0.333]                            % elements of B matrix
C=[1,0]                                % elements of C matrix
D=[0]                                  % elements of D matrix (a scalar)
```

```
v0=[1;0];                        % initial condition state vector
t=0:0.05:5;                      % start t:increment:stop t
X=[0*ones(size(t))]';            % need to transpose to make X a column
                                 % vector with one row for each time point
size (X)                         % checks the dimensions of the X matrix
[y,v]=kslsim(A,B,C,D,X,t,v0);    % call kslsim
%...plotting statements and pause
```

Output data

```
A =
     0     1
    -2    -3
B =
     0
  0.3330
C =
     1     0
D =
     0
ans =
   101  1
```

MATLAB Script

```
%F2_19b Unit step response
A=[0,1;-2,-3]
B=[0;0.333]
C=[1,0]
D=[0]
v0=[0;0];                        % initial condition state vector
t=0:0.05:5;                      % start t:increment:stop t
X=[1*ones(size(t))]';            % input row vector
size (X)                         % checks the dimensions of the X matrix
[y,v]=kslsim(A,B,C,D,X,t,v0);    % call kslsim
%...plotting statements and pause
```

CROSS-CHECK Are these plots reasonable? The IC response shows a stable system, which is the situation with characteristic roots of $s_{1,2} = -1, -2$. The step response settles out at about $y_{ss} = 0.16$. In steady state, for a constant input, all rates of change go to zero and the system DE becomes $2\theta_{ss} = 0.33\lambda = 0.33(1)$ and $\theta_{ss} = 0.167$, which agrees with the plot. ■

In each of the two preceding simulations we could also plot or look at other quantities. If, for example, we wanted to see the shaft angular velocity

$\omega(t) = v_2(t)$ as a function of time, we could do this by using the MATLAB command "plot (t, v(:, 2))." The result of doing this is given in Figure 2.19c.

SOLVED EXAMPLES AND MATLAB APPLICATIONS

EXAMPLE 2.1
An Oscillatory System

An electronic oscillator is modeled by the second-order differential equation

$$\ddot{y}(t) + \omega_0{}^2 y(t) = kx(t)$$

where $y(t)$ is the output, $x(t)$ is the input, and ω_0 and k are adjustable system constants.

a. Find the characteristic equation and the characteristic roots.

b. Find the equation for the response to initial conditions of $y(0) = 2$, $[dy(t)/dt]|_{t=0} = 0$ with the input $x(t) = 0$.

c. Write a state-space description and MATLAB script for this oscillator and an output equation if $y(t)$ is defined as the output. Assume that $f_0 = 1$ Hz or $\omega_0 = 2\pi$ rad/s and use the script to plot the IC response with initial conditions as in part (b).

d. Suppose we want the oscillator to produce the generic sinusoid

$$y_{\mathrm{IC}}(t) = A\cos(\omega t + \alpha)u(t)$$

where A, ω, and α can take on any desired values. Any of these three constants can be achieved by adjustment of the parameter ω_0 in the DE and by setting the ICs $y(0)$ and $[dy(t)/dt]|_{t=0}$. Determine general relationships for A, ω, and α in terms of ω_0, $y(0)$, and $[dy(t)/dt]|_{t=0}$. Test your results by choosing a sinusoid of your choice and using **lsim** or **kslsim** to plot the selected signal.

Solution

a. *Characteristic equation and roots.* The homogeneous DE

$$\ddot{y}(t) + \omega_0{}^2 y(t) = 0$$

yields the characteristic equation $s^2 + \omega_0{}^2 = 0$ and the characteristic roots $s_{1,2} = \pm j\omega_0$.

b. *Initial condition solution.* Assuming the solution $y_{\mathrm{IC}}(t) = C_1 e^{j\omega_0 t} + C_2 e^{-j\omega_0 t}$ and evaluating this equation at $t = 0$ gives

$$y_{\mathrm{IC}}(0) = 2 = C_1 + C_2. \tag{1}$$

Differentiating $y_{\mathrm{IC}}(t)$ yields $dy_{\mathrm{IC}}(t)/dt = j\omega_0 C_1 e^{j\omega_0 t} - j\omega_0 C_2 e^{-j\omega_0 t}$, and evaluating at $t = 0$ gives

$$\dot{y}_{\mathrm{IC}}(0) = 0 = j\omega_0 C_1 - j\omega_0 C_2. \tag{2}$$

Solving Equations (1) and (2) gives $C_1 = C_2 = 1$ and the initial condition solution

$$y_{\mathrm{IC}}(t) = 1e^{j\omega_0 t} + 1e^{-j\omega_0 t} = 2\cos(\omega_0 t)u(t).$$

c. *State-space description.* For $v_1(t) = y(t)$ and $v_2(t) = dv_1(t)/dt$, we have

$$\dot{\mathbf{v}}(t) = \begin{bmatrix} 0 & 1 \\ -39.478 & 0 \end{bmatrix}\mathbf{v}(t) + \begin{bmatrix} 0 \\ 1 \end{bmatrix}x(t) \quad \text{and} \quad y(t) = [1 \quad 0]\mathbf{v}(t).$$

To verify the analytical solution of part (b), we use the function **kslsim** from the California Functions file to plot the initial condition response in Figure E2.1a on page 74. The script is given next.

EXAMPLES

MATLAB Script

```
%E2_1 Oscillatory responses

%E2_1a IC response
A=[0,1;-39.478,0];            % A matrix ... 2 rows, 2 columns
B=[0;1];                      % B matrix ... 2 rows, 1 column
C=[1,0];                      % C matrix ... 1 row, 2 columns
D=[0];                        % D matrix ... a scalar
v0=[2;0];                     % IC vector ... 2 rows, 1 column
t=0:.01:4;                    % starting time:increment:final time
X=[0*ones(size(t))]';         % input vector, x(t)=0
[y,v]=kslsim(A,B,C,D,X,t,v0); % call kslsim
%...plotting statements and pause
```

d. From part (b) we see that the system parameter ω_0 determines the frequency of the IC response, so $\omega = \omega_0$. The desired signal $y_{\mathrm{IC}}(t)$ can be written as (dropping the $u(t)$ and the IC notation)

$$y(t) = A\cos(\omega t + \alpha) = A[\cos\omega t \cos\alpha - \sin\omega t \sin\alpha]$$

and its rate of change becomes

$$dy(t)/dt = -A\omega\sin(\omega t + \alpha) = -A\omega[\sin\omega t\cos\alpha + \cos\omega t\sin\alpha].$$

Evaluating these two equations for $t = 0$ yields the design relationships

$$y(0) = A\cos\alpha \quad \text{and} \quad [dy(t)/dt]|_{t=0} = -A\omega\sin\alpha.$$

To test these results we choose to generate the sinusoidal signal $y(t) = 0.001\sin(2\pi\cdot 10^3 t)$—that is, a zero-to-peak amplitude of 1 mV and a frequency of 1 kHz. Thus $\omega = \omega_0 = 2\pi\cdot 10^3$, $A = 0.001$, and $\alpha = -\pi/2$. The initial conditions become $y(0) = 10^{-3}\cos(-\pi/2) = 0$ and $[dy(t)/dt]|_{t=0} = (-10^{-3})(2\pi)(10^3) = 2\pi$. The matrix $\mathbf{A}$ changes from part (a) because ω_0 is now $2\pi\cdot 10^3$ rather than simply 2π. The script is next, with the plot in Figure E2.1b, where we see that the desired waveform has been generated.

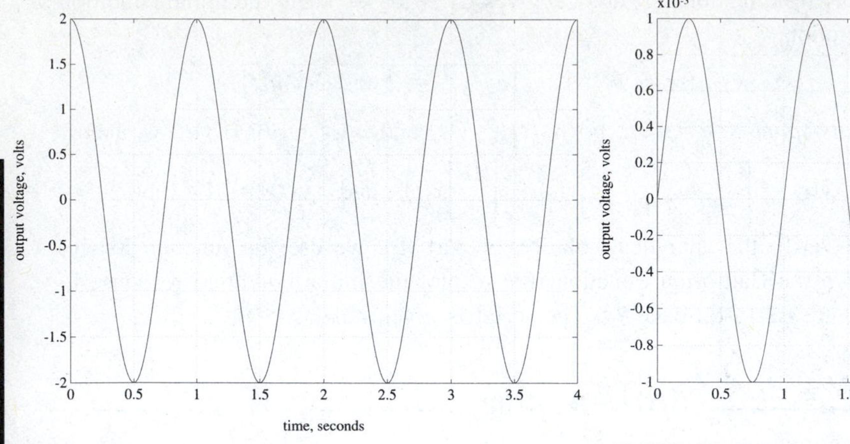

FIGURE E2.1a *IC response*

FIGURE E2.1b *Desired sinusoid*

MATLAB Script

```
%E2_1b Desired sinusoid
A=[0,1;-39.478e6,0];              % A matrix ... 2 rows, 2 columns
B=[0;1];                          % B matrix ... 2 rows, 1 column
C=[1,0];                          % C matrix ... 1 row, 2 columns
D=[0];                            % D matrix ... a scalar
v0=[0;6.2831];                    % IC vector ... 2 rows, 1 column
t=0:.00001:.004;                  % starting time:increment:final time
X=[0*ones(size(t))]';             % input vector, x(t)=0
[y,v]=kslsim(A,B,C,D,X,t,v0);     % call kslsim
%...plotting statements
```

EXAMPLE 2.2
Second-Order Systems

Many physical systems such as an *LRC* circuit or the equally traditional mass-spring-damper configuration can be modeled by the second-order DE

$$\ddot{y}(t) + 2\zeta\omega_n\dot{y}(t) + {\omega_n}^2 y(t) = Kx(t)$$

where ω_n = the undamped natural frequency in radians/time unit, ζ (zeta) = the damping ratio, K is a real constant, and $y(t)$ and $x(t)$ are the system's output and input, respectively.

a. The circuit of Figure E2.2a can be described by the simultaneous differential equations

$$(1)\ L\,di(t)/dt + Ri(t) + v(t) = e(t) \quad \text{and} \quad (2)\ C\,dv(t)/dt = i(t).$$

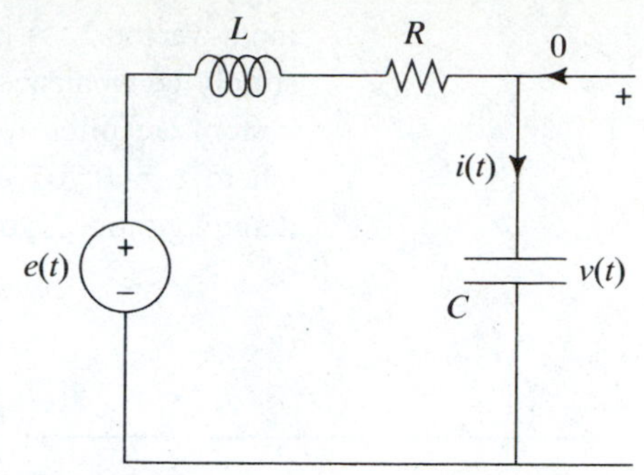

FIGURE E2.2a *Generic LRC circuit*

Combine these two first-order differential equations to obtain one second-order differential equation of the form

$$\ddot{v}(t) + \alpha\dot{v}(t) + \beta v(t) = \gamma e(t).$$

b. Find two varieties of the system's characteristic equation, one in terms of the circuit parameters L, R, and C and the other in terms of the more general parameters ω_n and ζ.

c. Assume that L and C are fixed and that $0 \leq R \leq \infty$. Find the range of values of R that will yield
 (i) Purely imaginary characteristic roots,
 (ii) Complex characteristic roots, and
 (iii) Real characteristic roots.
 Make a sketch of how you think the root locations vary for $0 \leq R \leq \infty$.

d. For a fixed value of ω_n and $0 \leq \zeta \leq \infty$, find the range of values of ζ that will yield
 (i) Purely imaginary characteristic roots,
 (ii) Complex characteristic roots, and
 (iii) Real characteristic roots.
 Make a sketch of how you think the root locations vary for $0 \leq \zeta \leq \infty$.

e. Put the second-order DE,

$$\ddot{v}(t) + 2\zeta\omega_n\dot{v}(t) + {\omega_n}^2 v(t) = kx(t),$$

into first-order or state-space form. Let $v_1(t) = y(t)$, $v_2(t) = dy(t)/dt$.

WHAT IF? Choose some different values of ζ and ω_n and investigate the unit step response of this system, making use of the MATLAB function **step** from either the Signals and Systems or the Control System Toolbox. Alternatively, the program **kslsim** can just as well be used with the input vector $\mathbf{X} = [\mathbf{1}^*\mathbf{ones}(\mathbf{size}(\mathbf{t}))]'$. The programs **step** (and **impulse**) are simply conveniences for the user, since the step and impulse responses of a system are often required. A script that you can use for **step** is given here, where $\zeta = 0.707$ and $K = \omega_n{}^2 = 100$. In selecting different values of the damping ratio ζ, you may want to include a negative value as well as a value of $\zeta > 1$.

■

MATLAB Script

```
%E2_2 Effect of roots on system response

%E2_2d Step response for ζ = 0.707
zeta=0.707;                            % damping ratio
wn = 10;                               % natural frequency
K = 100;                               % constant
wnsq = wn^2;
A=[0,1;-wnsq,-2*zeta*wn];              % A matrix
B=[0;K];                               % B matrix
C=[1,0];
D=[0];
t=0:.005:2;                            % start:increment:stop
y=step(A,B,C,D,1,t);                   % call step(the 1 designates the
                                       % number of the input to be used)
%...plotting statements and pause
```

Solution

a. *Combining equations.* We need to eliminate $i(t)$ and $di(t)/dt$ in the equation $L \cdot di(t)/dt + Ri(t) + v(t) = e(t)$. From Equation (2) we have $C \cdot dv(t)/dt = i(t)$ and $C \cdot d^2v(t)/dt^2 = di(t)/dt$. Substitution and some algebra give

$$\ddot{v}(t) + \frac{R}{L}\dot{v}(t) + \frac{1}{LC}v(t) = \frac{1}{LC}e(t).$$

b. *The characteristic equations.* The homogeneous DEs are

$$\ddot{v}(t) + \frac{R}{L}\dot{v}(t) + \frac{1}{LC}v(t) = 0 \quad \text{and} \quad \ddot{y}(t) + 2\zeta\omega_n\dot{y}(t) + \omega_n{}^2y(t) = 0$$

giving the two forms of the characteristic equation,

$$s^2 + \frac{R}{L}s + \frac{1}{LC} = 0 \quad \text{and} \quad s^2 + 2\zeta\omega_n s + \omega_n{}^2 = 0.$$

c. *Range of R for different kinds of roots.* Using the quadratic formula, the two roots are

$$s_{1,2} = \frac{1}{2}\left\{-\frac{R}{L} \pm \sqrt{\frac{R^2}{L^2} - \frac{4}{LC}}\right\}.$$

(i) $R = 0$ gives $s_{1,2} = \pm j1/\sqrt{LC}$.

(ii) For complex roots $(R^2/L^2 - 4/LC) < 0$, or $R < 2\sqrt{L/C}$.

(iii) For real roots $(R^2/L^2 - 4/LC) \geq 0$, or $R \geq 2\sqrt{L/C}$. Notice that when $R = 2\sqrt{L/C}$, the roots are real and equal, with $s_{1,2} = -1\sqrt{1/LC}$. The locus of the characteristic roots for $0 \leq R \leq \infty$ with $1/LC = 1$ is given in Figure E2.2b.[8]

EXAMPLES

d. *Range of ζ for different kinds of roots.* Using the quadratic formula, the two roots are

$$s_{1,2} = 0.5\left\{-2\zeta\omega_n \pm \sqrt{4\zeta^2\omega_n{}^2 - 4\omega_n{}^2}\right\} = -\zeta\omega_n \pm \omega_n\sqrt{\zeta^2 - 1}.$$

(i) For $\zeta = 0$, $s_{1,2} = \pm j\omega_n$

(ii) For $0 \leq \zeta < 1$, $s_{1,2} = \omega_n\left\{-\zeta \pm j\sqrt{1 - \zeta^2}\right\}$.

(iii) For $\zeta \geq 1$, the roots are negative real.

The locus of the characteristic roots for $0 \leq \zeta \leq \infty$ is given in Figure E2.2c for $\omega_n = 1$.

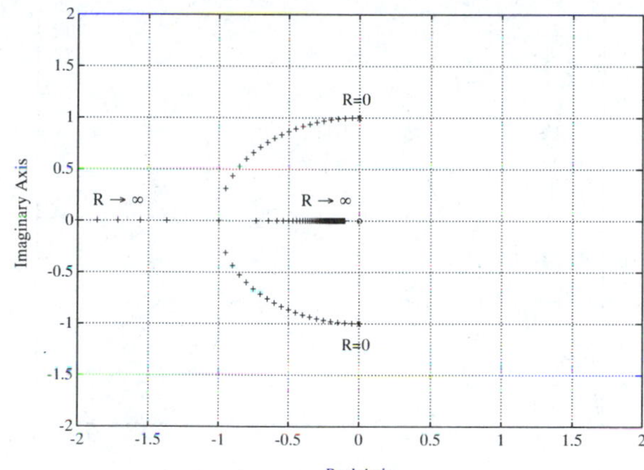

FIGURE E2.2b *Locus of roots,* $0 \leq R \leq \infty$

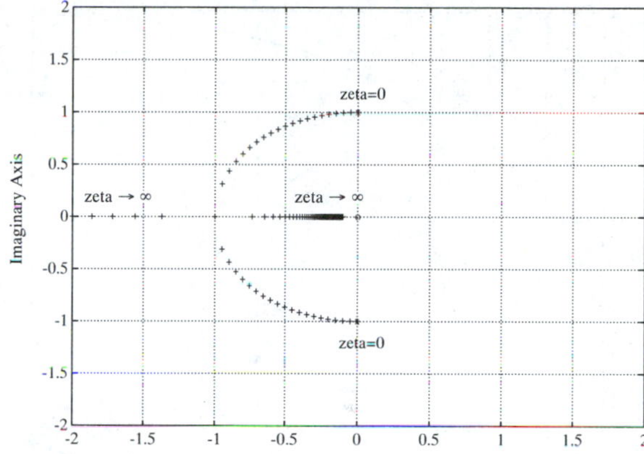

FIGURE E2.2c *Locus of roots,* $0 \leq \zeta \leq \infty$

e. *State-space, or first-order, model.* For $v_1(t) = y(t)$ we have

$$\dot{\mathbf{v}}(t) = \begin{bmatrix} 0 & 1 \\ -\omega_n{}^2 & -2\zeta\omega_n \end{bmatrix}\mathbf{v}(t) + \begin{bmatrix} 0 \\ K \end{bmatrix}x(t).$$

WHAT IF? See Figures E2.2d–f for three different values of ζ with $\omega_n = 10$ rad/time unit. Figures E2.2g and h show step responses for $\zeta = 0.1$ with $\omega_n = 10$ rad/time unit and $\omega_n = 100$ rad/time unit, respectively. ■

[8]There is a systematic procedure, called the root locus method, for showing the migration of a system's characteristic roots as a system parameter is varied. The method, developed in 1948 by Walter R. Evans, then an aerospace engineer in Southern California, is a graphical procedure that is available as the MATLAB function **rlocus**. Figure E2.7b was obtained by using **rlocus**. The root locus technique is a design method that is included in undergraduate courses in control systems.

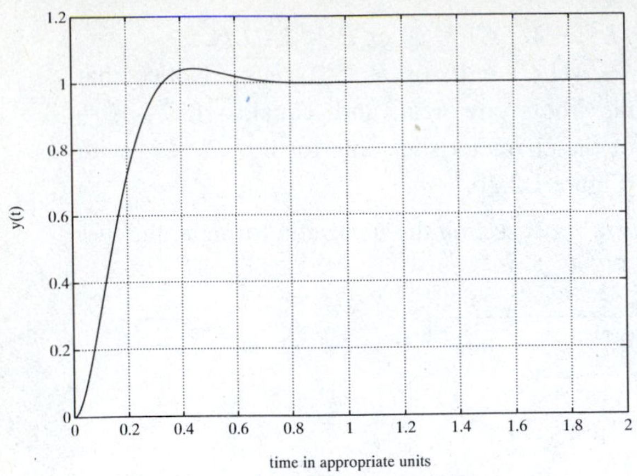

FIGURE E2.2d *Step response,* $\zeta = 0.707$

Comment: It is rumored that pilots like this value of ζ, because it yields a rapid response with just a slight overshoot that won't spill the coffee.

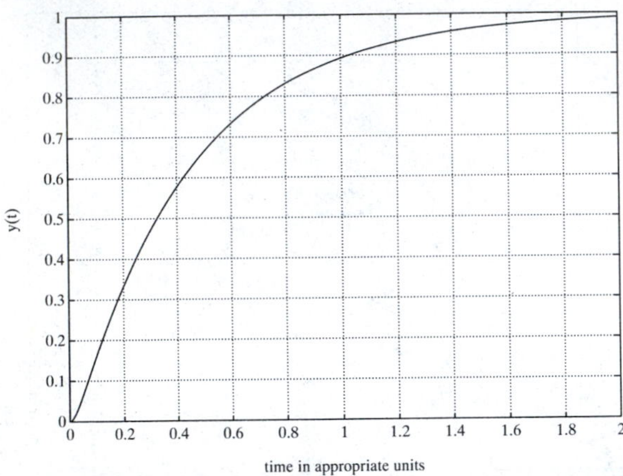

FIGURE E2.2f *Step response,* $\zeta = 2.3$

Comment: They (passengers and crew) like this better, but it's pretty slow.

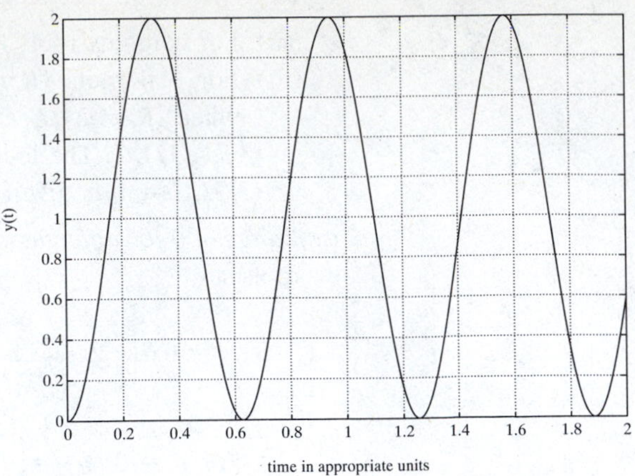

FIGURE E2.2e *Step response,* $\zeta = 0$

Comment: It is a fact that neither passengers nor crew like $\zeta = 0$.

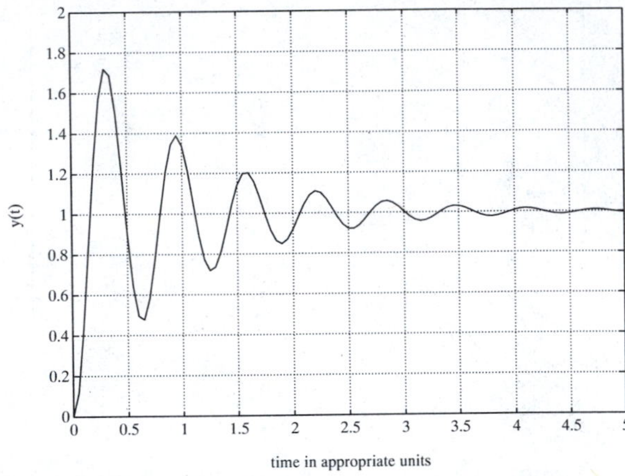

FIGURE E2.2g *Step response,* $\zeta = 0.1$, *natural frequency* $= 10$

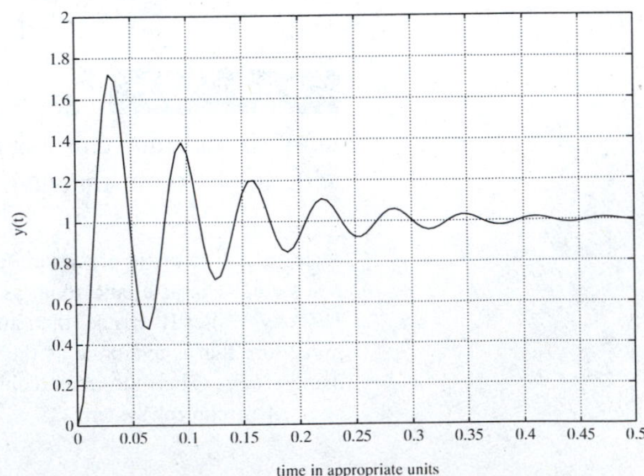

FIGURE E2.2h *Step response,* $\zeta = 0.1$, *natural frequency* $= 100$

Comment: The time required for a stable system to reach steady-state decreases as $\zeta\omega_n$ increases. For $\zeta\omega_n = 1$, it takes about 5 time units, whereas for $\zeta\omega_n = 10$, approximately 0.5 time unit elapses before reaching steady-state. This characteristic is shown in Figures E2.2g and h.

EXAMPLE 2.3
Unit Impulse Response of a Lowpass Filter

A lowpass RC filter can be modeled by the differential equation

$$\dot{v}(t) + \frac{1}{RC}v(t) = \frac{1}{RC}e(t)$$

where $e(t)$ and $v(t)$ are the input and output voltages, respectively. We want to determine an analytical expression for the unit impulse response for this filter. This will be accomplished by first finding the response to an input voltage of a pulse $p(t)$ with an "area" of 1.

a. What is the characteristic equation for this filter?

b. Find the output $v(t)$ for the step input $e_1(t) = (1/\Delta)u(t)$ with the initial voltage $v(0) = 0$. Your answer will be in terms of the literal quantity Δ, which will be a variable in this problem.

c. From the result in (b), deduce $v(t)$ if the input is the delayed step $e_2(t) = (1/\Delta)u(t - \Delta)$.

d. By letting $p(t) = e_1(t) - e_2(t)$, a voltage pulse of unit area is defined and is shown in Figure E2.3a. What is the output voltage $v(t)$ if the input is $e(t) = p(t)$?

e. Now take the limit of $v(t)$ as $\Delta \to 0$ and show that the unit impulse response of the filter is

$$h(t) = \frac{1}{RC}e^{-t/RC}u(t).$$

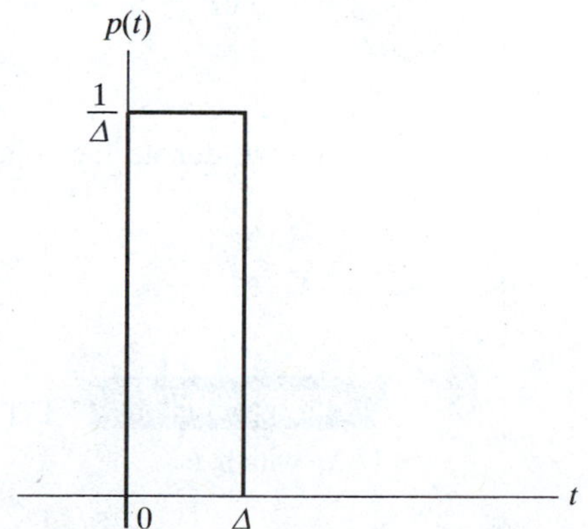

FIGURE E2.3a *Unit pulse input signal*

f. Let's look at the effect of the pulse width Δ on the output voltage $v(t)$ for the pulse input $p(t)$. Plot the output using **lsim** or **kslsim** for $\Delta = 1, 0.5, 0.1, 0.05$, and 0.01. For simplicity, let $1/RC = 1$. Does the output voltage $v(t)$ approach the shape of $h(t)$ as $\Delta \rightarrow 0$?

WHAT IF? Suppose that $1/RC = 10$ rather than 1. Estimate the pulse width Δ required for the unit pulse response to be a good approximation of the unit impulse response $h(t)$. Verify your answer with a computer-generated plot. ■

Solution

a. *Characteristic equation.* The homogeneous DE, $dv(t)/dt + (1/RC)v(t) = 0$, leads to the characteristic equation, $s + 1/(RC) = 0$.

b. *Output due a step input.* For the DE $dv(t)/dt + (1/RC)v(t) = (1/RC)e(t)$, we substitute $e(t) = (1/\Delta)u(t)$, which yields

$$\dot{v}(t) + \frac{1}{RC}v(t) = \frac{1}{\Delta RC} \quad \text{for } t \geq 0.$$

Since at this point we really don't know a general method to solve for $v(t)$, we'll use a bit of foreknowledge. As $t \rightarrow \infty$, it can be shown that the voltage $v(t)$ will reach a constant value. This condition is known as the steady-state and as $t \rightarrow \infty$, $dv(t)/dt \rightarrow 0$ and

$$\frac{1}{RC}v(t) = \frac{1}{\Delta RC} \quad \text{or} \quad v_p(t) = \frac{1}{\Delta},$$

where $v_p(t)$ is called the particular solution. The homogeneous solution is

$$v_h(t) = Ke^{-t/RC},$$

where K is a to-be-determined constant; consequently, the complete solution is

$$v(t) = v_h(t) + v_p(t) = Ke^{-t/RC} + \frac{1}{\Delta}.$$

With $v(0) = 0$, we find that

$$K = -\frac{1}{\Delta},$$

and we denote the solution as

$$v_1(t) = \frac{1}{\Delta}(1 - e^{-t/RC})u(t).$$

CROSS-CHECK Let's see if $v_1(t)$ satisfies the original DE with $e(t) = 1/\Delta$, which is

$$\dot{v}(t) + \frac{1}{RC}v(t) = \frac{1}{RC}e(t) = \frac{1}{\Delta RC} \quad \text{for } t \geq 0.$$

For $t \geq 0$ we have

$$\frac{dv_1(t)}{dt} = +\frac{1}{\Delta RC}e^{-t/RC} \quad \text{and}$$

$$\begin{aligned}\dot{v}_1(t) + \frac{1}{RC}v_1(t) &= \frac{1}{\Delta RC}e^{-t/RC} + \frac{1}{\Delta RC}[1 - e^{-t/RC}] \\ &= \frac{1}{RC}e(t) = \frac{1}{\Delta RC}.\end{aligned}$$

■

c. *Output due to a delayed step.* The circuit is time invariant, so the output voltage due to the delayed step voltage $e_2(t)$ is that of part (b) delayed by Δ seconds, or

$$v_2(t) = \frac{1}{\Delta}(1 - e^{-(t-\Delta)/RC})u(t - \Delta).$$

d. *Output due to a unit pulse input.* Because of linearity, the output due to $e_1(t) - e_2(t)$ will be $v(t) = v_1(t) - v_2(t)$, or

unit pulse response

$$v(t) = \frac{1}{\Delta}(1 - e^{-t/RC})u(t) - \frac{1}{\Delta}(1 - e^{-(t-\Delta)/RC})u(t - \Delta).$$

e. *Output due to an impulse input.* Using the series expansion $e^{-x} = 1 - x + x^2/2 - x^3/6 + \cdots$, the unit pulse response can be written as

$$\begin{aligned}v(t) =& \frac{1}{\Delta}\left[1 - \left(1 - \frac{t}{RC} + \frac{t^2}{2(RC)^2} - \ldots\right)\right]u(t) \\ &- \frac{1}{\Delta}\left[1 - \left(1 - \frac{t-\Delta}{RC} + \frac{(t-\Delta)^2}{2(RC)^2} - \ldots\right)\right]u(t - \Delta).\end{aligned}$$

Now we let $\Delta \to 0$ in the limit, which yields

unit impulse response $h(t)$

$$\begin{aligned}\lim_{\Delta\to 0} v(t) &= \lim_{\Delta\to 0} \frac{1}{\Delta}\left[\frac{t}{RC} - \frac{t^2}{2(RC)^2} + \cdots - \frac{t-\Delta}{RC} + \frac{(t-\Delta)^2}{2(RC)^2} - \cdots\right] \\ &= \lim_{\Delta\to 0} \frac{1}{\Delta}\left[\frac{\Delta}{RC} - \frac{2t\Delta}{2(RC)^2} + \frac{\Delta^2}{2(RC)^2} + \frac{3t^2\Delta}{6(RC)^3} - \frac{3t\Delta^2}{6(RC)^3} + \frac{\Delta^3}{6(RC)^3} - \cdots\right] \\ &= \frac{1}{RC}\left[1 - \frac{t}{RC} + \frac{t^2}{2(RC)^2} - \cdots\right] \\ &= \frac{1}{RC}e^{-t/RC}u(t).\end{aligned}$$

Note: The approach taken here is of limited use for more general systems because of the difficulty of dealing with the series expansions. (Can you imagine even a third-order system?) In Chapter 3 we'll see an alternative approach using Laplace transforms that simplifies the computations.

f. *Computer verification.* Shown in Figures E2.3b–f (pages 82–83) are the responses $v(t)$ for five different values of the pulse width Δ with the script for $\Delta = 0.05$ given next.

```
%E2_3 Pulse and impulse response of a lowpass filter

%E2_3e Unit pulse response for Δ=0.05
A=[-1];                          % A matrix
B=[1];                           % B matrix
C=[1];                           % C matrix
D=[0];                           % D matrix
v0=[0];                          % IC vector
t=0:.01:4;                       % starting time:increment:final time
x=zeros(1,401)';                 % sets x vector to zero
x(1:5)=20*ones(1,5);             % creates pulse 0.05 s. wide and
                                 % 20 v. tall
[y,v]=kslsim(A,B,C,D,x,t,v0);    % call kslsim
%...plotting statements and pause
```


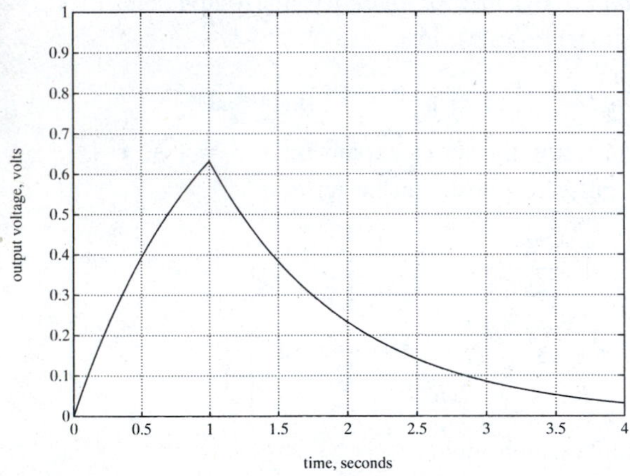

FIGURE E2.3b *Unit pulse response,* $\Delta = 1.0$

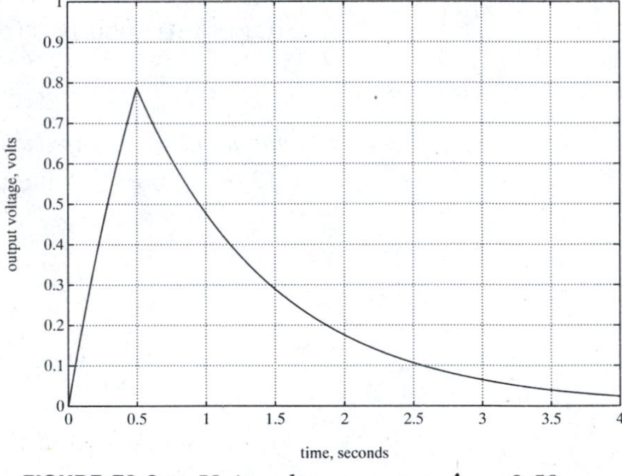

FIGURE E2.3c *Unit pulse response,* $\Delta = 0.50$

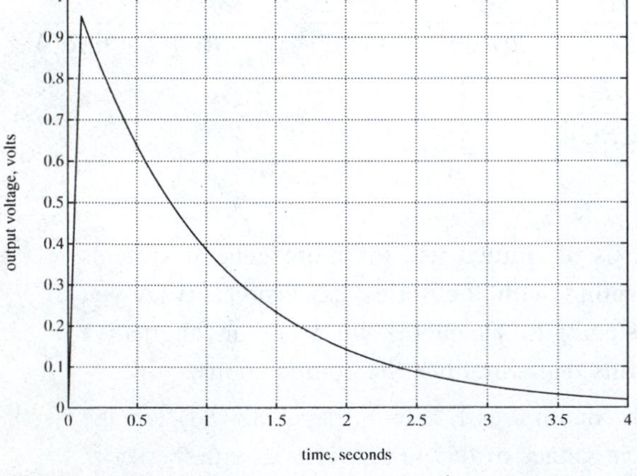

FIGURE E2.3d *Unit pulse response,* $\Delta = 0.10$

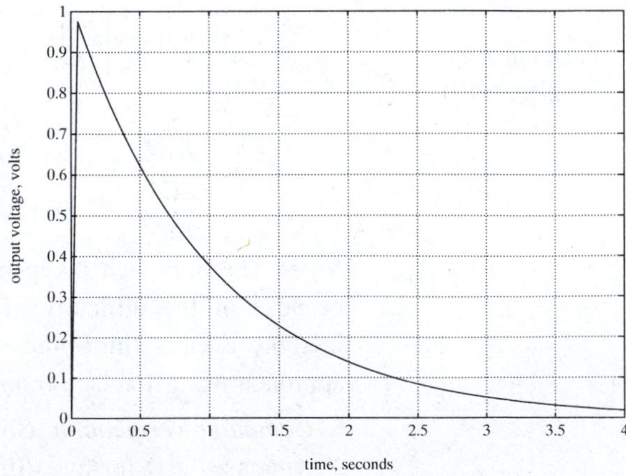

FIGURE E2.3e *Unit pulse response,* $\Delta = 0.05$

As the pulse width Δ becomes smaller, the pulse response approaches the impulse response $h(t)$ that is plotted in Figure E2.3g.

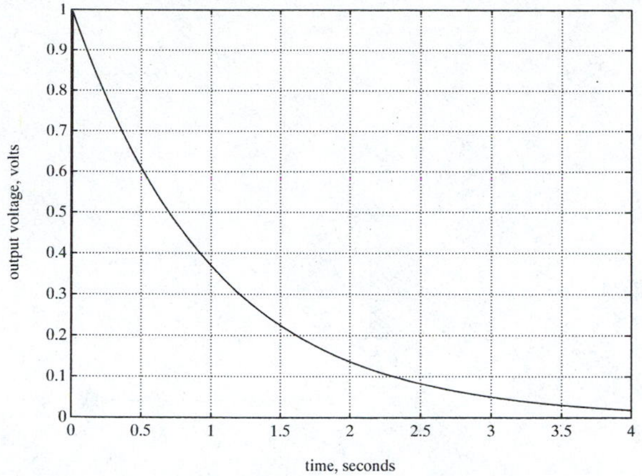

FIGURE E2.3f *Unit pulse response, $\Delta = 0.01$*

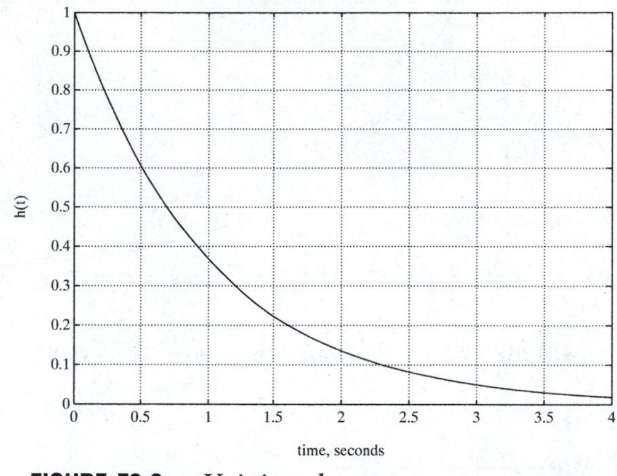

FIGURE E2.3g *Unit impulse response*

EXAMPLES

WHAT IF? The DE is now $dv(t)/dt + 10v(t) = 10e(t)$, making $\mathbf{A} = [-10]$ and $\mathbf{B} = [10]$. The unit impulse response is $h(t) = 10e^{-10t}u(t)$, a response that is 10 times "faster" than that for $1/RC = 1$, and we need to reduce the width of the pulse by a factor of 10. To keep its unit area, we make the pulse 10 times as high. Since $\Delta = 0.05$ gave a good approximation for $1/RC = 1$, we use $\Delta = 0.005$ (and the corresponding height of 200) for $1/RC = 10$. See Figure E2.3h on page 84. ■

MATLAB Script

```
%E2_3h Unit pulse response of a faster system
A=[-10];
B=[10];
C=[1];
D=[0];
v0=[0];                          % IC vector
t=0:.0005:0.4;                   % starting time:increment:final time
x=zeros(1,801);                  % sets x vector to zero
x(1:10)=200*ones(1,10);          % creates pulse 0.005 s. wide and
                                 % 200 v. tall
[y,v]=kslsim(A, B, C, D, x, t, v0);   % call kslsim
%...plotting statements
```

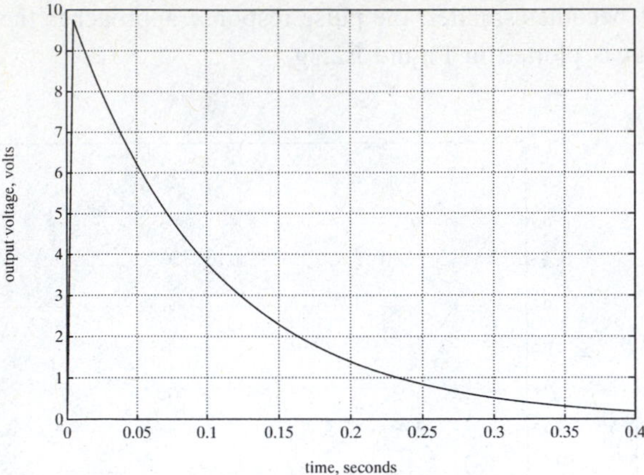

FIGURE E2.3h *Unit pulse response, $\Delta = 0.005$*

CHAPTER 2

EXAMPLE 2.4
Convolution

Here are several problems to provide practice in the different methods of evaluating convolution.

a. An LTI causal system is modeled by the unit impulse response $h(t)$. Show that the unit step response of this system—that is, the response to $x(t) = u(t)$—is given by

$$y(t) = \int_0^t h(\tau)\, d\tau\, .$$

b. A finite-duration integrator can be modeled by the unit impulse response $h(t) = u(t) - u(t - a)$, as in Figure E2.4a. If the input to this system is the exponential function $x(t) = Ae^{-bt}u(t)$, as in Figure E2.4b, use the graphical method for convolution to estimate the shape of the system output $y(t)$, where we have assumed $b > 0$.

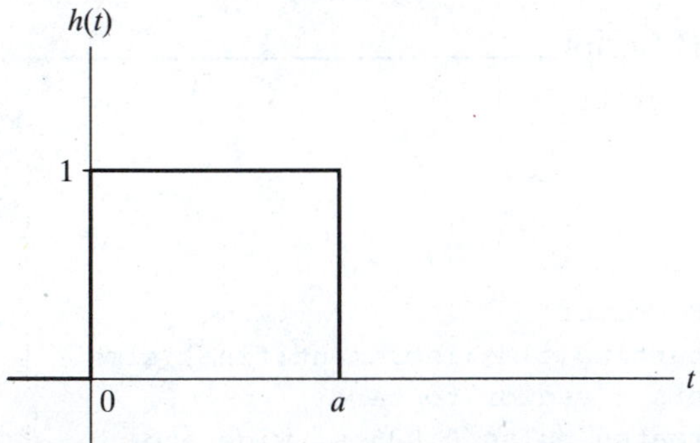

FIGURE E2.4a *Unit impulse response*

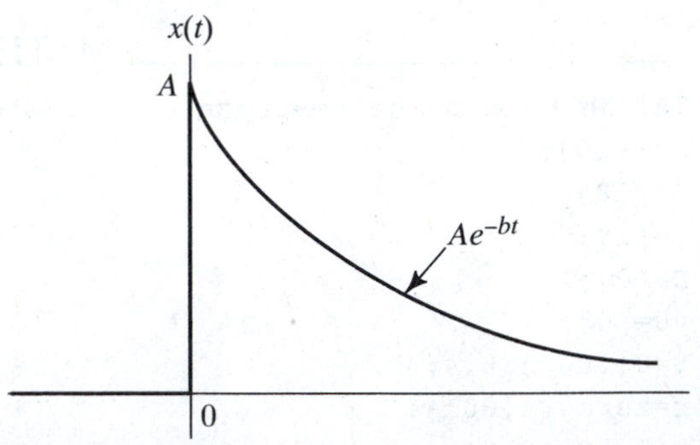

FIGURE E2.4b *System input*

c. Now use the analytical method to compute the exact equation for the convolution for the system and input of part (b).

d. Suppose that a hypothetical integrator is modeled by the noncausal unit impulse response $h(t) = u(t + a/2) - u(t - a/2)$. How will this affect the output $y(t)$ for the same exponential input of parts (b) and (c)?

Solution

a. *Relationship of impulse and step response.* Using the convolution integral

$$y(t) = \int_{-\infty}^{\infty} h(\tau)x(t - \tau)\,d\tau$$

the lower limit becomes $\tau = 0$ because $h(\tau) = 0$ for $\tau < 0$, a causal system. The folded function $u(t - \tau)$ is zero for $\tau > t$; consequently, the upper limit becomes $\tau = t$, and since $u(t - \tau)$ has a magnitude of unity, we can drop it from the integrand, leaving

$$y(t) = \int_{0}^{t} h(\tau)\,d\tau\,.$$

b. *Graphical method.* Using

$$y(t) = \int_{-\infty}^{\infty} h(\tau)x(t - \tau)\,d\tau$$

we have plotted $h(\tau)$ and $x(-\tau)$ in Figure E2.4c. For $t = 0$ we see that $h(\tau)x(0 - \tau) = 0$ for all values of τ, and so $y(0) = 0$. This situation will also hold for all values of $t < 0$; consequently, $y(t) = 0$ for all $t < 0$. For $0 \le t \le a$, the plot of $h(\tau)x(t - \tau)$ is shown in Figure E2.4d, with $y(t)$ being the area under that curve. This area, the value of $y(t)$, will increase until it reaches its maximum value at $t = a$ of

$$y(a) = \int_{0}^{a} Ae^{-b(a-\tau)}\,d\tau = Ae^{-ba}\int_{0}^{a} e^{b\tau}\,d\tau = \frac{A}{b}(1 - e^{-ba}).$$

For $t > a$, the plot of $h(\tau)x(t - \tau)$ will look like that in Figure E2.4e. The shape of $y(t)$ for $-\infty \le t \le \infty$ is given in Figure E2.4f.

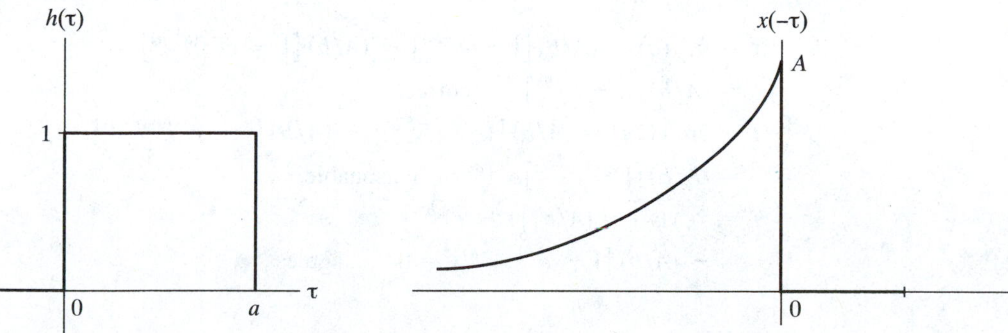

FIGURE E2.4c *Plots of $h(\tau)$ and a folded $x(-\tau)$.*

CHAPTER 2

c. *Analytical method.* For $x_1(t) = u(t)$,

$$y_1(t) = \int_0^t Ae^{-b\tau}\, d\tau = \frac{A}{b}(1 - e^{-bt})u(t).$$

For $x_2(t) = -u(t - a)$, $y_2(t) = -(A/b)\,[1 - e^{-b(t-a)}]u(t - a)$. Using time invariance and linearity, $y(t) = y_1(t) + y_2(t) = (A/b)[1 - e^{-bt}]u(t) - (A/b)\,[1 - e^{-b(t-a)}]u(t - a)$.

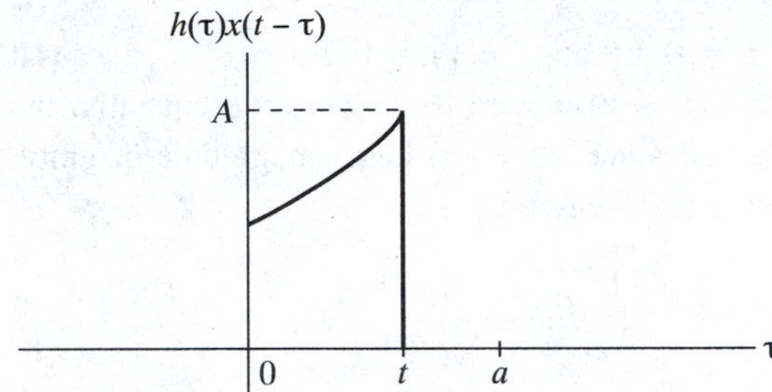

FIGURE E2.4d *Integrand for* $0 \le t \le a$

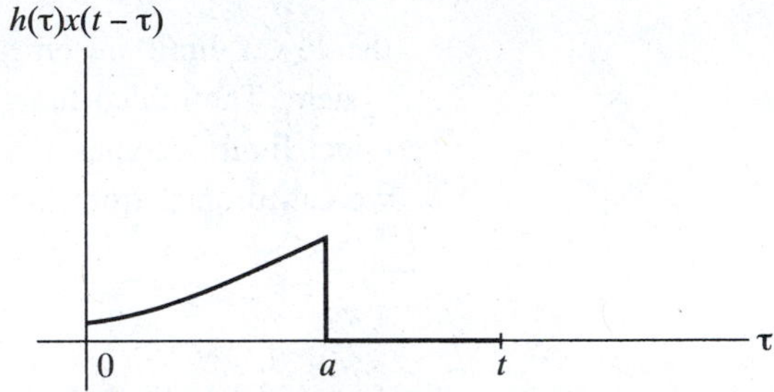

FIGURE E2.4e *Integrand for* $t > a$

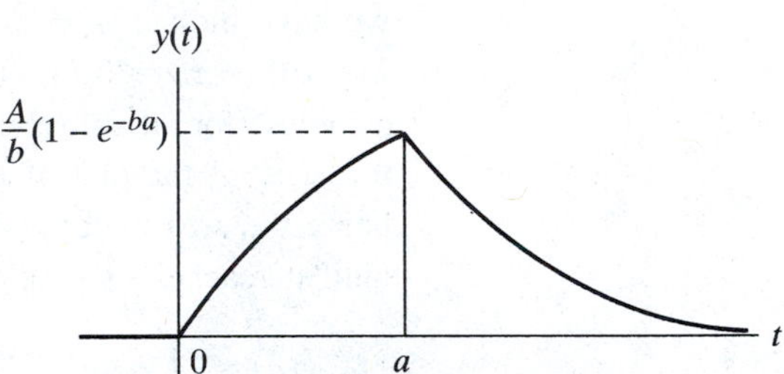

FIGURE E2.4f *Approximation of output* $y(t)$

CROSS-CHECK

$$\begin{aligned}
t &= a, y(a) = (A/b)\,[1 - e^{-ba}] - (A/b)\,[1 - e^{-b(a-a)}] \\
&= (A/b)\,[1 - e^{-ba}] \qquad \text{correct} \\
t &= 2a, y(2a) = (A/b)\,[1 - e^{-b2a}] - (A/b)\,[1 - e^{-b(2a-a)}] \\
&= (A/b)\,[1 - e^{-ba}]e^{-ba} \qquad \text{reasonable} \\
t &= \infty, y(\infty) = (A/b)\,[1 - e^{-b\infty}] \\
&\quad - (A/b)\,[1 - e^{-b(\infty-a)}] = 0 \qquad \text{agrees} \quad \blacksquare
\end{aligned}$$

d. Since this is a time-invariant system, the output will be the same as in part (c) but advanced in time by $a/2$ units. Replacing t by $t + a/2$ gives

$$y(t) = (A/b)\left[1 - e^{-b(t+a/2)}\right]u(t + a/2) - (A/b)\left[1 - e^{-b(t-a/2)}\right]u(t - a/2).$$

PROBLEMS

REINFORCEMENT PROBLEMS

P2.1 Characteristics of a linear system. An LTI system is described by the integro-differential equation

$$\frac{dy(t)}{dt} + Ky(t) + 25\int_0^t y(\tau)\,d\tau = x(t).$$

a. Find the characteristic equation in terms of K.

Hint: You can change the given integro-differential equation to a differential equation by differentiating both sides with respect to time.

b. Find an expression for the characteristic roots in terms of K.
c. For what range of positive values of K are the roots real?
d. Find the range of values of K for which the system is stable.
e. Determine the value of K for which the initial condition response would be a sustained oscillation of constant amplitude.

P2.2 Formulation of first-order or state models. Two simultaneous differential equations are used to model an LTI system. They are

$$3\ddot{p}(t) + 2\dot{p}(t) + p(t) - 2q(t) = 5f(t) - 7g(t)$$
$$2\ddot{q}(t) - 3\dot{q}(t) + 5p(t) = 3g(t)$$

where $f(t)$ and $g(t)$ are the system inputs. Choose $v_1(t) = p(t), v_3(t) = q(t), x_1(t) = f(t)$, and $x_2(t) = g(t)$ and write four first-order DEs that will model this system. Find the matrices **A** and **B** in the matrix representation $d\mathbf{v}(t)/dt = \mathbf{A}\mathbf{v}(t) + \mathbf{B}x(t)$.

P2.3 Initial condition response of a second-order system. A parallel LRC circuit is described by the differential equation

$$\ddot{v}(t) + 2\dot{v}(t) + 101v(t) = 0$$

when the input to the circuit is set to zero.

a. Find the characteristic roots, the natural frequency ω_n, and the damping ratio ζ for this circuit.
b. Find the initial condition solution $v(t)$ if it is known that $v(t)|_{t=0} = 10$ V and that $[dv(t)/dt]|_{t=0} = -10$ V/s.
c. Use **lsim** (or the California function **kslsim**) to plot the response $v(t)$.

Hint: The script F2_19 can be modified to fit this problem.

P2.4 System output by convolution. A causal, overdamped second-order system has the unit impulse response $h(t) = [e^{-t} - e^{-2t}]u(t)$. Use the analytical method of convolution to find the system output $y(t)$ for the following inputs: a. $x(t) = 10\delta(t)$; b. $x(t) = 2u(t)$; c. $x(t) = 3e^{-3t}u(t)$.

P2.5 Estimating the results of convolution. Use graphical convolution to sketch the following convolutions: a. $y(t) = e^{-t}u(t)*e^{-t}u(t)$; b. $z(t) = e^{t}u(-t)*e^{-t}u(t)$; c. $w(t) = \{u(t+1) - u(t)\}*2\{u(t) - u(t-3)\}$. Label as appropriate.

CHAPTER 2

P2.6 Characteristic equation and characteristic roots. Find the characteristic equation and the respective roots for the following scenarios. Remember that the MATLAB function **roots** can be useful for finding the roots of an algebraic equation.

a. $d^3p(t)/dt^3 - 2d^2p(t)/dt^2 + 3dp(t)/dt - 4p(t) = 5f(t)$, where $p(t)$ is the output and $f(t)$ is the input.
b. $y_{IC}(t) = [C_1 + C_2e^{-2t}]u(t)$
c. $d^3y/dt^3 + 3d^2y/dt^2 + 2dy/dt + 4y = 5dx/dt + 6x$
d. $d^2y(t)/dt^2 + 2dy(t)/dt + (1 - K)y(t) = dx(t)/dt$ for $K = 1, 3,$ and 5

P2.7 System stability. Determine the stability situation for each of the LTI systems described. In each case explain your answer.

a. $h(t) = [e^{-t} - e^{-2t}]u(t)$
b. $h(t) = [e^{t} - e^{-2t}]u(t)$
c. $d^2v(t)/dt^2 - 2dv(t)/dt + 101v(t) = x(t)$
d. $h(t) = e^{t}u(-t)$
e. $d^3y/dt^3 + 3d^2y/dt^2 + 2dy/dt + 4y = 5dx/dt + 6x$
f. $y_{IC}(t) = [C_1 + C_2e^{-2t}]u(t)$
g. $d^2y(t)/dt^2 + 2dy(t)/dt + (1 - K)y(t) = dx(t)/dt$ for $K = 1, 3,$ and 5

P2.8 Computer or state models. Find a state model for each of the following systems.

a. $d^3p(t)/dt^3 - 2d^2p(t)/dt^2 + 3dp(t)/dt - 4p(t) = 5f(t)$ with $p(t)$ the output and $f(t)$ the input
b. $d^3y(t)/dt^3 = x(t)$

P2.9 Unit impulse response. Find an analytical expression for the unit impulse response for each of the following causal systems.

a. $dy(t)/dt + 2y(t) = x(t)$
b. $dy(t)/dt - 2y(t) = 10x(t)$

Hint: Use the results of Example 2.3.

P2.10 Frequency responses from $h(t)$.

a. For $h(t) = e^{-5t}u(t)$ and $x(t) = 169\cos(12t), -\infty \le t \le \infty$, find $y(t), -\infty \le t \le \infty$.
b. Find the system frequency response $H(j\omega)$ for a system with unit impulse response $h(t) = [e^{-t} - e^{-2t}]u(t)$.
c. Repeat part (b) for $h(t) = te^{-t}u(t)$.

d. Repeat part (b) for $h(t) = 2e^{-5t}\sin(12t)u(t)$.
e. Repeat part (b) for $h(t) = \delta(t) - e^{-t}u(t)$.

P2.11 Frequency responses from the DE. Determine the system frequency response for the following differential equation models.

a. $dy(t)/dt + 2y(t) = x(t)$
b. $dy(t)/dt - 2y(t) = 10x(t)$
c. $dy(t)/dt + 2y(t) = dx(t)/dt$
d. $d^3y(t)/dt^3 = x(t)$
e. $d^3y(t)/dt^3 + 3d^2y(t)/dt^2 + 2dy(t)/dt + 6y(t) = 5f(t) + 10\,df(t)/dt$
f. $d^3y(t)/dt^3 + 3d^2y(t)/dt^2 + 2dy(t)/dt + 5y(t) = 10\,df(t)/dt + 5f(t)$

P2.12 Evaluation of convolution.

a. Given the continuous-time exponential signals $p(t) = e^{-t}u(t)$ and $r(t) = e^{t}u(-t)$, find an analytical expression for $y(t) = p(t)*r(t)$.
b. Find an equation for $y(t) = x(t)*h(t)$, where $x(t) = Bu(-t)$ and $h(t) = Ae^{at}u(t)$, $a < 0$.
c. In Illustrative Problem 2.4, $h(t) = e^{-t}u(t)$ and $x(t) = tu(t)$. Find $y(t) = h(t)*x(t)$.
d. For $x(t) = A\delta(t + a)$ and $y(t) = B\delta(t + b)$, find $z(t) = x(t)*y(t)$.

P2.13 Analytical and numerical convolution. Given the signals $f(t)$ and $g(t)$ as shown in Figure P2.13,

a. Plot $f(\tau) \cdot g(0.5 - \tau)$ versus τ and label carefully.
b. Now plot $g(\tau) \cdot f(0.5 - \tau)$ versus τ and label carefully.
c. Find the exact value of $y(t) = f(t)*g(t)$ at $t = 0.5$ from either (a) or (b).
d. Use **conv** to estimate the answer to (c).

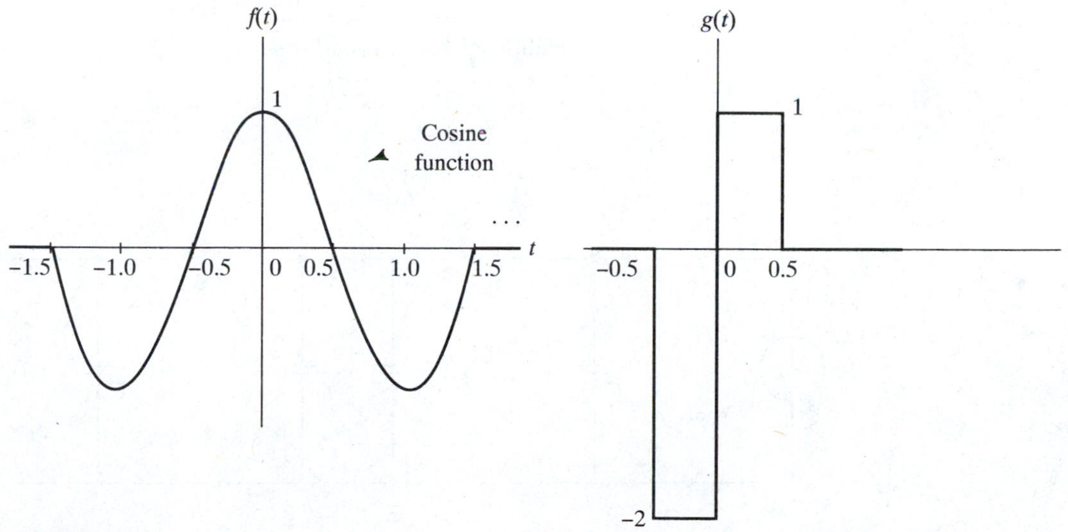

FIGURE P2.13

CHAPTER 2

EXPLORATION PROBLEMS

P2.14 Limits on convolution integral. Convolution is described by

$$(i)\ c_3(t) = \int_{-\infty}^{\infty} c_1(\tau)c_2(t-\tau)\,d\tau \quad \text{or} \quad (ii)\ c_3(t) = \int_{-\infty}^{\infty} c_2(\tau)c_1(t-\tau)\,d\tau\,.$$

Give the limits of integration for the following situations and for both forms of the convolution integral.

a. $c_1(t)$ nonzero for $-\infty \le t \le \infty$, $c_2(t)$ zero for $t < 0$.
b. $c_1(t)$ zero for $t < 0$, $c_2(t)$ nonzero for $-\infty \le t \le \infty$.
c. $c_1(t)$ zero for $t < 0$, $c_2(t)$ zero for $t < 0$.

P2.15 State models for *RC* filters. Write state and output equations that will model the *RC* filters of Figures P2.15a and b. Define the node-to-ground voltages and $v_j(t)$ as the states, and consider the filter output voltage $v(t)$ and the source current $i(t)$ as the outputs. The circuit in Figure P2.15b has *N* resistors and *N* capacitors arranged as shown.

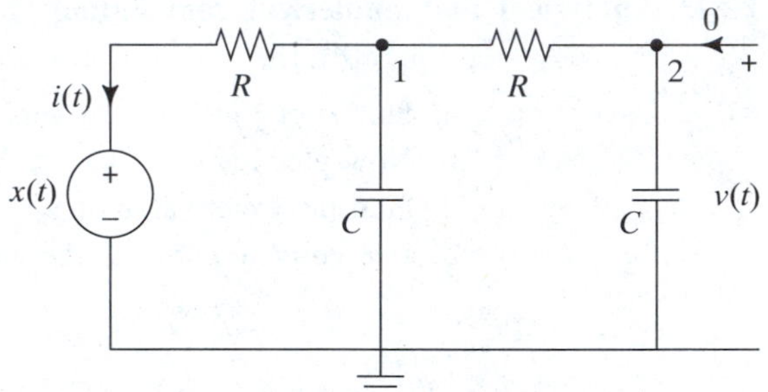

FIGURE P2.15a *Second-order RC filter*

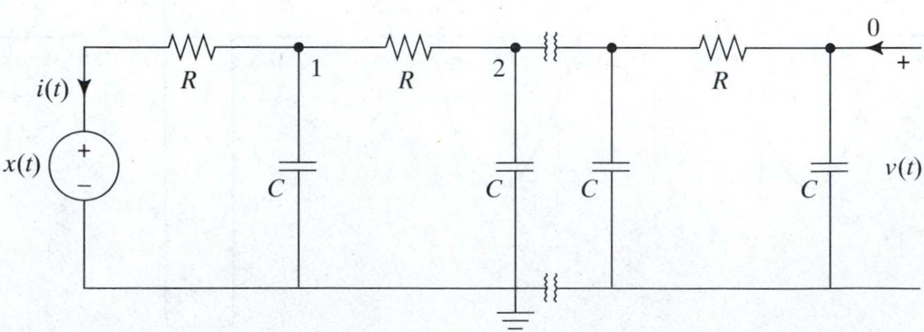

FIGURE P2.15b *Nth-order RC filter*

P2.16 Linearity. Determine whether the following systems are linear or nonlinear.

a. A system with inputs $x_1(t)$ and $x_2(t)$ whose output is

$$\text{(i) } y(t) = x_1(t) + x_2(t); \qquad \text{(ii) } y(t) = x_1(t)x_2(t);$$
$$\text{(iii) } y(t) = \min(x_1(t), x_2(t)).$$

b. A system with one input $x(t)$ whose output is

$$\text{(i) } y(t) = 1/x(t); \qquad \text{(ii) } y(t) = 10x(t).$$

P2.17 Lowpass and highpass filters. Two different analog filters are described by

$$(i)\ \ddot{y}(t) + 3\dot{y}(t) + 2y(t) = x(t)$$
$$(ii)\ \ddot{y}(t) + 3\dot{y}(t) + 2y(t) = \ddot{x}(t).$$

a. Find the characteristic roots for each filter.
b. Find the frequency response function for each filter.
c. Can you tell from the frequency responses of part (b) which filter describes a lowpass filter and which describes a highpass filter? Explain your reasoning.
d. The unit impulse responses of the two filters are known to be $h_a(t) = \delta(t) + [e^{-t} - 4e^{-2t}]u(t)$ and $h_b(t) = [e^{-t} - e^{-2t}]u(t)$. Can you determine which is for the lowpass filter and which is for the highpass?
e. Verify your answer to (c) by finding $y_{ss}(t)$ for $x(t) = A + B\cos(\omega_0 t) + C\cos(10\omega_0 t)$, where A is the amplitude of the input signal component at zero frequency, B is the amplitude of the component at midfrequency (ω_0) of the filter, and C is the amplitude of the component at the highest frequency ($10\omega_0$) applied to the filter.

Hint: Assume $A = B = C = 1$ and compare the output amplitudes at these frequencies with the amplitudes of the corresponding frequencies in $x(t)$.

P2.18 Impulse and step response using MATLAB. In the *RC* ladder network of Figure P2.15a, let $R = C = 1$ and, from the answer to P2.15, find the resulting matrices **A**, **B**, **C**, and **D**. Use **lsim** or **kslsim** to plot the unit impulse and the unit step responses for this network.

DEFINITIONS, TECHNIQUES, AND CONNECTIONS

DEFINITIONS TECHNIQUES

PROPERTIES

Linear system

$x_1(t) \to y_1(t)$
$x_2(t) \to y_2(t)$
$x_3(t) = C_1x_1(t) + C_2x_2(t) \to y_3(t) = C_1y_1(t) + C_2y_2(t)$
where C_1 and C_2 are arbitrary complex constants and $x_1(t)$ and $x_2(t)$ are arbitrary inputs.

Time-invariant system	$x_1(t) \rightarrow y_1(t)$ $x_2(t) = x_1(t - t_0) \rightarrow y_2(t) = y_1(t - t_0)$ for all t, t_0, $x_1(t)$.
Causal system	The system output does not precede the system input.
Stable system	1. All bounded inputs produce bounded outputs. 2. For a causal system the IC response decays to zero as $t \rightarrow \infty$, or, in terms of the characteristic roots, the roots (the r_k's) must be negative real or, if complex, they must have negative real parts. 3. $\int_{-\infty}^{\infty} \lvert h(\tau) \rvert \, d\tau < \infty$, the impulse response is absolutely integrable.

CHAPTER 2

*N*TH-ORDER DIFFERENTIAL EQUATION MODEL

$$\sum_{k=0}^{N} a_k \frac{d^k y(t)}{dt^k} = \sum_{k=0}^{L} b_k \frac{d^k x(t)}{dt^k}$$

Characteristic equation, CE	$\sum_{k=0}^{N} a_k s^k = 0$
Factored CE	$\sum_{k=0}^{N} a_k s^k = a_N(s - r_1)(s - r_2) \cdots (s - r_N) = 0$
Characteristic roots	$s_1 = r_1, \ldots, s_N = r_N$
Initial condition (IC) *solution*	$y_{\text{IC}}(t) = \sum_{k=1}^{N} C_k e^{r_k t}$, distinct roots

UNIT IMPULSE RESPONSE MODEL

an input $x(t) = \delta(t) \xrightarrow{\text{produces}}$ the output $y(t) = h(t)$

CONVOLUTION

Sifting property	$x(t) = \int_{-\infty}^{\infty} x(\tau)\delta(t - \tau)\, d\tau$
Convolution integral	$c_3(t) = \int_{-\infty}^{\infty} c_1(\tau)c_2(t - \tau)\, d\tau = \int_{-\infty}^{\infty} c_2(\tau)c_1(t - \tau)\, d\tau$
Operational notation	$c_3(t) = c_1(t) * c_2(t) = c_2(t) * c_1(t)$

Linear system forced output $y(t) = \int_{-\infty}^{\infty} h(\tau)x(t - \tau)\,d\tau = \int_{-\infty}^{\infty} x(\tau)h(t - \tau)\,d\tau$

Methods Analytical: Use the integral. Graphical: Use the integral graphically; that is, integrate by finding areas. Numerical: Use

$$y(nT) = T \sum_{m=-\infty}^{\infty} x(m)h(n - m) = T \sum_{m=-\infty}^{\infty} h(m)x(n - m).$$

DEFINITIONS TECHNIQUES

SINUSOIDAL STEADY-STATE RESPONSE

An input $x(t) = e^{j\omega t}$, $-\infty \le t \le \infty$ produces the output

$$y(t) = e^{j\omega t} \cdot H(j\omega), \qquad -\infty \le t \le \infty,$$

where

$$H(j\omega) = \int_{-\infty}^{\infty} h(\tau)e^{-j\omega\tau}\,d\tau$$

is the system frequency response and $h(t)$ is the unit impulse response of the stable system. When the sinusoidal input

$$x(t) = A\cos(\omega t + \alpha), \qquad -\infty \le t \le \infty$$

is applied to a stable linear time-invariant system, the steady-state output signal is

$$y_{ss}(t) = A|H(j\omega)|\cos(\omega t + \angle H(j\omega) + \alpha), \qquad -\infty \le t \le \infty.$$

An alternative method for obtaining $H(j\omega)$ is

$$\sum_{k=0}^{N} a_k \frac{d^k y(t)}{dt^k} = \sum_{k=0}^{L} b_k \frac{d^k x(t)}{dt^k} \rightarrow H(j\omega) = \frac{\sum_{k=0}^{L} b_k(j\omega)^k}{\sum_{k=0}^{N} a_k(j\omega)^k}$$

STATE-SPACE, OR FIRST-ORDER, MODEL FOR THE DE

$$\sum_{k=0}^{N} a_k \frac{d^k y(t)}{dt^k} = b_0 x(t) \quad \text{with } a_N \equiv 1.$$

General definition The state of a system is a minimum set of quantities $v_1(t), v_2(t), \ldots, v_N(t)$, which if known at $t = t_0$, are determined for $t \ge t_0$ by specifying the inputs to the system for $t \ge t_0$.

Matrix differential equation $\dot{\mathbf{v}}(t) = \mathbf{A}\mathbf{v}(t) + \mathbf{B}\mathbf{x}(t)$

Output equation $\mathbf{y}(t) = \mathbf{C}\mathbf{v}(t) + \mathbf{D}\mathbf{x}(t)$

SUMMARY OF RESULTS FOR A CAUSAL FIRST-ORDER SYSTEM

Differential equation	$\dot{y}(t) + a_1 y(t) = b_0 x(t)$
Unit sample response	$h(t) = b_0 e^{-a_1 t} u(t)$
State-space model	$\dot{v}(t) = -a_1 v(t) + b_0 x(t), \qquad y(t) = v(t)$
Frequency response $(a_1 > 0)$	$H(j\omega) = 1/(a_1 + j\omega)$

MATLAB FUNCTIONS USED

Function	*Purpose and Use*	*Toolbox*
conv	Given: unit sample response model of discrete system, input, **conv** returns system output.	MATLAB, Signal Processing
impulse	Given: state or TF model of continuous system, **impulse** returns impulse response.	Control System, Signals/Systems
kslsim	Given: state or TF model of continuous system, input, ICs, **kslsim** returns system output in continuous time.	California Functions
lsim	Given: state, or TF model of continuous system, input, ICs, **lsim** returns system output in continuous time.	Control System, Signals/Systems
rlocus	Given: an equation in the form $1 + K[\text{num}(\sigma)]/[\text{den}(\sigma)] = 0$, **rlocus** returns plot of locus of roots for K varying.	Signals/Systems, Control System
roots	Given: coefficients of polynomial p, **roots** returns zeros of $p = 0$.	MATLAB
step	Given: state or TF model of continuous system, **step** returns step response.	Signals/Systems, Control System

ANNOTATED BIBLIOGRAPHY

1. Braun, Martin, *Differential Equations and Their Applications,* Springer-Verlag, New York, 1975, 1978. *This mathematics text presents the important methods for solving differential equations as well as some key concepts of linear algebra. Many interesting physical systems are used as the vehicle for motivating the reader to learn the solution methods.*

2. Gabel, Robert A., and Richard A. Roberts, *Signals and Linear Systems, 3rd ed.,* John Wiley & Sons, New York, 1987. *Chapter Three treats convolution, impulse response, the solution of nonhomogeneous differential equations, frequency response, and the formulation of state-variable equations in the time domain for continuous-time systems.*

3. Kwakernaak, Huibert, and Raphael Sivan, *Modern Signals and Systems,* Prentice Hall, Inc., Englewood Cliffs, N.J., 1991. *Chapter Four, "Difference and Differential Systems," presents a parallel treatment of differential and difference equations that includes the general solution, stability, frequency response, and many problems and computer exercises.*

4. Oppenheim, Alan V., and Alan S. Willsky with Ian T. Young, *Signals and Systems,* Prentice Hall, Inc., Englewood Cliffs, N.J., 1983. *Chapter Three presents differential equations and the convolution integral.*

5. Reid, J. Gary, *Linear System Fundamentals,* McGraw-Hill Book Company, New York, 1983. *Chapters One and Two address the important continuous-time issues, including solution of differential equations, stability, and sinusoidal steady-state analysis. Chapter Six supports the derivation of the state model from the general linear network that can be electrical, mechanical, thermal, or whatever.*

6. Soliman, Samir S., and Mandyam D. Srinath, *Continuous and Discrete Signals and Systems,* Prentice Hall, Inc., Englewood Cliffs, N.J., 1991. *Chapter Three, "Continuous-Time Systems," covers classification and properties of continuous-time systems, LTI systems and the convolution integral, linear constant-coefficient differential equations, simulation diagrams, impulse response, and state-variable representation.*

ANSWERS

P2.1 a. $s^2 + Ks + 25 = 0$; b. $s_{1,2} = 0.5\{-K \pm \sqrt{K^2 - 4 \cdot 25}\}$; c. $K^2 \geq 100$, or $K \geq 10$ or $K \leq -10$; d. $K > 0$; e. $K = 0$.

P2.2 If you chose $v_2(t) = dp(t)/dt$ and $v_4(t) = dq(t)/dt$, the matrices are

$$\mathbf{A} = \begin{bmatrix} 0 & 1 & 0 & 0 \\ -0.333 & -0.667 & 0.667 & 0 \\ 0 & 0 & 0 & 1 \\ -2.500 & 0 & 0 & 1.500 \end{bmatrix} \quad \text{and} \quad \mathbf{B} = \begin{bmatrix} 0 & 0 \\ 1.667 & -2.333 \\ 0 & 0 \\ 0 & 1.500 \end{bmatrix}.$$

P2.3 a. CE: $s^2 + 2s + 101$; roots: $s_{1,2} = -1 \pm j10$; $\omega_n{}^2 = 101$ and $\omega_n = 10.05$; $2\zeta\omega_n = 2$ and $\zeta = 2/2\omega_n = 0.0995$; b. $v(t) = C_1e^{-t}\cos(10t + \phi)$; $v(0) = 10 = C_1\cos(\phi)$; $dv(t)/dt = -C_1e^{-t}\cos(10t + \phi) - 10C_1e^{-t}\sin(10t + \phi) = -10 = -C_1\cos(\phi) - 10C_1\sin(\phi)$; $\phi = 0$, $C_1 = 10$, and $v(t) = 10e^{-t}\cos(10t)$; c. See A2_3 on the CLS disk.

P2.4 a. $y(t) = 10h(t) = 10[e^{-t} - e^{-2t}]u(t)$; b. $y(t) = [1 - 2e^{-t} + e^{-2t}]u(t)$; c. $y(t) = [1.5e^{-t} - 3e^{-2t} + 1.5e^{-3t}]u(t)$

P2.5 See Figure A2.5a–c.

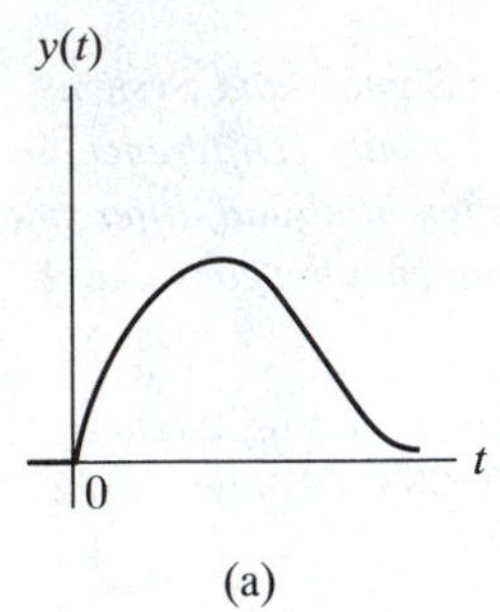

(a)

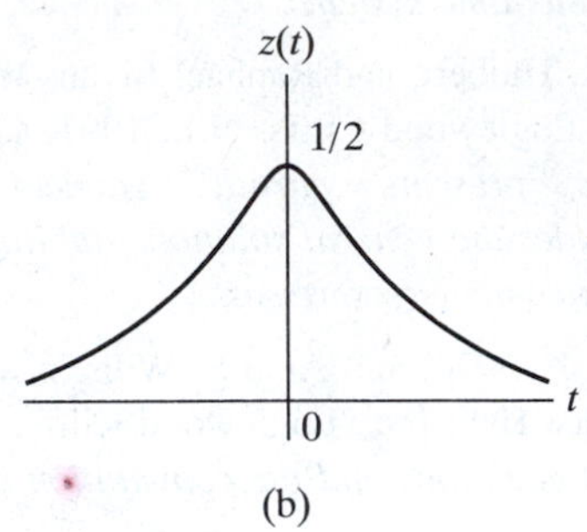

(b)

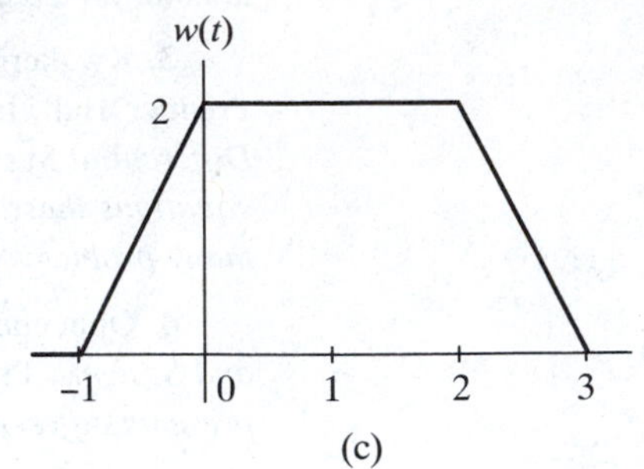

(c)

FIGURE A2.5

P2.6 a. $s^3 - 2s^2 + 3s - 4 = 0$, $s_1 = 1.651$, $s_{2,3} = 0.175 \pm j1.55$; b. $s^2 + 2s = 0$, $s_1 = 0$, $s_2 = -2$; c. $s^3 + 3s^2 + 2s + 4 = 0$, $s_1 = -2.796$, $s_{2,3} = -0.102 \pm j1.192$; d. $s^2 + 2s + (1 - K) = 0$, $K = 1$, $s_{1,2} = 0$ and -2; $K = 3$, $s_{1,2} = -2.732$ and 0.732; $K = 5$, $s_{1,2} = -3.236$ and 1.236

P2.7 a. Stable, $\int |h(\tau)|\, d\tau < \infty$; b. Unstable, $\int |h(\tau)|\, d\tau \to \infty$; c. Unstable, characteristic roots have positive real part; d. Stable, $\int |h(\tau)|\, d\tau < \infty$; e. Stable, see Problem P2.6c; f. Unstable, root at $s = 0$; g. Unstable for all three values of K; see Problem P2.6d.

P2.8 a. With $v_1(t) = p(t)$, $dv_1(t)/dt = v_2(t)$, $dv_2(t)/dt = v_3(t)$, and $x(t) = f(t)$, we have

$$\mathbf{A} = \begin{bmatrix} 0 & 1 & 0 \\ 0 & 0 & 1 \\ 4 & -3 & 2 \end{bmatrix}, \quad \mathbf{B} = \begin{bmatrix} 0 \\ 0 \\ 5 \end{bmatrix}, \quad \mathbf{C} = [1 \;\; 0 \;\; 0], \quad \mathbf{D} = \mathbf{0}.$$

b. With

$$v_1(t) = y(t), \quad \mathbf{A} = \begin{bmatrix} 0 & 1 & 0 \\ 0 & 0 & 1 \\ 0 & 0 & 0 \end{bmatrix}, \quad \mathbf{B} = \begin{bmatrix} 0 \\ 0 \\ 1 \end{bmatrix},$$

C, **D** as in (a).

P2.9 a. $h(t) = e^{-2t}u(t)$; b. $h(t) = 10e^{2t}u(t)$

P2.10 a. $y_{ss}(t) = 13\cos(12t - 1.18)$; b. $H(j\omega) = 1/(2 - \omega^2 + j3\omega)$; c. $H(j\omega) = 1/(1 + j\omega)^2$; d. $H(j\omega) = 24/(169 - \omega^2 + j10\omega)$; e. $H(j\omega) = j\omega/(1 + j\omega)$

P2.11 a. $H(j\omega) = 1/(2 + j\omega)$; b. Unstable; c. $H(j\omega) = j\omega/(2 + j\omega)$; d. Unstable; e. Unstable; f. $H(j\omega) = (5 + 10j\omega)/(-j\omega^3 - 3\omega^2 + j2\omega + 5)$

P2.12 a. $y(t) = 0.5[e^{t}u(-t) + e^{-t}u(t)]$; b. $y(t) = (-AB/a),\ t < 0$; $y(t) = (-AB/a)e^{at},\ t > 0$; $y(0) = -AB/a$; c. $y(t) = [t - 1 + e^{-t}]u(t)$; d. $z(t) = AB\delta(t + a + b)$

P2.13 a. See Figure A2.13a; b. See Figure A2.13b; c. Using the product in part (b), we have

$$y(0.5) = \int_{-0.5}^{0} -2\sin(\pi\tau)\,d\tau + \int_{0}^{0.5} \sin(\pi\tau)\,d\tau$$
$$= \frac{2\cos\pi\tau}{\pi}\Big|_{-0.5}^{0} + \frac{-\cos\pi\tau}{\pi}\Big|_{0}^{0.5} = \frac{3}{\pi} = 0.955\,.$$

d. Results depend on T used. For $T = 0.1$ (very large), from **conv** (discrete convolution) $y(n)|_{n=5} = 9.47$; after multiplying by T, $y(t)|_{t=0.5} \approx 0.947$.

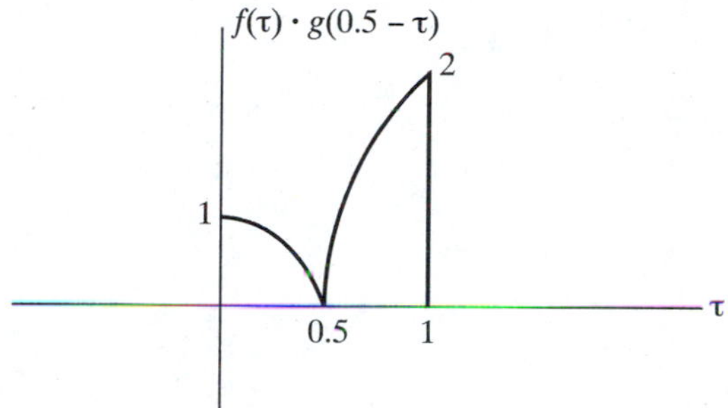

FIGURE A2.13a

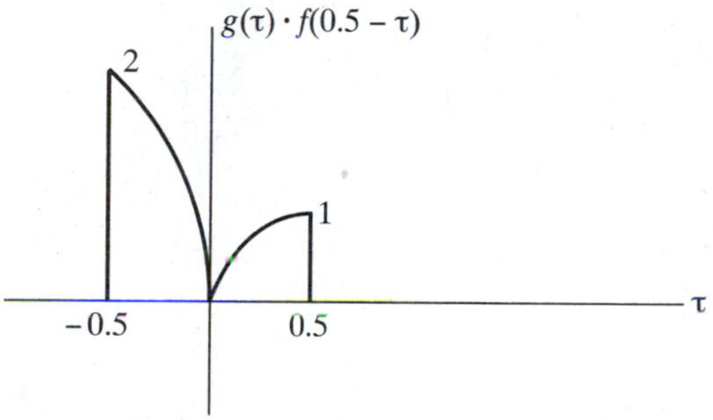

FIGURE A2.13b

P2.14

	(i)	(ii)
a.	$-\infty$ to t	0 to ∞
b.	0 to ∞	$-\infty$ to t
c.	0 to t	0 to t

P2.15 For Figure P2.15a with $v_1(t)$ and $v_2(t)$ at nodes 1 and 2, respectively, we have

$$\begin{bmatrix} \dot{v}_1(t) \\ \dot{v}_2(t) \end{bmatrix} = \begin{bmatrix} -2/RC & 1/RC \\ 1/RC & -1/RC \end{bmatrix}\begin{bmatrix} v_1(t) \\ v_2(t) \end{bmatrix} + \begin{bmatrix} 1/RC \\ 0 \end{bmatrix} x(t)\,.$$
$$\begin{bmatrix} y_1(t) \\ y_2(t) \end{bmatrix} = \begin{bmatrix} 0 & 1 \\ 1/R & 0 \end{bmatrix}\begin{bmatrix} v_1(t) \\ v_2(t) \end{bmatrix} + \begin{bmatrix} 0 \\ -1/R \end{bmatrix} x(t)\,.$$

For Figure P2.15b,

$$\dot{\mathbf{v}}(t) = \begin{bmatrix} -2/RC & 1/RC & 0 & \cdots & \cdots & \cdots \\ 1/RC & -2/RC & 1/RC & 0 & \cdots & \cdots \\ 0 & 1/RC & -2/RC & 1/RC & 0 & \cdots \\ \vdots & \vdots & \vdots & \vdots & \vdots & \vdots \\ \vdots & \vdots & \vdots & \vdots & \vdots & \vdots \\ 0 & \cdots & \cdots & 0 & 1/RC & -1/RC \end{bmatrix} \mathbf{v}(t) + \begin{bmatrix} 1/RC \\ 0 \\ 0 \\ \vdots \\ \vdots \\ 0 \end{bmatrix} x(t).$$

$$\mathbf{y}(t) = \begin{bmatrix} 0 & 0 & \cdots & \cdots & 1 \\ 1/R & 0 & \cdots & \cdots & \cdots \end{bmatrix} \mathbf{v}(t) + \begin{bmatrix} 0 \\ -1/R \end{bmatrix} x(t).$$

P2.16 a. (i) linear, (ii) nonlinear, (iii) nonlinear; b. (i) nonlinear, (ii) linear

P2.17 a. For both $s_{1,2} = -1, -2$; b. (i) $H(j\omega) = 1/(-\omega^2 + 2 + 3j\omega)$, (ii) $H(j\omega) = -\omega^2/(-\omega^2 + 2 + 3j\omega)$; c. (i) is lowpass because $|H(j0)| = 0.5$ and $|H(j\omega)| \to 0$ as $\omega \to \infty$, (ii) is highpass because $|H(j0)| = 0$ and $|H(j\omega)| \to 1$ as $\omega \to \infty$; d. $h_a(t)$ is highpass, $h_b(t)$ is lowpass; e. (i) $H(j0) = 1/2$, $H(j\omega_0) = 1/(2 - \omega_0{}^2 + j3\omega_0)$, $|H(j10\omega_0)| = M$, $y_{ss}(t) \approx 0.5A$; (ii) $H(j0) = 0$, $H(j\omega_0) = -\omega_0{}^2/(2 - \omega_0{}^2 + j3\omega_0)$, $|H(j10\omega_0)| = 100M$, $y_{ss}(t) \approx C\cos(10\omega_0 t)$.

P2.18 See A2_18 on the CLS disk.

Laplace Transforms and Applications

PREVIEW

In Chapter 2 we established a considerable amount of theory about and observed several applications of the analysis of continuous-time LTI systems. All this effort was accomplished in the time domain. It is also common practice to design and analyze LTI systems from the point of view of a transform or algebraic domain, which enables us to determine system response and characteristics using algebraic manipulations.

In this chapter we introduce the Laplace[1] transform and illustrate some of its applications in analyzing continuous-time systems. The development parallels

FIGURE 3.1 *Pierre Simon Laplace*

[1]Pierre Simon Laplace (1749–1827) of Normandy, a protégé of Jean d'Alembert, the leading mathematician in France, was appointed professor of mathematics at the Paris École Militaire

that of Chapter 8, where the *z* transform is presented, along with its applications in the study of discrete-time systems. The Laplace transform method for solving linear constant-coefficient DEs was introduced to graduate engineering students with a book by Gardner and Barnes [7] in the early 1940s, but Van Valkenburg's text [6], first published in 1953, made this method accessible to the masses, graduate and undergraduate.

Now, in the 1990s, Laplace methods are important for different reasons than solving linear constant-coefficient DEs. Computers, with the help of application software such as MATLAB with simulation functions such as **lsim** and **kslsim**, can solve DEs quite nicely and accurately where the order of the system is of little importance. So with a solution method for DEs in hand, why bother with Laplace transforms? There are several situations where transform methods can provide information and insight not easily discernible in the time domain. For example, (1) by transforming DEs into the Laplace or *s* domain as algebraic equations, we gain information about stability; (2) the powerful transfer function concept is developed, which includes a pole-zero model of a system; (3) we are able to use design procedures involving location of a system's characteristic roots which affect time response; (4) techniques are available that allow us to "see" the system as a diagram or structure; (5) with the aid of the MATLAB function **residue**, an analytical solution, if desired, can be determined quite easily even for systems of relatively high order.

The chapter begins by establishing the definition and some important properties of the Laplace transform. Some transform pairs for a few commonly occurring signals are derived with a more complete listing of pairs and properties appearing in Tables 3.1 and 3.2 at the end of the chapter. The role and fundamental importance of the region of convergence (ROC) in the study of causal, noncausal, and anticausal systems is emphasized. Next, system characteristics and models, such as stability and transfer functions, are investigated. Inverse transform techniques are then developed and applied to obtain procedures for solving linear differential equations. Special emphasis is

---

in 1769 and elected to the select (about 60 members) Académie Royale des Sciences in 1773. As an examiner of scholars of the royal artillery corps, Laplace examined and passed (probably just as well for Laplace) a 16-year-old sublieutenant named Napoleon Bonaparte. After a brief stint as minister of interior under Napoleon, Laplace became president of the senate and was made a count. When Napoleon was deposed in 1814 and the French monarchy was restored, Louis XVIII bestowed the title of marquis on Laplace and named him to the Chamber of Peers. When those around his deathbed were recalling to him the great discoveries he had made, Laplace replied, "What we know is but a little thing; what we are ignorant of is immense." (From David M. Burton, *The History of Mathematics,* Allyn & Bacon, Inc., 1985.)

given to the convolution property and to the steady-state response of a linear system to a sinusoidal input. Finding the system model in the form of a diagram, a block diagram or a signal flowgraph, is developed, and an algorithm called Mason's Gain Rule is applied to these diagrams to find system transfer functions.

Let us conclude with a short historical note. Operational calculus, invented by the British engineer Oliver Heaviside (1850–1925), was used to solve practical circuit problems before the Laplace transformation method was employed. Without careful justification of his method, Heaviside's heuristic point of view drew bitter criticism from the mathematicians of his time. The basis for substantiating the operational calculus was found in the writings of Laplace in 1780. Thus, the Laplace transformation has provided rigorous verification of the operational methods and no important errors have been discovered in Heaviside's results. So, in reality, Laplace transform methods and those of the operational calculus are much the same.

BASIC CONCEPTS

BASIC CONCEPTS

The *two-sided,* or bilateral, Laplace transform of a real signal $f(t)$ is defined as

bilateral $\mathcal{L}$ *transform*

$$\mathcal{L}[f(t)] = F(s) = \int_{-\infty}^{\infty} f(\tau)e^{-s\tau}\,d\tau$$

where $s = \sigma + j\omega$ is a complex variable. The region of the complex variable s for which the integral converges is called the existence region or the *region of convergence* and is designated ROC. The notation

signal ⇔ *transform*

$$f(t) \Leftrightarrow F(s)$$

is commonly used to denote a signal and its corresponding transform (and vice versa).

TRANSFORM PAIRS

(a) Impulse Function Using the definition of the bilateral transform, we have

$$\mathcal{L}[\delta(t)] = \int_{-\infty}^{\infty} \delta(\tau)e^{-s\tau}\,d\tau\,.$$

Restating the sifting property of the impulse function for $f(t)$ continuous at $t = t_0$,

sifting property

$$f(t_0) = \int_{-\infty}^{\infty} \delta(\tau - t_0) f(\tau)\, d\tau ,$$

and letting $f(t) = e^{-st}$ and $t_0 = 0$, the Laplace transform of a unit impulse function is

$$\mathcal{L}[\delta(t)] = \int_{-\infty}^{\infty} \delta(\tau) e^{-s\tau}\, d\tau = e^{-s(0)} = 1 .$$

We notice that this transform is independent of the complex variable s; consequently, we say that the transform converges for all s—that is, the ROC is the entire complex s plane. This result can be generalized for $f(t) = A\delta(t)$, an impulse function of strength A, to give

impulse function pair

$$A\delta(t) \Leftrightarrow A, \text{ all } s.$$

(b) Real Exponential Function, Causal[2] For $f(t) = Ae^{at}u(t)$, the transform is

$$F(s) = \int_{-\infty}^{\infty} Ae^{a\tau} u(\tau) e^{-s\tau}\, d\tau$$

and, since $u(\tau)$ is zero for $\tau < 0$, the lower limit of the integral becomes 0 and we have

$$F(s) = \int_{0}^{\infty} Ae^{a\tau} e^{-s\tau}\, d\tau = \left. \frac{Ae^{(a-s)\tau}}{a - s} \right|_0^{\infty} = \frac{A}{s - a}, \qquad \text{Re}[a - s] < 0 .$$

The region of convergence is $\text{Re}[a - s] < 0$ or $\text{Re}[s] > \text{Re}[a]$, and the transform pair is

causal exponential pair

$$Ae^{at}u(t) \Leftrightarrow \frac{A}{s - a}, \qquad \text{Re}[s] > \text{Re}[a].$$

Using the Symbolic Math Toolbox, we enter `f='A*exp(a*t)'` and **`F=laplace(f)`** with the function to be transformed `f=A*exp(a*t)` and the resulting transform `F=A/(s-a)` returned.

(c) Step Function For $f(t) = Au(t)$, we simply let $a = 0$ in the pair for the causal exponential and obtain

step function pair

$$Au(t) \Leftrightarrow \frac{A}{s}, \qquad \text{Re}[-s] < 0 \quad \text{or} \quad \text{Re}[s] > 0 .$$

---

[2] An LTI system is causal if and only if the unit impulse response $h(t)$ is zero for all $t < 0$. It is, therefore, convenient to refer to any signal $f(t)$ that is zero for all $t < 0$ as a causal signal, indicating that it could be the unit impulse response of a causal system.

CHAPTER 3

(d) Real Exponential Function, Anticausal For $f(t) = Ae^{at}u(-t)$, the transform is

$$F(s) = \int_{-\infty}^{\infty} Ae^{a\tau}u(-\tau)e^{-s\tau}\,d\tau$$

and, since $u(-\tau)$ is zero for $\tau > 0$, the upper limit of the integral becomes 0 and we have

$$F(s) = \int_{-\infty}^{0} Ae^{a\tau}e^{-s\tau}\,d\tau = \left.\frac{Ae^{(a-s)\tau}}{a-s}\right|_{-\infty}^{0} = -\frac{A}{s-a}, \qquad \mathrm{Re}[a-s] > 0$$

or, in the common notation,

anticausal exponential pair

$$Ae^{at}u(-t) \Leftrightarrow -\frac{A}{s-a}, \qquad \mathrm{Re}[s] < \mathrm{Re}[a].$$

Notice that the pairs for the anticausal and causal exponentials are quite similar, the differences being a sign and the region of convergence, as seen in Figure 3.2a and b. That is, for the Laplace transform $X(s) = 2/(s+3)$, the corresponding time signal is $x(t) = 2e^{-3t}u(t)$ or $x(t) = -2e^{-3t}u(-t)$, depending on the region of convergence. Since these are quite different signals, we see that it is essential that the ROC of a transform be specified. A listing of some common pairs with their ROCs is given in Table 3.1 at the end of the chapter.

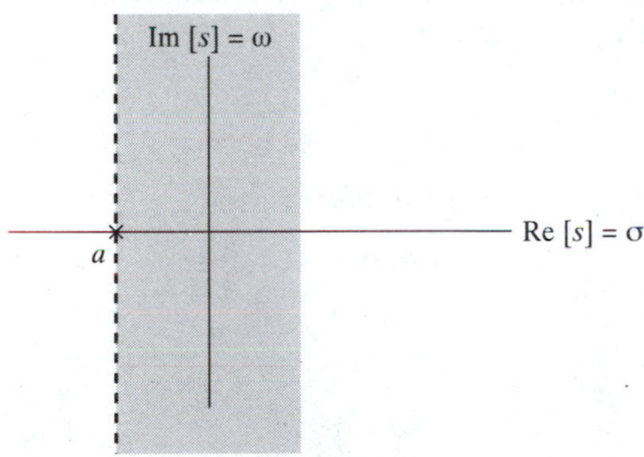

a. ROC for causal exponential $e^{at}u(t)$, $a < 0$

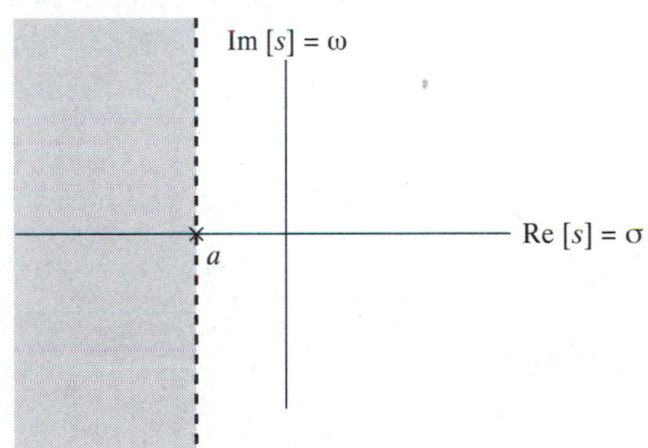

b. ROC for anticausal exponential $e^{at}u(-t)$, $a < 0$

FIGURE 3.2 *ROC for exponential signals*

ILLUSTRATIVE PROBLEM 3.1
Finding Transforms of Sinusoidal Signals

Use the pairs that were derived for the causal and anticausal exponentials to find the Laplace transforms of (a) the causal sinusoid $x_1(t) = \cos(\omega t)u(t)$ and (b) the anticausal sinusoid $x_2(t) = \sin(\omega t)u(-t)$.

Solution

a. Using the Euler relation $x_1(t) = 0.5\{e^{j\omega t} + e^{-j\omega t}\}u(t)$ and replacing a with $\pm j\omega$ in the causal pair gives

$$X_1(s) = 0.5\left\{\frac{1}{s - j\omega} + \frac{1}{s + j\omega}\right\} = \frac{s}{s^2 + \omega^2}, \qquad \text{Re}[\,j\omega - s] < 0\,.$$

Notice that this ROC may be written $\text{Re}[s] > 0$, since $\text{Re}[\,j\omega] = 0$.

b. In a similar manner with $x_2(t) = -0.5j\{e^{j\omega t} - e^{-j\omega t}\}u(-t)$, we use the anticausal exponential pair to obtain

$$X_2(s) = -0.5j\left\{\frac{-1}{s - j\omega} + \frac{1}{s + j\omega}\right\}$$
$$= \frac{-\omega}{s^2 + \omega^2}, \qquad \text{Re}[\,j\omega - s] > 0 \quad \text{or} \quad \text{Re}[s] < 0\,.$$

PROPERTIES AND RELATIONS FOR CAUSAL SIGNALS

CHAPTER 3

In this chapter we concentrate on signals that are causal (zero for $t < 0$) and causal systems ($h(t) = 0$ for $t < 0$) and normally use the unilateral transform

$$\mathcal{L}[f(t)u(t)] = F(s) = \int_0^t f(t)e^{-st}\,dt\,.$$

Some common properties and relations for this unilateral transform are given here and in Table 3.2 at the end of the chapter.

Property	*Signal*	*Transform*
1. Linearity	$af(t) + bg(t)$	$aF(s) + bG(s)$
2. Derivative	$\frac{df(t)}{dt}$	$sF(s) - f(t)\vert_{t=0} = sF(s) - f(0)$
3. Nth derivative	$\frac{d^N f(t)}{dt^N}$	$s^N F(s) - s^{N-1}f(t)\vert_{t=0} - s^{N-2}\frac{df(t)}{dt}\Big\vert_{t=0} - \cdots - \frac{d^{N-1}f(t)}{dt^{N-1}}\Big\vert_{t=0}$
4. Integral	$\int_0^t f(\tau)\,d\tau$	$\frac{F(s)}{s}$
5. Time shift ($a > 0$)	$f(t - a)u(t - a)$	$e^{-as}F(s)$
6. Frequency shift ($a > 0$)	$e^{at}f(t)$	$F(s - a)$

7. Convolution $\quad f_3(t) = \int_0^t f_1(\tau)f_2(t - \tau)\,d\tau \quad F_3(s) = F_1(s) \cdot F_2(s)$

$$= \int_0^t f_2(\tau)f_1(t - \tau)\,d\tau \qquad = F_2(s) \cdot F_1(s)$$

8. Final value theorem (FVT) $\quad \lim_{t\to\infty} f(t) \quad \lim_{s\to 0} sF(s)$, provided all poles of $sF(s)$ have negative real parts

9. Initial value theorem (IVT) $\quad \lim_{t\to 0} f(t) \quad \lim_{s\to\infty} sF(s)$

ILLUSTRATIVE PROBLEM 3.2
Use of Some Transform Properties

Certain transform properties can be used to take information available in the transform domain to determine time-domain behavior of a signal or system without evaluating the complete inverse transform. As an illustration, the starting value (initial) and the finishing value (final) of a signal are often of interest in the analysis and design of a linear system. For the following Laplace transforms of system signals, use the initial and final value theorems from Table 3.2 to determine the initial and final values of the corresponding causal time signals.

a. $I(s) = \dfrac{s - 2}{s(s + 2)}$

b. $H(s) = \dfrac{8}{s^2 + 10s + 169}$

c. $V(s) = \dfrac{2s^3 + 10}{s^3(s + 1)}$

Solution

a. For the initial value of $i(t)$, we have $i(0) = sI(s)|_{s\to\infty} = 1$. To find the final value, we first test the poles of $sI(s)$ to see that they are to the left of the imaginary axis. The only pole of $sI(s)$ is $s = -2$. We may apply the final value theorem (FVT), with the result $i(\infty) \to -1$.

b. For the value of $h(t)$ at $t = 0$, we see that $sH(s)|_{s\to\infty} = 0 = h(0)$. The poles of $sH(s)$ are to the left of the imaginary axis at $s = -5 \pm j12$, and so the final value theorem may be used, giving $h(\infty) \to 0$.

c. Using the initial value theorem (IVT), $v(0) = sV(s)|_{s\to\infty} = 2$. There are two poles of $sV(s)$ at $s = 0$ (not to the left of the imaginary axis) and so the FVT cannot be applied.

TRANSFER FUNCTIONS

The transfer function is defined as

output/input

$$\text{transfer function} = \frac{\text{Laplace transform of the output signal } y(t)}{\text{Laplace transform of the input signal } x(t)},$$

with initial conditions assumed to be zero. In symbols,

symbols

$$H(s) = \frac{Y(s)}{X(s)}.$$

Alternatively, taking the transform of both sides of a system's differential equation,

differential equation (DE)

$$\sum_{k=0}^{N} a_k \frac{d^k y(t)}{dt^k} = \sum_{k=0}^{L} b_k \frac{d^k x(t)}{dt^k}$$

and applying the linearity property along with the Nth derivative relationship for zero initial conditions gives the algebraic or s domain equation

transformed DE, zero ICs

$$Y(s)\left[\sum_{k=0}^{N} a_k s^k\right] = X(s)\left[\sum_{k=0}^{L} b_k s^k\right].$$

Solving for the transfer function $H(s)$, which is the ratio of two polynomials in s, we have

CHAPTER 3

rational function

$$H(s) = \frac{Y(s)}{X(s)} = \frac{\sum_{k=0}^{L} b_k s^k}{\sum_{k=0}^{N} a_k s^k}.$$

We notice that the coefficients of the input terms in the differential equation determine the coefficients of the numerator terms in the transfer function and that the coefficients of the output terms in the differential equation determine the denominator coefficients in the transfer function. Recall from the transform property involving derivatives that the Qth derivative in a differential equation translates to s^Q in the transformed differential equation.

ILLUSTRATIVE PROBLEM 3.3
Transfer Function from the DE

The differential equation of the rotating shaft pictured in Figure 3.3 (page 107) is

$$J\ddot{\theta}(t) + B\dot{\theta}(t) + K\theta(t) = \lambda(t).$$

Determine the transfer function $H(s) = \Theta(s)/\Lambda(s)$, where $\Theta(s) = \mathcal{L}[\theta(t)]$ and $\Lambda(s) = \mathcal{L}[\lambda(t)]$.

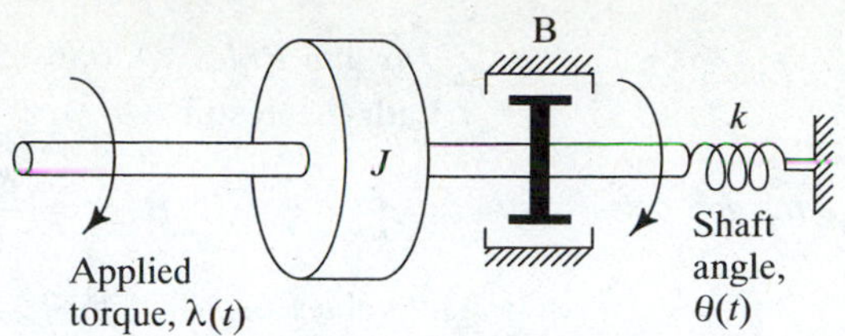

FIGURE 3.3 *Rotating shaft for Illustrative Problem 3.3*

Solution The transformed differential equation with zero initial conditions is $Js^2\Theta(s) + Bs\Theta(s) + K\Theta(s) = \Lambda(s)$, and the corresponding transfer function is

$$H(s) = \frac{\Theta(s)}{\Lambda(s)} = \frac{1}{Js^2 + Bs + K}.$$

If a unit impulse function $x(t) = \delta(t)$ is the input to an LTI system, $X(s) = 1$ and the transform of the output is $Y(s) = H(s) \cdot X(s) = H(s) \cdot 1$. Taking the inverse transform yields $h(t) = y(t)$. Thus the transfer function can be found by taking the Laplace transform of the system's unit impulse response $h(t)$ as

h(t) ⇔ H(s)

$$H(s) = \mathcal{L}[h(t)].$$

Conversely, the unit impulse response $h(t)$ can be found by evaluating the inverse transform of the transfer function $H(s)$.

ILLUSTRATIVE PROBLEM 3.4
Transfer Function from the Impulse Response

The unit impulse response of an oscillatory second-order *LRC* circuit is given by

$$h(t) = 2e^{-5t}\cos(12t - 1.57)u(t).$$

Find the transfer function $H(s)$.

Solution Using pair 8 in Table 3.1 with $A = 2$, $a = -5$, $\omega = 12$, and $\alpha = -1.57$ gives

$$H(s) = \frac{24}{s^2 + 10s + 169}.$$

Zeros and Poles A transfer function can be factored into first-order terms in s with the result

zeros and poles

$$H(s) = \frac{K(s - n_1)(s - n_2)\cdots(s - n_L)}{(s - d_1)(s - d_2)\cdots(s - d_N)}.$$

The values of s that make $H(s)$ go to zero and infinity are called the system zeros and system poles, respectively. There are zeros at $n_1, n_2, \ldots, n_L$ and poles at $d_1, d_2, \ldots, d_N$. Shown in Figure 3.4 is a plot for a typical $H(s)$ with the zeros indicated by O's, the poles by ×'s, and the scale factor by K. We are assuming that $N \geq L$, which means that there are also $N - L$ zeros at infinity.

Comment: There is a MATLAB function in the Control System Toolbox called **pzmap** that is often quite useful for finding and plotting the poles and zeros of a transform. The coefficients of the numerator and denominator polynomials are entered as the vectors **b** and **a**, respectively, and **pzmap(b, a)** returns a plot in the complex plane of the poles and zeros of the given transform.

CHAPTER 3

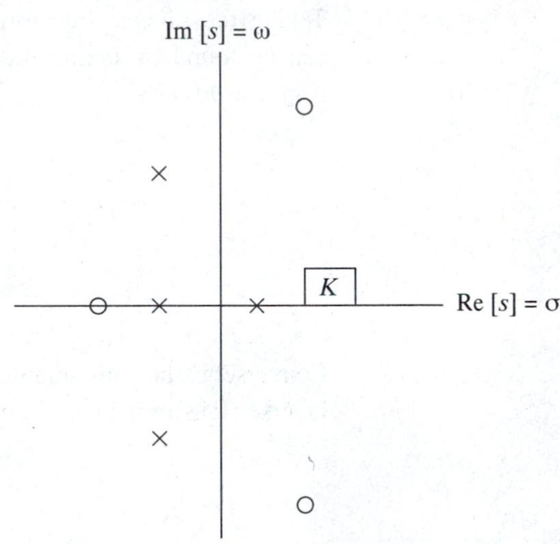

FIGURE 3.4 *Zeros and poles of $H(s)$*

ILLUSTRATIVE PROBLEM 3.5
Determining Zeros and Poles

Find the zeros and poles of the transfer function

$$H(s) = \frac{8}{s^2 + 10s + 169}.$$

Solution

There are no finite zeros, but with $N - L = 2$ we say that there are two zeros at infinity (i.e., $H(s) \to 0$ as $s \to \infty$). The poles are the roots of the equation $s^2 + 10s + 169 = 0$; using the quadratic formula gives $s_{1,2} = -5 \pm j12 = 13e^{\pm j1.966}$.

Region of Convergence The ROC must be specified for any Laplace transform, and the transfer function is not excepted from this rule. Let's look at the ROCs of transfer functions for causal, anticausal, and noncausal systems with nonrepeated poles.

(a) Causal Systems The unit impulse response of an Nth-order causal LTI system is

$$h_c(t) = C_1 e^{d_1 t} u(t) + \cdots + C_N e^{d_N t} u(t) = \sum_{k=1}^{N} C_k e^{d_k t} u(t)$$

and the corresponding transfer function $H_c(s)$, the Laplace transform of $h_c(t)$, is

$$H_c(s) = \frac{C_1}{s - d_1} + \cdots + \frac{C_N}{s - d_N} = \sum_{k=1}^{N} \frac{C_k}{s - d_k}, \qquad \text{Re}[s] > \max_k \{\text{Re}[d_k]\}$$

where the ROCs are $\text{Re}[s] > \text{Re}[d_1]$, $\text{Re}[s] > \text{Re}[d_2], \ldots, \text{Re}[s] > \text{Re}[d_N]$. Thus the ROC for the composite $H_c(s)$ is the half-plane lying to the right of the pole having the largest real part, as seen in Figure 3.5a (page 110).

(b) Anticausal Systems The unit impulse response of an Nth-order anticausal LTI system is

$$h_{ac}(t) = C_1 e^{d_1 t} u(-t) + \cdots + C_N e^{d_N t} u(-t) = \sum_{k=1}^{N} C_k e^{d_k t} u(-t).$$

From Table 3.1 its corresponding transfer function is

$$\begin{aligned} H_{ac}(s) &= -\frac{C_1}{s - d_1} - \cdots - \frac{C_N}{s - d_N} \\ &= -\sum_{k=1}^{N} \frac{C_k}{s - d_k}, \qquad \text{Re}[s] < \min_k \{\text{Re}[d_k]\} \end{aligned}$$

where the ROCs are $\text{Re}[s] < \text{Re}[d_1]$, $\text{Re}[s] < \text{Re}[d_2], \ldots, \text{Re}[s] < \text{Re}[d_N]$. Thus the ROC for the composite $H_{ac}(s)$ is the half-plane lying to the left of the pole having the smallest real part, as seen in Figure 3.5b.

(c) Noncausal Systems The unit impulse response of an Nth-order noncausal LTI system can be partitioned into its causal and anticausal parts:

$$h(t) = h_c(t) + h_{ac}(t) = \sum_{k=1}^{R} C_k e^{d_k t} u(t) + \sum_{k=R+1}^{N} C_k e^{d_k t} u(-t)$$

and the corresponding transfer function is

$$H(s) = H_c(s) + H_{ac}(s) = \sum_{k=1}^{R} \frac{C_k}{s - d_k} - \sum_{k=R+1}^{N} \frac{C_k}{s - d_k},$$
$$\sigma_1 < \text{Re}[s] < \sigma_2$$

where $H_c(s)$ contains all poles of the composite transfer function $H(s)$ having $\text{Re}[s] > \sigma_1$ and $H_{ac}(s)$ contains all the poles of $H(s)$ having $\text{Re}[s] < \sigma_2$, as in Figure 3.5c. Notice that if $\sigma_1 \geq \sigma_2$, the Laplace transform of $h(t)$ does not exist because there is no common region of convergence.

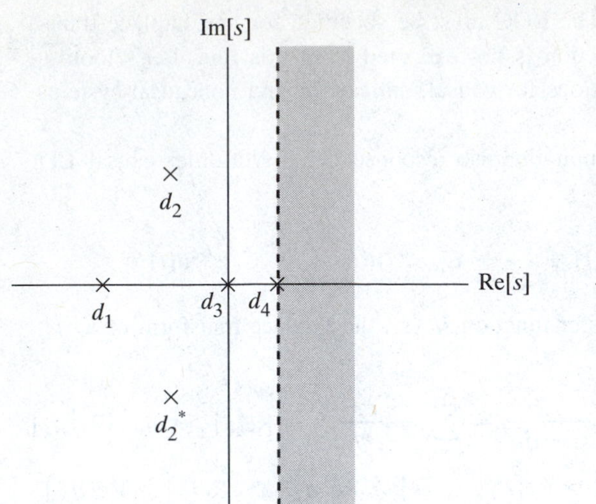

a. Causal system

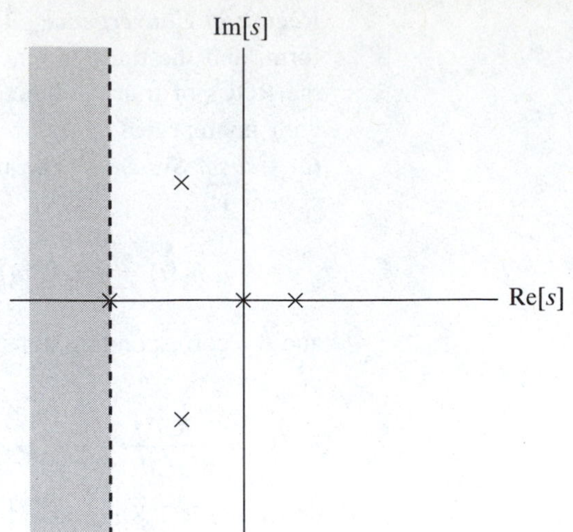

b. Anticausal system

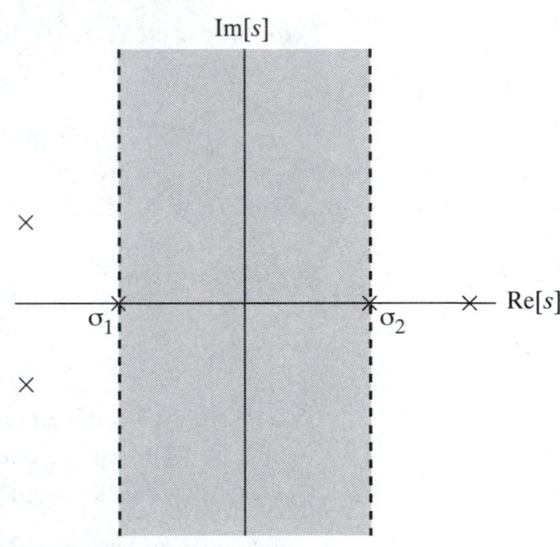

c. Noncausal system

FIGURE 3.5 *Causality, $H(s)$, and ROC*

CHAPTER 3

ILLUSTRATIVE PROBLEM 3.6
Region of Convergence for H(s)

Find the ROCs for the transfer functions of the following systems.

a. $h(t) = e^{-t}u(t) - e^{t}u(t)$

b. $h(t) = e^{-t}u(-t) - e^{t}u(-t)$

c. $h(t) = e^{-t}u(t) - e^{t}u(-t)$

d. $h(t) = e^{-t}u(t) + e^{-t}u(-t)$

Solution

a. $H(s) = \dfrac{1}{s+1} - \dfrac{1}{s-1}$, $\text{Re}[s] > -1$ and $\text{Re}[s] > 1$, so ROC is $\text{Re}[s] > 1$.

b. $H(s) = -\dfrac{1}{s+1} + \dfrac{1}{s-1}$, $\text{Re}[s] < -1$ and $\text{Re}[s] < 1$, so ROC is

$\text{Re}[s] < -1$.

c. $H(s) = \dfrac{1}{s+1} + \dfrac{1}{s-1}$, $\text{Re}[s] > -1$ and $\text{Re}[s] < 1$, so ROC is

$-1 < \text{Re}[s] < 1$.

d. $H(s) = \dfrac{1}{s+1} - \dfrac{1}{s+1}$, $\text{Re}[s] > -1$ and $\text{Re}[s] < -1$, so the ROC is empty and $H(s)$ does not exist. You can see that $H(s) = 0$, which is a good indication of the situation. See Figure 3.6 for a pictorial representation of these results.

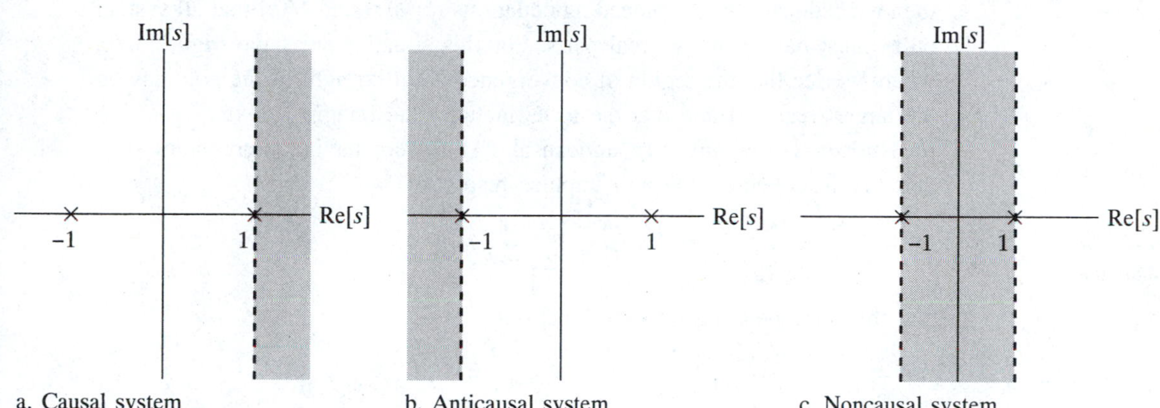

FIGURE 3.6 *ROCs for Illustrative Problem 3.6*

Stability System stability has been discussed in terms of characteristic roots, the unit impulse response, and the bounded-input–bounded-output criterion. Stability may also be deduced from transfer functions.

(a) Causal Systems A causal LTI system with nonrepeated poles and more poles than zeros is described by the unit impulse response

unit impulse response

$$h(t) = \sum_{k=1}^{N} C_k e^{d_k t} u(t)$$

with the corresponding transfer function

transfer function

$$H(s) = \sum_{k=1}^{N} \frac{C_k}{s - d_k}, \qquad \text{Re}[s] > \max_k \{\text{Re}[d_k]\}$$

where the region of convergence is shown in Figure 3.7a. If the system is stable, that is, if

stable system

$$\int_{-\infty}^{\infty} |h(t)|\, dt < \infty,$$

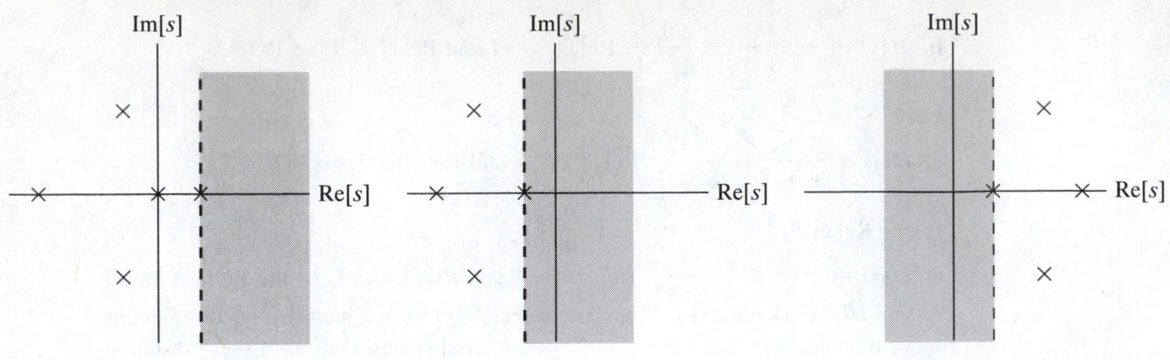

FIGURE 3.7 *Stability and ROCs*

then we deduce from the preceding equations for $h(t)$ and $H(s)$ that all system poles must have negative real parts. For this situation we have Figure 3.7b, where we see that the region of convergence is to the right of the pole having the largest real part and that the ROC includes the imaginary axis.[3]

(b) Anticausal Systems An anticausal LTI system having more poles than zeros is described by the unit impulse response

unit impulse response

$$h(t) = \sum_{k=1}^{N} C_k e^{d_k t} u(-t)$$

and the corresponding transfer function

transfer function

$$H(s) = -\sum_{k=1}^{N} \frac{C_k}{s - d_k}, \qquad \min_k \{\mathrm{Re}[d_k]\}.$$

If the system is stable, that is, if

$$\int_{-\infty}^{\infty} |h(t)|\, dt < \infty,$$

then we deduce from the equations for $h(t)$ and $H(s)$ that all system poles must have positive real parts. For this situation we have Figure 3.7c, where we see that the region of convergence is to the left of the pole having the smallest real part and that the ROC includes the imaginary axis.

Stable causal systems have all poles to the left of the imaginary axis, whereas stable anticausal systems have all poles to the right of the imaginary axis. For stable systems, consequently, the imaginary axis is included in the region of convergence.

CHAPTER 3

[3]Although we've restricted ourselves here to the predominant case of nonrepeated poles, the result that the ROC must include the imaginary axis applies as well to the multiple-pole case.

(c) Noncausal Systems The poles of the causal part of a noncausal system must lie to the left of the imaginary axis, whereas those of the anticausal part must lie to the right for the overall noncausal system to be stable.

ILLUSTRATIVE PROBLEM 3.7
Checking Stability

Given the system transfer function

$$H(s) = \frac{s^2 + 1}{(s + 1)(s + 2)}$$

determine stability for the following regions of convergence.

a. $\text{Re}[s] > -1$

b. $\text{Re}[s] < -2$

c. $-1 < \text{Re}[s] < -2$

Solution

The system poles are $s_1 = -1$ and $s_2 = -2$, both to the left of the imaginary axis.

a. The ROC includes the imaginary axis, so $H(s)$ describes a stable causal system.

b. The imaginary axis is not included in the ROC, so $H(s)$ describes an unstable anticausal system.

c. The imaginary axis is not included in the ROC, so $H(s)$ describes an unstable noncausal system.

Comment: Looking ahead to the next section, we can determine the inverse transform of $H(s)$, the unit impulse response $h(t)$, for each region of convergence. By substituting $t = \pm\infty$ in each of the resulting expressions for $h(t)$ that follows, the preceding stability determinations can be verified.

a. $h(t) = \delta(t) + 2e^{-t}u(t) - 5e^{-2t}u(t), h(\infty) \to 0$

b. $h(t) = \delta(t) - 2e^{-t}u(-t) + 5e^{-2t}u(-t),\ h(-\infty) \to \infty$

c. $h(t) = \delta(t) - 2e^{-t}u(-t) - 5e^{-2t}u(t),\ h(-\infty) \to \infty$

THE EVALUATION OF INVERSE TRANSFORMS

The inverse Laplace transform is given by

inverse transform

$$f(t) = \mathcal{L}^{-1}[F(s)] = \frac{1}{2\pi j}\int_{\sigma - j\infty}^{\sigma + j\infty} F(s)e^{st}\, ds\,,$$

a complex integral with σ a value that causes the line $\sigma - j\infty$ to $\sigma + j\infty$ to lie in the ROC. Fortunately, most of the signals we encounter allow us to evaluate the inverse transform without carrying out the integration.

(a) Partial Fraction Expansion (PFE), Simple Poles Consider the Laplace transform of a signal $f(t)$ in what is often called the *standard form:*

$$F(s) = K\frac{B(s)}{(s - d_1)(s - d_2)\cdots(s - d_N)}$$

where K is a scale factor, $B(s)$ is the numerator polynomial of order L, and all N poles are taken to be different (distinct). We also assume that the degree of the denominator is at least as large as the degree of the numerator—that is, $N \geq L$. If we are dealing with a realistic model of a physical system, the foregoing assumption of $N \geq L$ will always be valid. If $N = L$, we use long division to obtain a constant and a proper fraction.[4] Thus we have

constant + proper fraction

$$F(s) = C_0 + k\frac{Q(s)}{(s-d_1)(s-d_2)\cdots(s-d_N)} = C_0 + G(s)$$

where $Q(s)$ has degree $(N-1)$ or less. The proper fraction $G(s)$ is then expanded in partial fractions in the manner

$$\begin{aligned} G(s) &= k\frac{Q(s)}{(s-d_1)(s-d_2)\cdots(s-d_N)} \\ &= \frac{C_1}{s-d_1} + \frac{C_2}{s-d_2} + \cdots + \frac{C_N}{s-d_N} \end{aligned}$$

where for the pole at $s = d_j$ we have

$$\begin{aligned} C_j &= k\left.\frac{Q(s)\cdot(s-d_j)}{(s-d_1)(s-d_2)\cdots(s-d_j)\cdots(s-d_N)}\right|_{s=d_j,\ j=1,2,\ldots,N} \\ &= k\frac{Q(d_j)}{(d_j-d_1)(d_j-d_2)(d_j-d_{j-1})(d_j-d_{j+1})\cdots(d_j-d_N)}. \end{aligned}$$

Thus $F(s)$ may be expanded in partial fractions as

PF expansion of a transform

$$F(s) = C_0 + \sum_{k=1}^{N}\frac{C_k}{s-d_k}.$$

Now we need to face the region of convergence problem before evaluating the inverse transform of each term in the preceding summation. If, for example, the ROC is to the right of the pole in $F(s)$ with the largest real part, then the sequence is causal and we have

causal signal

$$f(t) = C_0\delta(t) + \sum_{k=1}^{N} C_k e^{d_k t}u(t).$$

If the ROC is to the left of the pole in $F(s)$ with the smallest real part, then the signal is anticausal and is given by

anticausal signal

$$f(t) = C_0\delta(t) - \sum_{k=1}^{N} C_k e^{d_k t}u(-t).$$

[4] A proper fraction exists when the degree of the denominator polynomial is greater than that of the numerator polynomial.

For a mixture of these two situations, where some poles are to the left of the ROC and some are to the right, the inverse transform will be a combination of the causal and noncausal signals given previously.

Comment: Notice that if $N = L$ in the transform $F(s)$, an impulse function $C_0\delta(t)$ appears in the time response $f(t)$. We have seen that an impulse is a mathematical function that does not represent a physical signal, which also makes the situation of $N = L$ somewhat unrealistic. Over certain ranges of frequencies, however, a transform having polynomials of equal degree in the numerator and denominator are realistic and yield reasonable results.

ILLUSTRATIVE PROBLEM 3.8
Finding Inverse Transforms

Given the transform

$$F(s) = \frac{s^2 - 9}{s^2 - 1},$$

find the inverse transform $f(t)$ for the following regions of convergence.

a. $\text{Re}[s] > 1$

b. $\text{Re}[s] < -1$

c. $-1 < \text{Re}[s] < 1$

Solution

First, we express $F(s)$ as a constant plus a proper fraction as

$$F(s) = 1 - \frac{8}{s^2 - 1} = 1 - \frac{8}{(s+1)(s-1)}.$$

Expansion in partial fractions gives

$$F(s) = 1 + \frac{C_1}{s+1} + \frac{C_2}{s-1} = 1 + \frac{4}{s+1} - \frac{4}{s-1}.$$

Notice that the PFE does not depend on the region of convergence.

a. ROC $\text{Re}[s] > 1$: ROC is to the right of both poles and

$$f(t) = \delta(t) + [4e^{-t} - 4e^{t}]u(t).$$

Using the Symbolic Math Toolbox, we enter `F=('(s^2-9)/(s^2-1')` and **`f=invlaplace(F)`** with the function to be inverse transformed `F=(s^2-9)/(s^2-1)` and its inverse transform `f=Dirac(t)-4*exp(t)+4*exp(-t)` returned.

b. ROC $\text{Re}[s] < -1$: ROC is to the left of both poles and

$$f(t) = \delta(t) + [-4e^{-t} + 4e^{t}]u(-t).$$

c. ROC $-1 < \text{Re}[s] < 1$: ROC is to the right of the pole at $s = -1$, which yields a causal signal, and to the left of the pole at $s = 1$, which produces an anticausal signal. The result is the noncausal signal

$$f(t) = \delta(t) + 4e^{-t}u(t) + 4e^{t}u(-t).$$

BASIC CONCEPTS

***(b) Inverse Transform Using the MATLAB Function* residue** The partial fraction expansion is easily accomplished by using the MATLAB function **residue**.[5] The coefficients of the numerator and denominator polynomials of the rational fraction to be expanded are the input data and the poles and PF constants are returned. The following annotated script (PFE) illustrates the use of **residue** on Illustrative Problem 3.8.

MATLAB Script

```
%PFE...computes PFE constants, poles, and the direct term of a rational
%function.
b=[1,0,-9]                   % numerator coefficients of F(s)
a=[1,0,-1]                   % denominator coefficients of F(s)
[r,p,k]=residue(b,a)         % call residue, r=PFE constant, p=pole or
                             % characteristic root, k=direct term (0 unless N=L)
```

Output

```
r=4                          % the two partial fraction constants
  -4
p=-1                         % the two poles of F(s)
   1
k=1                          % the direct term, nonzero because N=L.
```

CHAPTER 3

Thus matching the poles with the partial fraction constants (the r's), we have, as before,

$$F(s) = 1 + \frac{4}{s+1} - \frac{4}{s-1}.$$

(c) Partial Fraction Expansion, Multiple Poles We have concentrated our discussion of the partial fraction method to the practical situation of distinct or simple poles of the transform to be inverted. If a transform has multiple poles, a different strategy must be employed to find the PFE constants. We shall show one procedure (there are several) and illustrate the procedure by an example.

Suppose we have the transform

$$Y(s) = \frac{2s+12}{(s+1)^2}, \qquad \text{Re}[s] > 1$$

[5]The MATLAB function **residue**, which returns the poles of the transform along with the PFE constants, might be better named **pfe**. From complex analysis for a rational function, the partial fraction constants are known as the residues.

and we want the inverse transform $y(t)$. The transform is expanded as

expanded transform

$$Y(s) = \frac{2s + 12}{(s + 1)^2} = \frac{C_1}{(s + 1)^2} + \frac{C_2}{(s + 1)}$$

where, following the standard procedure,

$$C_1 = \left.\frac{(2s + 12)(s + 1)^2}{(s + 1)^2}\right|_{s=-1} = 10.$$

With $C_1 = 10$ we try to evaluate C_2 in the usual manner by multiplying through the expanded transform above by $s + 1$, obtaining

$$\frac{2s + 12}{s + 1} = \frac{10}{s + 1} + C_2;$$

we have a problem if we set $s = -1$. But if we subtract the term $10/(s + 1)$ from both sides of this equation before setting $s = -1$, we have

$$C_2 = \left\{\left.\frac{2s + 12}{(s + 1)} - \frac{10}{(s + 1)}\right|_{s=-1}\right\} = \left.\frac{2(s + 1)}{(s + 1)}\right|_{s=-1} = 2.$$

Using pairs 2 and 11 in Table 3.1, the inverse transform is

$$y(t) = [10te^{-t} + 2e^{-t}]u(t).$$

To use this approach, there is a specific procedure that must be followed. Given

$$P(s) = \frac{N(s)}{s^3(s + 1)} = \frac{C_1}{s^3} + \frac{C_2}{s^2} + \frac{C_3}{s} + \frac{C_4}{s + 1}$$

C_1, C_2, and C_3 must be determined *in that order.* That is, we must find the numerator associated with the *highest power* of the repeated root *first,* and so on down the list. The PF constant(s) for the simple root(s) (in this case $s = -1$) may be found at any time. In Exploration Problems P3.20 and P3.21, two other methods for handling multiple poles are presented.

BASIC CONCEPTS

ILLUSTRATIVE PROBLEM 3.9
Multiple Poles and the* residue *Function

Given the following transform, use **residue** to find the partial fraction expansion and determine the corresponding time signal from Table 3.1:

$$V(s) = \frac{s^4 + 2s^3 + 3s^2 + 2s + 1}{s^4 + 4s^3 + 7s^2 + 6s + 2}.$$

Solution

The data and the results from **residue** are given next.

MATLAB Script

```
%PFE
b=[1,2,3,2,1];            % numerator coefficients of V(s)
a=[1,4,7,6,2];            % denominator coefficients of V(s)
[r,p,k]=residue(b,a)      % call residue, r=PFE constant, p=pole or
                          % characteristic root, k=direct term (0 unless N=L)
```

Continues

Output

```
r=0.0000-0.5000i          the 4 PF constants
  0.0000+0.5000i
  -2.0000
  1.0000
p=-1.0000+1.0000i         the 4 poles
  -1.0000-1.0000i
  -1.0000
  -1.0000
  k=1                     the constant is nonzero because L=N
```

Matching the PF constants with the poles gives the expansion

$$V(s) = 1 + \frac{-j0.5}{s+1-j1} + \frac{j0.5}{s+1+j1} - \frac{2}{s+1} + \frac{1}{(s+1)^2}$$

and Table 3.1 (pairs 1, 8, 2, and 11) produces the time function

$$v(t) = \delta(t) + e^{-t}\cos(t - \pi/2) - 2e^{-t} + te^{-t}, \qquad t \geq 0.$$

CROSS-CHECK How did we know to match the PF constant -2 with the simple pole at $s = -1$ and the PF constant $+1$ with the double pole at $s = -1$? We simply tried some problems where we knew the result to determine the order of results given by the MATLAB algorithm. ■

CHAPTER 3

SOLUTION OF LINEAR DIFFERENTIAL EQUATIONS BY LAPLACE TRANSFORM

In this book we will consider the solution of differential equations that represent causal systems only and will typically be interested in the solutions for $t \geq 0$. Thus we need to consider the unilateral Laplace transform. Starting with the system's differential equation,

$$\sum_{k=0}^{N} a_k \frac{d^k y(t)}{dt^k} = \sum_{k=0}^{L} b_k \frac{d^k x(t)}{dt^k}$$

we take the Laplace transform term by term to obtain

$$\begin{aligned} a_0 Y(s) + a_1[sY(s) - y(0)] + a_2[s^2 Y(s) - sy(0) - \dot{y}(0)] + \cdots \\ + a_N[s^N Y(s) - s^{N-1}y(0) - \cdots - y^{N-1}(0)]^* \\ = b_0 X(s) + b_1[sX(s) - x(0)] \\ + b_2[s^2 X(s) - sx(0) - \dot{x}(0)] \\ + \cdots + b_L[s^L X(s) \\ - s^{L-1}x(0) - \cdots - x^{L-1}(0)]. \end{aligned}$$

*The notation $y^{N-1}(0)$ represents $\frac{d^{N-1}}{dt^{N-1}} y(t)|_{t=0}$.

Next we collect terms in the manner

$$\begin{aligned}[a_0 + a_1 s + a_2 s^2 + \cdots + a_N s^N]Y(s) = {} & [a_1 y(0) + a_2 \dot{y}(0) \\ & + \cdots + a_N y^{N-1}(0) - b_1 x(0) \\ & - b_2 \dot{x}(0) - \cdots - b_L x^{L-1}(0)] \\ & + [a_2 y(0) + a_3 \dot{y}(0) \\ & + \cdots + a_N y^{N-2}(0) - b_2 x(0) \\ & - b_3 \dot{x}(0) - \cdots - b_L x^{L-2}(0)]s \\ & + \cdots + a_N s^{N-1} y(0) \\ & + [b_0 + b_1 s + b_2 s^2 \\ & + \cdots + b_L s^L] X(s).\end{aligned}$$

The right side of this equation consists of terms due to the ICs and terms due to the transform of the input $X(s)$, and so we can write

$$Q(s)Y(s) = P_{\mathrm{IC}}(s) + P(s)X(s)$$

which, solved for $Y(s)$ (the Laplace transform of the output), yields

transform of y(t)

$$Y(s) = \frac{P_{\mathrm{IC}}(s)}{Q(s)} + \frac{P(s)}{Q(s)} \cdot X(s).$$

To complete the solution we need to know the initial conditions and $x(t)$ from which we obtain $X(s)$. Having this information we can then evaluate $Y(s)$ and expand it in partial fractions. Knowing the region of convergence, $y(t)$ can be determined using Table 3.1. A diagram of the procedure is given in Figure 3.8.

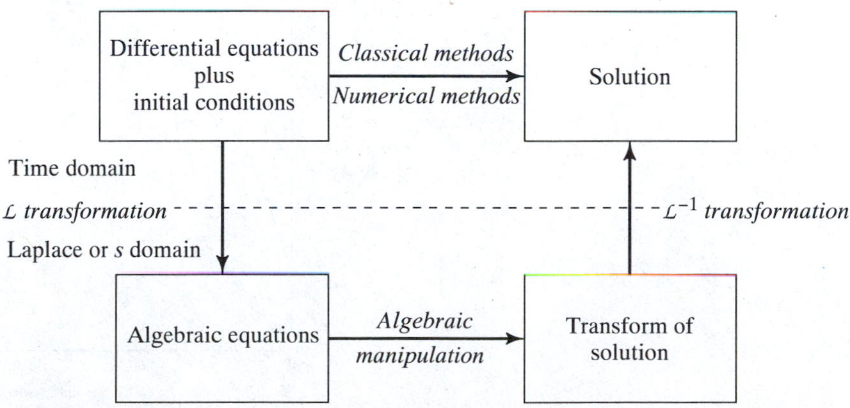

FIGURE 3.8 *A procedure for solving linear differential equations*

ILLUSTRATIVE PROBLEM 3.10 *Solving a Differential Equation Using Laplace Transforms*

Given the differential equation that describes a first-order circuit, $dy(t)/dt + 2y(t) = x(t)$, with $y(0) = 1$ and $x(t) = 10u(t)$.

a. Use the Laplace transform method to find $y(t)$ for $t \geq 0$.

b. Check your answer using the initial and final value theorems.

Solution

a. Using linearity, the derivative theorem, and the transform of a step function gives $sY(s) - 1 + 2Y(s) = 10/s$, $\text{Re}[s] > 0$. Solving for $Y(s)$ and its partial fraction expansion produces

$$Y(s) = \frac{s + 10}{s(s + 2)} = \frac{5}{s} - \frac{4}{s + 2}$$

and from Table 3.1, $y(t) = [5 - 4e^{-2t}]u(t)$.

b. Using the IVT, $sY(s) = (s + 10)/(s + 2) = 1$ as $s \to \infty$, which agrees with the given data and the time-domain solution. It is proper to use the FVT, since $sY(s)$ has a pole at $s = -2$, to the left of the imaginary axis. Thus $y(\infty) = sY(s)|_{s=0} = 5$, which agrees with the time-domain solution.

Comment: We can use the function **lsim** or **kslsim** to obtain a tabular solution and a graph for this problem. The script is given next and the plot is in Figure 3.9. To use either function, the data are entered in state-space form, which for this first-order system is straightforward as

$$dv(t)/dt = -2v(t) + x(t) \quad \text{with} \quad x(t) = 10u(t), \quad v(t) = y(t),$$
$$\text{and} \quad y(0) = v(0) = 1\,.$$

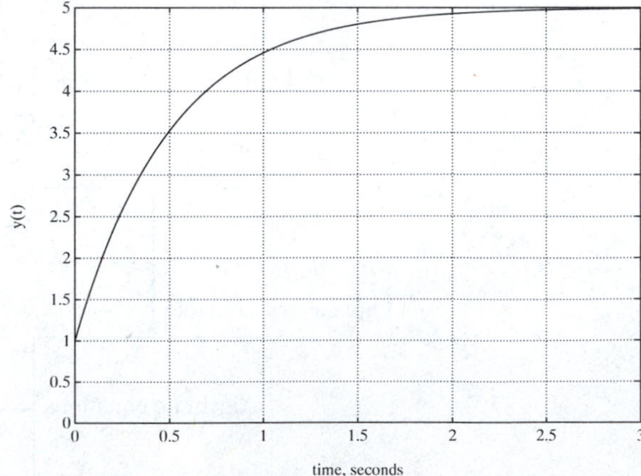

FIGURE 3.9 *Circuit output*

CHAPTER 3

MATLAB Script

```
%F3_9 Solution of first-order DE
A=[-2];
B=[1];
C=[1];
D=[0];
v0=[1];
t=0:0.01:3;                            % start time:time increment:final time
X=[10*ones(size(t))]';
[y,v]=kslsim(A,B,C,D,X,t,v0);          % call function kslsim
%...plotting statements
```

Linear systems with multiple inputs and multiple outputs are often modeled by simultaneous differential equations, such as

$$3\,\ddot{p}(t) + 2\dot{p}(t) + p(t) - 2q(t) = 5f(t) - 7g(t)$$
$$2\,\ddot{q}(t) - 3\dot{q}(t) + 5p(t) = 3g(t)$$

where $f(t)$ and $g(t)$ are the system inputs and $p(t)$ and $q(t)$ are its outputs. Taking the Laplace transform of these equations gives us the two simultaneous algebraic equations:

$$3\{s^2P(s) - sp(0) - \dot{p}(0)\} + 2\{sP(s) - p(0)\} + P(s) - 2Q(s) = 5F(s) - 7G(s)$$
$$2\{s^2Q(s) - sq(0) - \dot{q}(0)\} - 3\{sQ(s) - q(0)\} + 5P(s) = 3G(s).$$

These equations can be written in matrix form as

$$\begin{bmatrix} 3s^2 + 2s + 1 & -2 \\ 5 & 2s^2 - 3s \end{bmatrix}\begin{bmatrix} P(s) \\ Q(s) \end{bmatrix} = \begin{bmatrix} [3s + 2]p(0) + 3\dot{p}(0) + 5F(s) - 7G(s) \\ [2s - 3]q(0) + 2\dot{q}(0) + 3G(s) \end{bmatrix}$$

and given the two input signals and the four initial conditions, we could use matrix methods or determinants to find the transforms $P(s)$ and $Q(s)$. Then, with the help of partial fractions (**residue**) and Table 3.1 (and lots of luck!), the output signals $p(t)$ and $q(t)$ are found.

ILLUSTRATIVE PROBLEM 3.11 *Laplace Transform Solution of Simultaneous DEs*

A causal LTI system with the outputs $y(t)$ and $z(t)$ is described by

$$\dot{y}(t) - z(t) = -x_1(t) \quad \text{and} \quad \dot{z}(t) - y(t) = x_2(t),$$

with $y(0) = 0$, $z(0) = -1$, $x_1(t) = u(t)$, and $x_2(t) = tu(t)$. Find the solution for $z(t)$, $t \geq 0$.

Solution

Taking the Laplace transform gives us the algebraic equations

$$sY(s) - y(0) - Z(s) = -X_1(s) \quad \text{and} \quad sZ(s) - z(0) - Y(s) = X_2(s).$$

Substituting the two initial conditions and the transforms of the inputs $X_1(s)$ and $X_2(s)$ and then arranging in matrix form gives

$$\begin{bmatrix} s & -1 \\ -1 & s \end{bmatrix}\begin{bmatrix} Y(s) \\ Z(s) \end{bmatrix} = \begin{bmatrix} -1/s \\ -1 + 1/s^2 \end{bmatrix}$$

Solving the preceding matrix equation, we have

$$\begin{bmatrix} Y(s) \\ Z(s) \end{bmatrix} = \begin{bmatrix} s & -1 \\ -1 & s \end{bmatrix}^{-1}\begin{bmatrix} -1/s \\ -1 + 1/s^2 \end{bmatrix}$$
$$= \frac{1}{s^2 - 1}\begin{bmatrix} s & 1 \\ 1 & s \end{bmatrix}\begin{bmatrix} -1/s \\ -1 + 1/s^2 \end{bmatrix}$$
$$= \begin{bmatrix} (-2 + 1/s^2)/(s^2 - 1) \\ -s/(s^2 - 1) \end{bmatrix}.$$

Thus we have, from the second row above

$$Z(s) = \frac{-s}{s^2 - 1} = \frac{-0.5}{s + 1} + \frac{-0.5}{s - 1} \quad \text{and} \quad z(t) = -0.5[e^{-t} + e^{t}]u(t).$$

CONVOLUTION

In Chapter 2, "Continuous Systems," the forced response (ICs set to zero) of an LTI system was obtained from the convolution integral:

$$y(t) = \int_{-\infty}^{\infty} h(\tau)x(t - \tau)\,d\tau = \int_{-\infty}^{\infty} h(t - \tau)x(\tau)\,d\tau$$

where $h(t)$ is the unit impulse response of the system and $x(t)$ represents the system input. If the system is causal, $h(t) = 0$ for $t < 0$, and if the input is zero for $t < 0$, a causal signal, the convolution $y(t) = h(t)*x(t) = x(t)*h(t)$ is given by

$$y(t) = \int_{0}^{t} h(\tau)x(t - \tau)\,d\tau = \int_{0}^{t} h(t - \tau)x(\tau)\,d\tau.$$

Earlier in this chapter the system transfer function was defined as $H(s) = Y(s)/X(s)$. We see that if the transform $X(s)$ of the input $x(t)$ is known then the transform $Y(s)$ of the output signal $y(t)$ is

$$Y(s) = H(s) \cdot X(s) = X(s) \cdot H(s)$$

and the output $y(t)$ can be found by the inverse transform

$$y(t) = \mathcal{L}^{-1}[Y(s)] = \mathcal{L}^{-1}[H(s) \cdot X(s)] = \mathcal{L}^{-1}[X(s) \cdot H(s)].$$

Thus we have the convolution pair or property that allows the evaluation of a convolution in either the time domain directly or indirectly by the means of Laplace transforms,

convolution

$$y(t) = \int_{0}^{t} h(\tau)x(t - \tau)\,d\tau \Leftrightarrow Y(s) = H(s) \cdot X(s)$$

where the region of convergence for $Y(s) = H(s) \cdot X(s)$ will be greater than the largest real part among all of the poles of $H(s) \cdot X(s)$.

ILLUSTRATIVE PROBLEM 3.12

Convolution Using Transforms

An LRC circuit with the unit impulse response $h(t) = [e^{-t} - e^{-2t}]u(t)$ is subjected to a unit step input—that is, $x(t) = u(t)$.

a. Use Laplace transforms to find the circuit's output $y(t) = h(t)*x(t)$.

b. Verify your answer to part (a) by using the convolution integral.

c. Use the MATLAB function **step** or **lsim** (or the California function **kslsim**) to obtain a graphical solution to this problem.

Solution

a. We use the convolution pair or property $Y(s) = H(s) \cdot X(s)$. From Table 3.2,

$$H(s) = \frac{1}{s+1} - \frac{1}{s+2} = \frac{1}{(s+1)(s+2)}, \qquad \text{Re}[s] > -1$$

$$\text{and} \quad X(s) = \frac{1}{s}, \qquad \text{Re}[s] > 0.$$

Thus

$$Y(s) = \frac{1}{(s+1)(s+2)} \cdot \frac{1}{s} = \frac{0.5}{s} - \frac{1}{s+1} + \frac{0.5}{s+2}, \quad \text{Re}[s] > 0.$$

Table 3.1 gives the output $y(t) = [0.5 - e^{-t} + 0.5e^{-2t}]u(t)$.

b. Using the convolution integral, we have

$$y(t) = \int_0^t e^{-\tau}\, d\tau - \int_0^t e^{-2\tau}\, d\tau = [-e^{-t} + 1] + 0.5[e^{-2t} - 1]$$
$$= 0.5 - e^{-t} + 0.5e^{-2t}, \qquad t \geq 0.$$

c. The script that uses the MATLAB function **step** (Signals and Systems Toolbox) is next; the plot is in Figure 3.10. Recall that the transfer function is $H(s) = 1/(s^2 + 3s + 2)$. We use the option in **step** that allows describing the system in transfer function form rather than state-space form.

CROSS-CHECK We notice in Figure 3.10 that the steady-state value of $y(t)$ is 0.5, as verified in (a) and (b). ■

MATLAB Script

```
%F3_10 Step response of an LRC circuit
num=[1];              % numerator coefficients in descending powers of s
den=[1,3,2];          % denominator coefficients in descending powers of s
t=0:0.01:5;           % start time:time increment:final time
y=step(num,den,t);    % call step
%...plotting statements
```

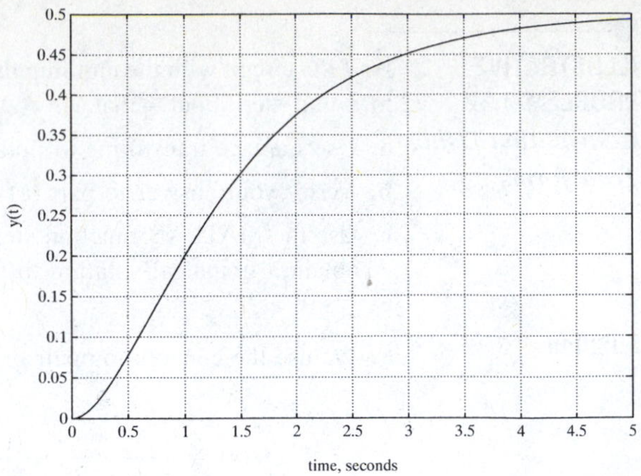

FIGURE 3.10 *Step response*

CHAPTER 3

SINUSOIDAL STEADY-STATE RESPONSE

A very important and practical situation occurs when a stable causal system described by the transfer function $H(s) = Y(s)/X(s)$ is subjected to the input signal $x(t) = A\cos(\omega t + \alpha)u(t)$. Let us show that the steady-state output response $y_{ss}(t)$ is given by the expression

$$y_{ss}(t) = A|H(j\omega)|\cos(\omega t + \angle H(j\omega) + \alpha).$$

The output caused by the input $x(t) = A\cos(\omega t + \alpha)u(t)$ has the Laplace transform

output transform

$$Y(s) = H(s)X(s) = H(s)\cdot\left\{\frac{0.5Ae^{j\alpha}}{s - j\omega} + \frac{0.5Ae^{-j\alpha}}{s + j\omega}\right\}.$$

Assuming nonrepeated poles and $N > L$, expanding $Y(s)$ in partial fractions produces

PFE

$$Y(s) = \frac{C_1}{s - p_1} + \frac{C_2}{s - p_2} + \cdots + \frac{C_N}{s - p_N} + \frac{C_\omega}{s - j\omega} + \frac{C_\omega{}^*}{s + j\omega}$$

where $C_\omega{}^*$ is the complex conjugate of C_ω. Taking the inverse transform gives us the solution $y(t)$ for $t \geq 0$:

$$y(t) = [C_1e^{p_1t} + C_2e^{p_2t} + \cdots + C_Ne^{p_Nt} + C_\omega e^{j\omega t} + C_\omega{}^*e^{-j\omega t}]u(t).$$

If the system is stable, terms such as $C_1 e^{p_1 t}, \ldots, C_N e^{p_N t}$ will decay to zero as $t \to \infty$, leaving only the steady-state solution

$$y_{ss}(t) = C_\omega e^{j\omega t} + C_\omega{}^* e^{-j\omega t} = 2|C_\omega| \cos(\omega t + \angle C_\omega).$$

The partial fraction constant C_ω is found in the usual way as

$$\begin{aligned} C_\omega &= Y(s)\,(s - j\omega)\big|_{s=j\omega} \\ &= H(s) \cdot \left\{ \frac{0.5Ae^{j\alpha}(s - j\omega)}{s - j\omega} + \frac{0.5Ae^{-j\alpha}(s - j\omega)}{s + j\omega} \right\}\Bigg|_{s=j\omega} \\ &= H(j\omega) \cdot 0.5Ae^{j\alpha} = 0.5A|H(j\omega)|e^{(\angle H(j\omega)+\alpha)}. \end{aligned}$$

Now we can write the steady-state response as

$$y_{ss}(t) = A|H(j\omega)| \cos(\omega t + \angle H(j\omega) + \alpha)$$

which is the result we set out to derive. This equation makes the evaluation of a system's steady-state response $y_{ss}(t)$ to a sinusoidal input very straightforward, and we will refer to the result as the sinusoidal steady-state formula. In words, it tells us that when a sinusoidal signal $x(t) = A\cos(\omega t + \alpha)u(t)$ is applied to a *stable* causal linear time-invariant system, then to obtain the steady-state output $y_{ss}(t)$,

1. The input amplitude A is multiplied by the gain $|H(j\omega)|$, and
2. The input phase α is shifted by the angle $\angle H(j\omega)$.

This is illustrated in Figure 3.11, where the input and the output signals of an LTI system subjected to a sinusoidal input are shown. In accordance with the problem statement, the input sinusoidal signal has been altered in both amplitude and phase, but not in frequency, in passing through the LTI system.

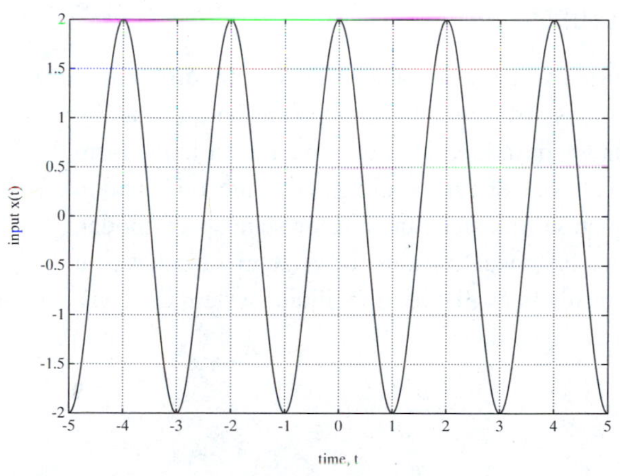

a. Input signal

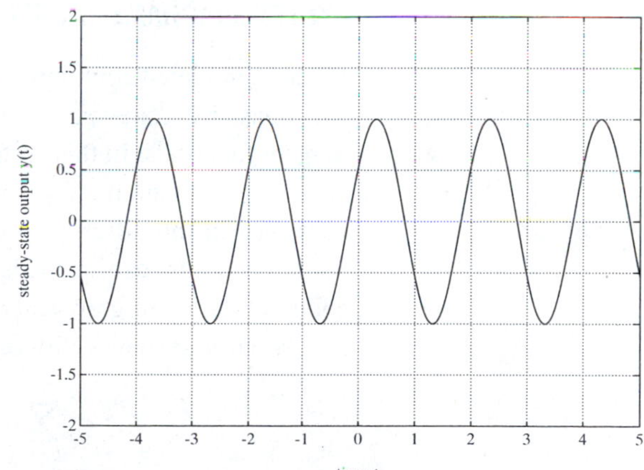

b. Steady-state output signal

FIGURE 3.11 *Illustration of sinusoidal steady-state formula*

ILLUSTRATIVE PROBLEM 3.13 *Using the Sinusoidal Steady-State Formula*

An analog notch filter that is realized with op-amps, resistors, and capacitors can be modeled by the differential equation

$$\ddot{v}(t) + B\dot{v}(t) + \omega_0^2 v(t) = K\ddot{e}(t) + K\omega_0^2 e(t)$$

where $v(t)$ is the output voltage, $e(t)$ is the input voltage, B is the bandwidth, K is an adjustable constant, and ω_0 is the notch frequency of the filter. An input signal $e(t) = A + C\cos(\omega_0 t)$ is applied to this filter. Find the equation for the steady-state output voltage $v_{ss}(t)$.

Solution

The characteristic equation is $s^2 + Bs + \omega_0^2 = 0$, which gives stable characteristic roots for the bandwidth $B > 0$, a natural assumption. Thus we can follow the procedure just outlined by first finding the transfer function

$$H(s) = \frac{V(s)}{E(s)} = \frac{K(s^2 + \omega_0^2)}{s^2 + Bs + \omega_0^2}.$$

Next $H(s)$ needs to be evaluated at the frequencies of the input signal, which are $s = j0$ (a constant such as A can be considered a sinusoid of zero frequency) and $s = j\omega_0$. Thus we have

$$H(j0) = \frac{V(s)}{E(s)} = \frac{K\{(j0)^2 + \omega_0^2\}}{(j0)^2 + B(j0) + \omega_0^2} = K$$

$$H(j\omega_0) = \frac{V(s)}{E(s)} = \frac{K\{(j\omega_0)^2 + \omega_0^2\}}{(j\omega_0)^2 + B(j\omega_0) + \omega_0^2} = 0.$$

Finally, applying the sinusoidal steady-state formula gives $y_{ss}(t) = KA$, and the filter does indeed notch out the unwanted frequency of ω_0.

SYSTEM DIAGRAMS OR STRUCTURES

We have modeled continuous systems by linear constant-coefficient differential equations, by state-space equations, by the unit impulse response, and by transfer functions. In this section we introduce an alternative model in the form of a system diagram, a graphical way of representing the same information contained in the differential equation, transfer function, or state-space model. Such diagrams can provide useful visualizations of system structure; in addition, there are graphical algorithms available to facilitate system analysis.

When a system's differential equation

$$\sum_{k=0}^{N} a_k \frac{d^k y(t)}{dt^k} = \sum_{k=0}^{L} b_k \frac{d^k x(t)}{dt^k}$$

is solved for $y(t)$, we have

$$y(t) = \frac{1}{a_0}\left\{-\sum_{k=1}^{N} a_k \frac{d^k y(t)}{dt^k} + \sum_{k=0}^{L} b_k \frac{d^k x(t)}{dt^k}\right\}$$

where it is seen that derivatives, multiplications, and additions are required to represent this differential equation. Although the use of differentiators is perfectly satisfactory from a mathematical point of view, an alternative is to recast this equation in the form of an integral equation and then use integrators in the implementation. For convenience we will let $N = L$ and start with a second-order differential equation

$$a_0 y(t) + a_1 \frac{dy(t)}{dt} + a_2 \frac{d^2 y(t)}{dt^2} = b_0 x(t) + b_1 \frac{dx(t)}{dt} + b_2 \frac{d^2 x(t)}{dt^2}.$$

The differentiation process is described by the operator D and integration by the operator D^{-1}; integrating 2 times then produces

$$\begin{aligned} D^{-2}[a_0 y(t)] + D^{-2}[a_1 D y(t)] + D^{-2}[a_2 D^2 y(t)] \\ = D^{-2}[b_0 x(t)] + D^{-2}[b_1 D x(t)] + D^{-2}[b_2 D^2 x(t)] \end{aligned}$$

where we have used the linearity property of the integration operation. We notice that

$$\begin{aligned} &D^{-2}[D^2 y(t)] = y(t), \qquad D^{-2}[D y(t)] = D^{-1} y(t), \\ &D^{-2}[D x(t)] = D^{-1} x(t), \quad \text{and} \quad D^{-2}[D^2 x(t)] = x(t) \end{aligned}$$

making it possible, again using linearity, to solve for $y(t)$ as

$$\begin{aligned} y(t) = \frac{1}{a_2}\{&-a_1 D^{-1} y(t) - a_0 D^{-2} y(t) + b_0 D^{-2} x(t) \\ &+ b_1 D^{-1} x(t) + b_2 x(t)\}. \end{aligned}$$

Thus to implement this integral equation we need integrators, multipliers, and adders. Figure 3.12 (page 128) shows two sets of commonly used symbols for these operators along with the describing relations. Normally, the sets of symbols are not mixed, and it is strictly a matter of preference as to the use of signal flowgraphs or block diagrams; we will use both.

Generalization to a Nth-order system by integrating N times (and assuming $L = N$) leads to

$$\begin{aligned} D^{-N}\{a_0 y(t) + a_1 D y(t) + \cdots + a_N D^N y(t)\} \\ = D^{-N}\{b_0 x(t) + b_1 D x(t) + \cdots + b_N D^N x(t)\} \end{aligned}$$

which—when solved for $y(t)$—may be described with the summations

$$y(t) = \frac{1}{a_N}\left\{-\sum_{k=0}^{N-1} a_k D^{-[N-k]} y(t) + \sum_{k=0}^{N} b_k D^{-[N-k]} x(t)\right\}.$$

Shown in Figure 3.13 is a diagram that represents the general differential equation, where we notice that the left half of the diagram represents (realizes) the input terms of the differential equation, whereas the right-hand part shows the output terms.[6] Unlabeled branches have gains of 1.

[6]Initial conditions can also be included in these diagrams. This topic is often included in a controls course, where system diagrams are constructed from a state-variable point of view.

Block diagram

Signal flowgraph

$q_1(t)$ → D^{-1} → $q_2(t) = D^{-1} q_1(t)$

$q_1(t)$ —D^{-1}→ $q_2(t) = D^{-1} q_1(t)$

$q_1(t)$ → b_1 → $q_2(t) = b_1 q_1(t)$

$q_1(t)$ —b_1→ $q_2(t) = b_1 q_1(t)$

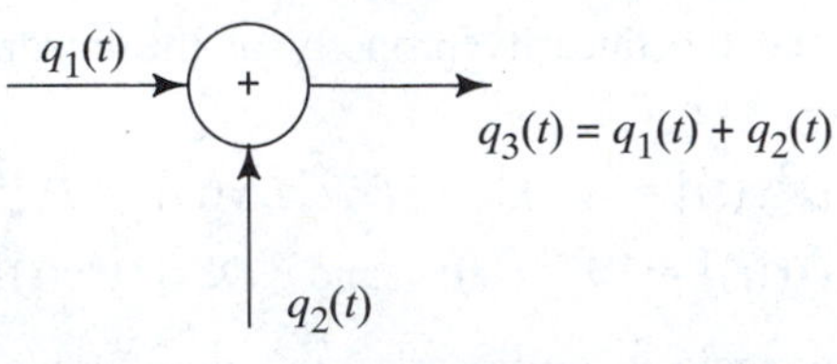

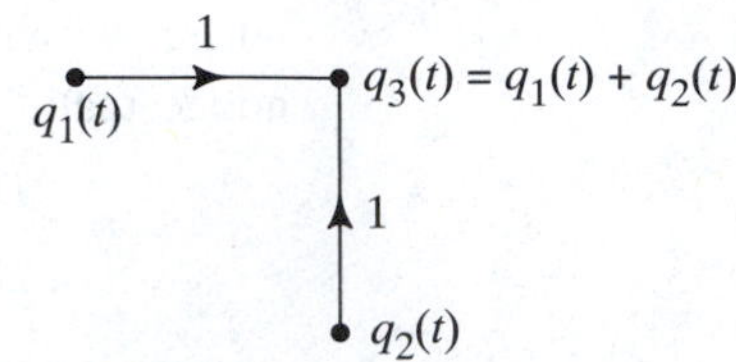

FIGURE 3.12 *System diagram symbols*

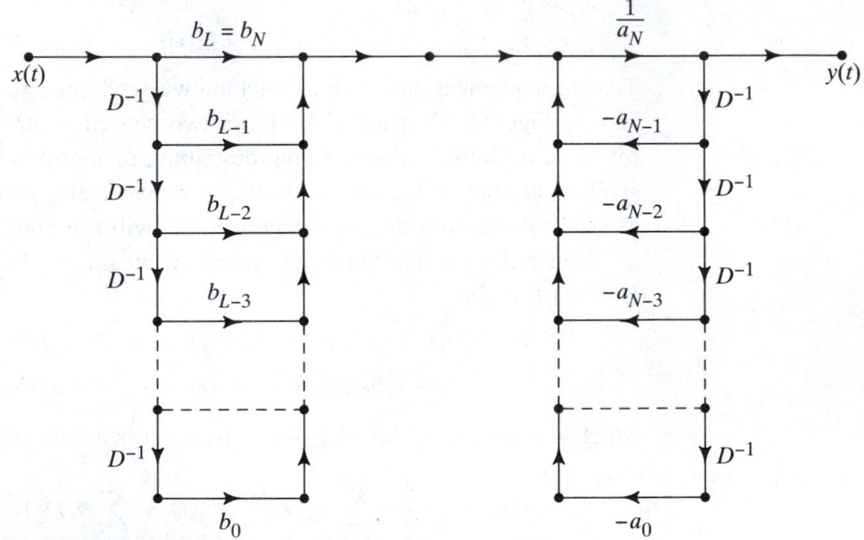

FIGURE 3.13 *Time-domain diagram of system DE*

ILLUSTRATIVE PROBLEM 3.14
Drawing a System Diagram

A bandstop continuous-time (analog) filter is described by the differential equation

$$\ddot{y}(t) + 50\dot{y}(t) + 10^4 y(t) = 20\{\ddot{x}(t) + 10^4 x(t)\}.$$

Draw a signal flowgraph in the manner of Figure 3.13 that represents this filter.

CHAPTER 3

Solution First, we solve the given equation for the highest derivative of $y(t)$, giving

$$\ddot{y}(t) = -50\dot{y}(t) - 10^4 y(t) + 20\{\ddot{x}(t) + 10^4 x(t)\}.$$

Then we integrate twice to obtain

$$y(t) = -50D^{-1}y(t) - 10^4 D^{-2}y(t) + 20x(t) + 20 \cdot 10^4 D^{-2}x(t).$$

The system diagram marked with the proper gains and necessary integrators (D^{-1}) is given in Figure 3.14. The multiplier showing zero gain can be eliminated, but this change is simply a matter of choice.

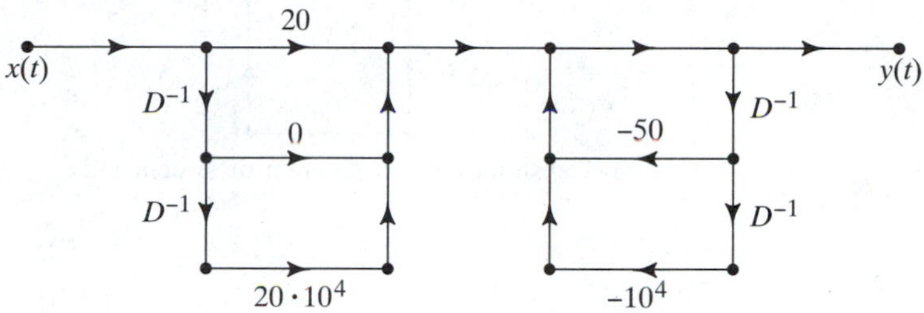

FIGURE 3.14 *Signal flowgraph for Illustrative Problem 3.14*

Figure 3.15a shows the general system diagram from the Laplace transform point of view, where we have replaced $x(t)$ with its transform $X(s)$, $y(t)$ with its transform $Y(s)$, the D^{-1} operators with the transform equivalent s^{-1}, and an integrated input $D^{-i}x(t)$ or output $D^{-j}y(t)$ with $s^{-i}X(s)$ and $s^{-j}Y(s)$, respectively. Consequently, the system's differential equation solved for the output $y(t)$,

$$y(t) = \frac{1}{a_N}\left\{-\sum_{k=0}^{N-1} a_k D^{-[N-k]}y(t) + \sum_{k=0}^{N} b_k D^{-[N-k]}x(t)\right\}$$

becomes the algebraic equation

$$Y(s) = \frac{1}{a_N}\left\{-\sum_{k=0}^{N-1} a_k s^{-[N-k]}Y(s) + \sum_{k=0}^{N} b_k s^{-[N-k]}X(s)\right\}.$$

A popular modification of Figure 3.15a is shown in Figure 3.15b, where the integrators (s^{-1} or D^{-1}) for the input and for the output are "shared," which yields a more compact diagram. This sort of diagram is seen frequently in Chapter 6, "State-Space Topics for Continuous Systems."

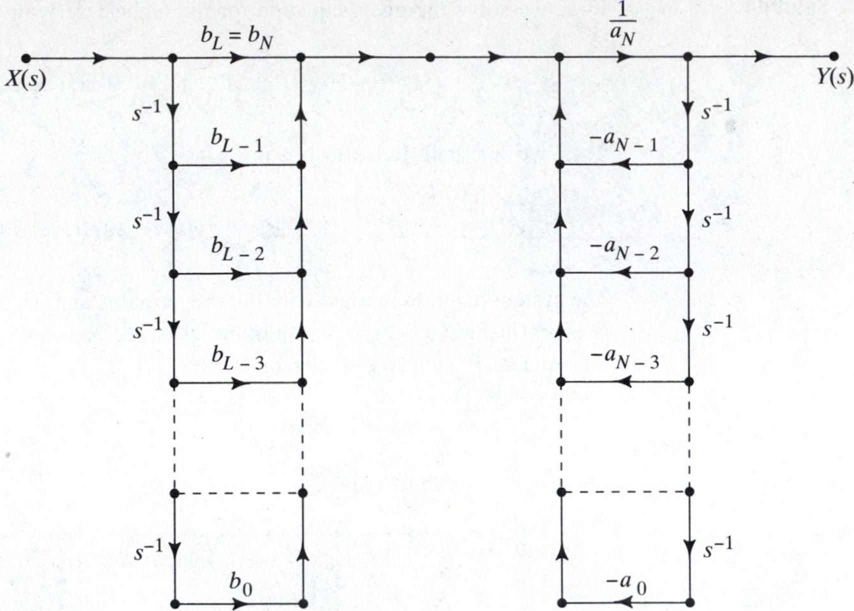

a. Transform-domain diagram of system DE

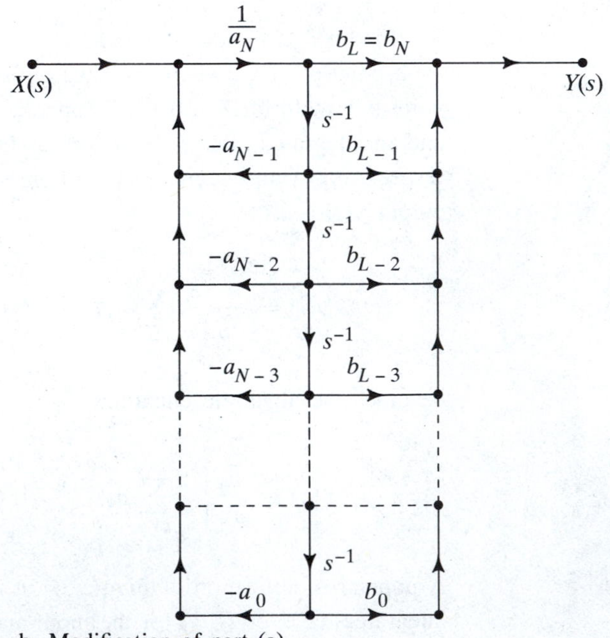

b. Modification of part (a)

FIGURE 3.15 *Transform-domain diagrams*

System Diagram from a Transfer Function To produce a system diagram from the transfer function, again assuming $L = N$, we begin by writing

$$H(s) = \frac{Y(s)}{X(s)} = \frac{\sum_{k=0}^{N} b_k s^k}{\sum_{k=0}^{N} a_k s^k}$$

or

$$Y(s)\left[\sum_{k=0}^{N} a_k s^k\right] = X(s)\left[\sum_{k=0}^{N} b_k s^k\right].$$

Looking ahead to the use of integrators (s^{-1}) rather than differentiators, we divide both sides of the equation by s^N, giving

$$Y(s)\left[\sum_{k=0}^{N} a_k s^{k-N}\right] = X(s)\left[\sum_{k=0}^{N} b_k s^{k-N}\right]$$

which can be written in the form

$$Y(s)\left[\sum_{k=0}^{N} a_k s^{-(N-k)}\right] = X(s)\left[\sum_{k=0}^{N} b_k s^{-(N-k)}\right].$$

Separating out the term on the left for $k = N$ and solving for $Y(s)$ produces

$$Y(s) = \frac{1}{a_N}\left[-\sum_{k=1}^{N-1} a_k s^{-(N-k)} Y(s) + \sum_{k=0}^{N} b_k s^{-(N-k)} X(s)\right]$$

which is the same expression that was stated earlier.

ILLUSTRATIVE PROBLEM 3.15 ***System Diagram from a Transfer Function***

For the transfer function

$$H(s) = \frac{Y(s)}{X(s)} = \frac{s^2 + 100}{s^3 + 20s^2 + 200s + 1000},$$

draw a system diagram in the manner of Figure 3.15b.

Solution

First, we write $Y(s)[s^3 + 20s^2 + 200s + 1000] = X(s)[s^2 + 100]$; after dividing through by s^3, we have $Y(s)[1 + 20s^{-1} + 200s^{-2} + 1000s^{-3}] = X(s) \cdot [s^{-1} + 100s^{-3}]$. Thus $Y(s) = -[20s^{-1} + 200s^{-2} + 1000s^{-3}]Y(s) + [s^{-1} + 100s^{-3}]X(s)$, giving the flowgraph of Figure 3.16a.

Comment: There are, of course, many equivalent system diagrams that can be drawn for a given transfer function. The choice depends on the application and, probably most important, the designer's preferences and prejudices. Shown in Figure 3.16b and c are two equivalent block diagrams that might be used by control system engineers.

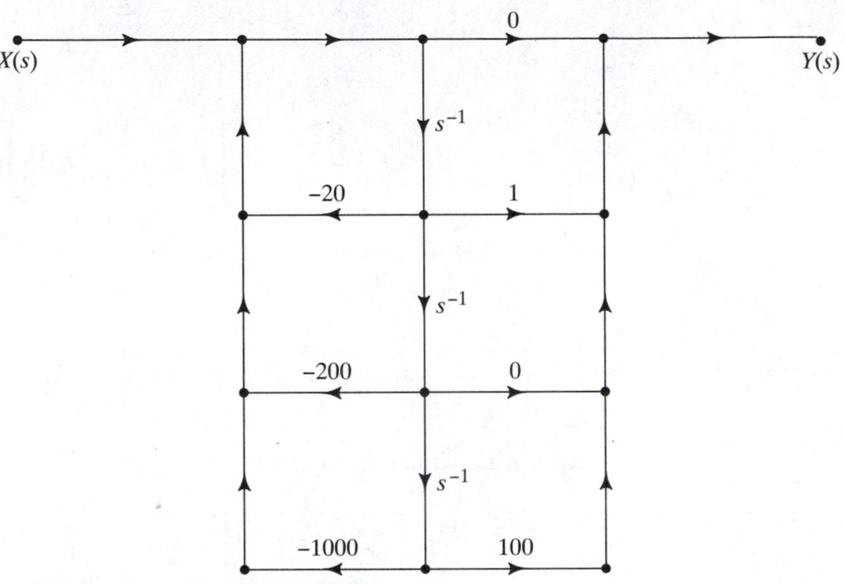

a. SFG for Illustrative Problem 3.15

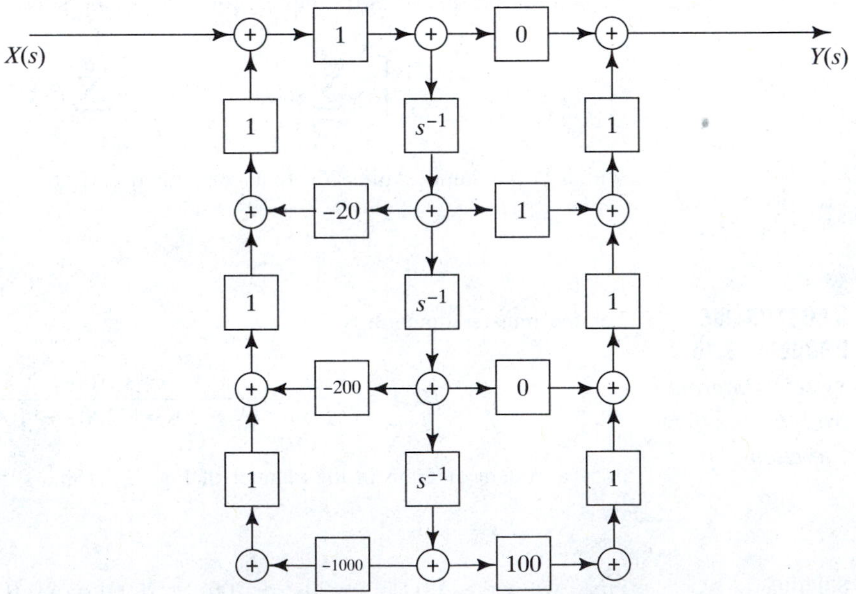

b. Block diagram for Illustrative Problem 3.15

FIGURE 3.16 *Three equivalent diagrams for Illustrative Problem 3.15*

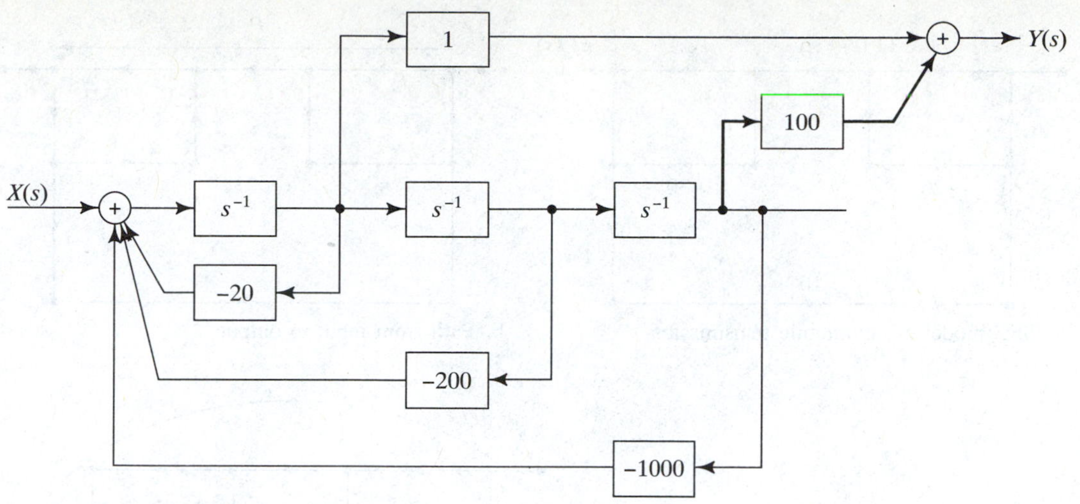

c. Another form of the block diagram for Illustrative Problem 3.15

FIGURE 3.16 *Continued*

BASIC CONCEPTS

MASON'S GAIN RULE AND APPLICATIONS

In the early 1950s as part of his doctoral studies at the Massachusetts Institute of Technology, Professor Samuel J. Mason developed a rule (or algorithm) that was based on a graphical procedure for calculating transfer functions between input and output points of a signal flowgraph or an equivalent block diagram. The proof of this rule is based on Cramer's method for solving a set of simultaneous algebraic equations; consequently, we apply the Mason Gain Rule (MGR) in the algebraic Laplace or s domain. We explain the rule by using as an example the SFG of Figure 3.17a (page 134) that models an automobile fluid-coupled transmission system.

Consider the following general definitions and their applicability to this signal flowgraph.

Nodes Points on a graph where the signals appear. All the heavy dots are nodes with the most important of these being $\Lambda(s)$, $\Omega_e(s)$, and $\Omega_d(s)$. At any node, the quantities associated with the incoming branches are summed, whereas the outgoing branches have no effect on the signal at the node. In other words, we can't apply Kirchhoff's law at the node. All outgoing branches carry the value of the node.

Branch A directed line segment, having an associated gain, that connects two nodes. An unmarked branch is assumed to have a gain of one. Two easily identified branches are the integrators with gains of s^{-1}.

Input node Has no incoming branches. Obviously $\Lambda(s)$ is the only input node.

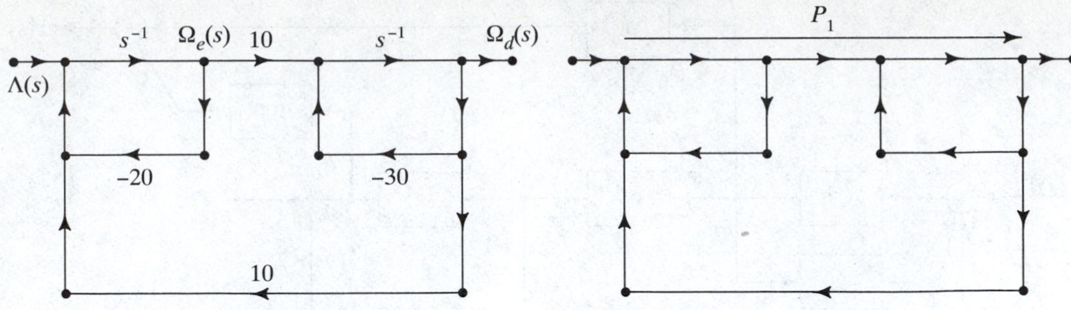

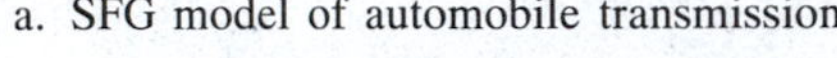

a. SFG model of automobile transmission

b. Path from input to output

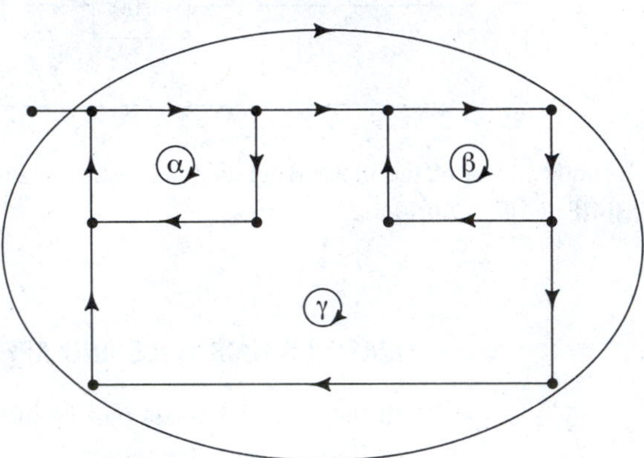

c. Three loops

FIGURE 3.17 *Mason's Gain Rule*

Output node Must have at least one incoming branch. All the rest of the nodes of this graph are output nodes. We use the node $\Omega_d(s)$ as the designated system output.

Path A continuous sequence of branches, traversed in the indicated branch directions, along which no node is encountered more than once. From $\Lambda(s)$ to $\Omega_d(s)$, there is only one path P_1, as indicated in Figure 3.17b.

Loop A continuous sequence of branches traversed in the indicated branch directions from one node around a closed path back to the same node, along which no node is encountered more than once. This graph has three loops: α, β, and γ, as indicated in Figure 3.17c.

The Mason Gain Rule is

MGR

$$H(s) = \frac{Y(s)}{X(s)} = \frac{\sum_{k=1}^{M} P_k(s)\Delta_k(s)}{\Delta(s)}$$

where

$H(s) =$ the transfer function relating an output node to an input node

$\Delta(s) =$ the graph determinant

$P_k(s) =$ the gain of the kth path from input to output

$\Delta_k(s) =$ cofactor of the kth path.

These terms will be defined as we find the transfer function $H(s) = Y(s)/X(s) = \Omega_d(s)/\Lambda(s)$ for the SFG of Figure 3.17a, where there is a particular order required for the calculation of the three quantities $\Delta(s)$, $P_k(s)$, and $\Delta_k(s)$.

1. The first quantity to determine is the graph determinant $\Delta(s)$, where

$$\begin{aligned}\Delta(s) = 1 &- \sum \text{loop gains} \\ &+ \sum \text{products of the gains of nontouching loops} \\ &\qquad \text{taken two at a time} \\ &- \sum \text{products of the gains of nontouching loops} \\ &\qquad \text{taken three at a time} \\ &+ \cdots,\end{aligned}$$

where the loop gain L_j is simply the product of the gains around the jth loop. For the example of Figure 3.17c, the loop gains are $L_1 = -20s^{-1}$ around the α loop, $L_2 = -30s^{-1}$ around the β loop, and $L_3 = (10)(10)s^{-1}s^{-1} = 100s^{-2}$ around the γ loop. Loops α and β do not touch. Two loops are nontouching if they have no branches or even nodes in common. Three or more loops are nontouching only if none of the loops touch one another. In this graph, we cannot find three loops that do not touch, so the calculation of $\Delta(s)$ terminates at this point, giving

$$\begin{aligned}\Delta(s) &= 1 - \underbrace{(L_1 + L_2 + L_3)}_{\textit{loop gains}} + \underbrace{(L_1 L_2)}_{\textit{gains of nontouching loops 2 at a time}} - 0 + 0 - \cdots \\ &= 1 - \{-20s^{-1} - 30s^{-1} + 100s^{-2}\} + \{(-20s^{-1})(-30s^{-1})\} \\ &= 1 + 50s^{-1} + 500s^{-2}\end{aligned}$$

2. Next, we take up the path gains $P_k(s)$ and their cofactors $\Delta_k(s)$, where

$$\begin{aligned}P_k(s) &= \text{gain of } k\text{th path from input to output} \\ &= \text{product of branch gains in } k\text{th path}\end{aligned}$$

and

$\Delta_k(s) =$ cofactor of the kth path, formed by striking out from $\Delta(s)$ all terms associated with loops that are touched by the kth path.

Consequently, for the SFG of Figure 3.17b, where there is only one path, we have

$$P_1(s) = (s^{-1})(s^{-1})(10) \quad \text{and} \quad \Delta_1(s) = 1$$

P_1 touches all the loops and all terms except the 1 are stricken from $\Delta(s)$.

3. Finally, the transfer function $H(s)$ is

transfer function

$$H(s) = \frac{Y(s)}{X(s)} = \frac{P_1(s) \cdot \Delta_1(s)}{\Delta(s)}$$

$$= \frac{10s^{-2}}{1 + 50s^{-1} + 500s^{-2}} = \frac{10}{s^2 + 50s + 500}.$$

ILLUSTRATIVE PROBLEM 3.16 ***Practice with Mason's Gain Rule***

For the LTI system modeled by Figure 3.17a, use the MGR to calculate another transfer function, $H(s) = \Omega_e(s)/\Lambda(s)$.

Solution

a. Using the three steps just outlined, we have the following:

(1) $\Delta(s) = 1 + 50s^{-1} + 500s^{-2}$, as before, because the graph determinant $\Delta(s)$ is a function of the graph and is the same for all input-output transmissions (transfer functions).

(2) There is only one path from $\Lambda(s)$ to $\Omega_e(s)$ and it does not touch the β loop. Thus, $P_1(s) = s^{-1}$ and $\Delta_1(s) = 1 + 30s^{-1}$.

(3) $H(s) = \dfrac{s^{-1}(1 + 30s^{-1})}{1 + 50s^{-1} + 500s^{-2}} = \dfrac{s + 30}{s^2 + 50s + 500}.$

Comment: Two block diagrams that are equivalent to the signal flowgraph of Figure 3.17a are given in Figure 3.18a and b. Mason's Gain Rule may be used in both situations, but the "signs" at the summations must be included when calculating loop and path gains in Figure 3.18b.

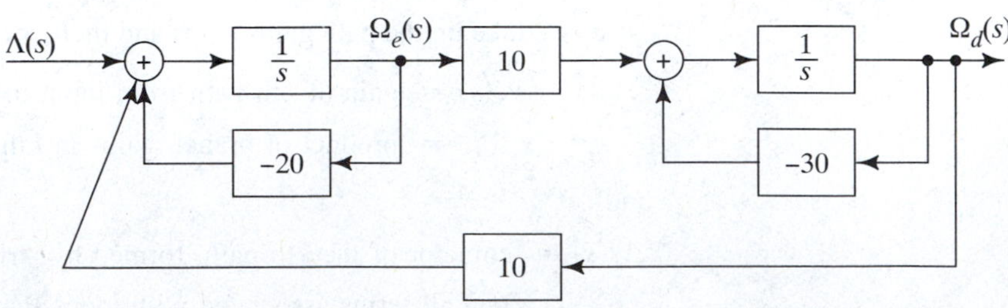

a. Block diagram with minus signs associated with block transmissions

FIGURE 3.18 *Block diagram equivalents to Figure 3.17a*

CHAPTER 3

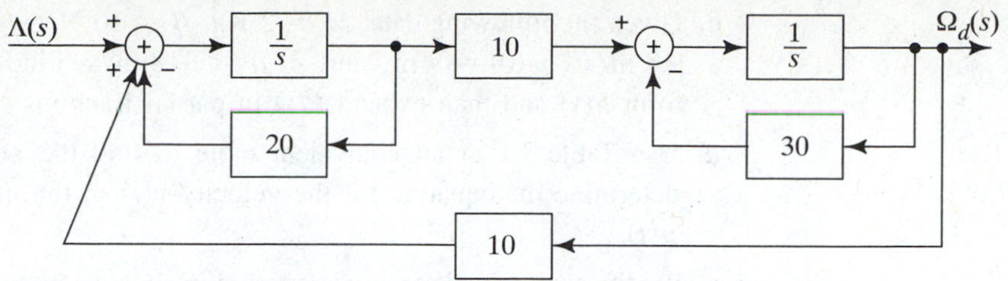

b. Block diagram with minus signs accounted for at summing junctions

FIGURE 3.18 *Continued*

SOLVED EXAMPLES AND MATLAB APPLICATIONS

EXAMPLE 3.1
Transform Solution of a Typical Differential Equation

Consider the simple situation in mechanics illustrated in Figure E3.1a, with a mass in a dashpot hanging on a spring. The dashpot is filled with SAE 50 oil to provide viscous damping. The reference position p is chosen so that the spring tension is zero when $p = 0$. Invoking Newton's law, we have the DE

$$M\ddot{p}(t) + B\dot{p}(t) + Kp(t) = Mg.$$

a. Transform this equation term by term (the gravitational force Mg is constant).

b. Solve for the transform $P(s)$ and put your answer in the form of

$$P(s) = \text{a rational function} = \text{polynomial/polynomial}.$$

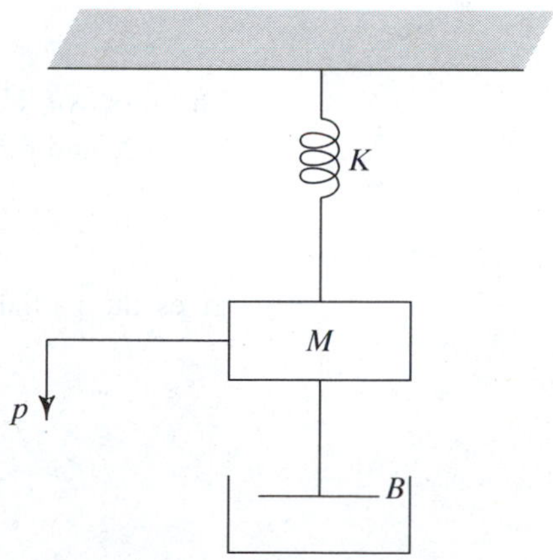

FIGURE E3.1a *Mass-spring-damper system*

EXAMPLES

c. Given the following data: $M = 2$ Kg, $B = 10$ N/m/s, $K = 12$ N/m, $g = 9.8$ m/s^2, $p(0) = 9$ m, and $\dot{p}(0) = -18$ m/s. Find the poles of the transform $P(s)$ and then expand $P(s)$ in partial fractions.

d. Use Table 3.1 or an equivalent table to find the position $p(t)$ and then determine the equation for the velocity $v(t)$ of the mass by differentiating $p(t)$.

e. Put the system's differential equation in state-space form and use MATLAB to plot $p(t)$ versus t and $v(t)$ versus t. Check a few points on the plots by substituting values of t into the expressions determined in part (d).

WHAT IF? Suppose that the type of viscous damping was changed (SAE 10 oil rather than SAE 50), making $B = 4$ N/m/s, with all other constants the same. What sort of change in the response would you expect? Use MATLAB to verify your reasoning. You might also try changing B to some other values, including a negative one. ■

CHAPTER 3

Solution

a. *Transform the DE.* Using the linearity property (1 in Table 3.2), the differentiation theorem (3 in Table 3.2), and the transform of a constant (3 in Table 3.1) yields

$$M\{s^2P(s) - sp(0) - \dot{p}(0)\} + B\{sP(s) - p(0)\} + KP(s) = Mg \cdot \frac{1}{s}.$$

b. *Solving for P(s).* Collecting terms and solving for $P(s)$ yields

$$P(s) = \frac{Mg/s + [B + Ms]p(0) + M\dot{p}(0)}{Ms^2 + Bs + K} = \frac{p(0)s^2 + \{\dot{p}(0) + (B/M)p(0)\}s + g}{s\{s^2 + (B/M)s + (K/M)\}}.$$

c. *The poles of P(s) and PFE.* Substituting the given data produces

$$P(s) = \frac{9s^2 + \{-18 + (5)(9)\}s + 9.8}{s(s^2 + 5s + 6)} = \frac{9s^2 + 27s + 9.8}{s(s^2 + 5s + 6)}.$$

The roots of $s^3 + 5s^2 + 6s = 0$ are the poles of $P(s)$. They are $s = 0$, $s = -2$, and $s = -3$. Expanding $P(s)$ as

$$\frac{9s^2 + 27s + 9.8}{s(s^2 + 5s + 6)} = \frac{C_1}{s} + \frac{C_2}{s + 2} + \frac{C_3}{s + 3}$$

gives the partial fraction constants

$$C_1 = \left.\frac{9s^2 + 27s + 9.8}{(s + 2)(s + 3)}\right|_{s=0} = 1.633,$$

$$C_2 = \left.\frac{9s^2 + 27s + 9.8}{s(s + 3)}\right|_{s=-2} = 4.100,$$

$$C_3 = \left.\frac{9s^2 + 27s + 9.8}{s(s + 2)}\right|_{s=-3} = 3.266.$$

d. *Find p(t) by table lookup and v(t) by calculus.* Knowing that $P(s)$ is

$$P(s) = \frac{1.633}{s} + \frac{4.100}{s+2} + \frac{3.266}{s+3}, \qquad \text{Re}[s] > 0$$

it follows from Table 3.1 (or an equivalent) that

$$p(t) = 1.633 + 4.100e^{-2t} + 3.266e^{-3t} \text{ meters}, \qquad t \geq 0$$

and by differentiating $p(t)$ we find that the velocity $v(t)$ is

$$v(t) = -8.2e^{-2t} - 9.8e^{-3t} \text{ meters/second}, \qquad t \geq 0.$$

e. *State-space representation.* The differential equation for the mass-spring-damper is

$$M\ddot{p}(t) + B\dot{p}(t) + Kp(t) = Mg$$

and defining $v_1(t) = p(t)$, $v_2(t) = dp(t)/dt = v(t)$, and $Mg = x(t)$, the state equation is

$$\dot{\mathbf{v}}(t) = \begin{bmatrix} 0 & 1 \\ -K/M & -B/M \end{bmatrix} \mathbf{v}(t) + \begin{bmatrix} 0 \\ 1/M \end{bmatrix} x(t)$$

with $\mathbf{v}(0)$ and $x(t)$ given. The script using **kslsim** is shown next and the plot is in Figure E3.1b on page 140.

EXAMPLES

MATLAB Script

```
%E3_1 Solution of DE using ksdlsim

%E3_1b Mass-spring-damper response
A=[0,1;-6,-5];                  % 2 rows, 1 column
B=[0;1/2];                      % 2 rows, 1 column
C=[1,0];                        % 1 row, 2 columns
D=[0];                          % a scalar
v0=[9,-18]';                    % 2 rows, 1 column
t=0:.01:2.5;                    % start time:increment:final time
X=[19.6*ones(size(t))]';        % a step input of 19.6 units in amplitude
[y,v]=kslsim(A,B,C,D,X,t,v0);  % call kslsim
%...plotting statements and pause
```

WHAT IF?

MATLAB Script

```
%E3_1c Mass-spring-damper response
A=[0,1;-6,-2];
B=[0;1/2];
C=[1,0];
D=[0];
v0=[9,-18]';                              % IC column vector
```

Continues

```
t=0:.01:4.5;                      % start time:increment:final time
X=[19.6*ones(size(t))]';          % input column vector
[y,v]=kslsim(A,B,C,D,X,t,v0);     % call kslsim
%...plotting statements
```

■

CROSS-CHECK The graph of Figure E3.1c shows an oscillatory response. The new CE is $s^2 + 2s + 6 = 0$, with complex conjugate characteristic roots of $s_{1,2} = -1 \pm j2.236$. From Chapter 2, where we had the generalized CE for second-order systems of $s^2 + 2\zeta\omega_n s + \omega_n^2 = 0$, we see that $\zeta = 0.408$, so the oscillatory response makes sense. We can use the FVT here and find that

$$p(t)|_{t\to\infty} = sP(s)|_{s=0} = \left.\frac{sMg/s + s[B + Ms]p(0) + sM\dot{p}(0)}{Ms^2 + Bs + K}\right|_{s=0}$$

$$= \frac{Mg}{K} = 1.63$$

which seems to agree with the plots. Notice that the value of the damping B does not change the final value of $p(t)$. ■

CHAPTER 3

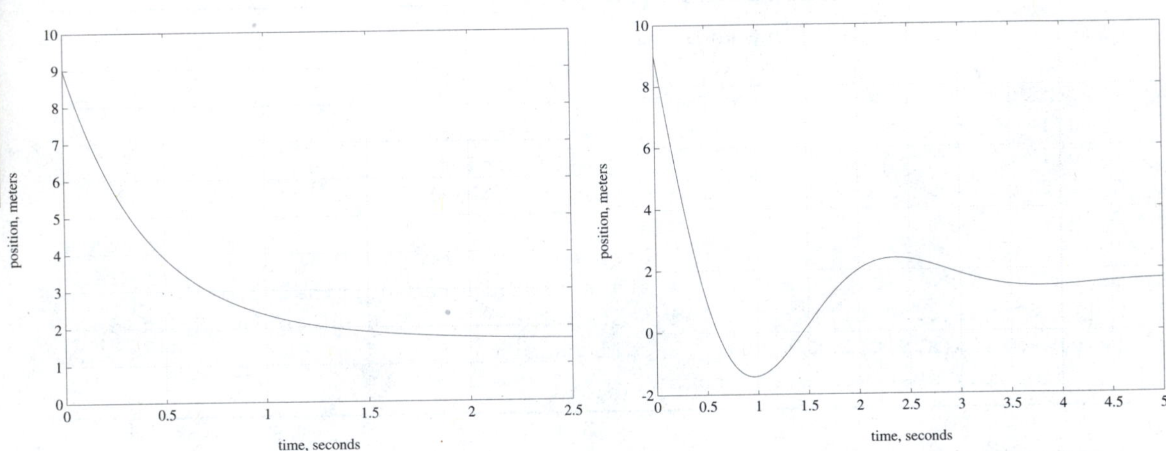

FIGURE E3.1b *Mass-spring-damper response*

FIGURE E3.1c *Mass-spring-damper response*

EXAMPLE 3.2 ***Partial Fraction Expansion with Complex Conjugate Roots***

Suppose we have an airplane whose ailerons are oscillated sinusoidally and we want to solve for the roll rate $p(t)$. The inertia torque is $J\,dp(t)/dt$, the damping torque is $Bp(t)$, and the aileron torque is $\lambda(t)$. Summing the moments about the roll axis gives the differential equation

$$J\dot{p}(t) + Bp(t) = \lambda(t).$$

a. Use $J = 0.5$, $B = 1.5$, $p(0) = 1$, and $\lambda(t) = \cos t$ and solve for the transform $P(s)$.

b. Write out the partial fraction expansion and use a transform table to find $p(t)$.

c. Use at least two MATLAB functions to plot $p(t)$ versus t. Check two values of $p(t)$ found by evaluating the expression in part (b) against these plots.

d. Use **residue** to verify your partial fraction calculations.

WHAT IF? In the system response, steady-state is reached at approximately $t = 1.7$ s. Suppose you want steady-state to be attained at $t = 0.5$ s and can adjust B arbitrarily. What value of B would you select? Use MATLAB to verify your choice. ■

Solution

a. *Solving for P(s).* Taking the transform gives the algebraic equation

$$J\{sP(s) - p(0)\} + BP(s) = \Lambda(s)$$

which—when solved for $P(s)$—produces

$$P(s) = \frac{Jp(0) + \Lambda(s)}{Js + B} = \frac{p(0) + (1/J)\Lambda(s)}{s + B/J}.$$

Substituting the given data with $\mathcal{L}[\cos t] = s/(s^2 + 1^2)$ and $p(0) = 1$ produces

$$P(s) = \frac{1 + 2s/(s^2 + 1)}{s + 3} = \frac{s^2 + 2s + 1}{(s^2 + 1)(s + 3)}.$$

b. *Partial fraction expansion of P(s) and table lookup for p(t).* We factor the denominator into first-order terms in order to use Table 3.1 or its equivalent. Thus

$$P(s) = \frac{s^2 + 2s + 1}{(s^2 + 1)(s + 3)} = \frac{s^2 + 2s + 1}{(s - j1)(s + j1)(s + 3)}$$
$$= \frac{C_1}{s - j1} + \frac{C_2}{s + j1} + \frac{C_3}{s + 3}$$

where

$$C_1 = \left.\frac{s^2 + 2s + 1}{(s + j1)(s + 3)}\right|_{s=j1} = \frac{2j}{-2 + j6} = \frac{2e^{j1.571}}{6.325e^{j1.893}} = 0.316e^{-j0.322}.$$

Comment: When evaluating complex numbers such as C_1, there are many golden opportunities for errors. We have found that the method used here reduces these opportunities somewhat:

1. Write both the numerator and denominator as complex numbers in exponential form.

EXAMPLES

2. The magnitude of the complex number is the product of numerator magnitudes divided by the product of denominator magnitudes. The angle (argument) of the complex number is the sum of numerator angles minus the sum of denominator angles.

In a similar manner we evaluate C_2 as

$$C_2 = \left.\frac{s^2 + 2s + 1}{(s - j1)(s + 3)}\right|_{s=-j1} = \frac{-2j}{-2 - j6}$$
$$= \frac{2e^{-j1.571}}{6.325e^{-j1.893}} = 0.316e^{j0.322}$$
$$= C_1^*(\text{complex conjugate of } C_1)\,.$$

The partial fraction constants associated with conjugate poles will always be conjugates. This is a consequence of the fact that the differential equations we are dealing with have *real* coefficients. In the future we will calculate only the first one and assume that we have done it correctly. For the real pole at $s = -3$,

$$C_3 = \left.\frac{s^2 + 2s + 1}{(s^2 + 1)}\right|_{s=-3} = 0.40\,.$$

Therefore, in summary,

$$P(s) = \frac{s^2 + 2s + 1}{(s^2 + 1)(s + 3)} = \frac{0.316e^{-j0.322}}{s - j1} + \frac{0.316e^{j0.322}}{s + j1} + \frac{0.40}{s + 3}$$

from which it follows that

$$p(t) = 0.632\, \boldsymbol{cos}(t - 0.322) + 0.40e^{-3t}, \qquad t \geq 0\,.$$

c. *Computer verification.* The state equation is $\dot{p}(t) + 3p(t) = 2\lambda(t)$, with

MATLAB Script

```
%E3_2 Solution using kslsim and ksimptf

%E3_2a Roll rate response using kslsim
A=[-3];
B=[2];
C=[1];
D=[0];
v0=[1]';
t=0:.05:25;
X=[1*cos(t)]';
[y,v]=kslsim(A,B,C,D,X,t,v0);            % call kslsim
%...plotting statements and pause
```

$p(0) = 1$ and $\lambda(t) = \cos(t)$. The following script makes use of the function **kslsim**. The plot for this response is given in Figure E3.2a. Another way to obtain a plot of the output $p(t)$ is to make use of the California function **ksimptf** or the MATLAB function **impulse**. To do this we think of the transform

$$P(s) = \frac{s^2 + 2s + 1}{s^3 + 3s^2 + s + 3}$$

as a transfer function $T(s)$ and then find the system output for an input $x(t) = \delta(t)$. That is, $Y(s) = T(s) \cdot X(s)$ with $X(s) = 1$. The script for finding the impulse response with the data in transfer function form is next. The plot is in Figure E3.2b.

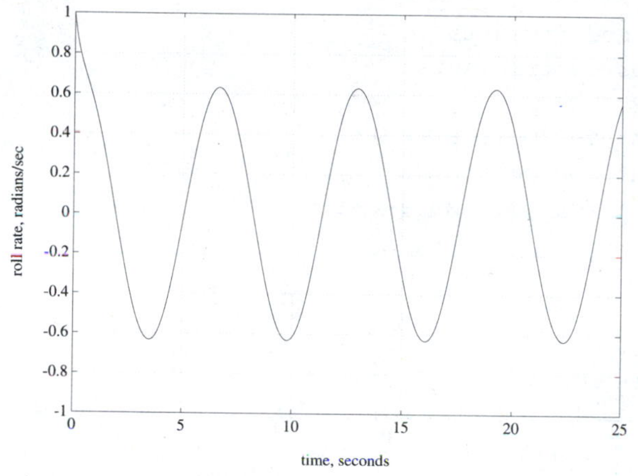

FIGURE E3.2a *Roll rate response*

FIGURE E3.2b *Roll rate response*

MATLAB Script

```
%E3_2b Roll rate response using ksimptf
num=[1,2,1];                    % numerator coefficients in descending
                                % powers of s
den=[1,3,1,3];                  % denominator coefficients in descending
                                % powers of s
t=0:.10:25;                     % start time:increment:finish
[y,v,t]=ksimptf(num,den,t);     % call ksimptf
%...plotting statements
```

CROSS-CHECK To check some values, we use the result $p(t) = 0.632\cos(t - 0.322) + 0.4e^{-3t}$, from which $p(5) = 0.632\cos(5 - 0.322) + 0.4e^{-15} \approx 0$ and $p(25) = 0.632\cos(25 - 0.322) + e^{-75} = 0.56$. Both of these values appear to agree with the plots. ■

d. *PFE verification.* Finally, we can verify our partial fraction constants by making use of **residue** with the script for PFE given next.

MATLAB Script

```
%PFE...computes poles and PF constants for a given transform
b=[0,1,2,1]
a=[1,3,1,3]
[r,p,k]=residue(b,a)
```

The program returns the following data. We have added the comments.

Output

```
b=  0  1  2  1               numerator coefficients
a=  1  3  1  3               denominator coefficients
r=
 0.4000                      PF constant for pole at s=-3
 0.3000-0.1000i              PF constant for pole at s=0+j1
 0.3000+0.1000i              PF constant for pole at s=0-j1
p=
 -3.0000                     s1=-3
 -0.0000+1.0000i             s2=0+j1
 -0.0000-1.0000i             s3=0-j1
k=  []                       k=0 because N>L
```

CHAPTER 3

WHAT IF? To observe the dynamics of the system alone, it is best to look at either the IC or the unit impulse response, and here we choose $h(t)$. The system transfer function is $H(s) = P(s)/\Lambda(s) = 1/(Js + B) = 2/(s + 2B)$, making $h(t) = 2e^{-2Bt}u(t)$, which we want to be 0 at about $t = 0.5$ s.[7] Thus $h(0.5) = 0$ for $B = 5$. This is not an exact calculation, of course; other values than -5 for the exponent will do quite well. (We like $e^{-4.56} = 0.01$, which is fairly close to zero.) The result of using $B = 5$ is shown in Figure E3.2c, where the steady-state requirement of $t = 0.5$ s has been achieved. ■

[7]A rule of thumb is that an exponential term $Ke^{-\alpha t}$, $\alpha > 0$, is approximately zero when $e^{-\alpha t} = e^{-5}$, or $t = 5/\alpha$. The quantity $t = 1/\alpha$ is often referred to as the time constant associated with this particular exponential.

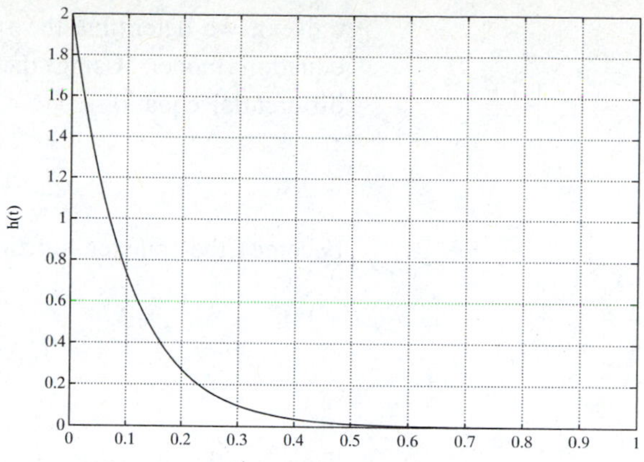

FIGURE E3.2c *Unit impulse response*

EXAMPLE 3.3
Sinusoidal Steady-State Response

An unstable spacecraft is stabilized by the addition of an analog filter, with the result that the relationship between the output position $\theta(t)$ and a reference input $r(t)$ can be described by the differential equation

system DE

$$\dddot{\theta}(t) + \ddot{\theta}(t) + 10\dot{\theta}(t) + 5\theta(t) = 5r(t).$$

a. Find an analytical expression for the steady-state output $\theta_{ss}(t)$ when the reference input is $r(t) = 10\cos(5t)$.

b. Estimate the time at which the steady-state solution is reached.

c. Use MATLAB to plot the output $\theta(t)$ for $t \geq 0$ and show where the steady-state solution is reached. Verify that the analytical solution and the computer simulation agree. Since the steady-state solution is unaffected by initial conditions, for simplicity we will assume them to be zero.

WHAT IF? Suppose we think that the system is described by the differential equation

$$\dddot{\theta}(t) + \ddot{\theta}(t) + 10\dot{\theta}(t) + 15\theta(t) = 5r(t) \text{ rather than}$$
$$\dddot{\theta}(t) + \ddot{\theta}(t) + 10\dot{\theta}(t) + 5\theta(t) = 5r(t).$$

Find the equation for the sinusoidal steady-state response and plot the result. ■

Solution

a. *Analytical solution.* First, we need to be sure that the system is stable even though the problem statement says that an unstable spacecraft has been stabilized. The characteristic equation is $s^3 + s^2 + 10s + 5 = 0$, and we can use MATLAB in a calculator mode to find the roots. We enter the coefficients of the characteristic equation as the vector $\mathbf{p} = [1, 1, 10, 5]$ and invoke a command such as **r = roots(p)**. The characteristic roots are $r_{1,2} = -0.244 \pm j3.311$ and $r_3 = -0.513$. These are stable roots, and it is safe (and legal) to proceed to find the steady-state solution.

EXAMPLES

Next, we determine the system's transfer function from the differential equation model. Using the algorithm from earlier in the chapter, the differential equation

$$\sum_{k=0}^{N} a_k \frac{d^k y(t)}{dt^k} = \sum_{k=0}^{L} b_k \frac{d^k x(t)}{dt^k}$$

becomes the transfer function

$$H(s) = \frac{Y(s)}{X(s)} = \frac{\sum_{k=0}^{L} b_k s^k}{\sum_{k=0}^{N} a_k s^k}.$$

Thus for the given system DE the transfer function is

$$H(s) = \frac{\Theta(s)}{R(s)} = \frac{5}{s^3 + s^2 + 10s + 5}.$$

Since the input frequency is $\omega = 5$ rad/s, we substitute $s = j5$ in $H(s)$, giving

$$H(j5) = \frac{5}{(j5)^3 + (j5)^2 + 10(j5) + 5} = 0.0644e^{j1.83}.$$

Using the sinusoidal steady-state formula, the steady-state output is

$$y_{ss}(t) = 10(0.0644)\cos(5t + 1.83) = 0.644\cos(5t + 1.83).$$

b. *Time at which steady-state is reached.* The characteristic roots of $-0.244 \pm j3.311$ cause a term in the response of the form $|K_1|e^{-0.244t}\cos(3.311t + \angle K_1)$, where K_1 is a complex number. This term will decay to approximately zero in 20 s because the exponential $e^{-0.244t}|_{t=20} = 0.008$, which is essentially zero. The other term in the response is of the form $K_2e^{-0.512t}$ and it will approach "zero" in less than 10 s. So we say that the steady-state condition is attained in approximately 20 s.[8]

c. *Computer simulation.* To plot this result, we use the California function **kslsim**, with the data entered in the form of the transfer function model of the spacecraft. The program is given next, with the resultant plot in Figure E3.3a. Notice that the steady-state zero-to-peak amplitude is reached in about 20 s, and the zero-to-peak value is 0.644, as predicted.

MATLAB Script

```
%E3_3 Spacecraft sinusoidal response

%E3_3a Spacecraft output position
num=[5];                                  % numerator coefficients
```

Continues

[8]To determine when steady-state is reached, we need to decide when the longest lasting of the transient terms has "died out."

```
den=[1,1,10,5];                    % denominator coefficients
t=0:.08:25;
x= [10*cos(5*t)]';                 % input column vector
[y,v]=kslsim(num,den,x,t);         % call kslsim
%...plotting statements and pause
```

WHAT IF? The transfer function is now

$$H(s) = \frac{\Theta(s)}{R(s)} = \frac{5}{s^3 + s^2 + 10s + 15}$$

and, again, for the input frequency of $\omega = 5$ rad/s, we have

$$H(j5) = \frac{5}{(j5)^3 + (j5)^2 + 10(j5) + 15} = 0.066e^{j1.70}.$$

Thus, using the sinusoidal steady-state algorithm, we have the predicted steady-state output

$$y_{ss}(t) = 0.066\cos(5t + 1.70)$$

which is almost the same as before. Using **lsim** or **kslsim** to verify the prediction gives the plot of Figure E3.3b, where it is seen that the response is diverging and will never reach the steady-state condition. ■

What is the problem? The answer is that we didn't check stability before using the sinusoidal steady-state formula. The characteristic roots are $s_{1,2} = 0.208 \pm j3.248$ and $s_3 = -1.416$, clearly an unstable situation because the complex roots have positive real parts.

EXAMPLES

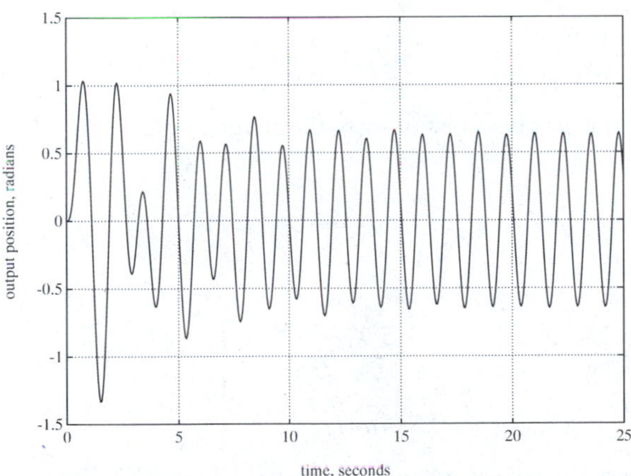

FIGURE E3.3a *Spacecraft output position*

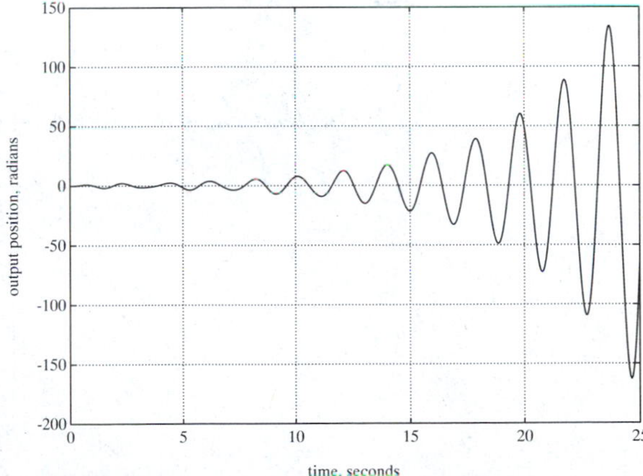

FIGURE E3.3b *Unstable spacecraft output position*

EXAMPLE 3.4 *Convolution by Transforms*

A bandpass analog filter is subjected to a step input $x(t) = u(t)$ and the response (plotted in Figure E3.4a) is $y(t) = 1 + 1.155e^{-5t}\cos(8.66t - 2.618)$, $t \geq 0$.

a. What is the transfer function model for the system?

b. Find an analytical expression for the unit impulse response $h(t)$. Use a MATLAB or California function (**ksimptf, impulse, kslsim**, or **lsim**) to plot this response.

c. Determine the output $y(t)$ for the input $x(t) = 10\cos(8.66t)u(t)$. Notice that this is the response for $t \geq 0$, not just the steady-state. Again, plot this response.

WHAT IF? Suppose that we wanted only the steady-state response to the sinusoid $x(t) = 10\cos(8.66t)u(t)$. ■

Solution

a. *Transfer function.* From convolution $y(t) = h(t) * x(t) = h(t) * u(t)$ for a step input. Thus, using the convolution property, we have $Y(s) = H(s) \cdot X(s) = H(s) \cdot (1/s)$, and we need to determine $Y(s)$ to get to $H(s)$. From pairs 3 and 8 in Table 3.1,

$$Y(s) = \mathcal{L}[y(t)] = \frac{1}{s} + \frac{1.155\{\cos[-2.618](s + 5) - 8.66\sin[-2.618]\}}{(s + 5)^2 + 8.66^2}$$

$$= \frac{1}{s} + \frac{-s}{s^2 + 10s + 100} = \frac{10(s + 10)}{s(s^2 + 10s + 100)}.$$

Now, dividing $Y(s)$ by $X(s) = 1/s$, we find $H(s)$ to be

$$H(s) = \frac{10(s + 10)}{s^2 + 10s + 100}, \qquad \text{Re}[s] > -5.$$

b. *Unit impulse response.* The unit impulse response is the inverse transform of $H(s)$, so

$$h(t) = \mathcal{L}^{-1}[H(s)]$$

$$= \mathcal{L}^{-1}\left[\frac{10(s + 10)}{s^2 + 10s + 100}\right]$$

$$= \mathcal{L}^{-1}\left[\frac{5.77e^{-j0.534}}{s + 5 - j8.66} + \frac{5.77e^{j0.534}}{s + 5 + j8.66}\right]$$

$$= 11.54e^{-5t}\cos(8.66t - 0.534)u(t).$$

The function **ksimptf** or **impulse** was used to plot $h(t)$ in Figure E3.4b with the data and script on the CLS disk.

CHAPTER 3

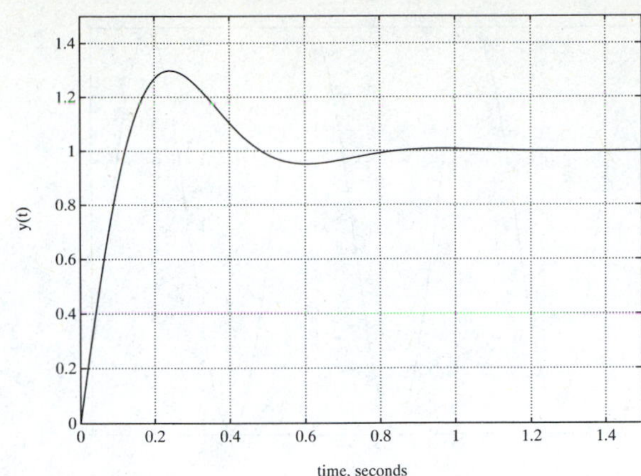

FIGURE E3.4a *Unit step response*

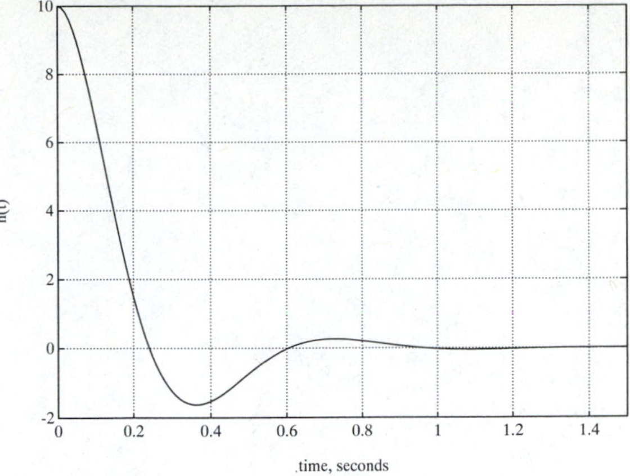

FIGURE E3.4b *Unit impulse response*

c. *Output for sinusoidal input.* Using $Y(s) = H(s) \cdot X(s)$ with $X(s) = \mathcal{L}[x(t)]$, we have

$$Y(s) = \frac{10(s + 10)}{s^2 + 10s + 100} \cdot \frac{10s}{s^2 + 8.66^2}$$
$$= \frac{C_1}{s + 5 - j8.66} + \frac{C_1{}^*}{s + 5 + j8.66}$$
$$+ \frac{C_2}{s - j8.66} + \frac{C_2{}^*}{s + j8.66}, \qquad \text{Re}[s] > -5.$$

This is a very good time to check your hand calculations of the partial fraction constants by using the function **residue**. The results are

$$C_1 = 6.41e^{j2.86} \quad \text{and} \quad C_2 = 7.34e^{-j0.576}$$

and using Table 3.1,

$$y(t) = 12.82e^{-5t}\cos(8.66t + 2.86) + 14.68\cos(8.66t - 0.576), \quad t \geq 0.$$

Notice that, using the rule of thumb stated in Example 3.3, the steady-state solution is reached in about 1 s. The plots in Figures E3.4a and b clearly exhibit this characteristic. It is also there, of course, in Figure E3.4c (page 150), but it is quite a bit harder to detect.

Comment: It is generally more convenient to use a function such as **lsim** or **kslsim** when simulating a system more than once with different inputs. The only change required in the script is the statement describing the input. For each of the three parts of this example, however, we thought of the output transform $Y(s)$ as a transfer function $T(s)$. Then we used this new "transfer function"

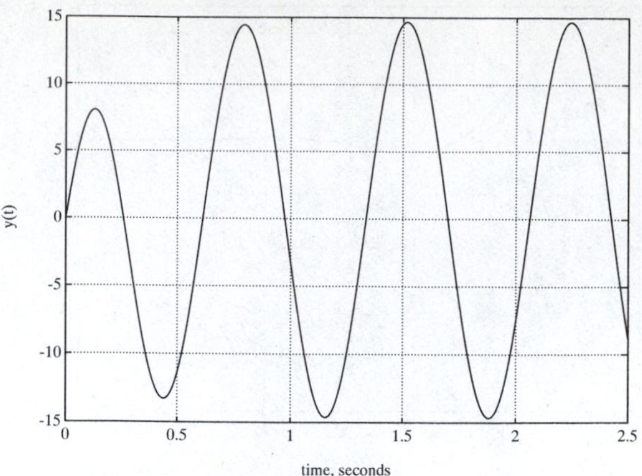

FIGURE E3.4c *Sinusoidal response*

to find the system output for an input $x(t) = \delta(t)$; that is, we reasoned that $Y(s) = T(s) \cdot X(s)$ with $X(s) = 1$. In part (c), for instance, we used

$$Y(s) = T(s) = \frac{10(s + 10)}{s^2 + 10s + 100} \cdot \frac{10s}{s^2 + 8.66^2}$$
$$= \frac{100s^2 + 100s}{s^4 + 10s^3 + 175s^2 + 750s + 7500}$$

and entered the numerator and denominator coefficients as data for **ksimptf** or **impulse**. However, in general, **lsim** or **kslsim** is recommended as a better procedure to follow.

WHAT IF? The most straightforward approach is to find the value of $H(s)$ at $s = j8.66$ and apply the sinusoidal steady-state formula. Thus

$$H(j8.66) = \frac{10(j8.66 + 10)}{(j8.66)^2 + 10(j8.66) + 100} = \frac{10(13.23e^{j0.71})}{90.19e^{j1.29}} = 1.47e^{-j0.58}$$

and applying the sinusoidal steady-state formula gives

$$y_{ss}(t) = (10)(1.47)\cos(8.66t - 0.58) = 14.7\cos(8.66t - 0.58).$$

Of course, having already found the response for $t \geq 0$ means that we could also simply discard those terms that vanish as $t \to \infty$; the result is the same as given by the sinusoidal steady-state formula. ■

EXAMPLE 3.5
DE, $h(t)$, $H(s)$, Poles and Zeros, SFG, and $H(j\omega)$ Models

A lowpass prototype (normalized) elliptic filter is described by the DE

$$\dddot{y}(t) + 1.024\ddot{y}(t) + 1.047\dot{y}(t) + 0.539y(t) = 0.306\ddot{x}(t) + 0.539x(t)$$

where $x(t)$ represents the input voltage to the filter and $y(t)$ its output.

a. Find the filter's transfer function.

b. Determine the zeros and poles for this lowpass filter.

c. What is the unit impulse response, $h(t)$?

d. Draw an SFG for this system using only gains of real numbers, summers, and integrators ($1/s$ or D^{-1}).

e. An input signal $x(t) = 1 + 1\cos(0.5t) + 1\cos(5t)$ is applied to this filter. Determine the steady-state output $y_{ss}(t)$.

WHAT IF? A friend suggests that a highpass filter can be derived from the lowpass transfer function by simply replacing s with its reciprocal, $1/s$. What do you think of this idea? ■

Solution

a. *Transfer function.* The input coefficients appear in the numerator as the coefficients of s^k, where k reflects the order of the derivative; a similar statement is true for the output in the denominator. Thus we have the transfer function

$$H(s) = \frac{Y(s)}{X(s)} = \frac{0.306s^2 + 0.539}{s^3 + 1.024s^2 + 1.047s + 0.539}.$$

b. *Zeros and poles.* The zeros are the roots of $0.306s^2 + 0.539 = 0$, or $s_{1,2} = \pm j1.327$. The poles are the roots of $s^3 + 1.024s^2 + 1.047s + 0.539 = 0$; using a calculator or a MATLAB function, $s_{1,2,3} = -0.667, -0.179 \pm j0.881$.

c. *Unit impulse response.* With $h(t) = \mathcal{L}^{-1}[H(s)]$, we expand $H(s)$ in PFE as

$$H(s) = \frac{0.665}{s + 0.667} + \frac{0.184e^{-j2.937}}{s + 0.179 - j0.881} + \text{conjugate term},$$
$$\text{Re}[s] > -0.179$$

and from Table 3.1,

$$h(t) = \left[0.665e^{-0.667t} + 0.368e^{-0.179t}\cos(0.881t - 2.937)\right]u(t).$$

d. *System diagram.* Writing the transfer function as

$$Y(s)\left[s^3 + 1.024s^2 + 1.047s + 0.539\right] = X(s)\left[0.306s^2 + 0.539\right]$$

we have, after dividing through by s^3 and separating out $Y(s)$,

$$Y(s) = \left[-1.024s^{-1} - 1.047s^{-2} - 0.539s^{-3}\right] + X(s)\left[0.306s^{-1} + 0.539s^{-3}\right]$$

and the SFG of Figure E3.5a (page 152) results.

EXAMPLES

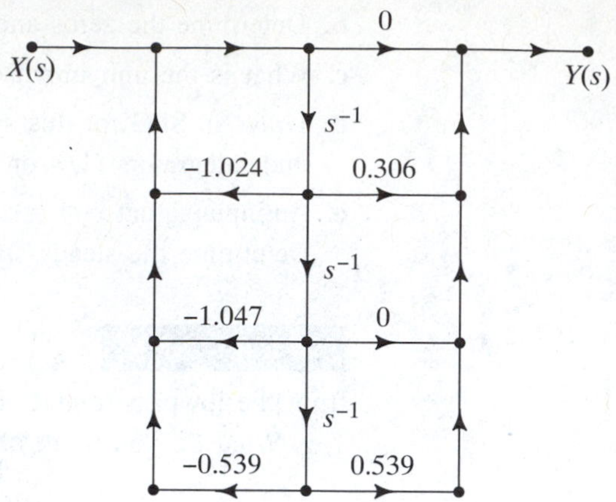

FIGURE E3.5a *SFG for filter*

e. *Sinusoidal steady state.* We need to evaluate $H(s)$ at the three input frequencies of $s = j0$ (a constant such as 1 can be considered a sinusoid of zero frequency from $1\cos(0t) = 1$), $s = j0.5$, and $s = j5$. Thus

$$H(j0) = \frac{0.306(j0)^2 + 0.539}{(j0)^3 + 1.024(j0)^2 + 1.047(j0) + 0.539} = 1$$

$$H(j0.5) = \frac{0.306(j0.5)^2 + 0.539}{(j0.5)^3 + 1.024(j0.5)^2 + 1.047(j0.5) + 0.539}$$

$$= \frac{0.463}{0.489e^{j0.953}} = 0.946e^{-j0.953}$$

$$H(j5) = \frac{0.306(j5)^2 + 0.539}{(j5)^3 + 1.024(j5)^2 + 1.047(j5) + 0.539} = 0.058e^{-j1.365}.$$

Applying the sinusoidal steady-state formula three times and adding the results gives the steady-state filter output

$$y_{ss}(t) = 1 + 0.946\cos(0.5t - 0.953) + 0.058\cos(5t - 1.365).$$

WHAT IF? Calling $H'(s)$ the transfer function of the possible high-pass filter, we have

$$H'(s) = H(s)|_{s=1/s}$$

$$= \frac{0.306(1/s)^2 + 0.539}{(1/s)^3 + 1.024(1/s)^2 + 1.047(1/s) + 0.539}$$

$$= \frac{0.306s + 0.539s^3}{1 + 1.024s + 1.047s^2 + 0.539s^3}.$$

At the three input frequencies of $s = j0$ ($\omega = 0$), $s = j0.5$ ($\omega = 0.5$), and $s = j5$ ($\omega = 5$) evaluating $H'(s)$ gives

$$H(j0) = 0, \qquad H(j0.5) = 0.10e^{j1.027}, \qquad \text{and} \qquad H(j5) = 0.981e^{j0.384}.$$

Your friend's idea seems a good one because the filter completely blocks the constant (DC) signal, attenuates 1 $\cos(0.5t)$ by a factor of 10, and passes 1 $\cos(5t)$ with a gain of about unity. ■

Comment: If we were to continue (heaven forbid) to compute the magnitude of $H(j\omega)$ for many different frequencies ω and then plot the results, a curve like that in Figure E3.5b would result. To obtain this plot, we used the MATLAB function **freqs**, which is introduced in Chapter 4, "Frequency Response of Continuous Systems."

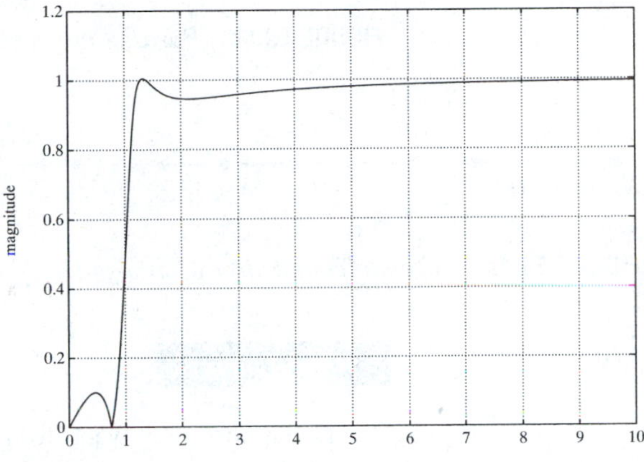

FIGURE E3.5b *Filter frequency response*

EXAMPLE 3.6
System Diagrams and Mason's Gain Rule

Given here are three different situations that illustrate some applications and uses of system diagrams and the MGR.

a. The unstable system of Figure E3.6a.1 can be stabilized by adding a feedback path with a gain of f, as in Figure E3.6a.2. For what values of f will the modified system be stable?

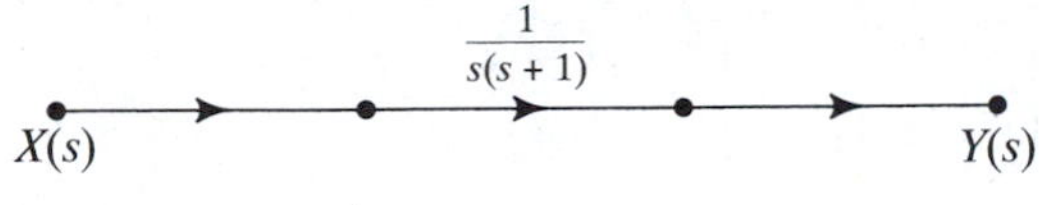

FIGURE E3.6a.1 *Unstable system*

FIGURE E3.6a.2 *Unstable system with feedback*

b. Three first-order systems are connected in parallel, as in Figure E3.6b (page 154). Use the MGR to find the transfer function $H(s) = Y(s)/X(s)$.

c. The same three first-order systems are connected in cascade, as in Figure E3.6c. Use the MGR to find the transfer function $H(s) = Y(s)/X(s)$.

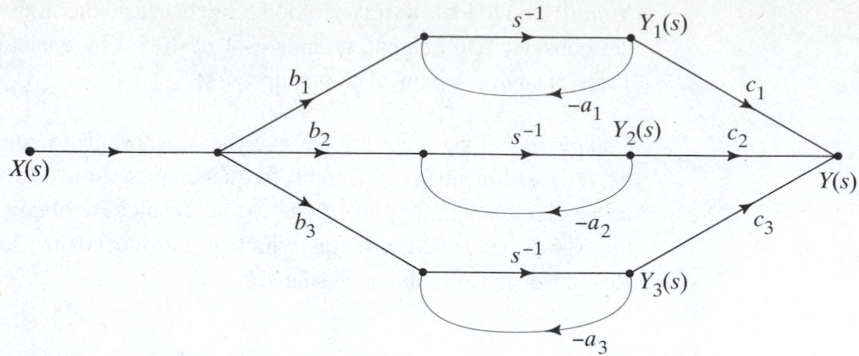

FIGURE E3.6b *Parallel connection of first-order systems*

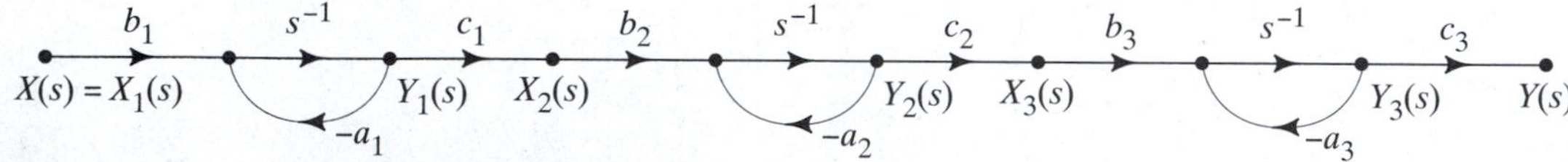

FIGURE E3.6c *Cascade connection of first-order systems*

CHAPTER 3

WHAT IF?

1. Suppose we want to generalize the SFG of Figure E3.6b to the situation of p first-order systems connected in parallel, with each individual system described by the transfer functions $H_1(s) = Y_1(s)/X(s), \ldots, H_p(s) = Y_p(s)/X(s)$. Find $H(s) = Y(s)/X(s)$ in terms of the first-order transfer functions and the gains $c_1, \ldots, c_p$.
2. Now generalize the SFG of Figure E3.6c to a connection of q first-order systems in cascade with each individual system described by the transfer functions $H_1(s) = Y_1(s)/X_1(s), \ldots, H_q(s) = Y_q(s)/X_q(s)$. Find $H(s) = Y(s)/X(s)$ in terms of the first-order transfer functions and the gains $c_1, \ldots, c_q$. ■

Solution

a. *Feedback.* For Figure E3.6a.2,

$$\Delta(s) = 1 - \left\{\frac{1}{s(s+1)} \cdot f\right\} = \frac{s^2 + s - f}{s(s+1)},$$

$$P_1(s) = \frac{1}{s(s+1)}, \quad \text{and} \quad \Delta_1(s) = 1$$

giving the transfer function

$$H(s) = \frac{P_1(s) \cdot \Delta_1(s)}{\Delta(s)} = \frac{1}{s^2 + s - f}.$$

The system poles are the roots of $s^2 + s - f = 0$, which are $s_{1,2} = -0.5 \pm \sqrt{0.25 + f}$. For positive values of f there will always be a positive real root, $\sqrt{0.25 + f} > 0.5$, and hence f must always be negative.

But are there any restrictions on negative values of f? No, because for $-0.25 \le f < 0$, the roots will be negative real and for $f < -0.25$, the roots will be complex conjugates with negative real parts. Thus we have stability for all negative values of f.

b. *Parallel systems.* There are three nontouching loops, so the graph determinant is

$$\begin{aligned}\Delta(s) = 1 &- \{-a_1 s^{-1} - a_2 s^{-1} - a_3 s^{-1}\} + \{(-a_1 s^{-1})(-a_2 s^{-1}) \\ &+ (-a_1 s^{-1})(-a_3 s^{-1}) + (-a_2 s^{-1})(-a_3 s^{-1})\} \\ &- (-a_1 s^{-1})(-a_2 s^{-1})(-a_3 s^{-1}) \\ = 1 &+ s^{-1}(a_1 + a_2 + a_3) + s^{-2}(a_1 a_2 + a_1 a_3 + a_2 a_3) + s^{-3}(a_1 a_2 a_3).\end{aligned}$$

There are three paths from $X(s)$ to $Y(s)$, and the path gains with their corresponding cofactors are

$$\begin{aligned}P_1(s) &= b_1 s^{-1}(c_1) \quad \text{and} \quad \Delta_1(s) = 1 + s^{-1}(a_2 + a_3) + s^{-2}(a_2 a_3) \\ P_2(s) &= b_2 s^{-1}(c_2) \quad \text{and} \quad \Delta_2(s) = 1 + s^{-1}(a_1 + a_3) + s^{-2}(a_1 a_3) \\ P_3(s) &= b_3 s^{-1}(c_3) \quad \text{and} \quad \Delta_3(s) = 1 + s^{-1}(a_1 + a_2) + s^{-2}(a_1 a_2).\end{aligned}$$

Putting all of this together, we have

$$\begin{aligned}H(s) = \frac{Y(s)}{X(s)} &= \frac{P_1(s)\Delta_1(s) + P_2(s)\Delta_2(s) + P_3(s)\Delta_3(s)}{\Delta(s)} \\ &= \frac{\beta_1 s^{-1} + \beta_2 s^{-2} + \beta_3 s^{-3}}{1 + (a_1 + a_2 + a_3)s^{-1} + (a_1 a_2 + a_1 a_3 + a_2 a_3)s^{-2} + (a_1 a_2 a_3)s^{-3}} \\ &= \frac{\beta_1 s^2 + \beta_2 s + \beta_3}{s^3 + (a_1 + a_2 + a_3)s^2 + (a_1 a_2 + a_1 a_3 + a_2 a_3)s + a_1 a_2 a_3}\end{aligned}$$

where

$$\begin{aligned}\beta_1 &= b_1 c_1 + b_2 c_2 + b_3 c_3, \\ \beta_2 &= b_1 c_1 a_2 + b_1 c_1 a_3 + b_2 c_2 a_1 + b_2 c_2 a_3 + b_3 c_3 a_1 + b_3 c_3 a_2, \quad \text{and} \\ \beta_3 &= b_1 c_1 a_2 a_3 + b_2 c_2 a_1 a_3 + b_3 c_3 a_1 a_2.\end{aligned}$$

c. *Cascade connection.* There is no change in the graph determinant with

$$\Delta(s) = 1 + s^{-1}(a_1 + a_2 + a_3) + s^{-2}(a_1 a_2 + a_1 a_3 + a_2 a_3) + s^{-3}(a_1 a_2 a_3).$$

There is only one path, and it touches all the loops, so we find

$$P_1(s)\Delta_1(s) = \{(b_1 s^{-1} c_1) \cdot (b_2 s^{-1} c_2) \cdot (b_3 s^{-1} c_3)\} \cdot \{1\}$$

and the new system transfer function is

$$\begin{aligned}H(s) = \frac{Y(s)}{X(s)} &= \frac{P_1(s)\Delta_1(s)}{\Delta(s)} \\ &= \frac{b_1 c_1 b_2 c_2 b_3 c_3}{s^3 + (a_1 + a_2 + a_3)s^2 + (a_1 a_2 + a_1 a_3 + a_2 a_3)s + a_1 a_2 a_3}.\end{aligned}$$

EXAMPLES

WHAT IF?

1. For p first-order systems connected in parallel, we have $Y(s) = Y_1(s) + Y_2(s) + \cdots + Y_p(s)$, which becomes

$$Y(s) = c_1H_1(s) \cdot X(s) + c_2H_2(s) \cdot X(s) + \cdots + c_pH_p(s) \cdot X(s).$$

Thus the transfer function of the newly created parallel system is

$$H(s) = \frac{Y(s)}{X(s)} = \frac{c_1b_1}{s + a_1} + \frac{c_2b_2}{s + a_2} + \cdots + \frac{c_pb_p}{s + a_p}$$
$$= \frac{N(s)}{(s + a_1)(s + a_2)\cdots(s + a_p)}$$

where we notice that the transfer functions of the "subsystems" *add* (with the appropriate gains) to form the transfer function of the overall system. Also, the poles of the new system are the poles of the original subsystems, whereas the system zeros are the roots of $N(s) = 0$.

2. For the cascade connection of q systems, we see that $X(s) = X_1(s)$, $X_2(s) = c_1Y_1(s), \ldots,$ and $Y(s) = c_qY_q(s)$, giving

$$Y_1(s) = \frac{b_1}{s + a_1} \cdot X(s), \; Y_2(s) = \frac{b_2}{s + a_2} \cdot X_2(s)$$
$$= \frac{b_2}{s + a_2} \cdot c_1\frac{b_1}{s + a_1} \cdot X(s), \ldots.$$

Continuing as before and with $Y(s) = C_qY_q(s)$, we have

$$H(s) = \frac{Y(s)}{X(s)} = \frac{b_1c_1b_2c_2\cdots b_qc_q}{(s + a_1)(s + a_2)\cdots(s + a_q)}$$

where we notice that the transfer functions of the "subsystems" *multiply* to form the transfer function of the overall system. Also, the poles of the new system are the poles of the original subsystems, and there are no finite zeros for this connection. ■

REINFORCEMENT PROBLEMS

P3.1 Initial condition solution for an electric circuit. An electric circuit is described by the differential equation

$$\ddot{i}(t) + 4\dot{i}(t) + 5i(t) = 5x(t) \quad \text{with the ICs } i(0) = 1 \quad \text{and} \quad \dot{i}(0) = 2,$$

where $i(t)$ is the current and $x(t)$ is the applied voltage. Find an analytical expression for $i(t)$ when the input $x(t) = 0$.

P3.2 Derivation of an important pair. For the oscillatory function

$$x(t) = Ae^{at}\cos(\omega t + \alpha)u(t),$$

use the Euler relation to express $x(t)$ as $x(t) = 0.5A\{e^{j\alpha}e^{(a+j\omega)t} + e^{-j\alpha}e^{(a-j\omega)t}\}$ and the transform pair

$$Ae^{bt}u(t) \Leftrightarrow \frac{A}{s-b}, \qquad \text{Re}[s] > \text{Re}[b],$$

to show that

$$X(s) = \frac{0.5Ae^{j\alpha}}{s-(a+j\omega)} + \frac{0.5Ae^{-j\alpha}}{s-(a-j\omega)}$$
$$= \frac{A\{(s-a)\cos\alpha - \omega\sin\alpha\}}{(s-a)^2+\omega^2}, \qquad \text{Re}[s] > a.$$

P3.3 Inverse transform practice.

a. For $F(s) = 1/((s^2+4)(s^2-4))$, $\text{Re}[s] > 2$, find $f(t)$.

b. If $G(s) = 1/((s+1)(s+2)^2)$, $\text{Re}[s] > -1$, what is the inverse transform $g(t)$?

c. Given the transform $Q(s) = 2/((s-1)(s+1))$, find three signals $q(t)$ that could have this transform.

P3.4 Impulse response $h(t)$ and transfer function $H(s)$.

a. For $Q(s) = (s+1)/(s+2)$, $\text{Re}[s] > -2$, find $q(t)$.

b. The response of a causal LTI system to a unit impulse is $y(t) = 3\delta(t) - 15e^{-10t}u(t)$. Find the system's transfer function.

c. If a system's impulse response is $h(t) = [6e^{-4t}\cos(5t + \pi/3)]u(t)$, find the poles and zeros of the transfer function $H(s)$.

P3.5 Final value theorem. For the following transforms, use the final value theorem as appropriate to find the final value of the corresponding time functions.

a. $X(s) = \dfrac{10s}{(s+1)(s+2)^2}$

b. $Y(s) = \dfrac{10s}{s^2+2^2}$

c. $Z(s) = \dfrac{5(s^2-2s+4)}{s(s+1)(s+2)(s+3)}$

P3.6 Impulse response of analog filters.

a. A second-order Butterworth highpass filter is described by the DE

$$\ddot{y}(t) + 1.41\dot{y}(t) + y(t) = \ddot{x}(t)$$

where $x(t)$ is the filter input and $y(t)$ its output. Find the unit impulse response $h(t)$.

b. The unit impulse response of an analog filter is given by

$$h(t) = Ae^{-at}\sin(bt)u(t)$$

with A, a, and b real and greater than zero. Find the unit step response of this filter. What is the unit ramp response—that is, the response to $x(t) = tu(t)$?

P3.7 Complete solution of a circuit's simultaneous DEs. An electric circuit is described by the simultaneous differential equations

$$(1)\ 0.50\frac{dv(t)}{dt} + i(t) = x(t) \quad \text{and} \quad (2)\ -v(t) + \frac{di(t)}{dt} + 2i(t) = 0,$$

with the initial conditions $v(0) = -2$ and $i(0) = 1$. If the circuit input is the exponential signal $x(t) = e^{-2t}u(t)$, use Laplace transform methods to find $v(t)$ for $t \geq 0$.

P3.8 Inverse transform of an improper fraction. Find the inverse transform of

$$Q(s) = \frac{s^3 + s^2 + 2s + 1}{s^3 + 3s^2 + s + 3}.$$

P3.9 Structures from other models. For the following system models, draw a system diagram in the manner of Figure 3.13 or Figure 3.15a or b.

a. $\ddot{y}(t) + 2\zeta\omega_n\dot{y}(t) + \omega_n{}^2y(t) = \omega_n{}^2x(t)$

b. $\ddot{y}(t) + 2\zeta\omega_n\dot{y}(t) + \omega_n{}^2y(t) = \omega_n{}^2x(t) + \omega_n{}^2\dot{x}(t)$

c. $H(s) = \dfrac{s + \beta_0}{s^4 - \alpha_3s^3 + \alpha_2s^2 - \alpha_1s + \alpha_0}$

d. $H(s) = \dfrac{s^2 + \omega_0{}^2}{(s^2 + a_1s + a_2)(s + p)}$

P3.10 Chebyshev filter. A bandpass Chebyshev filter has the transfer function

$$H_{BP}(s) = \frac{9.497(10^6)s^2}{s^4 + 3.671(10^3)s^3 + 7.197(10^8)s^2 + 1.303(10^{12})s + 1.259(10^{17})}.$$

a. What are the zeros and poles for this filter? (Use a MATLAB function to find the poles.)

b. Find the analytical expression for the unit impulse response $h(t)$ for this filter and estimate the time at which steady state is attained. Use **residue** for the PFE.

c. Plot this response using **impulse, ksimptf, kslsim,** or **lsim** and estimate the settling time (when steady-state has been reached) for this filter. Compare your steady-state estimate with the one obtained in part (b).

P3.11 Time scaling a filter. Suppose we want to slow down the unit impulse response of the filter in P3.10 by a factor of a thousand—that is, design for a settling time of about 5 s rather than 5 ms.

a. Find the zeros and poles of this new filter.

b. What is the new transfer function?

c. Without doing much work, determine $h(t)$ for the slower filter.

d. Verify (c) using **residue** and **impulse** or **ksimptf.**

P3.12 System models. An LTI causal system is modeled by the DE

$$\ddot{y}(t) - 6\dot{y}(t) + 25y(t) = \dot{x}(t) - 3x(t).$$

a. Find the characteristic equation and its characteristic roots.
b. Is the system stable?
c. Find the transfer function model.
d. Find the unit impulse response.
e. Find a system diagram based on Figure 3.15b.

P3.13 Practice with Mason's Gain Rule.

a. For the block diagram model of a diving submarine in Figure P3.13a, use the MGR to find the transfer functions $H_a(s) = C(s)/R(s)$ and $H_b(s) = E(s)/R(s)$.
b. What is the transfer function $H(s) = Y(s)/X(s)$ for the SFG of Figure P3.13b?
c. An SFG model of a steam turbine is shown in Figure P3.13c (page 160). The inputs are the turbine's angular velocity $\omega_1(t)$ and the load torque $\lambda(t)$; the outputs have been defined as the propeller's angular velocity and acceleration $\omega_3(t)$ and $\alpha_3(t)$, respectively. Find the four system transfer functions involving these inputs and outputs.

Note: $\mathcal{L}[\omega_1(t)] = \Omega_1(s)$, $\mathcal{L}[\omega_3(t)] = \Omega_3(s)$, $\mathcal{L}[\lambda(t)] = \Lambda(s)$, and $\mathcal{L}[\alpha_3(t)] = A_3(s)$.

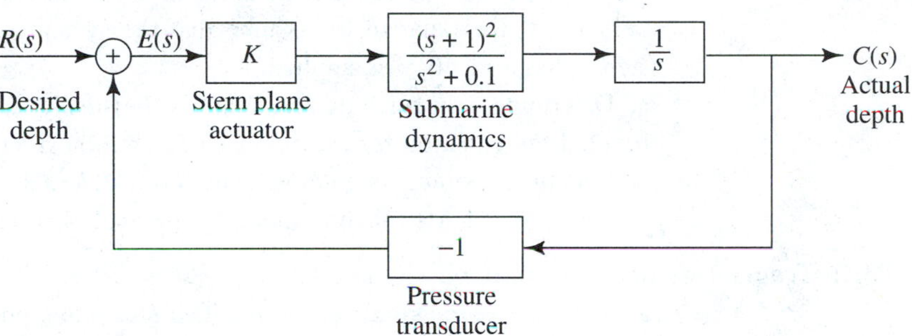

FIGURE P3.13a *Submarine block diagram*

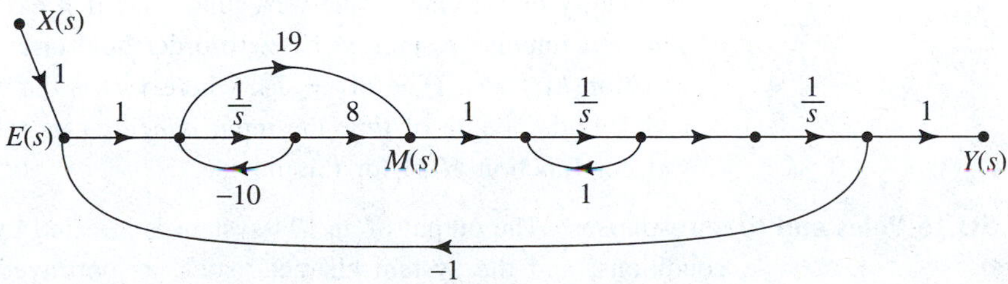

FIGURE P3.13b *SFG for a linear system*

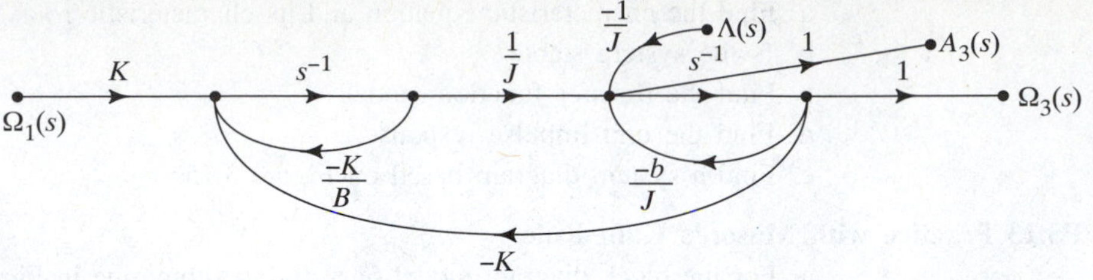

FIGURE P3.13c *SFG model of a steam turbine*

EXPLORATION PROBLEMS

P3.14 Radar antenna system. A radar antenna system can be described by the differential equation

$$\ddot{\omega}(t) + 65\dot{\omega}(t) + 3812\omega(t) = 154e(t)$$

where $e(t)$ is the applied voltage at the motor armature and $\omega(t)$ is the angular velocity of the radar dish. Assume that the system is inert (ICs = 0) and a step voltage of 40 V is applied.

a. Determine $\omega(t)$ analytically. What is the steady-state value?
b. Find the equation for the angular acceleration $\alpha(t) = d\omega(t)/dt$.
c. Find the maximum angular velocity, $\omega_{\max}(t)$ and the time at which it occurs.
d. Use a MATLAB function to verify the analytical results determined here.

P3.15 Transforms of some common signals.

a. What is the transform of the rectified sine-wave pulse $v(t) = \sin(t)u(t) - \sin(t - \pi)u(t - \pi)$?
b. Find $G(s)$ for the gate function $g(t) = Au(t) - Au(t - a)$, $A > 0$ and $a > 0$.
c. Can you find the Laplace transform of the constant $x(t) = A$?
d. The exponential pulse $x(t) = Ae^{-a|t|}$, $a > 0$ and $-\infty < t < \infty$, appears frequently in correlation analysis. Find $X(s)$ if it exists.
e. The unit impulse response of a zero-order hold circuit is described by the equation $h(t) = u(t) - u(t - T)$, where $y(t)$ is the output, $x(t)$ the input, and T is the length of time the input is held at its starting value. Find the transfer function $H(s)$ for this device.

P3.16 Poles and time response. The output of an LTI system is affected by the input, the initial conditions, and the system characteristics, as portrayed in Figure P3.16. In terms of transforms, we can describe the situation as $Y(s) = Y_{\mathrm{IC}}(s) + Y_{\mathrm{F}}(s)$, or in the time domain, as $y(t) = y_{\mathrm{IC}}(t) + y_{\mathrm{F}}(t)$. In this problem let's look at

CHAPTER 3

the effect the poles of a transform have on the time-domain response. In any case, we have, for typical terms in the PF expansion,

$$Z(s) = \sum_{k=1}^{N} \frac{C_k}{s - p_k} \quad \text{or} \quad z(t) = \sum_{k=1}^{N} C_k e^{p_k t}$$

where $z(t)$ is the time response of interest (IC, forced, combined, or whatever). We are assuming distinct poles and an ROC that yields causal terms. First, consider a term in $Z(s)$ such as $C/(s - p)$: (i) give the form of and (ii) sketch the corresponding time-domain term for (a) $p = 0$; (b) $p > 0$; (c) $p < 0$. Next, consider two poles such as $p = re^{\pm j\phi}$ and repeat (i) and (ii) for r real and (d) $\phi = \pi/2$; (e) $\pi/2 < \phi < \pi$; (f) $0 < \phi < \pi/2$.

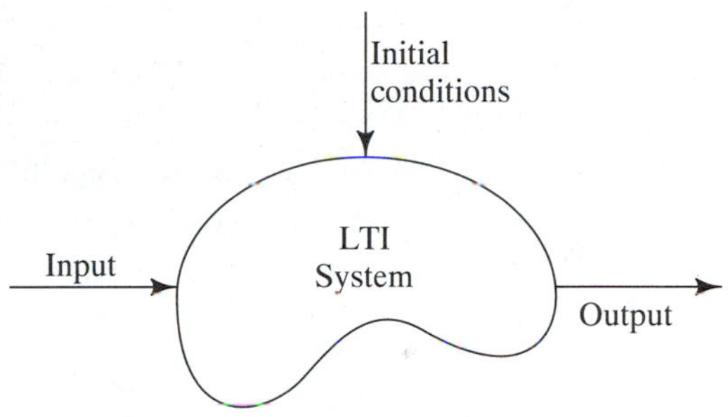

FIGURE P3.16 *LTI system*

P R O B L E M S

P3.17 Poles and time response with MATLAB. We can use a MATLAB function such as **impulse** or **ksimptf** to revisit Problem P3.16 and explore such items as the magnitude of the pole p, the magnitude of the radius r, and the size of the angle ϕ. We simply consider typical partial fraction terms such as

$$Z(s) = \frac{C}{s - p} \quad \text{or} \quad Z(s) = \frac{C}{s - re^{j\phi}} + \frac{C^*}{s - re^{-j\phi}}$$

as transfer functions and use an impulse input to observe the corresponding time response. For instance, for complex poles we can assume C to be real, which gives

$$Z(s) = \frac{C}{s - re^{j\phi}} + \frac{C}{s - re^{-j\phi}} = \frac{2C[s - r\cos(\phi)]}{s^2 - 2r\cos(\phi) + r^2}$$

and the numerator and denominator polynomials are in the required form for the application software. Investigate items (a)–(f) of P3.16 by plotting the time responses for different magnitudes of p, r, and ϕ, including the effect of a complex C in the oscillatory case.

P3.18 Locus of the roots of an equation with MATLAB. In Problems P3.16 and P3.17, we observed the effect of pole locations on the time response. In system design we often want to see the effect of varying a system's parameter, called G, on the values of the system's roots. Starting with the characteristic equation

$$\sum_{k=0}^{N} a_k s^k = 0$$

we collect all terms that involve the varying parameter G and form the polynomial $G \cdot \text{num}(s)$ and put the rest of the terms in the polynomial $\text{den}(s)$. Now the characteristic equation is in the form

$$G \cdot \text{num}(s) + \text{den}(s) = 0 \quad \text{or} \quad 1 + \frac{G \cdot \text{num}(s)}{\text{den}(s)} = 0.$$

For instance, the CE $\alpha_3 s^3 + \alpha_2 s^2 + G s^2 + \alpha_1 s + \alpha_0 + G = 0$ would be put in the form

$$G \cdot \{s^2 + 1\} + \alpha_3 s^3 + \alpha_2 s^2 + \alpha_1 s + \alpha_0 = 0$$

$$\text{or} \quad 1 + \frac{G \cdot \{s^2 + 1\}}{\alpha_3 s^3 + \alpha_2 s^2 + \alpha_1 s + \alpha_0} = 0$$

where, of course, some of the α's could be zero. A detailed discussion of this method is normally part of an undergraduate control systems course; we will not proceed further in the theory and ramifications of the root locus procedure except to point out that the MATLAB function **rlocus** is found in the Signals and Systems and Control System Toolboxes. Let's use this function to plot the locus of roots for some equations of interest to you. Some relatively simple ones for starters are (a) $s^2 + s + G = 0$, (b) $s^2 + Gs + 10 = 0$, and (c) $s^3 + 3s^2 + 2s + G = 0$. In each case you should check some easy-to-find roots to see if the computer-generated plot is reasonable.

P3.19 Zeros and time response with MATLAB. Before leaving the impression that the poles completely determine the corresponding time response, let's consider a few examples of the effect of the lowly zero. The unit step response of a second-order system described by the transfer function $H(s) = k/(s^2 + 2s + 100)$ is shown in Figure P3.19, where $k = 100$ was selected to yield a final value for $y(t)$ of 1.

In terms of the discussion of second-order systems in Chapter 2, the preceding system has a natural frequency of $\omega_n = 10$ and a damping ratio of $\zeta = 0.10$. If a real zero is added to this system, we have

$$H(s) = \frac{k(s - z)}{s^2 + 2s + 100}.$$

Use the MATLAB function **step** to plot the unit step response for several different values of z. In each case, adjust the gain k so that the final value of $y(t)$ is still 1. Some suggested values for the location of the negative axis zeros are $z = -5.0$ and -10.0. What do you observe about the change in the response due to the added zero? Investigate the effects of using zeros on the positive real axis.

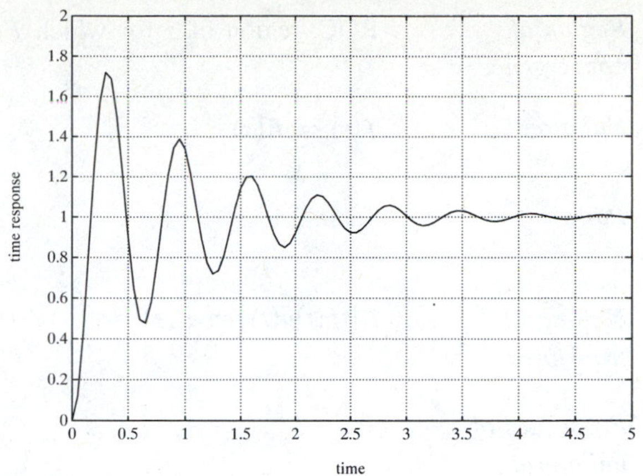

FIGURE P3.19

P3.20 Inverse transform with multiple poles.

a. Given the transfer function $P(s) = (2s^3 + 10)/s^3(s + 1)$, $\text{Re}[s] > -1$, find the analytical expression for $p(t)$.

b. Instead of having three poles of $P(s)$ at $s = 0$, keep one pole at $s = 0$ and move the other two to $s = -0.001$ and $s = -0.002$. With the simple pole still at $s = -1$, form a new transform $Q(s)$. Find the analytical expression for $q(t)$.

c. Compare the curves of parts (a) and (b) by plotting $p(t)$ and $q(t)$ for $0 \le t \le 5$. Use **impulse** or **ksimptf** with the input data for $P(s)$ and $Q(s)$ in transfer function form.

d. Suppose the curves were plotted for $0 \le t \le 25$. How would the results compare now?

P3.21 Another approach to multiple poles. In P3.20 we had $P(s) = A/s^3 + B/s^2 + C/s + D/(s + 1)$, where by the usual approach $A = 10$ and $D = -8$. By setting $s = s_k$, where s_k is any "nonpole" value of $P(s)$, the partial fraction constants B and C are easily evaluated. Try this with an s_k of your choice.

DEFINITIONS TECHNIQUES

DEFINITIONS, TECHNIQUES, AND CONNECTIONS

Bilateral transform $\qquad \mathcal{L}[f(t)] = F(s) = \int_{-\infty}^{\infty} f(t)e^{-st}\,dt$

Inverse transform $\qquad \mathcal{L}^{-1}[F(s)] = f(t) = \dfrac{1}{2\pi j}\int_{\sigma-j\infty}^{\sigma+j\infty} F(s)e^{st}\,ds$

Region of convergence ROC: region of s for which $F(s)$ exists

Notation $f(t) \Leftrightarrow F(s)$

Pairs Table 3.1

Unilateral transform $\mathcal{L}[f(t)u(t)] = F(s) = \int_0^\infty f(t)e^{-st}\,dt$

Properties of unilateral transforms Table 3.2

Transfer function $H(s) = \dfrac{Y(s)}{X(s)}$

1. $H(s) = \dfrac{\text{Laplace transform of the output signal } y(t)}{\text{Laplace transform of the input signal } x(t)}$, ICs $= 0$
2. $H(s) = \dfrac{Y(s)}{X(s)} = \dfrac{\sum_{k=0}^{L} b_k s^k}{\sum_{k=0}^{N} a_k s^k}$ given the DE $\sum_{k=0}^{N} a_k \dfrac{d^k y(t)}{dt^k} = \sum_{k=0}^{L} b_k \dfrac{d^k x(t)}{dt^k}$
3. $H(s) = \mathcal{L}[h(t)]$
4. *Zeros and poles* $H(s) = \dfrac{K(s - n_1)(s - n_2)\cdots(s - n_L)}{(s - d_1)(s - d_2)\cdots(s - d_N)}$.
5. *Stability from* $H(s)$
   a. Causal system—all poles to the left of the imaginary axis
   b. Anticausal system—all poles to the right of the imaginary axis
   c. General—ROC includes the imaginary axis

Convolution $y(t) = \int_0^t h(\tau)x(t - \tau)\,d\tau \Leftrightarrow Y(s) = H(s) \cdot X(s)$, causal system and $x(t) = 0, t < 0$

Sinusoidal steady state For $x(t) = A\cos(\omega t + \alpha)u(t)$
$y_{ss}(t) = A|H(j\omega)|\cos(\omega t + \angle H(j\omega) + \alpha)$
where $H(j\omega) = |H(j\omega)|e^{j\angle H(j\omega)}$

Structure with integrators

$$y(t) = \frac{1}{a_N}\left\{-\sum_{k=0}^{N-1} a_k D^{-[N-k]}y(t) + \sum_{k=0}^{L} b_k D^{-[N-k]}x(t)\right\}$$

$$Y(s) = \frac{1}{a_N}\left\{-\sum_{k=0}^{N-1} a_k s^{-[N-k]}Y(s) + \sum_{k=0}^{L} b_k s^{-[N-k]}X(s)\right\}$$

CHAPTER 3

Mason's Gain Rule

$$H(s) = \frac{Y(s)}{X(s)} = \frac{\sum_{k=1}^{M} P_k(s)\Delta_k(s)}{\Delta(s)}$$

MATLAB FUNCTIONS USED

Function	*Purpose and Use*	*Toolbox*
freqs	Given: TF model of continuous system, **freqs** returns magnitude and phase of frequency response.	Signals/Systems, Signal Processing
impulse	Given: state or TF model of continuous system, **impulse** returns impulse response.	Control System, Signals/Systems
invlaplace	Given: Laplace transform function F, **invlaplace** returns the inverse Laplace transform f.	Symbolic Math
ksimptf	Given: TF model of continuous system, **ksimptf** returns impulse response.	California Functions
kslsim	Given: state or TF model of continuous system, input, ICs, **kslsim** returns system output in continuous time.	California Functions
laplace	Given: The symbolic expression f, **laplace** returns the unilateral transform F.	Symbolic Math
lsim	Given: state or TF model of continuous system, input, ICs, **lsim** returns system output in continuous time.	Control System, Signals/Systems
residue	Given: rational function $T(\sigma) = N(\sigma)/D(\sigma)$, **residue** returns roots of $D(\sigma) = 0$ and PF constants of $T(\sigma)$.	MATLAB
rlocus	Given: an equation in the form $1 + K[\text{num}(\sigma)]/[\text{den}(\sigma)] = 0$, **rlocus** returns plot of locus of roots for K varying.	Signals/Systems, Control System
roots	Given: coefficients of polynomial p, **roots** returns roots of $p = 0$.	MATLAB
step	Given: state or TF model of continuous system, **step** returns step response.	Signals/Systems, Control System

ANNOTATED BIBLIOGRAPHY

1. Chua, Leon O., Charles A. Desoer, and Ernest S. Kuh, *Linear and Nonlinear Circuits,* McGraw-Hill Book Company, New York, 1987. *Chapters Ten and Eleven offer an excellent coverage of Laplace transforms, convolution, and network transfer functions.*

2. Gabel, Robert A., and Richard A. Roberts, *Signals and Linear Systems, 3rd ed.,* John Wiley & Sons, 1987. *Chapter Six, "The Laplace Transform," covers all the usual theorems and properties and includes a very subtle treatment of the inversion of two-sided Laplace transforms without the use of the residue theorem. Applications to the transient and steady-state analysis of linear systems are considered, and the chapter concludes with an explanation of the relationship of the Laplace transform to Fourier and z transforms.*

3. Karni, Shlomo, and William J. Byatt, *Mathematical Methods in Continuous and Discrete Systems,* Holt, Rinehart, and Winston, New York, 1982. *The objective of this concise text is to integrate the applications of linear algebra, complex variables, and transform methods in the analysis of continuous and discrete systems. Chapter Two presents a discussion of complex variables and contour integration that is then applied in the chapter on the Laplace transform and its inversion. Applications to periodic functions, to the diffusion equation, and to matrix differential equations provide interesting extensions. Several carefully constructed appendices contribute to the usefulness of this text.*

4. O'Flynn, Michael, and Eugene Moriarty, *Linear Systems, Time Domain and Transform Analysis,* John Wiley & Sons, New York, 1987. *This text contains material that is suitable for a first course in linear systems as well as for a more advanced course. Methods of modeling and analysis are developed for both continuous- and discrete-time systems. Chapter Five contains a thorough treatment of two-sided (bilateral) Laplace transforms, including applications in linear systems with random and signal plus noise inputs, where correlation and spectral functions are considered.*

5. Reid, J. Gary, *Linear System Fundamentals,* McGraw-Hill Book Company, New York, 1983. *The emphasis of the Laplace transform chapter is on examples taken from control theory and includes feedback, interconnection of systems, and root locus. Chapters Six and Seven on state-space analysis are very complete and include several interesting examples of modeling and analysis by this popular method.*

6. Van Valkenburg, M. E., *Network Analysis, 3rd ed.,* Prentice Hall, Inc., Englewood Cliffs, N.J., 1974. *For more than thirty years, Van Valkenburg's popular text,* Network Analysis, *has introduced Laplace transform methods for circuit analysis to hundreds of thousands of students. With a thorough knowledge of transform methods in hand, the student is introduced to sinusoidal analysis and frequency response in a mathematically rigorous fashion. Network functions, transfer functions, zeros and poles, and stability are included. Although advertised as a circuits text, this book also serves as a good introduction to the theory of linear systems.*

7. [Historical Reference] Gardner, Murray F., and John L. Barnes, *Transients in Linear Systems, Studied by the Laplace Transformation,* John Wiley & Sons, New York, 1942 (out of print). *This classic has been the source of "new and original ideas" for hundreds of authors for almost fifty years. This book evolved from lecture notes used at the Massachusetts Institute of Technology beginning in 1930. Before that a similar course was taught by the legendary Vannevar Bush. The first text to integrate the dynamics of electric and mechanical systems with classical transformation procedures, it includes a table containing more than two hundred operation-transform and function transform pairs. It is still worth looking at about fifty years after publication.*

ANSWERS

P3.1 $i(t) = 4.124e^{-2t}\cos(t - 1.326)u(t)$ or $i(t) = e^{-2t}(\cos t + 4\sin t)$.

P3.3 a. $f(t) = \{[1/16][\cos(2t - 3\pi/2)] + [1/32][e^{2t} - e^{-2t}]\}u(t)$; b. $g(t) = [e^{-t} - te^{-2t} - e^{-2t}]u(t)$; c. For $\text{Re}[s] > 1$, $q(t) = [e^{t} - e^{-t}]u(t)$; for $\text{Re}[s] < -1$, $q(t) = [-e^{t} + e^{-t}]u(-t)$; for $-1 < \text{Re}[s] < 1$, $q(t) = -e^{t}u(-t) - e^{-t}u(t)$

P3.4 a. $q(t) = \delta(t) - e^{-2t}u(t)$; b. $H(s) = (3s + 15)/(s + 10)$; c. Zero: $z = 4.66$, poles: $z_{1,2} = -4 \pm j5 = 6.40e^{\pm j2.25}$

P3.5 a. $\lim_{t\to\infty} x(t) = 0$; b. We cannot use FVT, $sY(s)$ has poles on the imaginary axis; steady state will be a sinusoidal oscillation. c. $\lim_{t\to\infty} z(t) = 10/3$

P3.6 a. $h(t) = \delta(t) - 1.414e^{-0.707t}\cos(0.707t)u(t)$; b. $s(t) = [Ab/(a^2 + b^2) + A/(\sqrt{a^2 + b^2})e^{-at}\cos(bt - \pi/2 - \tan^{-1}(b/-a)]u(t)$; ramp response, $y(t) = \{[A/(a^2 + b^2)][bt - 2ab + e^{-at}\cos(bt - 2\tan^{-1}[b/(-a)] - \pi/2)]\}u(t)$

P3.7 $v(t) = 2.82e^{-t}\cos(t + 3\pi/4)u(t)$

P3.8 $q(t) = 1\delta(t) - 2.3e^{-3t} + 0.316\cos(t - 0.322)$, $t \geq 0$

P3.9 See Figure A3.9.

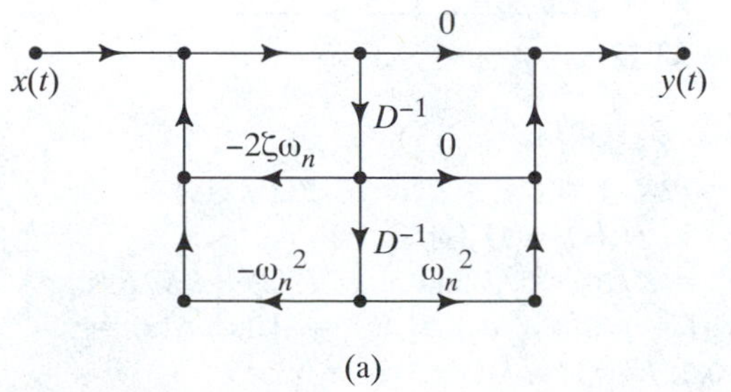

(a)

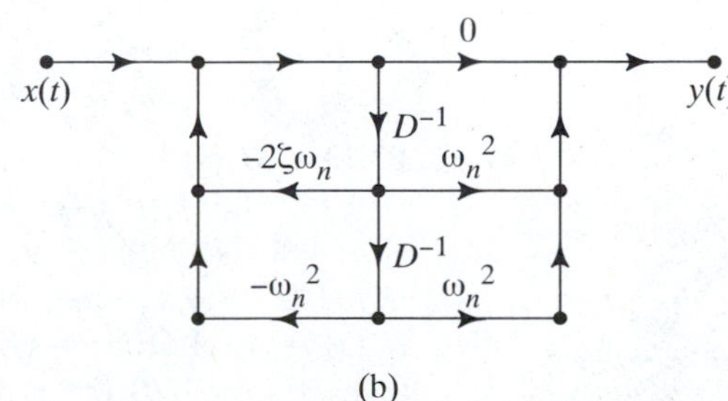

(b)

Figure A3.9 *Continues*

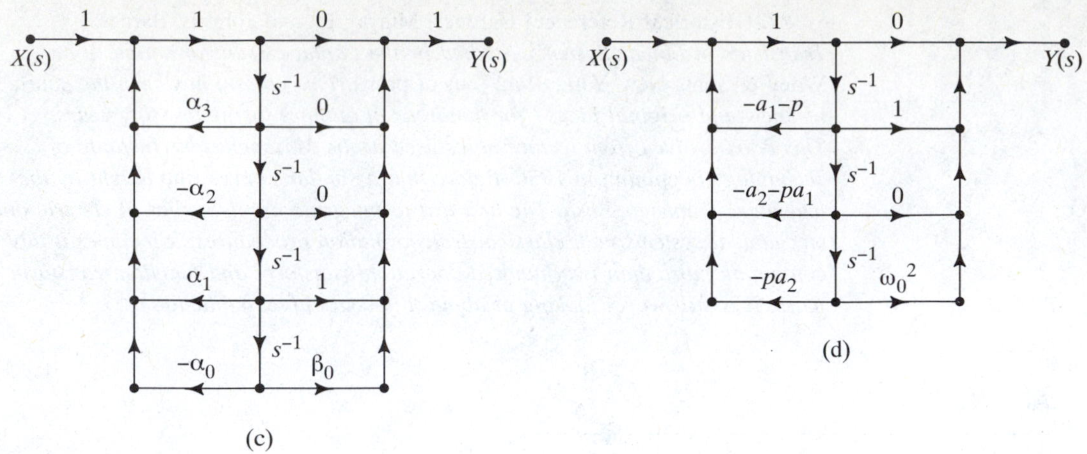

Figure A3.9 *Continued*

P3.10 a. Two finite zeros at $s = 0$, poles at $s_{1,2} = 2.018(10^4)e^{\pm j1.619}$, $s_{3,4} = 1.759(10^4)e^{\pm j1.620}$; b. $h(t) = [1965.2e^{-978t}\cos(20151t - 1.52) + 1713.1e^{-857t}\cos(17566t + 1.62)]u(t)$, steady state at $t \approx 6$ ms; c. See the plot of A3_10 on the CLS disk; the response settles out in about 6 ms.

P3.11 a. Change pole magnitudes by a factor of 10^3, giving $s_{1,2} = 2.018(10)e^{\pm j1.619}$, $s_{3,4} = 1.759(10)e^{\pm j1.620}$; b. $H(s) = 9.497s^2/(s^4 + 3.671s^3 + 7.197(10^2)s^2 + 1.303(10^3)s + 1.259(10^5))$; c. $h(t) = [1.96e^{-0.978t}\cos(20.1t - 1.52) + 1.7e^{-0.857t}\cos(17.5t + 1.62)]u(t)$; d. Settling time from A3_11 on the CLS disk is about 6 s.

P3.12 a. $s^2 - 6s + 25 = 0$, $s_{1,2} = 3 \pm j4$; b. unstable, causal system with positive real part for roots; c. $H(s) = Y(s)/X(s) = (s - 3)/(s^2 - 6s + 25)$; d. $h(t) = e^{3t}\cos(4t)u(t)$; e. See Figure A3.12.

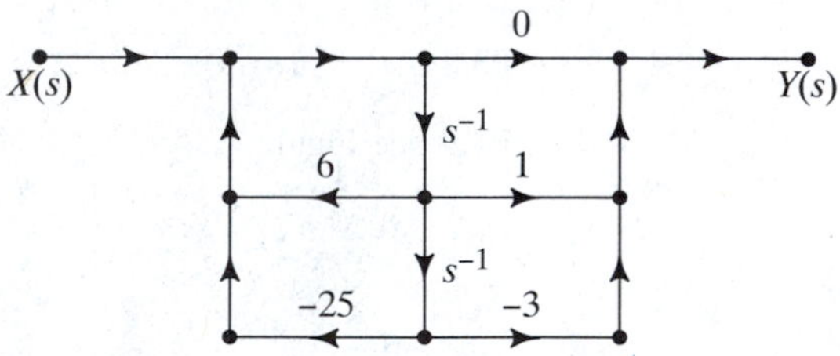

Figure A3.12

P3.13 a. $H_a(s) = K(s + 1)^2/(s^3 + Ks^2 + [2K + 0.1]s + K)$, $H_b(s) = s(s^2 + 0.1)/(s^3 + Ks^2 + [2K + 0.1]s + K)$; b. $Y(s)/X(s) = (19s + 8)/(s^3 + 9s^2 + 9s + 8)$; c. $H_a(s) = \Omega_3(s)/\Omega_1(s) = (K/J)/(s^2 + a_1 s + a_0)$, where $a_1 = K/B + b/J$ and $a_0 = K/J + bK/BJ$; $H_b(s) = A_3(s)/\Omega_1(s) = (K/J)(s)/(s^2 + a_1 s + a_0)$; $H_c(s) = \Omega_3(s)/\Lambda(s) = (-1/J)(s + K/B)/(s^2 + a_1 s + a_0)$; $H_d(s) = A_3(s)/\Lambda(s) = (-1/J)(s + K/B)(s)/(s^2 + a_1 s + a_0)$

P3.14 a. $\omega(t) = [1.62 + 1.90e^{-32.5t} \cos(52.5t + 2.587)]u(t)$, $\omega_{ss}(t) = 1.62$; b. $\alpha(t) = [117.34e^{-32.5t} \cos(52.5t - 1.57)]u(t)$; c. $t_{\max} = 0.06$, $\omega_{\max}(t) = 1.85$; d. See A3_14a–b on the CLS disk.

P3.15 a. $V(s) = (1 - e^{-\pi s})/(s^2 + 1)$, $\text{Re}[s] > 0$; b. $G(s) = [A/s](1 - e^{-as})$, $\text{Re}[s] > 0$; c. Does not exist, no ROC; d. $X(s) = -2Aa/(s^2 - a^2)$, $|a| > \text{Re}[s] < a$; e. $H(s) = [1/s](1 - e^{-Ts})$.

P3.16 (i) a. C; b. $Ce^{pt}u(t)$; c. $Ce^{-pt}u(t)$; d. $2|C| \cos(rt + \alpha)u(t)$; e. $2|C|e^{(r \cos \phi)t} \cos([r \sin \phi]t + \alpha)u(t)$; f. $2|C|e^{(r \cos \phi)t} \cos([r \sin \phi]t + \alpha)u(t)$ (ii) See the plots in the answers to Problem P3.17.

P3.17 We skipped (a) as not very interesting. b. $z(t) = 0.7e^{0.5t}u(t)$ and $z(t) = 0.7e^{t}u(t)$; c. $z(t) = 7e^{-t}u(t)$ and $z(t) = 7e^{-0.5t}u(t)$; d. $z(t) = 4 \cos(2.5t - \pi/3)u(t)$ and $z(t) = 4 \cos(5t)u(t)$; e. $z(t) = 4e^{-3t} \cos(4t - \pi/3)u(t)$ and $4e^{-0.5t} \cos(4.97t + \alpha)u(t)$; f. $z(t) = 4e^{t} \cos(4.88t + \alpha)u(t)$ and $4e^{0.5t} \cos(4.97t + \alpha)u(t)$; see A3_17 on the CLS disk.

P3.18 See A3_18 on the CLS disk.

P3.19 See A3_19 on the CLS disk. Adding the zero appears to change the overshoot radically. Zeros on the positive real axis cause some unusual effects in the response.

P3.20 a. $p(t) = [5t^2 - 10t + 10 - 8e^{-t}]u(t)$; b. $q(t) = \{[5 - 10.01e^{-0.001t} + 5.01e^{-0.002t}]10^6 - 8.02e^{-t}\}u(t)$; c. See A3_20 on the CLS disk for the scripts and plots. The time responses are identical. d. There is a slight difference between the two as t becomes very large. *Bonus:* See A3_20aa and A3_20bb for the results for poles at $s = 0$, -0.01, -0.02, and -1.

P3.21 For $s = 1$, $P(s) = 6 = 10 + B + C - 4$; for $s = 2$, $P(s) = 1.08 = 1.25 + 0.25B + 0.5C - 2.67$; solution gives $B = -10$ and $C = 10$, as before.

TABLE 3.1 *Laplace transform pairs*

$$F(s) = \int_{-\infty}^{\infty} f(t)e^{-st}\,dt$$

Signal	*Transform*	*ROC*
1. $A\,\delta(t)$	A	All s
2. $Ae^{at}u(t)$	$\dfrac{A}{s-a}$	$\mathrm{Re}[s] > a$
3. $Au(t)$	$\dfrac{A}{s}$	$\mathrm{Re}[s] > 0$
4. $Ae^{(a+j\omega)t}u(t)$	$\dfrac{A}{s-a-j\omega}$	$\mathrm{Re}[s] > a$
5. $Ae^{at}\cos(\omega t)u(t)$	$\dfrac{A(s-a)}{(s-a)^2+\omega^2}$	$\mathrm{Re}[s] > a$
6. $A\cos(\omega t)u(t)$	$\dfrac{As}{s^2+\omega^2}$	$\mathrm{Re}[s] > 0$
7. $A\sin(\omega t)u(t)$	$\dfrac{A\omega}{s^2+\omega^2}$	$\mathrm{Re}[s] > 0$
8. $Ae^{at}\cos(\omega t+\alpha)u(t)$	$\dfrac{A[(s-a)\cos\alpha - \omega\sin\alpha]}{(s-a)^2+\omega^2} = \dfrac{0.5Ae^{j\alpha}}{s-(a+j\omega)} + \dfrac{0.5Ae^{-j\alpha}}{s-(a-j\omega)}$	$\mathrm{Re}[s] > a$
9. $Atu(t)$	$\dfrac{A}{s^2}$	$\mathrm{Re}[s] > 0$
10. $At^nu(t)$	$\dfrac{An!}{s^{n+1}}$	$\mathrm{Re}[s] > 0$
11. $Ate^{at}u(t)$	$\dfrac{A}{(s-a)^2}$	$\mathrm{Re}[s] > a$
12. $Ae^{at}u(-t)$	$\dfrac{-A}{s-a}$	$\mathrm{Re}[s] < a$
13. $Au(-t)$	$\dfrac{-A}{s}$	$\mathrm{Re}[s] < 0$
14. $Ae^{at}\cos(\omega t)u(-t)$	$\dfrac{-A(s-a)}{(s-a)^2+\omega^2}$	$\mathrm{Re}[s] < a$

TABLE 3.2 *Unilateral Laplace transform properties and relations*

$$F(s) = \int_0^\infty f(t)e^{-st}\,dt$$

Property	*Signal*	*Transform*
1. Linearity	$af(t) + bg(t)$	$aF(s) + bG(s)$
2. Derivative	$\dfrac{df(t)}{dt}$	$sF(s) - f(t)\vert_{t=0} = sF(s) - f(0)$
3. Nth derivative	$\dfrac{d^N f(t)}{dt^N}$	$s^N F(s) - s^{N-1} f(t)\vert_{t=0} - s^{N-2}\dfrac{df(t)}{dt}\Big\vert_{t=0} - \cdots - \dfrac{d^{N-1}f(t)}{dt^{N-1}}\Big\vert_{t=0}$
4. Integral	$\int_0^t f(\tau)\,d\tau$	$\dfrac{F(s)}{s}$
5. Time shift ($a > 0$)	$f(t-a)u(t-a)$	$e^{-as}F(s)$
6. Frequency shift ($a > 0$)	$e^{at}f(t)$	$F(s-a)$
7. Convolution	$f_3(t) = \int_0^t f_1(\tau)f_2(t-\tau)\,d\tau = \int_0^t f_2(\tau)f_1(t-\tau)\,d\tau$	$F_3(s) = F_1(s) \cdot F_2(s) = F_2(s) \cdot F_1(s)$
8. Final value theorem (FVT)	$\lim_{t\to\infty} f(t)$	$\lim_{s\to 0} sF(s)$, provided all poles of $sF(s)$ have negative real parts
9. Initial value theorem (IVT)	$\lim_{t\to 0} f(t)$	$\lim_{s\to\infty} sF(s)$

A N S W E R S

CHAPTER 4

Frequency Response of Continuous Systems

PREVIEW

The concept of frequency response is used extensively in most areas of engineering and applied technology. We have all encountered the idea of filtering in audio systems, where—by adjusting the bass and treble controls—the frequency response of a CD or stereo system can be altered to suit better the fancy of the listener. Another example is the design of feedback control systems, where frequency response methods are often used. Mechanical engineers use frequency response methods in vibration and spectral analysis applications. And oceanographers model wave motion with frequency response techniques. Frequency methods are popular for several reasons, including the availability of sinusoidal sources, the ease of understanding the steady-state nature of the technique, the relative ease of performing tests on the system, and the vast array of accessible design rules and algorithms.

If we make the input to a system a unit amplitude sinusoid with zero phase and observe the steady-state output magnitude and phase as the frequency of the input varies, the result is called the *frequency response.* Thus, to perform a frequency response test on an electric circuit, only a variable-frequency oscillator and an oscilloscope are required to measure the input and output voltages. In other engineering applications the instrumentation problem is relatively similar. The frequency response of a high-fidelity CD system consisting of the magnitude plots of the two channels of the speaker voltage-input voltage ratio is given in Figure 4.1b. Notice that for frequencies in the audible range of 10 Hz to 20 kHz the responses are essentially flat.

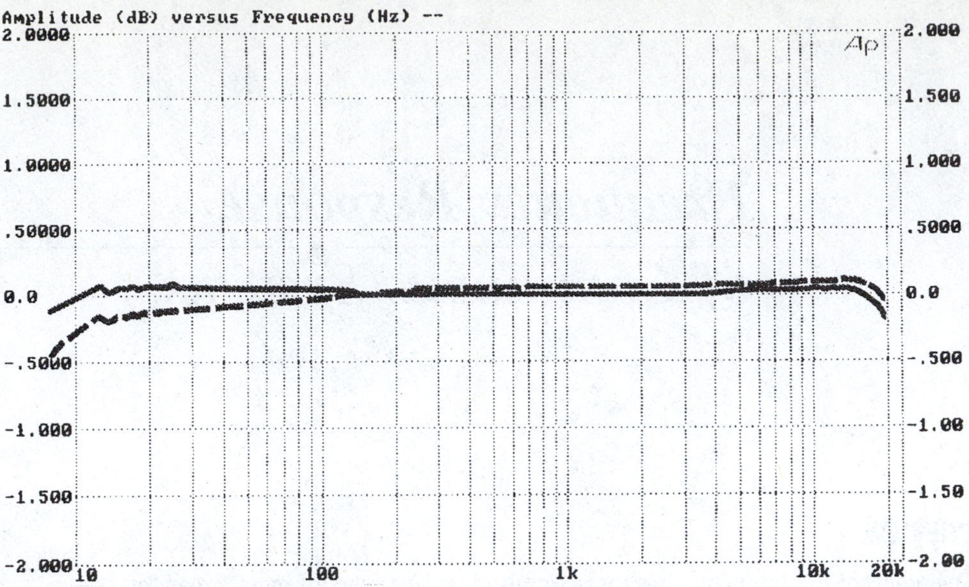

FIGURE 4.1 *CD frequency response*

In this chapter, we make use of several concepts developed earlier, including transforms, transfer functions, and stability. These results and ideas will be restated here, but if you have doubts about them you should refer to the appropriate sections in Chapters 2 and 3.

BASIC CONCEPTS

REVIEW

Transfer Function (See Chapter 3 for a detailed discussion.)

1. $$H(s) = \frac{Y(s)}{X(s)} = \frac{\text{Laplace transform of the output signal } y(t)}{\text{Laplace transform of the input signal } x(t)},$$
ICs $= 0$.

2. $H(s) = \dfrac{\sum_{k=0}^{L} b_k s^k}{\sum_{k=0}^{N} a_k s^k}$ given the DE $\sum_{k=0}^{N} a_k \dfrac{d^k y(t)}{dt^k} = \sum_{k=0}^{L} b_k \dfrac{d^k x(t)}{dt^k}$.
3. $H(s) = \mathcal{L}\,[h(t)]$.

Stability (See the appropriate sections in Chapters 2 and 3.)
Time Domain

1. All bounded inputs produce bounded outputs.
2. The IC response for arbitrary initial conditions of a causal system decays to zero as $t \to \infty$, or the characteristic roots (the r_k's) must have negative real parts.
3. The unit impulse response is absolutely integrable:

$$\int_{-\infty}^{\infty} |h(\tau)|\, d\tau < \infty.$$

Laplace Domain

1. Causal system: Real parts of all poles are less than zero.
2. Anticausal system: Real parts of all poles are greater than zero.
3. General: ROC includes the imaginary axis.

Sinusoidal Steady-State Response (See the appropriate sections in Chapters 2 and 3.)
From $h(t)$ For $x(t) = A\cos(\omega t + \alpha)$, $-\infty \le t \le \infty$,

$$y_{ss}(t) = A|H(j\omega)| \cos(\omega t + \alpha + \angle H(j\omega)),$$
$$\text{where } H(j\omega) = \int_{-\infty}^{\infty} h(\tau) e^{-j\omega\tau}\, d\tau.$$

From $H(s)$ For $x(t) = A\cos(\omega t + \alpha)u(t)$,

$$y_{ss}(t) = A|H(j\omega)| \cos(\omega t + \alpha + \angle H(j\omega)),$$
$$\text{where } H(j\omega) = H(s)|_{s=j\omega}.$$

FREQUENCY RESPONSE

The function $H(j\omega)$ is known as the frequency response of the system and it can be determined in several ways. The three following methods result in an analytical expression.

1. From the transfer function $H(s)$ of a stable system, the frequency response is

$$H(j\omega) = H(s)|_{s=j\omega} = \left.\frac{\sum_{k=0}^{L} b_k s^k}{\sum_{k=0}^{N} a_k s^k}\right|_{s=j\omega} = \frac{\sum_{k=0}^{L} b_k (j\omega)^k}{\sum_{k=0}^{N} a_k (j\omega)^k}.$$

2. Given the unit impulse response $h(t)$ of a stable LTI system, the frequency response is

$$H(j\omega) = \int_{-\infty}^{\infty} h(\tau)e^{-j\omega\tau}d\tau.$$

3. Starting with the DE for a stable system,

$$\sum_{k=0}^{N} a_k \frac{d^k y(t)}{dt^k} = \sum_{k=0}^{L} b_k \frac{d^k x(t)}{dt^k}, \qquad \text{gives } H(j\omega) = \frac{\sum_{k=0}^{L} b_k (j\omega)^k}{\sum_{k=0}^{N} a_k (j\omega)^k}.$$

Notice that for each value of the frequency ω in rad/s the frequency response function $H(j\omega)$ is a complex number that may be written in terms of its real and imaginary parts as $H(j\omega) = \alpha + j\beta$ or in exponential form as $H(j\omega) = Me^{jP}$, where M stands for the magnitude of the frequency response and P for its phase. Of course, both α and β, or M and P, depend on ω.

CHAPTER 4

ILLUSTRATIVE PROBLEM 4.1
Frequency Response Determination for Two Systems

a. A linear circuit is described by the differential equation

$$\dddot{v}(t) + 10\ddot{v}(t) + 8\dot{v}(t) + 5v(t) = 13\dot{e}(t) + 7e(t),$$

where $e(t)$ stands for the network's input voltage and $v(t)$ represents the output voltage. Find the frequency response function $H(j\omega)$.

b. A second-order network has the unit impulse response $h(t) = [e^{-t} - e^{-2t}]u(t)$. Find its frequency response $H(j\omega)$.

Solution

a. First, we need to check stability. We would assume that the system is stable (it wouldn't be of much use if not), but perhaps an error has been made in finding the system's DE. The transfer function of the causal circuit is

$$H(s) = \frac{13s + 7}{s^3 + 10s^2 + 8s + 5}$$

and its poles are $s_1 = -9.189$ and $s_{2,3} = -0.406 \pm j0.616$, clearly a stable situation. Thus we can safely say that the frequency response is

$$H(j\omega) = \frac{13j\omega + 7}{-j\omega^3 - 10\omega^2 + 8j\omega + 5}.$$

b. This system is stable because

$$\int_{-\infty}^{\infty} |h(\tau)|\, d\tau < \infty,$$

since both $e^{-t}u(t)$ and $e^{-2t}u(t)$ decay to zero as $t \longrightarrow \infty$. The frequency response is, therefore,

$$H(j\omega) = \int_{0}^{\infty} [e^{-\tau} - e^{-2\tau}]e^{-j\omega\tau}\, d\tau = \frac{1}{1 + j\omega} - \frac{1}{2 + j\omega}$$
$$= \frac{1}{-\omega^2 + 3j\omega + 2}.$$

Characteristics of the Frequency Response Function We have seen that the system frequency response function $H(j\omega)$ totally determines the steady-state system output signal for a known sinusoidal input signal. The basis of many continuous-system design procedures, therefore, is to determine an acceptable frequency response function that will satisfy both the magnitude and phase criteria of any particular design situation. The function $H(j\omega)$ is very important in system analysis and design, so let's look at its nature in greater detail.

Symmetry The magnitude M of $H(j\omega)$ is an even function of ω, and the phase P is an odd function. To show these characteristics we start with

$$H(j\omega) = \int_{-\infty}^{\infty} h(\tau)e^{-j\omega\tau}\, d\tau = \int_{-\infty}^{\infty} h(\tau)[\cos(\omega\tau) - j\sin(\omega\tau)]\, d\tau = a + jb.$$

Replacing ω with $-\omega$ gives

$$H(-j\omega) = \int_{-\infty}^{\infty} h(\tau)e^{j\omega\tau}\, d\tau = \int_{-\infty}^{\infty} h(\tau)[\cos(\omega\tau) + j\sin(\omega\tau)]\, d\tau = a - jb.$$

Thus we see that

$$|H(j\omega)| = |H(-j\omega)| = \sqrt{a^2 + b^2} = M$$

which satisfies the definition of an even function. The phase of $H(j\omega)$ is $\tan^{-1}(b/a)$, that of $H(-j\omega)$ is $\tan^{-1}(-b/a)$, and we find that

$$\angle H(j\omega) = -\angle H(-j\omega) = P$$

which satisfies the definition of an odd function.

Time and Frequency The unit impulse response $h(t)$ is a function of the continuous variable t. The frequency response function, $H(j\omega)$, is a function of the continuous (analog) frequency ω.

Decibels The values of the gain M may vary over a very wide range, and it is often useful to define M in decibels (dB) as $M_{dB} = 20 \log_{10} M$. We will use both versions of M in this chapter—that is, as a real number and in decibels.

Bandwidth If M_{max} is the maximum value of M, the range of frequencies $\omega_1 \le \omega \le \omega_2$ for which $0.707M_{max} \le M \le M_{max}$ is called the bandwidth or passband. Filters may have one or several passbands, and in terms of dB a passband is a region within -3 dB $= 20 \log_{10}(0.707)$ of the maximum value in dB.

Continuous-Time Fourier Transform $H(j\omega)$, which we have called the frequency response, is the Continuous-Time Fourier Transform (CTFT) of the unit impulse response $h(t)$. In a similar way, for the signals $x(t)$ and $y(t)$ we write

Fourier transform

$$X(j\omega) = \int_{-\infty}^{\infty} x(\tau)e^{-j\omega\tau}\,d\tau \quad \text{and} \quad Y(j\omega) = \int_{-\infty}^{\infty} y(\tau)e^{-j\omega\tau}\,d\tau$$

where $X(j\omega)$ and $Y(j\omega)$ are the CTFTs of the signals $x(t)$ and $y(t)$, respectively. This transform is considered in detail in the next chapter, "Continuous-Time Fourier Series and Transforms."

FREQUENCY RESPONSE PLOTS

There are many different varieties of frequency response plots, and their designations are sometimes confusing. Listed next are several different types of plots and their most common names.

Rectangular Plots Recalling that the frequency response can be expressed in terms of magnitude and phase as

$$H(j\omega) = |H(j\omega)|e^{j\angle H(j\omega)} = Me^{jP}$$

the most straightforward representation is to plot M versus ω and P versus ω on rectangular coordinates for the frequency range of $\omega_1 \le \omega \le \omega_2$. The range of frequencies must be determined by the user, which is in contrast with the frequency response of discrete systems (see Chapter 9), where $H(e^{j\theta})$ is normally plotted for $0 \le \theta \le \pi$. See Figure 4.2 for a rectangular plot of the frequency response for the stable system described by the transfer function $H(s) = 10/(s^2 + 11s + 10)$.

To plot this with MATLAB we use (invoke) the MATLAB function **freqs**, which stands for "frequency response in the s domain," or continuous-time domain. To use **freqs** we need to put the transfer function in the form

H(s) for **freqs**

$$H(s) = \frac{B(s)}{A(s)} = \frac{b(1)s^{(nb-1)} + b(2)s^{(nb-2)} + \cdots + b(nb)}{a(1)s^{(na-1)} + a(2)s^{(na-2)} + \cdots + a(na)}.$$

That is, the term with the highest power of s in both the numerator and denominator polynomials must be the first term, with the other terms following in decreasing powers of s. We enter the numerator and denominator coefficients in the vectors **b** and **a**, respectively. Running the program returns the frequency response vector **H** with a frequency range of ω_1 to ω_2 rad/s in the frequency vector **w**. For this example $H(s) = 10/(s^2 + 11s + 10)$, and we have specified a frequency range from -20 to $+20$ and an interval of 0.05 rad/s

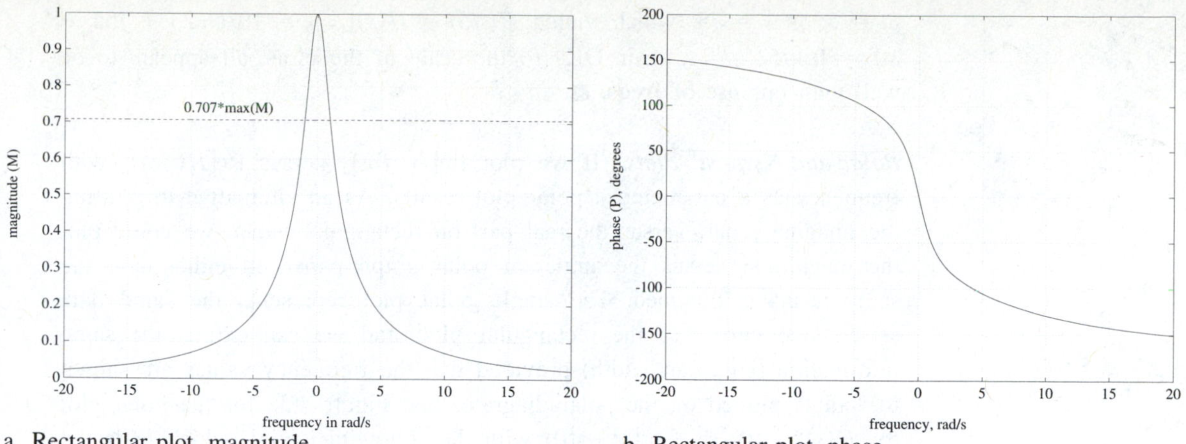

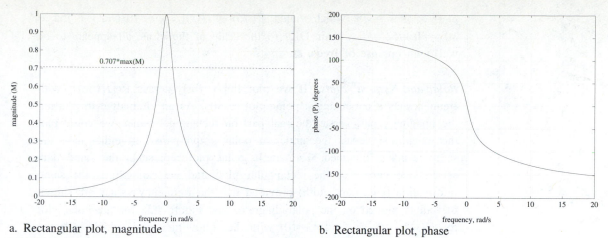

a. Rectangular plot, magnitude

b. Rectangular plot, phase

FIGURE 4.2 *Frequency response plots for a lowpass filter*

as seen in the following script. This system has a lowpass characteristic with the passband in the approximate range of positive frequencies of $0 \leq \omega \leq 1$ rad/s. If we want a more precise indication of the bandwidth, we could either look at the values obtained for **H** from MATLAB and the corresponding values of ω or we could create a second plot over the frequency range $0 \leq \omega \leq 3.0$, for example.

Notice the even symmetry of M and the odd symmetry of P about $\omega = 0$. The script for Figure 4.2 follows, along with some annotations and explanations.

MATLAB Script

```
%F4_2 Rectangular plots for Fig.4.2
a=[1,11,10];                        % denominator coefficients in
                                    % descending order
b=[10];                             % numerator coefficient
w=-20:.05:20;                       % initial frequency:increment in
                                    % rad/s:final frequency
H=freqs(b,a,w);                     % call freqs
mag=abs(H);                         % computes magnitude of H
phase=angle(H);                     % computes phase of H
phase=phase*180/pi;                 % changes phase to degrees
g=0.707*ones(size(w))*max(mag);     % line to show bandwidth
%...plotting statements
```

CROSS-CHECK Are these plots reasonable? First, for $\omega = 0$ or $s = j0$, $H(j0) = H(s)|_{s=j0} = 10/10 = 1e^{j0^\circ}$, which agrees with the plot. Next, try

$\omega = 5$ or $s = j5$, which yields $H(j5) = H(s)|_{s=j5} = 10/(-25 + j55 + 10) = 0.175e^{-j105^\circ}$, again OK. To the scale of the plots, all appears to be well with our use of **freqs**. ■

Polar and Nyquist[1] Plots If we plot $\text{Im}[H(j\omega)]$ versus $\text{Re}[H(j\omega)]$ with frequency as a parameter, a polar plot results. As an alternative to plotting the imaginary part versus the real part on rectangular paper, we could plot the magnitude versus the angle on polar graph paper. In either case the same result is obtained. The single polar plot represents the same data as the two curves of the rectangular plot, and we can extract the same information (e.g., bandwidth) provided that the frequency values are shown for points plotted on the polar diagram. See Figure 4.3a for the polar plot for $H(s) = 10/(s^2 + 11s + 10)$ with the following script, which calls the MATLAB function **nyquist**.

MATLAB Script

```
%F4_3 Polar and Nyquist diagrams

%F4_3a Polar plot for Figure 4.3a
num=[10];
den=[1,11,10];
w=logspace(-2,2);                        % generates 50 points between
                                         % 0.01 (10^-2) and 100 (10^2)
[re,im]=nyquist(num,den,w)               % call nyquist
plot(re,im),title('Fig. 4.3a Polar plot');
%...labeling statements and pause
```

Comment: The frequencies, $s = j0.01$, $s = j0.5$, . . . , were placed on the locus using the MATLAB function **text,** as seen in the file for F4_3a on the CLS disk.

If the polar diagram for $H(j\omega)$ is plotted for all ω, $-\infty \leq \omega \leq \infty$, the resulting curve is called the *Nyquist diagram.* Figure 4.3b shows a Nyquist diagram for the same transfer function, where the only script change is in the plotting statement.

[1] Harry Nyquist (1889–1976), a native of Sweden, was an electrical engineer with the Bell Telephone Laboratories who made a very important contribution concerning the stability of feedback amplifiers. His result is contained in the seminal paper "Regeneration Theory," *Bell System Technical Journal,* 11, 126–147 (1932).

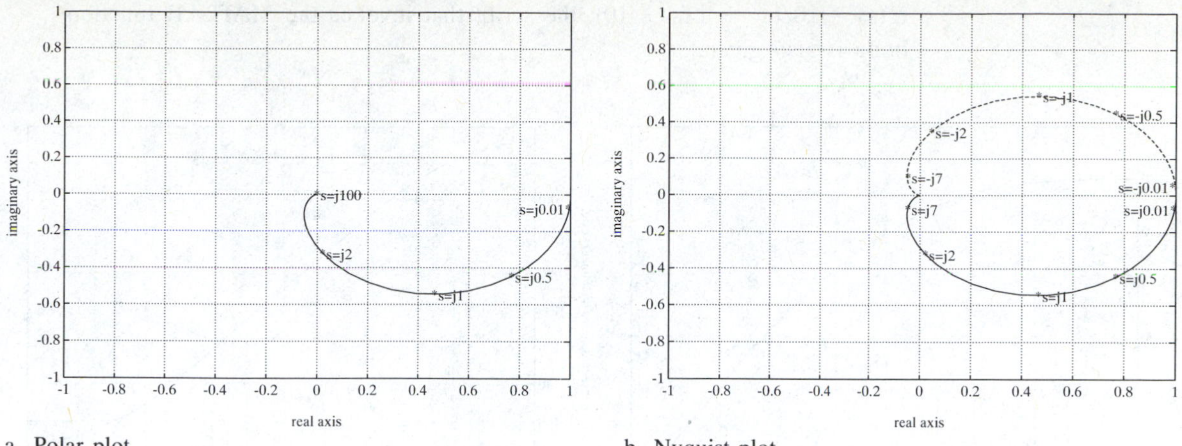

a. Polar plot

b. Nyquist plot

FIGURE 4.3 *Polar and Nyquist plots for* $H(s) = 10/(s^2 + 11s + 10)$

MATLAB Script

```
%F4_3b Nyquist diagram for Figure 4.3b
num=[10];
den=[1,11,10];
w=logspace(-2,2);
[re,im]=nyquist(num,den,w);
plot(re,im,re,-im),title('Fig. 4.3b Nyquist plot');
%...labeling statements
```

Logarithmic, or Bode,[2] Plots In practice the rectangular plots of Figure 4.2 are not particularly useful, since both $|H(j\omega)|$ and ω are likely to vary through several orders of magnitude. The obvious move is to make a log-log plot for $|H(j\omega)|$ versus ω and a semilog plot for $\angle H(j\omega)$ versus ω, since the phase is typically restricted to the range $-\pi$ to π. There are a number of equivalent possibilities, the most common being to plot $20 \log_{10}|H(j\omega)|$ versus ω and $\angle H(j\omega)$ versus ω with $20 \log_{10}|H(j\omega)|$ and $\angle H(j\omega)$ on linear scales and ω on a log scale. The quantity $20 \log_{10}|H(j\omega)|$ is called the *decibel,* or dB, gain of the frequency response $H(j\omega)$. See Figure 4.4 for the Bode plots for

[2]Logarithmic scales for both the magnitude and frequency scales were used extensively in the studies of Hendrik Wade Bode (1905–1984). Bode was a member of the technical staff at the Bell Telephone Laboratories and later joined the faculty at Harvard University. For further reading see *Network Analysis and Feedback Amplifier Design,* D. Van Nostrand Co., Princeton, N.J., 1945, or H. W. Bode, "Relations between Attenuation and Phase in Feedback Amplifier Design," *Bell Systems Technical Journal* 19, 421–454 (1940).

$H(s) = 10/(s^2 + 11s + 10)$; the script that invokes the MATLAB function **bode** is also given.

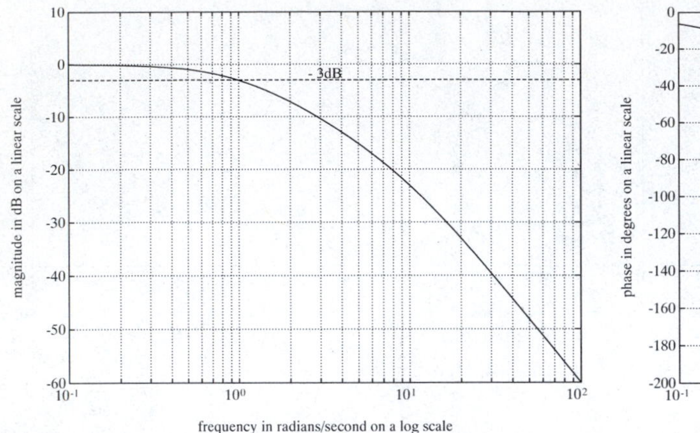

a. Bode plot magnitude

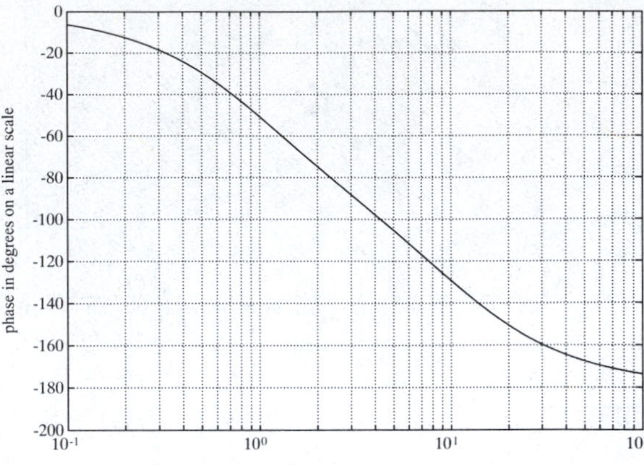

b. Bode plot phase

FIGURE 4.4 *Frequency response plots, semilog coordinates*

MATLAB Script

```
%F4_4 Semilog plots for Figure 4.4
num=[10];
den=[1,11,10];
w=logspace(-1,2);
[mag,phase]=bode(num,den,w);                     % call bode
mag=20*log10(mag);                               % change magnitude to dB
semilogx(w,phase),title('Fig.4.4 Bode plot');
%...labeling statements
semilogx(w,mag),title('Fig.4.4 Bode plot');
%...labeling statements
```

Graphical Estimation of Frequency Response The transfer function $H(s)$ of a continuous-time system can be described in terms of its zeros $n_1, n_2, \ldots, n_L$ and poles $d_1, d_2, \ldots, d_N$ as

H(s), factored form

$$H(s) = \frac{\pm k(s - n_1)(s - n_2)\cdots(s - n_L)}{(s - d_1)(s - d_2)\cdots(s - d_N)},$$

where k is a real constant. The frequency response $H(j\omega)$ is then

H(jω), factored form

exponential form

$$H(j\omega) = \frac{\pm k(j\omega - n_1)(j\omega - n_2)\cdots(j\omega - n_L)}{(j\omega - d_1)(j\omega - d_2)\cdots(j\omega - d_N)} = \frac{ke^{j(0 \text{ or } \pm\pi)}(N_1 e^{j\alpha_1})(N_2 e^{j\alpha_2})\cdots(N_L e^{j\alpha_L})}{(D_1 e^{j\beta_1})(D_2 e^{j\beta_2})\cdots(D_N e^{j\beta_N})}.$$

The N's, D's, α's, and β's can all be identified in the pole-zero plot of $H(s)$, as indicated in Figure 4.5, where $N_q e^{j\alpha_q}$ is a vector from the zero n_q to the tip of the $j\omega$ vector, $q = 1, 2, \ldots, L$, and $D_p e^{j\beta_p}$ is the vector from the pole d_p to the tip of the $j\omega$ vector, $p = 1, 2, \ldots, N$. It is customary to show the constant k on the plot, as in Figure 4.5. Remember that $s = j\omega$, where ω is the analog frequency of the input. Collecting the magnitude and phase components of the frequency response separately gives

$$\underbrace{M = \frac{kN_1 N_2 \cdots N_L}{D_1 D_2 \cdots D_N}}_{\text{magnitude}} \quad \text{and} \quad \underbrace{P = \{0 \quad \text{or} \quad \pm\pi\} + \sum_{q=1}^{L} \alpha_q - \sum_{p=1}^{N} \beta_p}_{\text{phase}}.$$

This graphical procedure for determining the frequency response is most useful as an estimation procedure, since a computer should always be used to obtain the final plot. Although possible, it is difficult to estimate phase information, so we will normally concern ourselves only with the magnitude plot.

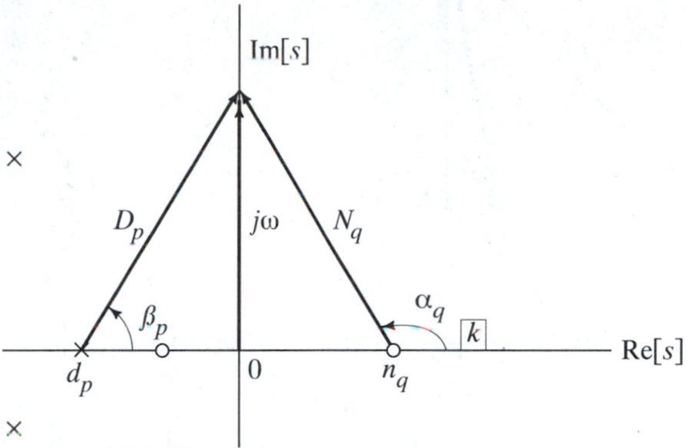

FIGURE 4.5 *Pole-zero plot for $H(s)$*

BASIC CONCEPTS

ILLUSTRATIVE PROBLEM 4.2 ***Estimating Filter Characteristics from a Zero-Pole Plot***

Two different filters are described by the zero-pole plots of Figure 4.6a and b (page 184). For each filter, estimate its type (i.e., lowpass, bandpass, highpass, or bandstop) and make a rough sketch of $|H(j\omega)|$ versus ω. What is the transfer function model? What is the DE model?

Solution

a. In Figure 4.6c $M = N/D$, and at any frequency N is always longer than D (they approach the same length as $\omega \to \infty$); thus this filter is a lowpass filter with a gain of 5 at $\omega = 0$ and a gain of unity as $\omega \to \infty$. The transfer function is

$$H(s) = \frac{Y(s)}{X(s)} = \frac{s + 5}{s + 1}$$

and the DE is $\dot{y}(t) + y(t) = \dot{x}(t) + 5x(t)$.

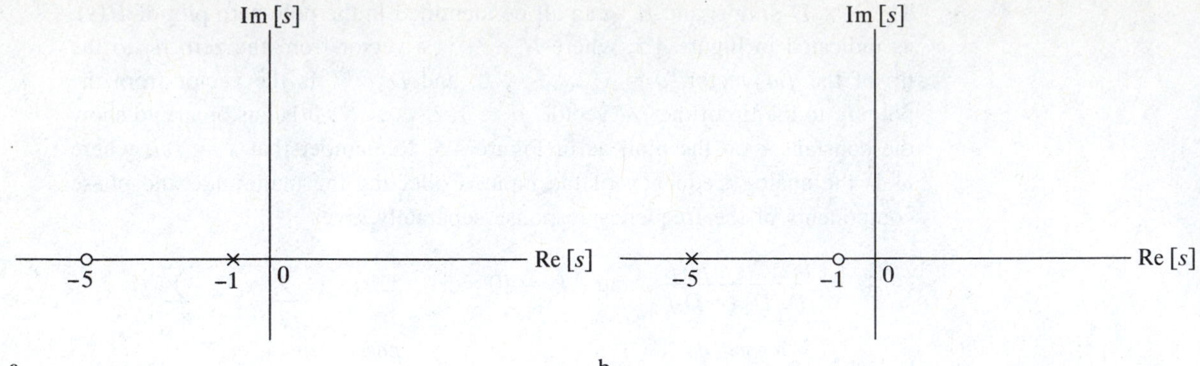

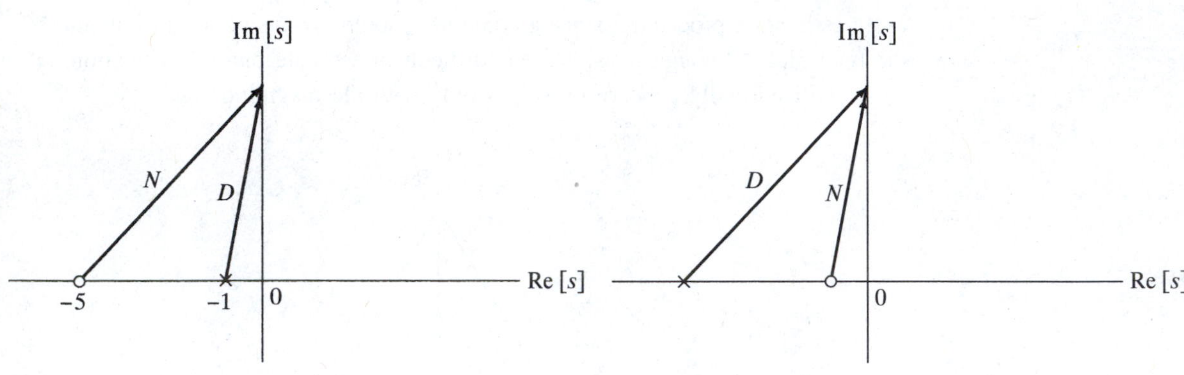

FIGURE 4.6 *Pole-zero plots and typical vectors used to estimate frequency responses*

b. From Figure 4.6d we see that N is always less than D, so the filter is highpass. The transfer function is

$$H(s) = \frac{Y(s)}{X(s)} = \frac{s+1}{s+5}$$

and the DE is $\dot{y}(t) + 5y(t) = \dot{x}(t) + x(t)$. ■

Asymptotic Approximation of $H(j\omega)$ Given a transfer function whose zeros and poles are real, such as

$$H(s) = \frac{k(s-n_1)(s-n_2)\cdots}{(s-d_1)(s-d_2)\cdots} = \frac{k(s+1/\tau_1)(s+1/\tau_2)\cdots}{(s+1/\tau_a)(s+1/\tau_b)\cdots}$$

the frequency response function can be written as

$$H(j\omega) = \frac{K(1+j\omega\tau_1)(1+j\omega\tau_2)\cdots}{(1+j\omega\tau_a)(1+j\omega\tau_b)\cdots},$$

where $K = k\tau_a\tau_b \ldots /\tau_1\tau_2 \ldots$. It follows that $H(j\omega)$ can be found at any frequency ω from *magnitude*

$$M = |H(j\omega)| = \frac{|K|\sqrt{1+\omega^2\tau_1{}^2}\sqrt{1+\omega^2\tau_2{}^2}\cdots}{\sqrt{1+\omega^2\tau_a{}^2}\sqrt{1+\omega^2\tau_b{}^2}\cdots}$$

and

phase

$$\begin{aligned} P = \angle H(j\omega) &= \{0 \quad \text{or} \quad \pm\pi\} + \tan^{-1}\omega\tau_1 + \tan^{-1}\omega\tau_2 \\ &\quad + \cdots - \{\tan^{-1}\omega\tau_a + \tan^{-1}\omega\tau_b + \cdots\} \\ &= \{0 \quad \text{or} \quad \pm\pi\} + \alpha_1 + \alpha_2 + \cdots - \{\beta_1 + \beta_2 + \cdots\} \end{aligned}$$

where the first term in the phase expression is 0 if $K > 0$ and $\pm\pi$ if $K < 0$.

Magnitude Let us consider more closely the magnitude of a typical numerator term in M, where we can write

$$N_i = |1 + j\omega\tau| = \sqrt{1+\omega^2\tau^2} = \begin{cases} 1 & \text{for } \omega^2\tau^2 << 1 \quad \text{or} \quad \omega << \frac{1}{\tau} \\ \omega\tau & \text{for } \omega^2\tau^2 >> 1 \quad \text{or} \quad \omega >> \frac{1}{\tau}. \end{cases}$$

Thus, the magnitude approaches the asymptote of 1 at low frequencies and the asymptote of $\omega\tau$ at high frequencies; the two asymptotes intersect at $\omega_b = 1/\tau$, which is known as the *break, or corner, frequency.* It is common practice to use decibels for these plots; in this situation, we have

dB approximation

$$\begin{aligned} N_{i_{dB}} &= 20\log_{10}|1 + j\omega\tau| = 20\log_{10}\sqrt{1+\omega^2\tau^2} \\ &= \begin{cases} 0 \text{ dB} & \text{for } \omega^2\tau^2 << 1 \quad \text{or} \quad \omega << \frac{1}{\tau} \\ 20\log_{10}\omega\tau \text{ dB} & \text{for } \omega^2\tau^2 >> 1 \quad \text{or} \quad \omega >> \frac{1}{\tau}. \end{cases} \end{aligned}$$

The exact plot of N_i in decibels versus ω on a log scale and its asymptotic approximation are shown in Figure 4.7 for $\tau = 1$. This plot will simply shift left or right, depending on the value of τ. Notice that the slope of the high-frequency asymptote is 20 dB per decade (a decade being a frequency ratio of 10:1) or 6 dB per octave (an octave being a frequency ratio of 2:1) and that

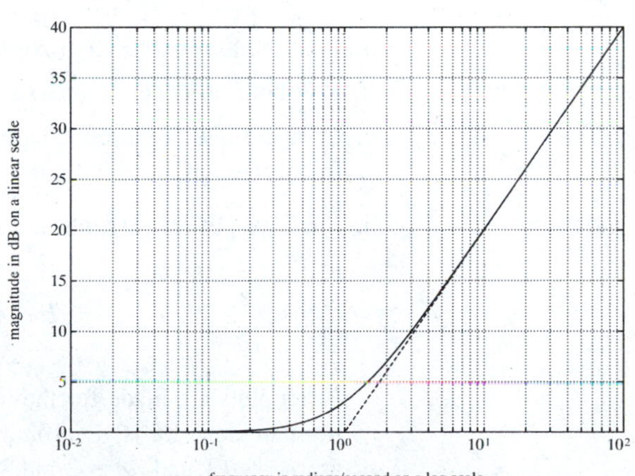

FIGURE 4.7 *Magnitude plot with asymptotes for a real zero*

BASIC CONCEPTS

the asymptotic approximation is reasonably good for values of ω remote from $\omega_b = 1/\tau$. Also observe that the actual plot is 3 dB above the asymptotic plot at the corner frequency.

There is one special case: The term $|j\omega|$ graphs as a straight line of slope 20 dB/decade through the point $\omega = 1$, $M_{\text{dB}} = 0$. And if any term appears with an exponent of R, all the corresponding slopes and magnitudes must be multiplied by R. For a denominator term—for example, $R = -1$—we see that $D_i = |1/(1 + j\omega\tau_i)|$ breaks *downward* at the break or corner frequency of $\omega_b = 1/\tau$ with an asymptotic slope of -20 dB/decade.

With the magnitude of the frequency response given as

$$M = \frac{KN_1 N_2 \cdots}{D_1 D_2 \cdots}$$

the decibel representation is

$$M_{\text{dB}} = 20 \log_{10}|K| + 20 \log_{10} N_1 + 20 \log_{10} N_2 + \cdots - (20 \log_{10} D_1 + 20 \log_{10} D_2 + \cdots).$$

The important result of this logarithmic operation is that the individual asymptotic plots for $N_1, N_2, \ldots, D_1, D_2, \ldots$ can be plotted *separately* and then *added* graphically to obtain the asymptotic approximation to $|H(j\omega)|_{\text{dB}}$.

CHAPTER 4

ILLUSTRATIVE PROBLEM 4.3
Asymptotic Plot of $|H(j\omega)|$

The transfer function of an analog filter is

$$H(s) = \frac{4(s + 0.1)}{s(s + 2)}.$$

a. Find the asymptotic approximation to $|H(j\omega)|$.

b. Use **bode** to generate an exact plot for $|H(j\omega)|$.

c. Compare the asymptotic approximation and the exact plots at the radian frequencies of $\omega = 0.01$, 0.1, 1, and 10 rad/s.

Solution

a. First, we put $H(j\omega)$ into the standard form for the asymptotic plots as

$$H(j\omega) = \frac{0.2(1 + 10j\omega)}{j\omega(1 + 0.5j\omega)}.$$

Exact plots are made for the terms 0.2 and $1/|j\omega|$ and asymptotic approximations are done for $|1 + j10\omega|$ and $1/|1 + j0.5\omega|$, where the break frequencies are $\omega = 0.1$ and $\omega = 2$ rad/s, respectively. The four separate plots and their graphical sum are shown in Figure 4.8a–e.

b. The Bode magnitude plot that was generated from **bode** is shown in Figure 4.9.

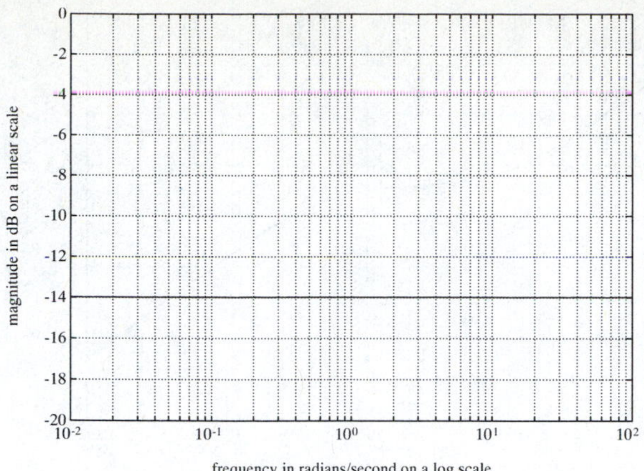

a. Exact plot for the factor 0.2

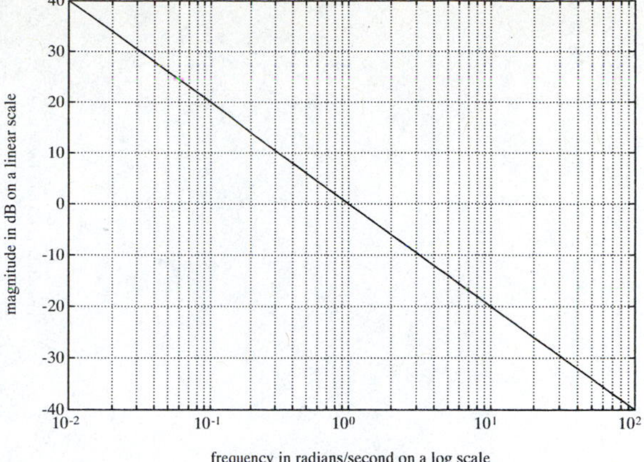

b. Exact plot for the factor $1/|j\omega|$

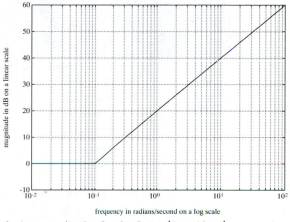

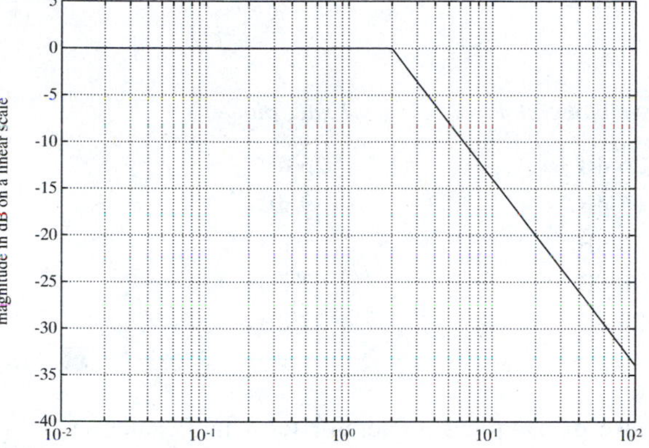

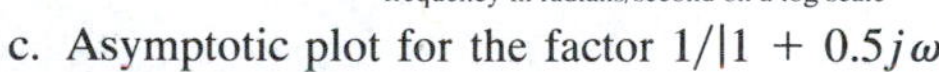

c. Asymptotic plot for the factor $1/|1 + 0.5j\omega|$

60
50
40
30
20
10
0
-10

magnitude in dB on a linear scale

10^{-2} 10^{-1} 10^{0} 10^{1} 10^{2}

frequency in radians/second on a log scale

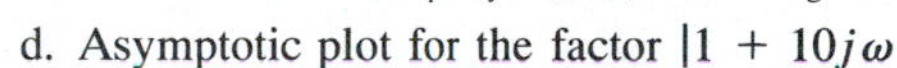

d. Asymptotic plot for the factor $|1 + 10j\omega|$

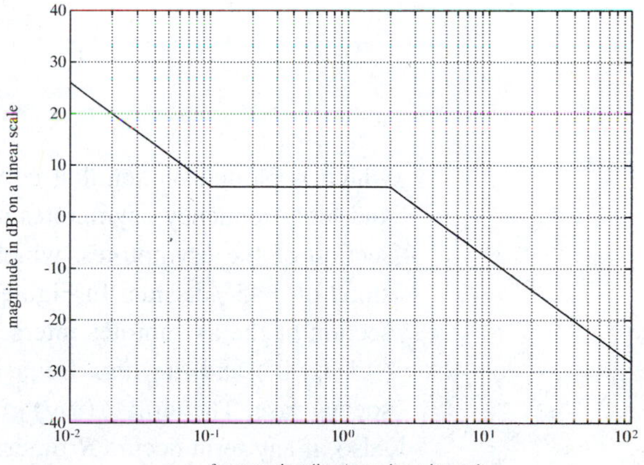

e. Asymptotic plot for $|H(j\omega)|$

FIGURE 4.8 *Exact and asymptotic plots for various pole-zero factors, and overall asymptotic magnitude plot*

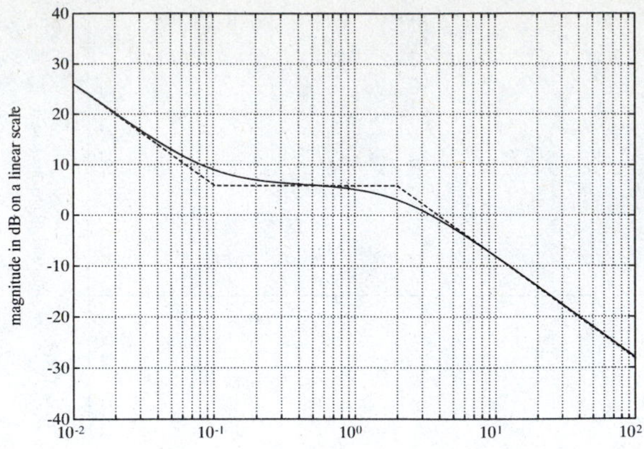

FIGURE 4.9 *Asymptotic and exact magnitude plots for $H(j\omega)$*

c.

Frequency	*Asymptotic plot*	*Exact plot*
$\omega = 0.01$ rad/s	26 dB	26 dB
$\omega = 0.1$ rad/s	6 dB	9 dB
$\omega = 1.0$ rad/s	6 dB	5 dB
$\omega = 2.0$ rad/s	6 dB	3 dB
$\omega = 10$ rad/s	−28 dB	−28 dB

Phase Turning now to the phase of a typical numerator term in P, such as α_1, we can write

$$\alpha_1 = \angle(1 + j\omega\tau) = \tan^{-1}\omega\tau = \begin{cases} 0° & \text{for } \omega\tau << 1 \quad \text{or} \quad \omega << \frac{1}{\tau} \\ 90° & \text{for } \omega\tau >> 1 \quad \text{or} \quad \omega >> \frac{1}{\tau}, \end{cases}$$

which is plotted on semilog coordinates in Figure 4.10 for $\tau = 1$. The low- and high-frequency asymptotes of 0° and 90° are shown as well as the center section of the asymptotes, which passes through the point $(\omega_b, 45°)$ with a slope of +45°/decade. In Figure 4.10 $\omega_b = \tau = 1$, but in general the center section of the asymptotes intersects the low-frequency asymptote at the point $(0.1\omega_b, 0°)$ and the high-frequency asymptote at $(10\omega_b, 90°)$. There is one special case: The term $\angle(j\omega)$ graphs as a constant of +90° at all frequencies. Also, if any term occurs R times, all the corresponding slopes and magnitudes must be multiplied by R. Setting $R = -1$, we see that the center section of the asymptotic plot for $\angle[1/(1 + j\omega\tau)]$ passes through the point $(\omega_b, -45°)$ with a slope of −45°/decade and with the low- and high-frequency asymptotes being 0° and −90°, respectively.

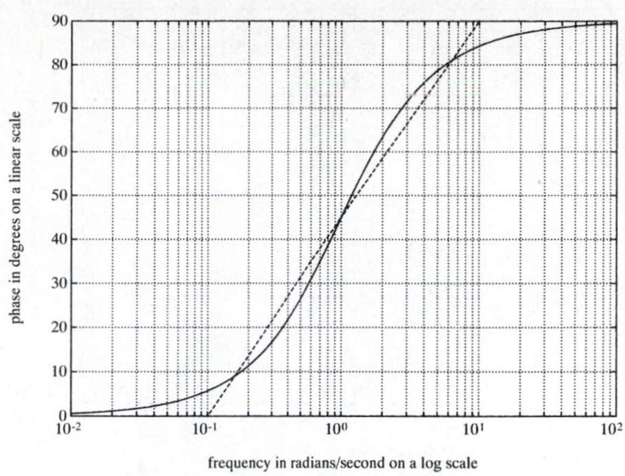

FIGURE 4.10 *Phase plot for the term* $(1 + j\omega)$

ILLUSTRATIVE PROBLEM 4.4

Asymptotic Plot of $\angle H(j\omega)$

Consider the transfer function

$$H(s) = \frac{4(s + 0.1)}{s(s + 2)}.$$

a. Find the asymptotic approximation to $\angle H(j\omega)$.

b. Use **bode** to generate an exact plot for $\angle H(j\omega)$.

c. Compare the asymptotic approximation and the exact plots at the radian frequencies of $\omega = 0.01$, 0.1, 1, and 10 rad/s.

Solution

a. First, we put $H(j\omega)$ into the standard form for the asymptotic plots:

$$H(j\omega) = \frac{0.2(1 + 10j\omega)}{j\omega(1 + 0.5j\omega)}.$$

An exact plot of $-90°$ is made for the phase of $1/j\omega$, and asymptotic approximations are made for $\angle(1 + j10\omega)$ and $\angle[1/(1 + j0.5\omega)]$, where the break frequencies are $\omega = 0.1$ and $\omega = 2$ rad/s, respectively. The three separate plots and their graphical sum are shown in Figure 4.11a–d (page 190).

b. The Bode plot (phase only) that was generated from **bode** is shown in Figure 4.11e, along with the asymptotic plot from part (a).

c.

Frequency	*Asymptotic plot*	*Exact plot*
$\omega = 0.01$ rad/s	$-90°$	$-84°$
$\omega = 0.1$ rad/s	$-45°$	$-48°$
$\omega = 1.0$ rad/s	$-31°$	$-32°$
$\omega = 10$ rad/s	$-77°$	$-80°$

BASIC CONCEPTS

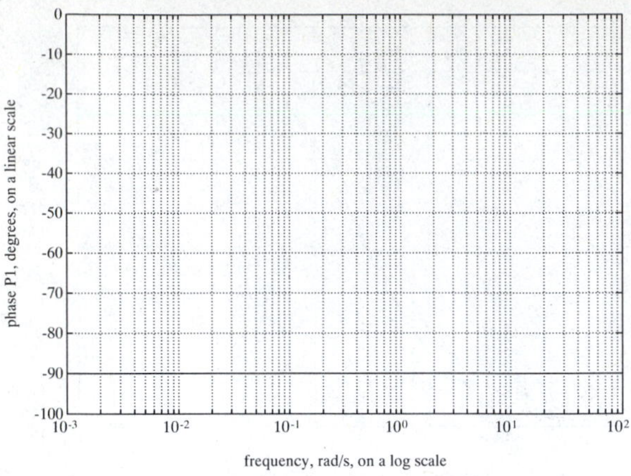

a. Exact phase due to the factor $1/j\omega$

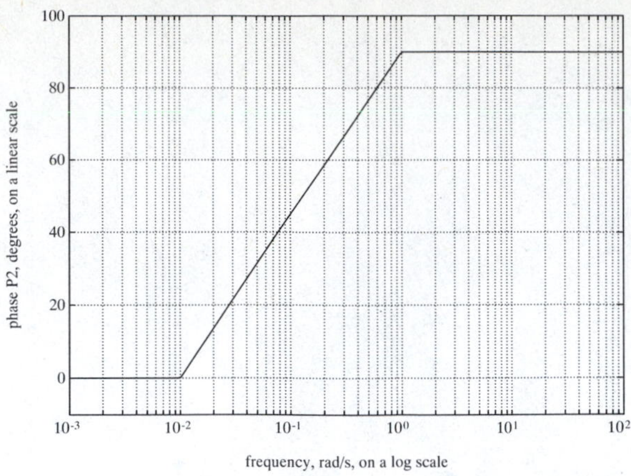

b. Asymptotic phase due to the factor $(1 + 10j\omega)$

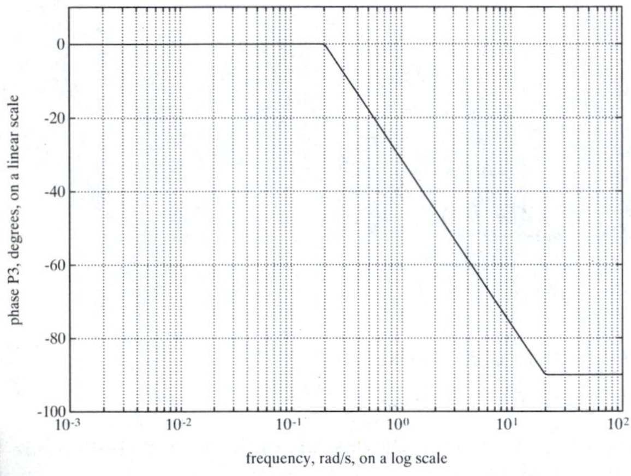

c. Asymptotic phase due to the factor $1/(1 + 0.5j\omega)$

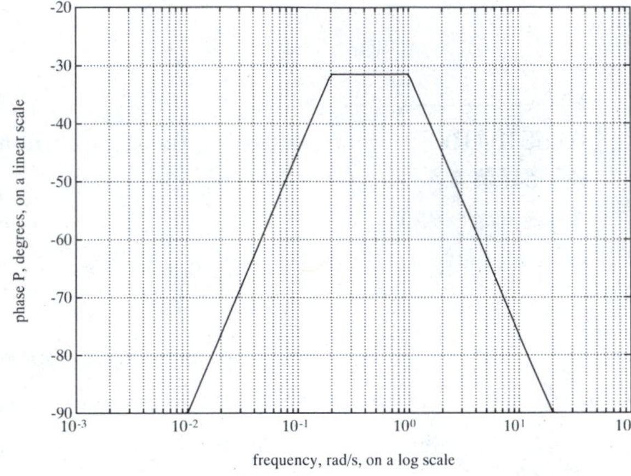

d. Overall asymptotic phase plot

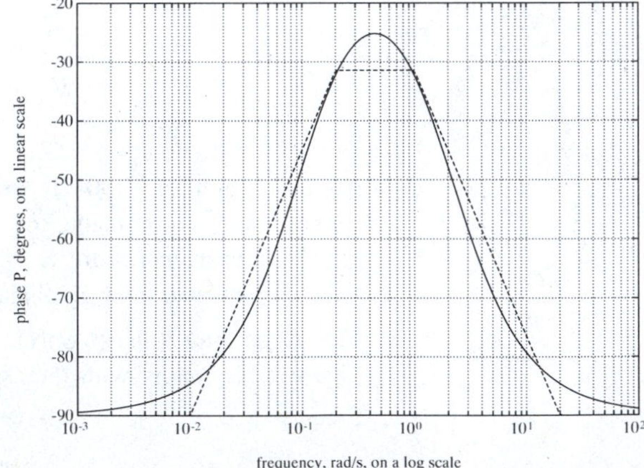

e. Bode phase plot and asymptotic phase plot

FIGURE 4.11 *Exact and asymptotic phase plots for various pole-zero factors, and the overall exact and asymptotic phase plots*

EXAMPLE 4.1
Frequency Response of a Simple Electric Circuit

Given the diagram for the lowpass filter shown in Figure E4.1a:

a. Find the transfer function $H(s) = V_o(s)/V_i(s)$.

b. Determine the frequency response function $H(j\omega)$.

c. Sketch the rectangular plot for this function using general values of R and C, labeling pertinent values.

d. Plot this frequency response (rectangular plot) for $R = 1\ \Omega$ and $C = 1$ F.

WHAT IF? Try some other combinations of R and C that are somewhat more realistic, such as $R = 1\ \text{M}\Omega$ with $C = 1$ pF. ■

Solution

a. *System transfer function from the model.* We need to find the Laplace transform of the output-input voltage ratio for the circuit of Figure E4.1a. Writing Kirchhoff's current law at the output gives

$$i_C(t) + i_R(t) = 0 \text{ and, consequently, } C\dot{v}_o(t) + \frac{1}{R}[v_o(t) - v_i(t)] = 0.$$

Taking the Laplace transform, setting the initial condition to zero, and solving for $V_o(s)/V_i(s)$ results in the transfer function

$$H(s) = \frac{V_o(s)}{V_i(s)} = \frac{1/RC}{s + 1/RC}.$$

b. *Evaluate the transfer function for all values of the frequency ω.* First, this is a stable system because the transfer function pole is at $s = -1/RC$, definitely negative real. Substituting $j\omega$ for s in $H(s)$ gives the frequency response

$$H(j\omega) = \frac{1/RC}{j\omega + 1/RC} = \frac{1/RC}{\sqrt{\omega^2 + 1/R^2C^2}} e^{-j\tan^{-1}(\omega/RC)}.$$

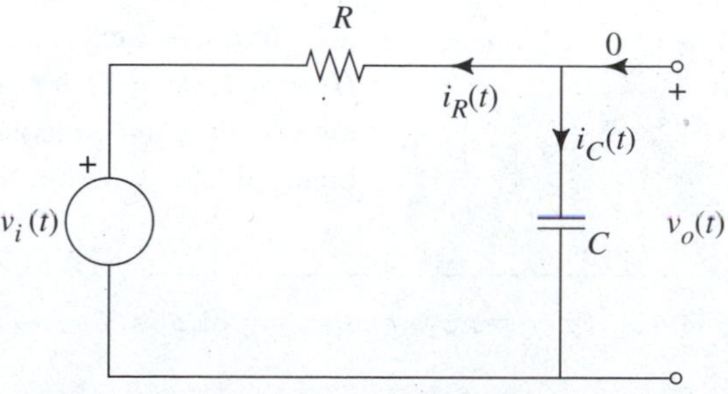

FIGURE E4.1a *Circuit model for a lowpass filter*

c. *Sketch the frequency response function.* With

$$M = \frac{1/RC}{\sqrt{\omega^2 + 1/R^2C^2}} \quad \text{and} \quad P = \tan^{-1}(\omega/RC)$$

we can quickly determine values for $\omega = 0$ ($M = 1$ and $P = 0°$) and for $\omega \to \infty$ ($M \to 0$ and $P \to -90°$). Another easy calculation is for $\omega = 1/RC$, where $M = 0.707$ and $P = -45°$. The frequency at which the magnitude of the frequency response is 0.707 times the maximum value is often called the *cutoff frequency* of the filter. In a Bode or dB plot, the gain at this "figure of merit frequency" is -3 dB. See Figure E4.1b for a sketch of the frequency response for this circuit, which passes lower frequencies relatively well while attenuating (blocking) higher frequencies. Consequently, this filter is known as a *lowpass* filter.

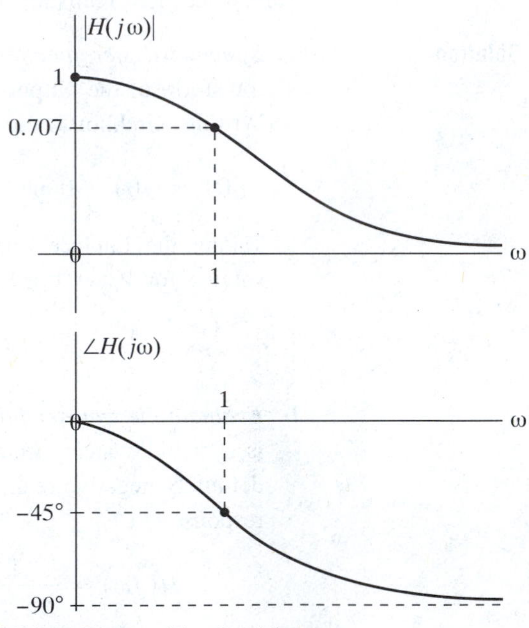

FIGURE E4.1b *Sketch of frequency response*

CHAPTER 4

d. *Plot the frequency response.* To plot this with MATLAB we use the function **freqs,** which requires that the transfer function be in the form $H(s) = 1/(s + 1)$ for $R = 1\ \Omega$, $C = 1$ F. The script in file E4_1c follows with some annotations and explanations. The plots are given in Figure E4.1c, where a dB plot is also included.

MATLAB Script

```
%E4_1 Frequency response plots using freqs

%E4_1c Frequency response plots for lowpass filter
```

```
a=[1,1];                  % denominator coefficients in descending order
b=[1];                    % numerator coefficient
w=0:.01:10;               % frequency range of 0 to 10 rad/s in increments
                          % of 0.01 rad/s
H=freqs(b,a,w);           % call freqs
mag=abs(H);
%...plotting statements
mag=20*log10(mag);        % change magnitude to magnitude in dB
%...plotting statements
phase=angle(H);
phase=phase*180/pi;       % change phase from radians to degrees
%...plotting statements and pause
```

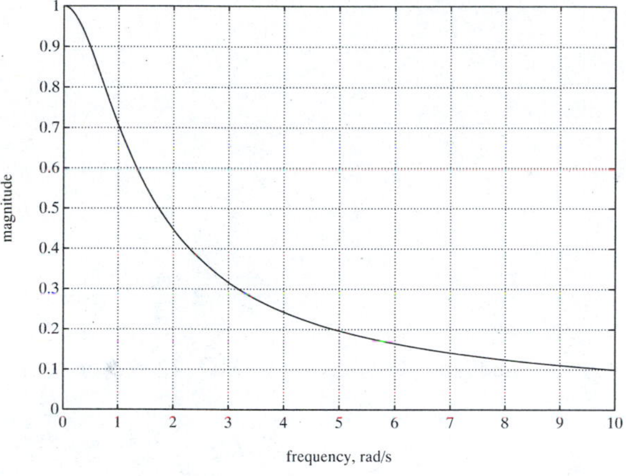

Lowpass filter magnitude

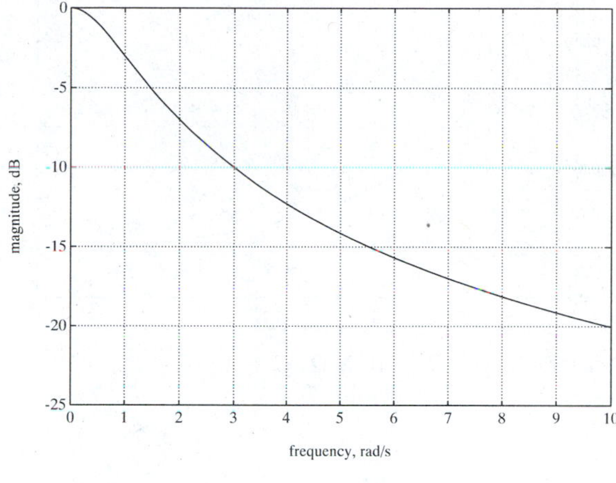

Lowpass filter magnitude in dB

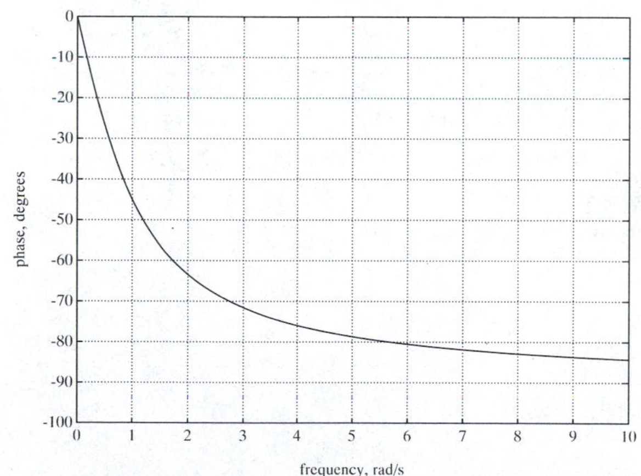

Lowpass filter phase

FIGURE E4.1c *Frequency response plots for low-pass filter*

WHAT IF? For the more practical values of $R = 1\ \text{M}\Omega = 10^6\ \Omega$ and $C = 1\ \text{pF} = 10^{-12}$ F, we need to change the vectors **a**, **b**, and **w** to **a** = [1, 1E6], **b** = [1E6], and **w** = 0:1E4:1E7. When the frequency response of this new filter is plotted in Figure E4.1d, we notice that the cutoff frequency is now $\omega = 10^6$ rad/s (gain = 0.707 times the maximum value) and the phase at this frequency is still $-45°$. You should try some combinations where $RC \neq 1$. ■

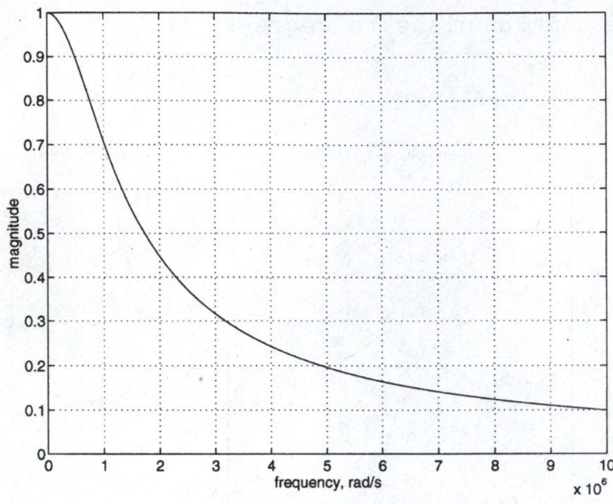

Magnitude with $R = 10\ \text{M}\Omega$, $C = 1$ pF

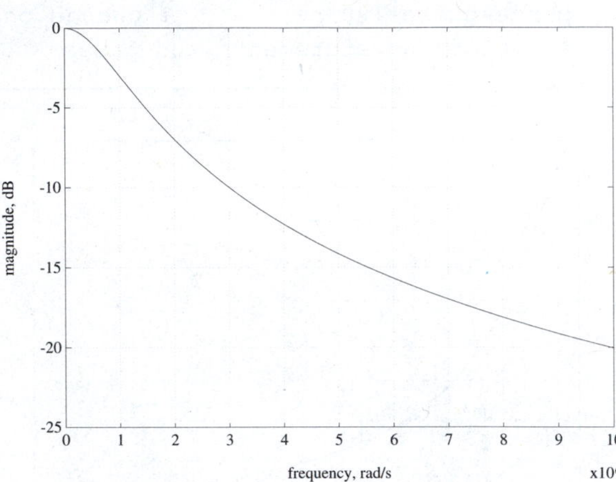

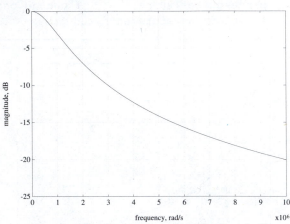

Magnitude in dB with $R = 10\ \text{M}\Omega$, $C = 1$ pF

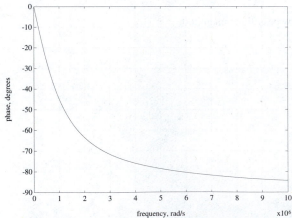

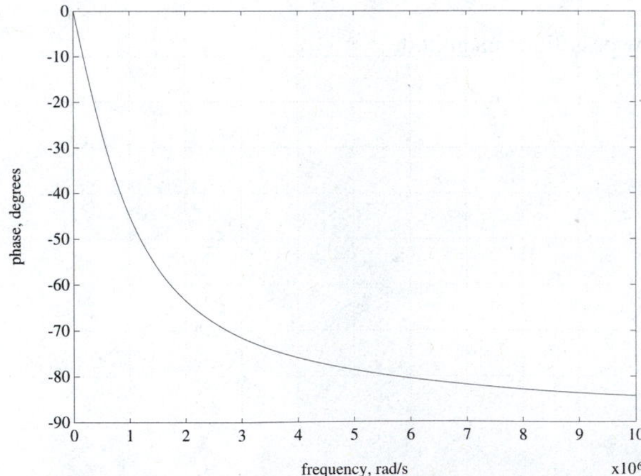

Phase with $R = 10\ \text{M}\Omega$, $C = 1$ pF

FIGURE E4.1d *Frequency response plots for a different lowpass filter*

CHAPTER 4

EXAMPLE 4.2
Finding H(s) from Frequency Response Data

A frequency response test is performed on a system that is assumed to be linear and time invariant. The resulting magnitude data are plotted on a dB (linear scale) versus ω (log scale), as shown in Figure E4.2a. The system transfer function is assumed to have simple poles and is of the form

$$H(s) = \frac{k(s - n_1)(s - n_2)\cdots}{(s - d_1)(s - d_2)\cdots} = \frac{k(s + 1/\tau_1)(s + 1/\tau_2)\cdots}{(s + 1/\tau_a)(s + 1/\tau_b)\cdots}.$$

Find an approximate expression for this transfer function.

WHAT IF? Use **freqs** to plot the phase for the transfer function you determined, and sketch the asymptotic phase plot. Can you find another transfer function that will have the same magnitude plot as given in Figure E4.2a? If there is one, sketch its asymptotic phase plot. ■

Solution

The low-frequency slope is -20 dB/decade and the high-frequency slope is also -20 dB/decade. The low-frequency slope tells us that there is a "free" s in the denominator and the high-frequency slope indicates a pole-zero excess of 1. The leveling off of the curve in its midrange indicates the presence of a zero, with a break frequency at approximately $\omega = 0.2$. Thus we have

$$H(s) = \frac{k(s - n_1)}{s(s - d_1)} = \frac{k(s + 1/\tau_1)}{s(s + 1/\tau_a)} = \frac{K(s\tau_1 + 1)}{s(s\tau_a + 1)}$$

where $K = k\tau_a/\tau_1$; substituting $s = j\omega$ gives the standard frequency response form

$$H(j\omega) = \frac{K(j\omega\tau_1 + 1)}{j\omega(j\omega\tau_a + 1)}.$$

Sketching the asymptotes, as in Figure E4.2b, we find slopes of -20, 0, and

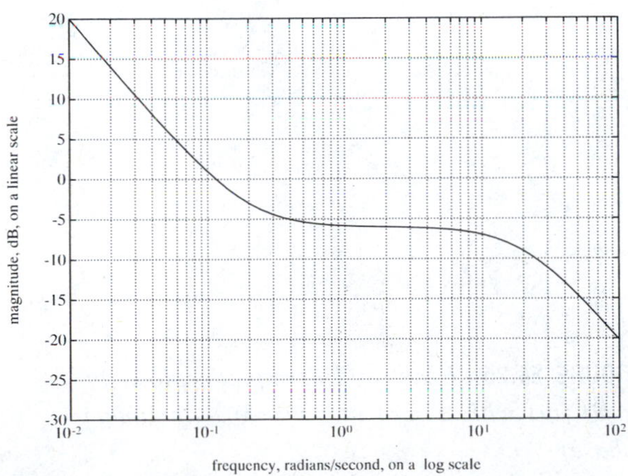

FIGURE E4.2a *Magnitude plot for unknown transfer function*

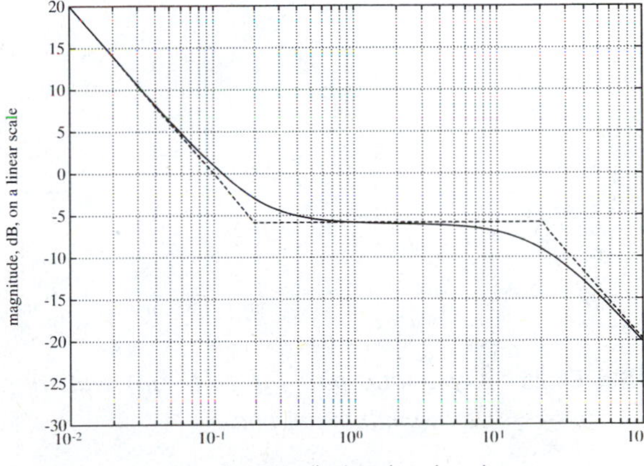

FIGURE E4.2b *Magnitude plot with asymptotes*

EXAMPLES

−20 from left to right. The asymptotes appear to intersect at about $\omega = 0.2$ rad/s and $\omega = 20$ rad/s, making $n_1 = -0.2$ or $\tau_1 = 5.0$ and $d_1 = -20$ or $\tau_a = 0.05$. Also notice that at $\omega = 0.2$, the actual plot is 3 dB above the asymptotic plot. Similarly, at $\omega = 20$ the actual plot is 3 dB below the asymptotic plot. At the low frequency of $\omega = 10^{-2}$, the gain is 20 dB, making $K = 0.10$. The estimated transfer function is

$$H(s) = \frac{10(s + 0.2)}{s(s + 20)}.$$

CROSS-CHECK At $\omega = 10^{-2}$ we have

$$H(j0.01) = \frac{10(j0.01 + 0.2)}{j0.01(j0.01 + 20)} \approx \frac{10(0.2)}{j0.01(20)} = 10e^{-j\pi/2}$$

and a magnitude of 20 dB. This seems to be OK. At $\omega = 10^2$ we have

$$H(j100) = \frac{10(j100 + 0.2)}{j100(j100 + 20)} \approx \frac{10(j100)}{j100(j100)} = 0.1e^{-j\pi/2}$$

and a magnitude of −20 dB, which also appears to be acceptable. ■

Comment: The ad hoc procedure works well here because there are only a few poles and zeros, their corner frequencies are well separated, and the poles and zeros are real. If these assumptions are not all true, such a procedure becomes much more difficult to use effectively. There are, however, other approaches, known as system identification procedures, that can be successfully applied. See the MATLAB System Identification Toolbox.

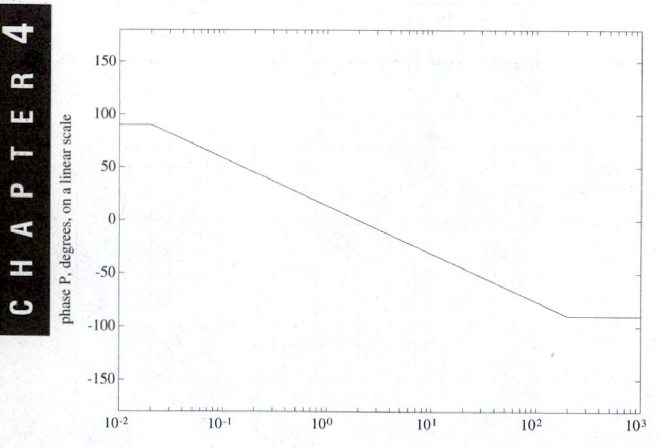

FIGURE E4.2c *Asymptotic plot for $\angle H(j\omega)$ with a zero at $s = +0.2$ (nonminimum phase)*

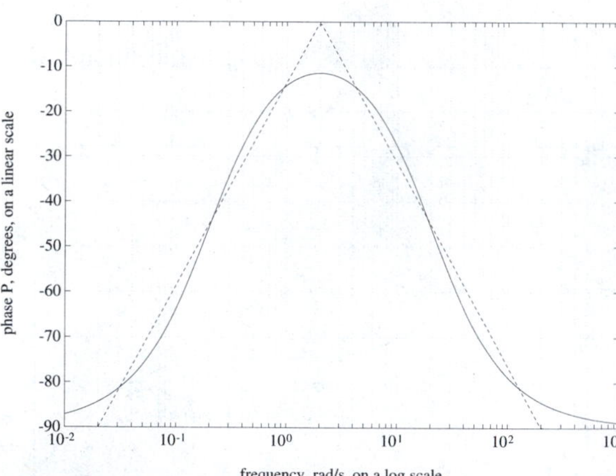

FIGURE E4.2d *Exact and asymptotic plots for $\angle H(j\omega)$ with a zero at $s = -0.2$ (minimum phase)*

WHAT IF? By placing the zero at $s = +0.2$ rather than at $s = -0.2$, we obtain the same magnitude plot as before, but with the asymptotic phase plot shown in Figure E4.2c where we observe phase shifts over the frequency range shown of between $+90°$ and $-90°$. In contrast, Figure E4.2d, drawn for the zero at $s = -0.2$, includes phase shifts in the range $-90°$ to $0°$. Thus, $H(s) = 10(s + 0.2)/(s(s + 20))$ is called a minimum phase filter and $H(s) = 10(s - 0.2)/(s(s + 20))$ is called a nonminimum phase filter. ■

EXAMPLE 4.3 *Graphical Estimation of* $|H(j\omega)|$

a. Given the transfer function models of several analog filters, make a pole-zero plot and use the graphical method to estimate the frequency response (magnitude only) for each of the following.

$$\text{(i) } H(s) = \frac{s+1}{s+5};\ \text{(ii) } H(s) = \frac{s-1}{s+5};\ \text{(iii) } H(s) = \frac{10s}{s^2 + 2s + 100};$$

$$\text{(iv) } H(s) = \frac{s^2 + 100}{s^2 + 2s + 100};\ \text{(v) } H(s) = \frac{0.306s^2 + 0.538}{s^3 + 1.026s^2 + 1.047s + 0.538}.$$

b. Verify your results in part (a) by using MATLAB.

Solution

a. *Graphical pole-zero estimates.*

(i) $H(s) = (s + 1)/(s + 5)$ has a zero at $s = -1$ and a pole at $s = -5$. From Figure E4.3a.1 the ratio of N/D will be 0.2 at $\omega = 0$ (DC) and will approach 1 as $\omega \to \infty$, a highpass filter.

(ii) The ratio N/D will follow exactly the same pattern as in part (a), as shown in Figure 4.3a.2; consequently, the magnitude plot will also be the same. This filter is a highpass filter with different phase characteristics.

(iii) $H(s) = 10s/(s^2 + 2s + 100)$ has a zero at $s = 0$ and poles at $s_{1,2} = -1 \pm j9.95$. The length of the zero vector N and the length of the pole vector D_2 will be more or less the same over most frequencies, with the length D_1 dominating the situation. D_1 will be very short near $\omega = 10$ rad/s, as seen in Figure 4.3a.3, and the filter is bandpass.

(iv) Zeros at $s_{1,2} = \pm j10 = 10e^{\pm j1.57}$ and poles at $s_{1,2} = -1 \pm j9.95 =$

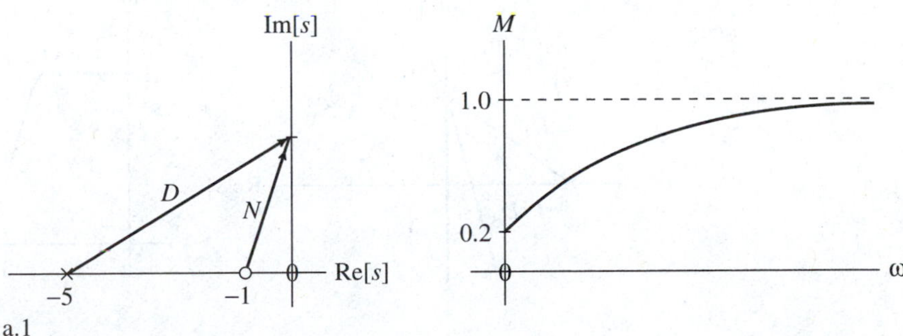

FIGURE E4.3a *Estimation of frequency response*

Continues

EXAMPLES

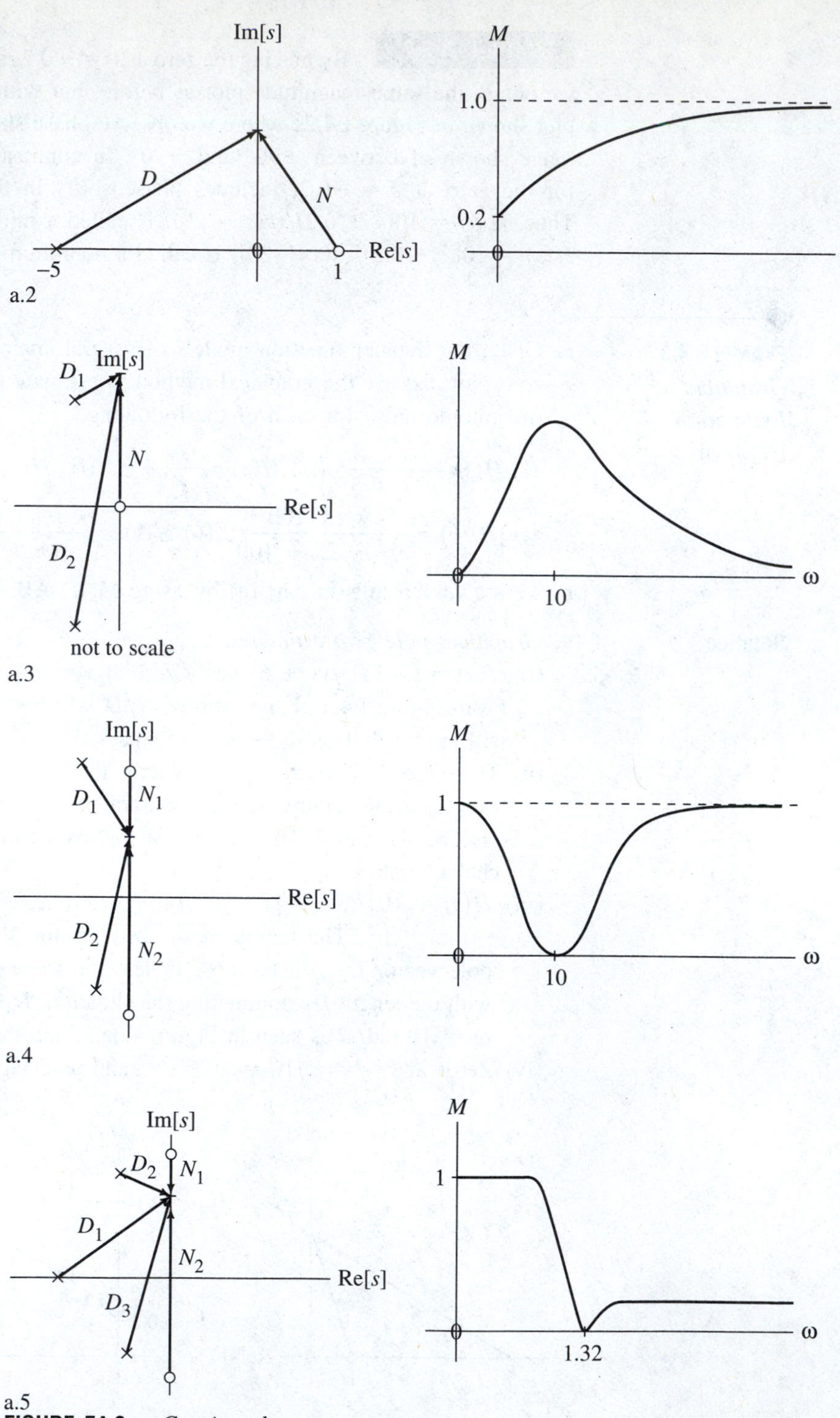

FIGURE E4.3a *Continued*

CHAPTER 4

$10e^{\pm j1.67}$. The filter gain is zero at $\omega = 10$; because of the close proximity of the poles and zeros in Figure E4.3a.4, the gain will be relatively constant at 1 for all other frequencies. This is a bandstop (or bandreject) filter.

(v) Using **roots** we find the zeros at $s_{1,2} = \pm j1.326$ and the poles at $s_1 = -0.667$, $s_{2,3} = -0.180 \pm j0.880$, as plotted in Figure E4.3a.5. The zeros drive the filter gain to 0 at $\omega = 1.326$ and the pole-zero excess of 1 makes the gain 0 as $\omega \to \infty$. The gain is 1 at $\omega = 0$ ($s = j0$), but not a great deal more can be determined except that the filter is lowpass. This happens to be the transfer function of an elliptic filter.

b. *MATLAB-generated plots.* The magnitude plots using **freqs** are given in Figure E4.3b.1–5 (page 200).

MATLAB Script

```
%E4_3 Filter frequency response plots

%E4_3b_1 Highpass filter
a=[1,5];                    % denominator coefficients
b=[1,1];                    % numerator coefficients
w=0:.1:50;                  % initial frequency:increment in rad/s:final
                            % frequency
H=freqs(b,a,w);             % call freqs
mag=abs(H);
%...plotting statements and pause

%E4_3b_2 Another highpass filter
a=[1,5];
b=[1,-1];
w=0:.1:50;
%...call freqs, plotting statements, and pause

%E4_3b_3 Bandpass filter
a=[1,2,100];
b=[10,0];
w=0:.1:50;
%...call freqs, plotting statements, and pause

%E4_3b_4 Bandstop filter
a=[1,2,100];
b=[1,0,100];
w=0:.1:50;
%...call freqs, plotting statements, and pause

%E4_3b_5 Elliptic lowpass filter
a=[1,1.026452,1.047986,0.539386];
b=[0.306214,0,0.537834];
w=0:.005:4;
%...call freqs and plotting statements
```

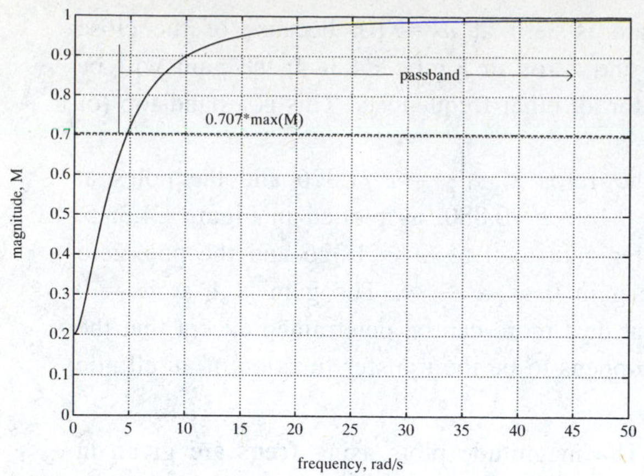

FIGURE E4.3b.1 *Highpass filter magnitude plot*

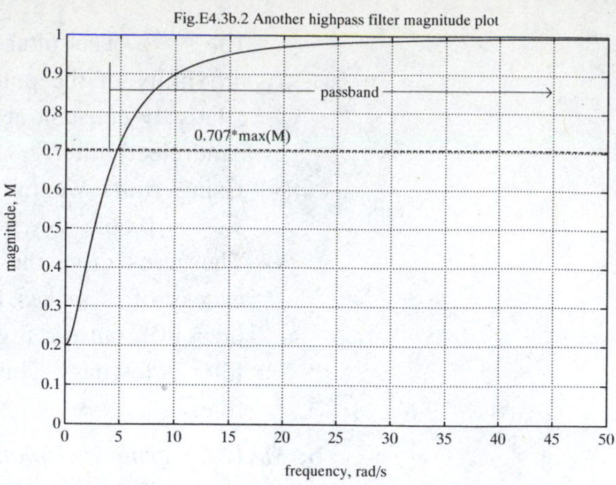

FIGURE E4.3b.2 *Another highpass filter magnitude plot*

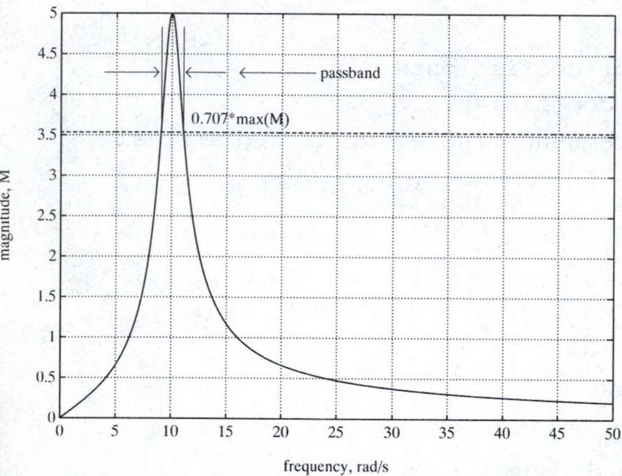

FIGURE E4.3b.3 *Bandpass filter magnitude plot*

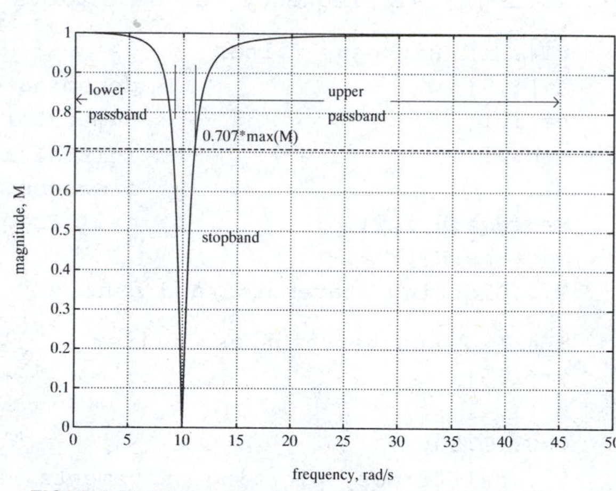

FIGURE E4.3b.4 *Bandstop filter magnitude plot*

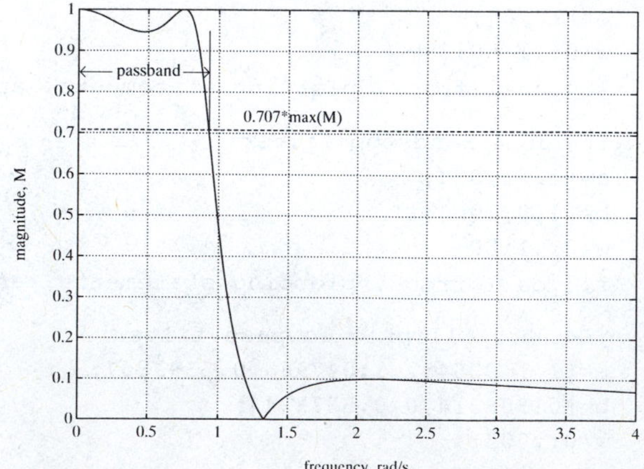

FIGURE E4.3b.5 *Elliptic lowpass filter magnitude plot*

CHAPTER 4

EXAMPLE 4.4

Active Circuits for Bandpass and Bandstop Filtering

Bandpass filters pass a band of frequencies, while attenuating those outside the band. The simplest such filter is second order with the transfer function

$$H(s) = \frac{Ks}{s^2 + Bs + \omega_0{}^2}$$

where ω_0 is called the resonant frequency in rad/s, and B, also in rad/s, is called the 3-dB bandwidth. The circuit's quality factor, $Q = \omega_0/B$, determines the height and sharpness of the resonant peak and the transfer function can be written in terms of Q as

bandpass filter in terms of Q

$$H(s) = \frac{Ks}{s^2 + (\omega_0/Q)s + \omega_0{}^2}.$$

Now consider the operational amplifier circuit of Figure E4.4a.

a. Write Kirchhoff's current law at nodes a and b, make the usual op-amp assumptions, and show that for zero initial conditions the following matrix equation is valid.

$$\begin{bmatrix} 1/R_1 + (C_1 + C_2)s & -C_2 s \\ C_1 s & 1/R_2 \end{bmatrix} \begin{bmatrix} V_b(s) \\ V(s) \end{bmatrix} = \begin{bmatrix} E(s)/R_1 \\ 0 \end{bmatrix}.$$

b. For $C_1 = C_2 = 1$, $R_1 = 1/(2Q)$ and $R_2 = 2Q$, find the resonant frequency ω_0, the bandwidth B, and the maximum value of $|H(j\omega)|$.

c. We want to design a bandpass filter that will yield a resonant frequency of 10 kHz, a maximum gain of 10, and a bandwidth of 500 Hz. Find the appropriate transfer function and verify your result with a frequency response plot.

WHAT IF? Find the transfer function of a bandstop filter with the same Q that blocks out 10 kHz. Verify your result with **freqs**. ■

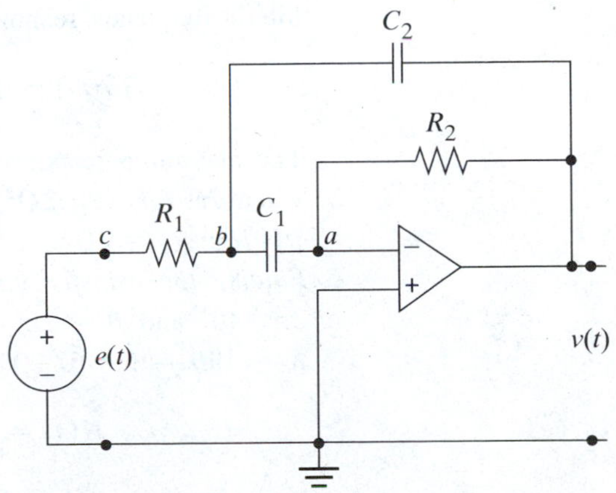

FIGURE E4.4a *Op-amp circuit*

EXAMPLES

Solution

a. *Verification of the given matrix equation.* Writing KCL at node b and substituting the current voltage (i-v) relationships gives

node b

$$\frac{1}{R_1}[v_b - e] + C_2[\dot{v}_b - \dot{v}] + C_1[\dot{v}_b - \dot{v}_a] = 0 \quad \text{where } v_a \approx 0.$$

Repeating KCL at node a, we obtain

node a

$$\frac{1}{R_2}[v - v_a] + C_1[\dot{v}_b - \dot{v}_a] = 0 \quad \text{where } v_a \approx 0.$$

Taking the Laplace transform of the preceding two equations with zero initial conditions and arranging into matrix form yields the desired result

$$\begin{bmatrix} 1/R_1 + (C_1 + C_2)s & -C_2 s \\ C_1 s & 1/R_2 \end{bmatrix} \begin{bmatrix} V_b(s) \\ V(s) \end{bmatrix} = \begin{bmatrix} E(s)/R_1 \\ 0 \end{bmatrix}.$$

b. *Finding ω_0, B, and $|H(j\omega)|_{\max}$ for the given data.* Substitution into the matrix equation gives

$$\begin{bmatrix} 2Q + 2s & -s \\ s & 1/(2Q) \end{bmatrix} \begin{bmatrix} V_b(s) \\ V(s) \end{bmatrix} = \begin{bmatrix} 2QE(s) \\ 0 \end{bmatrix},$$

which has the solution

solution by matrix inverse

$$\begin{bmatrix} V_b(s) \\ V(s) \end{bmatrix} = \frac{1}{s^2 + s/Q + 1} \begin{bmatrix} 1/(2Q) & s \\ -s & 2Q + 2s \end{bmatrix} \begin{bmatrix} 2QE(s) \\ 0 \end{bmatrix}.$$

Solving for the required transfer function produces

$$H(s) = \frac{V(s)}{E(s)} = \frac{-2Qs}{s^2 + s/Q + 1}.$$

Thus, by comparison with the general expression for a second-order bandpass filter given at the beginning of this example, the resonant frequency is $\omega_0 = 1$ rad/s and the bandwidth is $B = 1/Q$ rad/s. The filter's frequency response is

$$H(j\omega) = H(s)|_{s=j\omega} = \frac{-2Qj\omega}{-\omega^2 + j\omega/Q + 1}.$$

The maximum response occurs at the resonant frequency $\omega = \omega_0 = 1$, where $H(j\omega) = -2Q^2$ and we assign the negative sign to the phase, so $|H(j\omega)|_{\max} = 2Q^2$.

c. *Finding the transfer function and frequency response plot.* With $\omega_0 = 2\pi \cdot 10^4$ and $B = 2\pi \cdot 500$, we have $|H(j\omega)|_{\max} = K/B = 10$, giving $K = 10B = \pi \cdot 10^4$. Thus

$$H(s) = \frac{V(s)}{E(s)} = \frac{\pi 10^4 s}{s^2 + \pi 10^3 s + 4\pi^2 10^8}.$$

Using **freqs** to produce Figure E4.4b we see that all the specifications have been met.

CHAPTER 4

WHAT IF? Suppose zeros were placed at $s = \pm j\pi 10^4$ with the denominator coefficients remaining the same to maintain the same shape (Q). The result is shown in Figure E4.4c. ■

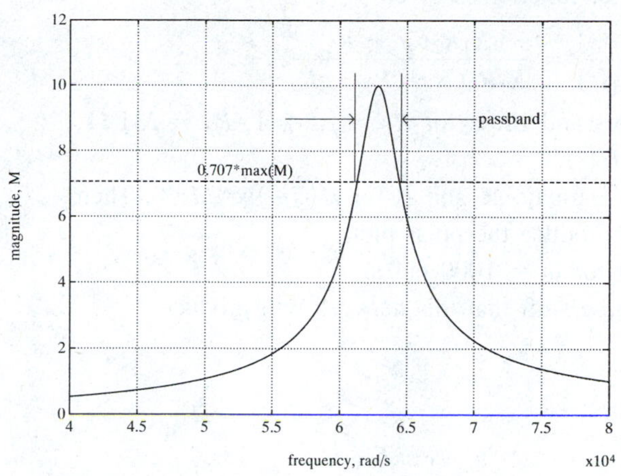

FIGURE E4.4b *Bandpass filter frequency response magnitude*

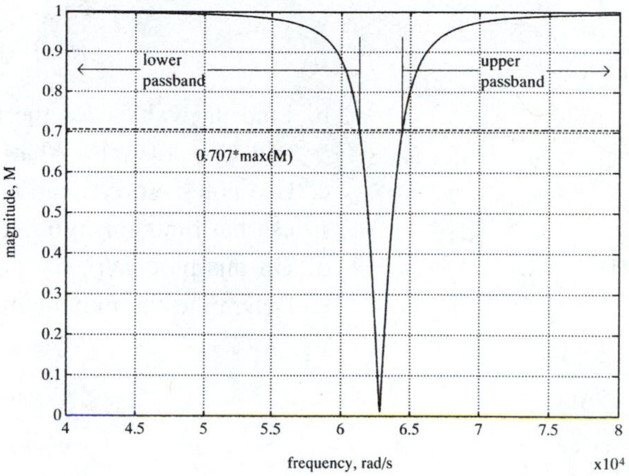

FIGURE E4.4c *Bandstop filter frequency response magnitude*

REINFORCEMENT PROBLEMS

P4.1 A second-order *RC* ladder network. For the network of Figure P4.1:

a. Find the transfer function $H(s) = V_o(s)/V(s)$ in terms of R and C.

b. For $R = 1 = C$, use the MATLAB function **freqs** to plot the frequency response for this network. Make a rectangular plot for both magnitude and phase and locate the cutoff frequency on these plots.

c. Plot the magnitude in dB (linear scale) versus ω (log scale) and the phase (linear scale) versus ω (log scale). Locate the cutoff frequency on these plots.

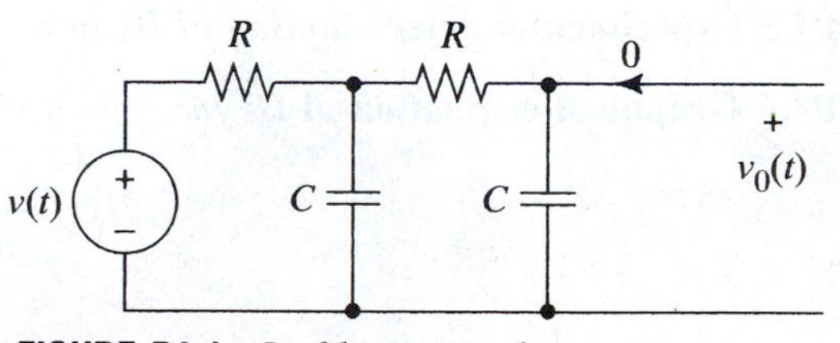

FIGURE P4.1 *Ladder network*

P4.2 A lead network. Control system engineers frequently use passive networks in design to make the overall system performance more satisfactory, an example being to speed up the response of an inherently sluggish system. A lead network as shown in Figure P4.2 produces "phase lead" for the output voltage when compared with the input voltage with sinusoidal excitation.

a. Show that the voltage transfer function is given by

$$H(s) = \frac{V_2(s)}{V_1(s)} = \frac{R_1R_2Cs + R_2}{R_1R_2Cs + R_1 + R_2}.$$

b. Find the values for the poles and zeros for $R_1 = 1.5\ \text{k}\Omega$, $R_2 = 231\ \Omega$, and $C = 1.0\ \mu\text{F}$. What is $H(s)$?
c. Use pole-zero estimation to estimate M and P for $H(j\omega) = Me^{jP}$. Then use the function **nyquist** to produce the polar plot.
d. On this plot mark the point for $\omega = 1000$ rad/s.
e. Determine the maximum phase shift that this network will produce.

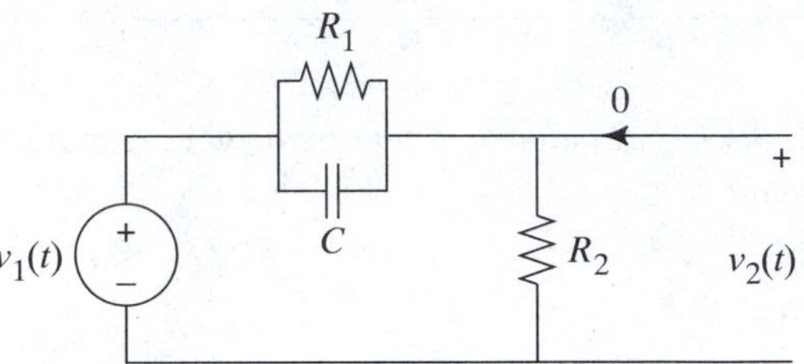

FIGURE P4.2 *Lead network*

P4.3 Sinusoidal steady-state. Given the following differential equations that describe different electric circuits, find the *steady-state* solution.

a. $dv(t)/dt + 2v(t) = \cos t$
b. $d^2v(t)/dt^2 + 2\,dv(t)/dt + 2v(t) = \cos(t + \pi/6)$
c. $dv(t)/dt + 2v(t) = \sin 2t + \cos t$

P4.4 Experimental determination of $H(j\omega)$. A frequency response test is performed on an analog control system that is assumed to be linear and time invariant. The resulting magnitude and phase data are plotted in Figure P4.4. Sketch the asymptotic approximations and find the approximate value of the transfer function for the system.

P4.5 Experimental determination of $H(j\omega)$. Repeat Problem P4.4 for the plots of Figure P4.5.

P4.6 Graphical evaluation of $H(j\omega)$. A stable LTI system is modeled by

$$H(s) = \frac{Y(s)}{X(s)} = \frac{10(s^2 + 4)}{s^3 + 3s^2 + 4s + 2}.$$

a. Make a zero-pole plot for $H(s)$.

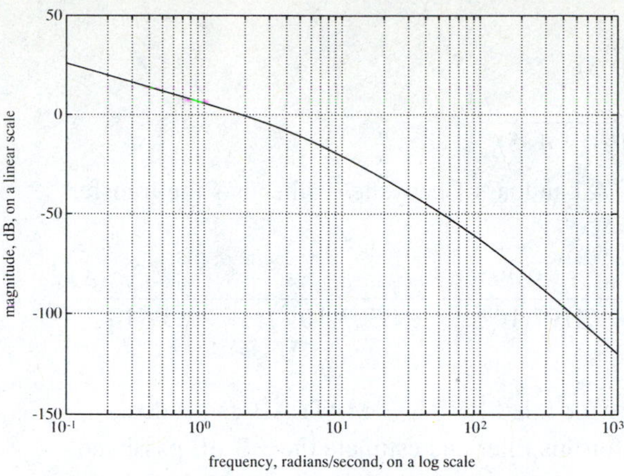

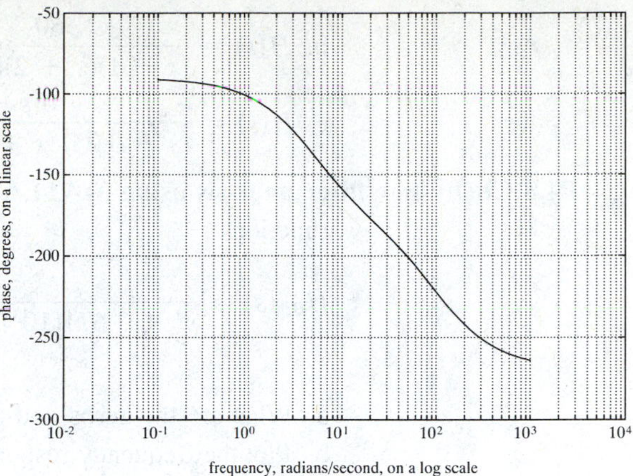

Bode magnitude plot — Bode phase plot

FIGURE P4.4 *Frequency response data for Problem P4.4*

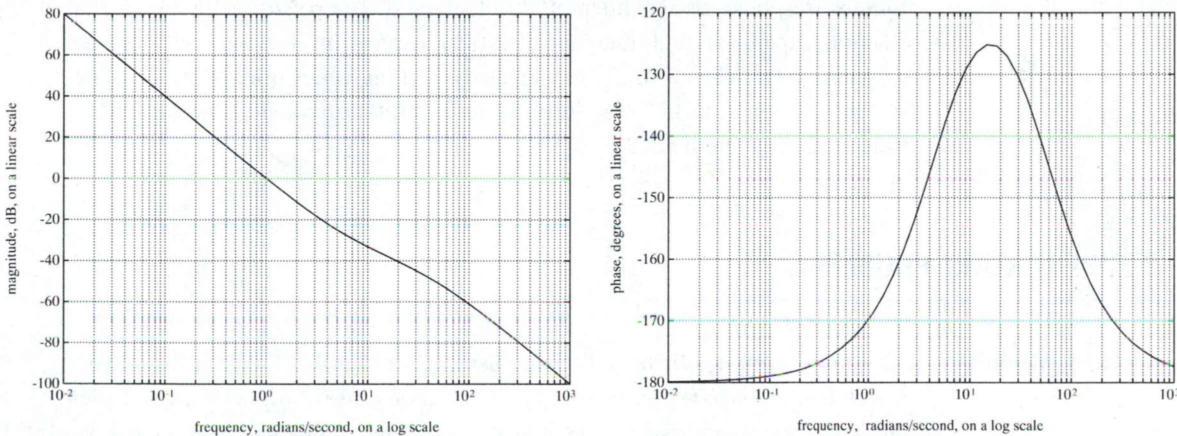

Bode magnitude plot — Bode phase plot

FIGURE P4.5 *Frequency response data for Problem P4.5*

b. Use this plot to evaluate $H(j1)$ graphically.

c. Check the results of part (b) by solving analytically for $H(j1)$. Then verify with **freqs.**

d. Find the steady-state output of this system for the sinusoidal input $x(t) = 4\sin(t - 1)$.

P4.7 Polar diagrams using MATLAB. Make computer-generated plots in polar form for the following transfer functions. In each case verify that the shape of the plot is correct by using the pole-zero frequency response estimation procedure.

a. $H(s) = \dfrac{10s}{(s + 1)(s + 10)}$

b. $H(s) = \dfrac{60}{(s + 1)(s + 2)(s + 3)}$

c. $H(s) = \dfrac{10(s + 1)}{(s + 0.05)(s + 0.10)(s + 5)}$

P4.8 Chebyshev filter analysis using MATLAB. A bandpass Chebyshev[3] filter has the transfer function

$$H_{BP}(s) = \frac{9.494(10^6)s^2}{s^4 + 3.671(10^3)s^3 + 7.197(10^8)s^2 + 1.303(10^{12})s + 1.259(10^{17})}.$$

a. What are the zeros and poles for this filter? Use MATLAB as needed.

b. Plot the frequency response for this filter and estimate the -3 dB passband. You may want to "blow up" certain sections of the plot by making a second run with different frequency limits, or you may want to use the MATLAB function **find.**

P4.9 Time-domain characteristics of a Chebyshev filter using MATLAB. Consider the unit impulse response of the filter of Problem P4.8. Use **residue** for the partial fraction expansion and find an analytical expression for the unit impulse response $h(t)$. What would you expect the settling time (time to steady state, in this case zero) to be? Use **impulse** or **ksimptf** to plot $h(t)$. Is the settling time what you predicted?

EXPLORATION PROBLEMS

CHAPTER 4

P4.10 Minimum and nonminimum phase networks using MATLAB. Given the network transfer function $H(s) = (s + z)/(s + p)$. For $z > 0$, the network is called *minimum phase,* and for $z < 0$, it is known as *nonminimum phase.*

a. For $z > 0$, find an analytical expression for the phase of this minimum phase network.

b. At what value of ω is the phase shift, P, maximum? Hint: $dP(\omega)/d\omega = 0$.

c. Show that $|H(j\omega)|$ for $z = +z_1$ is identical to $|H(j\omega)|$ for $z = -z_1$.

d. For $z < 0$, find an analytical expression for the phase of this network (nonminimum phase).

e. Make a Bode phase plot for both networks. Use $p = 10$ and $z = \pm 1$.

P4.11 Second-order systems and resonance using MATLAB. Shown in Figure P4.11 is a typical configuration for a second-order system that employs negative unity feedback.

[3]Pafnuti L. Chebyshev (1821–1894), a Russian mathematician.

a. Find the closed-loop frequency response $H(j\omega) = Y(j\omega)/R(j\omega) = Me^{jP}$. (Remember MGR from Chapter 3.) Give analytical expressions for M and P.

b. The frequency response of this system depends on the values of ζ and ω_n. Assume that $\omega_n = 1$ and $\zeta = 1$ and draw the *asymptotic* Bode magnitude and phase plots.

c. Now let's vary ζ. Consider the natural frequency ω_n to be constant at $\omega_n = 1$ rad/s and let the damping ratio ζ vary from 0.01 to 1, that is, $0 < \zeta \leq 1$. Use **freqs** or **bode** to obtain the Bode magnitude and phase plots and compare with part (b). You should also investigate the effect of changing the natural frequency ω_n.

d. It can be shown that the maximum value of M occurs at the frequency

$$\omega_{\text{peak}} = \omega_n\sqrt{1 - 2\zeta^2}, \qquad 0 \leq \zeta \leq 0.707 \quad \text{and that}$$

$$M_{\max} = \frac{1}{2\zeta}\sqrt{1/(1 - \zeta^2)}.$$

You might try deriving these formulas; you should at least check some of your results from part (c) against them.

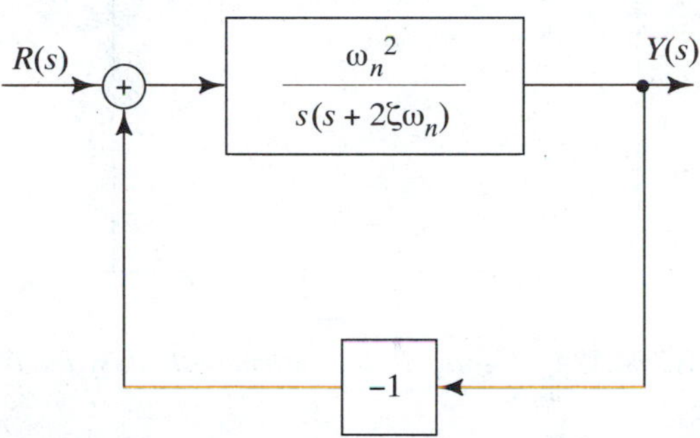

FIGURE P4.11 *Generic second-order unity feedback system*

P4.12 Gain margin using MATLAB. A plant described by the transfer function $G(s) = Y(s)/E(s) = k/(s(s + 1)(s + 2))$ is made part of a unity feedback system, as shown in Figure P4.12a (page 208).

a. Find the overall system transfer function $T(s) = Y(s)/R(s)$ and its characteristic equation.

b. Shown in Figure P4.12b is a Nyquist diagram of the plant frequency transfer function $G(j\omega)$ that was plotted for an original plant gain of $k = 3$. What are the roots of the CE from part (a) for this value of k?

c. Now increase the gain k to 6, find the system roots, and make a Nyquist plot for $G(j\omega)$. Locate the point on the negative real axis that intersects the Nyquist plot.

PROBLEMS

d. Finally, increase the gain k to 9, find the system roots, and make a Nyquist plot for $G(j\omega)$. Locate the point on the negative real axis that intersects the Nyquist plot.

Comment: The system is stable for $k = 3$, is borderline for $k = 6$, and is unstable for $k = 9$. The concepts of stability as determined from the frequency domain are an important part of an undergraduate control systems course; in that course you will learn that the *gain margin* of this system is $6/3 = 2$, or 6 dB. Consequently, if the gain k is increased by a factor of 2 (starting from $k = 3$), the Nyquist locus intersects the $(-1, 0)$ coordinate of the complex plane. Thus gain margin provides a measure of the relative stability of a system—that is, how much the gain can be increased before reaching an unstable condition. There is, of course, much more to the Nyquist stability criterion than shows up in this simple example, but we wanted to illustrate an application of a Nyquist plot.

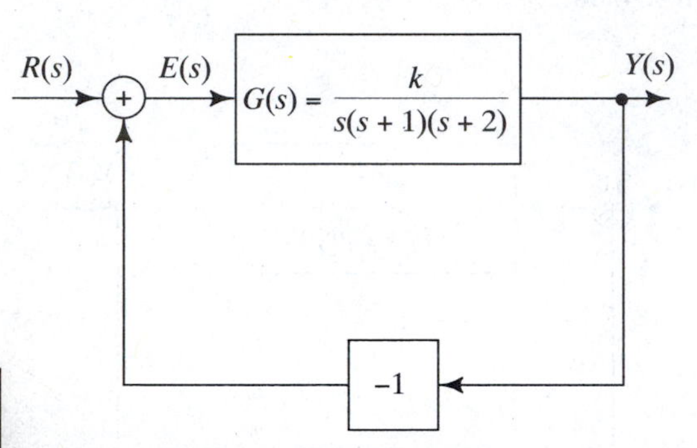

FIGURE P4.12a *A third-order system with unity feedback*

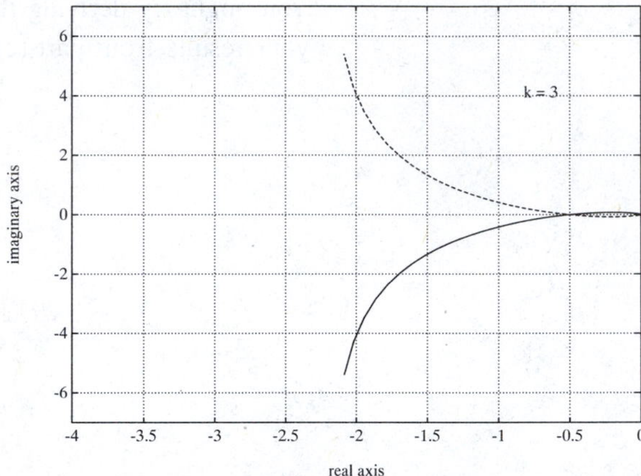

FIGURE P4.12b *Nyquist plot*

P4.13 Gain margin revisited.

a. Using a Nyquist diagram, find the gain margin of the system in Figure P4.13, where the plant is described by $G(s) = Y(s)/E(s) = k/((s+1)(s+2)(s+3))$. Assume that the original plant has $k = 20$.

b. Use the gain from part (a) that puts the Nyquist diagram through the $(-1, 0)$ coordinate to form the system's characteristic equation and then determine the roots. Two of the roots should be very close to pure imaginary.

P4.14 Time and frequency domain relationships for an LTI system. The California Function **zpresp** makes it easy to explore the characteristics of a continuous-time system in the time, frequency, and s domains. Running a script that specifies the numerator and denominator coefficients of a transfer function and invokes the California Function **zpresp** results in a display of the zero-pole diagram, the step and impulse responses, and the frequency response magnitude plot.

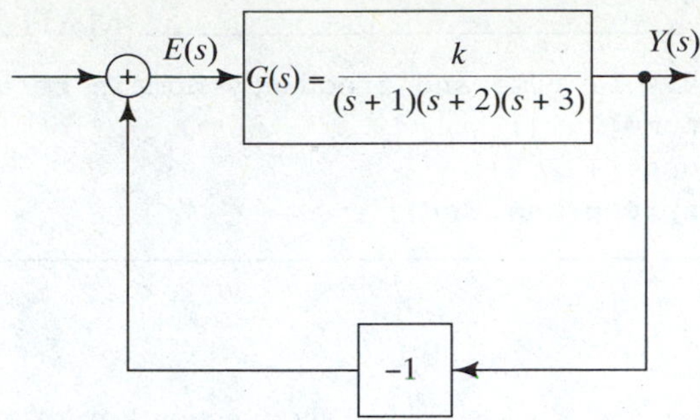

FIGURE P4.13 *Another third-order unity feedback system*

The script on page 210 illustrates the use of **zpresp** for the second-order bandpass analog filter $H(s) = (s + 1)/(s^2 + 2s + 2)$ with the resulting plot given in Figure P4.14.

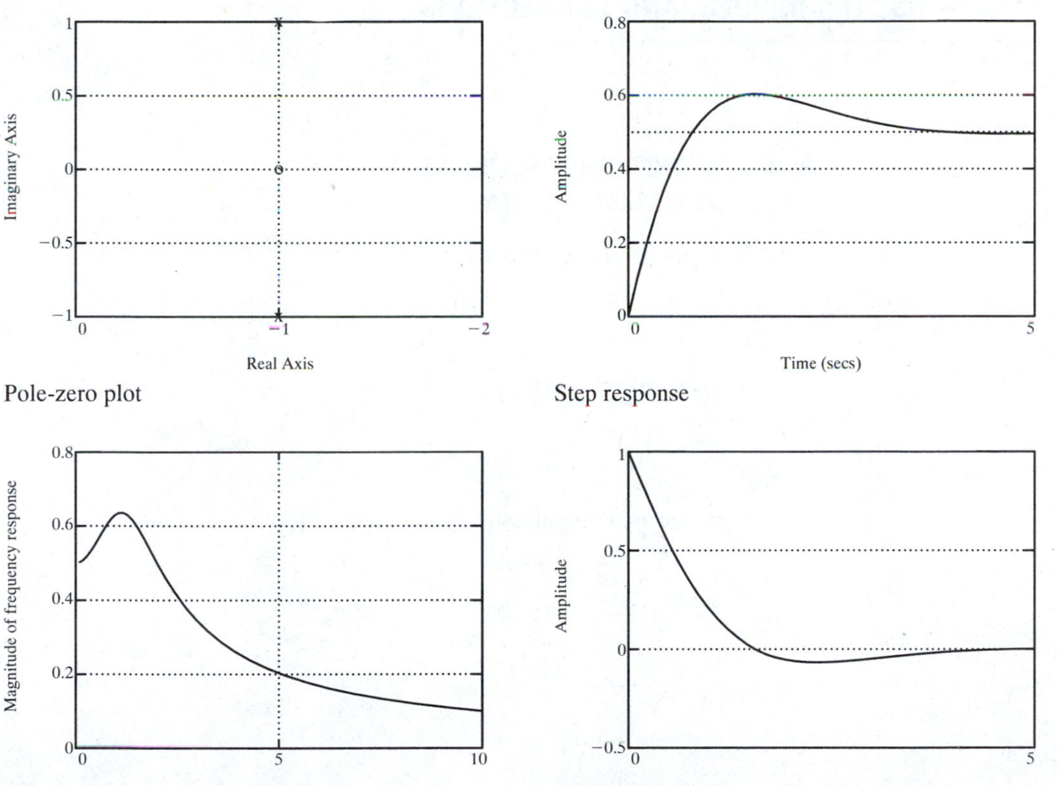

FIGURE P4.14 *Time and frequency relations for an LTI system*

MATLAB Script

```
%P4_14 Time and frequency domain relationships
num=[1,1];                          % numerator coefficients
den=[1,2,2];                        % denominator coefficients
zpresp(num, den);                   % call zpresp
```

Use the function **zpresp** with some transfer functions of your choice. You should verify your theoretical knowledge with the resulting graphical plots on items such as pole-zero locations and filter type, the effect of pole location on the settling time of the step and impulse responses, and the relationship between the step and impulse responses.

DEFINITIONS, TECHNIQUES, AND CONNECTIONS

FINDING THE FREQUENCY RESPONSE, $H(j\omega)$, FOR A STABLE SYSTEM

1. From the transfer function,

$$H(j\omega) = H(s)|_{s=j\omega} = \left.\frac{\sum_{k=0}^{L} b_k s^k}{\sum_{k=0}^{N} a_k s^k}\right|_{s=j\omega} = \frac{\sum_{k=0}^{L} b_k (j\omega)^k}{\sum_{k=0}^{N} a_k (j\omega)^k}.$$

2. From the unit impulse response $h(t)$,

$$H(j\omega) = \int_{-\infty}^{\infty} h(\tau) e^{-j\omega\tau}\, d\tau.$$

3. From the DE,

$$\sum_{k=0}^{N} a_k \frac{d^k y(t)}{dt^k} = \sum_{k=0}^{L} b_k \frac{d^k x(t)}{dt^k} \quad \text{gives } H(j\omega) = \frac{\sum_{k=0}^{L} b_k (j\omega)^k}{\sum_{k=0}^{N} a_k (j\omega)^k}.$$

CHAPTER 4

CHARACTERISTICS OF THE FREQUENCY RESPONSE FUNCTION

1. *Symmetry.* The magnitude M of $H(j\omega)$ is an even function of ω and the phase P is an odd function. Thus

$$|H(j\omega)| = |H(-j\omega)| = \sqrt{a^2 + b^2} = M \quad \text{and}$$

$$\angle H(j\omega) = -\angle H(-j\omega) = P.$$

2. *Time and frequency.* The unit impulse response $h(t)$ is a function of the continuous variable t and the frequency response $H(j\omega)$ is a function of the continuous variable ω.
3. *Decibels.* The values of the gain M may vary over a very wide range, and it is often useful to define M in decibels as $M_{\text{dB}} = 20\log_{10} M$.
4. *Bandwidth.* If $M_{\max}$ is the maximum value of M, the range of frequencies $\omega_1 \le \omega \le \omega_2$ for which $0.707M_{\max} \le M \le M_{\max}$ is called the bandwidth, or passband. Filters may have one or several passbands, and in terms of dB a passband is a region within $-3\text{ dB} \doteq 20\log_{10}(0.707)$ of the maximum value in dB.
5. *Continuous-Time Fourier Transform.* It is common practice to call $H(j\omega)$ the Continuous-Time Fourier Transform (CTFT) of the unit impulse response $h(t)$. In a similar way, for the sequences $x(t)$ and $y(t)$ we write

$$X(j\omega) = \int_{-\infty}^{\infty} x(\tau)e^{-j\omega\tau}\,d\tau \quad \text{and} \quad Y(j\omega) = \int_{-\infty}^{\infty} y(\tau)e^{-j\omega\tau}\,d\tau$$

where $X(j\omega)$ and $Y(j\omega)$ are the CTFTs of the signals $x(t)$ and $y(t)$, respectively.

DEFINITIONS TECHNIQUES

MAKING FREQUENCY RESPONSE PLOTS

1. *Rectangular.* $|H(j\omega)|$, linear scale, versus ω, linear scale, and $\angle H(j\omega)$, linear scale, versus ω, linear scale.
2. *Polar.* $\text{Im}[H(j\omega)]$ versus $\text{Re}[H(j\omega)]$ for $0 \le \omega \le \infty$ with frequencies marked on locus.
3. *Nyquist.* $\text{Im}[H(j\omega)]$ versus $\text{Re}[H(j\omega)]$ for $-\infty \le \omega \le \infty$ with frequencies marked on locus.
4. *Bode.* $20\log_{10}|H(j\omega)|$ versus ω and $\angle H(j\omega)$ versus ω with $20\log_{10}|H(j\omega)|$ and $\angle H(j\omega)$ on linear scales and ω on a log scale; $20\log_{10}|H(j\omega)|$ is called the decibel (dB) gain.
5. *Computer.* Use a program such as MATLAB's **freqs,** where the data are entered as

$$H(s) = \frac{B(s)}{A(s)} = \frac{b(1)s^{(nb-1)} + b(2)s^{(nb-2)} + \cdots + b(nb)}{a(1)s^{(na-1)} + a(2)s^{(na-2)} + \cdots + a(na)}.$$

ASYMPTOTIC APPROXIMATION

Given an experimental plot of $|H(j\omega)|_{\text{dB}}$ on a linear scale versus ω on a log scale, use asymptotic approximations to estimate $H(s)$. See Illustrative Problems 4.4 and 4.5.

GRAPHICAL ESTIMATION

$$M = \frac{kN_1N_2\cdots N_L}{D_1D_2\cdots D_N} \quad \text{and} \quad P = \{0 \quad \text{or} \quad \pm\pi\} + \sum_{k=1}^{L}\alpha_k - \sum_{k=1}^{N}\beta_k\,.$$

MATLAB FUNCTIONS USED

Function	*Purpose and Use*	*Toolbox*
bode	Given: state or TF model of continuous system, **bode** returns magnitude and phase of frequency response usually on semilog coordinates.	Signals/Systems, Control System
freqs	Given: TF model of continuous system, **freqs** returns frequency response.	Signals/Systems, Signal Processing
impulse	Given: state or TF model of continuous system, **impulse** returns impulse response.	Control System, Signals/Systems
ksimptf	Given: TF model of continuous system, **ksimptf** returns impulse response.	California Functions
nyquist	Given: state or TF model of continuous system, **nyquist** returns polar or Nyquist plot of frequency response.	Signals/Systems, Control System
roots	Given: coefficients of polynomial p, **roots** returns roots of $p = 0$.	MATLAB
zpresp	Given: TF model of continuous system, **zpresp** returns zero-pole plot, step and impulse responses, and frequency response magnitude plot.	California Functions

ANNOTATED BIBLIOGRAPHY

1. Dorf, Richard C., *Modern Control Systems, 6th ed.,* Addison-Wesley Publishing Company, Reading, Mass., 1992. *In Chapters Seven and Eight a practical discussion of the use of frequency response methods in the analysis and design of linear control systems is presented. Included are sections on the use of frequency response measurements, performance specifications in the frequency domain, and the application of the Nyquist criterion in stability determination.*

2. Houts, Ronald C. (with Oktay Atkin), *Signal Analysis in Linear Systems,* Saunders College Publishing, Philadelphia, Pa., 1991. *Chapter Three, "Linear System Considerations," contains a complete discussion of frequency response and filtering including realizability of passive filters.*

3. Oppenheim, Alan V., and Alan S. Willsky with Ian T. Young, *Signals and Systems,* Prentice Hall, Inc., Englewood Cliffs, N.J., 1983. *Frequency response from differential and difference equation characterizations of systems is considered in Chapters Four and Five, and Chapter Six presents a brief but insightful view of basic filtering principles.*

4. Van Valkenburg, M. E., *Network Analysis, 3rd ed.,* Prentice Hall, Inc., Englewood Cliffs, N.J., 1974. *Sinusoidal steady-state analysis is carefully developed from the point of view of transforms and rotating phasors. Frequency response plots are treated in Chapter Thirteen in many different ways, including a thorough discussion of asymptotic methods. The Nyquist stability criterion, a graphical method for determining system stability, is well documented, with several examples.*

ANSWERS

P4.1 a. $H(s) = \dfrac{1/(R^2C^2)}{s^2 + 3s/(RC) + 1/(R^2C^2)}$; b. See A4_1 on the CLS disk, cutoff at about $\omega = 0.375$ rad/s; c. See A4_1 on the CLS disk.

P4.2 a. Several approaches exist, including voltage divider; b. Zero at $s = -667$ and a pole at $s = -5000$; $H(s) = (s + 667)/(s + 5000)$; c. Graphical estimation gives $0.133 \leq M \leq 1$, $P > 0$, confirmed by Figure A4.2; d. $H(j1000) = 0.236e^{j\pi/4}$; e. By measurement, $P_{max} \approx 51°$.

MATLAB Script

```
%A4_2...Polar plot
num=[1,667];                    % numerator coefficients
den=[1,5000];                   % denominator coefficients
w=logspace(1,5);                % generates 50 points between 10 and 100,000
[re,im]=nyquist(num,den,w); % call nyquist
%...plotting statements
```

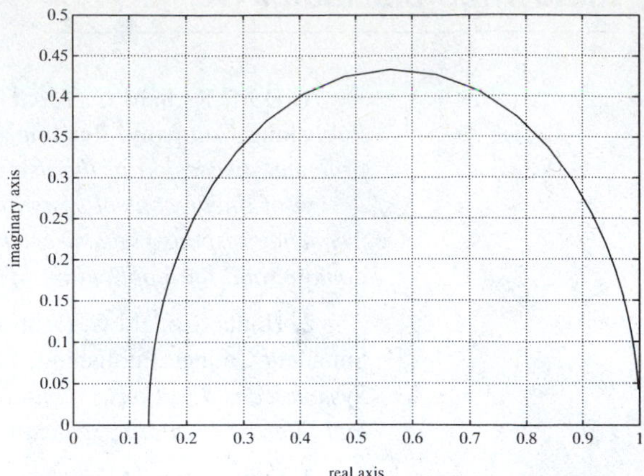

FIGURE A4.2

P4.3 a. $v_{ss}(t) = 0.447\cos(t - 0.464)$; b. $v_{ss}(t) = 0.447\cos(t - 0.583)$; c. $v_{ss}(t) = 0.354\sin(2t - 0.785) + 0.447\cos(t - 0.464)$

P4.4 a. $H(s) \approx 1000/(s(s + 5)(s + 100))$

Comment: Although this system is actually unstable due to the pole at $s = 0$, this instability caused no problem when evaluating the frequency response because $\omega = 0$ was not used as an input frequency.

P4.5 $H(s) \approx 10(s + 5)/(s^2(s + 50))$

P4.6 a. Zeros, $s_{1,2} = \pm j2$; poles, $s_1 = -1$, $s_{2,3} = -1 \pm j1$;
b. The answer depends on the scale used, but
$H(j1) \approx 10(1e^{-j1.57})(3e^{j1.57})/((1e^{j0})(\sqrt{2}e^{j0.785})(\sqrt{5}e^{j1.04})) \approx 9.5e^{-j1.8}$;
c. $H(j1) = 10(-1 + 4)/(-j1 - 3 + j4 + 2) = 9.49e^{-j1.89}$; see A4_6 on the CLS disk; d. $y_{ss}(t) = 37.96\sin(t - 2.89)$

P4.7 a. $\omega = \Delta$, N is small, $\alpha = \pi/2$, $\beta_1 = \beta_2 = \Delta$,
$H(j0^+) = \Delta e^{j\pi/2} : \omega \to \infty$; $M = N/(D_1D_2) \to 0$,
$P \to \pi/2 - (\pi/2 + \pi/2) \to (-\pi/2)$; the plot of A4_7 on the CLS disk appears to be all right; b. $\omega = \Delta$; $M = 60/(1 \cdot 2 \cdot 3) = 10$,
$\beta_1 = \beta_2 = \beta_3 = 0$, so $P = 0$; $\omega \to \infty$, $M = 60/(\infty \cdot \infty \cdot \infty) = 0$,
$P = -(\pi/2 + \pi/2 + \pi/2) = -3\pi/2$; see A4_7 on the CLS disk;
c. $\omega = \Delta = 0.01$, $M \to 10/0.025 = 400$, $\alpha = 0$, $\beta_1 = 0.2$, $\beta_2 = 0.1$, $\beta_3 = 0$, so $P = -0.3$; $\omega \to \infty$, $\omega = \infty/(\infty)^3 \to 0$, $P = -\pi$; see A4_7 on the CLS disk.

P4.8 a. Two finite zeros at $s = 0$, poles at $s_{1,2} = 2.018(10^4)e^{\pm j1.619}$, $s_{3,4} = 1.759(10^4)e^{\pm j1.619}$; b. See A4_8 on the CLS disk with the bandwidth about 3600 rad/s, or 573 Hz.

CHAPTER 4

P4.9 $h(t) = 1965e^{-978t}\cos(20151t - 1.524)u(t) + 1713e^{-857t}\cos(17566t + 1.625)u(t)$; for settling time $857t_s \approx 5$, or $t_s \approx 5.8$ ms, which agrees with the plot in A4_9 on the CLS disk.

P4.10 a. $P = \tan^{-1}(\omega/z) - \tan^{-1}(\omega/p)$; b. $\omega = \pm\sqrt{(p^2z - z^2p)/(p - z)} = \pm\sqrt{pz}$; c. $M = \sqrt{(\omega^2 + (\pm z_1)^2)/(\omega^2 + p^2)}$; d. $P = \tan^{-1}(\omega/(-z_1)) - \tan^{-1}(\omega/p)$; (e) See A4_10 on the CLS disk.

P4.11 a. $H(j\omega) = \omega_n{}^2/(-\omega^2 + 2j\zeta\omega_n\omega + \omega_n{}^2)$, where $M = \omega_n{}^2/\sqrt{[-\omega^2 + \omega_n{}^2]^2 + 4\zeta^2\omega_n{}^2\omega^2}$ and $P = -\tan^{-1}[2\zeta\omega_n\omega/(-\omega^2 + \omega_n{}^2)]$; b. For small ω, $M \rightarrow 1$ or 0 dB and $P \rightarrow 0°$; for large ω, $M \rightarrow 1/\omega^2$ or $[-40\log_{10}\omega]$ dB and $P \rightarrow -180°$; c. See A4_11 on the CLS disk for the magnitude and phase plots for $\zeta = 0.1$, 0.3, 0.5, 0.7, 0.9, and 1.0 with $\omega_n = 1$.

P4.12 a. $Y(s)/R(s) = k/(s^3 + 3s^2 + 2s + k)$; CE is $s^3 + 3s^2 + 2s + k = 0$; b. $s_1 = -2.672$, $s_{2,3} = -0.164 \pm j1.047$; intersects real axis at -0.5; c. $s_1 = -3$, $s_{2,3} = \pm j1.414$; intersects real axis at -1, as observed in A4_12 on the CLS disk; d. $s_1 = -3.240$, $s_{2,3} = 1.200 \pm j1.662$; from A4_12 on the CLS disk, intersects real axis at -1.5.

P4.13 a. See A4_13 on the CLS disk, where the negative axis intersection is at about $s = -0.333$, making the gain margin about $[1/0.333] \cdot 20 \approx 60$. Using this value of k, the CE is $s^3 + 6s^2 + 11s + 66 = 0$, with roots $s_1 = -6.0$, $s_{2,3} = \pm j3.317$.

RETROSPECTIVE

Chapters 2, 3, and 4

This is the first of four Retrospectives that have been designed to insert a break in the action to allow us to gather our thoughts and summarize what has gone before. Many different concepts and techniques have been presented so far in our study of linear systems, and some looking backward will help sort them out. One of the important issues is when to apply each of these approaches. To a great extent this depends on what the input signal is, how it is specified (sinusoidal, tabular data, analytical expression, etc.), the system model (frequency response, differential equations, transfer function, unit impulse response, etc.), and the desired form of the output (forced and/or initial condition response, steady-state response, analytical versus tabular or numerical form, etc.). Although it would be naive to think that the selection of an appropriate choice can be made in a mechanical fashion, the alternatives are summarized in the first two sections that follow. References to text material are given where appropriate.

SUMMARY OF CONTINUOUS-SYSTEM SOLUTION METHODS

INPUT	SYSTEM MODEL	SOLUTION METHOD	RESULT
$x(t)$ analytical	Differential equation	Classical solution	In analytical form (not covered in text)
$x(t)$ analytical	Differential equation	Laplace transforms	Total solution in analytical form $y(t) = y_{IC}(t) + y_F(t)$
$x(t)$ analytical	Transfer function	Laplace transforms	Forced response in analytical form $y(t) = y_F(t)$

INPUT	SYSTEM MODEL	SOLUTION METHOD	RESULT
$x(t)$ tabular	Unit impulse response $h(t)$, tabular form	Convolution numerical	Forced response in tabular form
$x(t)$ analytical	Unit impulse response $h(t)$, analytical form	Convolution analytical	Forced response in analytical form
$x(t) = X\cos(\omega t + \alpha)$	Frequency response $H(j\omega)$ stable system	Sinusoidal steady-state formula	Steady-state forced response, $y(t) = y_{ss}(t)$

SOLUTION CHOICES FOR SPECIFIED INPUT-OUTPUT CONDITIONS

INPUT	DESIRED OUTPUT	SOLUTION METHOD	SYSTEM MODEL
General $x(t)$	Total solution analytical	Classical; not covered	DE
General $x(t)$	Total solution analytical	Laplace transforms	DE
General $x(t)$	Forced response analytical	Laplace transforms	$H(s)$
General $x(t)$ tabular	Forced response tabular	Convolution, numerical	$h(t)$—numerical
General $x(t)$ analytical	Forced response analytical	Convolution, analytical	$h(t)$—analytical
$x(t) = X\cos(\omega t + \alpha)$	Forced response (steady state)	Sinusoidal steady-state formula	$H(j\omega)$

HOW TO FIND ONE MODEL FROM ANOTHER

DESIRED MODEL	STARTING POINT	LIKELY APPROACH
Unit impulse response $h(t)$	Differential equation	1. Classical; not covered 2. $\mathcal{L}^{-1}[H(s)]$
	System diagram	Obtain $H(s)$, find $\mathcal{L}^{-1}[H(s)]$.
	Transfer function	$\mathcal{L}^{-1}[H(s)]$

DESIRED MODEL	STARTING POINT	LIKELY APPROACH
	Frequency response, $H(j\omega)$	Depends on form given. If in rational form, find $H(s)$; then $\mathcal{L}^{-1}[H(s)] = h(t)$.
Transfer function $H(s)$	Differential equation	Laplace transforms
	System diagram	Via differential equation or MGR
	$h(t)$	$H(s) = \mathcal{L}[h(t)]$
	$H(j\omega)$	Depends on form. If rational, find $H(s)$ directly.
Differential equation	System diagram	By inspection.
	$h(t)$	Laplace transforms, i.e., $\mathcal{L}^{-1}[H(s) = Y(s)/X(s)]$
	$H(s)$	By inspection if rational, $s^n \sim d^n/dt^n$
	$H(j\omega)$	By inspection if rational, $(j\omega)^n \sim d^n/dt^n$
Frequency response	Differential equation	By inspection.
	System diagram	Via differential equation or MGR
	$h(t)$	Differential equation or $H(s)\|_{s=j\omega}$
	$H(s)$	$H(j\omega) = H(s)\|_{s=j\omega}$
System diagram	Differential equation	Solve for $y(t)$ and implement with integrators and gains.
	$h(t)$	Find $H(s)$ and implement as $H(s) = H_a(s) \cdot H_b(s)$.
	$H(s)$	See $h(t)$.
	$H(j\omega)$	Depends on form. If in rational form, find $H(s)$ directly and implement as before.

SUMMARY OF MATLAB FUNCTIONS USED

Function	*Purpose and Use*	*Toolbox*
bode	Given: state or TF model of continuous system, **bode** returns magnitude and phase of frequency response usually on semilog coordinates.	Signals/Systems, Control System
conv	Given: unit sample response model of continuous system, input, **conv** returns system output.	Signals/Systems, Signal Processing
displot	Plots discrete functions.	California Functions
freqs	Given: TF of continuous system, **freqs** returns frequency response.	Signals/Systems, Signal Processing
impulse	Given: state or TF model of continuous system, **impulse** returns impulse response.	Control System, Signals/Systems
invlaplace	Given: Laplace transform function F, **invlaplace** returns the inverse Laplace transform f.	Symbolic Math
ksimptf	Given: TF model of continuous system, **ksimptf** returns impulse response.	California Functions
kslsim	Given: state or TF model of continuous system, input, ICs, **kslsim** returns system output in continuous time.	California Functions
laplace	Given: the symbolic expression f, **laplace** returns the unilateral transform F.	Symbolic Math
lsim	Given: state or TF model of continuous system, input, ICs, **lsim** returns system output.	Control System, Signals/Systems
nyquist	Given: state or TF model of continuous system, **nyquist** returns polar or Nyquist plot of frequency response.	Signals/Systems, Control System
plot	Provides continuous or discrete plots.	MATLAB
residue	Given: rational function $T(\sigma) = N(\sigma)/D(\sigma)$, **residue** returns roots of $D(\sigma) = 0$ and PF constants of $T(\sigma)$.	MATLAB
rlocus	Given: an equation in the form, $1 + K[\text{num}(\sigma)]/[\text{den}(\sigma)] = 0$: **rlocus** returns plot of locus of roots for K varying.	Signals/Systems, Control System
roots	Given: coefficients of polynomial p, **roots** returns roots of $p = 0$.	MATLAB
step	Given: state or TF model of continuous system, **step** returns step response.	Signals/Systems, Control System
zpresp	Given: TF model of continuous system, **zpresp** returns zero-pole plot, step and impulse responses, and frequency response magnitude plot.	California Functions

Toolbox

MATLAB—Version 4.1, October 1, 1993
Signals/Systems, January 1, 1992
Control System—Version 3.06, October 13, 1993
Signal Processing—Version 3.06, February 7, 1994
California Functions, June 18, 1992

Continuous-Time Fourier Series and Transforms

PREVIEW

This is one of the two chapters of *Contemporary Linear Systems* devoted primarily to methods that use the theory proposed by Jean Baptiste Joseph Fourier.[1] The calculation of the steady-state response of a linear time-invariant system to a sinusoidal input signal, developed in Chapter 2 as an application of convolution, was repeated in Chapter 3 as an application of Laplace transforms

FIGURE 5.1 *Jean Baptiste Joseph Fourier*

[1]Baron Jean Baptiste Joseph Fourier (1768–1830) introduced the idea that an arbitrary function, even one defined by different analytic expressions in adjacent segments of its range, could be represented by a single analytic expression. This idea has proven to be most important in mathematics, science, and engineering. Not only was Fourier a professor of

and was the basis for Chapter 4, which was devoted exclusively to frequency response. Almost all practical periodic signals may be expressed as the sum of a series of cosine terms that are harmonically related, the so-called Fourier series. And since the response of an LTI system to a single sinusoidal term is easily calculated, it is equally easy, with the aid of superposition, to find the system response to a sum of sinusoidal inputs.

In this chapter the concept of the frequency spectrum of a periodic signal is introduced and applied. Through the vehicles of average power and Parseval's theorem, we will see that many practical engineering situations can be analyzed quite easily using only a few terms of the Fourier representation of a periodic signal. MATLAB functions are used to synthesize waveforms and to aid in the analysis of signal transmission through LTI systems.

Also presented in this chapter is a discussion of the Continuous-Time Fourier Transform (CTFT), including pairs, properties, and applications. Contrasted with the machine-oriented Discrete Fourier Transform (DFT) of Chapter 10, analysis using the CTFT (normally known as the $\mathcal{F}$ transform) relies on pairs and table lookup procedures similar to those used in the z and Laplace arenas. The main thrust of all Fourier methods, however, is that of using the frequency characteristics of a time-domain signal for analysis and design; consequently, much effort is devoted to the frequency spectrum of signals with its ramifications and utility. One of the important scientific ideas of the twentieth century, the sampling theorem of J. M. Whittaker, H. Nyquist, D. Gabor, and C. E. Shannon, is developed. Filtering (ideal and practical) principles, amplitude modulation, and signal transmission through a linear system are three of the engineering applications that are presented.

---

mathematics, but he had considerable success in several political and diplomatic assignments given to him by Napoleon and King Louis XVIII. For a succinct (three-page) story of the very interesting life of Fourier, see Chapter 24 of Ronald N. Bracewell, *The Fourier Transform and Its Applications, 2nd ed., rev.,* McGraw-Hill Book Company, New York, 1986.

FOURIER SERIES

Periodic Signals A continuous-time signal is periodic if

periodic signal

$$f(t) = f(t \pm T_0) \quad \text{for all } t$$

where the period is the smallest value of T_0 that satisfies this equation. If we have the signal $f(t) = A_1 \cos(\omega_1 t + \alpha_1) + A_2 \cos(\omega_2 t + \alpha_2) + A_3 \cos(\omega_3 t + \alpha_3)$, we need to test the ratio of ω_2/ω_1 and ω_3/ω_1 to see if the ratios are rational numbers. If they are, the signal is periodic and the period T_0 as well as the fundamental frequency $f_0 = 1/T_0$ can be determined. For instance, the voltage

$$v(t) = 1 \cos(628t + 1.23) + 2 \cos(1256t + 4.56) + 3 \cos(1884t + 7.89)$$

is periodic because $1256/628 = 2$ and $1884/628 = 3$, making $f_0 = 628/2\pi = 100$ Hz the fundamental frequency, with the period $T_0 = 1/f_0 = 0.01$ s. In terms of f_0,

$$v(t) = 1 \cos(2\pi f_0 t + 1.23) + 2 \cos(2 \cdot 2\pi f_0 t + 4.56) + 3 \cos(3 \cdot 2\pi f_0 t + 7.89).$$

The signal

$$p(t) = 1.23 \cos(2\pi t + 2) + 4.56 \cos(2.2\pi t + 2.5)$$

is also periodic because $2.2\pi/2\pi = 22/20$, a rational number. For these frequencies the greatest common factor is 0.2π, which is the fundamental radian frequency ω_0, with the fundamental hertz frequency being $f_0 = \omega_0/2\pi = 0.10$ Hz and the period, $1/f_0 = 10$ s. On the other hand, the current

$$i(t) = 2 \sin\left(\sqrt{2}t + 1\right) + 3 \cos(2.00t + 2)$$

is not periodic because the frequency ratio $2/\sqrt{2} = \sqrt{2}$ is an irrational number.

Complex Periodic Exponentials Let us now investigate how we might use complex exponentials to represent a periodic signal having a period of T_0. We can think of the complex exponential $e^{j\omega t}$ as a line of unit length rotating in a counterclockwise direction, as shown in Figure 5.2. We suspect that we need several (actually an infinity of) complex exponentials to represent a real

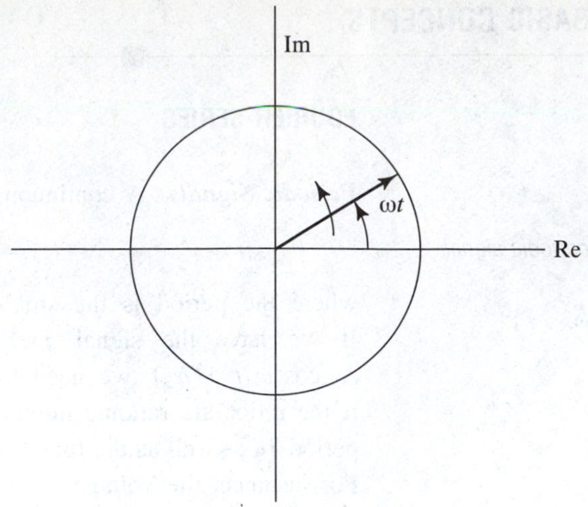

FIGURE 5.2 *Complex exponential*

periodic signal. Generally, one such exponential has the same period T_0 as the signal. This complex exponential is

$$p_1(t) = e^{j(2\pi/T_0)t} = e^{j(2\pi f_0)t} = e^{j\omega_0 t}.$$

A complex exponential having twice the frequency of $p_1(t)$ is

$$p_2(t) = e^{j(2\pi/T_0)2t} = e^{j(2\pi f_0)2t} = e^{j2\omega_0 t}$$

and, in general, for an exponential having k times the fundamental frequency of f_0, we have

$$p_k(t) = e^{j(2\pi/T_0)kt} = e^{j(2\pi f_0)kt} = e^{jk\omega_0 t}.$$

A unit line rotating in a *clockwise* direction is described by the exponential

$$p_{-1}(t) = e^{-j(2\pi/T_0)t} = e^{-j(2\pi f_0)t} = e^{-j\omega_0 t}$$

and for k times the fundamental frequency of f_0, we have

$$p_{-k}(t) = e^{-j(2\pi/T_0)kt} = e^{-j(2\pi f_0)kt} = e^{-jk\omega_0 t}.$$

Thus, a linear combination of all possible harmonically related complex exponentials can be described by the infinite summation

$$f(t) = \sum_{k=-\infty}^{\infty} F_k e^{jk\omega_0 t} = \sum_{k=-\infty}^{\infty} |F_k| e^{j(k\omega_0 t + \angle F_k)}$$

where we have considered the possibility that these exponentials can be of different amplitudes and can have different relative positions (phase) by assuming that the coefficients, the F_k's, are complex constants that need to be determined. The preceding equation is known as the *Fourier series*

CHAPTER 5

representation of a periodic signal and, of course, can be written in terms of the real frequency f_0 or the period T_0 as

alternative forms

$$f(t) = \sum_{k=-\infty}^{\infty} F_k e^{jk2\pi f_0 t} = \sum_{k=-\infty}^{\infty} F_k e^{j(k2\pi/T_0)t}.$$

The period of the signal is T_0 and the term for $k = 0$ is the constant, or DC, term. As we discussed earlier, the terms for $k = 1$ and $k = -1$ have the period T_0 (frequency $f_0 = 1/T_0$) and are commonly called the fundamental, or first-harmonic, components. In the same way the terms for $k = \pm 2$ constitute the second-harmonic components and the Qth-harmonic component is described by the $k = \pm Q$ terms.

ILLUSTRATIVE PROBLEM 5.1
Fundamental Concepts

Suppose we have a periodic signal $f(t)$ that is expressed as the Fourier series

$$f(t) = \sum_{k=-\infty}^{\infty} F_k e^{jkt}$$

where

$$F_0 = 0.25, \qquad F_1 = F_{-1} = 0.225, \qquad F_2 = F_{-2} = 0.159,$$
$$\text{and} \quad F_3 = F_{-3} = 0.075\,.$$

a. Find the period and the fundamental frequency.

b. Write out the seven terms in the exponential expression for $f(t)$ and then use the Euler relation to describe $f(t)$ as a sum of cosine terms.

Solution

a. Comparing the general expression,

$$f(t) = \sum_{k=-\infty}^{\infty} F_k e^{jk\omega_0 t},$$

with the specific expression of this example, we see that $\omega_0 = 1$ rad/s, making $f_0 = (1/2\pi)$ Hz and $T_0 = 2\pi$ s.

b. Writing out the Fourier series expression from $k = -3$ to $k = 3$, we have

$$f(t) = 0.075e^{-j3t} + 0.159e^{-j2t} + 0.225e^{-jt} + 0.250e^{j0t} + 0.225e^{jt} + 0.159e^{j2t} + 0.075e^{j3t}$$

and using the Euler relation produces the cosine series

$$f(t) = 0.250 + 0.450\cos(t) + 0.318\cos(2t) + 0.150\cos(3t)\,.$$

Figure 5.3a shows a plot of $f(t)$. Notice that the sum looks somewhat like a train of pulses, as pictured in Figure 5.3b. In the next section we show that the given coefficients for F_k are those that will be determined for a Fourier series representation of this pulse train.

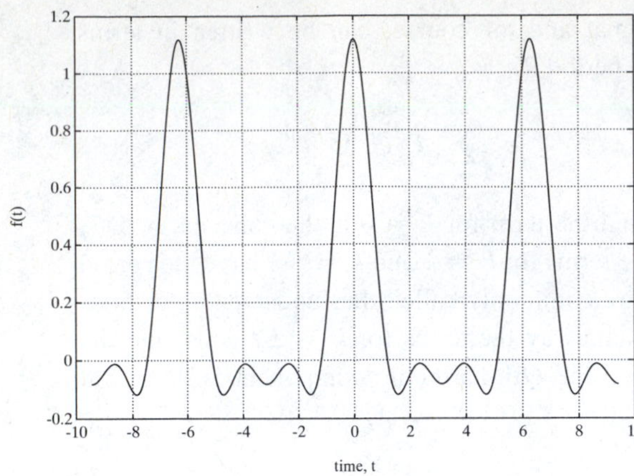

a. Fourier synthesis of a periodic signal

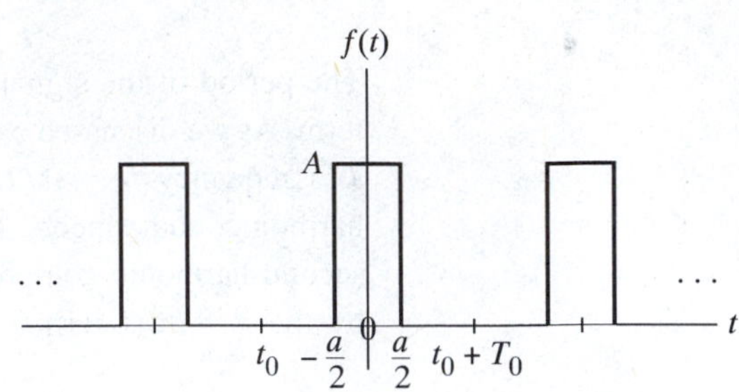

b. Periodic pulse train

FIGURE 5.3 *Fourier synthesis*

Fourier Coefficients The Fourier series representation for a periodic signal is

Fourier series

$$f(t) = \sum_{k=-\infty}^{\infty} F_k e^{jk\omega_0 t},$$

where it is assumed that we know the equation or have a graphical description for $f(t)$. This series can be written as a sum of complex exponentials, as before, or as sums of real cosine and/or sine functions. Notice that we need an infinite number of terms to do this exactly (even then some forms for $f(t)$ will cause problems!), so in practice we truncate the series at some practical value of k, such as $k = L$. The Fourier coefficients, the F_k's, are given by

Fourier coefficients

$$F_k = \frac{1}{T_0} \int_{t_0}^{t_0+T_0} f(t) e^{-jk\omega_0 t}\, dt,$$

where we see that the integral is evaluated over one period of the periodic signal $f(t)$. The preceding two equations—the first, which defines the exponential form of the Fourier series for $f(t)$, and the second, which allows the calculation of the coefficients of the terms in the series—are often known as a Fourier series pair and are written together as

Fourier series pair

$$f(t) = \sum_{k=-\infty}^{\infty} F_k e^{jk\omega_0 t} \Leftrightarrow F_k = \frac{1}{T_0} \int_{t_0}^{t_0+T_0} f(t) e^{-jk\omega_0 t}\, dt.$$

Sufficient, but not necessary, conditions to ensure that the Fourier series exists and converges to the function $f(t)$ are the Dirichlet conditions, which require that within a period the function $f(t)$ (1) has a finite number of maxima and minima; (2) has a finite number of finite discontinuities; and (3) is absolutely integrable, that is,

$$\int_{t_0}^{t_0+T_0} |f(t)|\, dt < \infty.$$

To write $f(t)$ as a cosine series, we first notice that

$$F_{-k} = \frac{1}{T_0} \int_{t_0}^{t_0+T_0} f(t) e^{-j(-k)\omega_0 t}\, dt .$$

which says that $F_{-k} = F_k^*$, that is, they are complex conjugates. Thus $|F_{-k}| = |F_k|$ and $\angle F_{-k} = -\angle F_k$. This allows us to write the Fourier series in the form

$$f(t) = \sum_{k=1}^{\infty} |F_k| e^{-j\angle F_k} e^{-jk\omega_0 t} + F_0 + \sum_{k=1}^{\infty} |F_k| e^{j\angle F_k} e^{jk\omega_0 t}.$$

Invoking the Euler identity once again, we have the cosine series

cosine series

$$f(t) = F_0 + 2 \sum_{k=1}^{\infty} |F_k| \cos(k\omega_0 t + \angle F_k).$$

If you are from a country other than California and prefer sine waves, we have

sine series

$$f(t) = F_0 + 2 \sum_{k=1}^{\infty} |F_k| \sin(k\omega_0 t + \angle F_k + \pi/2).$$

One other form of the series that is used is the sine-cosine series, where

$$f(t) = F_0 + \sum_{k=1}^{\infty} A_k \cos(k\omega_0 t) + \sum_{k=1}^{\infty} B_k \sin(k\omega_0 t).$$

In terms of the exponential coefficients, $A_k = 2\ \mathrm{Re}[F_k]$ and $B_k = -2\ \mathrm{Im}[F_k]$, $k = 1, 2, \ldots$. In all the preceding Fourier representations, the real constant F_0 is simply the average value of $f(t)$ over one period and can often be determined by inspection as the area under the curve $f(t)$ over one period divided by T_0.

ILLUSTRATIVE PROBLEM 5.2
Fourier Coefficients for a Pulse

Suppose we have a voltage $v(t)$ that is described by the periodic pulse train of Figure 5.4a on page 228 and we want to express this signal as (a) an exponential Fourier series and (b) a cosine series. Find the exponential Fourier coefficients V_k that are needed and then write out the two forms of the series.

Solution

a. As discussed earlier, we can choose t_0 arbitrarily, and it makes sense to use the symmetry of the voltage signal and choose t_0, as shown in Figure 5.4b. Then, from the Fourier coefficient equation, we have

$$V_k = \frac{1}{T_0} \int_{-a/2}^{a/2} A e^{-jk\omega_0 t}\, dt = \frac{A}{T_0} \left[\frac{e^{-jk\omega_0 (a/2)} - e^{-jk\omega_0(-a/2)}}{-jk\omega_0} \right].$$

But $e^{jk\omega_0 a/2} - e^{-jk\omega_0 a/2} = 2j \sin(k\omega_0 a/2)$ and $T_0 = 2\pi/\omega_0$, which allows us to write

$$V_k = \frac{A}{\pi k} \sin \frac{k\omega_0 a}{2}, \qquad -\infty \le k \le \infty;$$

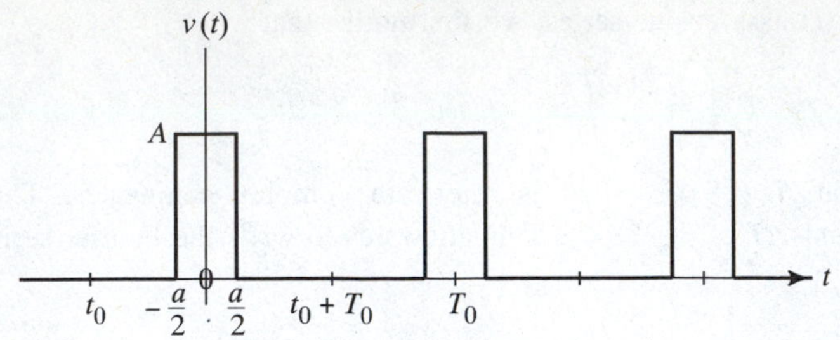

a. Pulse train for Illustrative Problem 5.2

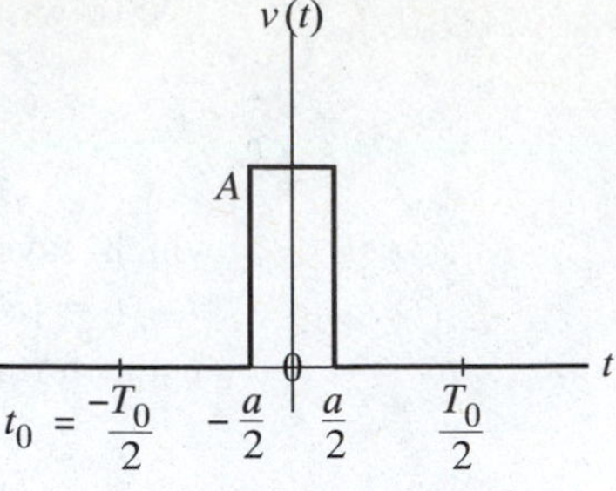

b. One period of the pulse train

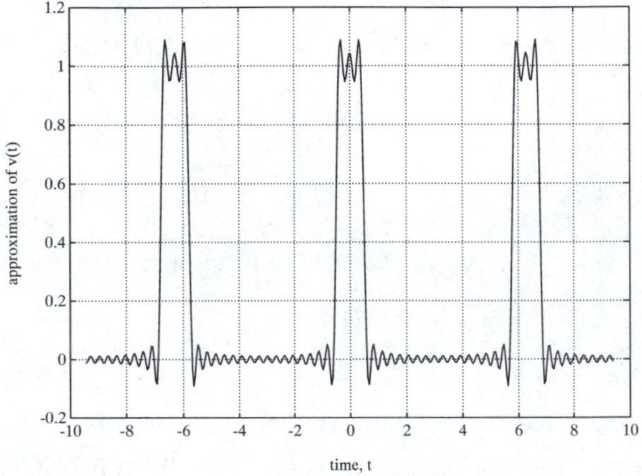

c. Synthesis of pulse waveform: 21 terms

FIGURE 5.4 *Periodic pulse train and Fourier synthesis*

in terms of the period T_0, the coefficients are given by

$$V_k = \frac{A}{\pi k} \sin \frac{k\pi a}{T_0}, \qquad -\infty \leq k \leq \infty.$$

Thus the exponential series is

$$v(t) = \sum_{k=-\infty}^{\infty} \left\{ \frac{A}{\pi k} \sin \frac{k\pi a}{T_0} \right\} e^{jk\omega_0 t}.$$

We should worry about evaluating V_k for $k = 0$, but using l'Hôpital's[2] rule, we find

$$V_0 = \frac{A}{\pi} \frac{\omega_0 a}{2} \cos\left(\frac{k\omega_0 a}{2} \right)\bigg|_{k=0} = \frac{Aa}{T_0},$$

which is also the average value of the pulse signal; that is,

$$V_0 = \frac{\text{area under } v(t) \text{ over one period}}{\text{the period}} = \frac{Aa}{T_0}.$$

[2]Guillaume François Antoine de l'Hôpital (1661–1704), a French nobleman (Marquis de St. Mesme), wrote the first calculus text, *Analyse des Infiniment Petits,* in 1696.

b. The cosine series is written from the Euler relation as

$$v(t) = \frac{Aa}{T_0} + \sum_{k=1}^{\infty} \left\{ \frac{2A}{\pi k} \sin \frac{k\pi a}{T_0} \right\} \cos(k\omega_0 t).$$

Comment: Figure 5.4c shows the synthesis of this waveform for $A = \omega_0 = a = 1$. The script for this plot is given next.

MATLAB Script

```
%F5_4 Fourier synthesis of sine waves
t=-3*pi:.05:3*pi;
x=zeros(size(t));
for k=1:1:20;
x=x+(2/(k*pi))*sin(0.5*k)*cos(k*t);
end
v=1/(2*pi)+x;                          % add in average value
%...plotting and labeling statements
```

Gibbs' Phenomenon In Figure 5.4c we notice two important effects that were first publicized by Josiah Willard Gibbs[3] in 1899: (1) There is a significant amount of ripple in the synthesized signal; and (2) there is clearly an overshoot at the points of discontinuity in the original waveform. This overshoot will always be about 8.95 percent when the synthesized signal consists of a large number of terms. The ripples at points other than the discontinuities, however, can be lessened with a larger number of terms.

Frequency Spectrum The coefficients, the F_k's, of the exponential Fourier series representation

$$f(t) = \sum_{k=-\infty}^{\infty} F_k e^{jk\omega_0 t}$$

are called the frequency or spectral components of $f(t)$. These complex components measure the portion of $f(t)$ that is present at DC and at each of the harmonic frequencies from $f = \pm f_0$, the fundamental frequency, to $f = \pm Lf_0$, the Lth harmonic. It is customary to plot these coefficients against k, f, or ω, depending on one's preference or habit. Since they are complex,

[3]Josiah Willard Gibbs (1839–1903), a Yale University physical chemist, wrote a very popular pamphlet, *Elements of Vector Analysis,* which was a great simplification and improvement of Hamilton's vector calculus. His work was so little appreciated by the authorities at Yale that he served without pay for 10 years, getting a reduced salary only after being invited to join the faculty of the new Johns Hopkins University. From *The History of Mathematics,* David M. Burton, Allyn & Bacon, Inc., 1985.

we normally plot two curves, one for magnitude and the other for phase. Figure 5.5a depicts the frequency spectrum for the pulse train of Figure 5.4a. Notice that in this case, because the V_k's are real numbers, we can simply plot V_k versus k, as in Figure 5.5b, and get by with only one plot.

If we use the cosine Fourier series with

$$v(t) = V_0 + \sum_{k=1}^{\infty} 2|V_k| \cos(k\omega_0 t + \angle V_k)$$

the magnitude and phase spectra of Figure 5.5c result.

Mean Square Error We know that for very practical reasons the Fourier series representation for $f(t)$ must be terminated or truncated at some value of k, say $k = \pm L$. This truncation will cause an error in the representation of $f(t)$, and we should ask if there is a better representation than the Fourier series. A widely used measure of "goodness," which is called a performance criterion, is that of mean square error, defined as

MSE

$$\text{MSE} = \frac{1}{T_0} \int_{t_0}^{t_0+T_0} e^2(t)\, dt$$

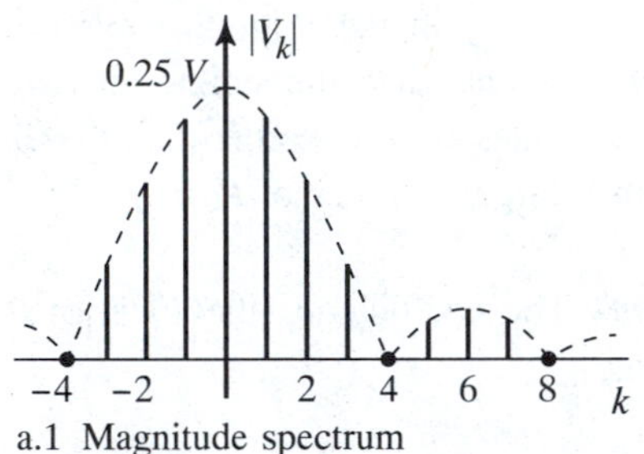

a.1 Magnitude spectrum

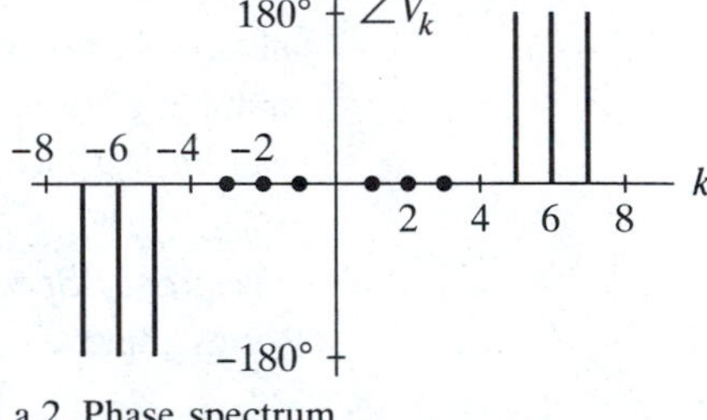

a.2 Phase spectrum

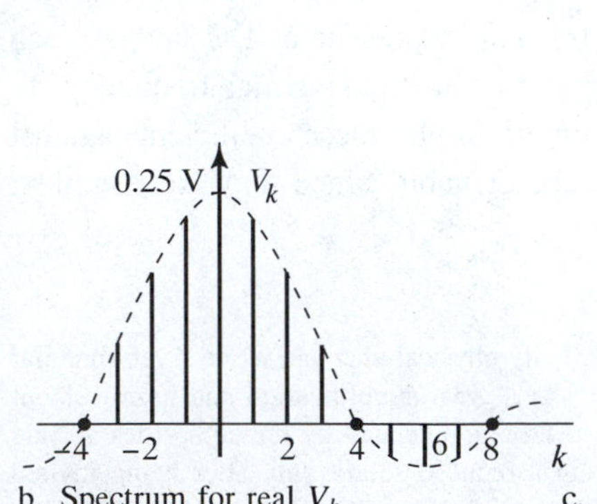

b. Spectrum for real V_k

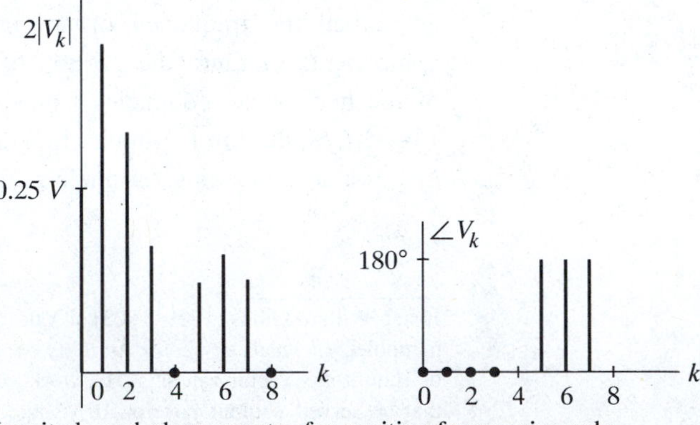

c. Magnitude and phase spectra for positive frequencies only

FIGURE 5.5 *Frequency spectrum*

CHAPTER 5

with the error $e(t)$ defined as

$$e(t) = f(t) - s_L(t).$$

In this expression for the error, $f(t)$ is the given periodic function and $s_L(t)$ is a series representation with a finite number of terms with coefficients S_k. Thus, $s_L(t)$ is given by

$$s_L(t) = \sum_{k=-L}^{L} S_k e^{jk\omega_0 t}.$$

The mean square error is, consequently, given by

$$\text{MSE} = \frac{1}{T_0}\int_{t_0}^{t_0+T_0} \left\{ f(t) - \sum_{k=-L}^{L} S_k e^{jk\omega_0 t} \right\}^2 dt$$

with the objective being to find the S_k's that will minimize this expression. The result of this minimization is that the mean square error is least when $S_k = F_k$. That is, the Fourier series representation is the best possible using the performance measure of mean square error.

Evaluation of the System Frequency Response H(jω) Reviewing the results from Chapters 2, 3, and 4, the frequency response can be determined analytically in any of the following ways.

1. From the transfer function of a stable system,

$$H(j\omega) = H(s)\big|_{s=j\omega} = \left. \frac{\sum_{k=0}^{L} b_k s^k}{\sum_{k=0}^{N} a_k s^k} \right|_{s=j\omega} = \frac{\sum_{k=0}^{L} b_k (j\omega)^k}{\sum_{k=0}^{N} a_k (j\omega)^k}.$$

2. Given the unit impulse response $h(t)$ of a stable system,

$$H(j\omega) = \int_{-\infty}^{\infty} h(\tau) e^{-j\omega\tau} d\tau.$$

3. Starting with the differential equation for a stable system,

$$\sum_{k=0}^{N} a_k \frac{d^k y(t)}{dt^k} = \sum_{k=0}^{L} b_k \frac{d^k x(t)}{dt^k} \quad \text{gives } H(j\omega) = \frac{\sum_{k=0}^{L} b_k (j\omega)^k}{\sum_{k=0}^{N} a_k (j\omega)^k}.$$

ILLUSTRATIVE PROBLEM 5.3 ***Determining the Frequency Response***

Find the frequency response of a causal LTI system that has the unit impulse response $h(t) = [e^{-t} - e^{-2t}]u(t)$. Plot $H(j\omega)$ for all ω—that is, $-\infty \le \omega \le \infty$.

Solution

From the preceding we have the frequency response

$$H(j\omega) = \int_{-\infty}^{\infty} h(\tau)e^{-j\omega\tau}\,d\tau = \int_{0}^{\infty} [e^{-\tau} - e^{-2\tau}]e^{-j\omega\tau}\,d\tau$$

$$= \frac{1}{1 + j\omega} - \frac{1}{2 + j\omega} = \frac{1}{-\omega^2 + 2 + j3\omega}.$$

To graph this frequency response we refer to Chapter 4 and use the MATLAB function **freqs.** The transfer function $H(s)$ is found from the Laplace transform of $h(t)$ as

$$H(s) = \frac{1}{1 + s} - \frac{1}{2 + s} = \frac{1}{s^2 + 3s + 2}.$$

The script is given next and the rectangular plot is in Figure 5.6. Some easy points to check are $\omega = 0$, where $H(j\omega) = 0.5$, and $\omega \to \pm\infty$, where $|H(j\omega)| \to 0$.

MATLAB Script

```
%F5_6 Rectangular plots for IP 5.3
a=[1,3,2];
b=[1];
w=-15:.05:15;              % initial freq.:increment in rad/s:final frequency
H=freqs(b,a,w);            % call freqs
mag=abs(H);
phase=angle(H)
%...statements that change phase to degrees and plot
```

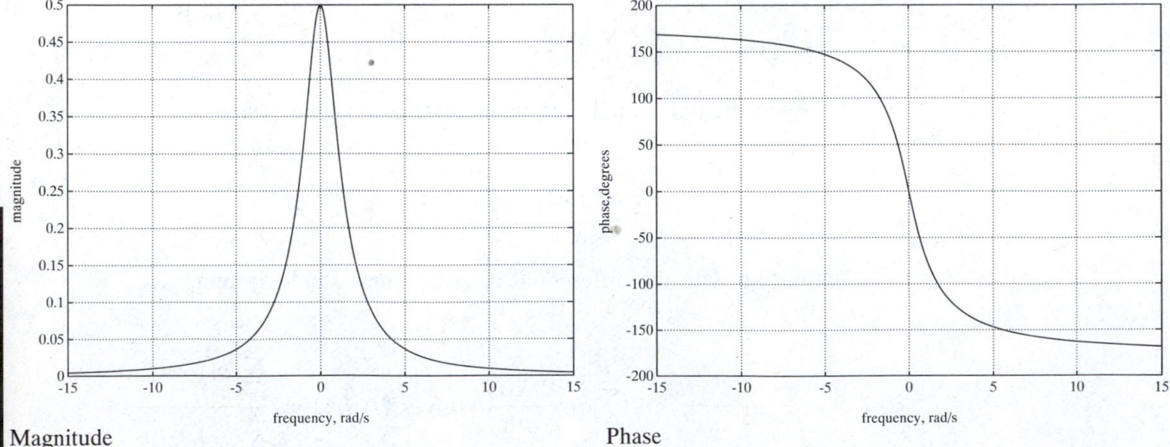

FIGURE 5.6 *Frequency response for a causal system*

Transmission of a Periodic Signal Through a Linear System Given a stable linear system with its frequency response $H(j\omega)$, we know that the steady-state response $y_{ss}(t)$ to a sinusoidal input $x(t) = A\cos(\omega t + \alpha)u(t)$ is

$$y_{ss}(t) = A|H(j\omega)|\cos(\omega t + \alpha + \angle H(j\omega)) = B\cos(\omega t + \beta).$$

For a periodic input described by the Fourier series

$$x(t) = \sum_{k=-\infty}^{\infty} X_k e^{jk\omega_0 t}$$

the steady-state output will be

$$y_{ss}(t) = \sum_{k=-\infty}^{\infty} Y_k e^{jk\omega_0 t}$$

where each output coefficient $Y_k = X_k \cdot H(jk\omega_0)$ is determined by multiplying the input coefficient by the value of the frequency response at the frequency of the input, $\omega = k\omega_0$. This is a direct application of the sinusoidal steady-state discussions of Chapters 2 and 3. With

$$H(jk\omega_0) = |H(jk\omega_0)|e^{j\angle H(jk\omega_0)} \quad \text{and} \quad X_k = |X_k|e^{j\angle X_k}$$

we have an exponential expression for the steady-state output

$$\begin{aligned} y_{ss}(t) &= \sum_{k=-\infty}^{\infty} X_k e^{jk\omega_0 t} H(jk\omega_0) \\ &= \sum_{k=-\infty}^{\infty} |X_k| \cdot |H(jk\omega_0)| e^{j(k\omega_0 t + \angle X_k + \angle H(jk\omega_0))} \end{aligned}$$

as in Figure 5.7. This equation can be written in the form of a cosine series as

$$\begin{aligned} y_{ss}(t) = X_0 H(j0) + \sum_{k=1}^{\infty} 2|X_k| \cdot |H(jk\omega_0)| \\ \cdot \cos(k\omega_0 t + \angle X_k + \angle H(jk\omega_0)), \end{aligned}$$

which can be considered to be the generalized form of the sinusoidal steady-state formula developed earlier in the book. Notice that the values of $H(jk\omega_0)$ are samples of the continuous frequency response $H(j\omega)$ at the frequencies $\omega_k = k\omega_0$.

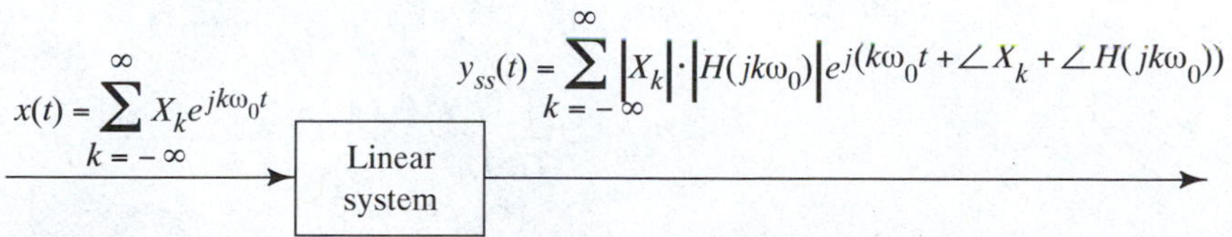

FIGURE 5.7 *Transmission of a periodic signal through a linear system*

ILLUSTRATIVE PROBLEM 5.4
Three Signals Applied to a Linear Circuit

Consider the frequency response of Illustrative Problem 5.3 to be that of a second-order *RC* circuit whose input voltage is the truncated Fourier series $x(t) = 10 + 5\cos(t) + 2\cos(10t)$ with $\omega_0 = 1$ rad/s. Find the equation for the steady-state output voltage $y_{ss}(t)$.

Solution

The circuit's frequency response is

$$H(j\omega) = \frac{1}{-\omega^2 + 2 + 3j\omega}$$

and the input frequencies are 0, 1, and 10 rad/s. Thus we calculate $H(j0) = 0.500$, $H(j1) = 0.316e^{-j1.249}$, and $H(j10) \approx 0$. Using the preceding cosine form gives

$$y_{ss}(t) = X_0 H(j0) + \sum_{k=1}^{\infty} 2|X_k| \cdot |H(jk\omega_0)| \cos(k\omega_0 t + \angle X_k + \angle H(jk\omega_0)).$$

With $X_0 = 10$, $X_1 = 2.5$, and $X_{10} = 1$, the steady-state output voltage is

$$\begin{aligned} y_{ss}(t) &= 10(0.500) + 5(0.316)\cos(t - 1.249) \\ &= 5 + 1.580\cos(t - 1.249). \end{aligned}$$

Power in a Periodic Signal Often we want to know the average power dissipated in a resistor or, more generally, the average power contained in a periodic signal given by

definition of average power

$$P_{\text{avg}} = \frac{1}{T_0}\int_{t_0}^{t_0+T_0} p(t)\,dt,$$

where $p(t)$ is the instantaneous power in the signal. For a resistor of R ohms having a voltage $v(t)$, we have $p(t) = v(t)^2/R$, which gives the average power

$$P_{\text{avg}} = \frac{1}{T_0}\int_{t_0}^{t_0+T_0} \frac{v^2(t)}{R}\,dt.$$

For the sinusoidal resistor voltage $v(t) = A\cos(\omega_0 t + \alpha)$, the average power is

$$\begin{aligned} P_{\text{avg}} &= \frac{1}{T_0}\int_{t_0}^{t_0+T_0} \frac{[A\cos(\omega_0 t + \alpha)]^2}{R}\,dt \\ &= \frac{A^2}{RT_0}\int_{t_0}^{t_0+T_0} \frac{1}{2}\{1 + \cos(2\omega_0 t + 2\alpha)\}\,dt. \end{aligned}$$

The integral of $\cos(2\omega_0 t + 2\alpha)$ over one period is clearly zero, and we have

power in a sine wave

$$P_{\text{avg}} = \frac{A^2}{RT_0}\int_0^{T_0} \frac{1}{2}(1)\,dt = \frac{A^2}{2R} \text{ watts}.$$

It is customary to talk about *normalized power* (or the power in a 1-Ω resistor if an electrical application) as

$$P_{\text{avgn}} = P_{\text{avg}} = \frac{A^2}{2}.$$

Extending this result to the case where $v(t)$ is a periodic signal and using the Fourier cosine representation

$$v(t) = V_0 + 2\sum_{k=1}^{\infty} |V_k| \cos(k\omega_0 t + \angle V_k),$$

the normalized average power is (see Problem P5.31)

$$\begin{aligned} P_{\text{avgn}} &= P_0 + P_1 + \cdots + P_k + \cdots \\ &= V_0{}^2 + \frac{4|V_1|^2}{2} + \frac{4|V_2|^2}{2} + \cdots + \frac{4|V_k|^2}{2} + \cdots = V_0{}^2 + 2\sum_{k=1}^{\infty} |V_k|^2. \end{aligned}$$

Comment: Notice that $2|V_k|$ is the zero-to-peak magnitude of the kth harmonic, and, from a circuits course, we recall that the root-mean-square (rms) value of the kth harmonic is $2|V_k|/\sqrt{2}$. With average power dependent on the square of the rms value, we have the same relationship as in the previous equation for P_{avgn}.

If one-half of the power is assigned to each exponential Fourier series component ($\pm k$), we have the equivalent expression

$$P_{\text{avgn}} = \sum_{k=-\infty}^{\infty} |V_k|^2.$$

Thus the two expressions for average power generate Parseval's relation for periodic signals:

$$P_{\text{avgn}} = \sum_{k=-\infty}^{\infty} |V_k|^2 = \frac{1}{T_0}\int_{t_0}^{t_0+T_0} p(t)\,dt.$$

The average power can be computed directly in the time domain or it may be approximated by using the Fourier coefficients of the exponential representation for a general periodic signal $f(t)$, as is shown next. The approximation results from terminating the Fourier series representation at $k = \pm L$ rather than using $k = \pm\infty$.

Parseval's theorem

$$P_{\text{avgn}} = \sum_{k=-\infty}^{\infty} |F_k|^2 = \frac{1}{T_0}\int_{t_0}^{t_0+T_0} p(t)\,dt$$

normalized power from k domain *normalized power from time domain*

An indication of the error introduced by truncating the Fourier series with $2L + 1$ terms can be obtained from these relationships. Defining

$$P_a = \sum_{k=-L}^{L} |F_k|^2$$

as the average power in the truncated series, then the smaller the quantity

$$\frac{P_{\text{avgn}} - P_a}{P_{\text{avgn}}}$$

the better the truncated Fourier approximation to $f(t)$. As expected, the more terms included in the Fourier expansion, the smaller the difference between P_{avgn} and P_a.

ILLUSTRATIVE PROBLEM 5.5 *Power Calculation in a Circuit*

In Illustrative Problem 5.4 the input voltage to the linear circuit was given as $x(t) = 10 + 5\cos(2t) + 2\cos(10t)$ and the steady-state output voltage was found to be $y_{ss}(t) = 5 + 1.50\cos(2t - 1.249)$. Find the normalized power in the input and output signals.

Solution

The normalized average power is given by

$$P_{\text{avgn}} = G_0{}^2 + \sum_{k=1}^{\infty} \frac{G_k{}^2}{2}$$

where G_k is the zero-to-peak value of the kth sinusoid. Thus for the input voltage $P_{\text{in}} = 10^2 + 5^2/2 + 2^2/2 = 114.5$ W, and for the output $P_{\text{out}} = 5^2 + 1.5^2/2 = 26.125$ W.

FOURIER TRANSFORM

Definition It is also possible and useful to analyze the frequency content of nonperiodic signals. The tool enabling us to do this is the Continuous-Time Fourier Transform of a signal $x(t)$, defined as

$$\mathcal{F}[x(t)] = X(\omega) = \int_{-\infty}^{\infty} x(t)e^{-j\omega t}\,dt.$$

$X(\omega)$ is a complex function of frequency. The inverse transform is given by

$$\mathcal{F}^{-1}[X(\omega)] = x(t) = \frac{1}{2\pi}\int_{-\infty}^{\infty} X(\omega)e^{j\omega t}\,d\omega\,.$$

We could use the notation $X(j\omega)$ to be consistent with $H(j\omega)$, which was used for frequency response in Chapters 2, 3, and 4, but $X(\omega)$ is widely accepted, so our Fourier transform pair is

$$x(t) = \frac{1}{2\pi}\int_{-\infty}^{\infty} X(\omega)e^{j\omega t}\,d\omega \Leftrightarrow X(\omega) = \int_{-\infty}^{\infty} x(t)e^{-j\omega t}\,dt\,.$$

Fourier transform pair ω (omega) form

Engineers (particularly electrical) often prefer to write the transform pair in terms of the hertz frequency f rather than the mathematical frequency ω, basing their prejudice, perhaps, on the fact that radio dials are labeled in kHz and MHz rather than krad or Mrad. With $\omega = 2\pi f$ we have the symmetrical pair

$$x(t) = \int_{-\infty}^{\infty} X(f)e^{j2\pi ft}\,df \Leftrightarrow X(f) = \int_{-\infty}^{\infty} x(t)e^{-j2\pi ft}\,dt.$$

Fourier transform pair f (frequency) form

The notation $\mathcal{F}[x(t)]$ denotes the Fourier transform of $x(t)$, where either the ω or f transform is assumed, depending on preference or custom. Irrespective of whether we are using f or ω, the preceding two equations for $x(t)$ are known as *synthesis equations,* since $x(t)$ is "synthesized" from an infinity of complex exponentials, $e^{j\omega t}$ or $e^{j2\pi ft}$, having amplitudes of $X(f)$ or $[1/2\pi]X(\omega)$. The equations for $X(\omega)$ and $X(f)$ are called the *analysis equations* because $x(t)$ is analyzed in terms of its frequency or spectral components. Notice that the frequency spectrum for $x(t)$ or $X(f)$ is *continuous*—that is, all frequencies are present—whereas the spectrum for the Fourier series coefficients, the X_k's, is *discrete*—only discrete frequencies are present. This is a very important distinction. Also notice that a Fourier transform *does not exist* for signals where the improper integral

$$X(\omega) = \int_{-\infty}^{\infty} x(t)e^{-j\omega t}\,dt$$

is undefined. Sufficient conditions (Dirichlet conditions) for the existence of the Fourier transform are that $x(t)$ (1) has a finite number of minima and maxima within any finite interval; (2) has a finite number of discontinuities within any finite interval (each of these discontinuities must be finite); and (3) is absolutely integrable; that is,

$$\int_{-\infty}^{\infty} |x(t)|\,dt < \infty.$$

Following is the development of the Fourier transform pairs for some important signals. A more complete listing is given in Table 5.1, Continuous-Time Fourier Transform Pairs, found at the end of the chapter.

Pairs for Finite-Energy Signals A signal $x(t)$ having finite energy is referred to as an *energy signal;* that is,

$$E = \int_{-\infty}^{\infty} x^2(t)\,dt < \infty.$$

Let's consider the Fourier transform pairs for several such signals.

Causal Exponential The Fourier transform[4] of the causal exponential $x(t) = Ae^{-at}u(t)$ with $a > 0$ is

$$X(\omega) = \int_{-\infty}^{\infty} Ae^{-at}e^{-j\omega t}u(t)\,dt = \int_{0}^{\infty} Ae^{-at}e^{-j\omega t}\,dt$$
$$= -\left.\frac{Ae^{-t(a+j\omega)}}{a + j\omega}\right|_{0}^{\infty} = \frac{A}{a + j\omega},$$

with the magnitude and phase of this transform plotted in Figure 5.8 for $a = A = 1$. To obtain these plots, **freqs** was used with $X(s) = 1/(s + 1)$. In the familiar notation we have

causal exponential

$$Ae^{-at}u(t) \Leftrightarrow \frac{A}{j\omega + a}, \quad a > 0.$$

Comment: Notice that the infinite energy signal $x(t) = Ae^{at}u(t)$ with $a > 0$ does not have a Fourier transform because

$$\int_{0}^{\infty} Ae^{at}e^{-j\omega t}\,dt \rightarrow \infty \quad \text{for } a > 0.$$

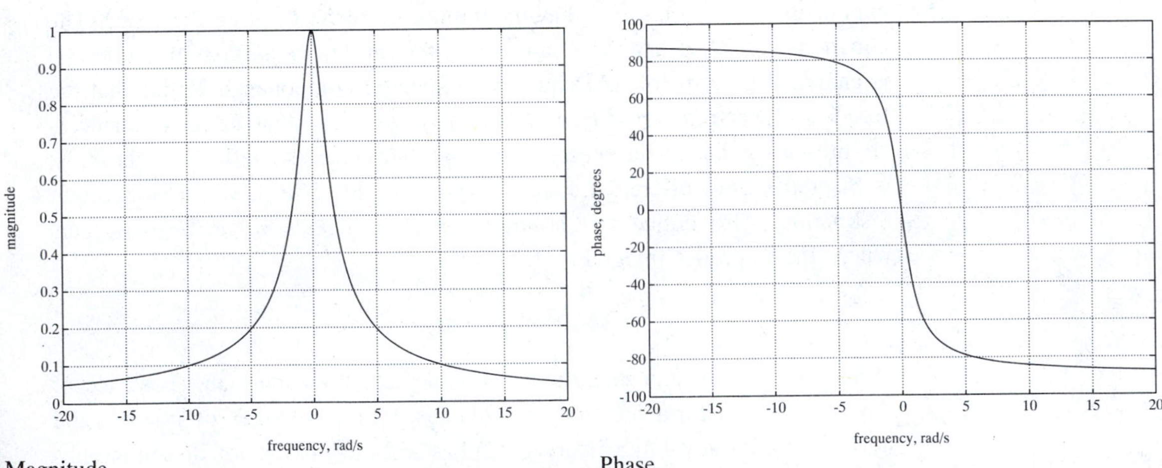

Magnitude Phase

FIGURE 5.8 *Fourier transform for the causal exponential* $x(t) = e^{-t}u(t)$

[4]For the most part we will use the "omega" Fourier transform pair,

$$f(t) = \frac{1}{2\pi}\int_{-\infty}^{\infty} F(\omega)e^{j\omega t}\,d\omega \Leftrightarrow F(\omega) = \int_{-\infty}^{\infty} f(t)e^{-j\omega t}\,dt$$

Anticausal Exponential For the exponential $x(t) = Ae^{at}u(-t)$ with $a > 0$, we have

$$X(\omega) = \int_{-\infty}^{\infty} Ae^{at}e^{-j\omega t}u(-t)\,dt = \int_{-\infty}^{0} Ae^{at}e^{-j\omega t}dt = \left.\frac{Ae^{t(a-j\omega)}}{a - j\omega}\right|_{-\infty}^{0}$$

$$= \frac{A}{a - j\omega} = \frac{-A}{j\omega - a}.$$

In standard notation the pair is

anticausal exponential

$$Ae^{at}u(-t) \Leftrightarrow \frac{-A}{j\omega - a}, \qquad a > 0,$$

with the magnitude and phase of this transform plotted in Figure 5.9 for $a = A = 1$. Again, the function **freqs** was used with $X(s) = -1/(s - 1)$.

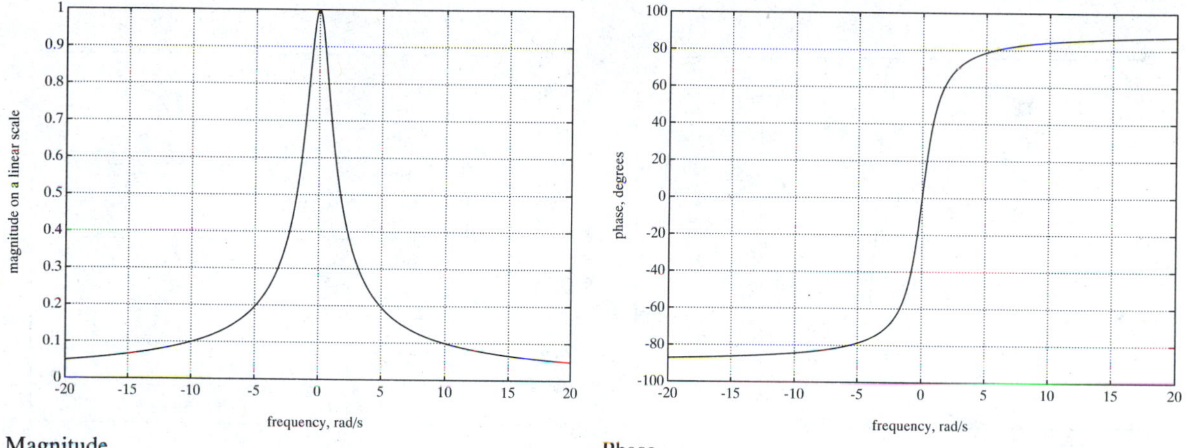

FIGURE 5.9 *Fourier transform for the anticausal exponential* $x(t) = e^{t}u(-t)$

Comment: Notice that the infinite energy signal $x(t) = Ae^{at}u(-t)$ with $a < 0$ does not have a Fourier transform because

$$\int_{-\infty}^{0} Ae^{at}e^{-j\omega t}dt \to \infty \quad \text{for } a < 0.$$

ILLUSTRATIVE PROBLEM 5.6
Transform of an Exponential Pulse

Now let's find the Fourier transform of the exponential pulse $x(t) = Ae^{-at}u(t) + Ae^{at}u(-t)$ for $a > 0$. This pulse is the sum of the anticausal and the causal exponentials discussed previously.

Solution Using linearity, a property whose general statement is given on page 247, we can write the transform as the sum of the transforms just derived, producing

$$X(\omega) = \frac{A}{a + j\omega} + \frac{A}{a - j\omega} = \frac{2Aa}{a^2 + \omega^2}$$

and the pair

exponential pulse

$$Ae^{-a|t|} \Leftrightarrow \frac{2Aa}{\omega^2 + a^2}.$$

To plot the magnitude and phase of this transform for $a = A = 1$, we use **freqs** with $X(s) = 1/(s + 1) - 1/(s - 1) = -2/(s^2 - 1)$, which gives the result in Figure 5.10.

CROSS-CHECK Notice that this transform is real and positive, as confirmed by both the mathematics and the plots. ■

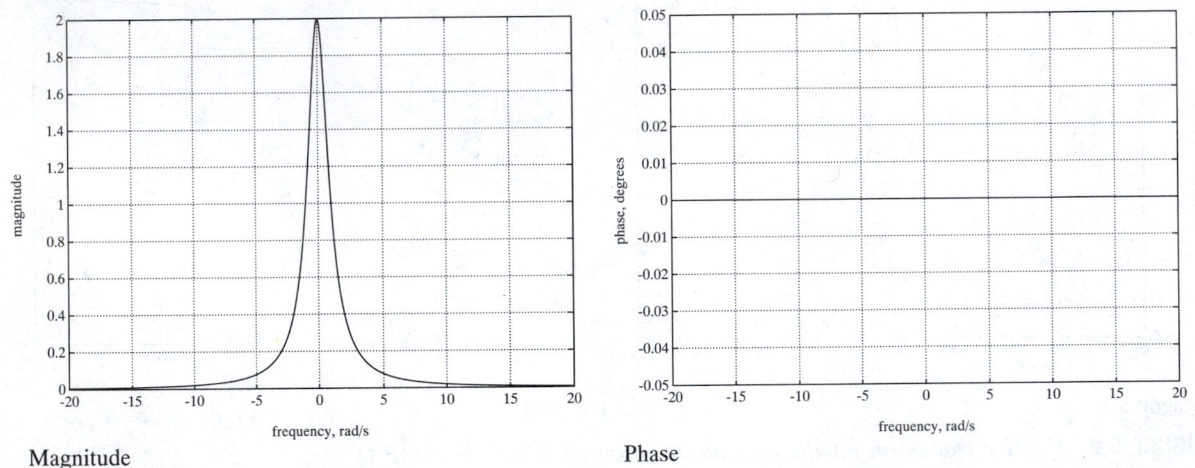

FIGURE 5.10 *Fourier transform for the exponential pulse* $x(t) = e^{-t}u(t) + e^{t}u(-t)$

Rectangular Pulse Now we'll determine the Fourier transform of a rectangular pulse of height A and width a that is described by $x(t) = A[u(t + a/2) - u(t - a/2)]$ and is shown in Figure 5.11a. Using the analysis equation with the limits of $t = -a/2$ and $t = a/2$ yields the Fourier transform

$$X(\omega) = \int_{-a/2}^{a/2} Ae^{-j\omega t}\, dt = A\left[\frac{e^{j\omega a/2} - e^{-j\omega a/2}}{j\omega}\right] = \frac{2A}{\omega} \sin \frac{\omega a}{2},$$

as shown in Figure 5.11b for $A = 1/2$ and $a = 2$, where we notice that the first zero crossings are found from $\omega a/2 = \pm\pi$ or $\omega = \pm 2\pi/a$. At $\omega = 0$, l'Hôpital's rule indicates that $X(0) = Aa = 1$.

As the time-domain pulse becomes narrower (the width a gets smaller), the frequency spectrum spreads out in the sense that the zero-crossing frequencies become greater. This is an indication that narrower pulses contain more significant high-frequency components than broader pulses. In the pair notation we have

rectangular pulse

$$A[u(t + a/2) - u(t - a/2)] \Leftrightarrow \frac{2A}{\omega} \sin \frac{\omega a}{2}.$$

Using the Symbolic Math Toolbox on the pulse of Figure 5.11a ($A = 0.5$ and $a = 2$), we have `x=('0.5*Heaviside(t+1)-0.5*Heaviside(t-1)')` and invoking **`X=fourier(x)`**, returns `X=-.5*i/w*exp(1.*i*w)+.5*i/w*exp(-1.*i*w)`, which can be written as `sinw/w`.

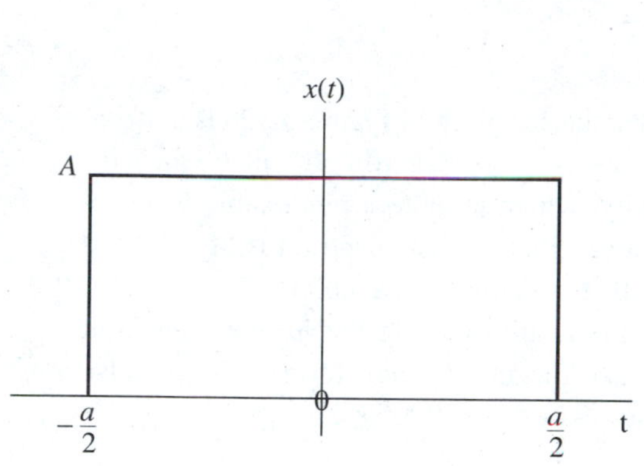

a. Rectangular pulse

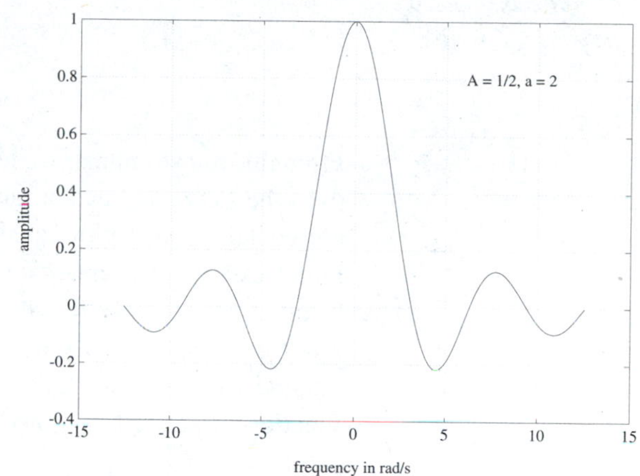

b. Fourier transform of rectangular pulse

FIGURE 5.11 *Rectangular pulse and its Fourier transform*

We have seen that when using the Fourier transform, a signal $x(t)$ is represented as an integral of complex exponential functions lying in the interval $\omega = -\infty$ to $\omega = \infty$. Looking at the inverse transform relation

$$x(t) = \frac{1}{2\pi} \int_{-\infty}^{\infty} X(\omega) e^{j\omega t}\, d\omega\,,$$

we notice that the relative amplitude of the component at any frequency ω is proportional to $X(\omega)$ and if, for example, $x(t)$ represents a voltage, $X(\omega)$ can be considered a voltage-density spectrum, or the *spectral-density function* of $x(t)$.

Pairs for Power Signals The procedure for finding the Fourier transform (spectral-density function) of signals having finite energy is simply a straightforward exercise in calculus. Now, what do we do about signals with infinite energy—that is, power signals? A power signal $x(t)$ has finite power, that is,

power signal

$$\lim_{T\to\infty} \frac{1}{T} \int_{-T/2}^{T/2} x^2(t)\, dt \quad \text{is finite.}$$

The Impulse Function Using the sifting property of the impulse function, the Fourier transform of the impulse $x(t) = A\delta(t)$ of Figure 5.12a is

$$X(\omega) = \int_{-\infty}^{\infty} A\delta(t)e^{-j\omega t}\,dt = Ae^{-j\omega 0} = A$$

as shown in Figure 5.12b. In the common notation, we have the pair

impulse function

$$A\delta(t) \Leftrightarrow A\,.$$

From this transform pair we notice that the spectrum of Figure 5.12b is uniform over the entire frequency range $-\infty \le \omega \le \infty$. Accordingly, all frequencies are contained and with equal intensity, which is quite a remarkable feat, but impulses are quite remarkable creatures. This type of spectrum is often called "white," in reference to white light that contains all the colors.

The Constant, A Let's investigate the result of using the inverse transform integral (the synthesis integral) to take the inverse transform of an impulse function of strength A—that is, $X(\omega) = A\,\delta(\omega)$. We have

using sifting property

$$x(t) = \frac{1}{2\pi}\int_{-\infty}^{\infty} X(\omega)e^{j\omega t}\,d\omega = \frac{1}{2\pi}\int_{-\infty}^{\infty} A\delta(\omega)e^{j\omega t}\,d\omega$$
$$= \frac{1}{2\pi}Ae^{j0t} = \frac{A}{2\pi}.$$

Thus $\mathcal{F}[A/2\pi] = A\delta(\omega)$ or $\mathcal{F}[A] = 2\pi A\delta(\omega)$. This makes sense from the spectral density point of view because a constant voltage of amplitude A is, in electrical engineering terminology, a DC signal, which certainly has all its

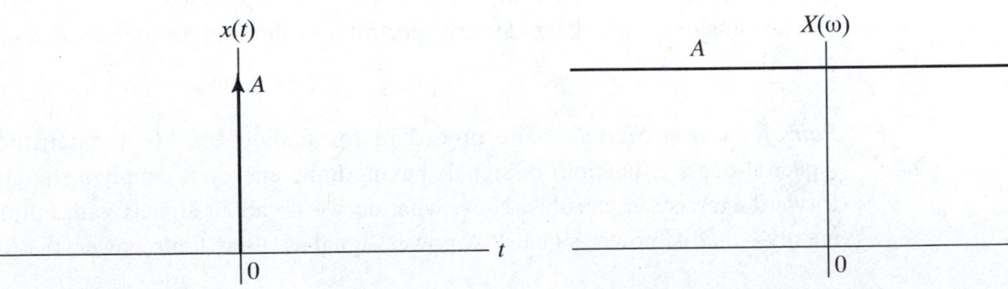

a. Impulse function

b. Transform of impulse function

FIGURE 5.12 *Impulse function and its Fourier transform*

power concentrated at $\omega = 0$. There are no other frequencies present. The constant and its transform are shown in Figure 5.13; in the usual manner

constant

$$A \Leftrightarrow 2\pi A\delta(\omega).$$

To verify this pair from another point of view, we use the Symbolic Math Toolbox with the transform `X='2*pi*A*Dirac(w)'` and invoke **`x=invfourier(X)`** to obtain $x=A$ as above.

The Complex Exponential, $Ae^{j\omega_1 t}$ A good guess for the transform of the complex exponential $Ae^{j\omega_1 t}$ would be a weighted impulse at $\omega = \omega_1$, since there is only one frequency (ω_1) present in the time-domain signal. Trying $X(\omega) = A\delta(\omega - \omega_1)$, we have

$$x(t) = \frac{1}{2\pi}\int_{-\infty}^{\infty} X(\omega)e^{j\omega t}\,d\omega = \frac{1}{2\pi}\int_{-\infty}^{\infty} A\delta(\omega - \omega_1)e^{j\omega t}\,d\omega = \frac{Ae^{j\omega_1 t}}{2\pi}.$$

Thus

$$\mathcal{F}\left[\frac{Ae^{j\omega_1 t}}{2\pi}\right] = A\delta(\omega - \omega_1) \quad \text{or} \quad \mathcal{F}[Ae^{j\omega_1 t}] = 2\pi A\delta(\omega - \omega_1).$$

And, of course, using the conventional pair notation, the result is

complex exponential

$$[Ae^{j\omega_1 t}] \Leftrightarrow 2\pi A\delta(\omega - \omega_1).$$

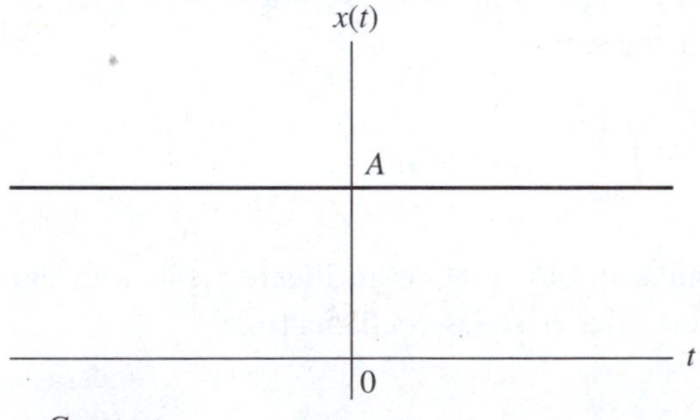

a. Constant

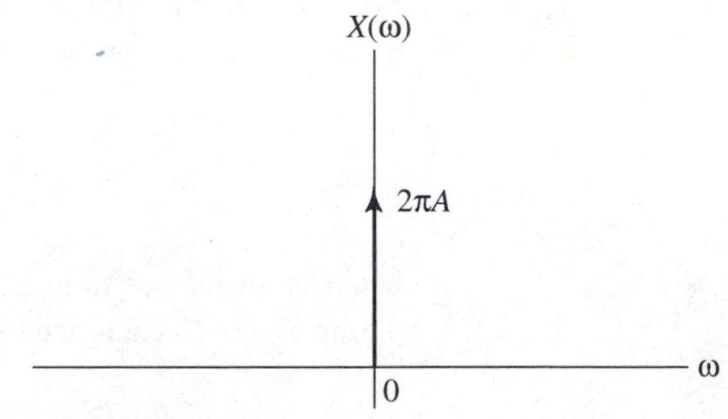

b. Transform of a constant

FIGURE 5.13 *Constant and its transform*

BASIC CONCEPTS

ILLUSTRATIVE PROBLEM 5.7 *Transform of a Sinusoid*

Determine the Fourier transform of the sinusoidal signal $x(t) = A\cos(\omega_a t)$, $-\infty \le t \le \infty$.

Solution

From the Euler relation, this sinusoid is

$$x(t) = 0.5A(e^{j\omega_a t} + e^{-j\omega_a t}).$$

Then, using the preceding results for the complex exponential, we find that

$$\mathcal{F}[0.5A(e^{j\omega_a t} + e^{-j\omega_a t})] = \pi A[\delta(\omega - \omega_a) + \delta(\omega + \omega_a)].$$

In the pair notation,

cosine

$$A\cos(\omega_a t) \Leftrightarrow \pi A[\delta(\omega - \omega_a) + \delta(\omega + \omega_a)].$$

Comments:

1. Notice that this is a two-sided infinite cosine signal, contrasted with the one-sided infinite signal $x(t) = \cos(\omega_a t)u(t)$.
2. Following a similar procedure we have the "sine pair"

sine

$$A\sin(\omega_a t) \Leftrightarrow -j\pi A[\delta(\omega - \omega_a) - \delta(\omega + \omega_a)].$$

Periodic Signals The exponential Fourier series for a periodic signal is

$$f(t) = \sum_{k=-\infty}^{\infty} F_k e^{jk\omega_0 t}$$

and since F_k is simply a complex constant, we can use the transform for the complex exponential $Ae^{j\omega_1 t}$ to form the important pair

periodic signal

$$\sum_{k=-\infty}^{\infty} F_k e^{jk\omega_0 t} \Leftrightarrow 2\pi \sum_{k=-\infty}^{\infty} F_k \delta(\omega - k\omega_0).$$

Fourier Series Coefficients from Fourier Transform A single pulse $f(t)$, as in Figure 5.14a, has the Fourier transform

$$F(\omega) = \int_{-a/2}^{a/2} f(t)e^{-j\omega t}\,dt,$$

whereas an infinite train of similar pulses, $g(t)$, as in Figure 5.14b with the period T_0, is characterized by the Fourier series coefficients

$$F_k = \frac{1}{T_0}\int_{-a/2}^{a/2} f(t)e^{-jk\omega_0 t}\,dt.$$

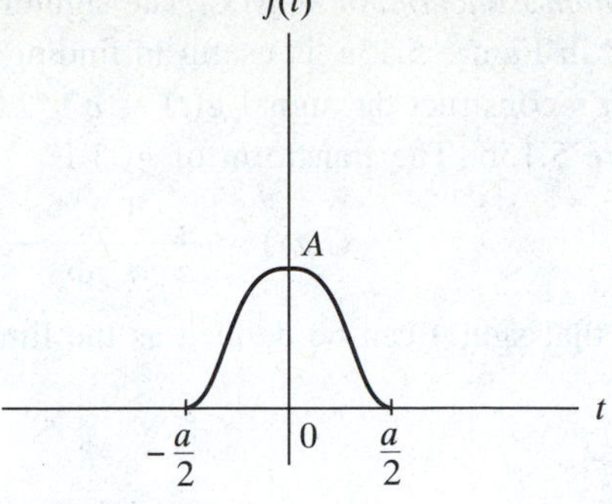

a. Single pulse

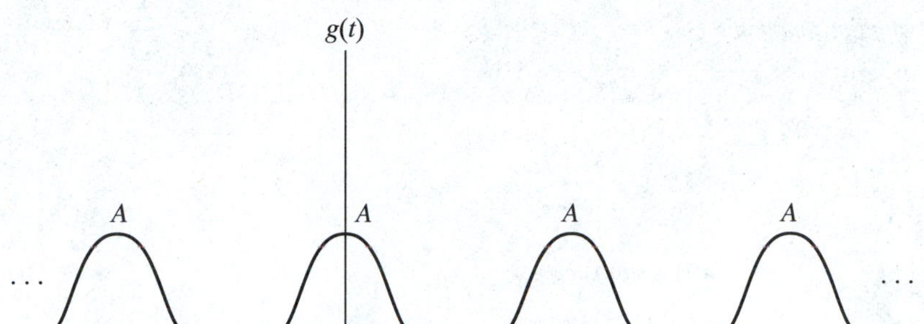

b. Pulse train

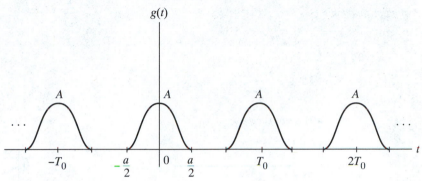

FIGURE 5.14 *Single pulse and periodic extension*

Comparing these two equations we see that

F_k from $F(\omega)$

$$F_k = \frac{1}{T_0} \cdot F(\omega)\big|_{\omega=k\omega_0}.$$

Recall that the single rectangular pulse of amplitude A and width a of Figure 5.11a on page 241 has the transform

$$X(\omega) = \frac{2A}{\omega} \sin \frac{\omega a}{2} = \frac{Aa}{\omega a/2} \sin \frac{\omega a}{2}$$

and if we want the Fourier coefficients for an infinite periodic train of similar pulses, such as in Figure 5.3b on page 226, we use the concept just developed with $\omega_0 = 2\pi/T_0$ to obtain

$$X_k = \frac{1}{T_0} \cdot X(j\omega)\big|_{\omega=k\omega_0} = \frac{1}{T_0} \cdot \frac{Aa}{\omega a/2} \sin \frac{\omega a}{2}\bigg|_{\omega=k\omega_0}$$
$$= \frac{1}{T_0} \cdot \frac{2A}{k\omega_0} \sin \frac{k\omega_0 a}{2} = \frac{A}{k\pi} \sin \frac{k\pi a}{T_0}$$

which is equivalent to the result we found in Illustrative Problem 5.2.

The Signum Function, or sgn(t) The signum, or sign, function $x(t) = \text{sgn}(t)$, pictured in Figure 5.15a, is useful in finding the transform of a step function. First, let's construct the signal $g(t) = e^{-at}u(t) - e^{at}u(-t)$, $a > 0$, as shown in Figure 5.15b. The transform of $g(t)$ is

$$G(\omega) = \frac{1}{a + j\omega} - \frac{1}{a - j\omega}.$$

We see that sgn(t) can be defined as the limit of $g(t)$ as $a \rightarrow 0$, yielding the pair

signum function

$$\text{sgn}(t) \Leftrightarrow \frac{2}{j\omega}.$$

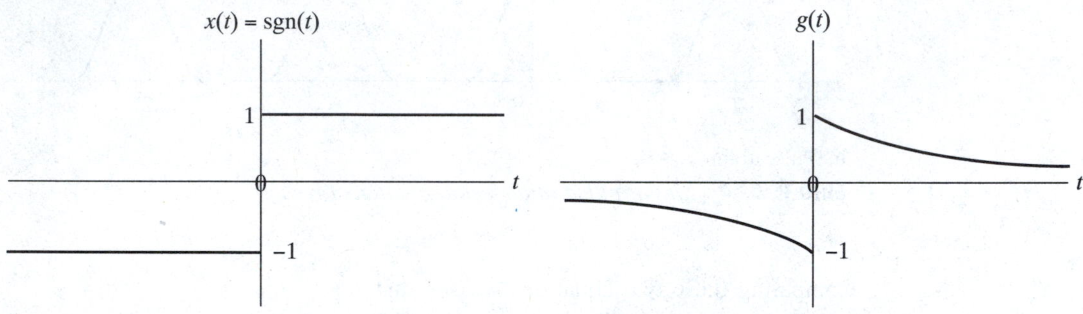

a. Signum function b. Approximation of sgn(t)

FIGURE 5.15 *Signum function and its approximation*

ILLUSTRATIVE PROBLEM 5.8 *Transform of a Step Function*

The unit step function $u(t)$ can be described as $u(t) = 0.5[1 + \text{sgn}(t)]$. Find its Fourier transform $U(\omega)$.

Solution Using the pairs

$$A \Leftrightarrow 2\pi A\delta(\omega) \quad \text{and} \quad \text{sgn}(t) \Leftrightarrow \frac{2}{j\omega}$$

we have $U(\omega) = \pi\delta(\omega) + 1/j\omega$, and for the general step function $Au(t)$ the pair is

$$Au(t) \Leftrightarrow A\pi\delta(\omega) + \frac{A}{j\omega}.$$

Properties It is useful to be able to work and analyze problems in both time and frequency domains without always having to use the defining integrals. Therefore, we here develop some useful Fourier transform properties that are included in Table 5.2 at the end of the chapter.

Linearity The linearity property comes directly from the defining integral as

linearity

$$\mathcal{F}[ag_1(t) + bg_2(t)] = \int_{-\infty}^{\infty} ag_1(t)e^{-j\omega t}dt + \int_{-\infty}^{\infty} bg_2(t)e^{-j\omega t}dt$$
$$= aG_1(\omega) + bG_2(\omega).$$

In words, the Fourier transform of the sum of two signals $g_1(t)$ and $g_2(t)$ that are multiplied by the arbitrary complex constants a and b, respectively, is a times the Fourier transform of $g_1(t)$ plus b times the Fourier transform of $g_2(t)$. Notice that this is a complex addition.

Time Shift Using the analysis integral

time shift

$$\mathcal{F}[g(t - \tau)] = \int_{-\infty}^{\infty} g(t - \tau)e^{-j\omega t}dt$$

and letting $m = t - \tau$ with $dm = dt$, we have

$$\mathcal{F}[g(t - \tau)] = \int_{-\infty}^{\infty} g(m)e^{-j\omega(m+\tau)}\,dm = e^{-j\omega\tau}\int_{-\infty}^{\infty} g(m)e^{-j\omega m}dm$$
$$= e^{-j\omega\tau}G(\omega) = |G(\omega)|e^{j(\angle G(\omega)-\omega\tau)}.$$

The term $e^{-j\omega\tau}$ simply adds a phase shift. Thus a time shift shows up in the frequency domain as a linear phase term.

Frequency Shift For the time function $g(t)e^{j\omega_0 t}$, the Fourier transform is

frequency shift

$$\mathcal{F}\left[g(t)e^{j\omega_0 t}\right] = \int_{-\infty}^{\infty} [g(t)e^{j\omega_0 t}]e^{-j\omega t}dt = \int_{-\infty}^{\infty} g(t)e^{-j(\omega-\omega_0)t}dt = G(\omega - \omega_0).$$

This property is quite useful in a situation where we have $p(t) = x(t)\cos(\omega_0 t)$ or something similar. In this case we have

$$p(t) = 0.5x(t)e^{j\omega_0 t} + 0.5x(t)e^{-j\omega_0 t} \quad \text{and}$$
$$P(\omega) = 0.5X(\omega - \omega_0) + 0.5X(\omega + \omega_0).$$

This is actually an example of what is known in communication theory as amplitude modulation. For example, if $x(t) = Ae^{-at}u(t) + Ae^{at}u(-t)$, $a > 0$, $X(\omega)$ is as shown in Figure 5.10 on page 240. Then $P(\omega) = \mathcal{F}[x(t)\cos(\omega_0 t)]$ is as represented in Figure 5.16, where we have assumed that ω_0 is much larger than the frequencies of $X(\omega)$ that have significant amplitudes. Notice that the transform of the signal $p(t)$, $P(\omega)$, has its significant frequency components around $\omega = \pm\omega_0$. This means that to transmit the signal $p(t)$ would require an antenna based on the higher frequency ω_0, which would require a smaller antenna than if we simply transmitted the signal $x(t)$.

Symmetry, or Duality This "two-for-one" property is useful for deriving or remembering pairs. For each Fourier transform pair, the duality property yields a second pair:

duality or symmetry

$$x(t) \Leftrightarrow X(\omega),$$
$$X(t) \Leftrightarrow 2\pi x(-\omega).$$

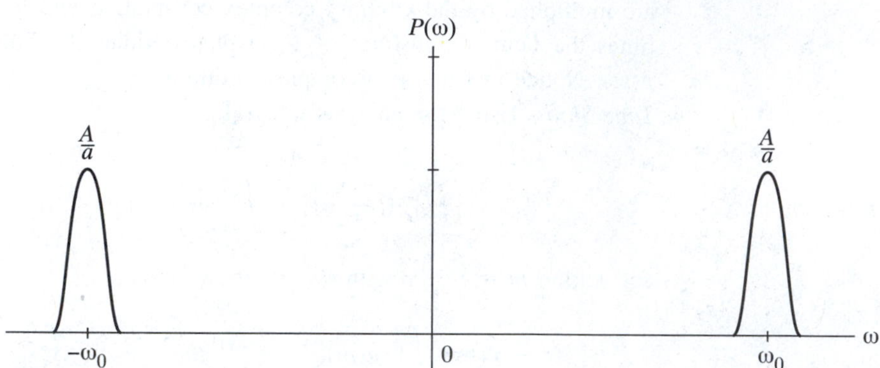

FIGURE 5.16 *Frequency shift (amplitude modulation)*

ILLUSTRATIVE PROBLEM 5.9 ***Using the Symmetry Property***

Starting from the pulse of height A and width a in the time domain and its transform $[Aa/(\omega a/2)]\sin(\omega a/2)$, use the symmetry property to find the transform of a time-domain function $[Aa/(ta/2)]\sin(ta/2)$.

Solution

Repeating the given transform pair (pair 3 in Table 5.1),

$$x(t) = A[u(t + a/2) - u(t - a/2)] \Leftrightarrow X(\omega) = \frac{Aa}{(\omega a/2)}\sin(\omega a/2)$$

we apply the symmetry property to write

$$\begin{aligned} X(t) = \frac{Aa}{(ta/2)}\sin(ta/2) &\Leftrightarrow 2\pi x(-\omega) \\ &= 2\pi A[u(-\omega + a/2) - u(-\omega - a/2)] \\ &= 2\pi A[u(\omega + a/2) - u(\omega - a/2)]. \end{aligned}$$

Comment: A pictorial view of this use of the symmetry property is given in Figure 5.17.

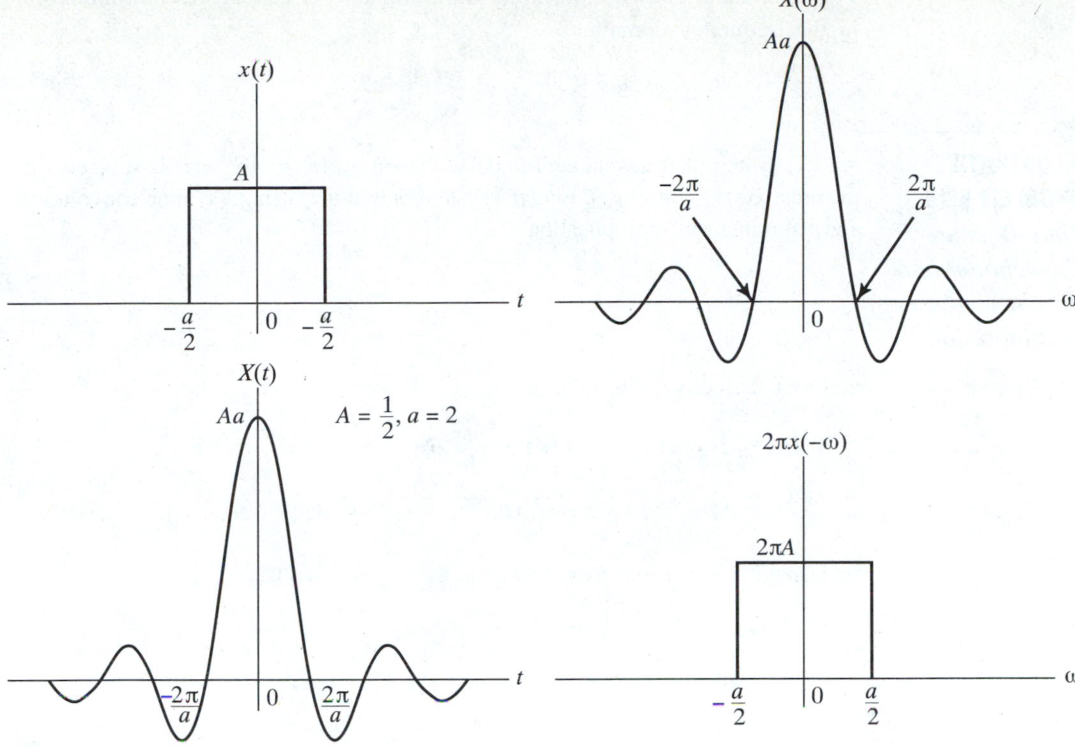

FIGURE 5.17 *Symmetry for Illustrative Problem 5.9*

Frequency Response Revisited Frequency response has been defined from the point of view of convolution (Chapter 2), differential equations (Chapter 2), and Laplace transforms (Chapter 3). If we evaluate the Fourier transform of the unit impulse response $h(t)$, the result is

frequency response

$$\mathcal{F}[h(t)] = H(\omega) = \int_{-\infty}^{\infty} h(\tau)e^{-j\omega\tau}\,d\tau$$

which is the same expression defined as the frequency response in Chapter 2. Recall that $h(t)$ must represent a stable system for $H(\omega)$ or $H(j\omega)$ to have any physical meaning.

Time-Domain Convolution Convolution was presented in Chapters 2 and 3; here we state the corresponding Fourier transform relation without proof:

$$\mathcal{F}\left[y(t) = \int_{-\infty}^{\infty} [h(\tau)x(t-\tau)\,d\tau\right] = Y(\omega) = H(\omega)\cdot X(\omega)$$

or, in operational notation,

time convolution

$$\mathcal{F}[y(t) = h(t)*x(t)] = Y(\omega) = H(\omega)\cdot X(\omega).$$

This states that convolution in the time domain corresponds to multiplication in the frequency domain.

ILLUSTRATIVE PROBLEM 5.10
Time-Domain Convolution and Frequency-Domain Multiplication

An LTI system that has the unit impulse response $h(t) = Ae^{-t}u(t)$ is subjected to the input $x(t) = Be^{-2t}u(t)$. Find the system output $y(t)$ using: (a) time convolution and (b) transform multiplication.

Solution

a. From the convolution integral

$$y(t) = \int_{-\infty}^{\infty} h(\tau)x(t-\tau)\,d\tau = \int_0^t Ae^{-\tau}Be^{-2(t-\tau)}\,d\tau$$
$$= ABe^{-2t}\int_0^t e^{\tau}\,d\tau = ABe^{-2t}[e^t - 1] = AB[e^{-t} - e^{-2t}], \qquad t \geq 0.$$

b. Using Table 5.1 and partial fraction expansion,[5] we have

$$Y(\omega) = H(\omega)\cdot X(\omega) = \frac{A}{1+j\omega}\cdot\frac{B}{2+j\omega} = \frac{AB}{1+j\omega} - \frac{AB}{2+j\omega}.$$

Using Table 5.1 once again, we have the same result as in part (a),

$$y(t) = AB[e^{-t} - e^{-2t}]u(t).$$

Frequency-Domain Convolution This property that is used extensively in communication applications can be derived from the duality property:

$$\mathcal{F}[z(t) = f(t)\cdot g(t)] = Z(\omega) = \frac{1}{2\pi}\int_{-\infty}^{\infty} F(u)G(\omega - u)\,du$$

or, in operational notation,

frequency convolution

$$\mathcal{F}[z(t) = f(t)\cdot g(t)] = \frac{1}{2\pi}F(\omega) * G(\omega).$$

Consequently, multiplication in the time domain corresponds to convolution in the frequency domain.

Frequency Convolution Involving Impulses The frequency convolution $C(\omega)$ of the arbitrary transform $F(\omega)$ with the shifted, weighted impulse $X(\omega) = D\,\delta(\omega - d)$ is given by the integral

$$C(\omega) = \int_{-\infty}^{\infty} F(u)D\,\delta(\omega - d - u)\,du.$$

[5]If you are unsure about partial fraction expansion (PFE), see the appropriate sections in Chapter 3. There we are dealing with the s domain, but the concepts and procedures transfer here.

Then, using the sifting property of the impulse function, namely,

$$\int_{-\infty}^{\infty} \delta(\sigma - a) f_1(\sigma)\, d\sigma = f_1(a)$$

we have the convolution

$$C(\omega) = \int_{-\infty}^{\infty} D\delta(\omega - d - u) F(u)\, du = DF(\omega - d).$$

This means that the spectrum $C(\omega)$ is simply $F(\omega)$ shifted by a frequency amount $\omega = d$ and scaled by D.

ILLUSTRATIVE PROBLEM 5.11
Time-Domain Multiplication and Frequency-Domain Convolution

Use frequency-domain convolution to find the Fourier transform of the causal time signal $y(t) = \cos(\omega_a t) \cdot u(t)$.

Solution

From Table 5.1

$$\mathcal{F}[\cos(\omega_a t)] = X(\omega) = \pi[\delta(\omega - \omega_a) + \delta(\omega + \omega_a)] \quad \text{and}$$

$$\mathcal{F}[u(t)] = U(\omega) = \pi\delta(\omega) + \frac{1}{j\omega}.$$

Thus, using frequency convolution as stated previously, we have

$$\begin{aligned}\mathcal{F}[y(t)] = Y(\omega) &= \frac{1}{2\pi}[X(\omega) * U(\omega)] \\ &= \frac{1}{2\pi}\left[\{\pi\delta(\omega - \omega_a) + \pi\delta(\omega + \omega_a)\} * \left\{\pi\delta(\omega) + \frac{1}{j\omega}\right\}\right].\end{aligned}$$

Thus the transform $Y(\omega)$ can be expressed in terms of the four convolutions

$$Y(\omega) = \frac{1}{2\pi}[C_1(\omega) + C_2(\omega) + C_3(\omega) + C_4(\omega)]$$

where

$$\begin{aligned}C_1(\omega) &= \pi\delta(\omega - \omega_a) * \pi\delta(\omega) = \pi^2\delta(\omega - \omega_a), \\ C_2(\omega) &= \pi\delta(\omega - \omega_a) * \frac{1}{j\omega} = \frac{\pi}{j(\omega - \omega_a)}, \\ C_3(\omega) &= \pi\delta(\omega + \omega_a) * \pi\delta(\omega) = \pi^2\delta(\omega + \omega_a), \quad \text{and} \\ C_4(\omega) &= \pi\delta(\omega + \omega_a) * \frac{1}{j\omega} = \frac{\pi}{j(\omega + \omega_a)}.\end{aligned}$$

Applying some algebra we have the transform of $y(t) = \cos(\omega_a t) \cdot u(t)$, namely,

$$Y(\omega) = \frac{\pi}{2}\delta(\omega - \omega_a) + \frac{\pi}{2}\delta(\omega + \omega_a) + \frac{j\omega}{-\omega^2 + \omega_a{}^2}.$$

Applications With Fourier transform pairs and properties defined and illustrated, let's turn to some important applications that occur in many engineering situations, including communications, signal processing, and control. Further treatment of these illustrations and discussion of many others will appear in later engineering courses.

Signal Transmission Through a Linear System The transmission of a signal through a linear system takes place, of course, in time but the system or filter is usually designed or analyzed in the frequency domain. So, knowing the input and the desired characteristics of the output, we then determine and implement the frequency response of the filter, $H(\omega)$. The sinusoidal transfer function may be defined as

$$H(\omega) = \frac{\text{Fourier transform of output}}{\text{Fourier transform of input}} = \frac{Y(\omega)}{X(\omega)}$$

which gives the output transform as

$$Y(\omega) = H(\omega) \cdot X(\omega) = |H(\omega)| \cdot |X(\omega)| e^{j(\angle H(\omega) + \angle X(\omega))}.$$

ILLUSTRATIVE PROBLEM 5.12 ***Signal Transmission Through a Lowpass Filter***

Consider the filter of Illustrative Problems 5.3 and 5.4 that is described by the frequency response $H(\omega) = 1/(-\omega^2 + 2 + j3\omega)$ and Figure 5.6 on page 232. If an exponential input $x(t) = 4e^{-4t}u(t)$ is applied to this filter, find the Fourier transform $Y(\omega)$ of the filter output signal $y(t)$ and use MATLAB to plot the frequency spectrum of the output.

Solution

Using pair 10, $X(\omega) = 4/(j\omega + 4)$, the output transform is

$$Y(\omega) = H(\omega) \cdot X(\omega) = \frac{4}{-j\omega^3 - 7\omega^2 + j14\omega + 8}.$$

To plot this spectrum (transform) with MATLAB's **freqs** we need

$$Y(s) = H(s) \cdot X(s) = \frac{1}{s^2 + 3s + 2} \cdot \frac{4}{s + 4} = \frac{4}{s^3 + 7s^2 + 14s + 8}$$

with the following script; the plot is given in Figure 5.18.

MATLAB Script

```
%F5_18 Output spectrum for IP 5.12
a=[1,7,14,8];
b=[4];
w=0:.025:15;
H=freqs(b,a,w);
mag=abs(H);
phase=angle(H);
%...changing phase to degrees and plotting statements
```

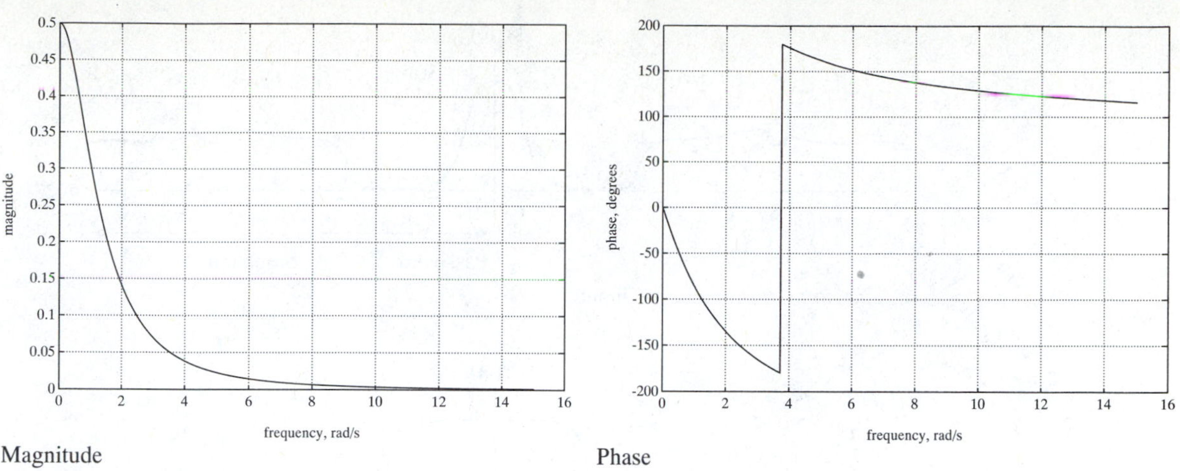

Magnitude Phase

FIGURE 5.18 *Output spectrum of a lowpass filter*

Ideal Filters Let's assume that an input signal to a filter consists of a desired signal $s(t)$ and some undesirable noise $n(t)$ and is represented as $x(t) = s(t) + n(t)$. Furthermore, we assume that their Fourier transforms $S(\omega)$ and $N(\omega)$ are bandlimited and, in an ideal situation, do not have any regions of overlap, as shown in Figure 5.19a (page 254). Here, all components of the signal $s(t)$ are concentrated at low frequencies. We would like to use a filter to pass the signal and attenuate—or even eliminate—the noise. Here it is clear that a lowpass filter (LPF) with the magnitude spectrum $|H(\omega)|$ of Figure 5.19b is desired. The resulting output spectrum is shown in Figure 5.19c.

If we could make an ideal filter, what would its characteristics be? For the input signal $s(t)$, we'd like the filter's output to be such that $y_{\text{ideal}}(t) = Ks(t - t_0)$ for all frequencies in the passband, where K is a gain constant and t_0 is a fixed delay constant. In other words, we'd like the output signal to have the same shape as the input. It is acceptable for the output to be scaled by a factor K and possibly delayed by t_0. Putting these time-domain specifications for the output signal in terms of desired frequency characteristics, we have

$$Y_{\text{ideal}}(\omega) = \begin{cases} KS(\omega)e^{-j\omega t_0}, & \text{for } \omega \text{ in passband} \\ 0, & \text{for } \omega \text{ in stopband.} \end{cases}$$

But in the passband, the Fourier transform of the input signal is $X(\omega) = S(\omega)$, while in the stopband, $X(\omega) = N(\omega)$. The ideal filter characteristics are, consequently, given by

$$H_{\text{ideal}}(\omega) = \begin{cases} Ke^{-j\omega t_0}, & \text{for } \omega \text{ in passband} \\ 0, & \text{for } \omega \text{ in stopband.} \end{cases}$$

Thus the magnitude M of the frequency response of an ideal filter is given by

$$M = |H_{\text{ideal}}(\omega)| = \begin{cases} K, & \text{for } \omega \text{ in passband} \\ 0, & \text{for } \omega \text{ in stopband} \end{cases}$$

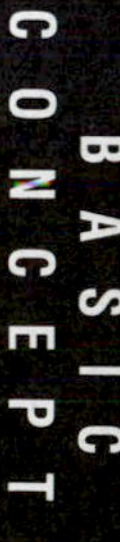

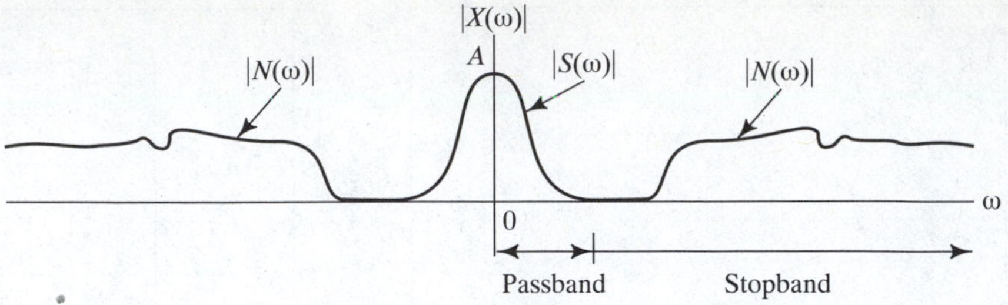

a. Transform magnitude for input

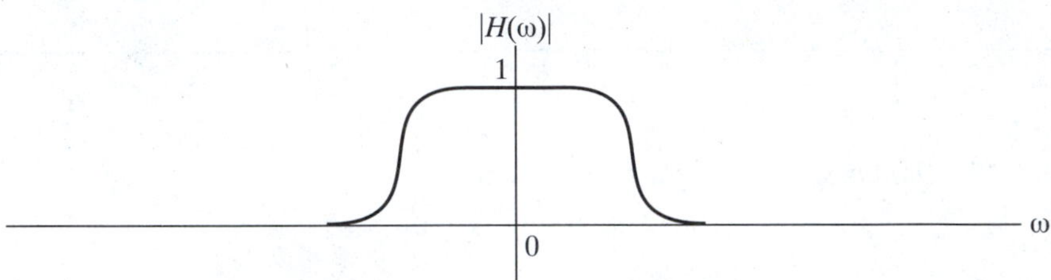

b. Transform magnitude for filter

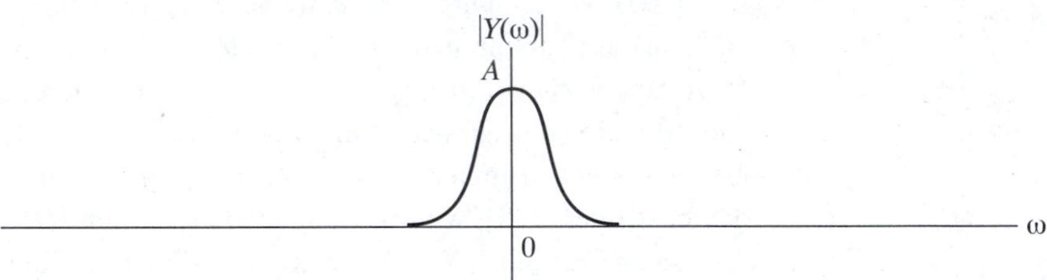

c. Transform magnitude for output

FIGURE 5.19 *Magnitude spectrum for $x(t)$, $h(t)$, and $y(t)$*

with the phase P of the frequency response described by

$$P = \angle H_{\text{ideal}}(\omega) = \begin{cases} -\omega t_0, & \text{for } \omega \text{ in passband} \\ \text{don't care}, & \text{for } \omega \text{ in stopband.} \end{cases}$$

We see that an ideal filter has a constant gain $M = K$ and linear phase $P = -\omega t_0$ in its passband and zero gain $M = 0$ in its stopband. (Phase has no significance when the gain is zero.) The frequency response characteristics of four types of ideal filters are given in Figure 5.20.

For implications and consequences of ideal and nonideal filters, see Example 5.6, "Impulse Response of an Ideal and a Butterworth Filter," and Example 5.7, "Distortionless Transmission." Design of practical analog filters, often based on the properties of ideal filters, is treated in detail in more advanced courses such as communications and signal processing systems, as well as in courses on filter design.

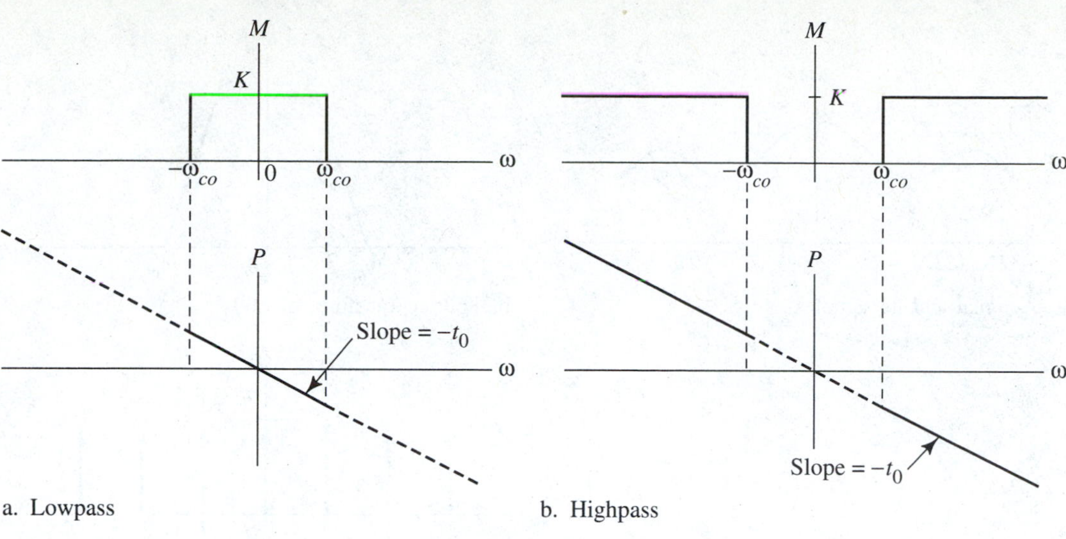

a. Lowpass

b. Highpass

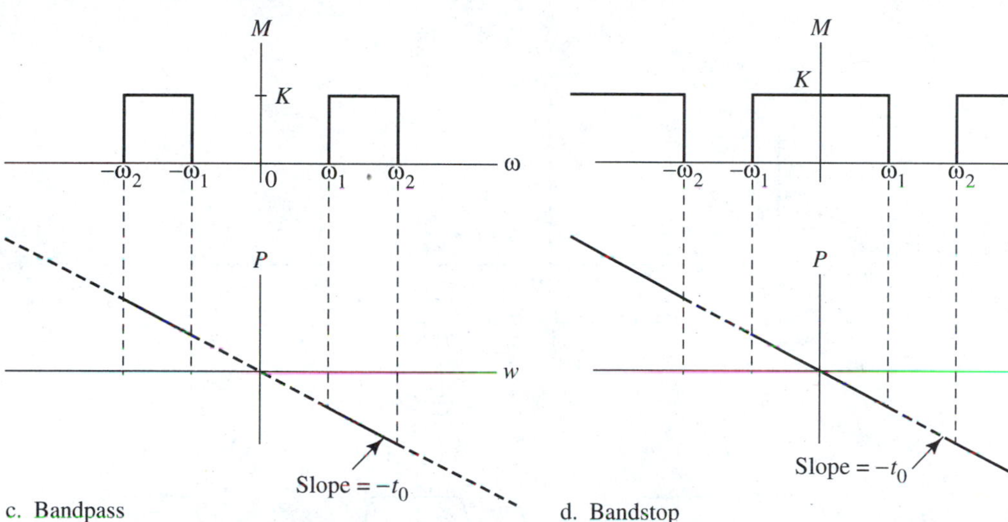

c. Bandpass

d. Bandstop

FIGURE 5.20 *Ideal filter frequency responses*

The Sampling Theorem The very important Sampling Theorem was discussed briefly in Chapter 1. Now let's consider sampling from the point of view of Fourier transforms. A signal $g(t)$ is shown in Figure 5.21a, along with the samples taken at equal intervals of T seconds. We assume that the Fourier transform of $g(t)$ is as shown in Figure 5.21b. The important feature of $G(\omega)$ is that it is bandlimited—that is, nonzero only in the frequency range $-\omega_m \leq \omega \leq \omega_m$. Its shape and the assumption that $G(\omega)$ is real are not important in the development that follows.

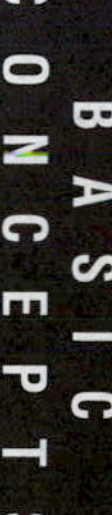

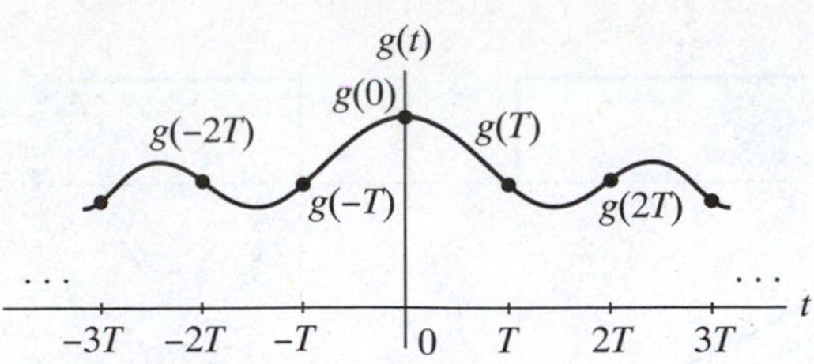

a. A signal and its samples

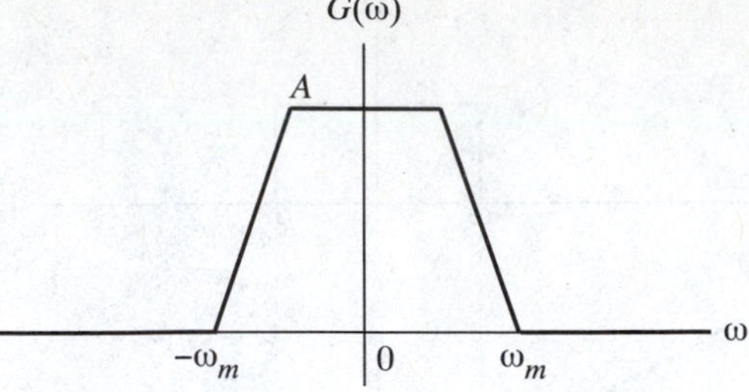

b. Fourier transform of $g(t)$

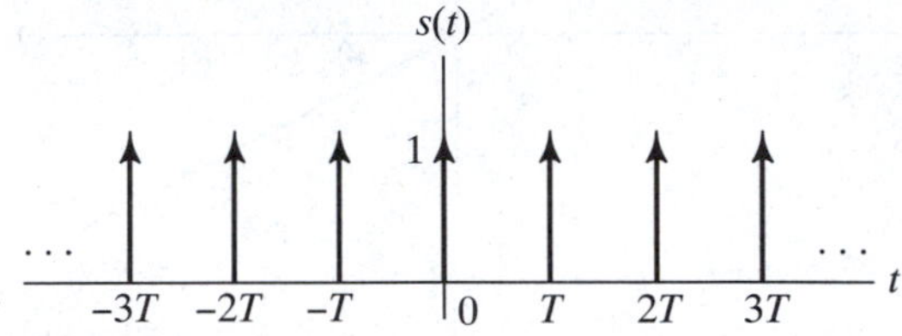

c. Impulse train representing the sampling operation

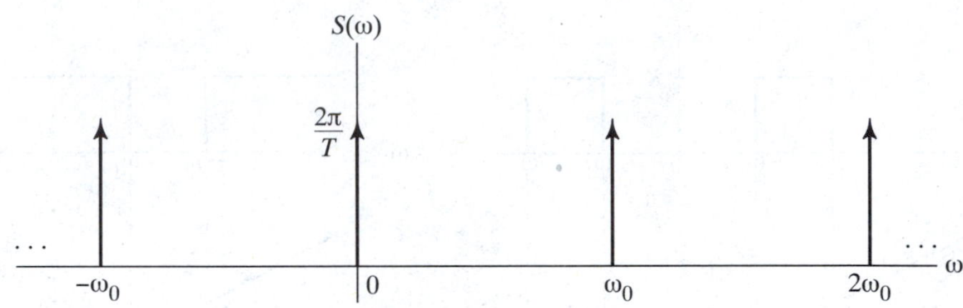

d. Fourier transform of impulse train

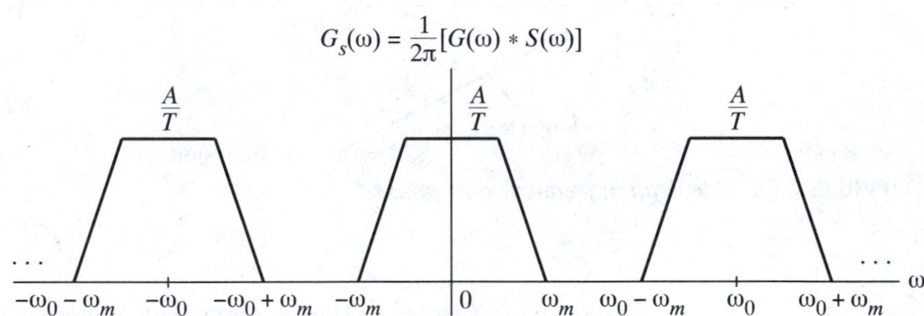

e. Fourier transform of sampled signal

FIGURE 5.21 *Steps in deriving the Sampling Theorem*

To write a mathematical expression for the sampled time signal $g_s(t)$, we model $g_s(t)$ as resulting from the multiplication of $g(t)$ with a train of impulses in the manner

sampled signal

$$g_s(t) = g(t) \cdot s(t) = g(t) \cdot \sum_{n=-\infty}^{\infty} \delta(t - nT),$$

where $s(t)$ is the impulse train, a periodic function, used to represent the sampling operation shown in Figure 5.21c described by

impulse train

$$s(t) = \sum_{n=-\infty}^{\infty} \delta(t - nT).$$

To derive the Fourier transform of the impulse train $s(t)$ shown in Figure 5.21c, we can make use of property 11 in Table 5.2. The Fourier coefficients are

Fourier coefficients

$$S_k = \frac{1}{T_0} \int_{t_0}^{t_0+T_0} s(t) e^{-jk\omega_0 t} dt = \frac{1}{T} \int_{t_0}^{t_0+T} \delta(t) e^{-jk\omega_0 t} dt = \frac{1}{T},$$

where we have used the sifting property of the impulse function and set $T_0 = T$. From transform pair 16 of Table 5.1 with $S_k = 1/T$, we have the Fourier transform of $s(t)$,

Fourier transform

$$S(\omega) = 2\pi \sum_{k=-\infty}^{\infty} S_k \delta(\omega - k\omega_0) = \frac{2\pi}{T} \sum_{k=-\infty}^{\infty} \delta(\omega - k\omega_0),$$

as pictured in Figure 5.21d with $\omega_0 = 2\pi/T$. This may be surprising because it indicates that an infinite impulse train in the time domain spaced T seconds apart has as its Fourier transform an infinite impulse train in the frequency domain separated by $\omega_0 = 2\pi/T$ rad/s. Notice that the closer together the time-domain impulses, the farther apart the frequency-domain impulses.

Using $G(\omega)$, $S(\omega)$, and the frequency-domain convolution property, we have

$$g_s(t) = g(t) \cdot s(t) \Leftrightarrow \frac{1}{2\pi}[G(\omega) * S(\omega)].$$

To find $G_s(\omega)$, the spectrum of $g_s(t)$, we perform the frequency domain convolution of $S(\omega)$ and $G(\omega)$. Since $S(\omega)$ is an impulse train in the frequency domain, this operation is particularly easy to carry out, with the result

$$\begin{aligned} G_s(\omega) &= \frac{1}{2\pi}\left\{[G(\omega)] * \left[\frac{2\pi}{T} \sum_{k=-\infty}^{\infty} \delta(\omega - k\omega_0)\right]\right\} \\ &= \frac{1}{T} \sum_{k=-\infty}^{\infty} [G(\omega)] * [\delta(\omega - k\omega_0)] = \frac{1}{T} \sum_{k=-\infty}^{\infty} G([\omega - k\omega_0]). \end{aligned}$$

We see that $G_s(\omega)$, the Fourier transform of the sampled signal, consists of periodic scaled replicas of $G(\omega)$ spaced ω_0 radians/second apart. A sketch of the spectrum for the sampled signal $g_s(t)$ with $\omega_0 >> \omega_m$ is shown in Figure 5.21e, where we notice that for no overlap in the spectrum of $G_s(\omega)$, the following condition must be arranged:

$$\omega_0 > 2\omega_m.$$

The sampling frequency ω_0 is often expressed in hertz as $f_s = \omega_0/2\pi$, and, of course, the maximum frequency ω_m can also be expressed in hertz as $f_m = \omega_m/2\pi$. The preceding inequality can also be written

Nyquist frequency

$$f_s > 2f_m \quad \text{or} \quad \frac{1}{T} > 2f_m .$$

The frequency $2f_m$ is referred to as the *Nyquist frequency.* Thus, to avoid overlap in the spectrum of $G_s(\omega)$, the sampling frequency f_s must be *at least* twice the highest frequency contained in $G(\omega)$.

To summarize, the preceding development leads to the following statement of the Sampling Theorem.

If an analog signal has no frequency components at frequencies greater than $f_{\max}$, the signal can be uniquely represented by equally spaced samples if the sampling frequency f_s is greater than $2 \cdot f_{\max}$. Furthermore, the original analog signal can be recovered from the samples by passing them through an ideal lowpass filter having an appropriate bandwidth.

The amazing and enormously useful characteristic of the Sampling Theorem is that if the sampling rate is sufficiently high $(f_s > 2f_m)$, $G_s(\omega)$ contains several separated replicas of $G(\omega)$, the spectrum of the original signal. This means that we can recover the original signal from its samples by, for example, lowpass filtering the sampled signal $g_s(t)$. Of course, to make this practical by not requiring an ideal analog filter, we must select a sampling rate more than the minimum specified by the sampling theorem. The trade-off is that the more gradual the recovery filter's roll-off is to be, the higher the required sampling rate. The implication of the ubiquitous Sampling Theorem is that we can sample an analog signal, process the samples with digital hardware and software, and then return to the analog domain with little or no loss of information by simply lowpass filtering the result. Thus the theorem provides the basis for the entire field of digital signal processing!

As a practical matter, no finite-duration analog signal has a bandlimited spectrum. To overcome this we generally lowpass filter the analog signal prior to sampling to limit its bandwidth. Again, the sharpness of the prefilter's transition from passband to stopband will have an impact on the sampling rate. The more gradual the roll-off, the higher the sampling rate should be. This prefilter is known as an antialiasing filter because it prevents the replicas of $G(\omega)$ shown in Figure 5.21e from overlapping, a condition known as aliasing.

Parseval's Theorem for Energy Signals The energy contained in a time-domain signal is often of interest, and it can be found by integrating the power in the signal over all time. By means of Parseval's theorem, this energy can also be computed in the frequency domain, so that we have the equivalence

Parseval's theorem

$$E_x = \int_{-\infty}^{\infty} |x(t)|^2 dt = \frac{1}{2\pi} \int_{-\infty}^{\infty} |X(\omega)|^2 d\omega ,$$

where $|X(\omega)|^2$ is called the *energy spectral density,* or the energy density of the signal $x(t)$. The energy E_y in the output signal $y(t)$ of a linear system subjected to an input signal of $x(t)$ is given by

$$E_y = \frac{1}{2\pi} \int_{-\infty}^{\infty} |H(\omega)|^2 |X(\omega)|^2 d\omega = \frac{1}{2\pi} \int_{-\infty}^{\infty} |Y(\omega)|^2 d\omega$$

where $H(\omega)$ is the frequency transfer function (frequency response) of the linear system.

ILLUSTRATIVE PROBLEM 5.13
Using the Parseval Relation

Use the Parseval relation to compute the energy in the signal described by

$$f(t) = \frac{16 \sin(4\pi t)}{4\pi(t)}, \qquad -\infty \le t \le \infty .$$

Solution

In the time domain we need to compute the integral of $|f(t)|^2$, which is not a simple task. However, if we find the transform of $f(t)$, we have (from Table 5.1) the expression

$$F(\omega) = 4[u(\omega + 4\pi) - u(\omega - 4\pi)]$$

which is a pulse of amplitude 4 extending over the range $|\omega| \le 4\pi$, as in Figure 5.22a. We need $|F(\omega)|^2$, as in Figure 5.22b, where the area is easily calculated as $16 \cdot 8\pi$; bringing in the $1/2\pi$ factor gives 64 J.

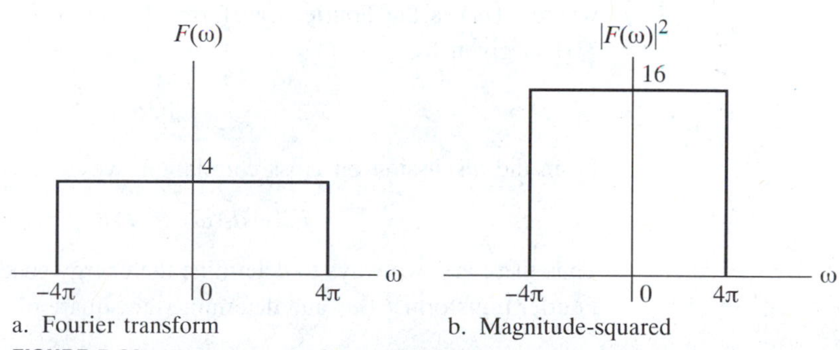

FIGURE 5.22 *Fourier transform and the magnitude-squared spectrum*

Cross Correlation for Energy Signals In an intuitive sense, the correlation of two quantities implies that knowledge of one quantity indicates knowledge of the other as, for example, in height–weight, driving under the influence–accidents, and so forth. Cross correlation of two signals provides a measure of the similarity (or difference) of the two. With the cross correlation of $x_1(t)$ and $x_2(t)$ defined by

time correlation

$$R_{x_1x_2}(\tau) = \int_{-\infty}^{\infty} x_1(t)x_2(t+\tau)\,dt,$$

the Fourier transform of the cross correlation[6] is given by

$$\mathcal{F}\left[R_{x_1x_2}(\tau) = \int_{-\infty}^{\infty} x_1(t)x_2(t+\tau)\,dt\right] = X_1(-\omega)\cdot X_2(\omega)$$
$$= X_1{}^*(\omega)\cdot X_2(\omega)\,.$$

If $x_1(t) = x_2(t) = x(t)$, we have $R_{xx}(\tau)$ and refer to the result as the autocorrelation. The autocorrelation can be used to detect the presence of periodic signals (limited to finite duration) embedded in noise. Cross correlation is useful in detecting the presence of a known signal in noise. An example of this is in radar, where a pulsed sinusoid of time-varying frequency is transmitted and, if a target is present, a delayed version of the transmitted signal corrupted by noise is received. The cross correlation enables the system to determine if a target is present and its range. For further discussion on a qualitative level and illustrative examples, see Robert W. Ramirez [5].

Notice the similarity of cross correlation to convolution, where we have

$$\mathcal{F}\left[x_1(t) * x_2(t) = \int_{-\infty}^{\infty} [x_1(\tau)x_2(t-\tau)\,d\tau\right] = X_1(\omega)\cdot X_2(\omega)\,.$$

The relationship between correlation and convolution is further explored in Problem P5.26.

Energy Spectral Density The spectral density of a finite energy signal $f(t)$ is given by

$$E(\omega)|F(\omega)|^2 = F^*(\omega)F(\omega)\,,$$

where $F(\omega)$ is the Fourier transform of $f(t)$. The total energy in the signal $f(t)$ is given by

$$E = \frac{1}{2\pi}\int_{-\infty}^{\infty} |F(\omega)|^2\,d\omega\,.$$

From the discussion on cross correlation, we observe that

$$F^*(\omega)F(\omega) = \mathcal{F}[R_{ff}(\tau)]\,.$$

Thus there are two ways to determine the energy spectral density: (1) Find the Fourier transform $F(\omega)$ and determine the square of the magnitude or (2) find

[6] Cross correlation can also be defined for power signals as

$$R_{x_1x_2}(\tau) = \lim_{T_0\to\infty}\frac{1}{2T_0}\int_{-T_0}^{T_0} x_1(t)x_2(t+\tau)\,dt\,.$$

As a practical matter, if numerical calculations are involved, the interval must be restricted to a finite duration.

the autocorrelation of $f(t)$ and then evaluate the Fourier transform of $R_{ff}(\tau)$. In practical situations where noise is present and the computations are performed using discrete-system techniques, the two methods do not generally provide the same results. For a more complete discussion see Alan V. Oppenheim and Ronald W. Schafer, *Discrete-Time Signal Processing,* Prentice Hall, Inc., Englewood Cliffs, N.J., 1989.

For a power signal $x(t)$ we define the *power spectral density* as $S_{xx}(\omega)$, which provides a measure of the power contributions of the various frequencies. Thus, the total power in $x(t)$ is

$$P = \frac{1}{2\pi}\int_{-\infty}^{\infty} S_{xx}(\omega)d\omega .$$

For additional information and a discussion of how to determine $S_{xx}(\omega)$, see Ferrel J. Stremler [6].

SOLVED EXAMPLES AND MATLAB APPLICATIONS

EXAMPLE 5.1 *An Impulse Train Applied to an Electric Circuit*

Let's consider the *RC* filter of Figure E5.1a on page 262 subjected to a voltage that is the periodic unit impulse train $x(t) = \delta_p(t)$ of Figure E5.1b, described by the infinite summation

impulse train

$$\delta_p(t) = \sum_{k=-\infty}^{\infty} \delta(t - kT_0) .$$

a. Find the Fourier series coefficients, the X_k's, for this unusual signal and then express the input $x(t)$ in terms of a cosine series.

b. This impulse train described by the exponential series

$$x(t) = \frac{1}{T_0}\sum_{k=-\infty}^{\infty} e^{jk\omega_0 t}$$

is applied to the circuit of Figure E5.1a that has the frequency response $H(j\omega) = 1/(1 + j\omega)$. Sketch the frequency response $H(j\omega)$ and plot the two-sided spectrum for the input $x(t)$ and the output $y_{ss}(t)$. Write the steady-state output in terms of a cosine series.

Solution

a. *Finding the X_k's.* The Fourier coefficients for the periodic impulse train are given by the analysis equation

$$X_k = \frac{1}{T_0}\int_{t_0}^{t_0+T_0} x(t)e^{-jk\omega_0 t}dt = \frac{1}{T_0}\int_{-T_0/2}^{T_0/2} \delta(t)e^{-jk\omega_0 t}dt.$$

Using the sifting property of the impulse function, we have

$$X_k = \frac{1}{T_0} \cdot e^{-jk\omega_0(0)} = \frac{1}{T_0}, \quad k = 0, \pm 1, 2, \ldots .$$

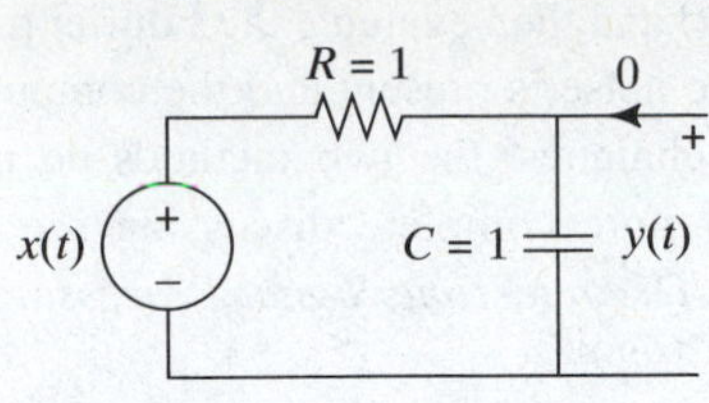

a. *RC* filter

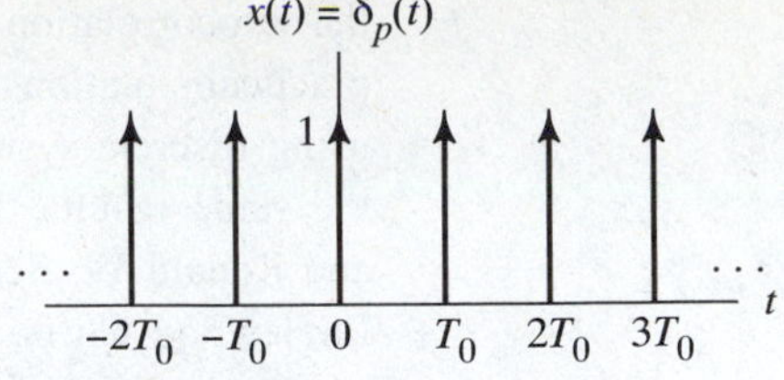

b. Voltage input, an impulse train

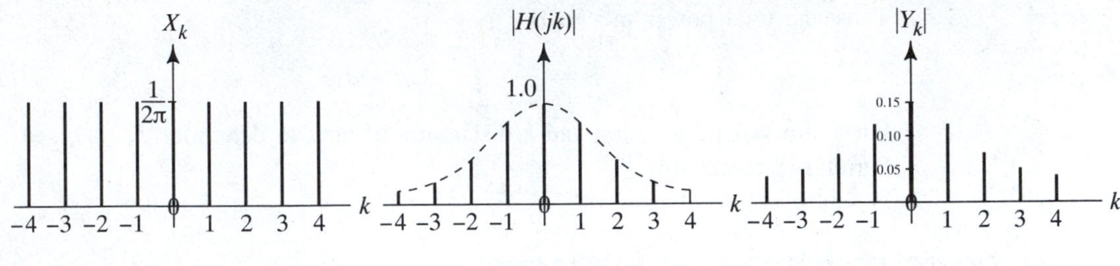

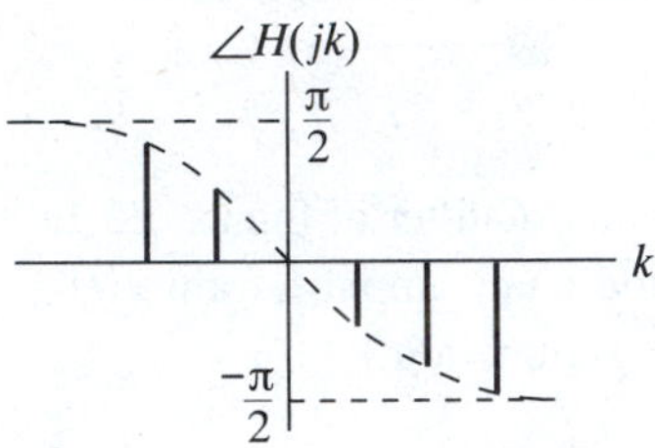

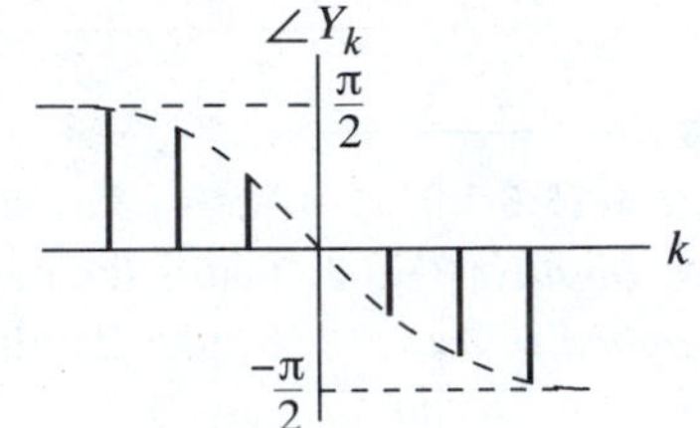

c. Input spectrum, system frequency response, and output spectrum

FIGURE E5.1 *Transmission through a linear network*

Notice that the frequencies ($\omega = k\omega_0$, $k = 0, \pm 1, \pm 2, \ldots$) are present in $x(t)$ with equal amplitudes of $1/T_0$. This is quite remarkable, but we have known for some time that the impulse function is quite something, and an infinite train of them is almost beyond belief. Using the Euler relation, the input can be put in the form of a cosine series as

cosine series

$$x(t) = \frac{1}{T_0} + \frac{2}{T_0} \sum_{k=1}^{\infty} \cos(k\omega_0 t)\,.$$

b. *Output spectrum.* The system's frequency response $H(j\omega) = 1/(1 + jk\omega_0)$ evaluated at integer multiples of the fundamental frequency ω_0 becomes

$$H(jk\omega_0) = \frac{1}{1 + jk\omega_0}, \qquad k = 0, \pm 1, \pm 2, \ldots.$$

Thus, in the output spectrum, the *magnitude* of each harmonic component is given by

$$|Y_k| = |X_k| \cdot |H(jk\omega_0)| = \frac{2}{T_0} \cdot \frac{1}{\sqrt{1 + k^2\omega_0{}^2}},$$

$$k = \pm 1, \pm 2, \ldots, \quad \text{and} \quad Y_o = \frac{1}{T_o} \cdot 1$$

$$= \text{(input magnitude)} \cdot \text{(magnitude of frequency response at } \omega = k\omega_0\text{)}$$

while the *phase* of each harmonic component is

$$\angle Y_k = \angle X_k + \angle H(jk\omega_0) = 0 + (-\tan^{-1} k\omega_0)$$

$$= \text{input phase} + \text{phase of frequency response at } \omega = k\omega_0.$$

Thus, in this particular example, the magnitude and phase spectra of the output will take the shapes of the magnitude and phase spectra of the system. Using these results, the steady-state output signal from this filter is given by

$$y_{ss}(t) = \frac{1}{T_0} + \frac{2}{T_0}\sum_{k=1}^{\infty} \frac{1}{\sqrt{1 + k^2\omega_0{}^2}} \cos(k\omega_0 t - \tan^{-1} k\omega_0).$$

The output spectrum is shown in Figure E5.1c for $\omega_0 = 1$, and the expression for the steady-state output is

$$y_{ss}(t) = 0.159 + 0.318\sum_{k=1}^{\infty} \frac{1}{\sqrt{1 + k^2}} \cos(kt - \tan^{-1} k).$$

EXAMPLE 5.2
Fourier Coefficients, Waveform Synthesis, and Power

Consider the sawtooth voltage waveform as shown in Figure E5.2a.

a. Show that the exponential Fourier series coefficients are given by

$$V_k = \frac{jV}{2\pi k}, \qquad k = \pm 1, \pm 2, \ldots, \quad \text{and} \quad V_0 = V/2.$$

b. Determine the cosine series that represents this voltage.

c. Use MATLAB to synthesize this periodic signal. Approximately how many terms are needed to get an adequate representation?

d. Determine the normalized average power in this signal.

e. Determine the percentage of the average power (P_{avgn}) contained in the DC and fundamental components of the Fourier series representation.

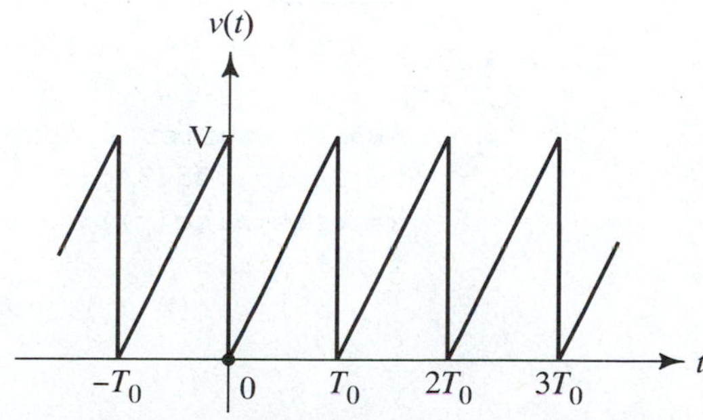

FIGURE E5.2a *Sawtooth waveform voltage*

EXAMPLES

WHAT IF? Investigate the effects in both the time and k domains of using this sawtooth wave as the input to the filter of Example 5.1. You also might want to consider the effect of a highpass filter on this sawtooth waveform—e.g., use $H(s) = s/(s + 10)$ rather than $H(s) = 1/(s + 1)$. ■

Solution

a. *Fourier coefficients.* For $0 \le t < T_0$, we have $v(t) = Vt/T_0$, and the coefficients for $k = \pm 1, \pm 2, \ldots$ are found from

$$\begin{aligned} V_k &= \frac{1}{T_0}\int_0^{T_0}\left\{\frac{V}{T_0}t\right\}e^{-jk\omega_0 t}\,dt = \frac{V}{T_0^{\,2}}\left[\frac{e^{-jk\omega_0 t}}{-k^2\omega_0^{\,2}}(-jk\omega_0 t - 1)\right]\Bigg|_0^{T_0} \\ &= \frac{-V}{T_0^{\,2}k^2\omega_0^{\,2}}[e^{-jk\omega_0 T_0}(-jk\omega_0 T_0 - 1) - e^{-jk\omega_0 0}(-jk\omega_0 0 - 1)] \\ &= \frac{-V}{T_0^{\,2}k^2\omega_0^{\,2}}[e^{-jk2\pi}(-jk2\pi - 1) - 1(-j0 - 1)] = \frac{-V}{T_0^{\,2}k^2\omega_0^{\,2}}[-jk2\pi] \\ &= \frac{jV}{2\pi k}. \end{aligned}$$

For $k = 0$ we have

$$V_0 = \frac{1}{T_0}\int_0^{T_0}\frac{V}{T_0}t\,dt = V/2.$$

b. *Cosine series and synthesis.* Using the results of part (a) gives

$$v(t) = \frac{V}{2} + \sum_{k=1}^{\infty}\frac{V}{k\pi}\cos(k\omega_0 t + \pi/2).$$

c. *MATLAB synthesis.* To synthesize the wave, we use $V = \omega_0 = 1$ with the following script. The result for the DC and 6 cosine terms is given in Figure E5.2b. The approximation is reasonably good for so few terms. (Remember, an infinity of terms is theoretically needed.) See Figure E5.2c for the result of the DC and 25 cosine terms. Notice, once again, the consequence of the Gibbs' effect.

MATLAB Script

```
%E5_2 Fourier coefficients and power

%E5_2b Fourier synthesis of a sawtooth
t=-3*pi:.025:3*pi;
x=zeros(size(t));
for k=1:1:6;
    x=x+(1/(k*pi))*cos(k*t+pi/2);
    end
v=.5+x;
%...plotting & labeling statements
```

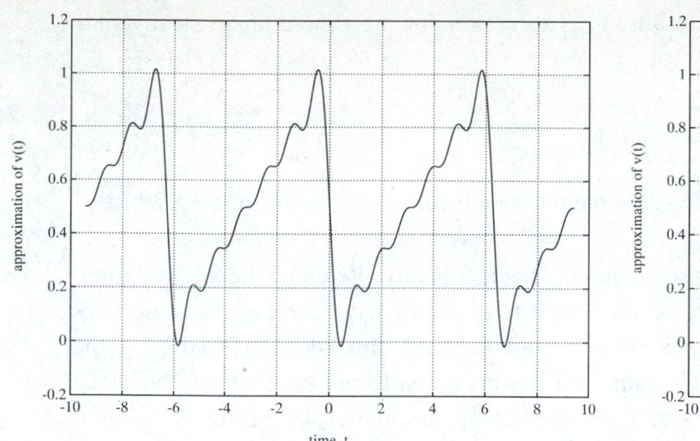

FIGURE E5.2b *Synthesis of sawtooth waveform: 7 terms*

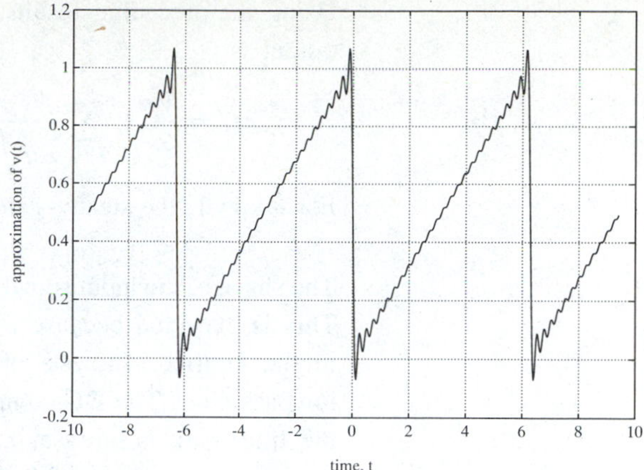

FIGURE E5.2c *Synthesis of sawtooth waveform: 26 terms*

d. *Average power in the given voltage waveform.* Here, over the period $0 \le t < T_0$, the voltage is given by the equation $v(t) = Vt/T_0$. The average normalized power is

$$P_{\text{avgn}} = \frac{1}{T_0}\int_0^{T_0} v(t)^2\,dt = \frac{1}{T_0}\int_0^{T_0}\left(\frac{Vt}{T_0}\right)^2 dt = \frac{V^2}{3}.$$

e. *Power in DC and fundamental.* Designating the average, or DC, value as V_0, the DC power is

$$P_0 = {V_0}^2 = (V/2)^2 = V^2/4\,.$$

From part (a), $|V_1| = |V_{-1}| = V/2\pi$ and the power in the fundamental is

$$P_1 = (V/2\pi)^2 + (V/2\pi)^2 = V^2/2\pi^2.$$

Thus the total average power in the DC and fundamental is

$$P_0 + P_1 = V^2\left\{\frac{1}{4} + \frac{1}{2\pi^2}\right\} = 0.301V^2.$$

Comparing this with the true average power of $0.333V^2$, we see that

$$\frac{0.301}{0.333}\cdot 100 \quad \text{or} \quad 90.4\%$$

of the power is contained in these two components. Thus most of the power of the sawtooth is at the lower frequencies.

WHAT IF? Calling the filter's steady-state output $y_{ss}(t)$, we find the output coefficients to be

$$Y_k = V_k \cdot H(jk\omega_0) = \frac{V}{2}\cdot 1 + \sum_{k=\pm 1}^{\pm\infty} \frac{jV}{2\pi k}\cdot\frac{1}{1 + jk\omega_0}.$$

EXAMPLES

Using the preceding results, the expression for the time-domain steady-state output is

$$y_{ss}(t) = \frac{V}{2} + \sum_{k=1}^{\infty} \frac{V}{k\pi\sqrt{1 + k^2\omega_0{}^2}} \cos(k\omega_0 t + \frac{\pi}{2} - \tan^{-1}(k\omega_0)).$$

For $\omega_0 = 1$, the steady-state output waveform is given in Figure E5.2d. ■

The shape of the input signal has not been radically altered by the lowpass filter. This is expected because the lower frequencies have the largest amplitudes in the Fourier synthesis of $v(t)$, and they are treated most kindly by the lowpass filter. The DC component has been unaffected (as it should be, since the filter gain is unity at $\omega = 0$) because the output waveform is centered about the DC value of 0.5. The highpass filter has altered the input waveform quite drastically, including the removal of the DC component as seen in Figure E5.2e.

EXAMPLE 5.3 *Steady-State Response of an LRC Circuit*

The Fourier cosine series for the sawtooth waveform of Figure E5.2a,

$$v(t) = \frac{V}{2} + \sum_{k=1}^{\infty} \frac{V}{\pi k} \cos(2\pi kt + \pi/2),$$

is the source voltage $v(t)$ for the circuit of Figure E5.3a. Notice that $T_0 = 1$ s.

a. Find the DC and first three sinusoidal terms in the cosine series for the current $i_{ss}(t)$ for $\omega_0 = 2\pi$ rad/s.

b. Use a function such as MATLAB's **freqs** to plot the frequency response of the circuit.

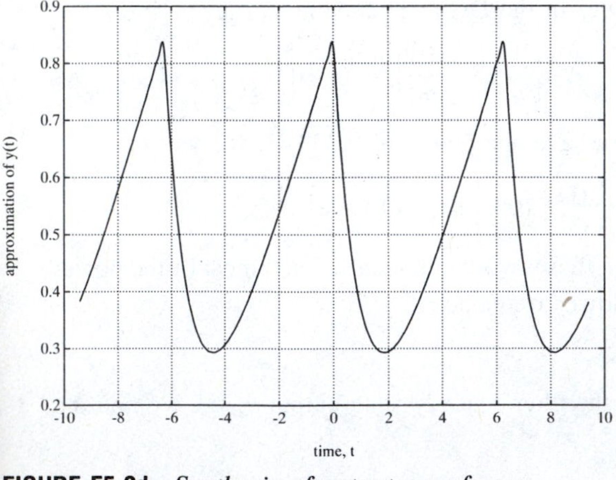

FIGURE E5.2d *Synthesis of output waveform: 26 terms*

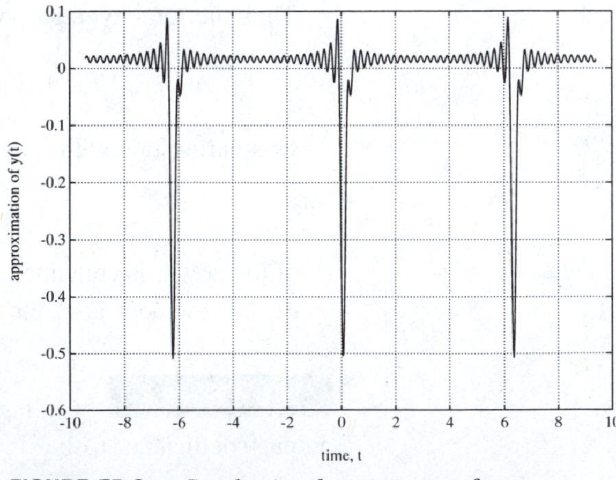

FIGURE E5.2e *Synthesis of output waveform: 26 terms, highpass filter*

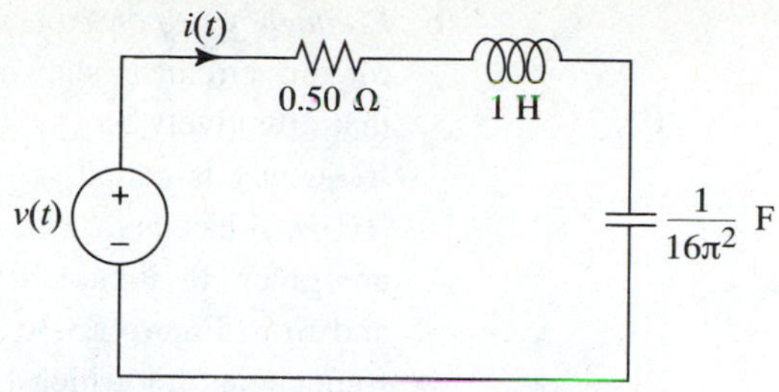

FIGURE E5.3a *Second-order circuit*

WHAT IF? Suppose we want to change the circuit's bandwidth without altering the resonant frequency. What circuit parameter will do this? Try a few different bandwidths and plot the frequency responses. ■

Solution

a. *Transmission through a linear system.* From circuit analysis we can write

$$H(s) = \frac{I(s)}{V(s)} = \frac{1}{R + Ls + 1/Cs} = \frac{Cs}{RCs + LCs^2 + 1}$$

and, after substituting the given data and applying some algebra, we have

$$H(s) = \frac{s}{s^2 + 0.5s + 16\pi^2}.$$

To determine the steady-state current, we need to apply the sinusoidal steady-state formula

$$i_{ss}(t) = I_0 + \sum_{k=1}^{\infty} 2|V_k| \cdot |H(jk\omega_0)| \cos(k\omega_0 t + \angle V_k + \angle H(jk\omega_0)).$$

With $\omega_0 = 2\pi$ rad/s, we need to evaluate

$$H(s)|_{s=jk\omega_0} = H(jk\omega_0) = \frac{jk\omega_0}{(jk\omega_0)^2 + 0.5(jk\omega_0) + 16\pi^2}$$

at the frequencies $k\omega_0 = 0,\ 2\pi,\ 4\pi$, and 6π, which gives us

$$H(j0) = 0,\quad H(j2\pi) \approx \frac{j}{6\pi},\quad H(j4\pi) = 2,\quad \text{and}\quad H(j6\pi) \approx \frac{-j6}{20\pi}.$$

Approximating the voltage input by the first few terms in its Fourier series,

$$v(t) = \frac{V}{2} + \frac{V}{\pi}\cos(2\pi t + \pi/2) + \frac{V}{2\pi}\cos(4\pi t + \pi/2) + \frac{V}{3\pi}\cos(6\pi t + \pi/2) + \cdots,$$

the steady-state current is then approximated by

$$\begin{aligned} i_{ss}(t) &= 0 + \frac{V}{6\pi^2}\cos(2\pi t + \pi) + \frac{V}{\pi}\cos(4\pi t + \pi/2) \\ &\quad + \frac{6V}{60\pi^2}\cos(6\pi t) + \cdots \\ &= V[0.0169\cos(2\pi t + \pi) + 0.318\cos(4\pi t + \pi/2) \\ &\quad + 0.010\cos(6\pi t) + \cdots] \\ &\approx 0.318V\cos(4\pi t + \pi/2). \end{aligned}$$

b. *Frequency response of circuit.* The script for plotting the frequency response for this circuit is shown next. Notice that this is a highly selective circuit that effectively blocks all frequencies except those close to $\omega = 4\pi$. This frequency is called the *resonance frequency.* It is the frequency at which $|H(j\omega)|$ has its maximum value. Two plots with different frequency ranges are given in Figures E5.3b and c. Notice that this is a bandpass filter, and from Figure E5.3c we see that the bandwidth, defined as the range of frequencies for which $|H(j\omega)| \geq 0.707|H_{\max}|$, is approximately equal to $12.8 - 12.3 = 0.50$ rad/s.

MATLAB Script

```
%E5_3 Frequency response

%E5_3b Bandpass circuit
a=[1,0.5,157.914];
b=[1,0];
w=-20:.05:20;
H=freqs(b,a,w);
mag=abs(H);
phase=angle(H)
%...changing phase to degrees and plotting statements
```

WHAT IF? For the same resonant frequency, the circuit resistance R will change the bandwidth, and we show in Figure E5.3d the effect of doubling the resistance to $R = 1\ \Omega$. Notice that $|H_{\max}|$ has been reduced by a factor of two and that the bandwidth is broadened. ■

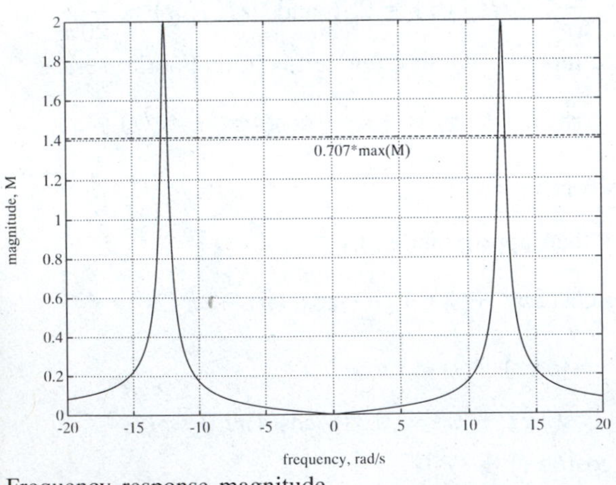

Frequency response magnitude

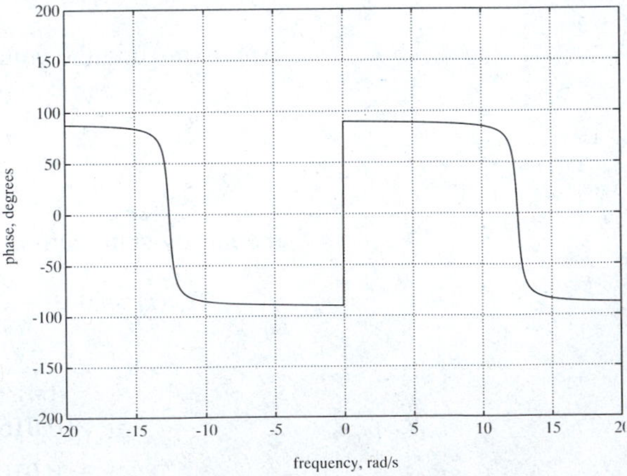

Frequency response phase

FIGURE E5.3b *Frequency response of circuit*

CHAPTER 5

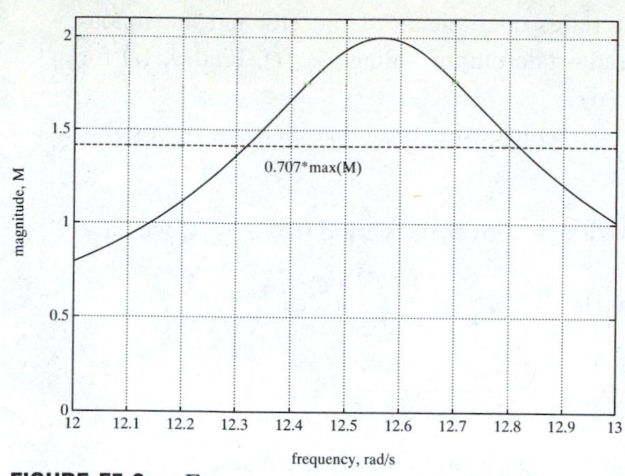

FIGURE E5.3c *Frequency response magnitude—zoom view*

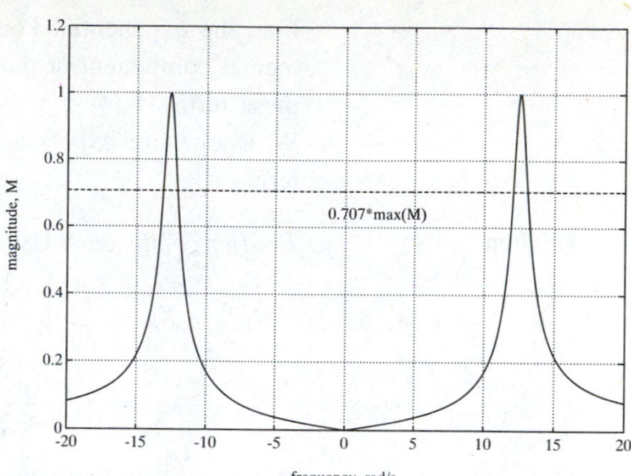

FIGURE E5.3d *Frequency response magnitude with larger bandwidth*

EXAMPLE 5.4
Fourier Coefficients, Power, and Steady-State Response

a. Show that the exponential Fourier series coefficients for the periodic voltage shown in Figure E5.4a are given by

$$X_k = \frac{1 - e^{-1}}{1 + j2\pi k}, \qquad k = 0, \pm 1, \pm 2, \ldots$$

b. If this voltage is impressed on a 1-Ω resistor, calculate the average power, $P_{\text{avg}} = P_{\text{avgn}}$, dissipated in the resistor.

c. Determine the average power contained in the DC ($k = 0$) and the first two harmonic components of the Fourier series representation of the signal.

d. This voltage is applied to the *RC* network shown in Figure E5.4b, which has the voltage transfer function

$$H(s) = \frac{Y(s)}{X(s)} = \frac{R_2}{R_1 + R_2 + R_1 R_2 C s}.$$

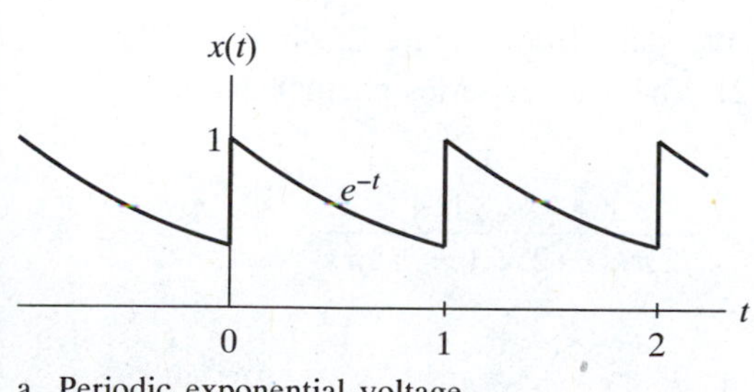

a. Periodic exponential voltage

$R_1 = 1$, $C = 1$, $R_2 = 1$, $x(t)$, $y(t)$

b. *RC* network

FIGURE E5.4

Find the exponential Fourier series coefficients of the DC and the fundamental component of the steady-state output voltage $y_{ss}(t)$. Find $y_{ss}(t)$ for these terms.

e. Write a general expression for $y_{ss}(t)$ in cosine form that includes all possible terms.

Solution

a. *Fourier coefficients.* Using $x(t) = e^{-t}$ over the period $0 \leq t < 1$, we have

$$X_k = \frac{1}{T_0}\int_{t_0}^{t_0+T_0} x(t)e^{-jk\omega_0 t}\,dt = \frac{1}{1}\int_0^1 e^{-t}e^{-jk\omega_0 t}\,dt$$

$$= -\left.\frac{e^{-t}e^{-jk\omega_0 t}}{1 + jk\omega_0}\right|_0^1 = \frac{1 - e^{-1}e^{-jk\omega_0}}{1 + jk\omega_0}.$$

But $\omega_0 = 2\pi/T_0 = 2\pi$, and this substitution yields the result

$$X_k = \frac{1 - e^{-1}}{1 + j2\pi k}, \qquad k = 0, \pm 1, \pm 2, \ldots.$$

b. *Average power in the resistor.* In this case,

$$P_{\text{avg}} = P_{\text{avgn}} = \frac{1}{T_0}\int_0^{T_0} x^2(t)\,dt = 1\int_0^1 (e^{-t})^2\,dt = \frac{1 - e^{-2}}{2} = 0.4323.$$

c. *Power in DC and the first two harmonics.* The specified input coefficients are

$$X_0 = \int_0^1 e^{-t}\,dt = 1 - e^{-1} = 0.6321,$$

$$|X_1| = |X_{-1}| = \frac{1 - e^{-1}}{|1 + j2\pi|} = 0.0994, \qquad \text{and}$$

$$|X_2| = |X_{-2}| = \frac{1 - e^{-1}}{|1 + j4\pi|} = 0.0500.$$

The average power in the DC and first two harmonics is given by

$$P = P_0 + P_1 + P_2 = X_0{}^2 + 2|X_1|^2 + 2|X_2|^2$$
$$= (0.6321)^2 + 2(0.0994)^2 + 2(0.0500)^2 = 0.4243.$$

Thus, the percentage in these components is $(0.4243/0.4323) \cdot 100 = 98.15\%$.

d. *Steady-state output.* Substituting the data from Figure E5.4b gives the transfer function $H(s) = 1/(s + 2)$. The Fourier series coefficients of the output are given by

$$Y_k = H(jk\omega_0) \cdot X_k = \frac{1}{j2\pi k + 2} \cdot \frac{1 - e^{-1}}{1 + j2\pi k}.$$

Thus,

$$Y_0 = H(j0) \cdot X_0 = \frac{1}{j2\pi(0) + 2} \cdot \frac{1 - e^{-1}}{1 + j2\pi(0)} = 0.3160$$

$$Y_1 = H(j2\pi) \cdot X_1 = \frac{1}{j2\pi(1) + 2} \cdot \frac{1 - e^{-1}}{1 + j2\pi(1)} = 0.015e^{-j2.676}$$

With $Y_{-1} = Y_1^*$, the steady-state output for these terms is

$$y_{ss}(t) = 0.316 + 0.030 \cos(2\pi t - 2.676).$$

e. *General expression.* If all terms are included, we have

$$y_{ss}(t) = 0.316 + \sum_{k=1}^{\infty} \frac{1.264}{\sqrt{(2 - 4\pi^2 k^2)^2 + (6\pi k)^2}} \cdot \cos\left(2\pi kt - \tan^{-1}\left[\frac{6\pi k}{2 - 4\pi^2 k^2}\right]\right).$$

EXAMPLE 5.5
Signal Transmission Through a Linear System

A periodic square-pulse waveform with a period T_0 is described by the Fourier series pair

$$x(t) = \sum_{k=-\infty}^{\infty} X_k e^{j2\pi kt/T_0} \Leftrightarrow X_k = \frac{\pi}{2} \frac{\sin(k\pi/2)}{k\pi/2}, \qquad k = 0, \pm 1, \pm 2, \ldots.$$

a. Find the Fourier transform $X(\omega)$ and plot the magnitude and phase spectra.

b. This signal $x(t)$ is applied to the ideal bandpass system whose frequency response $H_a(\omega)$ is shown in Figure E5.5a. Find the output transform $Y_a(\omega)$ and the output signal $y_a(t)$.

c. Find the output signal $y_b(t)$ if linear phase is now included in the system frequency response, as in Figure E5.5b.

Solution

a. *Input transform.* Using l'Hôpital's rule, $X_0 = \pi/2$. The other coefficients are $X_1 = X_{-1} = 1$, $X_2 = X_{-2} = 0$, $X_3 = X_{-3} = -\frac{1}{3}$, $X_4 = X_{-4} = 0$, $X_5 = X_{-5} = 1/5$, . . . and using the pair for periodic signals (16 in Table 5.1), we find that the Fourier transform of the input is

$$X(\omega) = \cdots + \frac{2\pi}{5}\delta(\omega + 5\omega_0) - \frac{2\pi}{3}\delta(\omega + 3\omega_0) + 2\pi\delta(\omega + \omega_0) + \pi^2\delta(\omega) + 2\pi\delta(\omega - \omega_0) - \frac{2\pi}{3}\delta(\omega - 3\omega_0) + \frac{2\pi}{5}\delta(\omega - 5\omega_0) - \cdots.$$

The magnitude spectrum is a series of shifted impulse functions, as shown in Figure E5.5c, where we have omitted the phase plot, since the transform is always real.

EXAMPLES

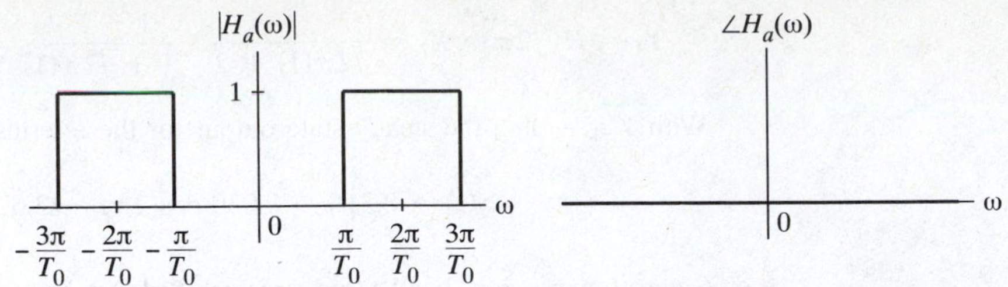

a. Bandpass system, zero phase

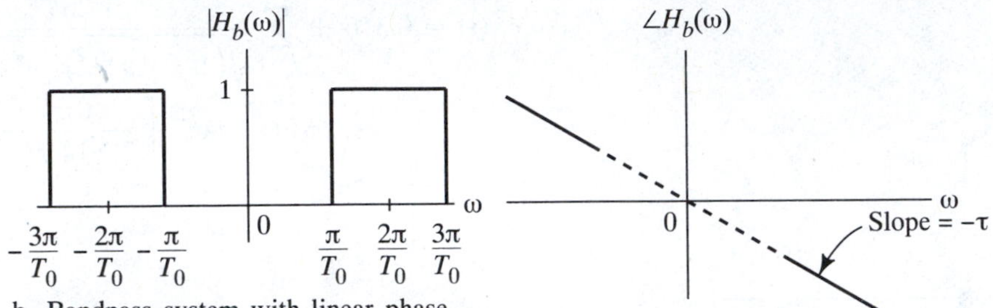

b. Bandpass system with linear phase

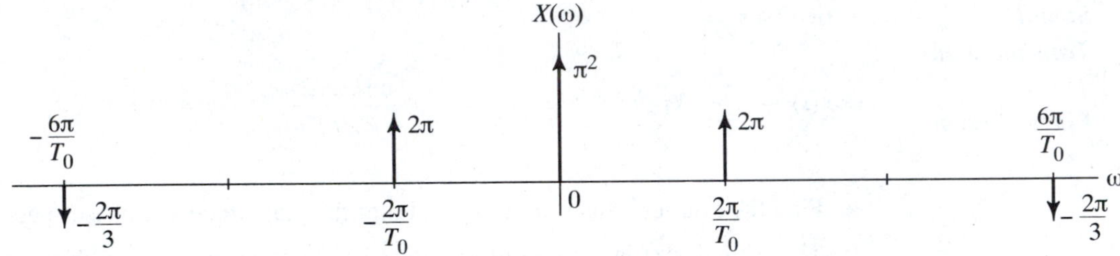

c. Input spectrum

FIGURE E5.5 *Spectrum and results for bandpass system*

b. *Output with system phase of zero.* With $\omega_0 = 2\pi/T_0$, the input transform can be written

$$\begin{aligned} X(\omega) = \cdots &+ \frac{2\pi}{5}\delta(\omega + 10\pi/T_0) - \frac{2\pi}{3}\delta(\omega + 6\pi/T_0) \\ &+ 2\pi\delta(\omega + 2\pi/T_0) + \pi^2\delta(\omega) + 2\pi\delta(\omega - 2\pi/T_0) \\ &- \frac{2\pi}{3}\delta(\omega - 6\pi/T_0) + \frac{2\pi}{5}\delta(\omega - 10\pi/T_0) - \cdots. \end{aligned}$$

From convolution we know that $Y_a(\omega) = H_a(\omega) \cdot X(\omega)$, and from the plots in Figure E5.5a and c, it is apparent that the input transform $X(\omega)$ is multiplied by 0 except at $\omega = \pm 2\pi/T_0$, where it is multiplied by 1; as a consequence,

$$\begin{aligned} Y_a(\omega) &= 2\pi\delta(\omega + 2\pi/T_0) + 2\pi\delta(\omega - 2\pi/T_0) \\ &= 2\pi\delta(\omega + \omega_0) + 2\pi\delta(\omega - \omega_0). \end{aligned}$$

Using pair 6 from Table 5.1,

$$y_a(t) = 2\cos(\omega_0 t).$$

c. *Output with linear phase.* In this situation the system frequency response is

$$H_b(\omega_0) = 1e^{-j\tau\omega_0}.$$

Using time-domain convolution (frequency-domain multiplication), the output transform is

$$Y_b(\omega) = 2\pi\delta(\omega + \omega_0) \cdot 1e^{j\tau\omega_0} + 2\pi\delta(\omega - \omega_0) \cdot 1e^{-j\tau\omega_0}.$$

The output signal is now time shifted, and from property 2 in Table 5.2 we have

$$y_b(t) = 2\cos(\omega_0[t - \tau]).$$

EXAMPLE 5.6
Impulse Response of an Ideal and a Butterworth[7] Lowpass Filter

The magnitude and phase characteristics of an ideal lowpass filter are given in Figure E5.6a (page 274). Notice that the gain $M = K$, the phase is linear with $P = -t_0\omega$, and the cutoff frequencies are $\omega_{co} = \pm B$ rad/s.

a. Describe the frequency response $H(\omega)$ analytically. Then, find an analytical expression for the unit impulse response of the filter $h(t)$ for zero phase—i.e., $P = 0$. Plot this response using MATLAB.

b. What is $h(t)$ if the phase $P = -t_0\omega$ is considered? Plot this response using MATLAB.

c. Let's look at the frequency response and the impulse response of a real-life lowpass filter such as Butterworth, Chebyshev, or elliptic. Look up the transfer function, plot the frequency response, and find an expression for and plot the impulse response $h(t)$ for a lowpass filter of your choice. Compare this filter with the ideal filter of parts (a) and (b).

[7] S. Butterworth published an early paper about analog filters, "On the Theory of Filter Amplifiers," *Wireless Engineer,* London, 1930.

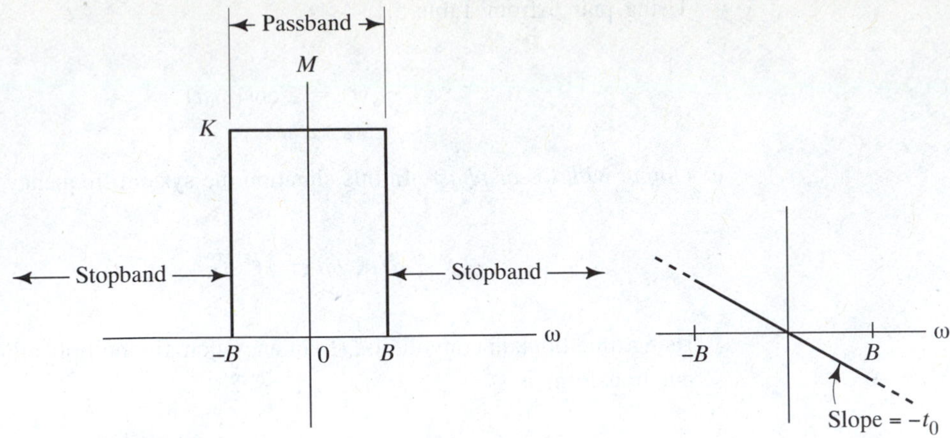

FIGURE E5.6a *Ideal filter frequency response characteristics*

Solution

a. *Finding the unit impulse response* $h(t)$. First the magnitude of $H(\omega)$ can be written in terms of step functions in the frequency domain as $M = K\{u(\omega + B) - u(\omega - B)\}$. The phase is $P = -t_0\omega$ and the frequency response is

$$H(\omega) = Me^{jP} = K\{u(\omega + B) - u(\omega - B)\}e^{-jt_0\omega}.$$

Using pair 4 in Table 5.1, we have the noncausal ($t_0 = 0$) response

$$h(t) = \frac{K}{\pi t}\sin(Bt) = \frac{KB}{\pi}\operatorname{sinc}(Bt), \qquad -\infty \leq t \leq \infty,$$

with the plot shown in Figure E5.6b with $B = 1$, $K = \pi$, and we have defined sinc $x = (\sin x)/x$.

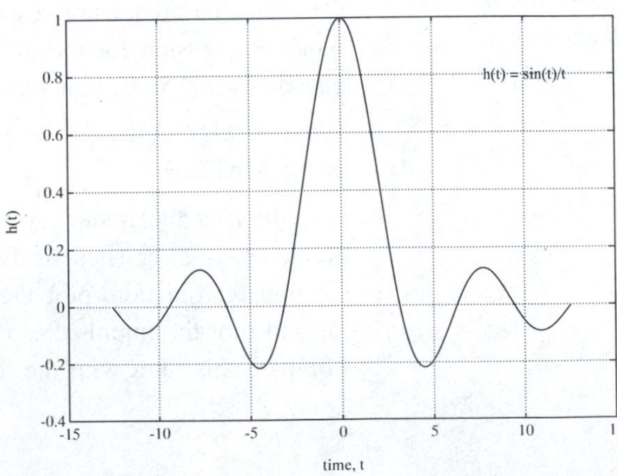

FIGURE E5.6b *Impulse response: ideal lowpass filter (zero phase)*

CHAPTER 5

b. For the phase we use the time-shift property of Table 5.2 to obtain the same $h(t)$ as in part (a), only delayed or shifted by t_0 time units:

$$h'(t) = \frac{K}{\pi[t - t_0]} \sin(B[t - t_0]) = \frac{KB}{\pi} \operatorname{sinc}(B[t - t_0]),$$
$$-\infty \leq t \leq \infty,$$

which is plotted in Figure E5.6c for $t_0 = 5$. From the result in part (b) we see that the effect of the linear phase, $P = -t_0\omega$, in the frequency domain is a time delay (time shift) of t_0 seconds in the time domain.

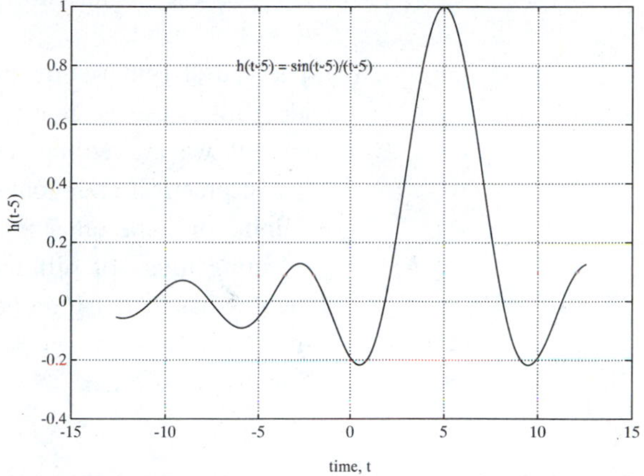

FIGURE E5.6c *Impulse response: ideal lowpass filter (linear phase)*

c. A third-order Butterworth lowpass filter with a -3-dB frequency of $\omega = 1000$ rad/s was chosen. Starting from the lowpass prototype with a cutoff frequency of 1 rad/s and the transfer function

$$H_{LP_p}(s) = \frac{1}{s^3 + 2s^2 + 2s + 1}, \qquad \operatorname{Re}[s] > -0.5,$$

we frequency scale by replacing s with $s/1000$ to obtain

$$H(s) = \frac{10^9}{s^3 + 2 \cdot 10^3 s^2 + 2 \cdot 10^6 s + 10^9}, \qquad \operatorname{Re}[s] > -500,$$

which has a cutoff of $\omega = 1000$ rad/s. The frequency response plot obtained using **freqs** is shown in Figure E5.6d. Comparing the ideal filter response of Figure E5.6a with the Butterworth characteristics of Figure E5.6d, we note the following:

(1) Unlike the ideal filter that has only a passband and a stopband, the Butterworth filter has a transition band between the passband and the stopband.

(2) An arbitrary definition of the passband is that it is the frequency range over which the magnitude response is within -3 dB of the filter's

largest frequency response magnitude. In this case the Butterworth filter has a bandwidth of 1000 rad/s. In the passband the Butterworth response is nearly flat, at least to about 500 rad/s, before beginning to "roll off."

(3) The definition of a filter's stopband is also arbitrary, with a typical definition being a frequency range over which the frequency response is at least a specified number of dB below the filter's maximum frequency response value. If this figure were 20 dB below maximum (not a great deal of attenuation as real filters go), which corresponds to $M = 0.1$, then this filter's stopband would begin at approximately 2000 rad/s and the frequency band $1000 < \omega < 2000$ rad/s would be considered the transition band.

(4) The causal Butterworth filter (see Figure E5.6d), unlike the noncausal ideal filter, can be implemented. The ideal filter can be approximated only if we are willing to introduce sufficient delay in its response, but in practical situations we generally use a Butterworth, Chebyshev, elliptic, or some other type of well-known filter. We can, however, by selecting filters of different types and orders, affect characteristics in the passband, transition band, and stopband. In addition, we can make the transition between passband and stopband steeper by varying the filter type or order.

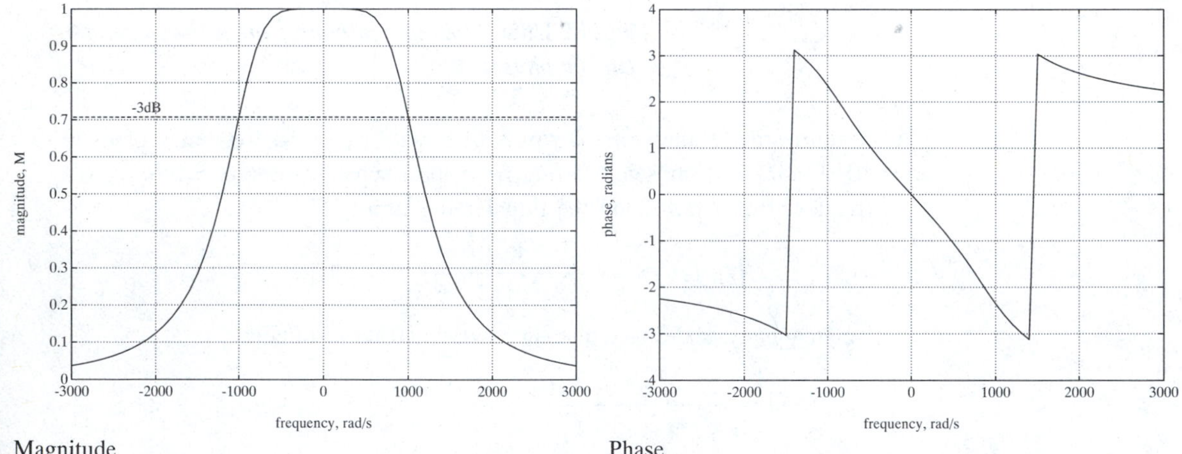

FIGURE E5.6d *Frequency response of Butterworth lowpass filter*

Using **residue** for the partial fraction expansion of $H(s)$, followed by the use of Table 3.1 on page 170, gives the unit impulse response

$$h(t) = 10^3[e^{-1000t} + 1.154e^{-500t}\cos(866t - 2.62)]u(t),$$

with the plot using **ksimptf** in Figure E5.6e.

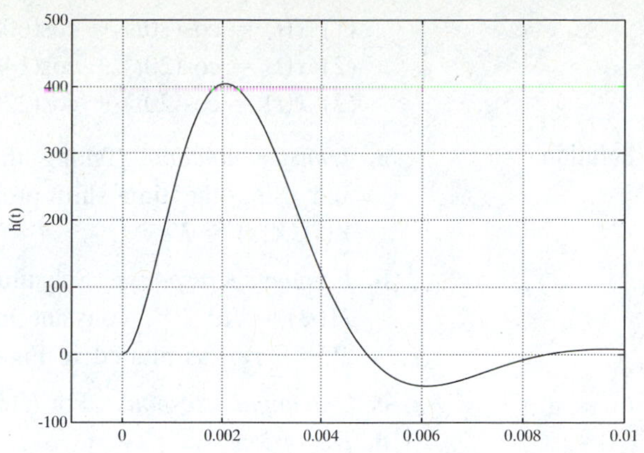

FIGURE E5.6e *Impulse response for LP Butterworth*

EXAMPLE 5.7
Distortionless Transmission

If $x(t)$ represents the input and $y(t)$ represents the output of an LTI system, distortionless transmission is described by

$$y(t) = Kx(t - \tau).$$

So, the input signal $x(t)$ is unchanged in shape and is delayed by an amount τ as it passes through the system, as pictured in Figure E5.7a.

a. Find the transfer function $H(s)$ for this system.

b. What is the frequency response $H(\omega) = Me^{jP}$? Sketch M and P versus frequency.

c. Find the unit impulse response, $h(t)$.

d. An analog filter has the magnitude and phase characteristics of Figure E5.7b. For the input signals given, determine if the filter distorts them. If there is distortion, is it magnitude distortion, phase distortion, or both?

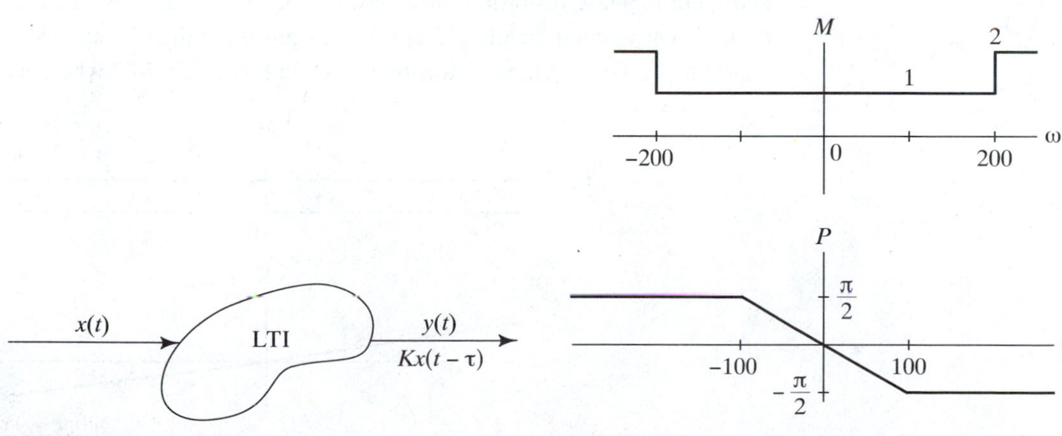

FIGURE E5.7a *System*

FIGURE E5.7b *Analog filter characteristics*

EXAMPLES

(1) $x(t) = \cos(20t) + \cos(60t)$
(2) $x(t) = \cos(20t) + \cos(140t)$
(3) $x(t) = \cos(20t) + \cos(220t)$

Solution

a. *Transfer function.* Taking the Laplace transform of $y(t) = Kx(t - \tau)$ and using the time-shift property gives $Y(s) = KX(s)e^{-\tau s}$, or $H(s) = Y(s)/X(s) = Ke^{-\tau s}$.

b. *Frequency response.* Substituting $s = j\omega$ in $H(s)$ produces $H(j\omega) = H(\omega) = Ke^{-j\tau\omega}$, constant magnitude of $M = K$, and linear phase of $P = -\tau\omega$, as plotted in Figure E5.7c.

c. *Unit impulse response.* For $H(s) = Ke^{-\tau s}$, $h(t) = \mathcal{L}^{-1}[H(s)] = K\delta(t - \tau)$.

d. *Distortion.*

(1) From the magnitude and phase plots we see that both sinusoids are multiplied by the same constant (1) and so there is no magnitude distortion. The phase in the linear region is given by

$$P = \frac{\text{rise}}{\text{run}} \cdot \omega = \frac{-\pi/2}{100} \cdot \omega = \frac{-\pi}{200} \cdot \omega.$$

Using these data we have the steady-state output signal

$$\begin{aligned} y_{ss}(t) &= \cos\left(20t - \frac{\pi}{10}\right) + \cos\left(60t - \frac{6\pi}{20}\right) \\ &= \cos\left[20\left(t - \frac{\pi}{200}\right)\right] + \cos\left[60\left(t - \frac{\pi}{200}\right)\right], \end{aligned}$$

and there is also no phase distortion, because each component of $x(t)$ is delayed by the same time, $t = \pi/200$ s. See Figure E5.7d.1.

(2) The steady-state output is

$$\begin{aligned} y_{ss}(t) &= \cos\left(20t - \frac{\pi}{10}\right) + \cos\left(140t - \frac{\pi}{2}\right) \\ &= \cos\left[20\left(t - \frac{\pi}{200}\right)\right] + \cos\left[140\left(t - \frac{\pi}{280}\right)\right], \end{aligned}$$

and there is phase distortion only because the components are delayed by different amounts while the amplitudes are multiplied by the same constant (1). The resulting distortion is seen in Figure E5.7d.2 where we

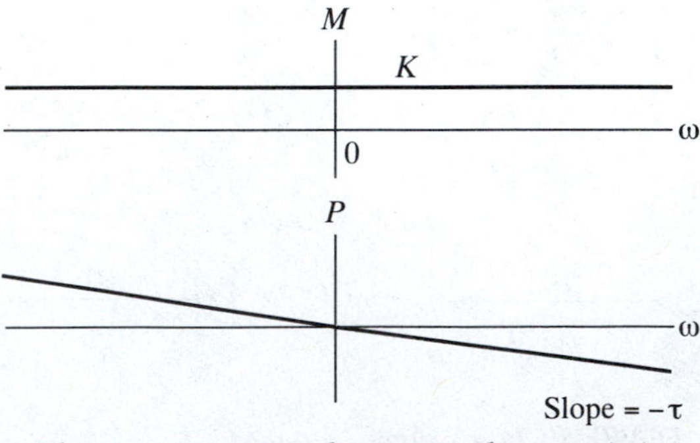

FIGURE E5.7c *Distortionless transmission-frequency characteristics*

observe in the output signal an asymmetry of peak values not present in the input signal.

(3) The steady-state output is

$$y_{ss}(t) = \cos\left(20t - \frac{\pi}{10}\right) + 2\cos\left(220t - \frac{\pi}{2}\right)$$
$$= \cos\left[20\left(t - \frac{\pi}{200}\right)\right] + 2\cos\left[220\left(t - \frac{\pi}{440}\right)\right]$$

and there is both magnitude and phase distortion because the amplitudes of the two components are multiplied by different constants (1 and 2) and the components are delayed by different amounts. The amplitude and phase distortion show up in the output signal of Figure E5.7d.3 (page 280), where we observe asymmetry in the peak values as well as different peak-to-peak values. For example, adjacent peaks of the output signal in the range of $0 \leq t \leq 0.05$ are from -1 to $+3$. The corresponding output peaks in Figure E5.7d.2 are from approximately 0 to 2.

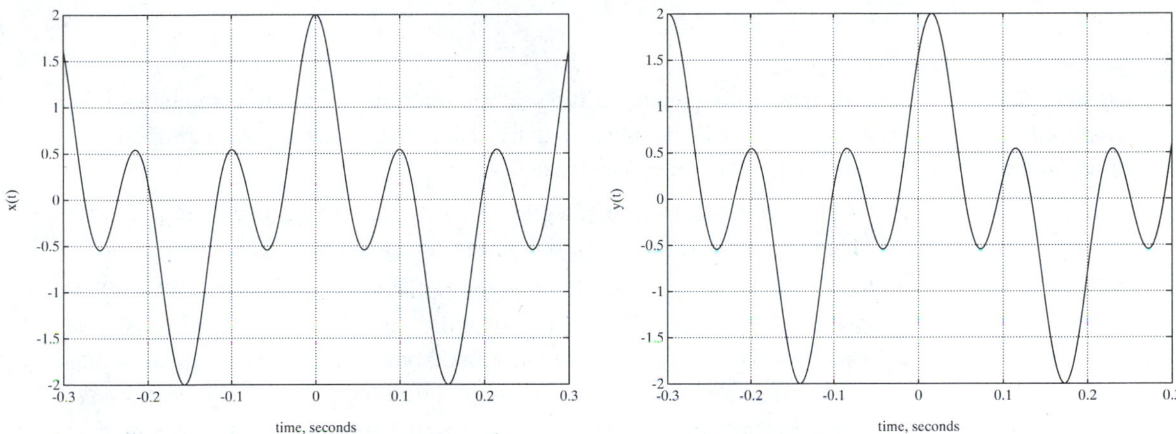

FIGURE E5.7d.1 *Input signal (left); output signal (right)*

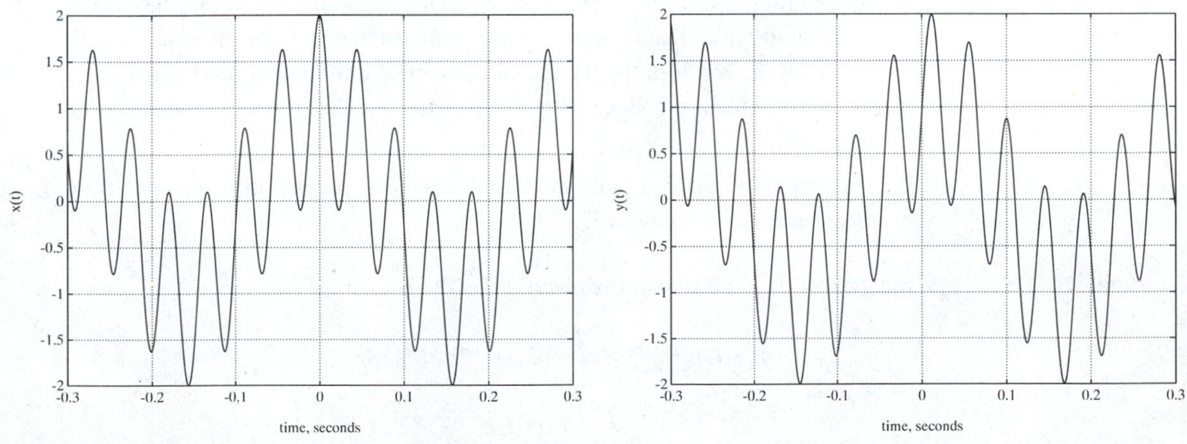

FIGURE E5.7d.2 *Input signal (left); output signal (right)*

EXAMPLES

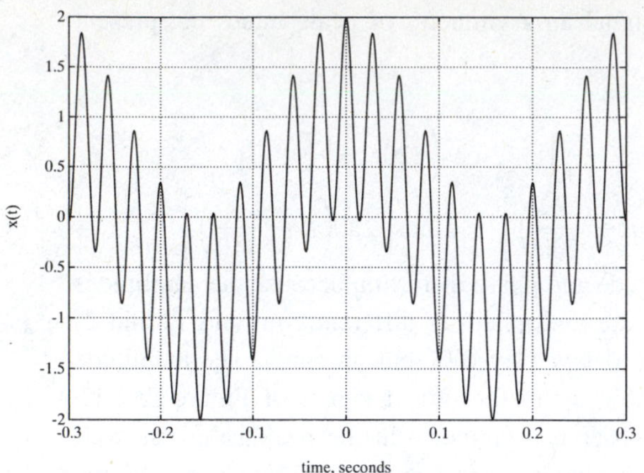

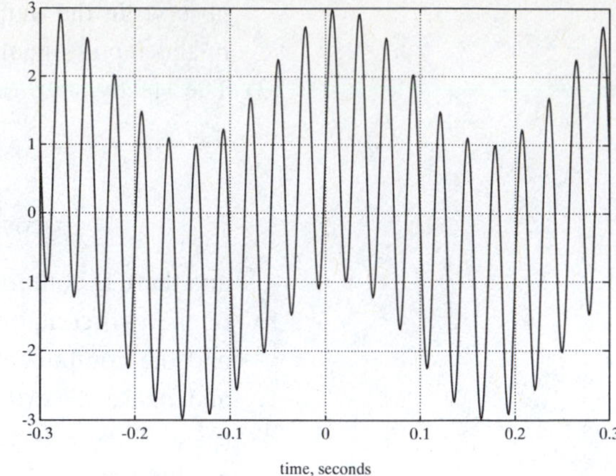

FIGURE E5.7d.3 *Input signal (left); output signal (right)*

EXAMPLE 5.8
Amplitude Modulation

Modulation or frequency shifting is an important part of communications systems, and it can be illustrated by using the frequency-shift property (3 in Table 5.2) of Fourier transforms.

a. To start, use the Fourier integral to derive this frequency-shift property. That is, given a signal $m(t)$, derive the transform of $m(t) \cdot e^{j\omega_0 t}$.

b. Now, for example, consider the multiplication of the sinusoid $c(t) = \cos \omega_c t$ by the signal $m(t)$, creating the modulated signal $m(t)\cos \omega_c t$. This process is called amplitude modulation because the original signal $m(t)$ modulates, or changes, the amplitude of the carrier signal $\cos \omega_c t$. Use the results of part (a) to find the Fourier transform of $m(t)\cos \omega_c t$.

c. In Figure E5.8a the modulation process is pictured; Figure E5.8b shows the resulting spectrum. Next, assume that the modulated signal has been transmitted to a distant site and we want to recover the original signal $s(t)$. Find the result in the transform domain of multiplying $m(t)\cos \omega_c t$ by $\cos \omega_c t$ to obtain the signal $r(t) = m(t)\cos^2 \omega_c t$. Frequency convolution is suggested.

d. The results of part (c) look promising. Can you suggest how the spectrum $M(\omega)$ might be recovered?

Solution

a. *Derivation.* The transform of $m(t) \cdot e^{j\omega_0 t}$ is

$$\mathcal{F}[m(t)e^{j\omega_0 t}] = \int_{-\infty}^{\infty} m(t)e^{j\omega_0 t}e^{-j\omega t}\,dt$$
$$= \int_{-\infty}^{\infty} m(t)e^{-j(\omega-\omega_0)}dt = M(\omega - \omega_0).$$

CHAPTER 5

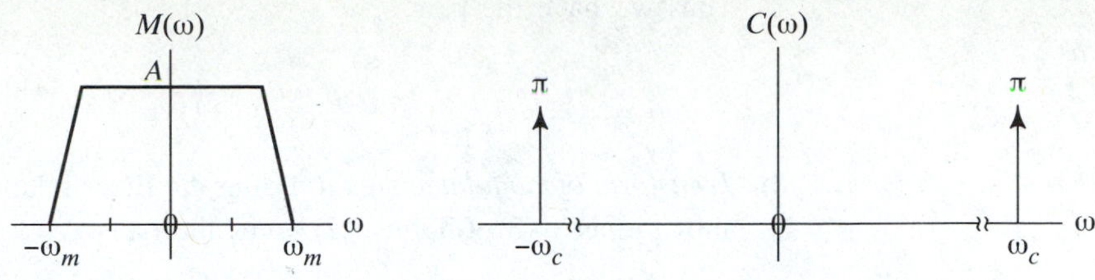

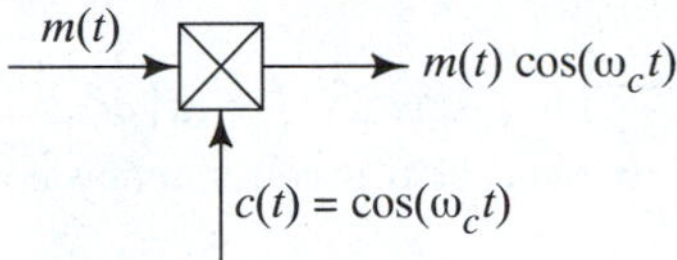

a. Modulation process and original spectrum

$$P(\omega) = \frac{1}{2\pi}\,[M(\omega) * C(\omega)]$$

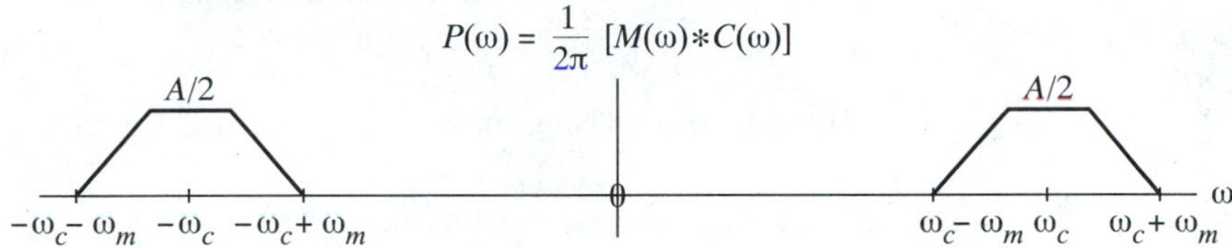

b. Resulting spectrum from modulation

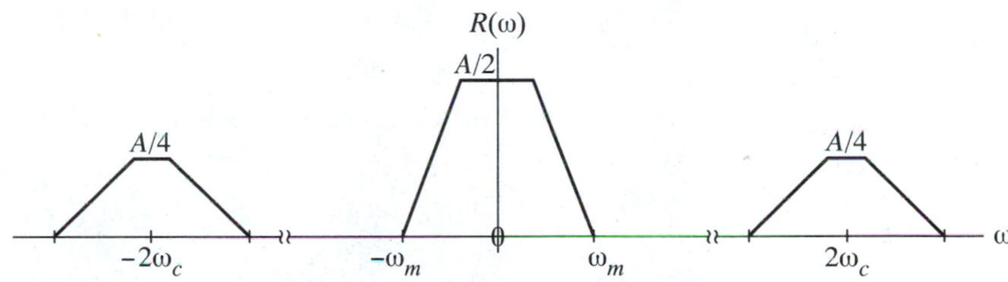

c. Spectrum after multiplying received wave by carrier

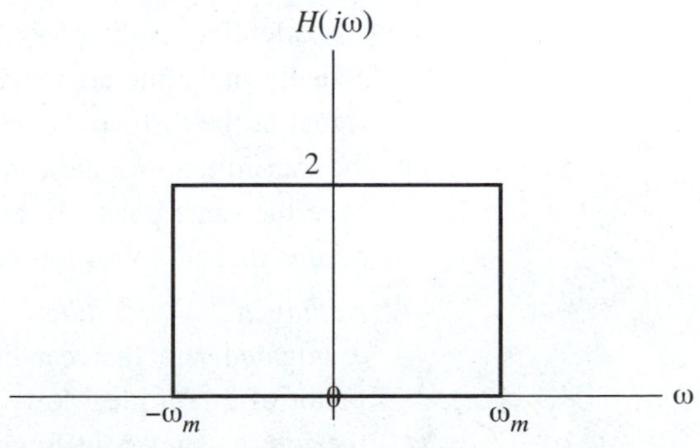

d. Lowpass filter gain characteristics

FIGURE E5.8 *Amplitude modulation*

EXAMPLES

Thus we have the pair

$$m(t)e^{j\omega_0 t} \Leftrightarrow M(\omega - \omega_0).$$

b. *Transform of modulated signal.* Using the Euler relation and the frequency-shift property (3 of Table 5.2) gives the pair

$$p(t) = m(t)\cos\omega_c t \Leftrightarrow \frac{M(\omega + \omega_c) + M(\omega - \omega_c)}{2}.$$

With $\mathcal{F}[\cos\omega_c t] = \pi[\delta(\omega + \omega_c) + \delta(\omega - \omega_c)]$ and $\mathcal{F}[m(t)] = M(\omega)$, we can also use frequency convolution to obtain the same result:

$$\begin{aligned}\mathcal{F}[m(t)\cos\omega_c t] &= \frac{1}{2\pi}\{[M(\omega)] * [\pi\delta(\omega + \omega_c) + \pi\delta(\omega - \omega_c)]\}\\ &= \frac{M(\omega + \omega_c) + M(\omega - \omega_c)}{2}.\end{aligned}$$

c. *Demodulation.* Using the results of part (b) we have

$$\begin{aligned}\mathcal{F}[m(t)\cos\omega_c t \cdot \cos\omega_c t] &= \mathcal{F}[r(t)] = R(\omega)\\ &= \left\{\frac{1}{2\pi}\left[\frac{M(\omega + \omega_c) + M(\omega - \omega_c)}{2}\right]\right.\\ &\qquad \left. * [\pi\delta(\omega + \omega_c) + \pi\delta(\omega - \omega_c)]\right\}\end{aligned}$$

The earlier section "Frequency Convolution Involving Impulses" gives us

$$\begin{aligned}R(\omega) &= \frac{1}{4}\{M(\omega + 2\omega_c) + M(\omega) + M(\omega) + M(\omega - 2\omega_c)\}\\ &= \frac{1}{2}M(\omega) + \frac{1}{4}\{M(\omega + 2\omega_c) + M(\omega - 2\omega_c)\}.\end{aligned}$$

The spectrum $R(\omega)$ is shown in Figure E5.8c. Notice that we haven't given any attention to the phase associated with the spectra $P(\omega)$ and $R(\omega)$. To actually make the approach we have described work properly, the $\cos\omega_c t$ signal at the distant site must have the same phase as the $\cos\omega_c t$ signal at the transmitter. We have implicitly made the assumption that both sinusoids have the same phase. In a practical situation, of course, one would have to ensure that this was the case.

d. *Adding a lowpass filter.* The spectra centered about $\pm 2\omega_c$ need to be eliminated and that centered about $\omega = 0$ needs to be multiplied by a factor of 2. An ideal lowpass filter with a low-frequency gain of 2 will do the job, as shown in Figure E5.8d.

REINFORCEMENT PROBLEMS

P5.1 Finding the frequency response.

a. Find the frequency response function $H(j\omega)$ for a filter that has the unit impulse response $h(t) = \delta(t) - e^{-t}u(t)$.
b. Repeat part (a) for a filter described by the transfer function $H(s) = s/(s + 1)$.
c. Repeat part (a) for a filter described by the DE $dy(t)/dt + y(t) = dx(t)/dt$.
d. Determine whether these filters are lowpass, highpass, bandpass, or bandstop filters.

P5.2 Fourier coefficients. Given the signal $x(t) = 2\cos(500\pi t) \cdot \cos(2500\pi t)$, find the period of $x(t)$ and the exponential Fourier coefficients, giving the magnitude, phase, and harmonic number for each.

P5.3 Fourier coefficients and MATLAB synthesis. Find the exponential Fourier coefficients for the periodic square wave $x(t)$ in Figure P5.3. Verify your results using MATLAB. About how many terms are needed to get a reasonable approximation to the original waveform?

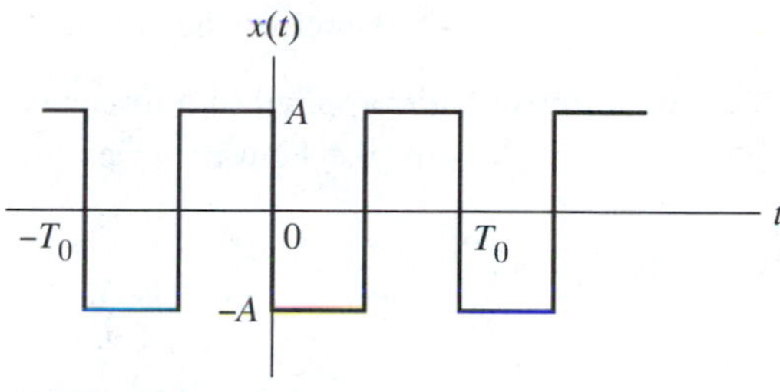

FIGURE P5.3

P5.4 Fourier coefficients and MATLAB synthesis. Repeat Problem P5.3 for the rectified cosine waveform $g(t)$ in Figure P5.4.

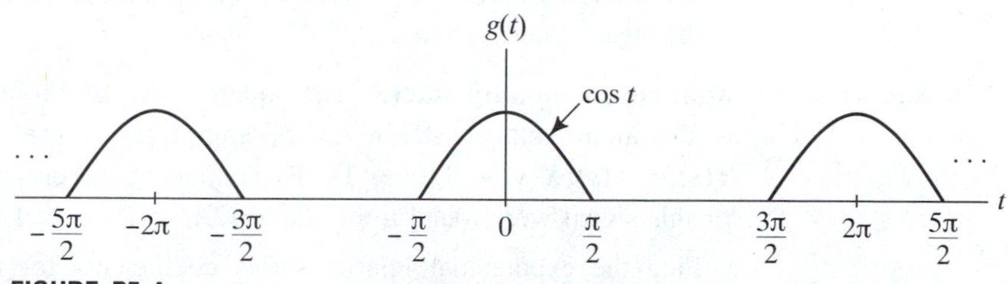

FIGURE P5.4

P5.5 Fourier cosine series and MATLAB synthesis. For the periodic signal $x(t)$ shown in Figure P5.5 (page 284), find the cosine Fourier series representation. Verify your results using MATLAB. About how many terms are needed to get a reasonable approximation to the original waveform?

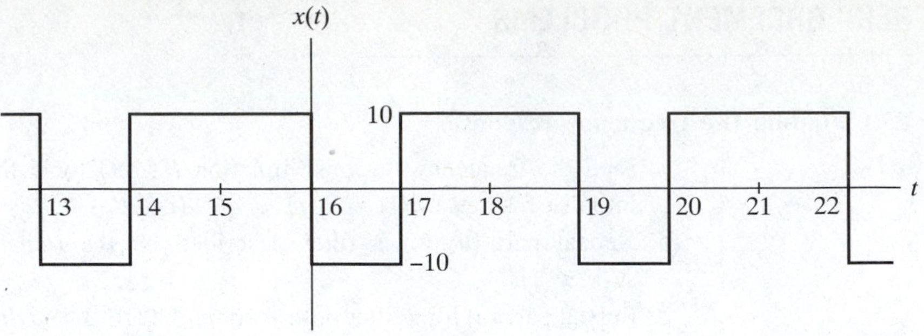

FIGURE P5.5

P5.6 Fourier series characteristics of an impulse train. A periodic train of impulse functions can be represented by the infinite summation

$$s(t) = \sum_{m=-\infty}^{\infty} \delta(t - mT_0)$$

where m and T_0 are real constants.

a. Determine the exponential Fourier coefficients and describe the frequency spectrum of this signal.
b. Describe the signal by a cosine Fourier series.

P5.7 An impulse train applied to a linear system. The impulse train of Problem P5.6 described by the Fourier series

$$s(t) = \frac{1}{T_0} \sum_{k=-\infty}^{\infty} e^{jk2\pi t/T_0}$$

is the input to a filter that has the unit impulse response $h(t) = \delta(t) - e^{-t}u(t)$.

a. What is the frequency response $H(j\omega)$ for this filter?
b. For $T_0 = 2\pi$, determine the exponential Fourier series representation of the steady-state output $y_{ss}(t)$ of this filter.
c. Write out the first three terms of the cosine Fourier series representation for the steady-state output $y_{ss}(t)$.

P5.8 Square-wave input to an op-amp filter. The square wave of Problem P5.3 is applied as the input voltage $x(t)$ to an op-amp filter whose transfer function is $H(s) = Y(s)/X(s) = 1/(s+1)$. The exponential Fourier series coefficients for this signal were found to be $X_k = j2A/\pi k$, $k = \pm1, \pm3, \pm5, \ldots$.

a. Find the exponential Fourier series coefficients for the output Y_k. Use $\omega_0 = 1$ rad/s and $A = \pi/4$.
b. Find a general expression for the cosine Fourier series representation of the steady-state output $y_{ss}(t)$.
c. Use MATLAB to plot this steady-state output. How did you decide on the number of terms to use?

P5.9 Power calculations in signal transmission. Consider the input signal of Problem P5.8 with the height of the square wave $A = \pi/4$.

a. Calculate the normalized power in the input signal.
b. Find the normalized power contained in the first two terms (first and third harmonics) of the input signal $x(t)$. What percentage of the input power is this?
c. This signal is applied to a filter whose transfer function is $H(s) = Y(s)/X(s) = 1/(s + 1)$. Find the normalized power contained in the first two terms (first and third harmonics) of the output signal $y_{ss}(t)$. What percentage of the input power is this?

P5.10 Attenuation by a filter. A first-order Butterworth filter is described by the unit impulse response $h(t) = e^{-t}u(t)$. Determine the attenuation in dB introduced by the filter to each of the first three terms of a Fourier series representation of a periodic input to the filter, where the period of the input is 2π seconds.

P5.11 Real and imaginary parts for a transform. For $x(t) = \delta(t - 0.5)$, find the real and imaginary parts of its transform $X(\omega)$.

P5.12 Fourier transform practice. For each of the following, find the Fourier transform and sketch its magnitude.

a. $a(t) = \delta(t + 1) + \delta(t + 2) + \delta(t - 2) + \delta(t - 1)$
b. $b(t) = [e^{-t} \cos(4t)]u(t)$
c. $c(t) = [\cos(4t)]u(t)$
d. $d(t) = \cos(4t), \ -\infty \leq t \leq \infty.$

P5.13 Frequency convolution to derive a pair. Given $\mathcal{F}[u(t)] = U(\omega)$ and $\mathcal{F}[x(t) = \sin \omega_0 t] = X(\omega)$, use frequency convolution to determine the Fourier transform of the time-domain product $y(t) = x(t) \cdot u(t) = \sin \omega_0 t \cdot u(t)$.

P5.14 Fourier transform via convolution and Fourier coefficients. Given $x(t)$ and $X(\omega)$ as in pair 3 in Table 5.1. Form a triangular pulse $p(t)$ from the convolution of $x(t)$ with itself—that is, $p(t) = x(t) * x(t)$.

a. Use the convolution property to find $P(\omega)$.
b. A periodic waveform $g_p(t)$ is shown in Figure P5.14. Use the result from part (a) to find the exponential Fourier series coefficients G_k.

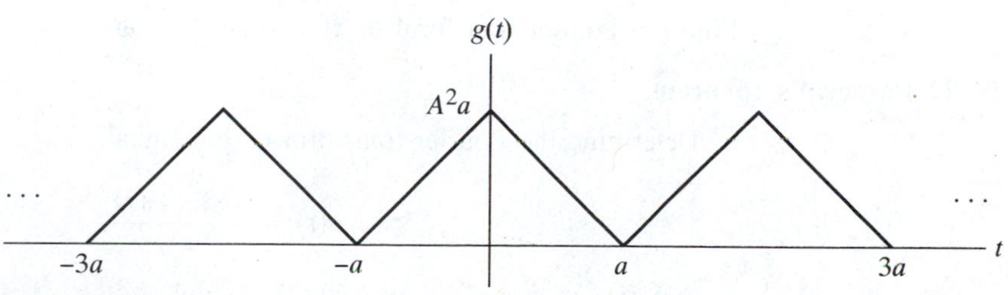

FIGURE P5.14

PROBLEMS

P5.15 Transform in terms of f rather than ω. Write the following in terms of f rather than ω:

a. the symmetry property;
b. $\mathcal{F}[\delta(t)]$;
c. $\mathcal{F}[A]$;
d. $\mathcal{F}[A \cos \omega_a t]$;
e. $H(\omega) = 1/(2 + j\omega)$.

P5.16 Response to an impulse train. A *noncausal* LTI system with the unit impulse response $h(t) = \sin(2\pi t)/\pi t$ is subjected to the periodic impulse train input

$$x(t) = \sum_{k=-\infty}^{\infty} \delta(t - 10k/3).$$

Determine the expression for the steady-state output $y_{ss}(t)$.

P5.17 Butterworth filter. A lowpass Butterworth filter that has a transfer function $H(s) = Y(s)/X(s) = (1/RC)/(s + 1/RC)$ is subjected to a single-pulse input signal of amplitude A and width of W seconds. Find the expression for the magnitude of the frequency content of the output signal $y(t)$—that is, find $|Y(\omega)|$ for $W = 5RC$.

P5.18 Transmission through an ideal filter. An ideal lowpass filter is described by $H(\omega) = 1\{u(\omega + 4) - u(\omega - 4)\}$ with zero phase shift. Calculate the normalized output energy (1-Ω resistor basis) if the input signal to the filter is (a) $x(t) = \delta(t)$; (b) $x(t) = e^{-4t}u(t)$.

P5.19 Transmission of a sampled signal. The signal $f(t) = \cos(2\pi t) + 0.1[\cos(5\pi t) + \cos(7\pi t)]$ is sampled (ideally) at an 8-Hz rate. The sampled waveform is then bandlimited to 2 Hz using an ideal lowpass filter. Write an expression for the steady-state output of the filter.

P5.20 Modulation. A rectangular pulse is described by $x_1(t) = u(t + 1) - u(t - 1)$. A new signal is formed as $x_3(t) = x_1(t) \cdot x_2(t)$, where $x_2(t) = \cos(20\pi t)$.

a. Find the equations for $X_1(f)$, $X_2(f)$, and $X_3(f)$.
b. Sketch these three transforms. Use MATLAB to plot $X_3(f)$.

P5.21 Amplitude modulation. Given the signal

$$K\{1 + 0.5 \cos \omega_m t\} \cos \omega_c t\,, \qquad \omega_c >> \omega_m\,.$$

Find the Fourier transform of $s(t)$—that is, $S(\omega)$.

P5.22 Parseval's theorem.

a. Determine the Fourier transform of the signal

$$v(t) = 4\,\frac{\sin(4\pi t)}{4\pi t}.$$

b. Use Parseval's theorem to find the energy in this signal.

CHAPTER 5

P5.23 Fourier transform from Laplace transform of causal signals. An important class of second-order analog filters can be described by the impulse response

$$h(t) = Ae^{-\zeta\omega_n t} \cos\left(\sqrt{1 - \zeta^2}\omega_n t + \alpha\right)u(t).$$

a. Find the transfer function $H(s)$. Write your answer as a rational function in terms of the variables ζ, ω_n, and α.
b. Use MATLAB to plot the Fourier transform of $h(t)$—i.e., the frequency response of the filter $H(j\omega)$, for $\omega_n = 100$ rad/s, $A = 10$, $\alpha = \pi/3$ rad, and $\zeta = 0.10$, 0.5, and 0.707.
c. Find the -3-dB bandwidth for each of these three filters.
d. Verify the shape of the phase plots for these filters.

P5.24 Unilateral Laplace transforms and Fourier transforms.

Case 1. If $F(s)$ has left half-plane poles only, then $F(\omega) = F(s)|_{s=j\omega}$.
Case 2. If $F(s)$ has poles in the left half-plane *and* on the ω-axis, then

$$F(\omega) = F(s)|_{s=j\omega} + \pi \sum_{k=1}^{N} C_k \delta(\omega - \omega_k)$$

where the C_k are the PF constants for the N ω-axis poles.
Case 3. If $F(s)$ has any poles in the right half-plane, then $F(\omega)$ does not exist.
Use information stated in Cases 1, 2, and 3 to find:

a. $H(\omega)$ for $H(s) = \dfrac{1}{(s+1)(s+2)}$, $\quad \text{Re}[s] > -1$.
b. $P(\omega)$ for $P(s) = \dfrac{2}{s(s+1)(s+2)}$, $\quad \text{Re}[s] > 0$.
c. $X(\omega)$ for $x(t) = [\sin \omega_0 t]u(t)$ with $X(s) = \dfrac{\omega_0}{s^2 + {\omega_0}^2}$.

P5.25 Solution of DE using Fourier transform. An LTI system is described by

$$\dddot{y}(t) + 6\ddot{y}(t) + 11\dot{y}(t) + 6y(t) = \dot{x}(t) + 5x(t).$$

a. Find the system frequency response $H(\omega)$.
b. Use Fourier transforms to find the unit impulse response $h(t)$.
c. Find the response to the input $x(t) = 6e^{-4t}u(t)$.

EXPLORATION PROBLEMS

P5.26 Correlation as convolution. Show that the correlation function

$$R_{xy}(\tau) = \int_{-\infty}^{\infty} x(t)y(t + \tau)\,dt$$

can be written as the convolution $x(-t) * y(t)$.

Hint: Make the change of variable $t = -\sigma$ in the correlation integral.

P5.27 Fourier transform of correlation. Show that the Fourier transform of the correlation function satisfies the relationships

$$\mathcal{F}[R_{xy}(\tau)] = X(-\omega) \cdot Y(\omega) = X^*(\omega) \cdot Y(\omega).$$

Hint: Use the definition of the Fourier transform, interchange the order of integrations, and perform a substitution of variables.

P5.28 Autocorrelation properties. Show the following properties of the autocorrelation function

$$R_{xx}(\tau) = \int_{-\infty}^{\infty} x(t)x(t + \tau)\, dt.$$

a. $R_{xx}(\tau)$ is an even function of τ, i.e., $R_{xx}(\tau) = R_{xx}(-\tau)$.
b. The maximum value of the autocorrelation function occurs at $\tau = 0$—i.e., $R_{xx}(0) \geq R_{xx}(\tau)$ for any τ.

Hint: Expand the inequality

$$\int_{-\infty}^{\infty} [x(t) - x(t + \tau)]^2 dt \geq 0.$$

P5.29 Cross correlation properties. Show the following properties of the cross correlation function

$$R_{xy}(\tau) = \int_{-\infty}^{\infty} x(t)y(t + \tau)\, dt.$$

a. $R_{xy}(\tau) = R_{yx}(-\tau)$.
b. $|R_{xy}(\tau)| \leq \sqrt{R_{xx}(0)R_{yy}(0)}$.

Hint: Expand the Cauchy-Schwartz inequality

$$\int_{-\infty}^{\infty} [x(t) - ky(t + \tau)]^2 dt \geq 0$$

for any real k to obtain a quadratic function of k, differentiate with respect to k to find where the quadratic is a minimum, and substitute this value of k to find the desired inequality.

P5.30 More Fourier transform properties. Show the following properties of Fourier transforms.

a. If $x(t)$ is a real and even function of t—i.e., $x(t) = x(-t)$—then $X(\omega)$ is a real and even function of ω.
b. If $x(t)$ is a real and odd function of t—i.e., $x(t) = -x(-t)$—then $X(\omega)$ is an imaginary and odd function of ω.

Hint: For both parts use the following properties:

1. $e^{-j\omega t} = \cos(\omega t) - j\sin(\omega t)$.
2. The product of two even functions or two odd functions is an even function, $f_e(t)$.
3. The product of an even function and an odd function is an odd function, $f_o(t)$.
4. $\int_{-\infty}^{\infty} f_o(t)\,dt = 0$.

P5.31 Average power for periodic signals. If

$$v(t) = V_0 + 2\sum_{k=1}^{\infty} |V_k| \cos(k\omega_0 t + \angle V_k)$$

show that the normalized average power is

$$\begin{aligned} P_{\text{avgn}} &= V_0^2 + \frac{4|V_1|^2}{2} + \frac{4|V_2|^2}{2} + \cdots + \frac{4|V_k|^2}{2} + \cdots \\ &= V_0^2 + 2\sum_{k=1}^{\infty} |V_k|^2. \end{aligned}$$

DEFINITIONS, TECHNIQUES, AND CONNECTIONS

FOURIER SERIES

Definitions and pairs

pair

$$f(t) = \sum_{k=-\infty}^{\infty} F_k e^{jk2\pi f_0 t} \Leftrightarrow F_k = \frac{1}{T_0}\int_{t_0}^{t_0+T_0} f(t)e^{-jk\omega_0 t}\,dt$$

cosine series

$$f(t) = F_0 + 2\sum_{k=1}^{\infty} |F_k| \cos(k\omega_0 t + \angle F_k)$$

sine series

$$f(t) = F_0 + 2\sum_{k=1}^{\infty} |F_k| \sin(k\omega_0 t + \angle F_k + \pi/2)$$

sine-cosine series

$$f(t) = F_0 + \sum_{k=1}^{\infty} A_k \cos(k\omega_0 t) + \sum_{k=1}^{\infty} B_k \sin(k\omega_0 t),$$

where $A_k = 2\,\mathrm{Re}[F_k]$ and $B_k = -2\,\mathrm{Im}[F_k]$

Transmission of a periodic signal through a linear time-invariant system

For a periodic input described by the Fourier series

$$x(t) = \sum_{k=-\infty}^{\infty} X_k e^{jk\omega_0 t}$$

we have the steady-state output in exponential and cosine forms as

$$y_{ss}(t) = \sum_{k=-\infty}^{\infty} X_k e^{jk\omega_0 t} H(jk\omega_0)$$

$$y_{ss}(t) = X_0 H(j0) + \sum_{k=1}^{\infty} 2|X_k| \cdot |H(jk\omega_0)| \cos(k\omega_0 t + \angle X_k + \angle H(jk\omega_0))$$

Frequency spectrum

Two-sided spectrum: Plots of the magnitude and phase of the exponential coefficients versus k—that is, $|F_k|$ versus k and $\angle F_k$ versus k for $k = 0, \pm 1, \pm 2, \ldots$.

One-sided spectrum: With $M_0 = F_0$ and $M_k = 2|F_k|$, plot M_k versus k for $k = 0, 1, 2, \ldots$. For the phase, plot $\angle F_k$ versus k for $k = 0, 1, 2, \ldots$.

Normalized power in a periodic signal

Parseval's theorem

$$P_{\text{avgn}} = \sum_{k=-\infty}^{\infty} |F_k|^2 \qquad P_{\text{avgn}} = \int_{t_0}^{t_0+T_0} p(t)\,dt$$

normalized power in k domain *normalized power in time domain*

FOURIER TRANSFORM

Definitions

ω pair

$$x(t) = \frac{1}{2\pi}\int_{-\infty}^{\infty} X(\omega)e^{j\omega t}\,d\omega \Leftrightarrow X(\omega) = \int_{-\infty}^{\infty} x(t)e^{-j\omega t}\,dt$$

f pair

$$x(t) = \int_{-\infty}^{\infty} X(f)e^{j2\pi ft}\,df \Leftrightarrow X(f) = \int_{-\infty}^{\infty} x(t)e^{-j2\pi ft}\,dt$$

Pairs and properties

See Tables 5.1 and 5.2.

Important applications

1. Time convolution:

$$y(t) = h(t) * x(t) = \mathcal{F}^{-1}[H(\omega)\cdot X(\omega)]$$

2. Frequency convolution:

$$y(t) = f(t)\cdot g(t) = \mathcal{F}^{-1}\left[\frac{1}{2\pi}F(\omega) * G(\omega)\right]$$

3. Transmission through a linear system:

$$Y(\omega) = H(\omega)X(\omega) = |H(\omega)|\,|X(\omega)|e^{j(\angle H(\omega)+\angle X(\omega))}$$

4. Parseval's theorem:

$$\int_{-\infty}^{\infty} |x(t)|^2 dt = \frac{1}{2\pi}\int_{-\infty}^{\infty} |X(\omega)|^2\, d\omega$$

5. Cross correlation:

$$\mathcal{F}[R_{xy}(p)] = X(-\omega)\cdot Y(\omega) = X^*(\omega)\cdot Y(\omega)$$

6. Ideal lowpass filter (linear phase):

$$H(j\omega) = \begin{cases} Ke^{-j\omega\tau}, & |\omega| \leq \omega_{co} \text{ (cutoff frequency)} \\ 0, & \text{elsewhere} \end{cases}$$

7. Sampling Theorem: If a signal contains no frequency components above a frequency $f_{\max}$, the signal can be uniquely represented by equally spaced samples if the sampling frequency f_s is greater than twice $f_{\max}$. Thus the sampling frequency must satisfy the inequality $f_s > 2f_{\max}$.

MATLAB FUNCTIONS USED

Function	*Purpose and Use*	*Toolbox*
bode	Given: state or TF model of continuous system, **bode** returns magnitude and phase of frequency response, usually on semilog coordinates.	Signals/Systems, Control System
fourier	Given: The symbolic expression *f*, **fourier** returns the Fourier transform *F*.	Symbolic Math
freqs	Given: TF model of continuous system, **freqs** returns frequency response.	Signals/Systems, Signal Processing
impulse	Given: state or TF model of continuous system, **impulse** returns impulse response.	Control System, Signals/Systems
invfourier	Given: Fourier transform function *F*, **invfourier** returns the inverse Fourier transform *f*.	Symbolic Math
ksimptf	Given: TF model of continuous system, **ksimptf** returns impulse response.	California Functions
nyquist	Given: state or TF model of continuous system, **nyquist** returns polar or Nyquist plot of frequency response.	Signals/Systems, Control System
residue	Given: rational function $T(\sigma) = N(\sigma)/D(\sigma)$, **residue** returns roots of $D(\sigma) = 0$ and PF constants of $T(\sigma)$.	MATLAB
roots	Given: coefficients of polynomial p, **roots** returns roots of $p = 0$.	MATLAB

ANNOTATED BIBLIOGRAPHY

1. Bracewell, Ronald N., *The Fourier Transform and Its Applications, 2nd ed. rev.,* McGraw-Hill Book Company, New York, 1986. *This very elegant book was written as a guide to understanding the use of transform methods when dealing with linear systems. Linear systems theory is approached through the Fourier transform; in fact, all the transforms that are discussed, which include Laplace, Hankel, Mullin, the DFT, and the discrete Hartley transform, are greatly illuminated by this approach. A useful and complete Pictorial Dictionary of Fourier Transforms is a major feature. Professor Bracewell's book is a must for anyone concerned with transform methods in linear systems. It merits a five-star rating and room in the book bag for desert island reading material.*

2. Brigham, E. Oran, *The Fast Fourier Transform and Its Applications,* Prentice Hall, Inc. Englewood Cliffs, N.J., 1988. *The title of this compact book is somewhat misleading, because the first five chapters include a tidy but thorough discussion of continuous-time Fourier procedures with an emphasis on convolution and correlation applications. All the important ideas are here, including graphical displays of Fourier transform pairs and properties. The last nine chapters are devoted to the theory and practice of machine computation of transforms and conclude with a 50-page bibliography of important articles and papers. This is a revision of the popular text* The Fast Fourier Transform *(Prentice Hall, Inc.) that was widely used for almost twenty years.*

3. Gabel, Robert A., and Richard A. Roberts, *Signals and Linear Systems, 3rd ed.,* John Wiley & Sons, New York, 1987. *Chapter Five treats Fourier methods for both discrete-time and continuous-time signals and systems. These Fourier methods are based on using real and complex sinusoids as basis functions beginning with a clear discussion of a generalized Fourier series and orthogonal functions that is followed by an introduction to the exponential Fourier series. Many interesting problems and examples, along with several tabular presentations of important results, are included.*

4. Oppenheim, Alan V., and Alan S. Willsky with Ian T. Young, *Signals and Systems,* Prentice Hall, Inc., Englewood Cliffs, N.J., 1983. *Chapter Four presents Fourier analysis for continuous-time signals and systems along with the appropriate applications, such as modulation, frequency response, and convolution. Chapters Six, Seven, and Eight on filtering, modulation, and sampling give numerous examples of the use of Fourier-inspired methods in system analysis, particularly communication applications.*

5. Ramirez, Robert W., *The FFT, Fundamentals and Concepts,* Prentice Hall, Inc., Englewood Cliffs, N.J., 1985. *This book is an interesting attempt to present sufficient Fourier concepts to bridge the gap between classroom theory and practical use. Written by Ramirez while he was an engineer with Tektronix, Inc., the book includes many diagrams and pictures that can be discussed in simple terms. A thin text (less than 200 pages), it is designed to eliminate any Fourier mysteries in the time and frequency domains.*

6. Stremler, Ferrel G., *Introduction to Communication Systems, 3rd ed.,* Addison-Wesley Publishing Company, Inc., Reading, Massachusetts, 1990.

This is a well-known text used in courses treating communications systems and engineering. Chapters Two, Three, Four, and Five provide insightful discussion on the Fourier series and transform with applications including correlation, convolution, power spectral density, and amplitude modulation.

7. [Historical Reference] Papoulis, Athanasios, *The Fourier Integral and Its Applications,* McGraw-Hill Book Company, New York, 1962 (Classic Textbook Reissue Series). *This text is probably the first popular book on Fourier methods for readers in applied science. All the material is presented on a uniform level of sophistication with, however, some variation in the details from topic to topic. Originally written for a graduate course in linear systems, in the book's preface the author states that he "hopes that it will be found to be* not too easy to read.*" Although thirty years old, this text is highly recommended for the serious reader.*

ANSWERS

P5.1 For all three, $H(j\omega) = j\omega/(1 + j\omega)$. It is a highpass filter with $H(j0) = 0$, $H(j1) = 0.5 + j0.5$, $H(j\infty) \to 1$.

P5.2 The period of $x(t)$ is 0.002 s; 2nd harmonic with $X_2 = X_{-2} = 1/2$, 3rd harmonic with $X_3 = X_{-3} = 1/2$, all other coefficients are zero.

P5.3 $X_k = 0$, k even; $X_k = j2A/\pi k$, $k = \pm1, \pm3, \pm5, \ldots$. See A5_3 on the CLS disk.

P5.4 $G_0 = 1/\pi$, $G_k = (1/2\pi)[\sin[(1-k)0.5\pi]/(1-k) + \sin[(1+k)0.5\pi]/(1+k)]$, k even; $G_1 = G_{-1} = 0.25$, $G_k = 0$ for all odd k other than $k = \pm1$. See A5_4 on the CLS disk for synthesis with 100 terms.

P5.5 $x(t) = 10/3 + \sum_1^\infty (20/k\pi)\{\sin(k2\pi/3) - (-1)^k \sin(k\pi/3)\}\cos(2\pi kt/3)$; see A5_5 on the CLS disk for synthesis with 50 terms.

P5.6 a. $S_k = 1/T_0$, same magnitude for all harmonics;
b. $s(t) = 1/T_0 + 2/T_0 \sum_{k=-\infty}^{\infty} \cos(k2\pi t/T_0)$

P5.7 a. $H(j\omega) = j\omega/(1 + j\omega)$; b. $y_{ss}(t) = [1/2\pi]\sum_{k=-\infty}^{\infty}[(jk)/(1 + jk)]e^{jkt}$; c. $y_{ss}(t) = [1/\pi][0.707\cos(t + 0.785) + 0.895\cos(2t + 0.462) + 0.949\cos(3t + 0.322)]$.

P5.8 a. $Y_k = [1/(1 + jk)] \cdot [j/2k]$, $k = \pm1, \pm3, \ldots$;
b. $y_{ss}(t) = [1/(k\sqrt{1 + k^2})]\cos(kt + \pi/2 - \tan^{-1}(k))$, $k = \pm1, \pm3, \ldots$;
c. See A5_8 on the CLS disk; increased number of terms used until change was minimal.

P5.9 a. $P_{\text{in}} = 0.617$; b. $P_{\text{in}} = P_1 + P_3 = 0.555$, 90% in these two components; c. $P_{\text{out}} = P_1 + P_3 = 0.2555$, 41.4% of input power

P5.10 $T_0 = 2\pi$ and $\omega_0 = 2\pi/T_0 = 1$ rad/s. Thus $|H(j1)| = 1/1.414 \to -3$-dB gain or 3-dB attenuation, $|H(j2)| = 1/2.236 \to -6.990$-dB gain or 6.990-dB attenuation, and $|H(j3)| = 1/3.162 \to -10$-dB gain or 10-dB attenuation.

P5.11 $\text{Re}[X(\omega)] = \cos(0.5\omega)$, $\text{Im}[X(\omega)] = -\sin(0.5\omega)$

P5.12 a. $A(\omega) = 2\cos(\omega) + 2\cos(2\omega)$; b. $B(\omega) = (1 + j\omega)/(-\omega^2 + 2j\omega + 17)$; c. $C(\omega) = j\omega/(-\omega^2 + 16) + \pi/2[\delta(\omega + 4) + \delta(\omega - 4)]$; d. $D(\omega) = \pi[\delta(\omega + 4) + \delta(\omega - 4)]$

P5.13 $Y(\omega) = \omega_0/(-\omega^2 + \omega_0{}^2) + 0.5j\pi[\delta(\omega + \omega_0) - \delta(\omega - \omega_0)]$

P5.14 a. $P(\omega) = [4A^2/\omega^2]\sin^2(\omega a/2)$; b. $G_k = [1/2a] \cdot P(\omega)$ with $\omega = k\pi/a$ gives $G_k = [2A^2 a/(\pi k)^2]\sin^2(\pi k/2)$

P5.15 a. $x(t) \Leftrightarrow X(f)$, $X(t) \Leftrightarrow x(-f)$; b. Using the sifting property, $\mathcal{F}[\delta(t)] = 1$; c. From symmetry, $A\,\delta(f)$; d. $0.5A[\delta(f - f_a) + \delta(f + f_a)]$; e. $H(f) = 1/(2 + j2\pi f)$

P5.16 From Table 5.1, $H(\omega) = u(\omega + 2\pi) - u(\omega - 2\pi)$. In the input $\omega_0 = 3\pi/5$. Thus the impulses for $k = 0$, $k = \pm 1$, $k = \pm 2$, and $k = \pm 3$ will pass through the filter (system). $y_{ss}(t) = 0.6[0.5 + \cos(0.6\pi t) + \cos(1.2\pi t) + \cos(1.8\pi t)]$.

P5.17 $X(\omega) = (2A/\omega)\sin(5RC\omega/2)$ and $H(\omega) = (1/RC)/(j\omega + 1/RC)$. Thus $|Y(\omega)| = |X(\omega)| \cdot |H(\omega)| = (2A/\omega)\sin(5RC\omega/2) \cdot (1/RC)\Big/\sqrt{\omega^2 + (1/RC)^2}$.

P5.18 a. $4/\pi$ J; b. $1/16$ J

P5.19 $y(t) = 8\cos(2\pi t)$

P5.20 a. $X_1(f) = \sin(2\pi f)/\pi f$, $X_2(f) = 0.5[\delta(f + 10) + \delta(f - 10)]$, and $X_3(f) = \text{sinc}\,2\pi(f - 10) + \text{sinc}\,2\pi(f + 10)$; b. A plot is given in A5_20 on the CLS disk, where MATLAB does hiccup at $f = \pm 10$ Hz, but this isn't a problem.

P5.21 $S(\omega) = K\pi\{\delta(\omega \pm \omega_c)\} + 0.25K\pi\{\delta(\omega \pm [\omega_c - \omega_m])\} + 0.25K\pi\{\delta(\omega \pm [\omega_c + \omega_m])\}$

P5.22 a. $V(\omega) = 1 \cdot [u(\omega + 4\pi) - u(\omega - 4\pi)]$; b. 4 J

P5.23 a. $H(s) = A\dfrac{\cos[\alpha](s + \zeta\omega_n) - \omega_n \sin[\alpha]\sqrt{1 - \zeta^2}}{s^2 + 2\zeta\omega_n s + \omega_n{}^2}$; b. The script is given for $\zeta = 0.707$. See A5_23 on the CLS disk for plots.

MATLAB Script

```
%A5_23c... Rectangular plots Problem 5.23
% zeta=0.707
a=[1,141.4,10^4];                      % denominator coefficients
b=[5,-258.8];                          % numerator coefficients
w=0:.5:1000;                           % initial freq.:increment in
                                       % rad/s:final freq
H=freqs(b,a,w);
bw=0.707*ones(size(w))*abs(max(H));    % marks 0.707 of abs(max(H))
plot(w,abs(H),w,bw),title('Fig.A5.23c Frequency Response');
%...changing phase to degrees and plotting statements
```

CHAPTER 5

c. The -3-dB bandwidth is indicated on each plot. d. Pole-zero plots show that the phase starts at π for $\omega = 0$ and approaches $-\pi/2$ as $\omega \to \infty$.

P5.24 a. $H(\omega) = 1/[(j\omega + 1)(j\omega + 2)]$;
b. $P(\omega) = 2/[(j\omega)(1 + j\omega)(2 + j\omega)] + \pi\delta(\omega)$;
c. $X(\omega) = \omega_0/(-\omega^2 + \omega_0^2) + 0.5j\pi[\delta(\omega + \omega_0) - \delta(\omega - \omega_0)]$

P5.25 a. $H(\omega) = (5 + j\omega)/(-j\omega^3 - 6\omega^2 + 11j\omega + 6)$; b. $h(t) = [2e^{-t} - 3e^{-2t} + e^{-3t}]u(t)$; c. $y(t) = [4e^{-t} - 9e^{-2t} + 6e^{-3t} - e^{-4t}]u(t)$

TABLE 5.1 *Continuous-time Fourier transform pairs*

$$x(t) = \frac{1}{2\pi}\int_{-\infty}^{\infty} X(\omega)e^{j\omega t}\,d\omega \quad \Leftrightarrow \quad X(\omega) = \int_{-\infty}^{\infty} x(t)e^{-j\omega t}\,dt$$

Signal	*Transform*
1. $A\delta(t)$	A
2. A	$2\pi A\delta(\omega)$
3. $A[u(t + a/2) - u(t - a/2)]$	$\dfrac{Aa}{(\omega a/2)}\sin(\omega a/2)$
4. $\dfrac{Aa}{(ta/2)}\sin(ta/2)$	$2\pi A[u(\omega + a/2) - u(\omega - a/2)]$
5. $Ae^{j\omega_a t}$	$2\pi A\delta(\omega - \omega_a)$
6. $A\cos(\omega_a t)$	$\pi A[\delta(\omega - \omega_a) + \delta(\omega + \omega_a)]$
7. $A\sin(\omega_a t)$	$-j\pi A[\delta(\omega - \omega_a) - \delta(\omega + \omega_a)]$
8. $A\,\mathrm{sgn}(t)$	$2A/j\omega$
9. $Au(t)$	$\pi A\delta(\omega) + A/j\omega$
10. $Ae^{-at}u(t),\ a > 0$	$A/(j\omega + a)$
11. $Ae^{at}u(-t),\ a > 0$	$A/(-j\omega + a)$
12. $Ae^{-a\|t\|}$	$2Aa/(\omega^2 + a^2)$
13. $Ae^{-bt}\cos(\omega_a t)u(t),\ b > 0$	$A\dfrac{b + j\omega}{b^2 + \omega_a{}^2 - \omega^2 + 2jb\omega}$
14. $Ae^{-bt}\sin(\omega_a t)u(t),\ b > 0$	$A\dfrac{\omega_a}{b^2 + \omega_a{}^2 - \omega^2 + 2jb\omega}$
15. $A\|t\|$	$-2A/\omega^2$
16. $\sum_{k=-\infty}^{\infty} F_k e^{jk\omega_0 t}$	$2\pi\sum_{k=-\infty}^{\infty} F_k\delta(\omega - k\omega_0)$
17. $\sum_{n=-\infty}^{\infty}\delta(t - nT)$	$\dfrac{2\pi}{T}\sum_{k=-\infty}^{\infty}\delta\left(\omega - \dfrac{2\pi k}{T}\right)$

TABLES

TABLE 5.2 *Continuous-time Fourier transform properties*

$x_1(t) \Leftrightarrow X_1(\omega)$ and $x_2(t) \Leftrightarrow X_2(\omega)$

Function	*Transform*	*Property*
1. $ax_1(t) + bx_2(t)$	$aX_1(\omega) + bX_2(\omega)$	*linearity*
2. $x_1(t - \tau)$	$e^{-j\omega\tau}X_1(\omega)$	*time shift*
3. $e^{-j\omega_0 t}x_1(t)$	$X_1(\omega - \omega_0)$	*frequency shift*
4. $x_3(t) = x_1(t) * x_2(t)$ $x_3(t) = \int_{-\infty}^{\infty} x_1(\tau)x_2(t - \tau)d\tau$	$X_3(\omega) = X_1(\omega) \cdot X_2(\omega)$	*time convolution*
5. $x_3(t) = x_1(t) \cdot x_2(t)$	$X_3(\omega) = \frac{1}{2\pi}[X_1(\omega) * X_2(\omega)]$ $X_3(\omega) =$ $\frac{1}{2\pi}\int_{-\infty}^{\infty} X_1(u)X_2(\omega - u)\,du$	*frequency convolution*
6. $\int_{-\infty}^{\infty} x_1(t)x_2(t + \tau)\,dt$	$X_1(-\omega) \cdot X_2(\omega) =$ $X_1{}^*(\omega) \cdot X_2(\omega)$	*correlation*
7. $\int_{-\infty}^{\infty} x^2(t)\,dt$	$[1/2\pi]\int_{-\infty}^{\infty} \lvert X(\omega)\rvert^2 d\omega$	*Parseval's theorem*
8. $\frac{d^N x(t)}{dt^N}$	$(j\omega)^N X(\omega)$	*time differentiation*
9. $\int_{-\infty}^{t} x(\tau)d\tau$	$\frac{1}{j\omega}X(\omega) + \pi X(0)\delta(\omega)$	*time integration*
10. $x(at)$	$\frac{1}{\lvert a\rvert} \times \left(\frac{\omega}{a}\right)$	*time-scaling*
11. $x(t)$ $X(t)$	$X(\omega)$ $2\pi x(-\omega)$	*symmetry or duality*
12. $x_p(t)$	$X_k = \frac{1}{T_0} \cdot X(\omega)\rvert_{\omega = k\omega_0}$	*Fourier coefficients*

13. If $x(t)$ is a real and even function of t, $X(\omega)$ is a real and even function of ω.

14. If $x(t)$ is a real and odd function of t, $X(\omega)$ is an odd and imaginary function of ω.

TABLE 5.3 *The forms of Fourier*

Fourier Series

$$x(t) = \sum_{k=-\infty}^{\infty} X_k e^{jk\omega_0 t} \Leftrightarrow X_k = \frac{1}{T_0} \int_{t_0}^{t_0+T_0} x(t)e^{-jk\omega_0 t}\, dt$$

Periodic in the continuous variable t — Nonperiodic in the discrete variable k

Continuous-Time Fourier Transform (CTFT)

$$x(t) = \frac{1}{2\pi} \int_{-\infty}^{\infty} X(\omega)e^{j\omega t} d\omega \Leftrightarrow X(\omega) = \int_{-\infty}^{\infty} x(t)e^{-j\omega t}\, dt$$

Nonperiodic in the continuous variable t — Nonperiodic in the continuous variable ω

Two additional forms of Fourier transforms that you will meet later follow.

Discrete-Time Fourier Transform (DTFT)

$$x(n) = \frac{1}{2\pi} \int_{\theta_0}^{\theta_0+2\pi} X(e^{j\theta})e^{jn\theta}\, d\theta \Leftrightarrow X(e^{j\theta}) = \sum_{n=-\infty}^{\infty} x(n)e^{-jn\theta}$$

Nonperiodic in the discrete variable n — Periodic in the continuous variable θ

Discrete Fourier Transform (DFT)

$$x(n) = \frac{1}{N} \sum_{k=0}^{N-1} X(k){W_N}^{-nk} \Leftrightarrow X(k) = \sum_{n=0}^{N-1} x(n){W_N}^{nk},\ W_N = e^{-j2\pi/N}$$

$n = 0, 1, \ldots, N-1$ — $k = 0, 1, \ldots, N-1$

Periodic in the discrete variable n — Periodic in the discrete variable k

TABLES

State-Space Topics for Continuous Systems

CHAPTER 6

PREVIEW

In previous chapters we've used state-space (state-variable) models such as **lsim** or **kslsim** to simulate linear time-invariant systems using MATLAB. Indeed, a strong motivation for the state form of differential equations is the ease with which such systems can be simulated. The only limits on the order of the system being simulated are imposed by the computer hardware and software, and the time available for solution. There are, however, other good reasons for state equations. We can effectively describe Multiple-Input–Multiple-Output (MIMO) systems using state variables. Also, the well-developed concepts of

FIGURE 6.1 *Computer-aided engineering*

linear algebra can be a valuable aid in analyzing and designing LTI systems. And finally, state equations allow a unified treatment of nonlinear and time-varying systems. These observations now lead us in this chapter to a more complete development of state-variable theory.

We begin by evaluating the time-domain response of LTI systems described in state-space form to find that the result is an initial condition solution and a forced response in the form of a convolution integral. Numerical solution of state equations can be accomplished by approximating the state differential equations by state difference equations. Closed-form analytical solutions of state equations can be obtained using Laplace transforms, although this is feasible only for systems of relatively low order and, consequently, is of limited value when studying practical systems that may be of fairly high order. Finally, we consider the manipulation of various forms of system models: transfer functions, state equations, and system diagrams. Throughout we will observe that although state equations are capable of describing high-order MIMO systems, they work just as well in characterizing the Single-Input–Single-Output system that we often encounter.

The state-space approach to analysis and design of continuous-time linear systems has its primary application in control theory. Its leading advocate was Rudolf E. Kalman, who, as a graduate student at Columbia University under the leadership of Professor John R. Ragazzini in the late 1950s, demonstrated the need for an alternative to the traditional frequency-domain methods that had their roots in communication systems using the results of Bode, Nyquist, and others. The application of state-variable methods broadened with the appearance of the landmark work by Zadeh and Desoer in 1963 [5]. Today, state-variable methods provide an important tool for system design.

BASIC CONCEPTS

STATE-SPACE MODEL

We begin by restating some results from Chapter 2. Starting with the Nth-order differential equation model

$$\sum_{k=0}^{N} a_k \frac{d^k y(t)}{dt^k} = b_0 x(t) \quad \text{with } a_N \equiv 1$$

this system can be represented by N first-order equations by defining N new variables $v_1(t), v_2(t), \ldots, v_N(t)$ as the states or state variables

$$v_1(t) = y(t),\ v_2(t) = \frac{dy(t)}{dt},\ v_3(t) = \frac{d^2y(t)}{dt^2}, \ldots,\ v_N(t) = \frac{d^{N-1}y(t)}{dt^{N-1}}.$$

Using these new variables the following set of state equations is established:

$$\begin{aligned}
\frac{dv_1(t)}{dt} &= \frac{dy(t)}{dt} = v_2(t) \\
\frac{dv_2(t)}{dt} &= \frac{d^2y(t)}{dt^2} = v_3(t) \\
&\vdots \\
\frac{dv_{N-1}(t)}{dt} &= \frac{d^{N-1}y(t)}{dt^{N-1}} = v_N(t) \\
\frac{dv_N(t)}{dt} &= \frac{d^Ny(t)}{dt^N} = -a_0v_1(t) - a_1v_2(t) - \cdots - a_{N-1}v_N(t) + b_0x(t).
\end{aligned}$$

In matrix form we have

state equation

$$\begin{aligned}
\frac{d\mathbf{v}(t)}{dt} &= \begin{bmatrix} 0 & 1 & 0 & 0 & \cdots & 0 \\ 0 & 0 & 1 & 0 & \cdots & 0 \\ \vdots & \vdots & \vdots & \vdots & \vdots & \vdots \\ 0 & 0 & 0 & 0 & \cdots & 1 \\ -a_0 & -a_1 & -a_2 & -a_3 & \cdots & -a_{N-1} \end{bmatrix} \mathbf{v}(t) + \begin{bmatrix} 0 \\ 0 \\ \vdots \\ 0 \\ b_0 \end{bmatrix} x(t) \\
&= \mathbf{A}\mathbf{v}(t) + \mathbf{B}x(t).
\end{aligned}$$

System outputs are defined by the $\mathbf{y}(t)$ vector as

output equation

$$\mathbf{y}(t) = \mathbf{C}\mathbf{v}(t) + \mathbf{D}x(t).$$

Notice that $x(t)$ represents a single or scalar input and, actually, many systems have multiple inputs. Writing state and output equations for Multiple-Input–Multiple-Output systems is discussed later in the chapter; for the ensuing discussion, both the inputs and the outputs will be described by the vectors $\mathbf{x}(t)$ and $\mathbf{y}(t)$, respectively.

INITIAL CONDITION SOLUTION OF THE STATE EQUATION IN THE TIME DOMAIN

First consider the initial condition (IC) solution[1] of the scalar equation

$$\dot{v}(t) = av(t)$$

which is

$$v_{\text{IC}}(t) = Ce^{at} = v(0)e^{at} = e^{at}v(0)$$

[1]The IC solution is also known as the zero-input solution.

where $v(0)$ is the initial value of $v(t)$—i.e., the value at $t = 0$. Now let's consider the matrix differential equation with the input vector $\mathbf{x}(t)$ set to zero, or

$$\dot{\mathbf{v}}(t) = \mathbf{A}\mathbf{v}(t).$$

In the same manner as for the scalar case, the IC solution of this matrix equation is

IC solution

$$\mathbf{v}_{\mathrm{IC}}(t) = \mathbf{e}^{\mathbf{A}t}\mathbf{v}(0).$$

The order of the matrix multiplication here is very important, since the matrix product $\mathbf{v}(0)\mathbf{e}^{\mathbf{A}t}$ does not exist. To show that $\mathbf{v}(t) = \mathbf{e}^{\mathbf{A}t}\mathbf{v}(0)$ is a solution of the homogeneous state equation $d\mathbf{v}(t)/dt = \mathbf{A}\mathbf{v}(t)$, we substitute it back in the differential equation and verify that equality results:

$$\frac{d\mathbf{v}(t)}{dt} = \frac{d}{dt}\{\mathbf{e}^{\mathbf{A}t}\mathbf{v}(0)\} = \mathbf{A}\{\mathbf{e}^{\mathbf{A}t}\mathbf{v}(0)\} = \mathbf{A}\mathbf{v}(t).$$

We need to know a little more about the matrix exponential $\mathbf{e}^{\mathbf{A}t}$. For the scalar case, e^{at} can be expanded as

expansion of e^{at}

$$e^{at} = 1 + at + \frac{a^2t^2}{2!} + \cdots + \frac{a^nt^n}{n!} + \cdots$$

and for the matrix case we can *define* the matrix exponential as

expansion of $\mathbf{e}^{\mathbf{A}t}$

$$\mathbf{e}^{\mathbf{A}t} = \mathbf{I} + \mathbf{A}t + \frac{\mathbf{A}^2t^2}{2!} + \cdots + \frac{\mathbf{A}^nt^n}{n!} + \cdots$$

where $\mathbf{I}$ is the identity matrix of dimension N, defined as the $N \times N$ matrix with ones on the main diagonal and zeros elsewhere. Suppose, for example, that the coefficient matrix $\mathbf{A}$ for an LTI system is

$$\mathbf{A} = \begin{bmatrix} -1 & 0 \\ 0 & -2 \end{bmatrix}$$

then the first three terms of the series will be

$$\mathbf{e}^{\mathbf{A}t} \approx \mathbf{I} + \mathbf{A}t + \frac{\mathbf{A}^2t^2}{2!} = \begin{bmatrix} 1 & 0 \\ 0 & 1 \end{bmatrix} + \begin{bmatrix} -1 & 0 \\ 0 & -2 \end{bmatrix}t + \begin{bmatrix} -1 & 0 \\ 0 & -2 \end{bmatrix}\begin{bmatrix} -1 & 0 \\ 0 & -2 \end{bmatrix}\frac{t^2}{2!}.$$

The next term is

$$\frac{\mathbf{A}^3t^3}{3!} \quad \text{or} \quad \frac{t}{3}\mathbf{A}\left\{\frac{\mathbf{A}^2t^2}{2!}\right\}$$

which means that it can be obtained by premultiplying the previous term by the matrix $\mathbf{A}$ and multiplying each of the resulting elements by the scalar $t/3$. This pattern continues, so the following term is $(t/4)\mathbf{A}\{\mathbf{A}^3(t^3/3!)\}$. If we add the first four terms (up to the terms involving t^3) in this series expansion, the result is

$$\mathbf{e}^{\mathbf{A}t} \approx \begin{bmatrix} 1 - t + 0.500t^2 - 0.133t^3 & 0 \\ 0 & 1 - 2t + 2t^2 - 1.333t^3 \end{bmatrix}.$$

We see that $\mathbf{e}^{\mathbf{A}t}$ can be written as a matrix with the same number of rows and columns as **A**. In general, each element in the $\mathbf{e}^{\mathbf{A}t}$ matrix will be an infinite series that will always converge for finite values of t, although slow convergence may make it desirable to use other methods[2] to actually compute $\mathbf{e}^{\mathbf{A}t}$.

ILLUSTRATIVE PROBLEM 6.1
Matrix Exponential and the IC Solution

a. For the preceding system, use $t = 0.1$ and do a paper-and-pencil calculation to evaluate the $\mathbf{e}^{\mathbf{A}t}$ matrix by using only, say, the first three terms.

b. Repeat part (a) using four terms.

c. Using the results of part (b) with $v_1(0) = 4.0$ and $v_2(0) = -2.5$, find the initial condition solution $\mathbf{v}_{\mathbf{IC}}(0.1)$.

Solution

a. Using three terms gives us

$$\mathbf{e}^{\mathbf{A}(0.1)} \approx \begin{bmatrix} 0.905 & 0 \\ 0 & 0.820 \end{bmatrix}.$$

b. If we sum four terms, the result is

$$\mathbf{e}^{\mathbf{A}(0.1)} \approx \begin{bmatrix} 0.9049 & 0 \\ 0 & 0.8187 \end{bmatrix}.$$

c. Thus, if the initial conditions are $v_1(0) = 4.0$ and $v_2(0) = -2.5$, we have

$$\mathbf{v}(0.1) = \begin{bmatrix} 0.9049 & 0 \\ 0 & 0.8187 \end{bmatrix}\begin{bmatrix} 4 \\ -2.5 \end{bmatrix} = \begin{bmatrix} 3.6196 \\ -2.0468 \end{bmatrix}.$$

Notice that in this example **A**—and hence $\mathbf{e}^{\mathbf{A}t}$—are diagonal matrices. In general, they are not diagonal.

Comment: We could (should!) use a computer to do this calculation. Following are the script that uses the function **kslsim** to do the task and the resulting output. Notice that the elements of the matrices **B**, **C**, and **D** are arbitrary (their dimensions are not) because we are interested in the IC solution only for the system states.

MATLAB Script

```
%IP6_1 IC solution of state equation
A=[-1,0;0,-2];                        % coefficients of A matrix
B=[0;10];                             % 2 × 1 arbitrary matrix
C=[1,0];                              % 1 × 2 arbitrary matrix
D=[0];                                % 1 × 1 arbitrary matrix
v0=[4;-2.5];                          % initial condition state vector
t=0:0.1:.1;                           % start t:increment:stop t
X=[0*ones(size(t))]';                 % input row vector of zeros
[y,v]=kslsim(A,B,C,D,X,t,v0)          % call kslsim
```

Continues

[2] See J. Gary Reid, *Linear System Fundamentals,* McGraw-Hill Book Company, New York, 1983, Chapters Seven and Eight.

Output

```
v= 4.0000 -2.5000            values of the states at t=0
   3.6194 -2.0468            values of the states at t=0.1
```

If we wanted to know the initial condition solution for several time instants, we would have to reevaluate $\mathbf{e}^{\mathbf{A}t}$ for the different values of t of interest (or use another approach; more on this later). We would often prefer to have a closed-form, or analytical, expression for $\mathbf{v}(t)$, and subsequently we will learn how this can be accomplished. At this point, we have learned a way to solve the matrix equation $d\mathbf{v}(t)/dt = \mathbf{A}\mathbf{v}(t)$ in the time domain. In other words, we have found the IC solution of the state equation

state equation

$$d\mathbf{v}(t)/dt = \mathbf{A}\mathbf{v}(t) + \mathbf{B}\mathbf{x}(t)\,.$$

Let us now consider the effects of inputs on the system response.

COMPLETE SOLUTION OF THE STATE EQUATION IN THE TIME DOMAIN

We want to obtain the complete solution that includes the effect of the inputs $\mathbf{x}(t)$ as well as that of the initial conditions $\mathbf{v}(0)$. First, we premultiply the state equation by the matrix integrating factor $\mathbf{e}^{-\mathbf{A}t}$ and move all terms containing $\mathbf{v}(t)$ and $d\mathbf{v}(t)/dt$ to the left side of the equation. This gives us

$$\mathbf{e}^{-\mathbf{A}t}d\mathbf{v}(t)/dt - \mathbf{e}^{-\mathbf{A}t}\mathbf{A}\mathbf{v}(t) = \mathbf{e}^{-\mathbf{A}t}\mathbf{B}\mathbf{x}(t)\,.$$

Notice that the left side of this equation is the time derivative of $\mathbf{e}^{-\mathbf{A}t}\mathbf{v}(t)$, which allows us to write the equation in the form

$$\frac{d}{dt}\{\mathbf{e}^{-\mathbf{A}t}\mathbf{v}(t)\} = \mathbf{e}^{-\mathbf{A}t}\mathbf{B}\mathbf{x}(t)\,.$$

Next, integrating both sides from zero to t we find

$$\mathbf{e}^{-\mathbf{A}t}\mathbf{v}(t) - \mathbf{e}^{-\mathbf{A}0}\mathbf{v}(0) = \int_0^t \mathbf{e}^{-\mathbf{A}\tau}\mathbf{B}\mathbf{x}(\tau)\,d\tau$$

where we are using the "dummy" variable τ in the definite integral on the right side. The matrix $\mathbf{e}^{-\mathbf{A}0}$ is simply the identity matrix $\mathbf{I}$, and if we now premultiply throughout by $\mathbf{e}^{\mathbf{A}t}$ and solve for $\mathbf{v}(t)$, we determine the complete solution to be

$$\mathbf{v}(t) = \mathbf{e}^{\mathbf{A}t}\mathbf{v}(0) + \mathbf{e}^{\mathbf{A}t}\int_0^t \mathbf{e}^{-\mathbf{A}\tau}\mathbf{B}\,d\tau = \mathbf{e}^{\mathbf{A}t}\mathbf{v}(0) + \int_0^t \mathbf{e}^{\mathbf{A}(t-\tau)}\mathbf{B}\,d\tau\,.$$

Looking at the solution we notice that the *initial condition response,* that part of the solution that is due to the initial conditions represented by the $\mathbf{v}(0)$ vector is given by

$$\mathbf{v}_{\mathbf{IC}}(t) = \mathbf{e}^{\mathbf{A}t}\mathbf{v}(0)$$

as found previously, while the *forced response*,[3] that part caused by the forcing function or input vector $\mathbf{x}(t)$, is

$$\mathbf{v}_\mathbf{F}(t) = \int_0^t \mathbf{e}^{\mathbf{A}(t-\tau)}\mathbf{B}\mathbf{x}(\tau)\,d\tau\,.$$

Recapitulating, if an LTI system is described by a set of first-order differential equations written as the matrix differential equation

matrix DE

$$\dot{\mathbf{v}}(t) = \mathbf{A}\mathbf{v}(t) + \mathbf{B}\mathbf{x}(t),$$

the complete solution is given by

solution of state equation

$$\begin{aligned}\mathbf{v}(t) &= \mathbf{e}^{\mathbf{A}t}\mathbf{v}(0) + \int_0^t \mathbf{e}^{\mathbf{A}(t-\tau)}\mathbf{B}\,d\tau \\ &= \mathbf{v}_{\mathbf{IC}}(t) + \mathbf{v}_\mathbf{F}(t)\,.\end{aligned}$$

ILLUSTRATIVE PROBLEM 6.2
Complete Solution of the State Equation

The rotational motion of a spacecraft about one axis can be described by the differential equation $d^2\theta(t)/dt^2 = k\lambda(t)$, where $\theta(t)$ is the roll angle and $\lambda(t)$ is the driving torque. Letting $v_1(t) = \theta(t)$ and $x(t) = \lambda(t)$, the state equation is

$$\dot{\mathbf{v}}(t) = \begin{bmatrix} 0 & 1 \\ 0 & 0 \end{bmatrix}\mathbf{v}(t) + \begin{bmatrix} 0 \\ 10 \end{bmatrix}x(t)\,.$$

a. Find an expression for the IC solution $\mathbf{v}_{\mathbf{IC}}(t)$ for any time t in terms of the initial condition vector $\mathbf{v}(0)$.

b. Determine the unit step response of this system if the output is defined as $y(t) = [1 \quad 0]\mathbf{v}(t)$ or $y(t) = v_1(t)$.

Solution

a. Using the series expansion for $\mathbf{e}^{\mathbf{A}t}$ gives

$$\begin{aligned}\mathbf{e}^{\mathbf{A}t} &= \mathbf{I} + \frac{t}{1}\mathbf{A} + \frac{t}{2}\mathbf{A}\{\mathbf{A}t\} + \frac{t}{3}\mathbf{A}\left\{\frac{\mathbf{A}^2t^2}{2}\right\} + \cdots \\ &= \begin{bmatrix} 1 & 0 \\ 0 & 1 \end{bmatrix} + \begin{bmatrix} 0 & 1 \\ 0 & 0 \end{bmatrix}t + \frac{t}{2}\begin{bmatrix} 0 & 1 \\ 0 & 0 \end{bmatrix}\begin{bmatrix} 0 & t \\ 0 & 0 \end{bmatrix} + \cdots = \begin{bmatrix} 1 & t \\ 0 & 1 \end{bmatrix}\end{aligned}$$

where we observe that this infinite series truncates after only two terms (not the usual situation). Thus the IC solution is given by

$$\mathbf{v}_{\mathbf{IC}}(t) = \begin{bmatrix} 1 & t \\ 0 & 1 \end{bmatrix}\mathbf{v}(0)\,.$$

From this matrix equation we see that due to the initial conditions $v_1(0)$ and $v_2(0)$, $v_1(t) = v_1(0) + tv_2(0)$ and $v_2(t) = v_2(0)$. We notice that the initial

[3]The forced response is also called the zero-state response. We prefer to define the response in terms of what causes it rather than what doesn't.

CHAPTER 6

condition response tends to infinity as t becomes large, which we saw in Chapter 2 to be a characteristic of an unstable system.

b. For the forced solution we have

$$\mathbf{v_F}(t) = \int_0^t \mathbf{e}^{\mathbf{A}(t-\tau)}\mathbf{B}\mathbf{x}(\tau)\,d\tau$$

and with $\mathbf{B}$, $\mathbf{e}^{\mathbf{A}t}$, and $x(t) = u(t)$ given, the result is

$$\mathbf{v_F}(t) = \int_0^t \begin{bmatrix} 1 & t-\tau \\ 0 & 1 \end{bmatrix}\begin{bmatrix} 0 \\ 10 \end{bmatrix} u(\tau)\,d\tau = \int_0^t \begin{bmatrix} 10(t-\tau) \\ 10 \end{bmatrix} d\tau = \begin{bmatrix} 5t^2 \\ 10t \end{bmatrix}.$$

The output is $y(t) = v_1(t) = 5t^2$, which indicates that the output is unbounded for a bounded input, also a characteristic of an unstable system.

STABILITY REVISITED

One definition of stability for a causal LTI system is the following:

The initial condition response of a stable causal system approaches zero as $t \to \infty$.

We investigate this from the point of view of the state-space equations. We have seen that the initial condition response is given by

$$\mathbf{v_{IC}}(t) = \mathbf{e}^{\mathbf{A}t}\mathbf{v}(0)$$

and in order for

$$\mathbf{v_{IC}}(t) \to \mathbf{0} \quad \text{as } t \to \infty$$

the matrix $\mathbf{e}^{\mathbf{A}t}$ must approach the null matrix $\mathbf{0}$ as $t \to \infty$. That is, all the elements of $\mathbf{e}^{\mathbf{A}t}$ must approach zero as t approaches infinity.

ILLUSTRATIVE PROBLEM 6.3
Stability from the Time Domain

Three $\mathbf{A}$ matrices for systems described by the state equation

$$\dot{\mathbf{v}}(t) = \mathbf{A}\mathbf{v}(t) + \mathbf{B}\mathbf{x}(t)$$

are listed next. Check for system stability by expanding $\mathbf{e}^{\mathbf{A}t}$.

a. $\mathbf{A} = \begin{bmatrix} -1 & 0 \\ 0 & -2 \end{bmatrix}$

b. $\mathbf{A} = \begin{bmatrix} -1 & 0 \\ 0 & 2 \end{bmatrix}$

c. $\mathbf{A} = \begin{bmatrix} 0 & 1 \\ 1 & 0 \end{bmatrix}$

Solution **a.** We will use only enough terms to find a pattern in the expansion.

$$\begin{aligned}\mathbf{e}^{\mathbf{A}t} &= \begin{bmatrix} 1 & 0 \\ 0 & 1 \end{bmatrix} + t\begin{bmatrix} -1 & 0 \\ 0 & -2 \end{bmatrix} + \frac{t}{2}\begin{bmatrix} -1 & 0 \\ 0 & -2 \end{bmatrix}\begin{bmatrix} -t & 0 \\ 0 & -2t \end{bmatrix} \\ &\quad + \frac{t}{3}\begin{bmatrix} -1 & 0 \\ 0 & -2 \end{bmatrix}\begin{bmatrix} t^2/2 & 0 \\ 0 & 4t^2/2 \end{bmatrix} \\ &\quad + \frac{t}{4}\begin{bmatrix} -1 & 0 \\ 0 & -2 \end{bmatrix}\begin{bmatrix} -t^3/6 & 0 \\ 0 & -8t^3/6 \end{bmatrix} + \cdots \\ &= \begin{bmatrix} 1 - t + t^2/2! - t^3/3! + t^4/4! - \cdots & 0 \\ 0 & 1 - 2t + 4t^2/2! - 8t^3/3! + 16t^4/4! - \cdots \end{bmatrix} \\ &= \begin{bmatrix} e^{-t} & 0 \\ 0 & e^{-2t} \end{bmatrix}.\end{aligned}$$

It is a stable system; both e^{-t} and e^{-2t} approach 0 as $t \to \infty$.

b. As in (a) this **A** matrix has only diagonal terms, so the expansion will be almost the same, with the only difference being the signs of the 2, 2 term. By deduction we have

$$\begin{aligned}\mathbf{e}^{\mathbf{A}t} &= \begin{bmatrix} 1 - t + t^2/2! - t^3/3! + t^4/4! - \cdots & 0 \\ 0 & 1 + 2t + 4t^2/2! + 8t^3/3! + 16t^4/4! + \cdots \end{bmatrix} \\ &= \begin{bmatrix} e^{-t} & 0 \\ 0 & e^{2t} \end{bmatrix}.\end{aligned}$$

This system is unstable; $e^{2t} \to \infty$ as $t \to \infty$.

c. Here we need to see what a few terms in the expansion tell us; we have

$$\begin{aligned}\mathbf{e}^{\mathbf{A}t} &= \begin{bmatrix} 1 & 0 \\ 0 & 1 \end{bmatrix} + t\begin{bmatrix} 0 & 1 \\ 1 & 0 \end{bmatrix} + \frac{t}{2}\begin{bmatrix} 0 & 1 \\ 1 & 0 \end{bmatrix}\begin{bmatrix} 0 & t \\ t & 0 \end{bmatrix} \\ &\quad + \frac{t}{3}\begin{bmatrix} 0 & 1 \\ 1 & 0 \end{bmatrix}\begin{bmatrix} t^2/2 & 0 \\ 0 & t^2/2 \end{bmatrix} + \cdots \\ &= \begin{bmatrix} 1 + t^2/2 + t^4/4! + \cdots & t + t^3/6 + \cdots \\ t + t^3/6 + \cdots & 1 + t^2/2 + t^4/4! + \cdots \end{bmatrix}.\end{aligned}$$

It is an unstable system; all four elements of the matrix will increase without bound as $t \to \infty$.

Comment: Later in the chapter we will state how stability can be determined in the time domain by finding the eigenvalues of the **A** matrix. The eigenvalue approach is a practical one compared to the tedious, unwieldly, and impractical approach of Illustrative Problem 6.3. As a bonus, the MATLAB function **eig** computes the eigenvalues.

NUMERICAL EVALUATION OF THE STATE EQUATION AND SYSTEMS WITH PIECEWISE-CONSTANT INPUTS

Now we need to turn the solution

$$\mathbf{v}(t) = \mathbf{e}^{\mathbf{A}T}\mathbf{v}(0) + \int_0^t \mathbf{e}^{\mathbf{A}(t-\tau)}\mathbf{B}\mathbf{x}(\tau)\, d\tau$$

CHAPTER 6

into a form that is amenable to numerical (computer) evaluation. The total time interval of interest, $0 \le t \le t_f$, is broken up into short segments of duration T, that is, $0, T, 2T, \ldots$; we start by solving for $\mathbf{v}(t)$ at $t = T$ with the initial condition $\mathbf{v}(0)$ and a known input $\mathbf{x}(t)$. Thus, for $t = T$,

$$\mathbf{v}(T) = \mathbf{e}^{\mathbf{A}T}\mathbf{v}(0) + \int_0^T \mathbf{e}^{\mathbf{A}(T-\tau)}\mathbf{B}\mathbf{x}(\tau)\,d\tau\,.$$

Further, suppose that the inputs are constants during the interval 0 to T—that is, $x_i(t) = x_i(0)$, $i = 1, 2, \ldots, M$—during the period $0 \le t < T$, as shown in Figure 6.2. This allows us to write the input vector as

$$\mathbf{x}(t) = \mathbf{x}(0) \qquad \text{for } 0 \le t < T\,.$$

If the inputs are not constant in the interval, then we must approximate them as constants. The accuracy of the approximation will be affected by the value of T. In general, the smaller the value of T, the more accurate the approximation. Since $\mathbf{x}(0)$ is a matrix of constants over the interval $[0, T)$,[4] we can take the constant term $\mathbf{x}(0)$ out of the integral. Thus the state vector at $t = T$ can be written

$$\mathbf{v}(T) = \mathbf{e}^{\mathbf{A}T}\mathbf{v}(0) + \left[\int_0^T \mathbf{e}^{\mathbf{A}(T-\tau)}\mathbf{B}\,d\tau\right]\mathbf{x}(0)\,.$$

Remembering that the order of matrix products must be preserved, $\mathbf{x}(0)$ must follow the integral. We introduce the notation

Phi and Del matrices

$$\mathbf{\Phi}(T) = \mathbf{e}^{\mathbf{A}T} \quad \text{and} \quad \mathbf{\Delta}(T) = \int_0^T \mathbf{e}^{\mathbf{A}(T-t)}\mathbf{B}\,d\tau$$

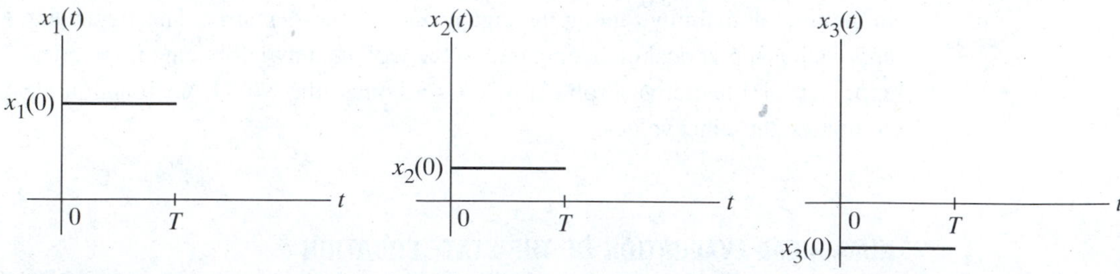

FIGURE 6.2 *Piecewise-constant inputs*

[4]The notation $[0, T)$ means the interval $0 \le t < T$; that is, the lower limit of the interval is 0 and it is contained in the interval, whereas the upper limit T is not.

where $\mathbf{\Phi}(T)$ is called the *Phi matrix* and $\mathbf{\Delta}(T)$ is the *Del matrix*. Using the notation of $\mathbf{\Phi}(T)$ and $\mathbf{\Delta}(T)$, the solution to the state equation at time T becomes

$$\mathbf{v}(T) = \mathbf{\Phi}(T)\mathbf{v}(0) + \mathbf{\Delta}(T)\mathbf{x}(0).$$

ILLUSTRATIVE PROBLEM 6.4

Evaluating the Phi and Del Matrices

Find the Phi matrix $\mathbf{\Phi}(T)$ and the Del matrix $\mathbf{\Delta}(T)$ for the unstable spacecraft described in Illustrative Problem 6.2 by the state equation

$$\dot{\mathbf{v}}(t) = \begin{bmatrix} 0 & 1 \\ 0 & 0 \end{bmatrix}\mathbf{v}(t) + \begin{bmatrix} 0 \\ 10 \end{bmatrix}x(t).$$

Solution

In Illustrative Problem 6.2, we found that

$$\mathbf{e}^{\mathbf{A}t} = \begin{bmatrix} 1 & t \\ 0 & 1 \end{bmatrix} \quad \text{and so } \mathbf{e}^{\mathbf{A}T} = \mathbf{\Phi}(T) = \mathbf{e}^{\mathbf{A}t}|_{t=T} = \begin{bmatrix} 1 & T \\ 0 & 1 \end{bmatrix}.$$

Taking advantage of the fact that the series for $\mathbf{e}^{\mathbf{A}T}$ truncates after two terms, it is relatively easy (which is very unusual) to find the Del matrix as

$$\mathbf{\Delta}(T) = \int_0^T \mathbf{\Phi}(T-\tau)\mathbf{B}\,d\tau = \int_0^T \begin{bmatrix} 1 & T-\tau \\ 0 & 1 \end{bmatrix}\begin{bmatrix} 0 \\ 10 \end{bmatrix}d\tau = \begin{bmatrix} 5T^2 \\ 10T \end{bmatrix}.$$

Comment: Finding the Phi and Del matrices is best done using the MATLAB function **c2d** from the Signals and Systems Toolbox. Shown here are the script and the results for the system described by the **A** and **B** matrices for two different values of T, 1, and 0.1.

MATLAB Script

```
%IP6_4 Phi, Del Calculation
A=[0,1;0,0];
B=[0;10];
[Phi,Del] = c2d(A,B,1)
A=[0,1;0,0];
B=[0;10];
[Phi,Del] = c2d (A,B,0.1)
```

Output

```
Phi=
        1   1
        0   1
Del=
        5.0000
        10.0000
Phi=
        1.0000   0.1000
        0        1.0000
Del=
        0.050
        1.0000
```

CHAPTER 6

Now, suppose we were to call the initial time t_0 rather than 0. This is no more than a change of notation, and $\mathbf{v}(0)$ and $\mathbf{x}(0)$ will become $\mathbf{v}(t_0)$ and $\mathbf{x}(t_0)$, respectively. In the same way, the solution after T seconds becomes $\mathbf{v}(t_0 + T)$. Merely changing the notation cannot, of course, change the solution, so the multipliers $\mathbf{\Phi}(T)$ and $\mathbf{\Delta}(T)$ are unchanged. We now have, therefore,

$$\mathbf{v}(t_0 + T) = \mathbf{\Phi}(T)\mathbf{v}(t_0) + \mathbf{\Delta}(T)\mathbf{x}(t_0).$$

It is instructive to look at some special values of t_0:

$$\begin{aligned}
t_0 &= 0 & \mathbf{v}(T) &= \mathbf{\Phi}(T)\mathbf{v}(0) + \mathbf{\Delta}(T)\mathbf{x}(0) \\
t_0 &= T & \mathbf{v}(2T) &= \mathbf{\Phi}(T)\mathbf{v}(T) + \mathbf{\Delta}(T)\mathbf{x}(T) \\
t_0 &= 2T & \mathbf{v}(3T) &= \mathbf{\Phi}(T)\mathbf{v}(2T) + \mathbf{\Delta}(T)\mathbf{x}(2T) \\
t_0 &= nT & \mathbf{v}(nT + T) &= \mathbf{\Phi}(T)\mathbf{v}(nT) + \mathbf{\Delta}(T)\mathbf{x}(nT).
\end{aligned}$$

Thus the general expression that allows us to calculate the values of the states at the time instants $T, 2T, 3T, \ldots, nT, \ldots$ is given by

numerical evaluation of the matrix DE

$$\mathbf{v}(nT + T) = \mathbf{\Phi}(T)\mathbf{v}(nT) + \mathbf{\Delta}(T)\mathbf{x}(nT) \quad \text{for } n = 0, 1, 2, \ldots$$
$$\text{with } \mathbf{x}(t) = \mathbf{x}(nT), \; nT \le t < nT + T$$

where

$$\mathbf{\Phi}(T) = \mathbf{e}^{\mathbf{A}T} \quad \text{and} \quad \mathbf{\Delta}(T) = \int_0^T \mathbf{e}^{\mathbf{A}(T-\tau)}\mathbf{B}\,d\tau = \int_0^T \mathbf{\Phi}(T - \tau)\mathbf{B}\,d\tau.$$

Notice that we have obtained a *state difference equation* by this approach. If $\mathbf{x}(t)$ is constant during the intervals $[0, T), [T, 2T), \ldots$, this equation can be used to find the exact values of the states, but only at the times $T, 2T, 3T, \ldots$. If $\mathbf{x}(t)$ is not piecewise constant, this state difference equation yields approximate values of $\mathbf{v}(t)$ at $t = T, 2T, 3T, \ldots$.

We can use this state equation "solution" in the following way. Assume that the initial condition $\mathbf{v}(0)$ is known and that the input vector $\mathbf{x}(t)$ is given for $t \ge 0$. Selecting T to be small enough so that it is reasonable to approximate $\mathbf{x}(t)$ as constant over the intervals $0 \to T, T \to 2T, \ldots, nT \to nT + T$, we can write, letting $n = 0$,

$$\mathbf{v}(T) = \mathbf{\Phi}(T)\mathbf{v}(0) + \mathbf{\Delta}(T)\mathbf{x}(0).$$

Then, with $n = 1$ in the equation $\mathbf{v}(nT + T) = \mathbf{\Phi}(T)\mathbf{v}(nT) + \mathbf{\Delta}(T)\mathbf{x}(nT)$, we find the values of the states at $t = 2T$ as

$$\mathbf{v}(2T) = \mathbf{\Phi}(T)\mathbf{v}(T) + \mathbf{\Delta}(T)\mathbf{x}(T)$$

and the value of $\mathbf{v}(T)$ found previously allows us to evaluate $\mathbf{v}(2T)$. One more step should be adequate to reinforce the emerging pattern, so letting $n = 2$ in the equation

$$\mathbf{v}(nT + T) = \mathbf{\Phi}(T)\mathbf{v}(nT) + \mathbf{\Delta}(T)\mathbf{x}(nT)$$

gives

$$\mathbf{v}(3T) = \mathbf{\Phi}(T)\mathbf{v}(2T) + \mathbf{\Delta}(T)\mathbf{x}(2T)$$

where the previously found value of $\mathbf{v}(2T)$ is again used on the right-hand side to evaluate $\mathbf{v}(3T)$. So we observe that we can calculate as many values of $\mathbf{v}(nT)$ as we wish. The accuracy of these values is determined by numerical factors, such as the word length of the computer used and the closeness of the piecewise approximation of $\mathbf{x}(t)$ by $\mathbf{x}(0)$, $\mathbf{x}(T)$, $\mathbf{x}(2T), \ldots$. This iterative procedure of solving the state difference equation will be encountered again in Chapter 11 in the discussion "Complete Solution of the State Difference Equation."

The system output at the time instants $t = 0, T, 2T, 3T, \ldots, nT, \ldots$ is given by the matrix equation

output equation

$$\mathbf{y}(nT) = \mathbf{C}\mathbf{v}(nT) + \mathbf{D}\mathbf{x}(nT).$$

ILLUSTRATIVE PROBLEM 6.5
Numerical Evaluation of the Matrix DE

For the system of Illustrative Problems 6.2 and 6.4, compute the state vector at $t = 0.5$ s for $\mathbf{v}(0) = [1 \quad 0]^T$ and $x(t) = 5tu(t)$. Use the fairly coarse value of $T = 0.1$ with the output defined as $y(t) = v_1(t)$, as before.

Solution

From Illustrative Problem 6.4,

$$\mathbf{\Phi}(T) = \begin{bmatrix} 1 & T \\ 0 & 1 \end{bmatrix} \quad \text{and} \quad \mathbf{\Delta}(T) = \begin{bmatrix} 5T^2 \\ 10T \end{bmatrix}.$$

Using $\mathbf{v}(nT + T) = \mathbf{\Phi}(T)\mathbf{v}(nT) + \mathbf{\Delta}(T)\mathbf{x}(nT)$ for $n = 0, 1, 2, \ldots$ with $T = 0.1$ gives

$$\mathbf{v}(0.1n + 0.1) = \begin{bmatrix} 1 & 0.10 \\ 0 & 1 \end{bmatrix}\begin{bmatrix} 1 \\ 0 \end{bmatrix} + \begin{bmatrix} 0.05 \\ 1 \end{bmatrix} 0.5n .$$

For $n = 0$,

$$\mathbf{v}(0.1(0) + 0.1) = \mathbf{v}(0.1) = \begin{bmatrix} 1 & 0.10 \\ 0 & 1 \end{bmatrix}\begin{bmatrix} 1 \\ 0 \end{bmatrix} + \begin{bmatrix} 0.05 \\ 1 \end{bmatrix} 0.5(0) = \begin{bmatrix} 1 \\ 0 \end{bmatrix}.$$

For $n = 1$,

$$\mathbf{v}(0.1(1) + 0.1) = \mathbf{v}(0.2) = \begin{bmatrix} 1 & 0.10 \\ 0 & 1 \end{bmatrix}\begin{bmatrix} 1 \\ 0 \end{bmatrix} + \begin{bmatrix} 0.05 \\ 1 \end{bmatrix} 0.5(1) = \begin{bmatrix} 1.025 \\ 0.500 \end{bmatrix}.$$

Continuing the iterative solution for $n = 2$, 3, and 4 gives

$$\mathbf{v}(0.1(2) + 0.1) = \mathbf{v}(0.3) = \begin{bmatrix} 1 & 0.10 \\ 0 & 1 \end{bmatrix}\begin{bmatrix} 1.025 \\ 0.500 \end{bmatrix} + \begin{bmatrix} 0.05 \\ 1 \end{bmatrix} 0.5(2) = \begin{bmatrix} 1.125 \\ 1.500 \end{bmatrix},$$

$$\mathbf{v}(0.1(3) + 0.1) = \mathbf{v}(0.4) = \begin{bmatrix} 1 & 0.10 \\ 0 & 1 \end{bmatrix}\begin{bmatrix} 1.125 \\ 1.500 \end{bmatrix} + \begin{bmatrix} 0.05 \\ 1 \end{bmatrix} 0.5(3) = \begin{bmatrix} 1.350 \\ 3.000 \end{bmatrix},$$

$$\mathbf{v}(0.1(4) + 0.1) = \mathbf{v}(0.5) = \begin{bmatrix} 1 & 0.10 \\ 0 & 1 \end{bmatrix}\begin{bmatrix} 1.350 \\ 3.000 \end{bmatrix} + \begin{bmatrix} 0.05 \\ 1 \end{bmatrix} 0.5(4) = \begin{bmatrix} 1.750 \\ 5.000 \end{bmatrix}.$$

If the value of T is reduced tenfold to $T = 0.01$, the result is

$$\mathbf{v}(0.5) = \begin{bmatrix} 2.011 \\ 6.125 \end{bmatrix}$$

with the exact value computed by calculus being

$$\mathbf{v}(0.5) = \begin{bmatrix} 2.042 \\ 6.250 \end{bmatrix}.$$

Computer Solution The functions **dlsim or ksdlsim**, which are designed for the solution of matrix difference equations, provide a sensible computer solution to this iterative process. These functions, designed for discrete-time systems, are similar to **lsim** and **kslsim**, which are used for continuous-time systems. Although the original data for this example were given in continuous form, we changed the description to discrete form for numerical computation. The jargon for this process is "digitizing the problem" for computer solution. For the output defined as $y(nT) = v_1(nT)$, the system data for Illustrative Problem 6.5 are entered in state form, as in the following script. Notice that we are using the given sampling interval of $T = 0.1$, with the output data appearing below the script and in Figure 6.3a.

MATLAB Script

```
%F6_3 Computer solution of the state equation

%F6_3a Output for IP6.5, T=0.1 using kslsim
Phi=[1,0.1;0,1];                   % Phi matrix
Del=[0.05;1];                      % Del matrix
C=[1,0];                           % C matrix
D=[0];                             % D matrix
v0=[1;0];                          % initial condition vector
n=0:1:5;                           % start:increment:stop
X=[0.5*n]';                        % input vector
sx=size(X);                        % check on dimensions of x
[y,v]=ksdlsim(Phi,Del,C,D,X,v0)    % call ksdlsim
%...plotting statements and pause
```

Continues

Output

```
v=                                  y=
   1.0000      0                          1.0000
   1.0000      0                          1.0000
   1.0250  0.5000                         1.0250
   1.1250  1.5000                         1.1250
   1.3500  3.0000                         1.3500
   1.7500  5.0000                         1.7500
```

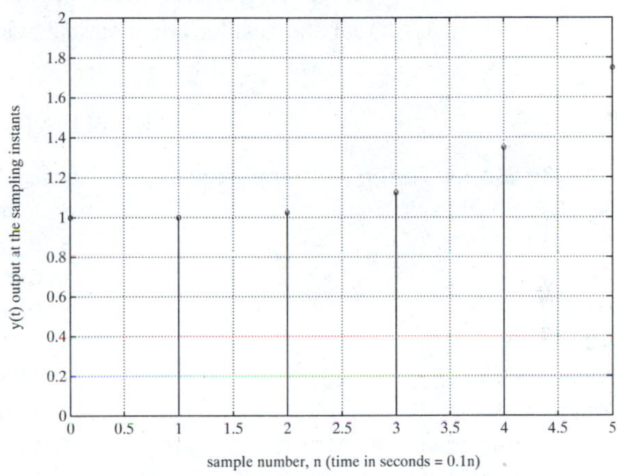

FIGURE 6.3a *Output for Illustrative Problem 6.5,* $T = 0.1$

A more accurate solution can, of course, be obtained by using a smaller value of T than $T = 0.1$, which was used to illustrate the iterative nature of the solution that could be done by hand. Shown next is a script for this same problem, where the Phi and Del matrices were calculated for $T = 0.01$ s rather than $T = 0.1$ s. The output $y(0.5) = 2.0106$ compares with $y(0.5) = 2.042$ from calculus. We could, of course, do better with a still-smaller value for T.

MATLAB Script

```
%F6_3b phidel and numde (Numerical Solution of Differential Equations)
A=[0,1;0,0];                       % continuous-time A matrix
B=[0;10];                          % continuous-time B matrix
[Phi,Del]=c2d(A,B,0.01)            % call c2d(A,B,T)
% At this stage Phi and Del are returned for the given A, B, and T.
C=[1,0];                           % C matrix
D=[0];                             % D matrix
v0=[1;0];                          % initial conditions at n=t=0
n=0:1:50;                          % start:increment:stop
```

Continues

```
X=[0.05*n]';                      % input vector
[y,v]=ksdlsim(Phi,Del,C,D,X,v0);  % call ksdlsim
%...plotting statements and pause
```

As in Chapters 2 and 3, the functions **lsim** or **kslsim** can always be used for the simulation of continuous-time systems with arbitrary inputs. Both these functions use algorithms that convert a continuous system to a discrete one using matrix exponentials and the interval between time points in the vector **t**, making the overall effect the same as above with **phidel** and **ksdlsim**. The script using **kslsim** with a time interval of 0.01 s is given next, where we see that the results are identical with Figure 6.3b.

MATLAB Script

```
%F6_3c Solution of state equation using kslsim
A=[0,1;0,0];                      % A matrix
B=[0;10];                         % B matrix
C=[1,0];                          % C matrix
D=[0];                            % D matrix
v0=[1;0];                         % initial condition state vector
t=0:0.01:0.5;                     % start t:increment:stop t
X=[5*t]';                         % input vector
[y,v]=kslsim(A,B,C,D,X,t,v0);     % call kslsim
%...plotting statements
```

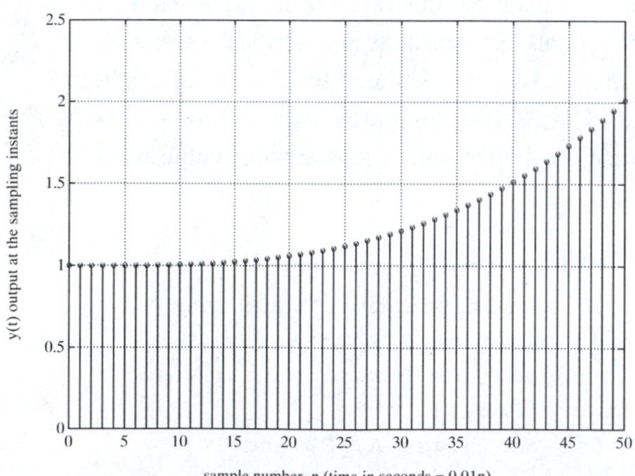

FIGURE 6.3b *Output for Illustrative Problem 6.5,* $T = 0.01$

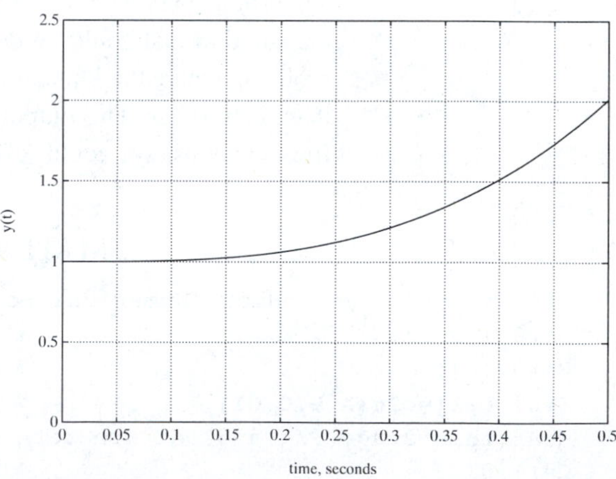

FIGURE 6.3c *Output for Illustrative Problem 6.5,* **kslsim**

We've shown the **lsim/kslsim** output as a discrete plot (Figure 6.3b) to emphasize that the values obtained correspond to discrete-time instants. Usually when we use **lsim**, the time increment is small enough for us to pretend the solution is obtained over the continuous-time interval as in Figure 6.3c.

Summary

1. For the simulation of continuous-time systems described by linear constant-coefficient differential equations, either of the following two procedures may be used:
   a. Use **lsim** or **kslsim** with the time increment t controlling the accuracy of the result.
   b. Use **c2d** and **dlsim/ksdlsim** with the value of T used to calculate Phi and Del, controlling the accuracy of the result.
2. If the system input is piecewise constant, i.e., $\mathbf{x}(t) = \mathbf{x}(nT)$, $nT \leq t < nT + T$ the **A** and **B** matrices of the continuous system should be digitized to the matrices $\mathbf{\Phi}(T)$ and $\mathbf{\Delta}(T)$ using **c2d**. Then **dlsim** or **ksdlsim** should be used to evaluate the difference equation

   $$\mathbf{v}(nT + T) = \mathbf{\Phi}(T)\mathbf{v}(nT) + \mathbf{\Delta}(T)\mathbf{x}(nT).$$

   If, however, we are interested in the output values between sampling instants, it is probably better to use **lsim/kslsim** and write a script to ensure that $\mathbf{x}(t) = \mathbf{x}(nT)$, $nT \leq t < nT + T$.

SOLUTION OF THE STATE EQUATION BY LAPLACE TRANSFORM

Given the state equation for a linear time-invariant system

state equation

$$\dot{\mathbf{v}}(t) = \mathbf{A}\mathbf{v}(t) + \mathbf{B}\mathbf{x}(t)$$

with the initial condition vector $\mathbf{v}(0)$. Taking the Laplace transform of this matrix equation gives us

transformed state equation

$$s\mathbf{V}(s) - \mathbf{v}(0) = \mathbf{A}\mathbf{V}(s) + \mathbf{B}\mathbf{X}(s).$$

The term $s\mathbf{V}(s)$, a column vector, is written as $s\mathbf{I}\mathbf{V}(s)$ in order that it may be combined with the **A** matrix to yield

$$(s\mathbf{I} - \mathbf{A})\mathbf{V}(s) = \mathbf{v}(0) + \mathbf{B}\mathbf{X}(s).$$

Premultiplying both sides of the equation by $(s\mathbf{I} - \mathbf{A})^{-1}$, the inverse of the matrix $(s\mathbf{I} - \mathbf{A})$, produces the $\mathbf{V}(s)$ vector

$$\mathbf{V}(s) = (s\mathbf{I} - \mathbf{A})^{-1}\mathbf{v}(0) + (s\mathbf{I} - \mathbf{A})^{-1}\mathbf{B}\mathbf{X}(s).$$

Defining $\mathbf{\Phi}(s) = (s\mathbf{I} - \mathbf{A})^{-1}$ gives us the more compact form

$$\mathbf{V}(s) = \mathbf{\Phi}(s)\mathbf{v}(0) + \mathbf{\Phi}(s)\mathbf{B}\mathbf{X}(s).$$

Taking the inverse Laplace transform we have

solution of state equation

$\mathbf{v}(t) = \mathbf{v}_{IC}(t) + \mathbf{v}_F(t)$

$$\begin{aligned}\mathbf{v}(t) &= \mathcal{L}^{-1}[\mathbf{V}(s)] = \mathcal{L}^{-1}[\mathbf{\Phi}(s)\mathbf{v}(0)] + \mathcal{L}^{-1}[\mathbf{\Phi}(s)\mathbf{B}\mathbf{X}(s)] \\ &= \boldsymbol{\phi}(t)\mathbf{v}(0) + \int_0^t \boldsymbol{\phi}(t-\tau)\mathbf{B}\mathbf{x}(\tau)\,d\tau \\ &= \text{IC response} + \text{forced response}\end{aligned}$$

Comment: The matrices $\boldsymbol{\phi}(t)$ and $\mathbf{e}^{\mathbf{A}t}$ are, of course, identical and we will use both descriptions.

THE TRANSITION MATRIX $\boldsymbol{\phi}(t)$

The matrix $\boldsymbol{\phi}(t)$ is called the state transition matrix (for a time-invariant system) and from the preceding we see that

$$\mathcal{L}[\boldsymbol{\phi}(t)] = \mathbf{\Phi}(s) = (s\mathbf{I} - \mathbf{A})^{-1}.$$

This equation implies the following steps to find $\boldsymbol{\phi}(t)$ using Laplace transforms:

1. Form $(s\mathbf{I} - \mathbf{A})$.
2. Obtain $\mathbf{\Phi}(s)$ by forming the matrix inverse $(s\mathbf{I} - \mathbf{A})^{-1}$.
3. Find the state transition matrix $\boldsymbol{\phi}(t)$ by taking the *inverse* Laplace transform of $\mathbf{\Phi}(s)$.

Notice that the transition matrix obtained by this method leads to closed-form analytical expressions for $\boldsymbol{\phi}(t)$, in contrast to the time-domain series expansion, which generally gives infinite series for the terms in $\boldsymbol{\phi}(t)$.

ILLUSTRATIVE PROBLEM 6.6
Calculating the State Transition Matrix $\boldsymbol{\phi}(t)$

Find the state transition matrix $\boldsymbol{\phi}(t)$ for the roll motion of a spacecraft (see Illustrative Problem 6.2) that is described by $d^2\theta(t)/dt^2 = k\lambda(t)$, where $\theta(t)$ is the roll angle and $\lambda(t)$ is the driving torque.

Solution

For $v_1(t) = \theta(t)$, $v_2(t) = d\theta(t)/dt$, and $x(t) = \lambda(t)$, the state equation is

$$\dot{\mathbf{v}}(t) = \begin{bmatrix} 0 & 1 \\ 0 & 0 \end{bmatrix}\mathbf{v}(t) + \begin{bmatrix} 0 \\ k \end{bmatrix}x(t).$$

Following the steps previously described:

1. $(s\mathbf{I} - \mathbf{A}) = \begin{bmatrix} s & 0 \\ 0 & s \end{bmatrix} - \begin{bmatrix} 0 & 1 \\ 0 & 0 \end{bmatrix} = \begin{bmatrix} s & -1 \\ 0 & s \end{bmatrix}.$
2. $(s\mathbf{I} - \mathbf{A})^{-1} = \mathbf{\Phi}(s) = \begin{bmatrix} 1/s & 1/s^2 \\ 0 & 1/s \end{bmatrix}.$
3. $\boldsymbol{\phi}(t) = \mathcal{L}^{-1}[\mathbf{\Phi}(s)] = \begin{bmatrix} 1 & t \\ 0 & 1 \end{bmatrix}.$

This is the same as the result of Illustrative Problem 6.2.

CHARACTERISTIC EQUATION AND EIGENVALUES

For a general Nth-order system, the matrix $s\mathbf{I} - \mathbf{A}$ has the following appearance:

$$s\mathbf{I} - \mathbf{A} = \begin{bmatrix} s - a_{11} & -a_{12} & -a_{13} & \cdots & -a_{1N} \\ -a_{21} & s - a_{22} & -a_{23} & \cdots & -a_{2N} \\ \vdots & \vdots & \vdots & \vdots & \vdots \\ \vdots & \vdots & \vdots & \vdots & \vdots \\ -a_{N1} & -a_{N2} & -a_{N3} & \cdots & s - a_{NN} \end{bmatrix}.$$

The inverse of any matrix $\mathbf{M}$ can be written as the adjoint matrix, **adj M**, divided by the determinant of $\mathbf{M}$, $|\mathbf{M}|$, or

matrix inverse

$$(s\mathbf{I} - \mathbf{A})^{-1} = \frac{\mathbf{adj}(s\mathbf{I} - \mathbf{A})}{|s\mathbf{I} - \mathbf{A}|}.$$

If we imagine calculating the determinant $|s\mathbf{I} - \mathbf{A}|$, we see that one of the terms will be the product of the diagonal elements of $s\mathbf{I} - \mathbf{A}$:

$$(s - a_{11})(s - a_{22})\cdots(s - a_{NN}) = s^N + c_{N-1}s^{N-1} + \cdots + c_0,$$

a polynomial of degree N with the leading coefficient of unity. There will also be other terms coming from the off-diagonal elements of $s\mathbf{I} - \mathbf{A}$, but none will have a degree in s as large as N. Thus we conclude that the form of the determinant is

characteristic polynomial

$$|s\mathbf{I} - \mathbf{A}| = s^N + \alpha_{N-1}s^{N-1} + \cdots + \alpha_1 s + \alpha_0.$$

This is known as the *characteristic polynomial* of the matrix **A**, and it plays a vital role in the dynamic behavior of the system. When equated to zero this characteristic polynomial becomes the *characteristic equation (CE):*

characteristic equation

$$s^N + \alpha_{N-1}s^{N-1} + \cdots + \alpha_1 s + \alpha_0 = 0.$$

This CE can be written in factored form as

$$(s - r_1)(s - r_2)\cdots(s - r_N) = 0$$

where $r_1, r_2, \ldots, r_N$ are the *characteristic roots* or *eigenvalues* of the system. These roots determine the essential features of the unforced (initial condition) behavior of the system, and for a stable system all of these roots must have negative real parts. The characteristic roots determine the terms in the partial fraction expansion of the matrix $\mathbf{\Phi}(s)$, whose inverse Laplace transform is the transition matrix $\boldsymbol{\phi}(t)$.

The eigenvalues, usually denoted as the λ_k's, of the matrix **A** are the same as the characteristic roots of the system. These eigenvalues are the roots of the equation

characteristic equation

$$|\lambda\mathbf{I} - \mathbf{A}| = 0\,.$$

Thus, what engineers often call characteristic roots, mathematicians call eigenvalues of the matrix **A**. According to Shakespeare, however, "that which we call a rose by any other name would smell as sweet."

ILLUSTRATIVE PROBLEM 6.7
Solution by Laplace Transform

An overdamped second-order system is described by the state-space and output equations

$$\dot{\mathbf{v}}(t) = \begin{bmatrix} 0 & 1 \\ -2 & -3 \end{bmatrix}\mathbf{v}(t) + \begin{bmatrix} 0 \\ 2 \end{bmatrix}x(t) \quad \text{and} \quad y(t) = [0 \quad 1]\mathbf{v}(t)\,.$$

a. Find the system's characteristic equation.

b. What are the characteristic roots or eigenvalues? Use the MATLAB function **eig** to check your answer. Is the system stable?

c. For the initial condition $\mathbf{v}(0) = [4 \quad -5]^T$ and the exponential input $x(t) = 3e^{-4t}u(t)$, use the Laplace transform method to find $\mathbf{V}(s) = \mathbf{V}_{\mathbf{IC}}(s) + \mathbf{V}_{\mathbf{F}}(s)$.

d. Using **residue** to help with the PFE, find $\mathbf{v}_{\mathbf{IC}}(t)$ and $\mathbf{v}_{\mathbf{F}}(t)$.

e. What is the system output $y(t)$?

Solution

a. First, we form the matrix

$$s\mathbf{I} - \mathbf{A} = \begin{bmatrix} s & -1 \\ 2 & s+3 \end{bmatrix}$$

from which we obtain the CE as $|s\mathbf{I} - \mathbf{A}| = 0$, or $s^2 + 3s + 2 = 0$.

b. The characteristic roots are $s_{1,2} = -1, -2$; since these roots are both negative real, the system is stable. To use **eig** (MATLAB Toolbox) the **A** matrix is entered as $\mathbf{A} = [0, 1; -2, -3]$ and **eig**(A) returns the eigenvalues -1 and -2.

c. Now we need the inverse of the matrix $s\mathbf{I} - \mathbf{A}$, which can be found from

$$(s\mathbf{I} - \mathbf{A})^{-1} = \frac{\mathbf{adj}(s\mathbf{I} - \mathbf{A})}{|s\mathbf{I} - \mathbf{A}|} = \frac{1}{s^2 + 3s + 2} \cdot \begin{bmatrix} s + 3 & 1 \\ -2 & s \end{bmatrix}$$

$$= \begin{bmatrix} \dfrac{s+3}{s^2+3s+2} & \dfrac{1}{s^2+3s+2} \\ \dfrac{-2}{s^2+3s+2} & \dfrac{s}{s^2+3s+2} \end{bmatrix}.$$

The Laplace transform of the state vector $\mathbf{v}(t)$ is given by

$$\mathbf{V}(s) = (s\mathbf{I} - \mathbf{A})^{-1}\mathbf{v}(0) + (s\mathbf{I} - \mathbf{A})^{-1}\mathbf{B}\mathbf{X}(s)$$
$$= \mathbf{V}_{\mathbf{IC}}(s) + \mathbf{V}_{\mathbf{F}}(s).$$

The Laplace transform of the response due to the initial conditions $\mathbf{v}(0)$ is given by

$$\mathbf{V}_{\mathbf{IC}}(s) = \begin{bmatrix} \dfrac{s+3}{s^2+3s+2} & \dfrac{1}{s^2+3s+2} \\ \dfrac{-2}{s^2+3s+2} & \dfrac{s}{s^2+3s+2} \end{bmatrix} \begin{bmatrix} 4 \\ -5 \end{bmatrix} = \begin{bmatrix} \dfrac{4s+7}{s^2+3s+2} \\ \dfrac{-5s-8}{s^2+3s+2} \end{bmatrix}$$

and the transform of the response due to the input $x(t) = 3e^{-4t}u(t)$ is

$$\mathbf{V}_{\mathbf{F}}(s) = \begin{bmatrix} \dfrac{s+3}{s^2+3s+2} & \dfrac{1}{s^2+3s+2} \\ \dfrac{-2}{s^2+3s+2} & \dfrac{s}{s^2+3s+2} \end{bmatrix} \begin{bmatrix} 0 \\ 2 \end{bmatrix} \cdot \frac{3}{s+4}$$

$$= \begin{bmatrix} \dfrac{6}{(s+4)(s^2+3s+2)} \\ \dfrac{6s}{(s+4)(s^2+3s+2)} \end{bmatrix}.$$

d. With the help of **residue** we find

$$\mathbf{V}_{\mathbf{IC}}(s) = \begin{bmatrix} \dfrac{1}{s+2} + \dfrac{3}{s+1} \\ \dfrac{-2}{s+2} + \dfrac{-3}{s+1} \end{bmatrix} \quad \text{and} \quad \mathbf{v}_{\mathbf{IC}}(t) = \begin{bmatrix} e^{-2t} + 3e^{-t} \\ -2e^{-2t} - 3e^{-t} \end{bmatrix}, \quad t \geq 0.$$

$$\mathbf{V}_{\mathbf{F}}(s) = \begin{bmatrix} \dfrac{1}{s+4} - \dfrac{3}{s+2} + \dfrac{2}{s+1} \\ \dfrac{-4}{s+4} + \dfrac{6}{s+2} + \dfrac{-2}{s+1} \end{bmatrix} \quad \text{and} \quad \mathbf{v}_{\mathbf{F}}(t) = \begin{bmatrix} e^{-4t} - 3e^{-2t} + 2e^{-t} \\ -4e^{-4t} + 6e^{-2t} - 2e^{-t} \end{bmatrix}, \quad t \geq 0.$$

CROSS-CHECK

1. In the given state equation we had $dv_1(t)/dt = v_2(t)$; this relation should hold true in the preceding solution, where we see that for both $\mathbf{v}_{\mathbf{IC}}(t)$ and $\mathbf{v}_{\mathbf{F}}(t)$ the second row, $v_2(t)$ is the derivative of the first row, $v_1(t)$. The chances are good that this solution may actually be correct.
2. Notice that the IC response has terms due to the system only, C_1e^{-2t} and C_2e^{-t}, whereas the forced response contains terms due to the system,

C_3e^{-2t} and C_4e^{-t}, plus a term due to the input, C_5e^{-4t}. This appears to be reasonable from a physical point of view. ■

e. With $y(t) = v_2(t)$, we have from part (d),

$$\begin{aligned} y(t) &= -2e^{-2t} - 3e^{-t} - 4e^{-4t} + 6e^{-2t} - 2e^{-t} \\ &= -4e^{-4t} + 4e^{-2t} - 5e^{-t}, \qquad t \geq 0. \end{aligned}$$

■

MATRIX OF TRANSFER FUNCTIONS

For an initial state $\mathbf{v}(0)$ assumed to be zero, the Laplace transform of the state vector $\mathbf{v}(t)$ is

$$\mathbf{V}(s) = (s\mathbf{I} - \mathbf{A})^{-1}\mathbf{B}\mathbf{X}(s).$$

If the output vector $\mathbf{y}(t)$ is defined by

output vector

$$\mathbf{y}(t) = \mathbf{C}\mathbf{v}(t) + \mathbf{D}\mathbf{x}(t)$$

then the Laplace transform of the output vector is

$$\begin{aligned} \mathbf{Y}(s) &= \mathbf{C}\mathbf{V}(s) + \mathbf{D}\mathbf{X}(s) = \mathbf{C}\boldsymbol{\Phi}(s)\mathbf{B}\mathbf{X}(s) + \mathbf{D}\mathbf{X}(s) = [\mathbf{C}\boldsymbol{\Phi}(s)\mathbf{B} + \mathbf{D}]\mathbf{X}(s) \\ &= \mathbf{H}(s) \cdot \mathbf{X}(s) \end{aligned}$$

The matrix of transforms,

transfer function matrix

$$\mathbf{H}(s) = \mathbf{C}\boldsymbol{\Phi}(s)\mathbf{B} + \mathbf{D}$$

relating $\mathbf{Y}(s)$, the Laplace transform of the outputs, to $\mathbf{X}(s)$, the Laplace transform of the inputs, is known as the *transfer function matrix.* With $\mathbf{Y}(s) = \mathbf{H}(s)\mathbf{X}(s)$, the transfer function matrix $\mathbf{H}(s)$ is of the form

$$\mathbf{H}(s) = \begin{bmatrix} H_{11}(s) & H_{12}(s) & \cdots & H_{1M}(s) \\ H_{21}(s) & H_{22}(s) & \cdots & H_{2M}(s) \\ \vdots & \vdots & \vdots & \vdots \\ H_{P1}(s) & H_{P2}(s) & \cdots & H_{PM}(s) \end{bmatrix}$$

where the transforms of the M by 1 input vector $\mathbf{x}(t)$ and the P by 1 output vector $\mathbf{y}(t)$ are

$$\begin{aligned} \mathbf{X}(s) &= [X_1(s) \quad X_2(s) \quad \cdots \quad X_M(s)]^T \quad \text{and} \\ \mathbf{Y}(s) &= [Y_1(s) \quad Y_2(s) \quad \cdots \quad Y_P(s)]^T. \end{aligned}$$

Thus the particular transfer function that relates the ith *output* to the jth *input* is $H_{ij}(s)$, and for M system inputs and P designated outputs, we see that a system has $M \cdot P$ different transfer functions. In reality, therefore, the more complete transfer function model for an LTI system is the transfer function matrix $\mathbf{H}(s)$ rather than a specific scalar transfer function $H(s)$ as discussed in Chapter 3.

FREQUENCY RESPONSE MATRIX

In Chapter 4, "Frequency Response for Continuous Systems," one method that was used to find a system's frequency response was to substitute $j\omega$ for s in the transfer function $H(s)$. The same idea transfers to a system's state-space model, and we have the frequency response matrix

frequency response matrix

$$\mathbf{H}(j\omega) = \mathbf{C\Phi}(j\omega)\mathbf{B} + \mathbf{D}.$$

ILLUSTRATIVE PROBLEM 6.8 ***Transfer Functions and Frequency Response***

A bandpass analog filter is described by the state equation

$$\dot{\mathbf{v}}(t) = \begin{bmatrix} 0 & 1 \\ -1 & -1 \end{bmatrix}\mathbf{v}(t) + \begin{bmatrix} 0 \\ 1 \end{bmatrix}x(t)$$

and by the output equation $y(t) = v_2(t)$.

a. Find the transfer function matrix $\mathbf{H}(s)$.

b. What is the frequency response?

Solution

a. With $y(t) = v_2(t)$, we have $\mathbf{C} = [0 \quad 1]$ and $\mathbf{D} = \mathbf{0}$, which when used with the previously given expression for the transfer function matrix, gives

$$\mathbf{H}(s) = \mathbf{C\Phi}(s)\mathbf{B} = [0 \quad 1][s\mathbf{I} - \mathbf{A}]^{-1}\begin{bmatrix} 0 \\ 1 \end{bmatrix}$$

where

$$(s\mathbf{I} - \mathbf{A})^{-1} = \frac{\mathbf{adj}(s\mathbf{I} - \mathbf{A})}{|s\mathbf{I} - \mathbf{A}|} = \frac{1}{s^2 + s + 1} \cdot \begin{bmatrix} s+1 & 1 \\ -1 & s \end{bmatrix}.$$

Forming the product $\mathbf{C}(s\mathbf{I} - \mathbf{A})^{-1}\mathbf{B}$ produces the single transfer function $H(s) = s/(s^2 + s + 1)$. In this situation, with a single input and a single output, the matrix of transfer functions $\mathbf{H}(s)$ is simply a scalar $H(s)$, as shown before.

b. The frequency response is the rational function $H(j\omega) = j\omega/((j\omega)^2 + j\omega + 1)$.

MANIPULATION OF SYSTEM MODELS

In Chapter 3 we started with an SISO system's differential equation (DE) and determined the corresponding scalar transfer function $H(s)$ and/or a system diagram (SFG) that related the output transform $Y(s)$ to the input transform $X(s)$. In this same chapter on Laplace transform applications, we moved from the unit impulse response model $h(t)$ to the system transfer function model $H(s)$ and vice versa. Now, with a background in state-space concepts, let's look at a few other possibilities for moving about or manipulating these system models that can now include the state model. Frequently we may know or be

CHAPTER 6

supplied with one particular system description and want to convert it to a different one for easier design or analysis computations. In general, in contrast with Chapter 3, we concentrate on MIMO systems.

State and Output Equations from the System Diagram Often we are supplied with a system diagram for an LTI system and we need to write a set of state equations as well as the system output equations. The state model is useful for simulating MIMO systems with nonzero initial conditions. We can derive this model directly from the signal flowgraph or block diagram. To do this, the states are designated as the outputs of the integrators, and the signals with their associated branch gains are summed at the inputs of the integrators producing, say, the jth state equation

$$\dot{v}_j(t) = \sum_{k=1}^{N} a_{jk} v_k(t) + \sum_{k=1}^{M} b_{jk} x_k(t).$$

This process is repeated at the input of all N integrators, yielding the matrix state equation

$$\dot{\mathbf{v}}(t) = \mathbf{A}\mathbf{v}(t) + \mathbf{B}\mathbf{x}(t).$$

By using the graph to obtain the indicated linear combination of states $\mathbf{v}(t)$ and the inputs $\mathbf{x}(t)$, we can write the output equation

$$\mathbf{y}(t) = \mathbf{C}\mathbf{v}(t) + \mathbf{D}\mathbf{x}(t).$$

ILLUSTRATIVE PROBLEM 6.9
State Model from an SFG

A highly simplified model for the power unit of a ship's propulsion system is shown in Figure 6.4a, and its SFG representation is given in Figure 6.4b. The system inputs are the input velocity $\omega_1(t)$ from the driving turbine and a load torque $\lambda(t)$ on the ship's propeller. We define the outputs to be the output velocity $\omega_3(t)$ and the output acceleration $d\omega_3(t)/dt$. Write a set of state equations and an output equation and put them into matrix form.

Solution

Summing the input signals at the $dv_1(t)/dt$ and $dv_2(t)/dt$ nodes gives

$$\dot{v}_1(t) = -\frac{b}{J} v_1(t) + \frac{1}{J} v_2(t) - \frac{1}{J} \lambda(t) \quad \text{and}$$

$$\dot{v}_2(t) = -K v_1(t) - \frac{K}{B} v_2(t) + K\omega_1(t).$$

With the input and output vectors defined as

$$\mathbf{x}(t) = [\omega_1(t) \quad \lambda(t)]^T \quad \text{and} \quad \mathbf{y}(t) = [v_1(t) \quad \dot{v}_1(t)]^T,$$

respectively, the state and output equations are (notice that $dv_1(t)/dt$ is replaced by its expression in $\mathbf{y}(t)$ in terms of states and inputs)

$$\dot{\mathbf{v}}(t) = \begin{bmatrix} -b/J & 1/J \\ -K & -K/B \end{bmatrix} \mathbf{v}(t) + \begin{bmatrix} 0 & -1/J \\ K & 0 \end{bmatrix} \mathbf{x}(t) \quad \text{and}$$

$$\mathbf{y}(t) = \begin{bmatrix} 1 & 0 \\ -b/J & 1/J \end{bmatrix} \mathbf{v}(t) + \begin{bmatrix} 0 & 0 \\ 0 & -1/J \end{bmatrix} \mathbf{x}(t).$$

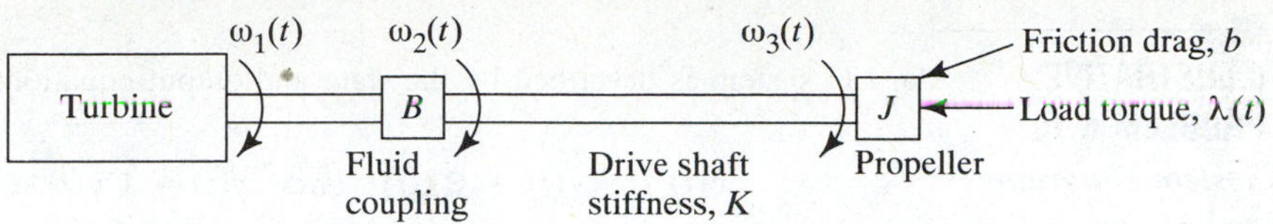

a. Model of power unit

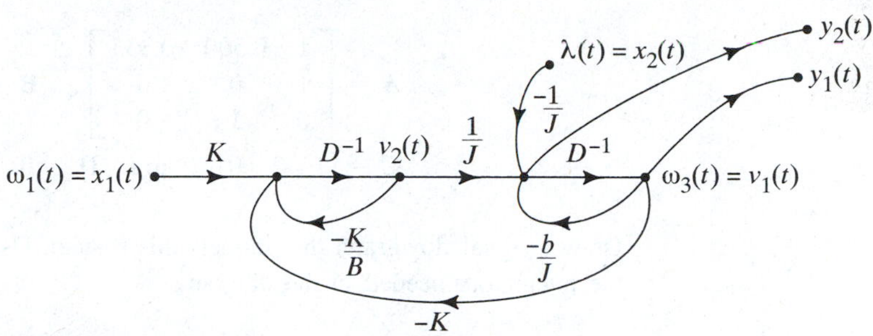

b. SFG for power unit

FIGURE 6.4 *State model from SFG for Illustrative Problem 6.9*

System Diagram from the State Equation We may want to reverse the preceding process and obtain a pictorial representation of a system from a set of state and output equations. To develop an SFG or block diagram, we start with the matrix state equation

$$\dot{\mathbf{v}}(t) = \mathbf{A}\mathbf{v}(t) + \mathbf{B}\mathbf{x}(t)$$

and write out the jth state equation, giving

$$\dot{v}_j(t) = \sum_{k=1}^{N} a_{jk} v_k(t) + \sum_{k=1}^{M} b_{jk} x_k(t) .$$

Now $v_j(t)$ can be found by integrating $dv_j(t)/dt$, as pictured in Figure 6.5a in the time domain and in Figure 6.5b for its transform domain counterpart. Thus we need $N + M$ (N for the states and M for the inputs) branches in the SFG for the inputs to each integrator unit. (Of course, some of these branches usually have transmission gains of zero and are omitted from the SFG.) This process needs to be repeated for each of the N states. The final step is to add the graphical representation of the output equation $\mathbf{y}(t) = \mathbf{C}\mathbf{v}(t) + \mathbf{D}\mathbf{x}(t)$ to the signal flowgraph.

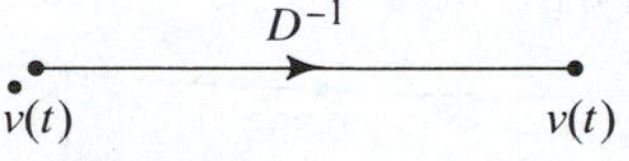

a. Time domain

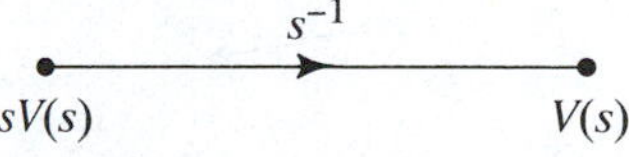

b. Transform domain

FIGURE 6.5 *Integrator models*

ILLUSTRATIVE PROBLEM 6.10
System Diagram from State and Output Equations

An LTI system is described by the state and output equations

$$\dot{\mathbf{v}}(t) = \mathbf{A}\mathbf{v}(t) + \mathbf{B}\mathbf{x}(t) \quad \text{and} \quad \mathbf{y}(t) = \mathbf{C}\mathbf{v}(t) + \mathbf{D}\mathbf{x}(t)$$

where

$$\mathbf{A} = \begin{bmatrix} 1 & 0.500 & 0.333 \\ 1 & 0 & 0 \\ 0 & 1 & 0 \end{bmatrix}, \quad \mathbf{B} = \begin{bmatrix} 1 \\ 0 \\ 0 \end{bmatrix},$$
$$\mathbf{C} = [1 \quad 2 \quad 0], \quad \text{and} \quad \mathbf{D} = \mathbf{0}.$$

Draw a signal flowgraph that models this system. Use D^{-1} to denote the gain of the integrators needed in the diagram.

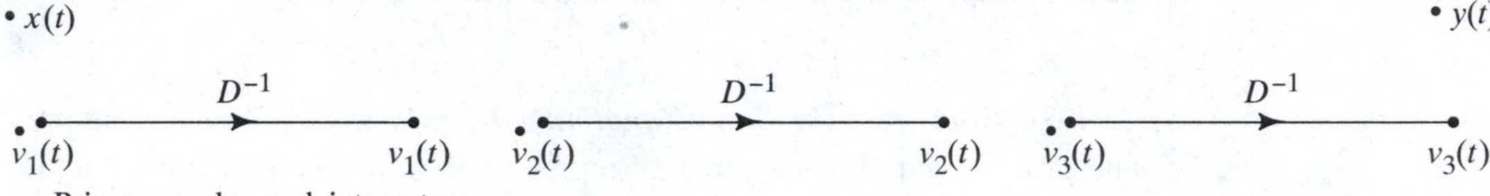

a. Primary nodes and integrators

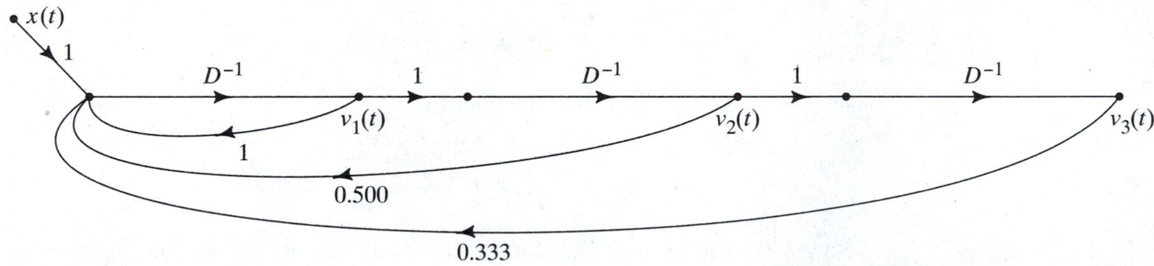

b. Implementing the state equation

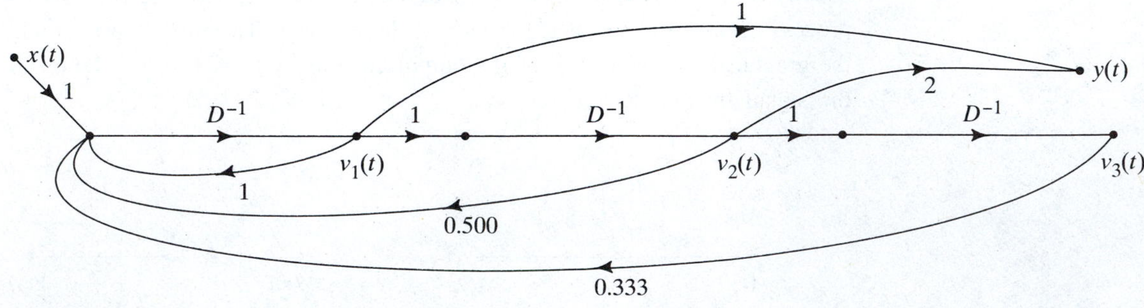

c. Complete SFG

FIGURE 6.6 *SFG for Illustrative Problem 6.10*

Solution The primary nodes established for the states are

$$v_1(t),\ \dot{v}_1(t),\ v_2(t),\ \dot{v}_2(t),\ v_3(t),\quad \text{and}\quad \dot{v}_3(t);$$

$x(t)$ for the input; $y(t)$ for the output. See Figure 6.6a for this realization, where $v_1(t)$ and $dv_1(t)/dt$ are connected by D^{-1}, and so forth. Then the state equations are used to obtain Figure 6.6b; finally, the output equation is used to establish the complete diagram of Figure 6.6c.

Transfer Function to State Equations

Single-Input–Single-Output (SISO) system In general, a system is characterized by a matrix of transfer functions, and it may be useful to represent the system in state-space form. First, we will consider an SISO system whose transfer function description is

$$H(s) = \frac{Y(s)}{X(s)} = \frac{\sum_{k=0}^{L} b_k s^k}{\sum_{k=0}^{N} a_k s^k}$$

where we make $a_N \equiv 1$ and $N \geq L$.[5] Writing $H(s)$ as the product of two transfer functions gives

$$H(s) = \frac{Y(s)}{X(s)} = \frac{1}{\sum_{k=0}^{N} a_k s^k} \cdot \frac{\sum_{k=0}^{L} b_k s^k}{1} = H_a(s) \cdot H_b(s)\,.$$

Next we define a new variable $Q(s)$ such that

$$H(s) = \frac{Y(s)}{X(s)} = \frac{Q(s)}{X(s)} \cdot \frac{Y(s)}{Q(s)} = \frac{1}{\sum_{k=0}^{N} a_k s^k} \cdot \frac{\sum_{k=0}^{L} b_k s^k}{1}$$

where

$$H_a(s) = \frac{Q(s)}{X(s)} = \frac{1}{\sum_{k=0}^{N} a_k s^k} \quad \text{and} \quad H_b(s) = \frac{Y(s)}{Q(s)} = \frac{\sum_{k=0}^{L} b_k s^k}{1}\,.$$

We notice that $H_a(s)$ implements the poles of $H(s)$ and $H_b(s)$ implements the zeros of $H(s)$. Rearranging the equation for $H_a(s)$ and taking the inverse transform gives

$$Q(s) \sum_{k=0}^{N} a_k s^k = X(s) \quad \text{or} \quad \sum_{k=0}^{N} a_k \frac{d^k q(t)}{dt^k} = x(t)\,.$$

[5]This is the case for a practical system. We will not allow $N < L$.

Referring to the section "The State-Space Model for Nth-Order Systems" in Chapter 2, we define the state variables as $v_1(t), v_2(t), \ldots, v_N(t)$ and write

$$v_1(t) = q(t), \quad v_2(t) = dq(t)/dt,$$
$$v_3(t) = d^2q(t)/dt^2, \ldots, v_N(t) = d^{N-1}q(t)/dt^{N-1}.$$

Then, recalling that $a_N \equiv 1$, we can establish the following relationships:

$$\begin{aligned} dv_1(t)/dt &= v_2(t) \\ dv_2(t)/dt &= v_3(t) \\ &\vdots \\ dv_N(t)/dt &= -a_0v_1(t) - a_1v_2(t) - a_2v_3(t) - \cdots \\ &\quad - a_{N-1}v_N(t) + x(t). \end{aligned}$$

Putting all this into matrix form, we define the state vector as

state vector

$$\mathbf{v}(t) = [v_1(t) \quad v_2(t) \quad \cdots \quad v_N(t)]^T$$

and the matrix state equation follows as

state equation

$$\begin{aligned} \frac{d\mathbf{v}(t)}{dt} &= \begin{bmatrix} 0 & 1 & 0 & 0 & \cdots & 0 \\ 0 & 0 & 1 & 0 & \cdots & 0 \\ \vdots & \vdots & \vdots & \vdots & \cdots & \vdots \\ 0 & 0 & 0 & 0 & \cdots & 1 \\ -a_0 & -a_1 & -a_2 & -a_3 & \cdots & -a_{N-1} \end{bmatrix} \mathbf{v}(t) + \begin{bmatrix} 0 \\ 0 \\ \vdots \\ 0 \\ 1 \end{bmatrix} x(t) \\ &= \mathbf{Av}(t) + \mathbf{Bx}(t). \end{aligned}$$

Now we need to concern ourselves with obtaining an equation for the system output $y(t)$. From the transfer function $H_b(s) = Y(s)/Q(s)$, we can solve for $Y(s)$ and take the inverse transform to obtain

$$Y(s) = H_b(s) \cdot Q(s) = \frac{\sum_{k=0}^{L} b_k s^k}{1} \cdot Q(s) \quad \text{or} \quad y(t) = \sum_{k=0}^{L} b_k \frac{d^kq(t)}{dt^k}.$$

For the most usual situation, where the number of derivatives of the output is greater than the number of derivatives of the input (i.e., $N > L$), the output equation is

$$\mathbf{y}(t) = [b_0 \quad b_1 \quad \cdots \quad b_{L-1}]\mathbf{v}(t).$$

If $N = L$ (remember we are not allowing $N < L$), we have

output equation

$$\begin{aligned} \mathbf{y}(t) &= [\, b_0 - a_0b_N \quad b_1 - a_1b_N \quad \cdots \quad b_{N-1} - a_{N-1}b_N \,]\mathbf{v}(t) + b_Nx(t). \\ &= \mathbf{Cv}(t) + b_Nx(t) \end{aligned}$$

ILLUSTRATIVE PROBLEM 6.11
State Equations from a Scalar Transfer Function

Given the transfer function $H(s) = Y(s)/X(s) = s/(s^2 + 4)$, find a state-space representation.

Solution

We write

$$H(s) = H_a(s) \cdot H_b(s) = \frac{1}{s^2 + 4} \cdot \frac{s}{1}$$

where $H_a(s) = Q(s)/X(s) = 1/(s^2 + 4)$ can be written in state-space form as

$$q(t) = v_1(t), \quad \dot{v}_1(t) = v_2(t), \quad \text{and} \quad \dot{v}_2(t) = -4v_1 + x(t).$$

From $H_b(s) = s$ we have $Y(s) = s \cdot Q(s)$, or $y(t) = v_2(t)$. In matrix form the results are

$$\dot{\mathbf{v}}(t) = \begin{bmatrix} 0 & 1 \\ -4 & 0 \end{bmatrix}\mathbf{v}(t) + \begin{bmatrix} 0 \\ 1 \end{bmatrix}x(t) \quad \text{and} \quad y(t) = [0 \quad 1]\mathbf{v}(t).$$

Single-Input–Multiple-Output (SIMO) system A system in which there is a single input, but several designated outputs is called an SIMO system. The system output vector (P by 1) is given by

$$\mathbf{Y}(s) = \mathbf{H}(s)X(s)$$

and the transfer function matrix will be the column vector of the form

$$\mathbf{H}(s) = \begin{bmatrix} H_1(s) \\ H_2(s) \\ \vdots \\ H_P(s) \end{bmatrix} = \begin{bmatrix} N_1(s)/D(s) \\ N_2(s)/D(s) \\ \vdots \\ N_P(s)/D(s) \end{bmatrix}$$

where the denominator polynomial of each transfer function is denoted as $D(s)$ and the numerator polynomials are $N_1(s), N_2(s), \ldots, N_P(s)$, respectively. Consider the first transfer function $H_1(s) = N_1(s)/D(s) = H_a(s) \cdot H_b(s)$, where we let $H_a(s) = 1/D(s)$ and $H_b(s) = N_1(s)$, and then follow the procedure of the previous section to create the state and output equations for $H_1(s)$. We repeat this procedure for each of the P transfer functions, and the result is one state equation, $d\mathbf{v}(t)/dt = \mathbf{A}\mathbf{v}(t) + \mathbf{B}x(t)$, and P scalar output equations, each of the form $y_i(t) = \mathbf{C_i}\mathbf{v}(t) + d_i x(t)$. The P output equations are then simply grouped into the matrix equation

$$\mathbf{y}(t) = \mathbf{C}\mathbf{v}(t) + \mathbf{D}x(t)$$

where $\mathbf{y}(t)$ is a P by 1 output vector, $\mathbf{C}$ is a P by N matrix, $\mathbf{v}(t)$ is the N by 1 state vector, $\mathbf{D}$ is a P by 1 column vector, and $x(t)$ is the scalar or single input.

CHAPTER 6

ILLUSTRATIVE PROBLEM 6.12
State and Output Equations for an SIMO System

Find a state and output equation representation for the transform equation $\mathbf{Y}(s) = \mathbf{H}(s) \cdot X(s)$ where $\mathbf{H}(s)$ is the transfer function matrix

$$\mathbf{H}(s) = \begin{bmatrix} 10/(s^2 + 2s + 3) \\ (s^2 + 1)/(s^2 + 2s + 3) \\ 7s/(s^2 + 2s + 3) \end{bmatrix}.$$

Solution

Using the technique of the previous section, the common state equation for all three transfer functions is

$$\dot{\mathbf{v}}(t) = \begin{bmatrix} 0 & 1 \\ -3 & -2 \end{bmatrix}\mathbf{v}(t) + \begin{bmatrix} 0 \\ 1 \end{bmatrix}x(t).$$

The output equation for the first transfer function is $y_1(t) = 10v_1(t)$, for the second the output equation is $y_2(t) = dv_2(t)/dt + v_1(t)$ or $y_2(t) = -2v_1(t) - 2v_2(t) + x(t)$, and for the third it is $y_3(t) = 7v_2(t)$. Putting these three output equations together into one matrix equation gives

$$\mathbf{y}(t) = \begin{bmatrix} 10 & 0 \\ -2 & -2 \\ 0 & 7 \end{bmatrix}\mathbf{v}(t) + \begin{bmatrix} 0 \\ 1 \\ 0 \end{bmatrix}x(t).$$

Comment: An SFG for this SIMO system is shown in Figure 6.7, and the MGR could be used three times to verify that the given transfer functions are represented.

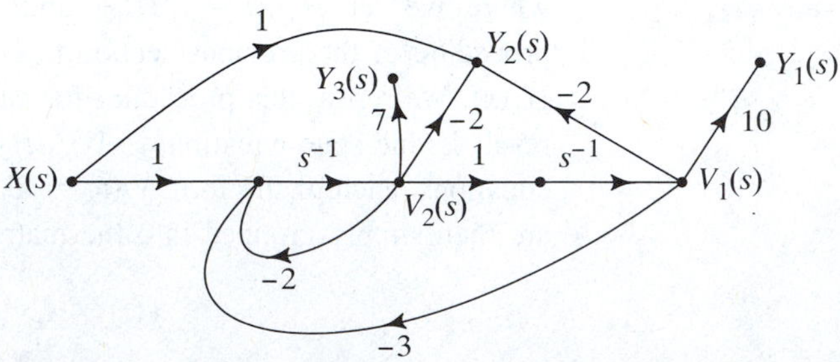

FIGURE 6.7 *SFG for SIMO system of Illustrative Problem 6.12*

SOLVED EXAMPLES AND MATLAB APPLICATIONS

EXAMPLE 6.1
Complete Solution of the State Equation by Laplace Transform

A causal LTI system is described by the state and output equations

$$\dot{\mathbf{v}}(t) = \begin{bmatrix} -2 & -2 \\ 1 & 0 \end{bmatrix}\mathbf{v}(t) + \begin{bmatrix} 10 \\ 0 \end{bmatrix}x(t) \quad \text{and} \quad y(t) = [1 \quad 0]\mathbf{v}(t)$$

with $x(t) = tu(t)$, $v_1(0) = 5$, and $v_2(0) = 0$.

a. Find the analytical expression for $y(t)$ due to the *initial conditions only.*

b. Find the transfer function $H(s) = Y(s)/X(s)$.

c. What are the system poles? Is the system stable?

d. Find the analytical expression for $y(t)$ due to the *input only.*

e. Use **lsim** or **kslsim** with the system described in state-space form to plot $y(t)$ in parts (a) and (d) that were solved earlier. Do your results agree?

f. Check your results to part (b) using the MATLAB function **ss2tf**.

WHAT IF? Suppose the output equation is changed to $y(t) = [0 \quad 1]\mathbf{v}(t)$. Is the stability of the system affected? Explain. ■

Solution

a. *Initial condition solution.* We form $\mathbf{V_{IC}}(s) = \mathbf{\Phi}(s)\mathbf{v}(0) = (s\mathbf{I} - \mathbf{A})^{-1}\mathbf{v}(0)$. First,

$$(s\mathbf{I} - \mathbf{A}) = \begin{bmatrix} s+2 & 2 \\ -1 & s \end{bmatrix}$$

and the inverse of $(s\mathbf{I} - \mathbf{A})$ is found as

matrix inverse

$$(s\mathbf{I} - \mathbf{A})^{-1} = \frac{\mathbf{adj}(s\mathbf{I} - \mathbf{A})}{|s\mathbf{I} - \mathbf{A}|} = \frac{1}{s^2+2s+2} \cdot \begin{bmatrix} s & -2 \\ 1 & s+2 \end{bmatrix}$$

$$= \begin{bmatrix} \dfrac{s}{s^2+2s+2} & \dfrac{-2}{s^2+2s+2} \\ \dfrac{1}{s^2+2s+2} & \dfrac{s+2}{s^2+2s+2} \end{bmatrix}, \qquad \text{Re}[s] > -1.$$

Each of the four elements of the $(s\mathbf{I} - \mathbf{A})^{-1}$ or $\mathbf{\Phi}(s)$ matrix must be expanded in partial fractions. We use the function **residue** that was introduced in Chapter 3, with the result

$$(s\mathbf{I} - \mathbf{A})^{-1} = \boldsymbol{\phi}(s) = \begin{bmatrix} \dfrac{0.707e^{j0.785}}{s+1-j1} + \text{conjugate} & \dfrac{1.000e^{j1.571}}{s+1-j1} + \text{conjugate} \\ \dfrac{0.500e^{-j1.571}}{s+1-j1} + \text{conjugate} & \dfrac{0.707e^{-j0.785}}{s+1-j1} + \text{conjugate} \end{bmatrix}.$$

Then using Table 3.1, we have the state transition matrix

$$\boldsymbol{\phi}(t) = \begin{bmatrix} 1.41e^{-t}\cos(t+0.78) & 2.00e^{-t}\cos(t+1.57) \\ 1.00e^{-t}\cos(t-1.57) & 1.41e^{-t}\cos(t-0.78) \end{bmatrix}, \qquad t \geq 0.$$

Now we can finally answer the question, What is the output $y(t)$ due to the initial condition $\mathbf{v}(0) = [5 \quad 0]^T$? We have

IC response

$$\begin{aligned} y_{IC}(t) &= \mathbf{C}\mathtt{f}(t)\mathbf{v}(0) \\ &= [1 \quad 0]\begin{bmatrix} 1.41e^{-t}\cos(t+0.78) & 2.00e^{-t}\cos(t+1.57) \\ 1.00e^{-t}\cos(t-1.57) & 1.41e^{-t}\cos(t-0.78) \end{bmatrix}\begin{bmatrix} 5 \\ 0 \end{bmatrix} \\ &= 7.07e^{-t}\cos(t+0.785). \end{aligned}$$

b. *Transfer function* $H(s) = Y(s)/X(s)$. The transfer function matrix is

$$\mathbf{H}(s) = \mathbf{C\Phi}(s)\mathbf{B} + \mathbf{D} = [1 \quad 0]\mathbf{\Phi}(s)\begin{bmatrix} 10 \\ 0 \end{bmatrix} = 10\phi_{11}(s) = \frac{10s}{s^2+2s+2}.$$

c. *System poles and stability.* The system poles are the roots of $s^2 + 2s + 2 = 0$, or $s_{1,2} = -1 \pm j1$. The causal system is *stable* because the roots have *negative real* parts (the imaginary axis is included in the ROC). We also notice that the initial condition response decays to zero as $t \to \infty$, another widely accepted definition of stability.

d. *Forced response, that is, response due to input only.* This is a good time to use the transfer function that we have just found. The transform of the output is

$$\begin{aligned} Y(s) = H(s)X(s) &= \frac{10s}{s^2+2s+2}\cdot\frac{1}{s^2}, \qquad \mathrm{Re}[s] > 0 \\ &= \frac{5}{s} + \frac{3.54e^{-j5\pi/4}}{s+1-j1} + \text{conjugate term} \end{aligned}$$

and the forced response is $y_F(t) = 5 + 7.07e^{-t}\cos(t - 5\pi/4)$, $t \geq 0$.

e. *MATLAB.* Using **kslsim** or **lsim** for verification of the IC response gives the plot of Figure E6.1a with the following script.

MATLAB Script

```
%E6_1 Solution of state equation

%E6_1a Initial condition response
A=[-2,-2;1,0];
B=[10;0];
C=[1,0];
D=[0];
v0=[5;0];
t=0:.01:4.5;
X=[0*ones(size(t))]';                   % makes the input zero
[y,v]=kslsim(A,B,C,D,X,t,v0);
%...plotting statements and pause
```

CROSS-CHECK From the analytical expression found in part (a), $y_{IC}(t) = 7.07e^{-t}\cos(t + 0.785)$, we find that $y_{IC}(0) = 5.001$, $y_{IC}(4.0) =$

0.00938, and $y_{IC}(1.5) = -1.033$, which compare favorably with Figure E6.1a. For the forced response we simply set the initial condition vector to zero and make the input equal to $x(t) = 1 * t$, as indicated in the following script, with the output plot in Figure E6.1b. Notice, here, that the steady-state value is 5, which agrees with the analytical solution, and that $y_F(1.5) = 5 + 7.07e^{-1.5}\cos(1.5 - 5\pi/4) = 3.808$, which is consistent with the plot of Figure E6.1b. ■

EXAMPLES

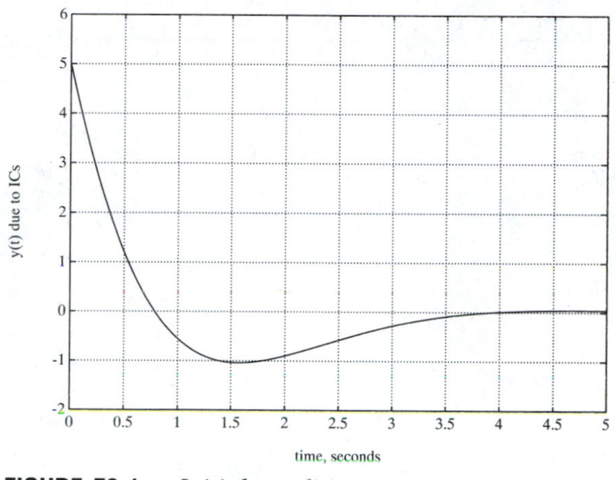

FIGURE E6.1a *Initial condition response*

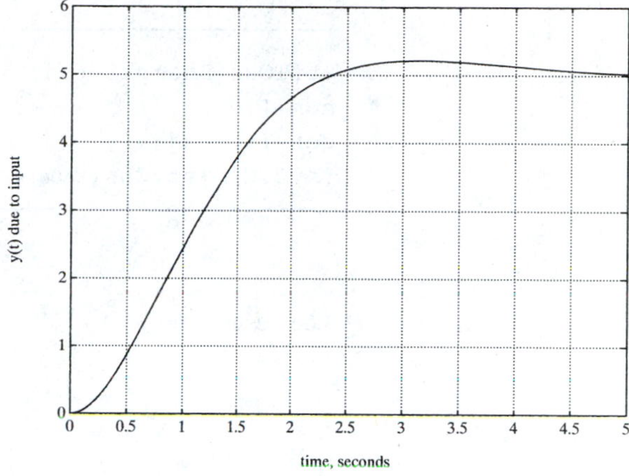

FIGURE E6.1b *Forced response*

f. The MATLAB function **ss2tf** stands for state-space to transfer function, and the script using this function is given next.

MATLAB Script

```
%E6_1b Forced response
A=[-2,-2;1,0];
B=[10;0];
C=[1,0];
D=[0];
v0=[0,0]';
t=0:.01:4.5;
X=[1*t]';
[y,v]=kslsim(A,B,C,D,X,t,v0);
%...plotting statements
```

MATLAB Script

```
%ssTOtf state-space to transfer function
A=[-2,-2;1,0]
B=[10;0]
```

Continues

```
C=[1,0]
D=[0]
[num,den]=ss2tf(A,B,C,D,1)          % call ss2tf
```

The program returns the numerator vector **num** = [0 10 0] and the denominator vector **den** = [1 2 2]. These results agree with the calculated $H(s) = 10s/(s^2 + 2s + 2)$. We can go the other way and use the program **tf2ss**, as shown next.

MATLAB Script

```
%tfTOss Example 6.1
num=[10,0];
den=[1,2,2];
[A,B,C,D]=tf2ss(num,den)
```

The results are

A = −2 −2

1 0

B = 1

0

C = 10 0 Notice that the constant 10 appears in the **C** matrix rather than the **B** matrix.

D = 0

WHAT IF? Changing the output equation has no effect on the matrix **A** or its eigenvalues, so stability is unaffected. ■

EXAMPLE 6.2
Complete Solution of the State Equation in the Time Domain with the Help of MATLAB

Two tanks, both of cross-sectional area α, are connected by a narrow pipe, as shown in Figure E6.2a. Fluid is introduced to the system in the form of three inputs, $x_1(t)$, $x_2(t)$, and $x_3(t)$. The fluid level in the tanks is given by $v_1(t)$ and $v_2(t)$, with the flow between them described by $q(t) = k\{v_1(t) - v_2(t)\}$. The pertinent linearized equations are

$$\alpha \dot{v}_1(t) = x_1(t) + x_2(t) - q(t);\ \alpha \dot{v}_2(t) = x_3(t) + q(t);$$
$$q(t) = k\{v_1(t) - v_2(t)\}.$$

a. Put these equations in state-space form and include an output equation with the flow $q(t)$ between the tanks defined as the system output $y(t)$. Give your answer as the **A**, **B**, **C**, and **D** matrices.

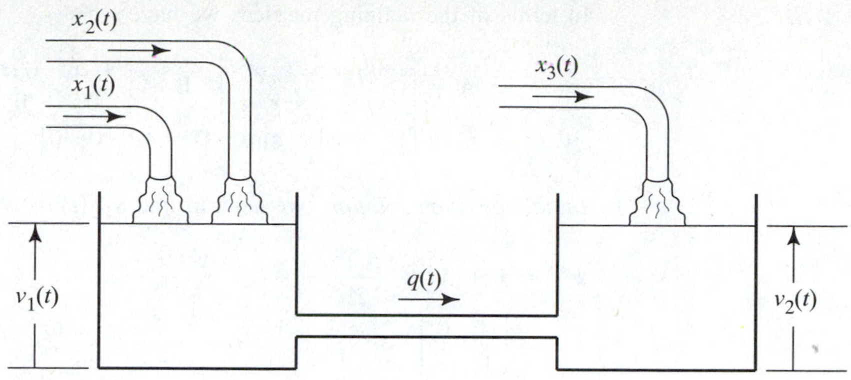

FIGURE E6.2a *Two mixing tanks*

b. Given that $k = \alpha = 1$, $v_1(0) = 2$, and $v_2(0) = 3$, solve for $\mathbf{v_{IC}}(t)$. Carry out the expansion of $\mathbf{e}^{\mathbf{A}t}$ through the t^4 term. The units of time applicable depend on the units for the constants k and α as well as the units for the fluid levels $v_1(t)$ and $v_2(t)$ and for the flow $q(t)$. Assume hours to be the unit of time.

c. From the tedious work involved in part (b), it is very apparent that we should turn to a numerical solution as quickly as possible. Use $T = 0.1$ and determine the Phi matrix; that is, find $\mathbf{\Phi}(0.1)$. In this case, include terms through T^3. What do you get for $\mathbf{\Phi}(0.1)$ using the **c2d** function from MATLAB?

d. For the given initial conditions and all inputs zero, about how long will it take for the fluid level in the two tanks to equalize? Use MATLAB along with the results of (c).

e. Now consider that $x_1(t) = x_3(t) = 0$ and that the input $x_2(t) = 2u(t)$. Use MATLAB to find and plot $q(t) = y(t)$ for the initial conditions of part (b).

WHAT IF? There are several things that are interesting to do here.

1. Try some inputs for $x_1(t)$ and $x_3(t)$ other than zero. Are the responses to these inputs reasonable?
2. Use **lsim** or **kslsim** for the various parts of the problem. Are the results similar to those from **dlsim** or **ksdlsim**? ■

Solution

a. *State and output equations.* The total system order is two, first-order in $v_1(t)$, and first-order in $v_2(t)$. We therefore need two state equations, which are found by substituting for $q(t)$ in the first two equations:

$$\alpha\dot{v}_1(t) = x_1(t) + x_2(t) - k\{v_1(t) - v_2(t)\} \quad \text{and}$$
$$\alpha\dot{v}_2(t) = x_3(t) + k\{v_1(t) - v_2(t)\}.$$

In terms of the defining matrices we have

$$\mathbf{A} = \begin{bmatrix} -k/\alpha & k/\alpha \\ k/\alpha & -k/\alpha \end{bmatrix}, \quad \mathbf{B} = \begin{bmatrix} 1/\alpha & 1/\alpha & 0 \\ 0 & 0 & 1/\alpha \end{bmatrix},$$
$$\mathbf{C} = [k \quad -k], \quad \text{and} \quad \mathbf{D} = [0 \quad 0 \quad 0].$$

b. *Initial condition solution.* We need to find $\mathbf{v_{IC}}(t) = \mathbf{e}^{\mathbf{A}t}\mathbf{v}(0)$, where

$$\begin{aligned}
\mathbf{e}^{\mathbf{A}t} &= \mathbf{I} + \mathbf{A}t + \frac{\mathbf{A}^2t^2}{2!} + \cdots + \frac{\mathbf{A}^n t^n}{n!} + \cdots \\
&= \begin{bmatrix} 1 & 0 \\ 0 & 1 \end{bmatrix} + \begin{bmatrix} -1 & 1 \\ 1 & -1 \end{bmatrix} t + \begin{bmatrix} 2 & -2 \\ -2 & 2 \end{bmatrix} \frac{t^2}{2!} \\
&\quad + \begin{bmatrix} -4 & 4 \\ 4 & -4 \end{bmatrix} \frac{t^3}{3!} + \begin{bmatrix} 8 & -8 \\ -8 & 8 \end{bmatrix} \frac{t^4}{4!} + \cdots \\
&= \begin{bmatrix} 1 - t + t^2 - \frac{2t^3}{3} + \frac{t^4}{3} - \cdots & t - t^2 + \frac{2t^3}{3} - \frac{t^4}{3} + \cdots \\ t - t^2 + \frac{2t^3}{3} - \frac{t^4}{3} + \cdots & 1 - t + t^2 - \frac{2t^3}{3} + \frac{t^4}{3} - \cdots \end{bmatrix}.
\end{aligned}$$

Thus the states $\mathbf{v}(t)$ due to their initial values $\mathbf{v}(0)$ are given by

$$\begin{aligned}
\mathbf{v_{IC}}(t) &= \mathbf{e}^{\mathbf{A}t}\mathbf{v}(0) \\
&= \begin{bmatrix} 1 - t + t^2 - \frac{2t^3}{3} + \frac{t^4}{3} - \cdots & t - t^2 + \frac{2t^3}{3} - \frac{t^4}{3} + \cdots \\ t - t^2 + \frac{2t^3}{3} - \frac{t^4}{3} + \cdots & 1 - t + t^2 - \frac{2t^3}{3} + \frac{t^4}{3} - \cdots \end{bmatrix} \\
&\quad \times \begin{bmatrix} 2 \\ 3 \end{bmatrix} = \begin{bmatrix} 2 + t - t^2 + \frac{2t^3}{3} - \frac{t^4}{3} + \cdots \\ 3 - t + t^2 - \frac{2t^3}{3} + \frac{t^4}{3} - \cdots \end{bmatrix}.
\end{aligned}$$

c. *Computation of the Phi matrix for* $T = 0.10$. Using the results of part (b),

$$\begin{aligned}
\mathbf{\Phi}(0.10) &= \mathbf{e}^{\mathbf{A}T} = \mathbf{e}^{\mathbf{A}t}\big|_{t=0.10} \\
&= \begin{bmatrix} 1 - 0.1 + (0.1)^2 - 2(0.1)^3/3 + \cdots & 0.1 - (0.1)^2 + 2(0.1)^3/3 - \cdots \\ 0.1 - (0.1)^2 + 2(0.1)^3/3 - \cdots & 1 - 0.1 + (0.1)^2 - 2(0.1)^3/3 + \cdots \end{bmatrix} \\
&= \begin{bmatrix} 0.9093 & 0.0907 \\ 0.0907 & 0.9093 \end{bmatrix}.
\end{aligned}$$

Using **c2d** we find that

$$\mathbf{\Phi}(0.10) = \begin{bmatrix} 0.9094 & 0.0906 \\ 0.0906 & 0.9094 \end{bmatrix},$$

and it appears that a sufficient number of terms was used in the $\mathbf{e}^{\mathbf{A}T}$ expansion.

d. *Time to steady-state.* When the flow $q(t)$ between the tanks reaches zero, the fluid levels are the same. We modified the script from Illustrative Problem 6.5 on page 313; after trying a few different values of n, we settled on the plot in Figure E6.2b, where we see that a little more than 2 hours is needed for things to settle to steady-state. The negative values for $q(t)$ indicate that the flow is from tank 2 to tank 1.

MATLAB Script

```
%E6_2 Solution of state equation using ksdlsim and kslsim

%E6_2b phidel and numde (Numerical Solution of Differential Equations)
A=[-1,1;1,-1]                          % continuous-time A matrix
B=[1,1,0;0,0,1]                        % continuous-time B matrix
[Phi,Del]=c2d(A,B,0.1)                 % call c2d(A,B,T)
% At this stage Phi and Del are returned for the given A, B, and T.
C=[1,-1];                              % C matrix
D=[0,0,0];                             % D matrix
v0=[2;3];                              % initial conditions at n=t=0
n=0:1:30;                              % start:increment:stop
X=[0*n;0*n;0*n]';                      % input column vector
[y,v]=ksdlsim(Phi,Del,C,D,X,v0);       % call ksdlsim
%...plotting statements and pause
```

e. *Total response.* Using **c2d**, $\mathbf{\Delta}(0.10)$ is computed as

$$\mathbf{\Delta}(0.10) = \begin{bmatrix} 0.0953 & 0.0953 & 0.0047 \\ 0.0047 & 0.0047 & 0.0953 \end{bmatrix}.$$

The input vector in the script for E6_2b is changed to $\mathbf{X} = [0 * \text{n}; 2 * \text{ones(n)}; 0 * \text{n}]'$ to obtain the script E6_2c, with the results in Figure E6.2c.

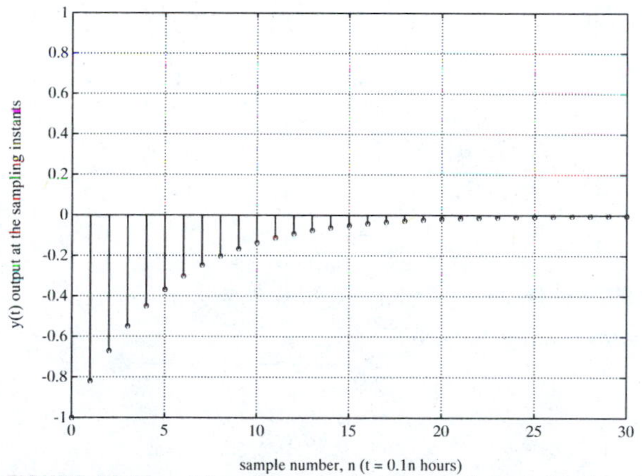

FIGURE E6.2b *Flow from tank 1 to tank 2: initial condition response*

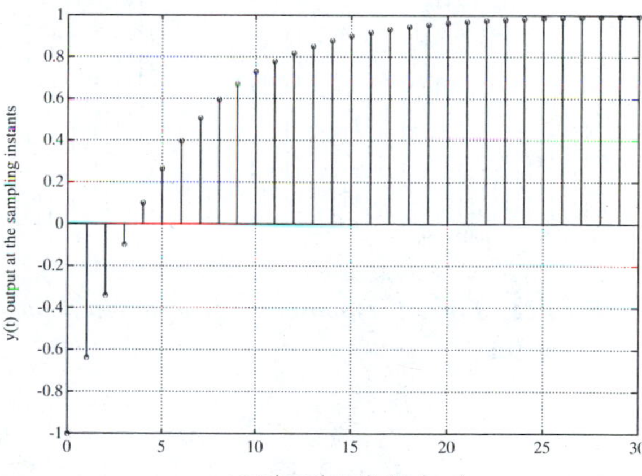

FIGURE E6.2c *Flow from tank 1 to tank 2: total response*

CROSS-CHECK How can we check the results in Figure E6.2c without doing a lot of computations?

1. The value at $t = n = 0$ is correct for $y(0) = v_1(0) - v_2(0) = 2 - 3 = -1$.

2. A bounded-input $x_2(t) = 2u(t)$ produced the preceding bounded-output, meaning that the system could be stable. We have a strong indication that it is from the IC response in Figure E6.2b, although to be sure we'd have to try several other sets of ICs or use analytical methods.

3. The transfer function $H_2(s) = Y(s)/X_2(s)$ is easily computed as $Y(s)/X_2(s) = 1/(s+2)$, giving $Y(s) = (1/(s+2)) \cdot 2/s$ and using the final value theorem $y(\infty) \longrightarrow sY(s)|_{s=0} = 1$. (It is OK to use the FVT because the pole of $sY(s)$ is negative real.) From these three checks it appears that the simulation in Figure E6.2c is correct. ■

WHAT IF?

1. We used some sinusoidal signals with $x_1(t) = 3\cos(5t)$ and $x_3(t) = -3 \cdot \cos(10t)$, keeping $x_2(t) = 2u(t)$, with the result in Figure E6.2d.
2. Using the same inputs as in (1), the results using **kslsim** are in Figure E6.2e.

■

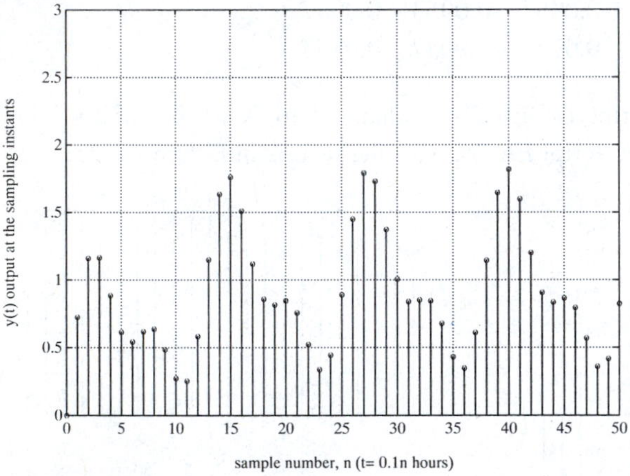

FIGURE E6.2d *Forced response to a generalized input using* **ksdlsim**

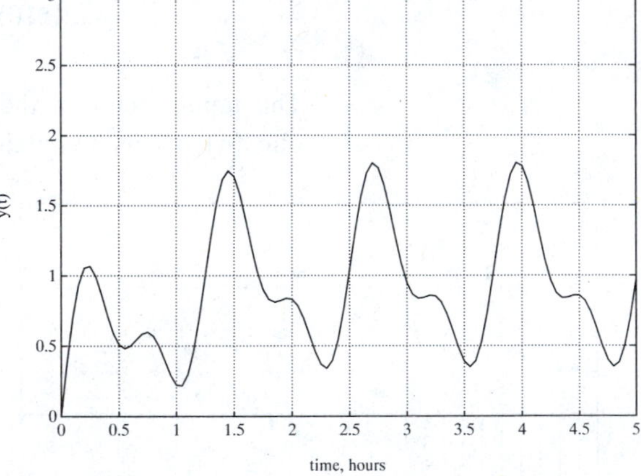

FIGURE E6.2e *Forced response to a generalized input using* **kslsim**

EXAMPLE 6.3 ***System Models and Time Response Characteristics***

An unstable plant described by the transfer function $H_1(s) = Y(s)/M(s) = 1/(s(s-1))$ is made part of a new feedback system shown in Figure E6.3a that includes a filter $H_2(s) = W(s)/E(s) = (ks+8)/(s+10)$ in the forward path and a sensor with a gain of negative one in the feedback path.

a. Use Mason's Gain Rule to find the overall transfer function $H(s) = Y(s)/R(s)$.

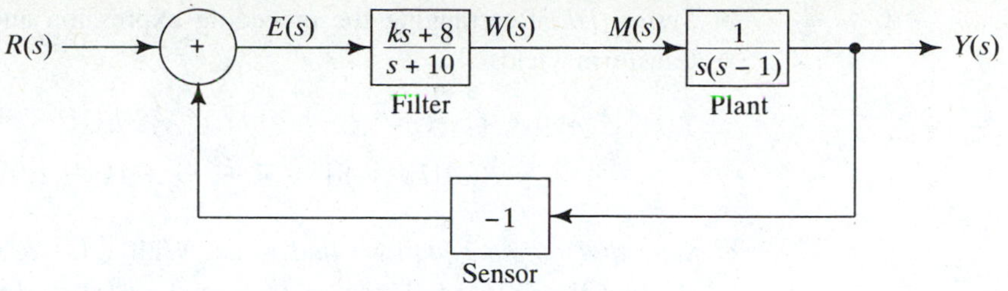

FIGURE E6.3a *System block diagram*

b. Determine the system differential equation relating the input $r(t)$ and the output $y(t)$.

c. Write state and output equations for this system.

d. Your supervisor decides that characteristic roots (eigenvalues) of $s_1 = -8$ and $s_{2,3} = -0.5 \pm j0.866$ will yield a decent step response. Determine the value of the filter gain k that will produce these roots.

e. Simulate this system and plot the response to a step input of $r(t) = u(t)$ with all initial conditions set to zero. What is the approximate settling time?

WHAT IF?

1. Suppose it is decided that the overshoot (the maximum value of $y(t)$) is too great. Find a value of k that will reduce the overshoot and, at the same time, keep the settling time more or less the same as for the value of k found in part (d).

2. We often want to design a system that will attempt to follow a linearly increasing reference signal such as $r(t) = Btu(t)$, where B is a positive real constant. With this input, the system error $e(t) = r(t) - y(t)$ is of interest. Use MATLAB to plot the error $e(t)$ for $k = 19$ and $B = 1$. Compare the steady-state error, $e(t)$, as $t \to \infty$ with the steady-state error for $r(t) = u(t)$.

■

Solution

a. *System transfer function.* The graph determinant is

$$\Delta(s) = 1 + \frac{ks + 8}{s + 10} \cdot \frac{1}{s(s - 1)} = \frac{s^3 + 9s^2 + s(k - 10) + 8}{(s + 10)(s)(s - 1)}.$$

The path gain and its cofactor are

$$P_1(s)\Delta_1(s) = \frac{ks + 8}{s + 10} \cdot \frac{1}{s(s - 1)} \cdot 1$$

which leads us to the overall transfer function of

$$H(s) = \frac{Y(s)}{R(s)} = \frac{P_1(s)\Delta_1(s)}{\Delta(s)} = \frac{ks + 8}{s^3 + 9s^2 + s(k - 10) + 8}.$$

b. *System DE.* Rearranging the preceding expression and taking the inverse transform yields

$$Y(s)\left[s^3 + 9s^2 + s(k - 10) + 8\right] = R(s)[ks + 8] \Leftrightarrow \dddot{y} + 9\ddot{y} + (k - 10)\dot{y} + 8y = k\dot{r} + 8r\,.$$

c. *State and output equations and roots.* With $Q(s)/R(s) = 1/(s^3 + 9s^2 + s(k - 10) + 8)$, we define $v_1(t) = q(t)$, $v_2(t) = dq(t)/dt$, and $v_3(t) = d^2q(t)/dt^2$. The state equation is

$$\dot{\mathbf{v}}(t) = \begin{bmatrix} 0 & 1 & 0 \\ 0 & 0 & 1 \\ -8 & -(k - 10) & -9 \end{bmatrix} \mathbf{v}(t) + \begin{bmatrix} 0 \\ 0 \\ 1 \end{bmatrix} r(t)\,.$$

For $Y(s)/Q(s) = ks + 8$, the output equation is

$$y(t) = [8 \quad k \quad 0]\mathbf{v}(t)\,.$$

d. *Selection of k.* The desired characteristic equation is $(s + 8)(s + 0.5 - j0.866)(s + 0.5 + j0.866) = 0$, or $s^3 + 9s^2 + 9s + 8 = 0$. The system CE in terms of the variable k is $s^3 + 9s^2 + (k - 10)s + 8 = 0$, and equating the coefficients of s^1 in both equations yields $k = 19$.

e. *Computer simulation.* The settling time of the response will be determined by the real part of $s_{2,3} = -0.5 \pm j0.866$, or -0.5. Thus the output term $Ke^{-0.5t}\cos(0.866t + \beta)$ will decay to approximately zero in about 10 time units. We used **kslsim** with the following script; the graph obtained is in Figure E6.3b on page 339.

MATLAB Script[6]

```
%E6_3 System models and time response characteristics

%E6_3b Step response, k=19
A=[0,1,0;0,0,1;-8,-9,-9];
B=[0;0;1];
C=[8,19,0];
D=[0];
v0=[0,0,0]';
t=0:.02:10;
R=[1*ones(size(t))]';
[y,v]=kslsim(A,B,C,D,R,t,v0);
%...plotting statements and pause
```

[6]Notice that here we use `a, b, c,` and `d` for the system matrices `A, B, C,` and `D`. The point is that either is acceptable but it's best to be consistent within a script. If you want to use a and A to mean the same thing within a script, use the MATLAB function **casesen**.

WHAT IF?

1. After experimenting a bit using **kslsim**, we selected $k = 38$, with the results given in Figure E6.3c. Observe that the peak value here is about 1.4, rather than the value of about 1.9 obtained with $k = 19$.
2. Figures E6.3d and e show the error responses for the ramp and step inputs, respectively. We see that the system has a steady-state error of about -1.25 for the ramp, with zero steady-state error for the step input. In Figure E6.3f we observe that the output tracks, or follows, the ramp but with a constant misalignment. In a later course, you will learn how to reduce this steady-state error. ■

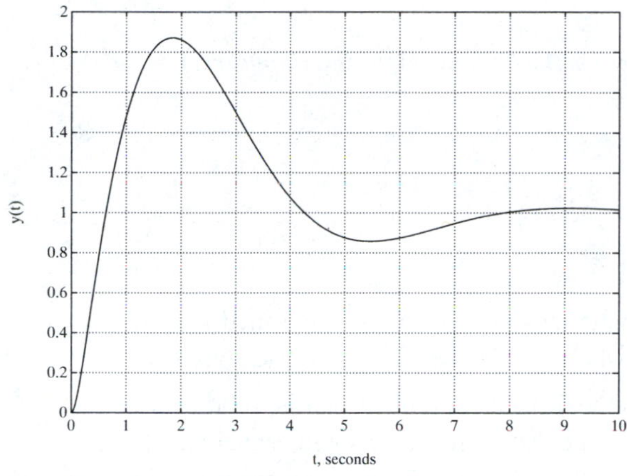

FIGURE E6.3b *Step response, $k = 19$*

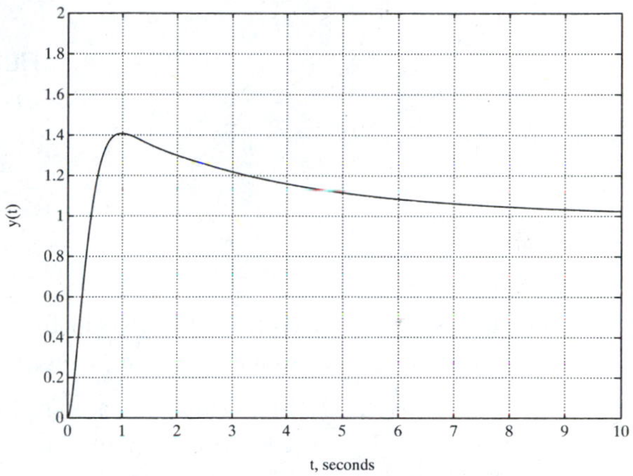

FIGURE E6.3c *Step response, $k = 38$*

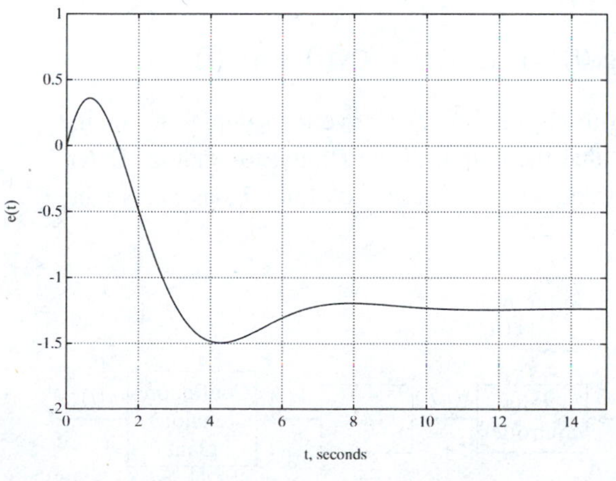

FIGURE E6.3d *Error response to $r(t) = tu(t)$, $k = 19$*

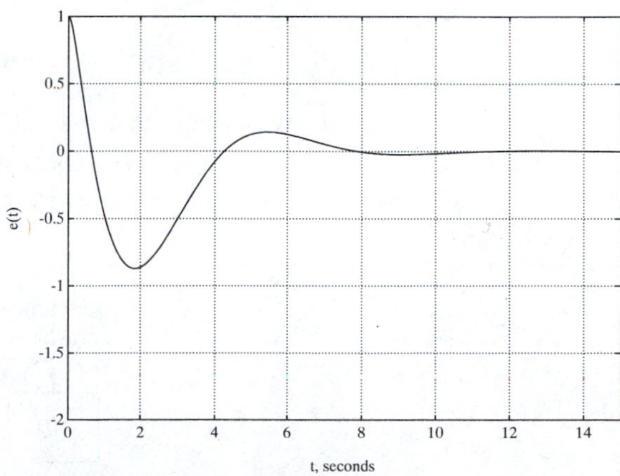

FIGURE E6.3e *Error response to $r(t) = u(t)$, $k = 19$*

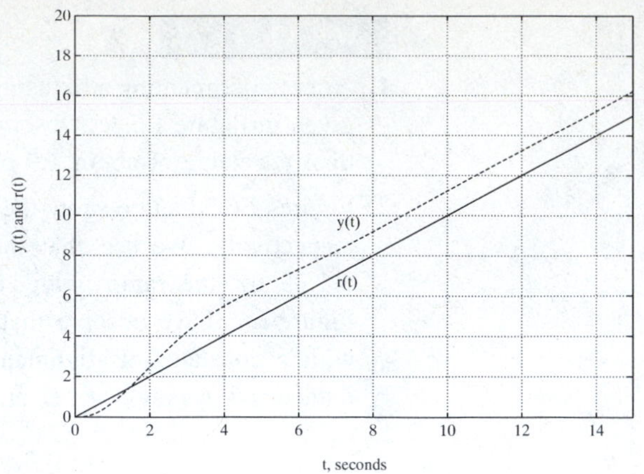

FIGURE E6.3f *Output $y(t)$ and input $r(t) = tu(t)$, $k = 19$*

EXAMPLE 6.4 ***Continuous Systems with Piecewise-Constant Inputs (Digital Control)***

A typical arrangement for a continuous system that is controlled by a digital computer is given in Figure E6.4a, with the following observations.

1. State equations are used to model the analog plant that is to be controlled, where the "pipelike" lines indicate that more than a single-state variable may be represented. In the configuration of Figure E6.4a, a Single-Input–Single-Output (SISO) plant is assumed, but a Multiple-Input–Multiple-Output (MIMO) plant is, of course, equally easy to describe with state equations. Thus we have, for the "analog" plant, the continuous-time state and output equations

$$\dot{\mathbf{v}}(t) = \mathbf{A}\mathbf{v}(t) + \mathbf{B}x(t) \quad \text{and} \quad y(t) = \mathbf{C}\mathbf{v}(t) + \mathbf{D}x(t)\,.$$

2. We assume that the Analog-to-Digital (A/D) Converter samples its input signals every T seconds and that the output of the Digital-to-Analog (D/A) Converter is piecewise constant over the period of time T, as shown in

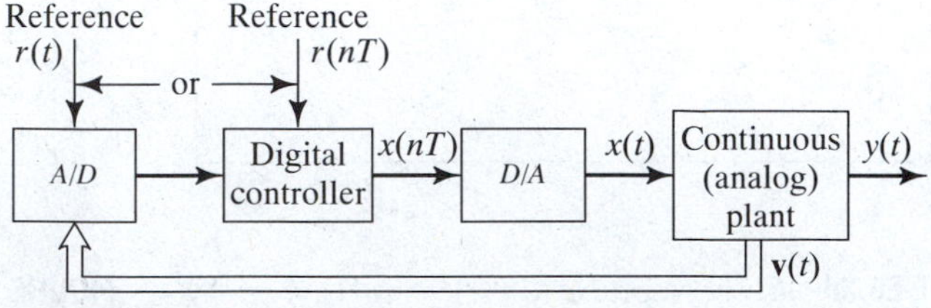

FIGURE E6.4a *Control by computer*

Figure E6.4b. Then the continuous system can be described at the *sampling instants only* by the state and output difference equations

$$\mathbf{v}(nT + T) = \mathbf{\Phi}(T)\mathbf{v}(nT) + \mathbf{\Delta}(T)x(nT) \quad \text{with} \quad nT \le t < nT + T$$
$$\text{and} \quad y(nT) = \mathbf{C}\mathbf{v}(nT) + \mathbf{D}x(nT).$$

As an application of digital control, we consider the highly unstable spacecraft of Illustrative Problem 6.2 modeled by the state equation

$$\dot{\mathbf{v}}(t) = \begin{bmatrix} 0 & 1 \\ 0 & 0 \end{bmatrix}\mathbf{v}(t) + \begin{bmatrix} 0 \\ 10 \end{bmatrix}x(t).$$

For a piecewise-constant input $x(t)$, we saw in Illustrative Problem 6.4 that the system is described at the sampling instants by the state difference equation

$$\mathbf{v}(nT + T) = \begin{bmatrix} 1 & T \\ 0 & 1 \end{bmatrix}\mathbf{v}(nT) + \begin{bmatrix} 5T^2 \\ 10T \end{bmatrix}x(nT) \quad \text{for } n = 0, 1, 2, \ldots.$$

For simplicity, we define the output as $v_1(nT)$, giving the output equation

$$y(nT) = [1 \quad 0]\mathbf{v}(nT).$$

a. To illustrate the unstable nature of this system, we make the reference input $r(t) = u(t)$ and disconnect the feedback "pipe," so that $x(nT) = r(nT) = u(nT)$. Choosing $T = 0.1$, make a computer generated plot of the system output $y(nT)$ versus the sample number n (or time t). Assume zero initial conditions.

b. Now suppose that we simply feed back the output signal $v_1(t)$ in a negative sense, making $x(nT) = r(nT) - v_1(nT)$ or, dropping the T, $x(n) = r(n) - v_1(n)$. Thus we have

$$\begin{aligned}\mathbf{v}(n + 1) &= \mathbf{\Phi}\mathbf{v}(n) + \mathbf{\Delta}x(n) \\ &= \mathbf{\Phi}\mathbf{v}(n) + \mathbf{\Delta}\{r(n) - [1 \quad 0]\mathbf{v}(n)\} = \mathbf{\Phi}_{\text{cl}}\mathbf{v}(n) + \mathbf{\Delta}r(n),\end{aligned}$$

where the modified or closed loop Phi matrix is $\mathbf{\Phi}_{\text{cl}} = \mathbf{\Phi} - \mathbf{\Delta}\mathbf{F}$, with $\mathbf{F} = [f_1 \quad f_2] = [1 \quad 0]$. Has this helped the situation?

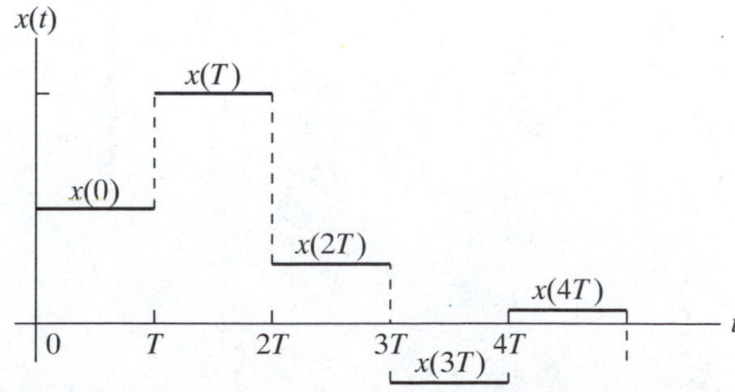

FIGURE E6.4b *Piecewise-constant input signal*

c. Now let's try feeding back the second state $v_2(t)$, which from the original set of equations is $dv_1(t)/dt$. That is, if $v_1(t)$ is the output position of the spacecraft, then $v_2(t)$ will be the output rate or output velocity, which will be relatively easy to measure. For a start, try a gain of unity on both states and plot the output $y(t)$.

WHAT IF? Maybe the response of part (c) is too sluggish and some overshoot is desired. Try some different feedback gains acting on $v_2(t)$ and plot $y(t)$. Do you find any that you like better? Also, notice the effect of having a gain on $v_1(t)$ of other than unity. Does this bother you at all? ■

Solution

a. *Step response without feedback.* For $T = 0.1$ we have

$$\mathbf{v}(0.1n + 0.1) = \begin{bmatrix} 1 & 0.1 \\ 0 & 1 \end{bmatrix}\begin{bmatrix} 1 \\ 0 \end{bmatrix} + \begin{bmatrix} 0.05 \\ 1 \end{bmatrix}1 \quad \text{and}$$

$$y(0.1n) = [1 \quad 0]\mathbf{v}(0.1n);$$

using **ksdlsim** we have the following script and the plot in Figure E6.4c, which shows the inherent instability of this spacecraft.

MATLAB Script

```
%E6_4 A system with a piecewise-constant input

%E6_4c Step response with no feedback-unstable
Phi=[1,0.1;0,1];
Del=[0.05;1];
C=[1,0];
D=[0];
v0=[0;0];                          % initial condition vector
n=0:1:20;                          % start n:1:stop n
R=[ones(size(n))]';                % input vector
[y,v]=ksdlsim(Phi,Del,C,D,R,v0);   % call ksdlsim
%...plotting statements and pause
```

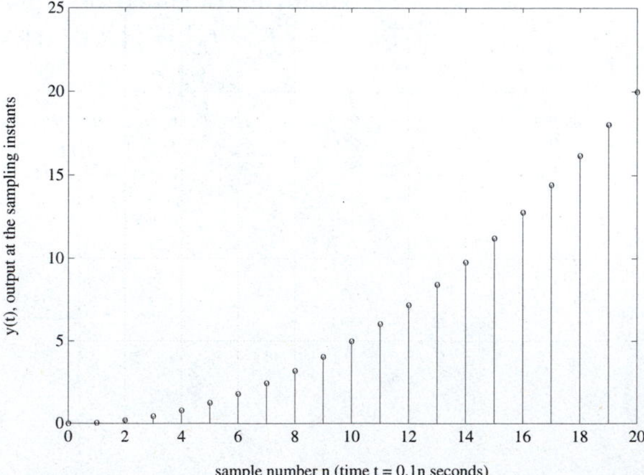

FIGURE E6.4c *Step response of unstable system*

b. *Step response with one state being fed back.* The script that implements this modification is given next, with the feedback gains entered in the new matrix **F**.

MATLAB Script

```
%E6_4d Step response with position feedback-unstable
Phi=[1,0.1;0,1];
Del=[0.05;1];
C=[1,0];
D=[0];
F=[1,0];                            % feedback gain matrix
v0=[0;0];                           % initial condition vector
n=0:1:100;                          % start n:1:stop n
R=[ones(size(n))]';                 % system input vector
Phicl=Phi-Del*F
[y,v]=ksdlsim(Phicl,Del,C,D,R,v0); % call ksdlsim
%...plotting statements and pause
```

The result of this single-state feedback is given in Figure E6.4d, where we see that the situation is somewhat better, but the system is still unstable, a highly unsatisfactory situation.

c. *Feeding back both states with unity gain.* We use the same program as in part (b), changing the **F** matrix to $\mathbf{F} = [1 \quad 1]$ rather than $\mathbf{F} = [1 \quad 0]$. The results are given in Figure E6.4e. Hooray, the spacecraft is now stable and locks onto the input reference signal of $r(n) = 1$ in about 40 samples, or 4 s.

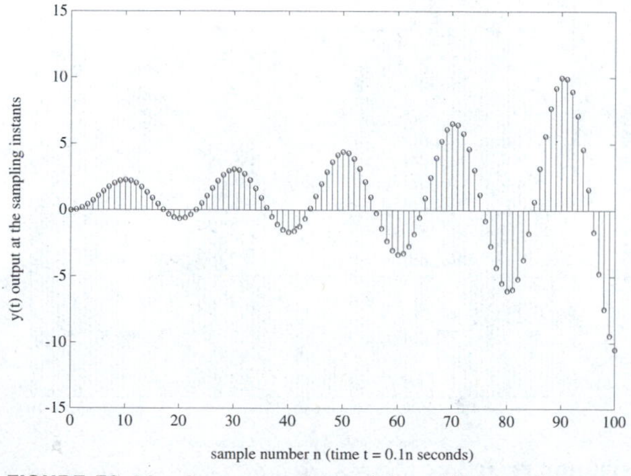

FIGURE E6.4d *Step response of the still-unstable system*

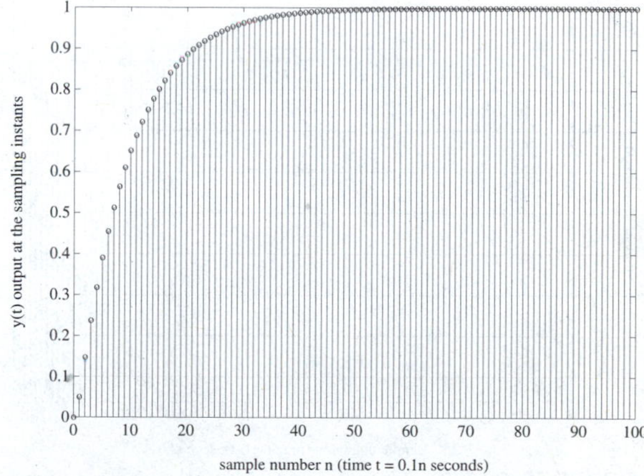

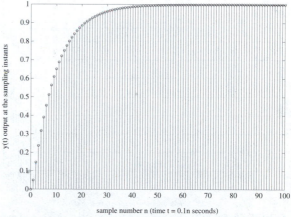

FIGURE E6.4e *Step response of stable system*

WHAT IF? A small sample of the effects of changing the gain on $v_2(t)$ is shown in the three plots of Figures E6.4f–h. The choice depends on your (or, more likely, your supervisor's) taste or prejudices. ■

Making the gain on $v_1(t)$ other than unity results in the plot of Figure E6.4i, where we see that the output does not return exactly to the reference input value $r(t) = 1u(t)$. It actually goes beyond unity. This may or may not be important in any particular situation, but topics such as these are normally considered in an undergraduate control course.

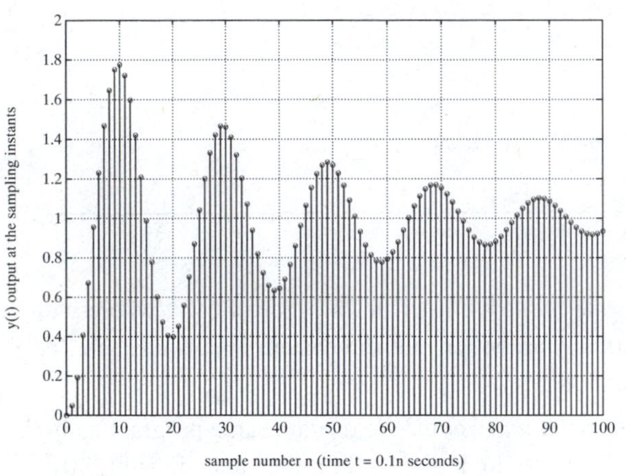

FIGURE E6.4f *Step response with feedback gain matrix* $\mathbf{F} = [1 \quad 0.1]$

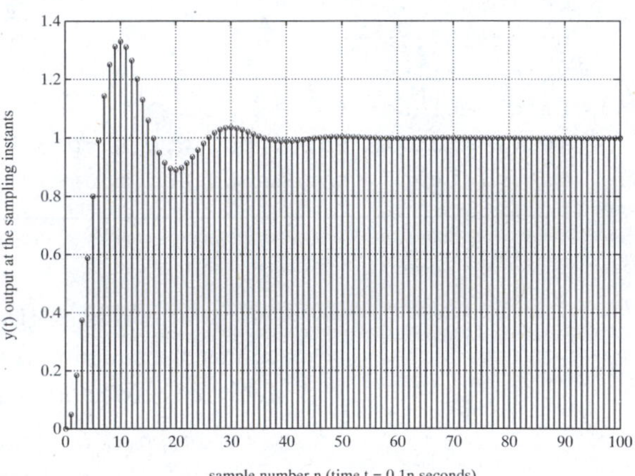

FIGURE E6.4g *Step response with feedback gain matrix* $\mathbf{F} = [1 \quad 0.25]$

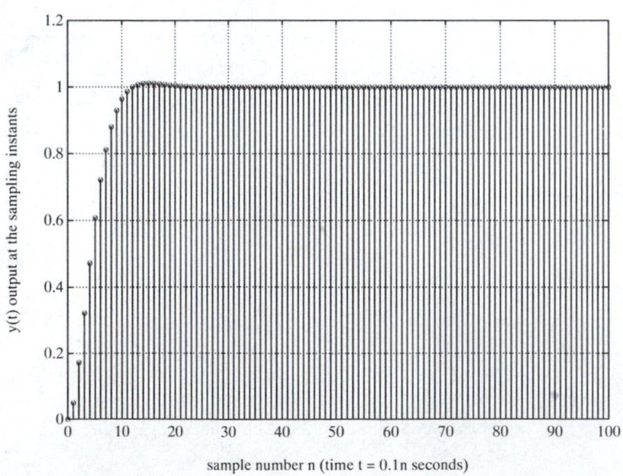

FIGURE E6.4h *Step response with feedback gain matrix* $\mathbf{F} = [1 \quad 0.5]$

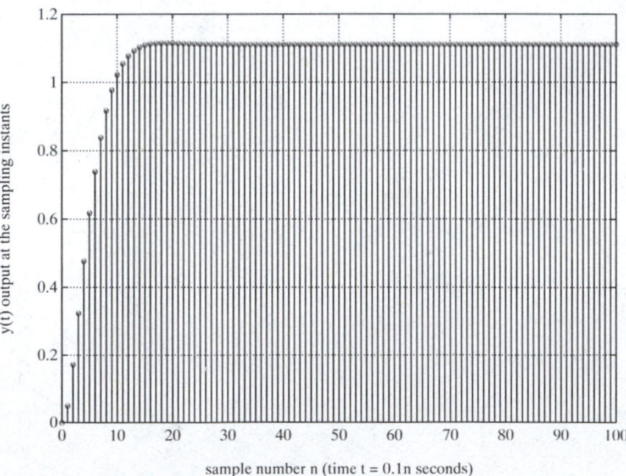

FIGURE E6.4i *Step response with feedback gain matrix* $\mathbf{F} = [0.9 \quad 0.5]$

REINFORCEMENT PROBLEMS

P6.1 Initial condition solution of a state equation. Given the matrix equation

$$\dot{\mathbf{v}}(t) = \begin{bmatrix} 1 & 0 \\ 0 & -2 \end{bmatrix}\mathbf{v}(t) + \begin{bmatrix} 3 \\ -4 \end{bmatrix}x(t).$$

a. Find the state transition matrix $\mathbf{e}^{\mathbf{A}t} = \boldsymbol{\phi}(t)$ by expanding $\mathbf{e}^{\mathbf{A}t}$. In this case, you should be able to recognize a closed form for $\boldsymbol{\phi}(t)$ from the expansion.

b. Determine $\mathbf{v}_{\mathbf{IC}}(t)$ for $\mathbf{v}(0) = [5 \quad -6]^T$.

P6.2 Forced solution of the state equation. In P6.1 we saw that $\boldsymbol{\phi}(t)$ can be written in the closed form

$$\boldsymbol{\phi}(t) = \begin{bmatrix} e^t & 0 \\ 0 & e^{-2t} \end{bmatrix}.$$

Use this result for $\boldsymbol{\phi}(t)$ to find a closed-form expression for the forced response $\mathbf{v}_{\mathbf{F}}(t)$ if $x(t) = u(t)$ in the system of Problem P6.1. Work in the time domain; i.e., do not use transforms.

P6.3 Characteristic roots (eigenvalues) and stability. Several **A** matrices for systems described by the state equation $d\mathbf{v}(t)/dt = \mathbf{A}\mathbf{v}(t) + \mathbf{B}\mathbf{x}(t)$ are listed here. For each system, find the characteristic equation, find the characteristic roots or eigenvalues, and check for stability.

a. $\mathbf{A} = \begin{bmatrix} -1 & 0 \\ 0 & -2 \end{bmatrix}$ b. $\mathbf{A} = \begin{bmatrix} -1 & 0 \\ 0 & 2 \end{bmatrix}$

c. $\mathbf{A} = \begin{bmatrix} 0 & 1 \\ 0 & -2 \end{bmatrix}$ d. $\mathbf{A} = \begin{bmatrix} 0 & 1 \\ -2 & 0 \end{bmatrix}$

P6.4 Transform solution. For the two tanks of Example 6.2, use the Laplace transform method to find an analytical expression for the output flow $q(t)$ and evaluate your result at $t = 1.0$. Assume $k = \alpha = 1$, $v_1(0) = 2$, $v_2(0) = 3$, $x_1(t) = x_3(t) = 0$, and $x_2(t) = 2u(t)$. Compare this result with the computer solution of Example 6.2e in Figure E6.2c.

P6.5 State and output equations from transfer functions. Find a set of state and output equations for each of the systems described here.

a. $H(s) = \dfrac{Y(s)}{X(s)} = \dfrac{5}{s + 25}$

b. $H(s) = \dfrac{Y(s)}{X(s)} = \dfrac{s + 5}{s + 25}$

c. $H(s) = \dfrac{Y(s)}{X(s)} = \dfrac{s}{s^3 - 2s^2 + 3s - 4}$

d. $H(s) = \dfrac{Y(s)}{X(s)} = \dfrac{s^2}{s^3 - 2s^2 + 3s - 4}$

e. $H(s) = \dfrac{Y(s)}{X(s)} = \dfrac{s^3}{s^3 - 2s^2 + 3s - 4}$

f. $H(s) = \dfrac{Y(s)}{X(s)} = \dfrac{s^3 - 2s + 1}{s^3 - 2s^2 + 3s - 4}$

P6.6 State and output equations to SFGs and then transfer functions. Two different LTI systems are described by the state and output equations $d\mathbf{v}(t)/dt = \mathbf{A}\mathbf{v}(t) + \mathbf{B}\mathbf{x}(t)$ and $\mathbf{y}(t) = \mathbf{C}\mathbf{v}(t) + \mathbf{D}\mathbf{x}(t)$.

a. Draw an SFG that represents these equations.

b. Use Mason's Gain Rule to find the appropriate transfer function $H(s)$ or transfer function matrix $\mathbf{H}(s)$ as appropriate.

$$\text{(i) } \mathbf{A} = \begin{bmatrix} 0 & 1 & 0 \\ 0 & 0 & 1 \\ 4 & -3 & -2 \end{bmatrix}, \quad \mathbf{B} = \begin{bmatrix} 0 \\ 0 \\ 1 \end{bmatrix}, \quad \mathbf{C} = [0 \quad 1 \quad 0], \quad \mathbf{D} = \mathbf{0}.$$

$$\text{(ii) } \mathbf{A} = \begin{bmatrix} 0 & 1 & 0 \\ 0 & 0 & 1 \\ 4 & -3 & -2 \end{bmatrix}, \quad \mathbf{B} = \begin{bmatrix} 0 \\ 0 \\ 1 \end{bmatrix}, \quad \mathbf{C} = [-3 \quad 1 \quad -2],$$

$$\mathbf{D} = d = 1.$$

P6.7 Solved Example 6.3 revisited. The block diagram of Figure E6.3a has been redrawn with $k = 19$ in Figure P6.7. Notice that this SFG is based on three "integrators" with gains of s^{-1}, or $1/s$.

a. Write a set of state equations where the output of each integrator is defined as a state.

b. From $|s\mathbf{I} - \mathbf{A}| = 0$, show that the characteristic equation is the same as in Example 6.3.

c. Three system outputs have been designated as $y_1(t) = c(t)$, $y_2(t) = e(t)$, and $y_3(t) = m(t)$. Find the corresponding output equation.

d. Use MGR to find the transfer functions $H_1(s) = C(s)/X(s)$ and $H_2(s) = E(s)/X(s)$.

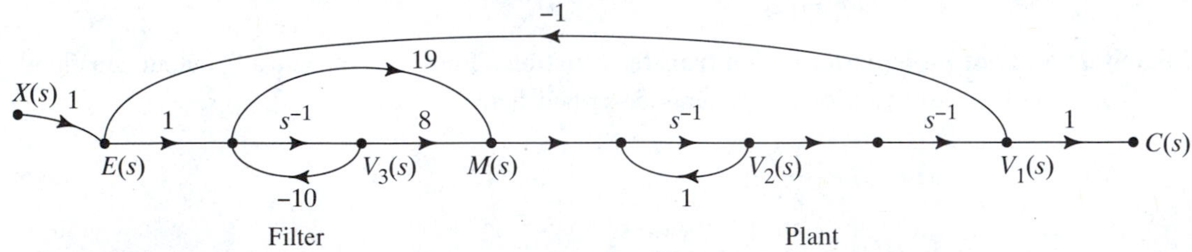

FIGURE P6.7 *Unstable plant with filter*

P6.8 SFG, state model, and transfer functions. The SFG of Figure P6.8 models an LTI system in the manner of Figure P6.7.

a. Write a set of state equations where the output of each integrator is defined as a state.

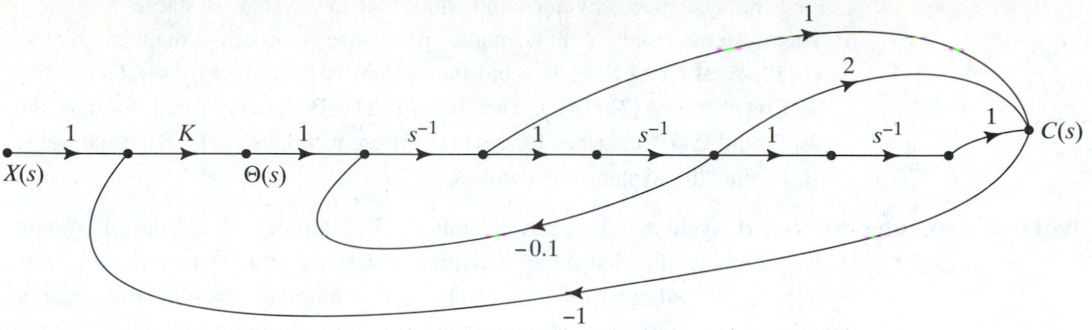

FIGURE P6.8 *LTI system*

b. From $|s\mathbf{I} - \mathbf{A}| = 0$, find the characteristic equation.

c. Two system outputs have been designated as $y_1(t) = c(t)$ and $y_2(t) = \theta(t)$. Find the corresponding output equation.

d. Use MGR to find the transfer functions $H_a(s) = C(s)/X(s)$ and $H_b(s) = \Theta(s)/X(s)$.

P6.9 Numerical solution of a DE. A causal LTI system is described by the DE

$$\dddot{y}(t) + 2\ddot{y}(t) = x(t) \quad \text{with } y(0) = 1, \qquad \dot{y}(0) = 2, \qquad \ddot{y}(0) = 3,$$
$$\text{and} \quad x(t) = 4u(t).$$

a. Use the Laplace transform method to find an analytical expression for $y(t)$.

b. Now solve this differential equation by the numerical method based on the Phi and Del matrices. For $v_1 = y$, $v_2 = dy/dt$, and $v_3 = d^2y/dt^2$, show that $\mathbf{\Phi}(T)$ and $\mathbf{\Delta}(T)$ are given by

$$\mathbf{\Phi}(T) = \begin{bmatrix} 1 & T & 0.5T - 0.25 + 0.25e^{-2T} \\ 0 & 1 & 0.5 - 0.5e^{-2T} \\ 0 & 0 & e^{-2T} \end{bmatrix} \quad \text{and}$$

$$\mathbf{\Delta}(T) = \begin{bmatrix} 0.25T^2 - 0.25T + 0.125 - 0.125e^{-2T} \\ 0.5T - 0.25 + 0.25e^{-2T} \\ 0.5 - 0.5e^{-2T} \end{bmatrix}.$$

c. Use the script developed in Illustrative Problem 6.4 to find $\mathbf{\Phi}(T)$ and $\mathbf{\Delta}(T)$ for several different values of T. (We have given the results for $T = 0.1$.)

d. Complete the solution using the appropriate MATLAB function(s). Compare the numerical results with the analytical result of part (a). For convenience, use $T = 0.1$ and $0 \leq t \leq 2.5$.

P6.10 Systems with piecewise-constant inputs. A continuous-time system is described by the state equation

$$\dot{\mathbf{v}}(t) = \begin{bmatrix} 0 & 1 \\ -2 & -3 \end{bmatrix}\mathbf{v}(t) + \begin{bmatrix} 0 \\ 1 \end{bmatrix}x(t).$$

a. Find the system eigenvalues and show that the system is stable.

b. The system input is now made piecewise constant—that is, $x(t) = x(nT)$, $T \leq t < nT + T$—and the system is described by $\mathbf{v}(nT + T) = \mathbf{\Phi}(T)\mathbf{v}(nT) + \mathbf{\Delta}(T)x(nT)$. Use the MATLAB function **c2d** to find the $\mathbf{\Phi}(T)$ and $\mathbf{\Delta}(T)$ matrices for several different values of T. For each value of T, find the system eigenvalues.

P6.11 A computer-controlled system. The power unit (a DC motor) of a control system is described by the first-order equations $d\theta(t)/dt = \omega(t)$ and $d\omega(t)/dt = -\omega(t) + x(t)$, where $\theta(t)$ and $\omega(t)$ are the angular position and angular velocity of the motor and $x(t)$ represents the motor's input voltage.

a. Find the transition matrix $\boldsymbol{\phi}(t)$ for this system. Let $\theta(t) = v_1(t)$ and $\omega(t) = v_2(t)$.

b. The input voltage to the motor is piecewise constant; that is, $x(t) = x(nT)$, $T \leq t < nT + T$. Determine the $\mathbf{\Phi}(T)$ and $\mathbf{\Delta}(T)$ matrices.

c. What is the state difference equation that will give the values of $\theta(t)$ and $\omega(t)$ every T seconds? Use $T = 1$ s and verify your hand computations with the script, developed in Illustrative Problem 6.4, that uses the MATLAB function **c2d**.

P6.12 Another computer-controlled system. Consider the same situation as Problem P6.11 with a different power unit described by $\dot{\theta}(t) = \omega(t)$ and $\dot{\omega}(t) = -2\theta(t) - 3\omega(t) + x(t)$. Repeat (a), (b), and (c) of Problem P6.11.

P6.13 Radar antenna system revisited. A radar antenna system was modeled in Chapter 3 (Problem P3.14) by the differential equation

$$\ddot{\omega}(t) + 65\dot{\omega}(t) + 3812\omega(t) = 154e(t),$$

where $e(t)$ is the applied voltage at the motor armature and $\omega(t)$ is the angular velocity of the radar dish.

a. Find the transfer function model $H(s) = \Omega(s)/E(s)$.

b. Find a state-space model with $v_1(t) = \omega(t)$, $v_2(t) = d\omega(t)/dt = \alpha(t)$, and $x(t) = e(t)$. For an initial velocity of $\omega(0) = -2$ rad/min, an initial acceleration of $\alpha(0) = 0$ rad/min^2, and a step motor voltage of 40 V, use **kslsim** or its equivalent to plot $\omega(t)$ and $\alpha(t)$ versus t.

c. Draw an SFG model that has $\Omega(s)$, $A(s)$, and $E(s)$ as the primary nodes. Use MGR on this graph to verify your answer to part (a).

d. What is the unit impulse response model $h(t)$? Use a MATLAB function to plot this result.

EXPLORATION PROBLEMS

P6.14 Submarine depth control—a design problem. Automatic depth control of a submarine is an interesting and important problem. It is also worth noting that the U.S.

PROBLEMS

Navy is not into this idea. Figure P6.14a illustrates a representative situation, where the actual depth of the submarine is denoted by $c(t)$. This depth $c(t)$ is measured by a pressure transducer and is compared with the desired depth $r(t)$. An equivalent block diagram of this system is given in Figure P6.14b, where

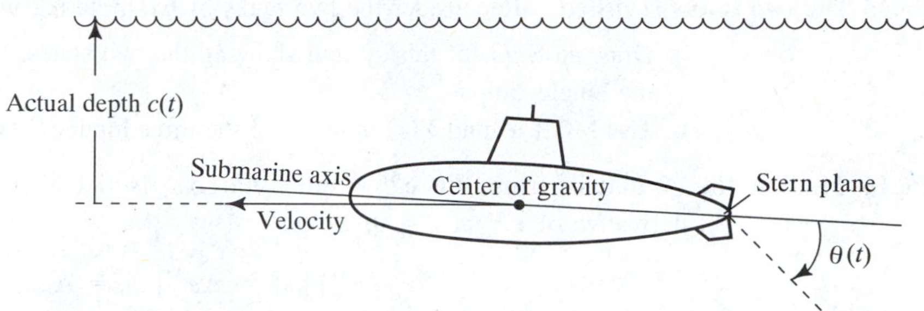

FIGURE P6.14a *Submarine*

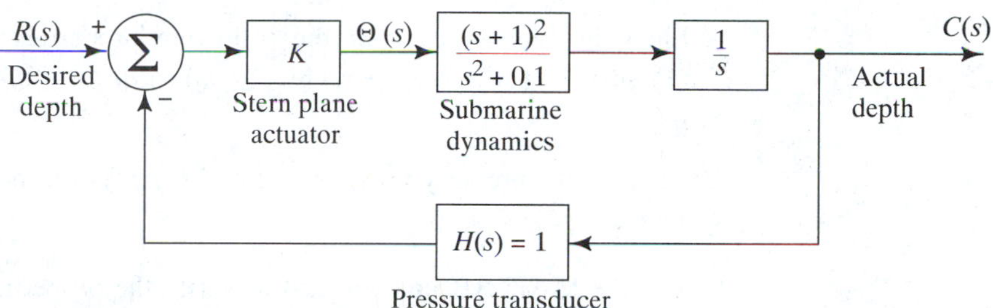

FIGURE P6.14b *Block diagram of depth-control system*

it is seen that any difference between $c(t)$ and $r(t)$ is fed back to the stern plane actuator, which appropriately adjusts the stern plane actuator angle $\theta(t)$.

a. Use Mason's Gain Rule to find the transfer function $H(s) = C(s)/R(s)$. This transfer function will be in terms of the gain K of the stern plane actuator. In this model, time is measured in minutes, making radians/minute the units for the complex variable s.
b. Write state and output equations for this system with the output defined as $c(t)$.
c. Suppose the Captain of the submarine issues a command to dive to 200 ft—that is, $r(t) = -200u(t)$. Simulate this dive and plot $c(t)$ for the following values of K:
(1) $K = 0.45$ (Is the period of oscillation about 2π minutes?)
(2) $K = 2$
(3) $K = 10$. Assume that the initial conditions are zero.

d. For each value of K in (c), answer the following:
   (i) Does the submarine reach the desired depth of -200 ft in steady-state?
   (ii) If so, how long does it take?
   (iii) What are the system eigenvalues?
   (iv) Which of the three values of K do you recommend? Pick another if you don't like any of these. In any case, justify your choice for K.

P6.15 The two tanks revisited. Reconsider the two tanks of Example 6.2 with $k = \alpha = 1$.

a. Draw an SFG for this system showing the two states, the three inputs, and the single output.
b. Use MGR to find $Y(s)$ in terms of the three inputs $X_1(s)$, $X_2(s)$, and $X_3(s)$.

P6.16 Finding the A matrix from the $\mathbf{e}^{\mathbf{A}t} = \boldsymbol{\phi}(t)$ matrix. Notice that if we evaluate the derivative of $\mathbf{e}^{\mathbf{A}t}$ at $t = 0$, we get

$$\frac{d}{dt}\{e^{\mathbf{A}t}\}\,|_{t=0} = \mathbf{A}e^{\mathbf{A}t}|_{t=0} = \mathbf{A}\,.$$

Use this idea to find the $\mathbf{A}$ matrix for the system whose transition matrix is

$$\boldsymbol{\phi}(t) = \begin{bmatrix} 1.414e^{-t}\cos(t + 0.785) & 2.000e^{-t}\cos(t + 1.571) \\ 1.000e^{-t}\cos(t - 1.571) & 1.414e^{-t}\cos(t - 0.785) \end{bmatrix}.$$

P6.17 Some properties of the $\boldsymbol{\phi}(t)$ matrix.

a. Show in general, each of the following properties of the $\boldsymbol{\phi}(t)$ matrix: (1) $\boldsymbol{\phi}(0) = \mathbf{I}$; (2) $\boldsymbol{\phi}(t_2 - t_1)\boldsymbol{\phi}(t_1 - t_0) = \boldsymbol{\phi}(t_2 - t_0)$; (3) $\boldsymbol{\phi}^{-1}(t_2 - t_1) = \boldsymbol{\phi}(t_1 - t_2)$.

Hint: Use the property $\mathbf{v}(t_\beta) = \boldsymbol{\phi}(t_\beta - t_\alpha)\mathbf{v}(t_\alpha)$ for the unforced ($x(t) = 0$) system.

b. Use the MATLAB function **c2d** to verify the properties of part (a) for

$$(1)\ \mathbf{A} = \begin{bmatrix} -2 & -2 \\ 1 & 0 \end{bmatrix} \quad (2)\ \mathbf{A} = \begin{bmatrix} 0 & 1 \\ -2 & -3 \end{bmatrix}.$$

P6.18 Another way to evaluate $\boldsymbol{\Delta}(T)$. In numerical evaluation of the state equation, we found that

$$\mathbf{v}(nT + T) = \boldsymbol{\Phi}(T)\mathbf{v}(nT) + \Delta(T)\mathbf{x}(nT)$$

where

$$\boldsymbol{\Phi}(T) = \mathbf{e}^{\mathbf{A}T} = \{\mathcal{L}^{-1}[(s\mathbf{I} - \mathbf{A})^{-1}]\}\,|_{t=T}$$

and

$$\boldsymbol{\Delta}(T) = \int_0^T \boldsymbol{\phi}(t - \tau)\mathbf{B}\mathbf{x}(\tau)\,d\tau\,.$$

Starting from the general relationship

$$\mathbf{v}(t) = \boldsymbol{\phi}(t)\mathbf{v}(0) + \int_0^t \boldsymbol{\phi}(t - \tau)\mathbf{B}\mathbf{x}(\tau)\,d\tau$$

assume that $t = T$ and that $\mathbf{x}(t) = \mathbf{x}(0)$, $0 \le t < T$, and show that

$$\boldsymbol{\Delta}(T) = \{\mathcal{L}^{-1}[\boldsymbol{\Phi}(s)\mathbf{B}/s]\}|_{t=T}\,.$$

DEFINITIONS, TECHNIQUES, AND CONNECTIONS

State equation	$\dot{\mathbf{v}}(t) = \mathbf{A}\mathbf{v}(t) + \mathbf{B}\mathbf{x}(t)$
IC solution	$\mathbf{v}_{\mathbf{IC}}(t) = \mathbf{e}^{\mathbf{A}t}\mathbf{v}(0) = \boldsymbol{\phi}(t)\mathbf{v}(0)$
Expansion of $\mathbf{e}^{\mathbf{A}t}$	$\mathbf{e}^{\mathbf{A}t} = \mathbf{I} + \mathbf{A}t + \dfrac{\mathbf{A}^2t^2}{2!} + \cdots + \dfrac{\mathbf{A}^nt^n}{n!} + \cdots$
Forced solution	$\mathbf{v}_{\mathbf{F}}(t) = \int_0^t \mathbf{e}^{\mathbf{A}(t-\tau)}\mathbf{B}\mathbf{x}(\tau)\,d\tau = \int_0^t \boldsymbol{\phi}(t-\tau)\mathbf{B}\mathbf{x}(\tau)\,d\tau$
Phi matrix	$\boldsymbol{\Phi}(T) = \mathbf{e}^{\mathbf{A}T} = \mathbf{e}^{\mathbf{A}t}\vert_{t=T}$
Del matrix	$\boldsymbol{\Delta}(T) = \int_0^T \boldsymbol{\Phi}(T-\tau)\mathbf{B}\,d\tau$
Numerical evaluation	$\mathbf{v}(nT + T) = \boldsymbol{\Phi}(T)\mathbf{v}(nT) + \boldsymbol{\Delta}(T)\mathbf{x}(nT)$, for $n = 0, 1, 2, \ldots$
Piecewise-constant input	$\mathbf{v}(nT + T) = \boldsymbol{\Phi}(T)\mathbf{v}(nT) + \boldsymbol{\Delta}(T)\mathbf{x}(nT)$ $\mathbf{x}(t) = \mathbf{x}(nT),\; nT \le t < nT + T$
Laplace transform solution	$\mathbf{V}(s) = \boldsymbol{\Phi}(s)\mathbf{v}(0) + \boldsymbol{\Phi}(s)\mathbf{B}\mathbf{X}(s)$
Matrix inverse	$\boldsymbol{\Phi}(s) = (s\mathbf{I} - \mathbf{A})^{-1} = \dfrac{\mathbf{adj}(s\mathbf{I} - \mathbf{A})}{\vert s\mathbf{I} - \mathbf{A}\vert}$
Characteristic equation	$\vert s\mathbf{I} - \mathbf{A}\vert = 0$
Characteristic roots	Zeros of $\vert s\mathbf{I} - \mathbf{A}\vert = 0$ or eigenvalues of the $\mathbf{A}$ matrix

Transition matrix	$\mathrm{f}(t) = \mathrm{L}^{-1}[\mathbf{\Phi}(s)]$
Output equation	$\mathbf{y}(t) = \mathbf{Cv}(t) + \mathbf{Dx}(t)$
Transfer function matrix	$\mathbf{H}(s) = \mathbf{C\Phi}(s)\mathbf{B} + \mathbf{D}$
Frequency response matrix	$\mathbf{H}(j\omega) = \mathbf{C\Phi}(j\omega)\mathbf{B} + \mathbf{D}$

MATLAB FUNCTIONS USED

Function	*Purpose and Use*	*Toolbox*
c2d	Given: state model of continuous system and sampling interval, **c2d** returns discrete state model for a sampling interval of T seconds.	Signals/Systems, Control System
dlsim	Given: state or TF model of discrete system, input, ICs, **dlsim** returns system output at the sampling instants.	Control System, Signals/Systems
eig	Given: an $N \times N$ matrix **A**, **eig** returns the eigenvalues of **A**.	MATLAB, Signals/Systems
ksdlsim	Given: state or TF model of discrete system, input, ICs, **ksdlsim** returns system output at sampling instants.	California Functions
kslsim	Given: state or TF model of continuous system, input, ICs, **kslsim** returns system output in continuous time.	California Functions
lsim	Given: state or TF model of continuous system, input, ICs, **lsim** returns system output in continuous time.	Control System, Signals/Systems
residue	Given: rational function $T(\sigma) = N(\sigma)/D(\sigma)$, **residue** returns roots of $D(\sigma) = 0$ and PF constants of $T(\sigma)$.	MATLAB
rlocus	Given: an equation in the form $1 + K[\text{num}(\sigma)]/[\text{den}(\sigma)] = 0$, **rlocus** returns plot of locus of roots for K varying.	Signals/Systems, Control System
roots	Given: coefficients of polynomial p, **roots** returns roots of $p = 0$.	MATLAB
ss2tf	Given: State model of continuous system, **ss2tf** returns TF model.	Control System, Signals/Systems
tf2ss	Given: TF model of continuous system, **tf2ss** returns equivalent state model.	Signals/Systems, Control System

ANNOTATED BIBLIOGRAPHY

1. Friedland, Bernard, *Control System Design, An Introduction to State Space Methods,* McGraw-Hill Book Company, New York, 1986. *This sophisticated, yet practical, treatise presents the use of state-space methods in the design of control systems. Freidland's emphasis on applications has motivated the selection of topics to include those that have the most practical utility. The first four chapters provide a thorough discussion of the role of state-space methods in linear system theory. It is an important book authored by an academician/practicing engineer.*

2. Kailath, Thomas, *Linear Systems,* Prentice Hall, Inc., Englewood Cliffs, N.J., 1980. *This comprehensive text on linear systems is normally used in a senior/graduate course. According to the author, the presentation is deliberately loosely organized in the early chapters with emphasis on discussion and motivation rather than formal development. Many good problems are included for both discrete and continuous systems. This well-conceived book should be studied by all interested in a clear exposition on linear systems from the state-space point of view.*

3. Kwakernaak, Huibert, and Raphael Sivan, *Modern Signals and Systems,* Prentice Hall, Inc., Englewood Cliffs, N.J., 1991. *This text provides a thorough treatment of state-space techniques and procedures. In Chapter Five, "State Description of Systems," the authors present a parallel treatment of state differential and difference equations that includes state description of systems, realization, solution of state equations, stability, frequency response, and many interesting problems and computer exercises.*

4. Reid, J. Gary, *Linear System Fundamentals,* McGraw-Hill Book Company, New York, 1983. *Just about all that is important in the study of linear systems is covered in this book. This characteristic is indicated by the subtitle,* Continuous and Discrete, Classic and Modern. *Chapters Six and Seven on state-space analysis are very complete and include several interesting examples of modeling and analysis by this method.*

5. [*Historical Reference*] Zadeh, Lotfi A., and Charles A. Desoer, *Linear System Theory, The State Space Approach,* Krieger Publishing Company, Melbourne, Florida, 1979. Reprint of original, McGraw-Hill Book Company, New York, 1963. *This prescient book, along with the beginnings of accessibility to relatively high-speed digital computers, kindled an interest among engineers, mathematicians, and scientists in the state variable method. Written at a graduate level,* Linear System Theory *includes a complete and rigorous treatment of linear systems, continuous and discrete, from the state-variable point of view. Although difficult to read, this text is a "bible" for state-space enthusiasts.*

ANSWERS

P6.1 a. $\mathbf{e}^{\mathbf{A}t} = \begin{bmatrix} 1 + t + t^2/2 + t^3/6 + \cdots & 0 \\ 0 & 1 - 2t + 2t^2 - 8t^3/6 + \cdots \end{bmatrix} = \begin{bmatrix} e^t & 0 \\ 0 & e^{-2t} \end{bmatrix}$; b. $\mathbf{v}_{\mathbf{IC}}(t) = [5e^t \quad -6e^{-2t}]^T$, $t \geq 0$

P6.2 $\mathbf{v}_{\mathbf{F}}(t) = \int_0^t \begin{bmatrix} e^{t-\tau} & 0 \\ 0 & e^{-2(t-\tau)} \end{bmatrix} \begin{bmatrix} 3 \\ -4 \end{bmatrix} u(\tau)\, d\tau = \begin{bmatrix} -3 + 3e^t \\ -2 + 2e^{-2t} \end{bmatrix}$, $t \geq 0$

P6.3 a. $s^2 + 3s + 2 = 0$, $s_{1,2} = -1, -2$, stable; b. $s^2 - s - 2 = 0$, $s_{1,2} = -1, 2$, unstable; c. $s^2 + 2s = 0$, $s_{1,2} = 0, -2$, unstable; d. $s^2 + 2 = 0$, $s_{1,2} = \pm j1.41$, unstable.

P6.4 $q(t) = y(t) = 1 - 2e^{-2t}$, $t \geq 0$; $q(1) = 1 - 2e^{-2} = 0.7293$, which agrees with the plot.

P6.5 *Note:* There are several correct state and output equations that will represent the transfer functions correctly. a. $\dot{v} = -25v + 5x$, $y = v$; b. $\dot{v} = -25v + x$, $y = -20v + x$;

$$\text{c. } \mathbf{A} = \begin{bmatrix} 0 & 1 & 0 \\ 0 & 0 & 1 \\ 4 & -3 & 2 \end{bmatrix}, \quad \mathbf{B} = \begin{bmatrix} 0 \\ 0 \\ 1 \end{bmatrix}, \quad \mathbf{C} = [0 \quad 1 \quad 0], \quad \mathbf{D} = \mathbf{0};$$

d. **A**, **B**, and **D** as in part (c), $\mathbf{C} = [0 \quad 0 \quad 1]$; e. **A** and **B** as in part (c), $\mathbf{C} = [4 \quad -3 \quad 2]$, $\mathbf{D} = 1$; f. **A** and **B** as in part (c), $\mathbf{C} = [5 \quad -5 \quad 2]$, $\mathbf{D} = 1$

P6.6 a. See Figure A6.6. b. $H_1(s) = s/(s^3 + 2s^2 + 3s - 4)$, $H_2(s) = (s^3 + 4s - 7)/(s^3 + 2s^2 + 3s - 4)$.

P6.7 a. Defining the states as the integrator outputs from right to left: $dv_1(t)/dt = v_2(t)$, $dv_2(t)/dt = -19v_1(t) + v_2(t) - 182v_3(t) + 19x(t)$, $dv_3(t)/dt = -v_1(t) - 10v_3(t) + x(t)$; b. $|s\mathbf{I} - \mathbf{A}| = s^3 + 9s^2 + 9s + 8$ as before; c. $y_1(t) = v_1(t)$, $y_2(t) = -v_1(t) + x(t)$, $y_3(t) = -19v_1(t) - 182v_3(t) + 19x(t)$; d. $H_1(s) = (19s + 8)/(s^3 + 9s^2 + 9s + 8)$, as before; $H_2(s) = (s^3 + 9s^2 - 10s)/(s^3 + 9s^2 + 9s + 8)$

P6.8 a. $dv_1(t)/dt = v_2(t)$, $dv_2(t)/dt = v_3(t)$, $dv_3(t)/dt = -Kv_1(t) - (2K + 0.1)v_2(t) - Kv_3(t) + Kx(t)$; b. $s^3 + Ks^2 + (2K + 0.1)s + K = 0$; c. $y_1(t) = [1 \quad 2 \quad 1]\mathbf{v}(t)$, $y_2(t) = [-K \quad -2K \quad -K]\mathbf{v}(t) + Kx(t)$; d. $H_a(s) = K(s^2 + 2s + 1)/(s^3 + Ks^2 + [2K + 0.1]s + K)$, $H_b(s) = Ks(s^2 + 0.1)/(s^3 + Ks^2 + [2K + 0.1]s + K)$

P6.9 a. $y(t) = [0.25e^{-2t} + t^2 + 2.5t + 0.75]u(t)$;

$$\text{c. } \mathbf{\Phi}(0.1) = \begin{bmatrix} 1 & 0.1 & 0.0047 \\ 0 & 1 & 0.0906 \\ 0 & 0 & 0.8187 \end{bmatrix} \quad \text{and} \quad \mathbf{\Delta}(0.1) = \begin{bmatrix} 0.00016 \\ 0.00468 \\ 0.09063 \end{bmatrix};$$

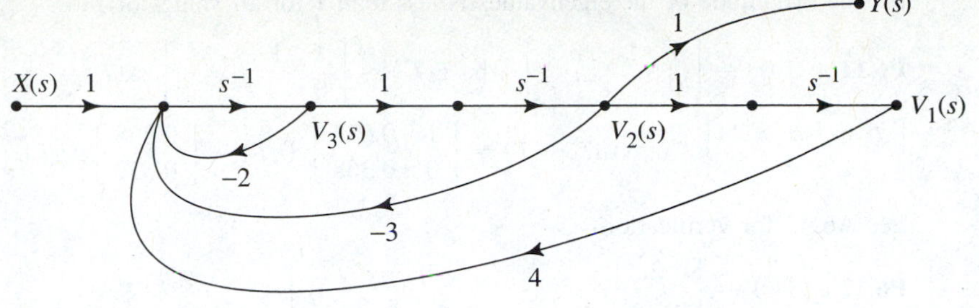

System 1

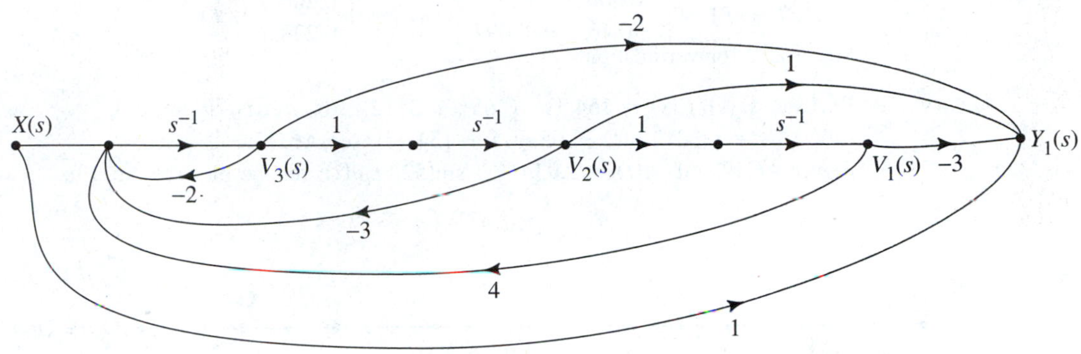

FIGURE A6.6

System 2

d. From the Laplace solution, $y(2) = 0.25e^{-4} + 2^2 + 2.5(2) + 0.75 = 9.7546$. From the output of file A6_9, we have $y = 1.0000$ at $n = 0$ or $t = 0$, 4.2838 at $n = 10$ or $t = 1$, 9.7546 at $n = 20$ or $t = 2$, 13.2517 at $n = 25$ or $t = 2.5$.

P6.10 a. $\lambda_{1,2} = -1, -2$, causal system with left half-plane roots.

Output

```
From A6_10
T=1; Phi=0.6004     0.2325
        -0.4651    -0.0972
     Del=0.1998
           0.2325
     lambda=0.3679
            0.1353
```

Output

```
From A6_10
T=0.1; Phi=0.9909  0.0861
          -0.1722  0.7326
       Del=0.0045
           0.0861
       lambda=0.9048
              =0.8187
```

b. The magnitude of the eigenvalues is less than 1 for all values of T.

P6.11 a. $\mathrm{f}(t) = \begin{bmatrix} 1 & 1 - e^{-t} \\ 0 & e^{-t} \end{bmatrix}$; b. $\mathrm{f}(T) = \begin{bmatrix} 1 & 1 - e^{-T} \\ 0 & e^{-T} \end{bmatrix}$, $\mathbf{\Delta}(T) = \begin{bmatrix} T - 1 + e^{-T} \\ 1 - e^{-T} \end{bmatrix}$; c. $\mathbf{v}(nT + T) = \begin{bmatrix} 1 & 0.632 \\ 0 & 0.368 \end{bmatrix}\mathbf{v}(nT) + \begin{bmatrix} 0.368 \\ 0.632 \end{bmatrix}x(nT)$.

See A6_11 for verification.

P6.12 a. $\mathrm{f}(t) = \begin{bmatrix} 2e^{-t} - e^{-2t} & e^{-t} - e^{-2t} \\ -2e^{-t} + 2e^{-2t} & -e^{-t} + 2e^{-2t} \end{bmatrix}$;

b. $\mathrm{f}(T) = \mathrm{f}(t)|_{t=T}$, $\mathbf{\Delta}(T) = \begin{bmatrix} 0.5 - e^{-T} + 0.5e^{-2T} \\ e^{-T} - e^{-2T} \end{bmatrix}$;

c. $\mathbf{v}(nT + T) = \begin{bmatrix} 0.600 & 0.233 \\ -0.465 & -0.097 \end{bmatrix}\mathbf{v}(nT) + \begin{bmatrix} 0.200 \\ 0.233 \end{bmatrix}x(nT)$.
See A6_12 for verification.

P6.13 a. $\Omega(s)/E(s) = 154/(s^2 + 65s + 3812)$; b. $dv_1(t)/dt = v_2(t)$, $dv_2(t)/dt = -3812v_1(t) - 65v_2(t) + 154x(t)$; see A6_13a. c. See Figure A6.13. d. $h(t) = 2.93e^{-32.5t}\sin(52.5t)u(t)$; for the plot see A6_13b.

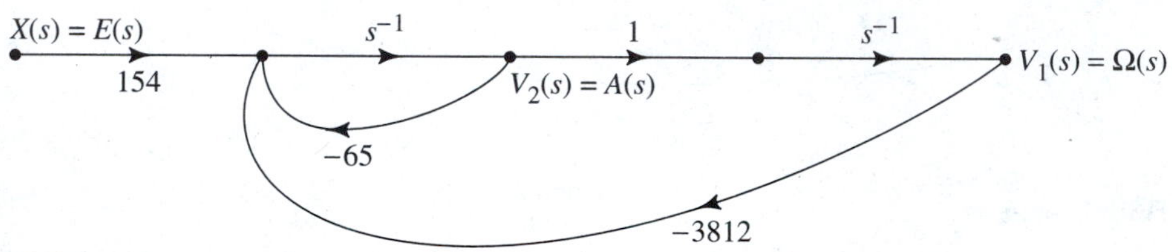

FIGURE A6.13

P6.14 a. $\dfrac{C(s)}{R(s)} = \dfrac{K(s + 1)^2}{s^3 + Ks^2 + (2K + 0.1)s + K}$; b. One possibility:

$$\mathbf{A} = \begin{bmatrix} 0 & 1 & 0 \\ 0 & 0 & 1 \\ -K & -(2K + 0.1) & -K \end{bmatrix}, \quad \mathbf{B} = \begin{bmatrix} 0 \\ 0 \\ 1 \end{bmatrix}, \quad \mathbf{C} = [K \quad 2K \quad K], \quad \mathbf{D} = \mathbf{0}$$

c. See A6_14 on the CLS disk. d. $K = 0.45$: (i) no, oscillation about -200 ft, very disturbing to the crew members; (ii) never; (iii) $s_{1,2} = \pm j1$, $s_3 = -0.45$; $K = 2$: (i) yes; (ii) about 6 min; (iii) $s_{1,2} = -0.696 \pm j1.664$, $s_3 = -0.616$; $K = 10$: (i) yes; (ii) about 3 min, but actually the response is -200 ft and remains within 2 ft of -200 at about $t = 1.5$; (iii) $s_1 = -0.769$, $s_2 = -1.734$, $s_3 = -7.497$; (iv) $K = 10$ is the best among the three. We would probably want to look at the solution more closely to determine the maximum acceleration required.

P6.15 a. See Figure A6.15; b. $Y(s) = [1/(s + 2)]\{X_1(s) + X_2(s) - X_3(s)\}$.

Comment: You may wonder why $Y(s)$ has a denominator of degree 1. This denominator occurs because the system is not completely observable, which, in this case, means that knowing $y(t) = v_1(t) - v_2(t)$ does not enable us to "know" both $v_1(t)$ and $v_2(t)$.

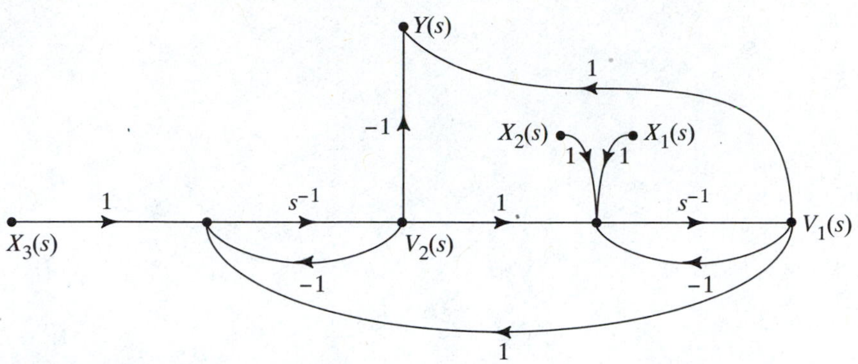

FIGURE A6.15

P6.16 A $= [-2 \quad -2; 1 \quad 0]$

P6.17 b. See A6_17 on the CLS disk.

RETROSPECTIVE

Continuous Systems

Having completed the chapters devoted to continuous-time LTI systems, it's a good time to take our second look backward to see where we've been. The retrospective following Chapter 4 emphasized analytical methods and when they may be used. An important factor is deciding what approach to apply. In Chapter 5 we encountered the first of two chapters in the book that discuss Fourier methods; Chapter 6 presented solution methods for the state-space description for linear systems. Thus we have added to our repertoire of models, and at this point we can use a system's differential equation(s), transfer function, unit impulse response, frequency response, block diagram or signal flowgraph, and state and output equations. Furthermore, we now are able to convert from one model to another.

SYSTEM MODELS

Differential equation	$\sum_{k=0}^{N} a_k \frac{d^k y(t)}{dt^k} = \sum_{k=0}^{L} b_k \frac{d^k x(t)}{dt^k}$
Transfer function	$H(s) = \frac{Y(s)}{X(s)}, \quad \text{ICs} = 0$
Unit impulse response	$h(t)$
Frequency response	$H(j\omega) \quad \text{or} \quad H(\omega)$
State space	$\dot{\mathbf{v}}(t) = \mathbf{A}\mathbf{v}(t) + \mathbf{B}\mathbf{x}(t)$ $\mathbf{y}(t) = \mathbf{C}\mathbf{v}(t) + \mathbf{D}\mathbf{x}(t)$
Block diagram or SFG	

FINDING ONE MODEL FROM ANOTHER

See Figure R.1 for a pictorial view of the possibilities and procedures.

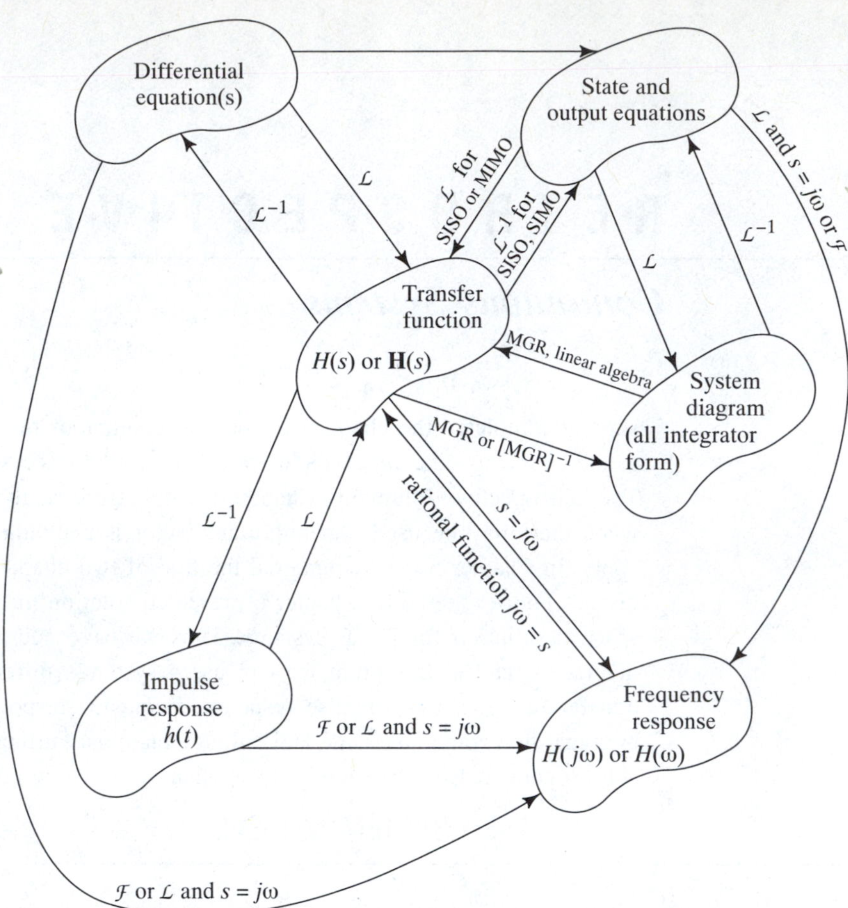

FIGURE R.1 *Finding one model from another*

PROPERTIES AND PERTINENT CHARACTERISTICS

Linearity	Homogeneity and additivity
Causality	Causal, anticausal, and noncausal
Time invariance	Fixed system parameters
Characteristic roots (eigenvalues)	$(s - r_1)(s - r_2)\cdots(s - r_N) = 0$
Characteristic equation	$\sum_{k=0}^{N} a_k \frac{d^k y(t)}{dt^k} = 0 \rightarrow \sum_{k=0}^{N} a_k s^k = 0$ by assuming $y(t) = Ce^{st}$

Stability (causal system)	$y_{IC}(t) \to 0$ as $t \to \infty$ $\int_0^\infty \lvert h(t)\rvert dt < \infty$, all characteristic roots have negative real parts

TECHNIQUES FOR FINDING SYSTEM OUTPUT

Convolution	$y(t) = \int_{-\infty}^{\infty} h(\tau)x(t-\tau)\,d\tau, \quad \text{ICs} = 0$
Laplace transform	$Y(s)$ found from DE(s) with ICs $\neq 0$ or $Y(s) = H(s)X(s)$ with ICs $= 0$
(for distinct poles of $Y(s)$)	$Y(s) = \sum_{k=1}^{R} \frac{C_k}{s - d_k}, \quad y(t) = \sum_{k=1}^{N} C_k e^{d_k t}$
Sinusoidal steady state	$y_{ss}(t) = A\lvert H(j\omega)\rvert \cos(\omega t + \alpha + \angle H(j\omega))$ for $x(t) = A\cos(\omega t + \alpha)$
Fourier series of output	$y_{ss}(t) = X_0 H(j0) + \sum_{k=1}^{\infty} 2\lvert X_k\rvert\cdot$ $\lvert H(jk\omega_0)\rvert \cos(k\omega_0 t + \angle X_k)$ for the input $x(t) = \sum_{k=-\infty}^{\infty} X_k e^{jk\omega_0 t}, \; H(jk\omega_0) = H(j\omega)\rvert_{\omega = k\omega_0}$
Fourier transform	$Y(\omega) = H(\omega)X(\omega)$

MISCELLANEOUS

Alternative path to transfer function using Mason's Gain Rule	$H(s) = \frac{\sum_{k=1}^{M} P_k(s)\Delta_k(s)}{\Delta(s)}$

ADDITIONAL MATLAB FUNCTIONS USED

The Retrospective at the end of Chapter 4 contains a list (page 220) of the MATLAB functions used in the Chapters 1, 2, 3, and 4. Descriptions of additional functions that appear in Chapters 5 and 6 follow.

Function	*Purpose and Use*	*Toolbox*
c2d	Given: state model of continuous system and sampling interval, **c2d** returns discrete state model for a sampling of interval T seconds.	Signals/Systems, Control System
fourier	Given: The symbolic expression f, **fourier** returns the Fourier transform F.	Symbolic Math
dlsim	Given: state or TF model of discrete system, input, ICs, **dlsim** returns system output at the sampling instants.	Control System, Signals/Systems
eig	Given: an $N \times N$ matrix **A**, **eig** returns the eigenvalues of **A**.	MATLAB, Signals/Systems
invfourier	Given: Fourier transform function F, **invfourier** returns the inverse Fourier transform f.	Symbolic Math
ksdlsim	Given: state or TF model of discrete system, input, ICs, **ksdlsim** returns system output at the sampling instants.	California Functions
ss2tf	Given: State model of continuous system, **ss2tf** returns TF model.	Control System, Signals/Systems
tf2ss	Given: TF model of continuous system, **tf2ss** returns equivalent state model.	Signals/Systems, Control System

Discrete Systems

PREVIEW

Here we begin a treatment of discrete systems that parallels the presentation for continuous systems in Chapters 2–6. It may be helpful to review briefly the material in Chapter 1 on sequences.

A discrete-time system receives an input sequence $x(n)$ and, using a rule or an algorithm, generates an output sequence $y(n)$. This concept is illustrated in Figure 7.2 (page 364), where the block labeled "discrete-time system" could represent a digital filter to eliminate unwanted noise in a telecommunications system, a speech-processing system for use as voice input to a computer, or a seismic-sounding system used to help gain information about a reservoir that

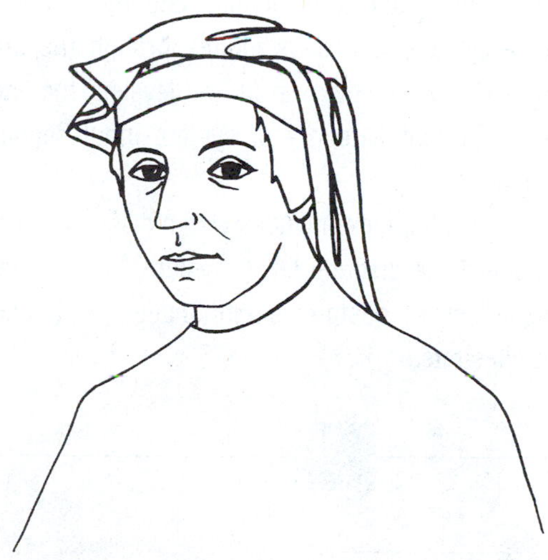

FIGURE 7.1 *Leonardo of Pisa (Fibonacci)*

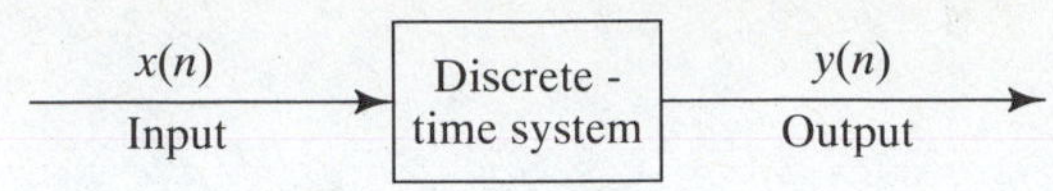

FIGURE 7.2 *Discrete-time system*

holds oil and gas in porous formations. On a more familiar level, perhaps, is your savings account in the local banking institution. According to one dictionary, a system is defined as a "complex whole set of connected things or parts, organized body of material or immaterial things."[1] We will generally live with this definition.

In this chapter we introduce several models for discrete systems and discuss techniques for finding system response using these models. This analysis is all achieved in the time domain—that is, without the use of transform methods. We begin by investigating the importance and application of system properties such as linearity, time invariance, causality, and stability. Linear time-invariant (LTI) systems are defined and will be used throughout the following chapters.

Next, LTI systems are described by linear constant-coefficient difference equations (DEs) with examples to illustrate the general nature of these systems. The characteristic equation and its roots are then determined and used in the initial condition (IC) solution. The forced solution of one basic system is calculated, but since this is normally a difficult task when using time-domain procedures, a thorough treatment of this topic is postponed until Chapter 8, "z Transforms and Applications."

The unit sample response, which plays a significant role in the analysis of linear systems, is defined and then applied in determining the response of a system to various inputs through the operation of discrete convolution. Convolution is then used to develop the steady-state system response to an important category of system input sequences, the ubiquitous sinusoidal sequence.

The chapter concludes with a discussion of the state-space or state-variable model for a discrete linear system. Formal methods of solution of these state equations are postponed until Chapter 11, but solution using a MATLAB program is illustrated.

[1] *The Concise Oxford Dictionary, 1st ed.,* Oxford Press, 1911.

BASIC CONCEPTS

PROPERTIES

Let us now investigate some properties and characteristics of discrete systems.

Linearity One of the most important concepts in system theory is linearity. Many accepted mathematical techniques can be applied to linear systems, which allow for easier and more convenient methods of analysis and design. In addition, many practical engineering systems may be modeled as being linear. Linear systems possess the property of superposition, which leads to the following definition: *A system is linear if and only if it satisfies the principle of homogeneity and the principle of additivity.*

1. *Homogeneity.* In Figure 7.3a we see that an input sequence $x_1(n)$ applied to a linear system produces the output sequence $y_1(n)$. When a new input $x(n) = C_1x_1(n)$ is applied to the same linear system, the output is $y(n) = C_1y_1(n)$. That is, if the input is scaled by the

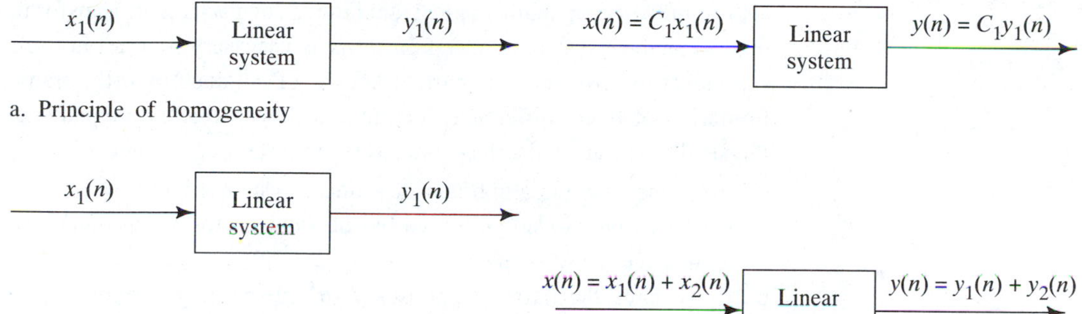

a. Principle of homogeneity

$x_2(n)$ Linear system $y_2(n)$

b. Principle of additivity

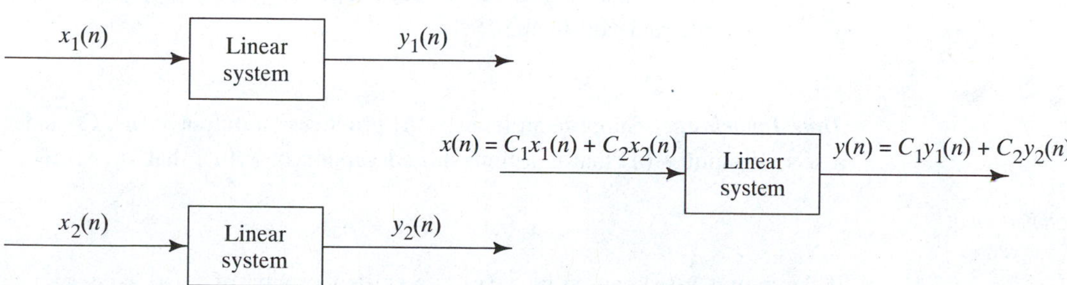

c. Principle of superposition

FIGURE 7.3 *Linearity*

complex constant C_1, the output will be scaled by the same constant. This characteristic is the homogeneity property.

2. *Additivity.* In Figure 7.3b we see that $x_1(n)$ again produces $y_1(n)$ and another input $x_2(n)$ produces $y_2(n)$ in our same linear system. If a new input $x(n) = x_1(n) + x_2(n)$ is now applied to the linear system, the output is $y(n) = y_1(n) + y_2(n)$. That is, the output due to the sum of two (or several inputs) is equal to the sum of the outputs produced by each separate input.
3. *Homogeneity and additivity.* Again with $x_1(n) \rightarrow y_1(n)$ and $x_2(n) \rightarrow y_2(n)$, as in Figure 7.3c, we make the input $x(n) = C_1x_1(n) + C_2x_2(n)$, combining the preceding items 1 and 2. In a linear system the output will be $y(n) = C_1y_1(n) + C_2y_2(n)$. Thus linear discrete systems satisfy homogeneity and additivity for all possible inputs $x_1(n)$ and $x_2(n)$ and for arbitrary complex constants C_1 and C_2. When both of these principles are satisfied, a system is said to satisfy the principle of superposition. A discrete system, therefore, is linear if and only if it satisfies the principle of superposition. A system that does not satisfy the principle of superposition is called *nonlinear.*
4. *Observations on linearity.* Linearity is a property of systems, but it is also a property of mathematical operators, such as the z transform, used to analyze systems. Indeed, linearity is necessary (but not in itself sufficient) to provide an important benefit of z transforms: the transformation of linear difference equations to linear algebraic equations. Physically, linearity also has some very important consequences:
   a. If we increase the amplitude of a linear system's input by a factor F, the output is also amplified by the same factor F but otherwise retains the same shape.
   b. If we have the output responses $y_1(n)$, $y_2(n)$ of a system to two inputs $x_1(n)$, $x_2(n)$, linearity guarantees that exactly the same terms will appear in the output, perhaps in different amounts, if we apply an input $x(n) = C_1x_1(n) + C_2x_2(n)$.
   c. Linearity also implies that if a system has several inputs or if it has one input that can be represented as a weighted sum of signals, we can find the outputs due to each input separately and then simply scale and add them.

Time Invariance Suppose an input $x_1(n)$ produces an output $y_1(n)$. Consider a second input $x_2(n)$ that is a time-shifted version of $x_1(n)$; that is,

shifted input

$$x_2(n) = x_1(n - n_0).$$

If the output $y_2(n)$ caused by $x_2(n)$ is a shifted version of $y_1(n)$—i.e.,

shifted output

$$y_2(n) = y_1(n - n_0)$$

for arbitrary $x_1(n)$ and n_0 and for all n—then the system is said to be a *time-invariant system.* Loosely speaking, the system parameters do not change with time. That is, the same input applied at different times will produce outputs that are identical in shape and size but shifted in time. A pictorial description of time invariance is given in Figure 7.4.

Linear Time-Invariant Systems When a system is both linear and time invariant, it is called a linear time-invariant (LTI) system, and it is amenable to analysis using many techniques, in particular, z transforms. LTI systems are typically described in the time domain by linear difference equations with constant coefficients. In the transform domain the equations are linear and *algebraic* in form. This means that we can use our (and the computer's) facility for doing algebra to study LTI systems.

Causality A causal (not casual!) system is a nonpredictive system in that the output does not precede, or anticipate, the input. See Figure 7.5 on page 368 for the output sequences of a causal, noncausal, and anticausal system, respectively, to a unit sample input $\delta(n)$. More is said about these three different kinds of systems in the next chapter.

Stability We will define system stability from many points of view, with the first being in terms of input-output behavior. If the system input is bounded (the input magnitude does not approach $\pm\infty$) and if the system is *stable,* then the output must also be bounded. This stability test is known as the Bounded-Input–Bounded-Output (BIBO) criterion. This is a difficult test to perform practically, for it implies that to be really certain that a system is stable, all possible bounded inputs must be applied if the output continues to meet the bounded test. It is quite difficult to have sufficient time to apply all possible inputs. Fortunately, there are other tests that can be applied.

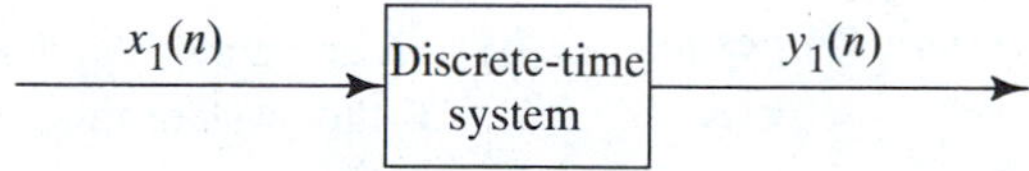

FIGURE 7.4 *Time-invariant system*

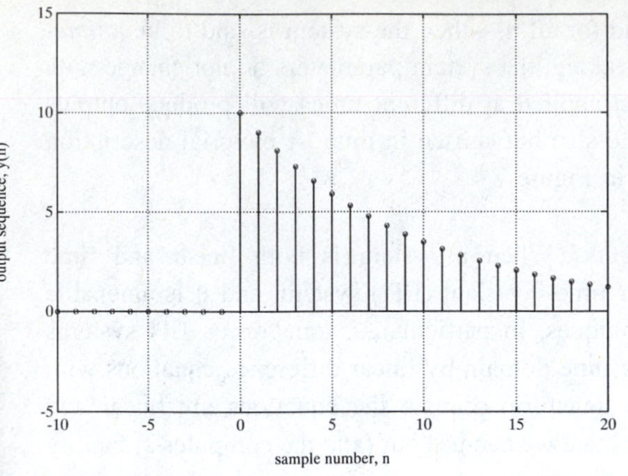

a. Causal system

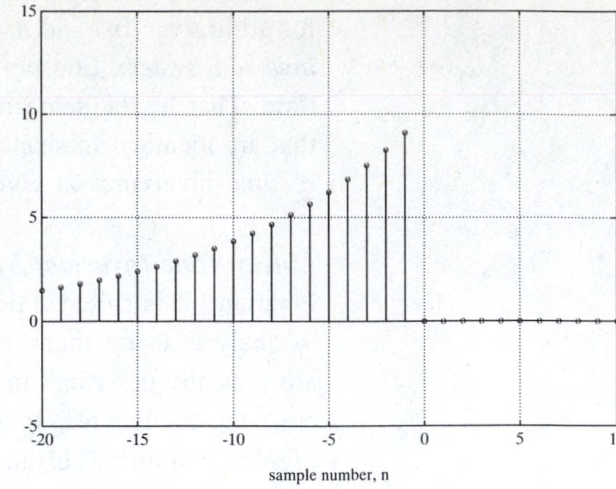

b. Anticausal system

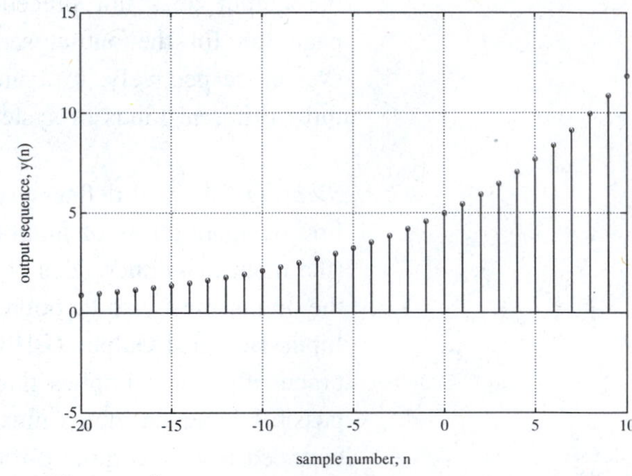

c. Noncausal system

FIGURE 7.5 *Responses to the input sequence $x(n) = \delta(n)$ for causal, anticausal, and noncausal systems*

ILLUSTRATIVE PROBLEM 7.1
System Properties

a. A discrete system is described by the input-output algorithm $y(n) = Kx(n) + A$, where K and A are real constants.

(i) Investigate this system for linearity using the tests illustrated in Figure 7.3.

(ii) Is this system time invariant or time varying?

b. A unit delay element (a shift register) in a discrete system is described by the equation $y(n) = x(n - 1)$. Is such a unit delay element causal or noncausal?

c. The relationship $y(n) = n|x(n)|^2$ describes the input-output relationship of a discrete system. Is this a stable system?

Suggestion: A step function $x(n) = Au(n)$ is a bounded input that you might want to use.

Solution

a. (i) For the input $x_1(n)$, the output is $y_1(n) = Kx_1(n) + A$, and for the input $x_2(n)$, the output is $y_2(n) = Kx_2(n) + A$. For the input $x_3(n) = \alpha_1 x_1(n) + \alpha_2 x_2(n)$, the output is $y_3(n) = Kx_3(n) + A = K\{\alpha_1 x_1(n) + \alpha_2 x_2(n)\} + A$. But $\alpha_1 y_1(n) + \alpha_2 y_2(n) = \alpha_1\{Kx_1(n) + A\} + \alpha_2\{Kx_2(n) + A\}$ and $y_3(n)$ does not equal $\alpha_1 y_1(n) + \alpha_2 y_2(n)$. The system is nonlinear.

Comment: It would be linear if $A = 0$.

(ii) An input $x_1(n)$ produces an output $y_1(n) = Kx_1(n) + A$ and a shifted input $x_2(n) = x_1(n - n_0)$ produces an output $Kx_1(n - n_0) + A$. But $y_1(n - n_0) = Kx_1(n - n_0) + A$, the same output for all n_0 so the system is time invariant (or shift invariant).

b. With $y(n) = x(n - 1)$, the output of the delay depends only on the immediate past input, and so the system is causal.

c. Using $x(n) = Au(n)$, the output sequence is $y(n) = n|x(n)|^2 = \{0, A^2, 2A^2, 3A^2, \ldots\}$, a sequence that is increasing without bound. The system is unstable.

BASIC CONCEPTS

*N*th-ORDER DIFFERENCE EQUATION MODEL

Discrete-time systems can be described by means of linear constant-coefficient difference equations. A compound-interest calculation is described by the difference equation

$$y(nT) = y(nT - T) + \frac{p}{100}y(nT - T) + x(nT)$$

where $y(nT)$ represents the money in an account at time $t = nT$, $y(nT - T)$ represents the money in the account at the previous time of computation, p is the percent interest paid in the time interval T, and $x(nT)$ represents the money deposited or withdrawn at the time sample $t = nT$. The time interval T, of course, may be measured in days, weeks, months, or whatever the policy of the financial institution.

In general, linear, constant-coefficient Nth-order difference equations (DEs)[2] for Single-Input–Single-Output (SISO) systems are written as

$$y(n) + a_1 y(n - 1) + \cdots + a_N y(n - N) = b_0 x(n) + b_1 x(n - 1) + \cdots + b_L x(n - L)$$

where we have accounted for L possible delays of the input and N possible delays of the output and there is no restriction on the relative sizes of the finite integers L and N. The notation $y(n)$ means the value of the sequence y at the

[2]Here, we use DE for difference equation, in contrast to DE for differential equation in Chapter 2, which deals with continuous-time systems. The expression "Diff Eqs" that has been around for a long time must be interpreted more carefully in these digital days, since we are getting more and more interested in difference equations at the expense of differential equations.

nth sample or at $t = nT$, but we drop the T to simplify the notation. A more concise representation of the Nth-order DE using finite summations is

recursive DE of order N

$$y(n) + \sum_{k=1}^{N} a_k y(n-k) = \sum_{k=0}^{L} b_k x(n-k).$$

The Nth-order difference equation just described represents an important class of discrete systems known as *recursive systems,* because the output depends on previous values of the output as well as on the input.

Another class of discrete systems is described by

nonrecursive DE

$$y(n) = \sum_{k=0}^{L} b_k x(n-k)$$

and is known as *nonrecursive* because the previous values of the output do not enter the calculations.

The *order* of the difference equation is equal to the number of delays N of the output sequence irrespective of the number of delays L of the input sequence. Thus the savings and loan account is described by a first-order equation, while a recursive filter given by $y(n) + a_1 y(n-1) + a_2 y(n-2) = b_0 x(n) + b_3 x(n-3)$ is second order because there are two delays of the output. Normally, we don't describe nonrecursive systems in terms of order but simply indicate the number of input delays.

Initial Condition Solution of a Difference Equation The *characteristic equation* for a recursive system is found by substituting a trial solution $y(n) = Cz^n$ into the homogeneous (input terms set to zero) difference equation

$$y(n) + \sum_{k=1}^{N} a_k y(n-k) = 0$$

with the result

$$Cz^n + a_1 Cz^{n-1} + a_2 Cz^{n-2} + \cdots + a_N Cz^{n-N} = 0 \quad \text{or}$$
$$Cz^n (1 + a_1 z^{-1} + a_2 z^{-2} + \cdots + a_N z^{-N}) = 0.$$

But Cz^n cannot be zero (this corresponds to the quite uninteresting "solution" $y(n) = 0$, $n = 0, 1, \ldots$) and so the equation to be satisfied is

$$1 + a_1 z^{-1} + a_2 z^{-2} + \cdots + a_N z^{-N} = 0.$$

This is known as the *characteristic equation* (CE) and can also be written

characteristic equation, CE

$$1 + \sum_{k=1}^{N} a_k z^{-k} = 0$$

where the N roots of this equation are called the *characteristic roots.* The CE in factored form becomes

$$(1 - r_1 z^{-1})(1 - r_2 z^{-1}) \cdots (1 - r_N z^{-1}) = 0$$

or in positive powers of z

factored CE

$$(z - r_1)(z - r_2)\cdots(z - r_N) = 0$$

where $z_1 = r_1,\ z_2 = r_2, \ldots, z_N = r_N$ are the characteristic roots that may be real or complex. If we are dealing with systems described by difference equations with real coefficients, complex roots must appear in conjugate pairs.

The solution of the homogeneous equation

$$y(n) + \sum_{k=1}^{N} a_k y(n - k) = 0$$

for a given set of initial conditions $y(-1), y(-2), \ldots, y(-N)$ is called the *initial condition* (IC) *solution* and for *distinct roots* has the form

$$y_{\text{IC}}(n) = C_1(r_1)^n + C_2(r_2)^n + \cdots + C_N(r_N)^n$$

for an Nth-order system. Thus for a system with no input(s), the output is given by

IC solution

$$y_{\text{IC}}(n) = \sum_{k=1}^{N} C_k(r_k)^n .$$

If the CE contains multiple roots, indicated by the factor $(z - r_k)^q$, terms of the form

multiple roots

$$C_1 n^{q-1}(r_k)^n + C_2 n^{q-2}(r_k)^n + \cdots + C_q(r_k)^n$$

will appear in the IC solution.

System stability is often defined in terms of this initial condition (IC) response; a causal LTI system is said to be stable if and only if the IC response decays to zero as $n \to \infty$. This happens if and only if all the r_k's have magnitudes *less than 1.*

ILLUSTRATIVE PROBLEM 7.2
Finding the Characteristic Equation, the Form of the Initial Condition Solution, and Checking Stability

A causal LTI system is described by the difference equation

$$y(n) - 0.25y(n - 1) - 0.125y(n - 2) = 3x(n) .$$

a. Find the characteristic equation.

b. Find the form of the IC response.

c. Determine if the system is stable.

Solution

a. The characteristic equation $1 - 0.25z^{-1} - 0.125z^{-2} = 0$ can be factored as

$$(1 - 0.5z^{-1})(1 + 0.25z^{-1}) = 0$$

and so the roots are $r_1 = 0.5$ and $r_2 = -0.25$.

BASIC CONCEPTS

b. The initial condition solution has the form

$$y_{IC}(n) = C_1(0.5)^n + C_2(-0.25)^n$$

where C_1 and C_2 can be found if $y(-1)$ and $y(-2)$ are given (known).

c. Since both characteristic roots have magnitudes less than 1.0, the causal system is stable.

The initial condition response can also be determined by an iterative procedure. For a given set of initial conditions, $y(-1), y(-2), \dots, y(-N)$, we start by computing $y(0)$; then we find $y(1)$ from $y(0), y(-1), \dots, y(-N+1)$; $y(2)$ from $y(1), y(0), \dots, y(-N+2)$; and so forth. This solution will not be in analytical form but will simply be a sequence of numbers $\{y(n)\}$ for $n \geq 0$.

CHAPTER 7

ILLUSTRATIVE PROBLEM 7.3
Two Methods for the Initial Condition Solution

Solve the homogeneous difference equation

$$y(n) - 0.25y(n-1) - 0.125y(n-2) = 0$$

with $y(-1) = 1$ and $y(-2) = 0$ by (a) iteration and (b) by finding an analytical solution of the form

$$y_{IC}(n) = \sum_{k=1}^{N} C_k(r_k)^n.$$

Solution

a. From the given difference equation we compute

$$y(0) = 0.25y(-1) + 0.125y(-2) = 0.25(1) + 0.125(0) = 0.25$$
$$y(1) = 0.25y(0) + 0.125y(-1) = 0.25(0.25) + 0.125(1) = 0.188$$
$$y(2) = 0.25y(1) + 0.125y(0) = 0.25(0.188) + 0.125(0.25) = 0.078$$

and so forth.

b. From Illustrative Problem 7.2 the IC solution is of the form

$$y_{IC}(n) = C_1(0.5)^n + C_2(-0.25)^n.$$

Using $y(-1) = 1$ and $y(-2) = 0$ yields the two simultaneous equations

$$\text{(i) } C_1(0.5)^{-1} + C_2(-0.25)^{-1} = 1 \quad \text{and}$$
$$\text{(ii) } C_1(0.5)^{-2} + C_2(-0.25)^{-2} = 0$$

or, after some algebra, $1 = 2C_1 - 4C_2$ and $0 = 4C_1 + 16C_2$. Solution by any number of ways gives $C_1 = 0.333$ and $C_2 = -0.083$. Thus

$$y_{IC}(n) = 0.333(0.5)^n - 0.083(-0.25)^n$$

and substituting $n = 0, 1, 2$ gives

$$y(0) = 0.333(0.50)^0 - 0.083(-0.25)^0 = 0.250$$
$$y(1) = 0.333(0.5)^1 - 0.083(-0.25)^1 = 0.188$$
$$y(2) = 0.333(0.5)^2 - 0.083(-0.25)^2 = 0.078$$

and so forth. These values agree with those in part (a).

Comment: The initial condition solution is decaying toward zero as the number of samples becomes large. That's good, because this system was found to be stable in Illustrative Problem 7.2.

Complete Solution of a Difference Equation Given an input sequence $x(n)$ and known initial conditions, $y(-1), y(-2), \ldots, y(-N)$, a difference equation can always be solved by iteration to obtain a tabulation of output values. There is, however, one common and important situation that can be solved easily in an analytical fashion. Consider the first-order system

$$y(n) + a_1 y(n-1) = b_0 x(n)$$

with the initial condition $y(-1) = 0$ and the constant input $x(n) = D$, $n \geq 0$. By iteration

$$\begin{aligned}
y(0) &= b_0 D, \\
y(1) &= b_0 D - a_1 b_0 D = b_0 D(1 - a_1) = b_0 D(1 + \alpha) \quad \text{with } -a_1 = \alpha, \\
y(2) &= b_0 D - a_1(b_0 D - a_1 b_0 D) = b_0 D(1 - a_1 + a_1{}^2) \\
&= b_0 D(1 + \alpha + \alpha^2), \\
&\vdots \\
y(q) &= b_0 D(1 + \alpha + \alpha^2 + \cdots + \alpha^q) = b_0 D \cdot \sum_{k=0}^{q} \alpha^k.
\end{aligned}$$

This finite geometric series can be written in closed form (see Appendix B) as

$$S = \sum_{k=0}^{q} \alpha^k = \frac{1 - \alpha^{q+1}}{1 - \alpha}, \qquad \alpha \neq 1$$

and, consequently, the solution to the difference equation at any sample n is

$$y(n) = b_0 D \, \frac{1 - \alpha^{n+1}}{1 - \alpha}.$$

Comment: If $\alpha = 1$, the finite geometric series is

$$S = \sum_{k=0}^{q} 1^k = q + 1,$$

that is, 1 added $q + 1$ times. This result can also be determined by applying l'Hôpital's rule to the closed form of the summation, giving

$$S = \frac{1 - \alpha^{q+1}}{1 - \alpha} \rightarrow \left. \frac{-(q+1)}{-1} \right|_{\alpha=1} = q + 1.$$

BASIC CONCEPTS

ILLUSTRATIVE PROBLEM 7.4
A Savings Account

Suppose Tom Robbins' daughter opens a savings account on her third birthday and deposits \$25 on the first of each month thereafter. Without using an iterative approach, determine the amount of money in the account on her sixth birthday if the interest rate is 9% per year, or 0.75% per month.

Solution

This savings account system is described by the first-order difference equation $y(n) = y(n-1) + ry(n-1) + b_0x(n) = [1 + r]y(n-1) + b_0x(n)$, where the system parameters are $r = 0.75/100 = 0.0075$ and $b_0 = 1$. Thus, in comparison with the general form of the difference equation $y(n) + a_1y(n-1) = b_0x(n)$, we have $a_1 = -1.0075$ ($\alpha = +1.0075$). The input is $x(n) = D = 25$ and $n = 35$ (we'll compute the savings just before she makes her sixth-birthday deposit). Using the closed form for the summation, we have

$$y(n) = b_0\, D\, \frac{1 - \alpha^{n+1}}{1 - \alpha}, \quad \text{with } y(35) = (1)\,(25)\,\frac{1 - (1.0075)^{36}}{1 - 1.0075} = \$1028.81\,.$$

From iterative methods we would obtain $y(0) = \$25$, $y(1) = (1.0075)\,(25) + 25 = \$50.19$, $y(2) = (1.0075) \cdot y(1) + 25 = \$75.54$, and so forth, until $y(35) = \$1028.80$.

Comment: There is a MATLAB function called **filter** (Signals and Systems or Signal Processing Toolbox) that can be used to verify (and plot) these results. To use this program the DE must be put in the form

$$\sum_{k=0}^{N} a_k y(n-k) = \sum_{k=0}^{L} b_k x(n-k) \quad \text{with } a_0 = 1\,.$$

The script is given below and the plot (Figure 7.6) on page 375, where we have also shown the cumulative deposit $d(n)$.

MATLAB Script

```
%F7_6 Solution to IP7.4 using filter
n=0:1:35;                      % evaluate for 36 samples
b=[1];                         % coefficients of input terms, the x(n)'s
a=[1,-1.0075];                 % coefficients of output terms, the y(n)'s
x=[25*ones(size(n))];          % constant input of 25
y=filter(b,a,x);               % call filter
d=25+n*25                      % cumulative deposit d(n)
%...plotting statements
```

For a more comprehensive discussion of the complete solution of difference equations, see Chapter 8, "z Transforms and Applications."

THE UNIT SAMPLE (IMPULSE) RESPONSE MODEL

Difference equations are one way of modeling linear, time-invariant, discrete-time systems. A second model is the unit sample (impulse) response, the output

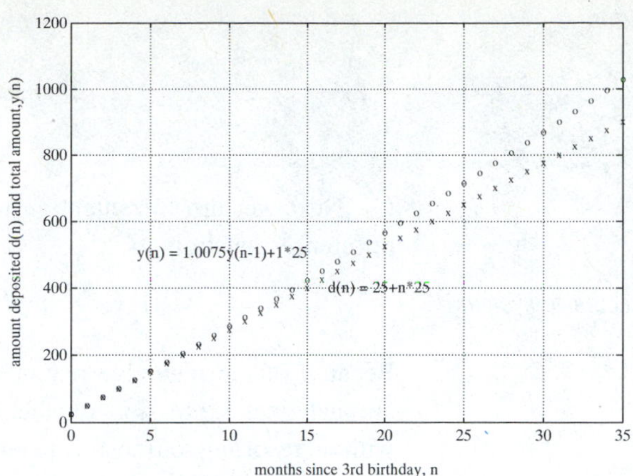

FIGURE 7.6 *Savings account history*

sequence $y(n)$ produced by applying an input of $x(n) = \delta(n)$ to the system. This output sequence is usually denoted as $h(n)$. That is, if $x(n) = \delta(n)$, then $y(n) = h(n)$.

For a causal, nonrecursive system described by the difference equation

nonrecursive system

$$y(n) = \sum_{k=0}^{L} b_k x(n - k)$$

subjected to the unit sample sequence of $x(n) = \delta(n)$, we see that the values of the output sequence are

$$\begin{aligned} y(n) &= 0, \; n < 0 \\ y(0) &= h(0) = b_0 \\ y(1) &= h(1) = b_1 \\ y(2) &= h(2) = b_2 \\ &\vdots \\ y(L) &= h(L) = b_L \\ y(L+1),\; y(L+2), \ldots &= 0. \end{aligned}$$

Notice that the values of the unit sample response $h(n)$ are the same as the values of the coefficients of the system's difference equation, namely,

$$y(n) = h(n) = \begin{cases} b_n, & 0 \le n \le L \\ 0, & n < 0 \text{ and } n > L. \end{cases}$$

This impulse response has a finite number of nonzero sample values; the general expression for the unit sample (impulse) response of this Finite Impulse Response (FIR) system is *h(n) for FIR system*

$$h(n) = \sum_{k=0}^{L} b_k \delta(n-k) = \sum_{k=0}^{L} h(k)\delta(n-k).$$

Next we turn our attention to causal recursive systems described by the difference equation

recursive system

$$y(n) + \sum_{k=1}^{N} a_k y(n-k) = \sum_{k=0}^{L} b_k x(n-k).$$

CHAPTER 7

Because of the recursive nature of the solution, it is very difficult to obtain an analytical expression for the unit sample response $h(n)$ for most systems without resorting to transform methods. Consequently, we postpone any serious discussion of this topic until Chapter 8 and consider only a simple solvable time-domain example at this time. For a first-order system with no delays of the input, given by

$$y(n) + a_1 y(n-1) = b_0 x(n),$$

we make $x(n) = \delta(n)$ and set $y(-1) = 0$. Then the values of the output sequence are

$$y(0) = h(0) = b_0,$$
$$y(1) = h(1) = -b_0 a_1,$$
$$y(2) = h(2) = b_0(-a_1)^2,$$
$$\vdots$$
$$y(q) = h(q) = b_0(-a_1)^q.$$

Thus, for this simple first-order system, the unit sample response $h(n)$ is

$$h(n) = b_0(-a_1)^n, \quad n \geq 0, \quad \text{or} \quad h(n) = b_0(-a_1)^n u(n).$$

In general, the unit sample response $h(n)$ of recursive, or Infinite Impulse Response (IIR), systems can be described by

h(n) for IIR system

$$h(n) = \sum_{k=1}^{N} C_k (r_k)^n + \sum_{k=0}^{L-N} A_k \delta(n-k),$$

where N is the order of the system, L is the number of delays of the input, and the r_k's are the system's characteristic roots, which are assumed to be different (distinct). The shifted samples appear in the second summation if L is equal to or exceeds the order of the system N. This result is developed in the next chapter, "z Transforms and Applications."

Earlier in this chapter system stability was defined in terms of the system's input and output sequences—that is, in a stable system a bounded input sequence produces a bounded output sequence. Turning to the unit sample response $h(n)$, we say that the unit sample response is absolutely summable in a stable system. That is,

stability definition

$$\sum_{n=-\infty}^{\infty} |h(n)| < \infty .$$

Since the unit sample response of a FIR system is given by

$$h(n) = \sum_{k=0}^{L} b_k \delta(n-k) = \sum_{k=0}^{L} h(k)\delta(n-k)$$

we see that FIR systems are always stable because

$$\sum_{n=-\infty}^{\infty} |h(n)| = \sum_{n=0}^{L} |h(n)| = \sum_{n=0}^{L} |b_n| = S, \qquad \text{a finite sum.}$$

For causal recursive systems it can be shown that the unit sample response is absolutely summable if the magnitudes of all of the characteristic roots are less than 1; that is, for a stable causal recursive system

$$|r_k| < 1, \quad k = 1, 2, \ldots, N .$$

This, of course, is the same result that was derived earlier in the discussion of the initial condition response.

ILLUSTRATIVE PROBLEM 7.5
Unit Sample Response and Stability

Find the unit sample response $h(n)$ for the two following systems. In each case determine if the system is stable.

a. A comb filter that can be used in radar signal processing applications described by the difference equation $y(n) = x(n) - x(n-8)$.

b. A causal first-order highpass filter modeled by $y(n) + 0.9y(n-1) = 2x(n)$.

Solution

a. The comb filter is a nonrecursive (FIR) system where $h(n) = b_n$ and so $h(n) = 1\delta(n) - 1\delta(n-8)$. This FIR system is stable, since

$$\sum_{n=-\infty}^{\infty} |h(n)| = 2 .$$

b. For the recursive filter $a_1 = 0.9$ and $b_0 = 2$ and by comparison with the expression derived earlier for a first-order system, $h(n) = 2(-0.9)^n u(n)$. Here, using the infinite geometric sum formula

$$\sum_{n=-\infty}^{\infty} |h(n)| = 2\sum_{n=0}^{\infty} |(-0.9)^n| = 2 \cdot 10 .$$

Hence it is a stable system. Or, from another point of view, the characteristic root is $r_1 = -0.9$, its magnitude is less than 1, and the causal system is stable. A plot, using **filter**, of the unit sample response $h(n) = 2(-0.9)^n u(n)$ is given in Figure 7.7, where it is evident that this is a summable response—that is, a stable system.

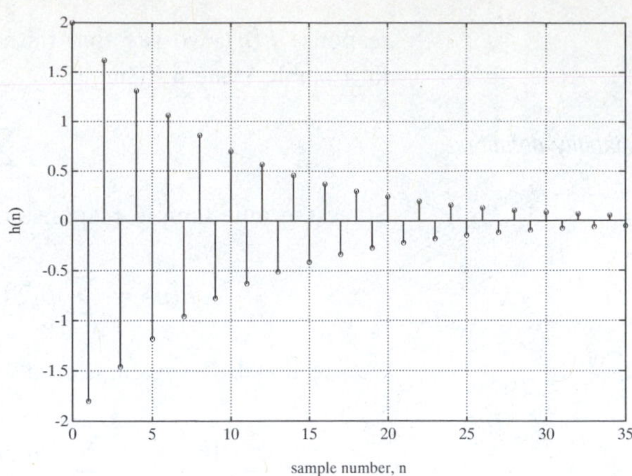

FIGURE 7.7 *Unit sample response for a first-order highpass filter*

CONVOLUTION

In Chapter 1 it was shown that an arbitrary sequence $f(n)$ can be represented as

sequence definition

$$f(n) = \sum_{m=-\infty}^{\infty} f(m)\delta(n - m)$$

which means that any sequence is equivalent to a sum of an appropriate number of sequences each consisting of a single sample function. The size (weight) of the sample function is given by $f(m)$ and its location, by the m in $\delta(n - m)$.

The Convolution Sum We will show that the output $y(n)$ of an LTI system due to a known input sequence $x(n)$ can be found in terms of the unit sample response $h(n)$. We will consider only the system's response due to the input $x(n)$ (its forced response), so all initial conditions are set to zero.

1. An LTI discrete system has the unit sample response $h(n)$. Using symbols,

unit sample response

$$\delta(n) \xrightarrow{\text{produces}} h(n)\,.$$

2. The system is time invariant, which means that a shifted input $\delta(n - m)$ will produce a shifted output $h(n - m)$, or

time or shift invariance

$$\delta(n - m) \xrightarrow{\text{produces}} h(n - m)\,.$$

3. Any sequence may be described in terms of shifted weighted samples, so the input $x(n)$ can be described by

input description

$$x(n) = \sum_{m=-\infty}^{\infty} x(m)\delta(n - m)\,.$$

4. Linearity consists of the two properties, (a) homogeneity and (b) additivity. Using the homogeneity property, a shifted sample of size $x(m)$ will produce a proportionate change in the size of the output; that is,

homogeneity

$$x(m)\delta(n - m) \xrightarrow{\text{produces}} x(m)h(n - m).$$

5. Finally, if we sum all the shifted samples that are the inputs, we have $x(n)$, and summing all the shifted samples that are the outputs caused by these inputs, we have the output $y(n)$. Described by symbols this becomes

additivity

$$x(n) = \sum_{m=-\infty}^{\infty} x(m)\delta(n - m) \xrightarrow{\text{produces}} y(n) = \sum_{m=-\infty}^{\infty} x(m)h(n - m).$$

Thus the forced output of an LTI discrete system with unit sample response $h(n)$ is given by the *convolution sum*

convolution sum

$$y(n) = \sum_{m=-\infty}^{\infty} x(m)h(n - m).$$

Consequently, if the unit sample response of an LTI system is known, the forced response caused by an arbitrary input $x(n)$ can be determined by evaluating the convolution sum. Furthermore, as we shall see, this summation is easily evaluated by computer, even for the usual situation in which inputs and unit sample responses cannot be described by analytical expressions. It is important to note that the convolution sum must be evaluated once for each value of n for which $y(n)$ is to be determined.

ILLUSTRATIVE PROBLEM 7.6
Evaluating the Convolution Sum

A noncausal LTI discrete system has the unit sample response $h(n)$ shown in Figure 7.8a (page 380). Find the output in response to the input $x(n)$ shown in Figure 7.8b (page 380).

Solution

Let $n = 0$. Then

$$\begin{aligned} y(0) &= \sum_{m=-\infty}^{\infty} x(m)h(0 - m) \\ &= x(-1)h(1) + x(0)h(0) + x(1)h(-1) \\ &= (3)(1) + (4)(2) + (-2)(-1) = 13 \end{aligned}$$

where we have ignored the values of m for which either $x(m)$ or $h(0 - m)$ or both are 0. Graphically, the evaluation we have just performed for $y(0)$ corresponds to

BASIC CONCEPTS

summing the product sequence $x(m)h(-m)$, known as the summand and shown in Figure 7.8c, over all values of m, which gives $0 + 3 + 8 + 2 + 0 + \cdots = 13$. Now let $n = 1$, in which case

$$\begin{aligned} y(1) &= \sum_{m=-\infty}^{\infty} x(m)h(1 - m) \\ &= x(0)h(1) + x(1)h(0) + x(2)h(-1) \\ &= (4)(1) + (-2)(2) + (1)(-1) = -1. \end{aligned}$$

Continuing this process leads to the final result shown in Figure 7.8d, which was obtained using the MATLAB function **conv** with the script on page 381.

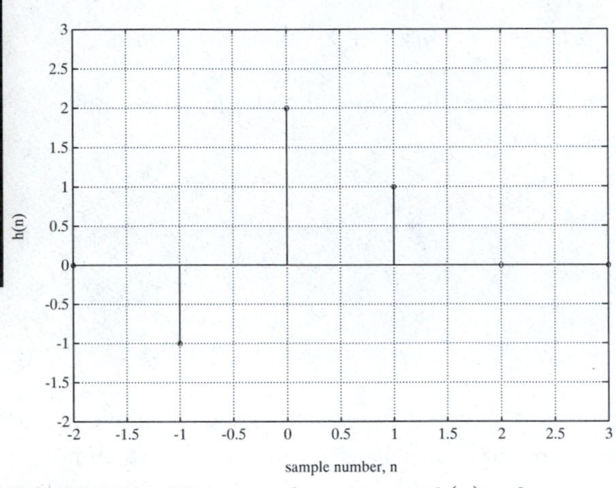

FIGURE 7.8a *Unit sample response, $h(n)$, of a system*

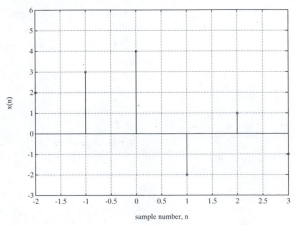

FIGURE 7.8b *Input sequence, $x(n)$*

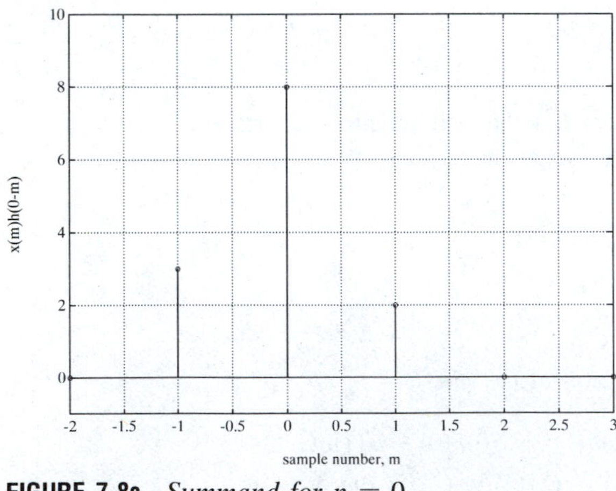

FIGURE 7.8c *Summand for $n = 0$*

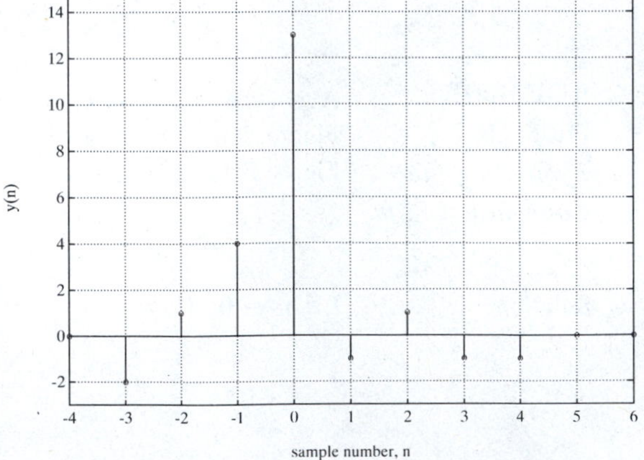

FIGURE 7.8d *Convolution of $h(n)$ and $x(n)$*

MATLAB Script

```
%F7_8 Evaluation of convolution for IP7.6

%F7_8d Convolution of h(n) and x(n)
n=-4:1:6;
h=[0,-1,2,1,0,0];
x=[2,3,4,-2,1,-1];
y=conv(h,x);
%...plotting statements
```


Important Comment: In Illustrative Problem 7.6, $h(n)$ contained $N_1 = 6$ samples and the input sequence $x(n)$ was also of length $N_2 = 6$ samples. The resultant convolution was 11 samples in length, or $(N_1 + N_2 - 1)$ samples, which is a general result.

An alternative form of the convolution sum can be obtained by the change of variable $i = n - m$, which gives

alternative form

$$y(n) = \sum_{i=-\infty}^{\infty} x(n-i)h(i) = \sum_{m=-\infty}^{\infty} x(n-m)h(m).$$

Convolution is a mathematical operation that applies to arbitrary sequences; that is, if $c_3(n)$ is the convolution of the sequences $c_1(n)$ and $c_2(n)$, then

$$c_3(n) = \sum_{m=-\infty}^{\infty} c_1(m)c_2(n-m) = \sum_{m=-\infty}^{\infty} c_2(m)c_1(n-m).$$

Notice that to perform the convolution of two sequences, one of the sequences must be written with a negative time index of $-m$—that is, $c_1(n-m)$ or $c_2(n-m)$. This process is also known as *folding* the sequence about the $m = 0$ index. Since it is immaterial which sequence is reversed or folded, we normally choose to fold the simpler sequence. To shorten the notation, the convolution operation is denoted using a "star" as

operational notation

$$c_3(n) = c_1(n) * c_2(n) = c_2(n) * c_1(n).$$

There are two ways to evaluate the convolution sum:

1. *Numerically.* If the sequences to be convolved are available as lists or tables of values, the evaluation of the convolution proceeds numerically as in Illustrative Problem 7.6. It is also relatively easy to write a computer program to do this.
2. *Analytically.* If the sequences are given as analytical expressions, it may be possible to find an analytical expression for the convolution result. This approach is feasible, however, in only a limited number of situations.

ILLUSTRATIVE PROBLEM 7.7
Analytical Convolution

a. Suppose the constant sequence $x(n) = A$, $-\infty \le n \le \infty$, is applied to an LTI system whose unit sample response is $h(n) = B\beta^n u(n)$ with $|\beta| < 1$. Find the equation for the system output $y(n)$.

b. Now find $y(n)$ for the input $x(n) = Au(n)$, a step sequence.

Solution

a. First, the better sequence to fold is $x(n)$, since $x(m) = x(-m)$, a characteristic of an even sequence. Then we see that for any value of n that shifts $x(-m)$ to the left or to the right, the product $x(n - m)h(m)$ is zero for $m < 0$ and nonzero for $m \ge 0$. Thus we have

$$y(n) = \sum_{m=0}^{\infty} AB\beta^m = \frac{AB}{1 - \beta}, \quad -\infty \le n \le \infty,$$

as in Figure 7.9a.

b. With $x(n) = Au(n)$, we again fold and shift $x(m)$ to obtain $x(n - m) = Au(n - m)$, as shown in Figure 7.9b for an arbitrary value of n. Shifting $x(-m)$ to the left for $n < 0$ makes $x(n - m)h(m)$ zero for all negative m. For $n > 0$ the product $x(n - m)h(m)$ is zero for $m < 0$ and for $m > n$. This gives

$$\begin{aligned} y(n) &= \sum_{m=0}^{n} AB\beta^m \\ &= AB\frac{1 - \beta^{n+1}}{1 - \beta}, \quad \text{for } n \ge 0 \quad \text{or} \quad y(n) = \frac{AB}{1 - \beta}[1 - \beta^{n+1}]u(n). \end{aligned}$$

This result is shown in Figure 7.9c.

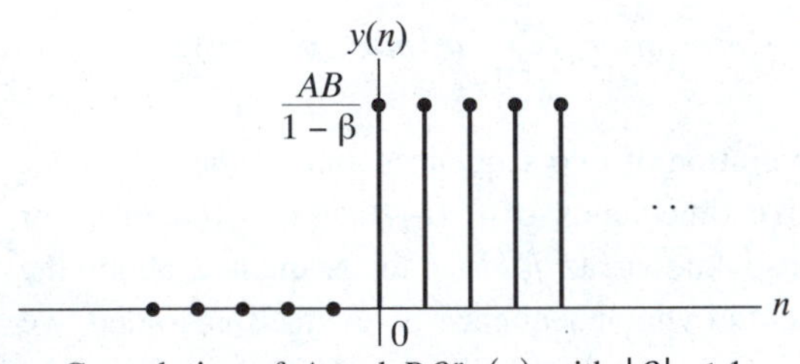

a. Convolution of A and $B\beta^n u(n)$ with $|\beta| < 1$

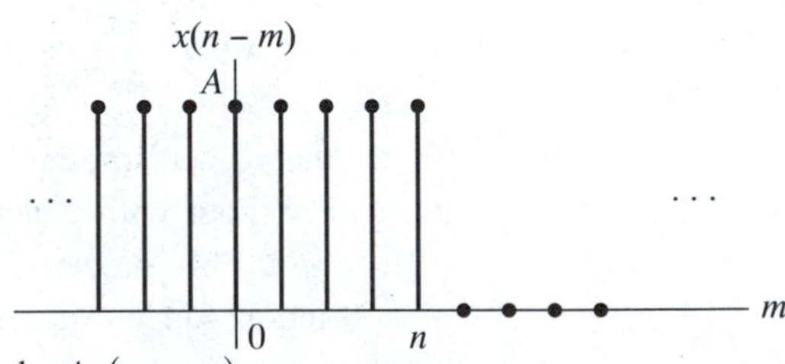

b. $Au(n - m)$ versus m

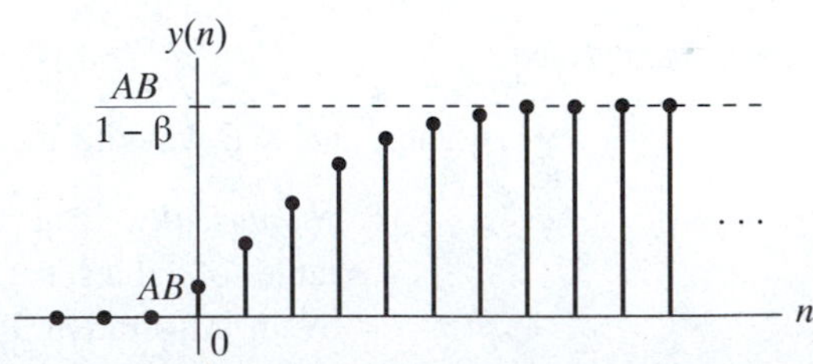

c. Convolution of $Au(n)$ and $B\beta^n u(n)$ with $|\beta| < 1$

FIGURE 7.9 *Illustrative Problem 7.7*

SINUSOIDAL STEADY-STATE RESPONSE

Let's use the convolution summation to develop an important characteristic of an LTI system, its steady-state response to a sinusoidal input sequence. Knowing that a sinusoidal sequence can be synthesized from two complex exponentials as $A\cos(n\theta) = 0.5A(e^{jn\theta} + e^{-jn\theta})$, we'll find the steady-state response to each exponential and then invoke linearity and add the results.[3] Given the unit sample response $h(n)$, the system output $y(n)$ is found from the convolution sum as

$$y(n) = \sum_{m=-\infty}^{\infty} h(m)x(n-m).$$

For the complex exponential input $x_1(n) = e^{jn\theta}$, $-\infty \le n \le \infty$, the steady-state output is

$$y_{1ss}(n) = \sum_{m=-\infty}^{\infty} h(m)e^{j\theta(n-m)}, \qquad -\infty \le n \le \infty$$

and $e^{jn\theta}$ can be taken outside the summation, giving the steady-state solution for a stable system of

$$y_{1ss}(n) = e^{jn\theta} \sum_{m=-\infty}^{\infty} h(m)e^{-jm\theta} = e^{jn\theta}H(e^{j\theta})$$

where the summation, which is a complex constant, is defined as

frequency response

$$\sum_{m=-\infty}^{\infty} h(m)e^{-jm\theta} = H(e^{j\theta}).$$

This complex constant $H(e^{j\theta})$ is known as the *frequency response function* or simply the *frequency response* of the system. For the complex exponential input $x_2(n) = e^{-jn\theta}$, $-\infty \le n \le \infty$, the steady-state output is found in the same manner as

$$y_{2ss}(n) = e^{-jn\theta} \sum_{m=-\infty}^{\infty} h(m)e^{jm\theta} = e^{-jn\theta}H(e^{-j\theta}).$$

Using linearity to sum the two responses gives

$$y_{ss}(n) = y_{1ss}(n) + y_{2ss}(n) = e^{jn\theta}H(e^{j\theta}) + e^{-jn\theta}H(e^{-j\theta}).$$

But $H(e^{j\theta}) = |H(e^{j\theta})|e^{j\angle H(e^{j\theta})}$ and $H(e^{-j\theta}) = |H(e^{j\theta})|e^{-j\angle H(e^{j\theta})}$, which gives

$$\begin{aligned} y_{ss}(n) &= e^{jn\theta}|H(e^{j\theta})|e^{j\angle H(e^{j\theta})} + e^{-jn\theta}|H(e^{j\theta})|e^{-j\angle H(e^{j\theta})} \\ &= |H(e^{j\theta})|\{e^{j(n\theta+\angle H(e^{j\theta}))} + e^{-j(n\theta+\angle H(e^{j\theta}))}\} \\ &= 2|H(e^{j\theta})|\cos(n\theta + \angle H(e^{j\theta})), \quad -\infty \le n \le \infty. \end{aligned}$$

[3]See "Sampling Theorem" on pp. 16–17 in Chapter 1 for a review of discrete frequency θ and sinusoidal sequences.

This is the output resulting from the system input

$$x(n) = x_1(n) + x_2(n) = e^{jn\theta} + e^{-jn\theta} = 2\cos(n\theta), \qquad -\infty \le n \le \infty.$$

Generalizing to a sinusoidal input amplitude of A and a phase of α, we see that when a sinusoidal sequence such as

sinusoidal input

$$x(n) = A\cos(n\theta + \alpha), \qquad -\infty \le n \le \infty,$$

is applied to a stable LTI system, the steady-state output sequence is

$$y_{ss}(n) = A|H(e^{j\theta})|\cos(n\theta + \alpha + \angle H(e^{j\theta})), \qquad -\infty \le n \le \infty.$$

This result makes the evaluation of a discrete system's steady-state response to a sinusoidal input sequence very straightforward, and we refer to the result as the sinusoidal steady-state formula. In other words, it tells us that when a sinusoidal sequence

$$x(n) = A\cos(n\theta + \alpha), \qquad -\infty \le n \le \infty,$$

is applied to a *stable* linear time-invariant system, then (1) the input amplitude A is multiplied by the gain $|H(e^{j\theta})|$ (magnitude of frequency response) and (2) the input phase α is shifted by the angle $\angle H(e^{j\theta})$ (phase of frequency response). This result also holds for a constant, or DC, input, since using $\theta = 0$ in the Euler[4] relation, $A\cos(n\theta) = 0.5A(e^{jn\theta} + e^{-jn\theta})$, yields $A\cos(n0) = 0.5A(e^{jn0} + e^{-jn0}) = A$.

ILLUSTRATIVE PROBLEM 7.8
Using the Sinusoidal Steady-State Formula

a. A nonrecursive digital comb filter is described by the unit sample response $h(n) = \delta(n) - \delta(n - 10)$. Find the response to the input sequence

$$x(n) = \cos(n\pi/10) + \cos(n\pi/5), \qquad -\infty \le n \le \infty.$$

b. A highpass recursive filter has the unit sample response $h(n) = (-0.9)^n u(n)$. Find the gain of this filter at $\theta = 0$ and $\theta = \pi$.

[4]Leonhard Euler (1707–1783), a key figure in eighteenth-century mathematics, was the son of a Lutheran pastor who lived near Basel, Switzerland. Euler received a master's degree in mathematics at the age of 16, and when only 19 he won a prize from the Academie des Sciences for a treatise on the most efficient arrangement of ship masts. He held appointments at the Academy of St. Petersburg and at the Royal Academy of Berlin, and for a while was a lieutenant in the Russian Navy. Euler, who became blind early in life, is accepted as being the most versatile and prolific writer in the history of mathematics. He wrote or dictated over 700 books and papers in his lifetime and was an inspiration to generations of younger mathematicians, including P. S. Laplace whose advice was, "Read Euler, he is our master in all."

Solution **a.** The frequency response is

$$H(e^{j\theta}) = \sum_{m=-\infty}^{\infty} [\delta(m) - \delta(m-10)]e^{-jm\theta} = 1 - e^{-j10\theta}.$$

For $\theta = \pi/10$ we have $H(e^{j\pi/10}) = 1 - e^{-j10\pi/10} = 2$ and for $\theta = \pi/5$ the frequency response takes on the value $H(e^{j\pi/5}) = 1 - e^{-j10\pi/5} = 0$. Thus $y_{ss}(n) = 2\cos(n\pi/10) + 0$. The input and output sequences are plotted in Figure 7.10, where it is seen that the filter blocks out the sinusoid of digital frequency $\theta = \pi/5$.

b. The system frequency response is

$$H(e^{j\theta}) = \sum_{m=0}^{\infty} (-0.9)^m e^{-jm\theta} = \frac{1}{1 + 0.9e^{-j\theta}}.$$

For the low frequency of $\theta = 0$, $H(e^{j0}) = 1/(1 + 0.9e^{-j0}) = 0.526$, and for the high frequency of $\theta = \pi$, $H(e^{j\pi}) = 1/(1 + 0.9e^{-j\pi}) = 10$. The filter is indeed highpass, as announced.

BASIC CONCEPTS

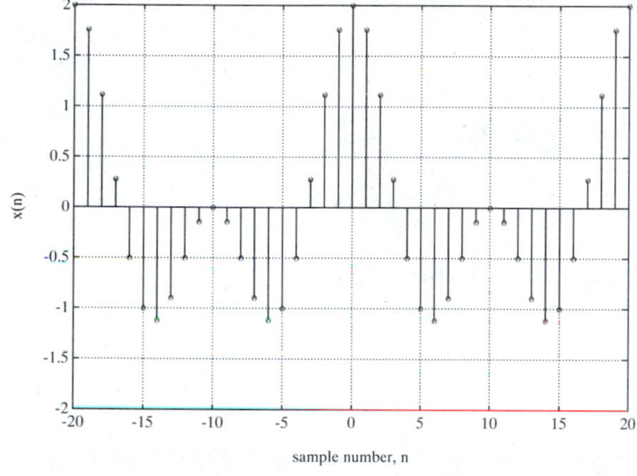

a. Input sequence

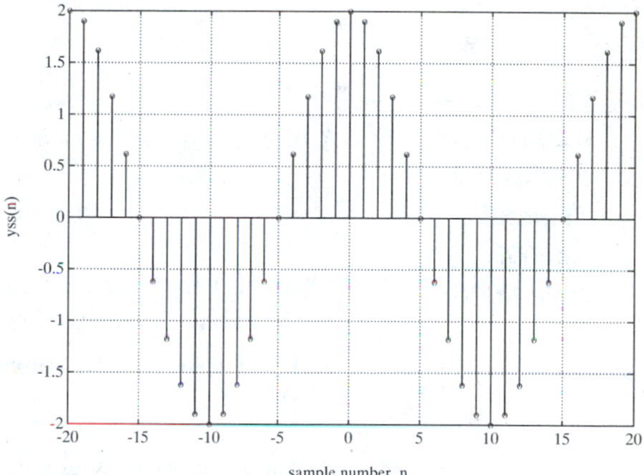

b. Steady-state output sequence

FIGURE 7.10 *Input sequence and steady-state output sequence for a highpass filter*

Alternative path to $H(e^{j\theta})$ Frequently we would like to find the frequency response directly from a system's difference equation model

$$y(n) + \sum_{k=1}^{N} a_k y(n-k) = \sum_{k=0}^{L} b_k x(n-k).$$

For the complex exponential input $x(n) = e^{j\theta n}$, the input delayed by k samples is $e^{j\theta(n-k)}$. Knowing that the steady-state solution for $x(n) = e^{j\theta n}$

is $y_{ss}(n) = e^{j\theta n}H(e^{j\theta})$, it follows that the output delayed by k samples is $e^{j\theta(n-k)}H(e^{j\theta})$. Substituting these expressions for $x(n)$ and $y(n)$ into the general difference equation gives

$$e^{j\theta n}H(e^{j\theta}) + \sum_{k=1}^{N} a_k e^{j\theta(n-k)}H(e^{j\theta}) = \sum_{k=0}^{L} b_k e^{j\theta(n-k)}$$

which can be solved for the frequency response $H(e^{j\theta})$ as

$$H(e^{j\theta}) = \frac{\sum_{k=0}^{L} b_k e^{-j\theta k}}{1 + \sum_{k=1}^{N} a_k e^{-j\theta k}}.$$

Thus the frequency response can be found from the DE coefficients and the appropriate powers of $e^{-j\theta}$. Remember that the system must always be checked for stability before applying the frequency response to obtain the steady-state response.

CHAPTER 7

ILLUSTRATIVE PROBLEM 7.9
Sinusoidal Analysis from the DE

A bandpass digital filter is described by the DE

$$y(n) + 0.81y(n-2) = x(n) - x(n-2)$$

where $y(n)$ represents the output sequence and $x(n)$ its input.

a. Determine the filter's frequency response.

b. Find the steady-state output for the input sequence $x(n) = 10 + 10\cos(n\pi/2) + 10\cos(n\pi)$.

Solution

a. The system characteristic roots are $z_{1,2} = \pm j0.90$, indicating a stable system. The input coefficients are $b_0 = 1$ and $b_2 = -1$, with the output coefficients being $a_1 = 0$ and $a_2 = 0.81$, giving the frequency response

$$H(e^{j\theta}) = \frac{1 - e^{-j2\theta}}{1 + 0.81e^{-j2\theta}}.$$

b. $H(e^{j0}) = 0$, $H(e^{j\pi/2}) = 10.53$, and $H(e^{j\pi}) = 0$, making the steady-state output $y_{ss}(n) = 105.3\cos(n\pi/2)$, where it is seen that the midfrequency of $\pi/2$ rad is amplified quite nicely.

Comment: In Chapter 9 we use the MATLAB function called **freqz** to plot a system's frequency response. Shown in Figure 7.11 is the frequency response for this filter, and we see very clearly its bandpass nature.

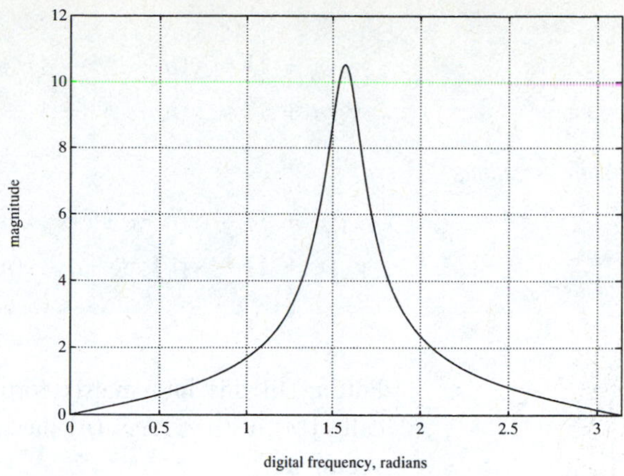

FIGURE 7.11 *Frequency response for a bandpass filter*

THE STATE-SPACE MODEL

A third description of linear discrete systems treated in this chapter is called the state-space, or first-order, model. Used extensively in computer simulation of systems, this model represents difference equations of arbitrary order as a set of first-order difference equations. Assume an Nth-order linear, constant-coefficient difference equation

$$y(n) + a_1 y(n-1) + a_2 y(n-2) + \cdots + a_N y(n-N) = b_0 x(n)$$

where $y(n)$ is the output, $x(n)$ is the input, and there are N delays of the system output and no (zero) delays of the system input.[5] To represent this system by N first-order equations, we define N new variables $v_1(n), v_2(n), \ldots, v_N(n)$ as

$$v_1(n) = y(n-N), \quad v_2(n) = y(n-N+1),$$
$$v_3(n) = y(n-N+2), \ldots, v_N(n) = y(n-1).$$

Then we can establish the following set of first-order difference equations, which are commonly called *state equations.*

[5] In Chapter 11 on state-space topics for discrete systems we will consider the general situation where the difference equation contains terms involving delays of the input, that is, $x(n-1)$, $x(n-2), \ldots, x(n-L)$.

state equations

$$\begin{aligned}
v_1(n+1) &= y(n-N+1) = v_2(n) \\
v_2(n+1) &= y(n-N+1+1) = v_3(n) \\
&\vdots \\
v_{N-1}(n+1) &= y(n-1) = v_N(n) \\
v_N(n+1) &= y(n) = -a_N v_1(n) - a_{N-1}v_2(n) - \cdots - a_1 v_N(n) + b_0 x(n).
\end{aligned}$$

Putting all this into matrix form, which is well suited for computation, we call $v_1(n)$, $v_2(n), \ldots, v_N(n)$ the *states,* or *state variables,*[6] and define the state vector as

state vector

$$\mathbf{v}(n) = [\, v_1(n) \quad v_2(n) \quad \ldots \quad v_N(n)\,]^T$$

where the superscript T stands for the matrix transpose operator. The matrix state equation is

matrix state difference equation

$$\begin{aligned}
\mathbf{v}(n+1) &= \begin{bmatrix} 0 & 1 & 0 & \cdots & 0 \\ 0 & 0 & 1 & 0 & \cdots \\ \vdots & \vdots & \vdots & \vdots & \vdots \\ 0 & 0 & 0 & \cdots & 1 \\ -a_N & -a_{N-1} & \cdots & -a_2 & -a_1 \end{bmatrix} \mathbf{v}(n) + \begin{bmatrix} 0 \\ 0 \\ \vdots \\ 0 \\ b_0 \end{bmatrix} x(n) \\
&= \mathbf{A}\mathbf{v}(n) + \mathbf{B}x(n).
\end{aligned}$$

A common definition for the state of a system is as follows:

The state of a system is a minimum set of quantities $v_1(n)$, $v_2(n), \ldots, v_N(n)$, which if known at $n = n_0$, are determined for $n > n_0$ by specifying the inputs to the system for $n \geq n_0$.

The *output* of this Single-Input–Single-Output system is related to the state vector $\mathbf{v}(n)$ and the single (scalar) input $x(n)$ by the *output equation*

$$y(n) = \mathbf{C}\mathbf{v}(n) + \mathbf{D}x(n)$$

where $\mathbf{C}$ is 1 by N and $\mathbf{D}$ is a scalar multiplier that may be written simply as d.

[6] Some disciplines prefer $x_1(n)$, $x_2(n), \ldots, x_N(n)$ rather than $v_1(n)$, $v_2(n), \ldots, v_N(n)$ to define the state variables. See the section "The State-Space Model" in Chapter 2 for our view of the notation situation.

ILLUSTRATIVE PROBLEM 7.10
Finding the State and Output Equations

Find the state and output equations for the third-order system

$$y(n) - 0.25y(n-1) - 0.125y(n-2) + 0.5y(n-3) = 3x(n)$$

where $y(n)$ is the output and $x(n)$ is the input.

Solution

Defining $v_1(n) = y(n-3)$, $v_2(n) = y(n-2)$, and $v_3(n) = y(n-1)$, we have $v_1(n+1) = v_2(n)$, $v_2(n+1) = v_3(n)$, and $v_3(n+1) = 0.25v_3(n) + 0.125v_2(n) - 0.5v_1(n) + 3x(n)$. In matrix form the state and output equations are

$$\mathbf{v}(n+1) = \begin{bmatrix} 0 & 1 & 0 \\ 0 & 0 & 1 \\ -0.5 & 0.125 & 0.25 \end{bmatrix} \mathbf{v}(n) + \begin{bmatrix} 0 \\ 0 \\ 3 \end{bmatrix} x(n) \quad \text{and}$$

$$y(n) = [\,-0.5 \quad 0.125 \quad 0.25\,]\mathbf{v}(n) + 3x(n).$$

BASIC CONCEPTS

SYSTEM SIMULATION

An important application of the state-space model is its use for the simulation of linear systems subjected to given input sequences and known initial conditions. Two functions that are available are the MATLAB function **dlsim** and the California function **ksdlsim**. The function **filter**, which was introduced in Illustrative Problem 7.4, may also be employed, but its use with nonzero initial conditions is best explained from a system diagram and, consequently, is considered in Chapter 8, where these diagrams are discussed. To illustrate the use of the state model, consider the system of Illustrative Problem 7.10 subjected to a step input with nonzero initial conditions; that is, we want to find $y(n)$ for $x(n) = u(n)$ with some arbitrary initial conditions, say $y(-1) = -1$, $y(-2) = -2$, and $y(-3) = -3$. Using the California function **ksdlsim** we have

$$[\mathbf{y}, \mathbf{v}] = \mathbf{ksdlsim}(\mathbf{A}, \mathbf{B}, \mathbf{C}, \mathbf{D}, \mathbf{X}, \mathbf{v0})$$

where it is seen that if the **A**, **B**, **C**, **D**, and **X** matrices are supplied along with the initial condition vector **v0**, **ksdlsim** will return the time histories **y** of the output and **v** of the states. The matrix **X** must have as many columns as there are inputs, where each row of **X** corresponds to a new time point. Following is a script that might be used (along with some comments), the input data as a check, and a plot of the results in Figure 7.12. Notice that this script returns the system data, the initial conditions, and a check of the size of **X**, a common place for errors. The initial values for the states are computed from $v_1(0) = y(-3) = -3$, $v_2(0) = y(-2) = -2$, and $v_3(0) = y(-1) = -1$.

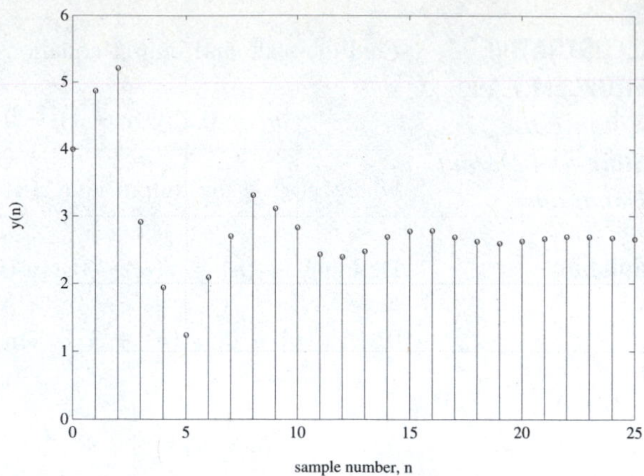

FIGURE 7.12 *Step response of a third-order system with nonzero initial conditions*

CHAPTER 7

MATLAB Script

```
%F7_12 Step response of a third-order system
A=[0,1,0;0,0,1;-0.5,0.125,0.25] % A matrix
B=[0;0;3]                        % B matrix
C=[-0.5,0.125,0.25]              % C matrix
D=[3]                            % D matrix
v0=[-3;-2;-1]                    % initial condition state vector
n=0:1:25;                        % start n:increment:stop n
X=[1*ones(size(n))]';            % input column vector
sx=size (X)                      % shows the dimensions of the X matrix
[y,v]=ksdlsim(A,B,C,D,X,v0);     % call ksdlsim
xlab='sample number, n';
ylab='y(n)';
ptitle='Fig.7.12 Step response';
response';
displot(n,y,xlab, ylab, ptitle);
axis([0,25,0,6]);                % sets graph scales [left, right, lower,
                                 % upper]
grid;                            % reset axis scaling to automatic
```

Output

```
A=   0    1.0000    0
     0         0    1.0000
 -0.5000  0.1250  0.2500
B=   0
     0
     3
C=-0.5000  0.1250  0.2500
D=   3
v0= -3
    -2
    -1
sx=       Note: The size of X is correct: 26 rows(26 time points) and
    26 1          1 column (1 input)
```

CROSS-CHECK

1. Let's see if the plot in Figure 7.12 seems reasonable by calculating a few points by iteration. From the given DE, $y(0) = 0.25y(-1) + 0.125y(-2) - 0.5y(-3) + 3x(0) = 0.25(-1) + 0.125(-2) - 0.5(-3) + 3 = 4.0$, $y(1) = 0.25y(0) + 0.125y(-1) - 0.5y(-2) + 3x(1) = 4.88$, and $y(2) = 0.25y(1) + 0.125y(0) - 0.5y(-1) + 3x(2) = 5.22$. These appear to agree with the plot, so the simulation is apparently correct.
2. Is this system stable? The system characteristic equation is $1 - 0.25z^{-1} - 0.125z^{-2} + 0.5z^{-3} = 0$, and the MATLAB function **roots** can be used to find the characteristic roots. The coefficients of the characteristic polynomial are entered as the vector $\mathbf{p} = [1 \quad -0.25 \quad -0.125 \quad 0.5]$. Invoking the function **r=roots(p)** returns the following data:

$$
\begin{aligned}
\mathbf{p} &= 1.0000 \quad -0.2500 \quad -0.1250 \quad 0.5000 \\
\mathbf{r} &= 0.5079 + 0.6284i \\
&\quad\; 0.5079 - 0.6284i \\
&\quad\; -0.7658
\end{aligned}
$$

With $|0.5079 \pm 0.6284i| = 0.808$ for the complex roots, we see that all the roots have magnitudes less than 1, and the causal system is stable. This appeared to be possible from the plot in Figure 7.12, because a bounded input gave a bounded output, but we can't be certain using the BIBO stability check unless all possible inputs are applied. That takes awhile. ■

EXAMPLE 7.1
Unit Sample Response of First- and Second-Order Systems

a. An LTI system is described by

$$y(n) + a_1 y(n-1) = b_0 x(n) + b_2 x(n-2).$$

(1) Find an analytical expression for the unit sample response.
(2) Check your answer by computing a few samples of $y(n) = h(n)$ directly from the given difference equation and comparing these values with those calculated from the solution for $h(n)$.

b. An oscillatory system is described by the difference equation

$$y(n) - y(n-1) + y(n-2) = 10x(n) - 5x(n-1).$$

Use iterative methods to find the first 10 values of the unit sample response $h(n)$ and show that your results fit the pattern $h(n) = A\cos(n\theta + \alpha)$.

WHAT IF? The system of part (a) can describe a filter of various types depending on its parameter values. In particular, for $b_0 = 1$, $b_1 = 0.01$, and $a_1 = -0.95$, it is a lowpass filter, whereas changing a_1 to 0.95 makes it a highpass filter. Find the first five values of the unit sample response in each case and relate the outputs to the lowpass and highpass nature described. ■

Solution

a. *First-order recursive system.*

(1) This is a good place to use the linearity and time-invariance properties discussed early in the chapter. We consider the system to be driven by two inputs, $x_1(n) = x(n)$ and $x_2(n) = x(n-2)$, find the response to each, and add the results. For the input

$$x_1(n) = \delta(n) \rightarrow y_1(n) = h_1(n) = b_0(-a_1)^n u(n)$$

and for the input

$$x_2(n) = \delta(n-2) \rightarrow y_2(n) = h_2(n) = b_2(-a_1)^{n-2} u(n-2)$$

where we have used the time-invariant property. Then, from linearity

$$h(n) = h_1(n) + h_2(n) = b_0(-a_1)^n u(n) + b_2(-a_1)^{n-2} u(n-2).$$

(2) Check:

From the difference equation	*From $h(n)$*
$y(0) = b_0$	$h(0) = b_0$
$y(1) = -a_1 b_0$	$h(1) = b_0(-a_1)$
$y(2) = a_1^2 b_0 + b_2$	$h(2) = b_0 a_1^2 + b_2$
$y(3) = -a_1^3 b_0 - a_1 b_2$	$h(3) = b_0(-a_1^3) + b_2(-a_1)$

b. *Iterative solution.* With $x(n) = \delta(n)$ and assuming $y(-1) = y(-2) = 0$, we have

$$y(0) = h(0) = 10$$
$$y(1) = h(1) = 10 - 5 = 5$$
$$y(2) = h(2) = 5 - 10 = -5$$
$$y(3) = h(3) = -5 - 5 = -10$$
$$y(4) = h(4) = -10 + 5 = -5$$
$$y(5) = h(5) = -5 + 10 = 5$$
$$y(6) = h(6) = 5 + 5 = 10$$
$$y(7) = h(7) = 10 - 5 = 5$$
$$y(8) = h(8) = 5 - 10 = -5$$
$$y(9) = h(9) = -5 - 5 = -10$$

The sequence is periodic with $N = 6$; trying $h(n) = 10\cos(n2\pi/6 + 0)$, we see that this expression will yield the correct sequence values.

E X A M P L E S

WHAT IF? By iterative solution we obtain the following:

Lowpass filter (LPF)	*Highpass filter (HPF)*
$y(0) = 1$	$y(0) = 1$
$y(1) = 0.95$	$y(1) = -0.95$
$y(2) = 0.9025$	$y(2) = 0.9025$
$y(3) = 0.8674$	$y(3) = -0.8674$
$y(4) = 0.8240$	$y(4) = 0.8240$

A unit sample $\delta(n)$ (it will be shown later) contains "all" frequencies in equal amplitudes. Lowpass filters allow only "slowly varying" signals to pass, whereas highpass filters allow only "rapidly varying" signals to pass. The sequences of $y(n) = h(n)$ just computed illustrate this characteristic. ■

EXAMPLE 7.2 ***Designing a Digital Oscillator***

We want to design a digital oscillator that will generate the sequence

$$y(n) = A\cos\left(\frac{n2\pi}{N} + \alpha\right), \quad n \geq 0$$

from the initial condition solution of the second-order difference equation

$$y(n) + a_1 y(n-1) + a_2 y(n-2) = 0.$$

a. Determine the required values for a_1 and a_2.

b. Using the values of a_1 and a_2 from part (a) find the initial conditions $y(-1)$ and $y(-2)$ that will produce arbitrary values of amplitude and phase, A and α, respectively. Your answers will be in terms of the period N, amplitude A, and phase α.

c. Sketch the sequence $y(n)$ for $N = 16$, $A = 10$, and $\alpha = 0$.

d. Find the initial conditions $y(-1)$ and $y(-2)$ to generate the sequence of part (c). Use the MATLAB function **dlsim** or the California function **ksdlsim** to verify your sketch from part (c).

CHAPTER 7

WHAT IF? Select appropriate values of a_1, a_2, $y(-1)$, and $y(-2)$ to generate some oscillatory sequences of your choice. For sine-wave zealots, one choice could be $y(n) = 20\sin(2\pi n/20)u(n)$. ■

Solution

a. *Calculation of a_1 and a_2.* From the difference equation, the characteristic equation is

CE from DE

$$1 + a_1 z^{-1} + a_2 z^{-2} = 0 \quad \text{or} \quad z^2 + a_1 z + a_2 = 0$$

and the desired time-domain solution is

$$y(n) = A\cos([n2\pi/N] + \alpha) = 0.5A\{e^{j[n2\pi/N]}e^{j\alpha} + e^{-j[n2\pi/N]}e^{-j\alpha}\}$$

which has the characteristic roots $r_{1,2} = 1e^{\pm j2\pi/N}$. This makes the desired characteristic equation

desired CE

$$(1 - e^{j[2\pi/N]}z^{-1})(1 - e^{-j[2\pi/N]}z^{-1}) = 0 \quad \text{or} \quad z^2 - 2\cos[2\pi/N]z + 1 = 0$$

and by equating coefficients of like powers of z, we have $a_1 = -2\cos[2\pi/N]$ and $a_2 = 1$.

b. *General expression for the initial conditions.* The ICs required to produce the arbitrary values of A and α may be found by first expanding

$$y(n) = A\cos([n2\pi/N] + \alpha) = A\{\cos[n2\pi/N]\cos\alpha - \sin[n2\pi/N]\sin\alpha\}.$$

Thus general expressions for the ICs of a sinusoidal sequence are given by

values of ICs

$$y(-1) = A\{\cos[(-1)2\pi/N]\cos\alpha - \sin[(-1)2\pi/N]\sin\alpha\}$$
$$y(-2) = A\{\cos[(-2)2\pi/N]\cos\alpha - \sin[(-2)2\pi/N]\sin\alpha\}.$$

c. *Plot for $N = 16$.* See Figure E7.2a for a sketch of $y(n) = 10\cos(n2\pi/16)$.

d. *Computer verification.* For $A = 10$, $\alpha = 0$, and a period of $N = 16$,

$$y(-1) = 10\{\cos[(-1)2\pi/16]\cos 0 - \sin[(-1)2\pi/16]\sin 0\} = 9.239$$
$$y(-2) = 10\{\cos[(-2)2\pi/16]\cos 0 - \sin[(-2)2\pi/16]\sin 0\} = 7.071\,.$$

The coefficients of the characteristic equation are $a_1 = -2\cos[2\pi/16] = -1.848$ and $a_2 = 1$. For the first-order model we let $v_1(n) = y(n-2)$ and $v_2(n) = y(n-1)$, which gives the matrix equation

$$\mathbf{v}(n+1) = \begin{bmatrix} 0 & 1 \\ -1 & 1.848 \end{bmatrix}\mathbf{v}(n) + \begin{bmatrix} 0 \\ 0 \end{bmatrix}x(n) \quad \text{with } \mathbf{v}(0) = \begin{bmatrix} 7.071 \\ 9.239 \end{bmatrix}$$

and the output equation $y(n) = [-1 \quad 1.848]\mathbf{v}(n)$.

A script that uses **ksdlsim** is given below, with the resulting plot in Figure E7.2b on page 396. Notice that the period is 16 samples and we have a cosine sequence of amplitude 10 with zero phase shift.

EXAMPLES

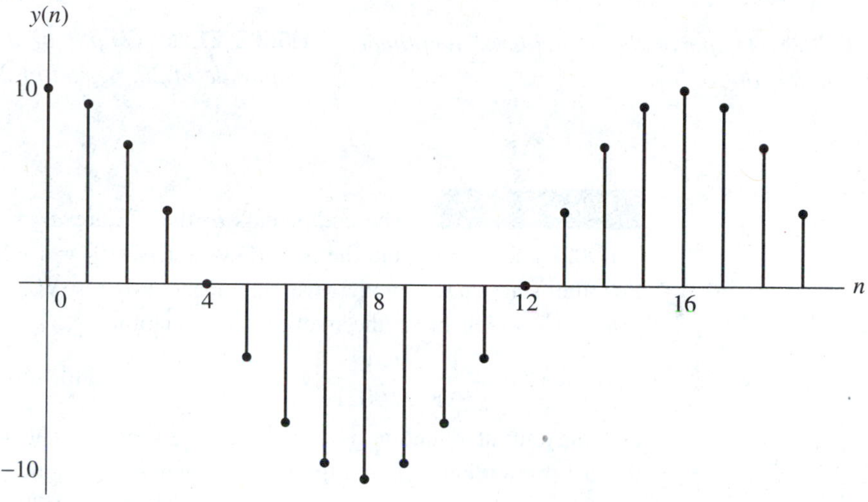

FIGURE E7.2a $y(n) = 10\cos(2\pi n/16)$

MATLAB Script

```
%E7_2 Designing a digital oscillator

%E7_2b Output of digital oscillator
A=[0,1;-1,1.848];                % A matrix
B=[0;0];                         % B matrix
C=[-1,1.848];                    % C matrix
D=[0];                           % D matrix
v0=[7.071;9.239];                % initial state vector
n=0:1:32;                        % start n:increment:stop n
X=[0*n]';                        % input column vector
[y,v]=ksdlsim(A,B,C,D,X,v0);     % call ksdlsim
%...plotting statements and pause
```

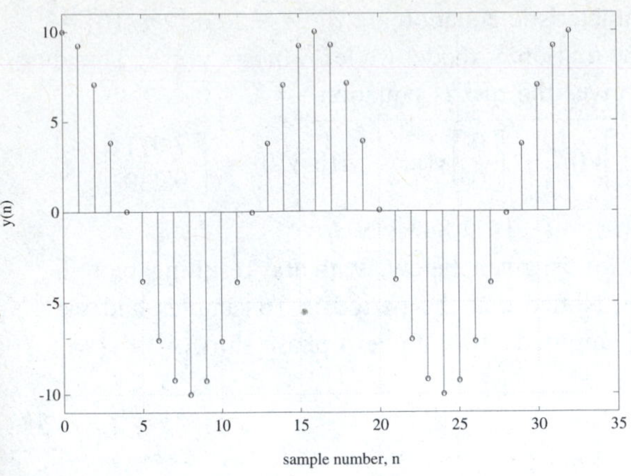

FIGURE E7.2b *Output of digital oscillator: amplitude of 10, period of 16*

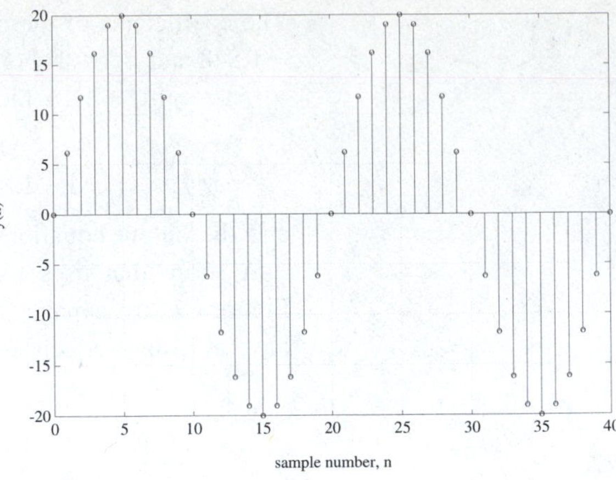

FIGURE E7.2c *Output of a digital oscillator: amplitude of 20, period of 20*

CHAPTER 7

WHAT IF? The coefficients of the CE are $a_1 = -2\cos[2\pi/20] = -1.9021$ and $a_2 = 1$ and the new IC vector is $\mathbf{v}(0) = [-11.7557, -6.1803]'$. For the state-space model we again let $v_1(n) = y(n-2)$ and $v_2(n) = y(n-1)$, which gives the matrix state equation

$$\mathbf{v}(n+1) = \begin{bmatrix} 0 & 1 \\ -1 & 1.9021 \end{bmatrix}\mathbf{v}(n) + \begin{bmatrix} 0 \\ 0 \end{bmatrix}x(n) \quad \text{with } \mathbf{v}(0) = \begin{bmatrix} -11.7557 \\ -6.1803 \end{bmatrix}$$

and the output equation $y(n) = [-1 \quad 1.848]\mathbf{v}(n)$. The plot resulting from the use of **ksdlsim** is in Figure E7.2c, where we notice that the period is 20 samples and the sine sequence has an amplitude of 20 with no phase shift. ■

MATLAB Script

```
%E7_2c Sinusoidal sequence
A=[0,1;-1,1.9021];               % A matrix
B=[0;0];                         % B matrix
C=[-1,1.9021];                   % C matrix
D=[0];                           % D matrix
v0=[-11.7557;-6.1803];           % initial state
n=0:1:40;                        % start n:increment:stop n
X=[0*n]';                        % input column vector
[y,v]=ksdlsim(A,B,C,D,X,v0);     % call ksdlsim
%...plotting statements
```

EXAMPLE 7.3
A Moving Average Filter

A very popular nonrecursive system known as a moving average filter comes from the study of time-series data generated by physical, economic, or biological processes. A running or moving average is obtained by accumulating the average over the previous L samples; then as each successive group of samples is taken, the new samples are included in the average and the oldest samples are removed. In this manner, the accumulated signal is always the average of the L most recent samples. The nonrecursive difference equation description is

$$
\begin{aligned}
y(n) &= \frac{1}{L}\{x(n) + x(n-1) + \cdots + x(n-[L-1])\} \\
&= \frac{1}{L}\sum_{k=0}^{L-1} x(n-k).
\end{aligned}
$$

EXAMPLES

a. Find the unit sample response $h(n)$ for this filter.

b. Suppose $L = 5$ and $x(n)$ is the 10-point sequence $x(n) = \{10, 11, 12, 11, 10, 9, 8, 8, 9, 10\}$ representing the winter midday temperature in degrees Celsius (C) on successive days in San Francisco, California as shown in Figure E7.3a on page 398. Find $y(n)$ for $n = 0$ to 9, assuming that $x(n) = 0$ for $n < 0$. Which values of $y(n)$ least accurately represent the average value of the input and why?

WHAT IF? You can try out the moving average concept by designing a filter that measures something important in your life, such as the number of hours that you study per day, the number of hours that you sleep per night, the number of glasses of water you drink per day, or whatever. ■

Solution

a. *Unit sample response.*

$$
x(n) = \delta(n), \quad y(0) = h(0) = \frac{1}{L},
$$

$$
y(1) = h(1) = \frac{1}{L}, \ldots, y(L-1) = h(L-1) = \frac{1}{L}
$$

and, in general,

$$
y(n) = h(n) = \begin{cases} \dfrac{1}{L}, & 0 \le n \le L-1 \\ 0, & n < 0 \quad \text{and} \quad n \ge L. \end{cases}
$$

b. *Calculating a moving average.* With

$$y(n) = \frac{1}{5}\sum_{k=0}^{4} x(n-k),$$

$$y(0) = \frac{1}{5}[x(0) + x(-1) + x(-2) + x(-3) + x(-4)]$$

$$= \frac{1}{5}[10 + 0 + 0 + 0 + 0] = 2$$

$$y(1) = \frac{1}{5}[x(1) + x(0) + x(-1) + x(-2) + x(-3)]$$

$$= \frac{1}{5}[11 + 10 + 0 + 0 + 0] = 4.2\,.$$

Completing the calculations gives the output sequence

$$y(n) = \{2, 4.2, 6.6, 8.8, 10.8, 10.6, 10.0, 9.2, 8.8, 8.8\}$$

which is plotted in Figure E7.3b using the MATLAB function **filter.** The first four values of the output $y(0)$ through $y(3)$, which are start-up values of $y(n)$, are not very good estimates of the average because of the assumed values of $x(n) = 0$ for $n < 0$.

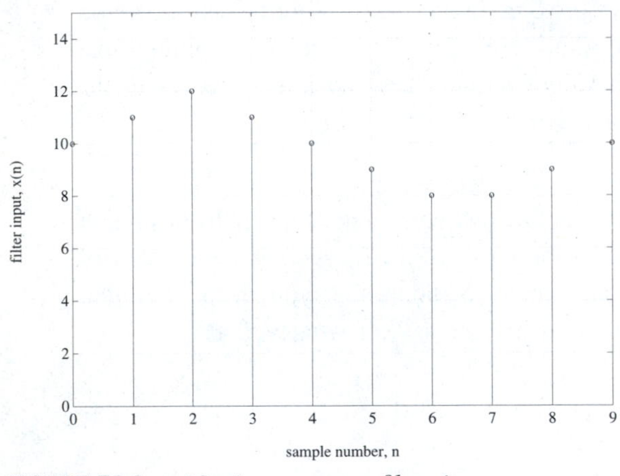

FIGURE E7.3a *Moving average filter input*

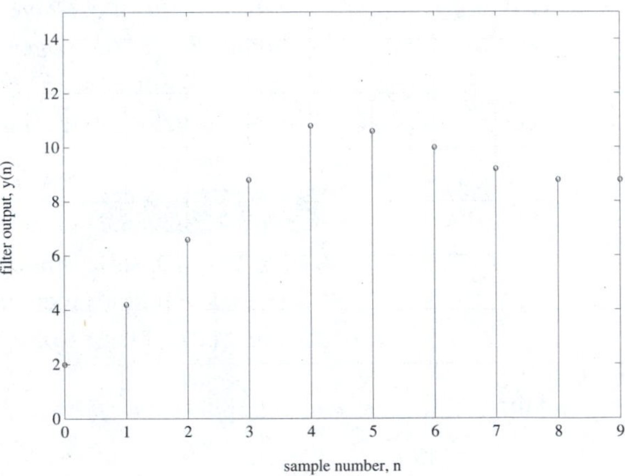

FIGURE E7.3b *Moving average filter output*

WHAT IF? A moving average filter was designed to obtain the running average of Professor Ed Thornton's golf scores. A filter length of 20 was decided upon; Ed's scores for the last 6 months (66 scores) were:

[96, 93, 92, 95, 89, 90, 94, 98, 101, 103, 98, 97, 94, 95, 96, 93, 88, 89, 96, 92, 97, 91, 93, 93, 91, 95, 96, 94, 93, 87, 88, 87, 86, 89, 87, 85, 88, 88, 87, 89, 91, 88, 86, 90, 84, 83, 88, 89, 95, 96, 95, 93, 91, 97, 95, 96, 96, 93, 95, 94, 92, 92, 95, 98, 100, 95]

1. A plot of Thornton's scores is given in Figure E7.3c, and the 20-point moving average is shown in Figure E7.3d. The filter smooths the input data but does not give reasonable answers until it "fills up" at the 20th sample. Notice that the vertical axis has been limited to values in the range 80 to 110 to emphasize the smoothing effect of the filter.

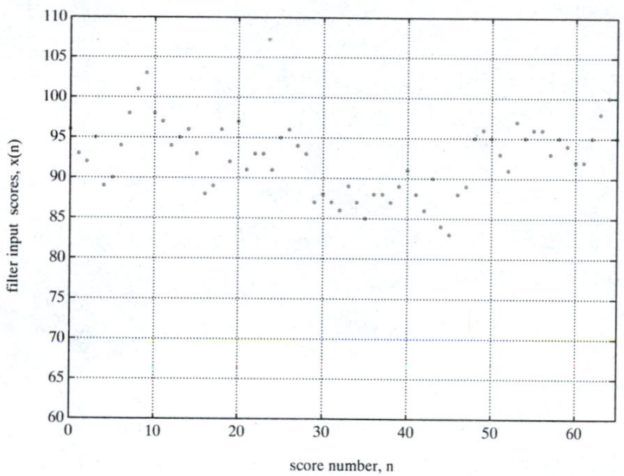

FIGURE E7.3c *Golf scores input to moving average filter*

FIGURE E7.3d *Moving average filter output with* $w(k) = 1$

2. In the filter just designed, each score was treated equally; that is, they weren't weighted. In order to make the average reflect the recent scores more than the earlier ones, it was decided to use the weighted average

$$y(n) = \frac{G}{L}\sum_{k=0}^{L-1} w(k)x(n-k)$$

where $L = 20$, $w(k)$ is a weighting factor and G is a gain factor used to normalize the calculation. The first choice was the linear sequence $w(k) = 1 - 0.05k$, which weights the current score $x(n)$ with a weight of $w(0) = 1$ and $x(n - 19)$ with $w(19) = 0.050$. The weights sum to 10.5, giving $10.5G = 20$ or $G/L = 0.0952$, with the resulting average scores in Figure E7.3e. In the figure, the wiggles reflect the usual ups and downs of a very amateur golfer.

3. Finally, an exponential weighting sequence was chosen with $w(k) = (0.86)^k$, giving $w(0) = 1$ and $w(19) = 0.057$, close to the values for the linear weights. Adding the weights again, we now have the scaling factor $G/L = 0.147$, with the weighted scores in Figure E7.3f. ■

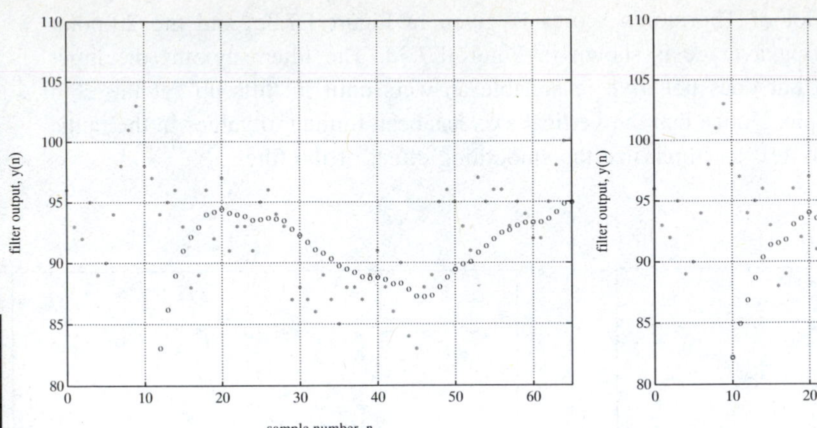

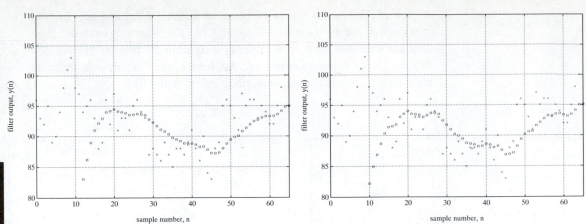

FIGURE E7.3e *Moving average filter output with* $w(k) = 1 - 0.05k$

FIGURE E7.3f *Moving average filter output with* $w(k) = (0.86)^k$

EXAMPLE 7.4 *Convolution Filters*

A digital filter is often modeled by its unit sample response $h(n)$. Another important characteristic of a filter is its response to a unit step input $u(n)$. For each of the following filters, use convolution to determine an analytical expression for the step response. Plot a few samples of these responses using a computer program such as MATLAB's **conv.**

a. A lowpass filter described by $h(n) = (0.90)^n u(n)$

b. A highpass filter described by $h(n) = (-0.90)^n u(n)$

c. A bandpass filter with $h(n) = 2(0.90)^n \cos(n\pi/2)u(n)$

WHAT IF? Suppose your supervisor suggests that you use **filter** rather than **conv** for this problem. Can this request be accommodated? ■

Solution

a. *Lowpass filter.* Using

$$y(n) = \sum_{m=-\infty}^{\infty} h(m)x(n-m)$$

we see that both the input $x(n) = u(n)$ and the unit sample response $h(n) = (0.90)^n u(n)$ equal zero for $n < 0$; consequently, the limits of the convolution summation become $m = 0$ and $m = n$ rather than $m = -\infty$ and $m = \infty$. The filter output is

$$y(n) = \sum_{m=0}^{n} (0.90)^m 1^{n-m} = \frac{1 - (0.90)^{n+1}}{1 - 0.90} = [10 - 9(0.90)^n]u(n).$$

See Figure E7.4a.

MATLAB Script

```
%E7_4 Filter outputs using convolution

%E7_4a Lowpass filter
n=0:1:30;
a=(0.9).^(n);
b=1*ones(size(n));
y=conv(a,b);
%...plotting statements and pause
```

EXAMPLES

b. *Highpass filter.* The limits are the same as in (a), so we have

$$y(n) = \sum_{m=0}^{n} (-0.90)^m 1^{n-m} = \frac{1 - (-0.90)^{n+1}}{1 - (-0.90)}$$
$$= [0.526 + 0.474(-0.90)^n]u(n).$$

See Figure E7.4b.

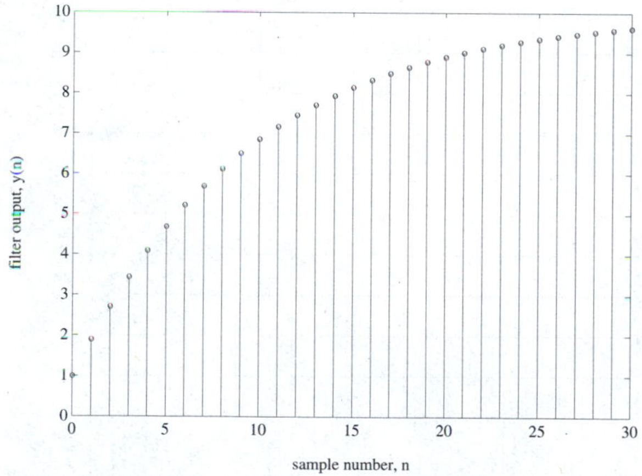

FIGURE E7.4a *Step response of a lowpass filter*

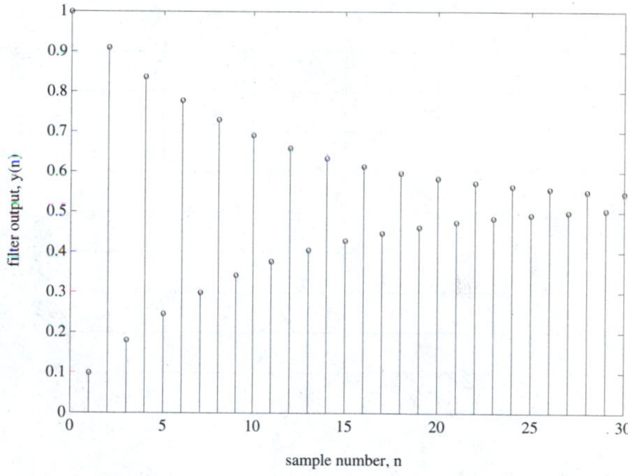

FIGURE E7.4b *Step response of a highpass filter*

MATLAB Script

```
%E7_4b Highpass filter
n=0:1:30;
a=(-0.9).^(n);
b=1*ones(size(n));
y=conv(a,b);
%...plotting statements and pause
```

c. *Bandpass filter.* With the same limits as in (a) we have

$$y(n) = \sum_{m=0}^{n} 2(0.90)^m \cos\left(\frac{m\pi}{2}\right)1^{n-m}$$

$$y(n) = \sum_{m=0}^{n} (0.90e^{j\pi/2})^m + \sum_{m=0}^{n} (0.90e^{-j\pi/2})^m$$

$$= \frac{1 - (0.90e^{j\pi/2})^{n+1}}{1 - 0.90e^{j\pi/2}} + \frac{1 - (0.90e^{-j\pi/2})^{n+1}}{1 - 0.90e^{-j\pi/2}}$$

$$= \{1.105 + 1.486(0.90)^{n+1} \cos(0.5\pi[n + 1] - 2.409)\}u(n).$$

See Figure E7.4c.

CHAPTER 7

MATLAB Script

```
%E7_4c Bandpass filter
n=0:1:50;
p=[(0.9).^n];
q=[2*cos(n*pi/2)];
a=p.*q;
b=1*ones(size(n));
y=conv(a,b);
%...plotting statements
```

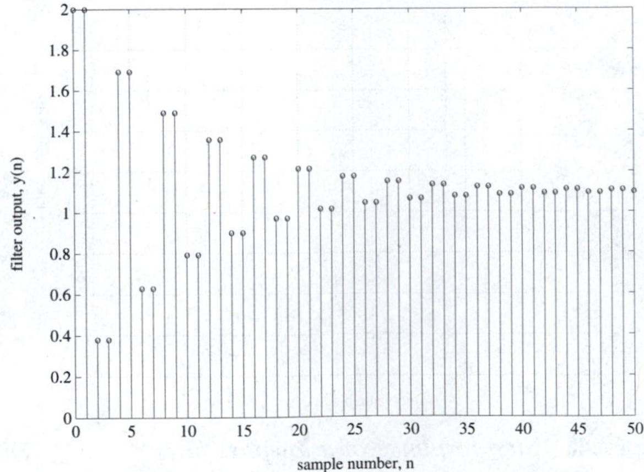

FIGURE E7.4c *Step response of a bandpass filter*

CROSS-CHECK There are plenty of opportunities for errors in the preceding convolution, so let's compare the calculations with the plot. First, for $n = 0$ we have $y(0) = 1.105 + 1.486(0.90)^1 \cos(0.5\pi - 2.409) = 1.105 + 1.337 \cos(-0.838) = 1.999$, which checks. Also, $y(1) = 1.105 +$

$1.486(0.90)^2 \cos(\pi - 2.409) = 2.000$; still OK. Finally, $y(\infty) = 1.105$, which appears to be reasonable from the plot; a data printout confirms the calculation. ■

WHAT IF? MATLAB's **filter** requires the DE model for the system. For the first-order systems of parts (a) and (b), we can guess (with help from Example 7.1) a DE of the form $y(n) + ay(n-1) = bx(n)$ and for $x(n) = \delta(n)$, the iterative solution is $y(0) = h(0) = b$, $h(1) = -ab$, $h(2) = a^2b$, $h(3) = -a^3b, \ldots.$ Thus for part (a) the DE is $y(n) - 0.9y(n-1) = x(n)$; for part (b) we have $y(n) + 0.9y(n-1) = x(n)$. A wizard might guess the DE for part (c) as $y(n) + 0.81y(n-2) = 2x(n)$ and be correct, but the rest of us will have to wait until Chapter 8 to find a logical and direct path to this result. ■

EXAMPLES

EXAMPLE 7.5
Cascade and Parallel Connections of Systems; Evaluation of the Impulse, Step, and Sinusoidal Steady-State Responses of the Resulting Systems

Shown in Figure E7.5a is a cascade connection of two linear systems. The input to the first system is $x(n)$, and its output is $w(n)$; the input to the second system is $w(n)$, and its output is $y(n)$. System 1 is modeled by $h_1(n) = a^n u(n)$ and system 2, by $h_2(n) = b^n u(n)$. Assume that a and b satisfy the condition $0 < a < b < 1$, which will ensure stability for each system and for the overall system.

a. For $x(n) = A\cos(n\pi/2)u(n)$, $-\infty \le n \le \infty$, what are the responses $w_{ss}(n)$ and $y_{ss}(n)$?

b. Determine the unit sample response $h(n)$ of the overall system. Plot $h(n)$ for $a = 0.80$ and $b = 0.90$.

c. Find the unit step response $y(n)$ of the overall system. Plot $y(n)$.

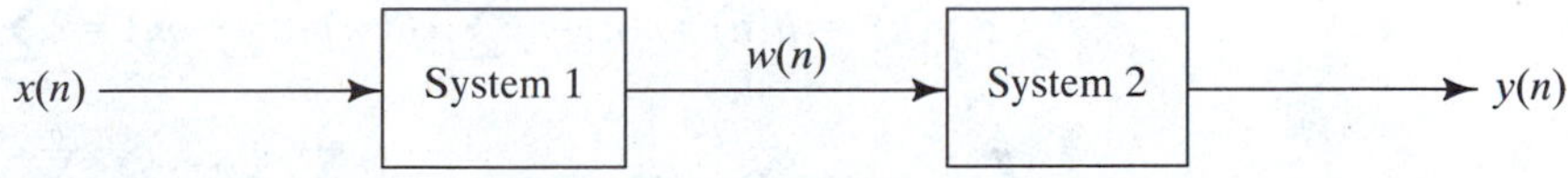

FIGURE E7.5a *Cascade connection*

WHAT IF? Investigate the parallel connection of these two systems. ■

Solution **a.** *Sinusoidal steady-state.* The frequency response for system 1 is

$$H_1(e^{j\theta}) = \sum_{m=0}^{\infty} a^m e^{-jm\theta} = \frac{1}{1 - ae^{-j\theta}} \quad \text{and, with } \theta = \pi/2,$$

$$H_1(e^{j\pi/2}) = \frac{1}{1 - ae^{-j\pi/2}} = \frac{1}{\sqrt{1 + a^2}} e^{-j \tan^{-1} a}.$$

Thus, using the sinusoidal steady-state formula

$$w_{ss}(n) = A \cdot \frac{1}{\sqrt{1 + a^2}} \cos(n\pi/2 - \tan^{-1} a), \quad -\infty \leq n \leq \infty.$$

Using the same method of analysis on system 2, we have

$$y_{ss}(n) = A \cdot \frac{1}{\sqrt{1 + a^2}} \cdot \frac{1}{\sqrt{1 + b^2}} \cos(n\pi/2 - \tan^{-1} a - \tan^{-1} b), \quad -\infty \leq n \leq \infty.$$

b. *Overall unit sample response* $h(n)$. The fact that $w(n) = x(n) * h_1(n)$ and that $y(n) = w(n) * h_2(n)$ yields $y(n) = [h_1(n) * h_2(n)] * x(n)$, and the overall unit sample response is $h(n) = h_1(n) * h_2(n)$. Because $h_1(n)$ and $h_2(n)$ are both causal, the convolution limits are $m = 0$ and $m = n$, giving

$$h(n) = \sum_{m=0}^{n} h_1(m) h_2(n - m) = \sum_{m=0}^{n} a^m b^{n-m} = b^n \sum_{m=0}^{n} (ab^{-1})^m$$

$$= \left[b^n \frac{1 - (ab^{-1})^{n+1}}{1 - ab^{-1}} \right] u(n) = \frac{1}{b - a} [b^{n+1} - a^{n+1}] u(n).$$

See Figure E7.5b for the plot of $h(n)$, the convolution of $h_1(n)$ and $h_2(n)$.

CROSS-CHECK From the convolution,

$$h(n) = \frac{1}{0.9 - 0.8} [(0.9)^{n+1} - (0.8)^{n+1}] u(n)$$

giving $h(0) = 1$ and $h(5) = 2.693$. Both values agree with the plot (to the scale of the plot), so there is a good chance that both the analytical and MATLAB solutions are correct. ■

c. *Step response of the total system.* Once again,

$$y(n) = h(n) * x(n) = \sum_{m=0}^{n} h(m) u(n - m) = \sum_{m=0}^{n} h(m) 1^{n-m}$$

$$= \sum_{m=0}^{n} h(m) = \frac{1}{b - a} \left[\sum_{m=0}^{n} b^{m+1} - \sum_{m=0}^{n} a^{m+1} \right]$$

$$= \frac{1}{b - a} \left[b \frac{1 - b^{n+1}}{1 - b} - a \frac{1 - a^{n+1}}{1 - a} \right] u(n)$$

$$= [C_1 + C_2 b^n + C_3 a^n] u(n).$$

See Figure E7.5c for $y(n)$, the convolution of $h(n)$ and $u(n)$.

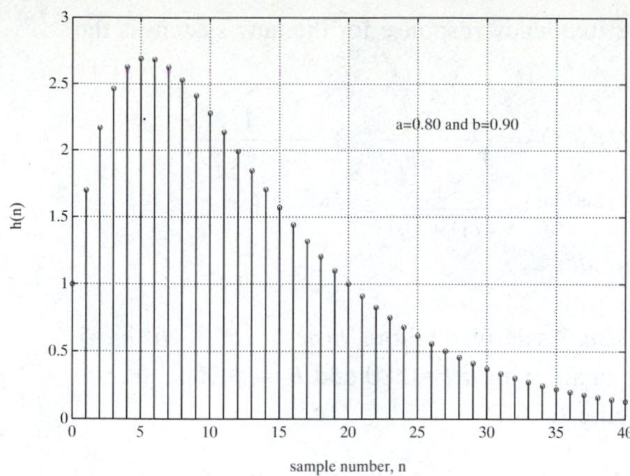

FIGURE E7.5b *Unit sample response for a cascade connection*

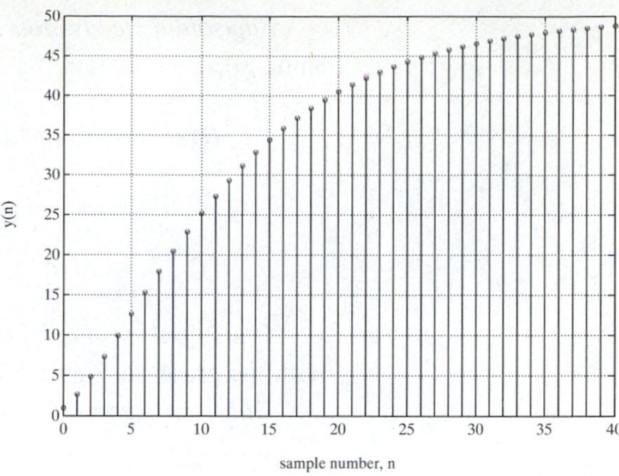

FIGURE E7.5c *Step response for a cascade connection*

EXAMPLES

CROSS-CHECK From the analytical solution with $a = 0.80$ and $b = 0.90$,

$$\begin{aligned} y(20) &= \frac{1}{0.9 - 0.8}\left[0.9\frac{1 - (0.9)^{21}}{1 - 0.1} - 0.8\frac{1 - (0.8)^{21}}{1 - 0.8}\right] \\ &= 90(1 - 0.109) - 41(1 - 0.009) \\ &= 40.550 \end{aligned}$$

which agrees with the plot in Figure E7.5c. ■

WHAT IF? See Figure E7.5d for the parallel connection of the two systems. Now $y(n) = y_1(n) + y_2(n) = h_1(n) * x(n) + h_2(n) * x(n) = [h_1(n) + h_2(n)] * x(n)$ and the system unit sample response for the parallel connection is $h(n) = h_1(n) + h_2(n)$. ■

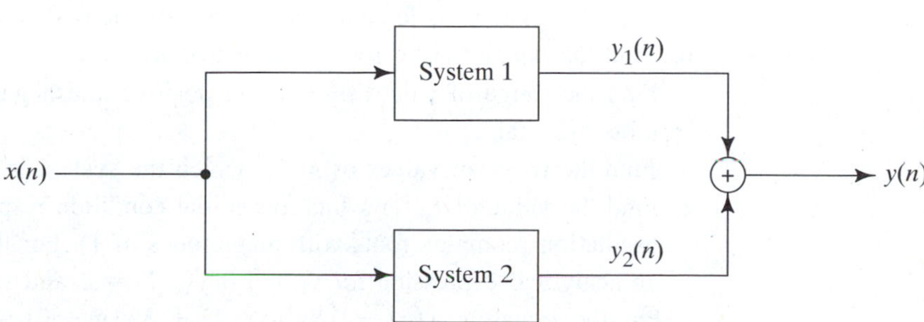

FIGURE E7.5d *Parallel connection*

Sinusoidal steady-state. The frequency response for the new system is the sum

$$H(e^{j\theta}) = H_1(e^{j\theta}) + H_2(e^{j\theta}) = \frac{1}{1 - ae^{-j\theta}} + \frac{1}{1 - be^{-j\theta}}$$
$$= \frac{1}{\sqrt{1 + a^2}} e^{-j\tan^{-1} a} + \frac{1}{\sqrt{1 + b^2}} e^{-j\tan^{-1} b}$$
$$= M_1 e^{jP_1} + M_2 e^{jP_2}.$$

Unit impulse (sample) response. As derived before, $h(n) = [a^n + b^n]u(n)$, with the graph in Figure E7.5e, again with $a = 0.80$ and $b = 0.90$.

Unit step response. See Figure E7.5f.

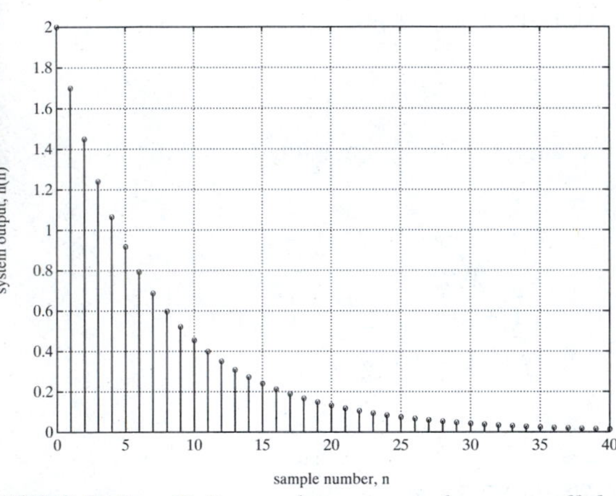

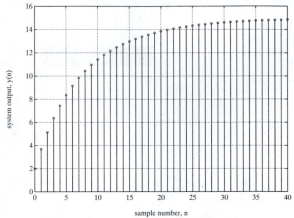

FIGURE E7.5e *Unit sample response for a parallel connection*

FIGURE E7.5f *Unit step response for a parallel connection*

REINFORCEMENT PROBLEMS

P7.1 Characteristics of a linear system. An LTI system is described by the difference equation

$$y(n) - 0.8y(n - 1) + Ky(n - 2) = 2x(n) + 3x(n - 2).$$

a. Find the characteristic equation in terms of the real constant K.

b. Find the characteristic roots in terms of K.

c. Find the range of values of K (both positive and negative) for which the roots are real.

d. Find the range of values of K for which the system is stable.

e. Find the value of K for which the initial condition response is a sustained oscillation (complex roots with magnitudes of 1). For this value of K find an analytical expression for $y_{IC}(n)$ if $y(-1) = 1$ and $y(-2) = 0$.

f. Put the equation $y(n) - 0.8y(n - 1) + Ky(n - 2) = 0$ into first-order (state-space) form.

g. Use **dlsim or ksdlsim** to verify the results of part (e) using the value of K determined for sustained oscillation.

P7.2 Initial condition response of a second-order system. An LTI system is described by the difference equation $y(n) - 0.9y(n-1) + 0.81y(n-2) = x(n) + x(n-2)$.

a. Is the system stable?

b. The initial condition response is of the form $y_{IC}(n) = Aa^n \cos(n\theta + \alpha)$. From the given data find θ and a.

Hint: Write $y_{IC}(n) = Aa^n \cos(n\theta + \alpha) = 0.5Ae^{j\alpha}(ae^{j\theta})^n + 0.5Ae^{-j\alpha}(ae^{-j\theta})^n$.

c. Find the values of $y(-1)$ and $y(-2)$ to make $A = 10$ and $\alpha = 0$ in the initial condition response of part (b).

d. Use MATLAB with the initial conditions from part (c) to plot the initial condition response. From the plot estimate the values of a and θ. Compare with the values found and used in parts (b) and (c).

P7.3 Initial condition response of a lowpass filter. A lowpass digital filter is modeled by the DE $y(n) - 1.7y(n-1) + 0.72y(n-2) = x(n) + 2x(n-1) + x(n-2)$.

a. Find the algebraic form of the initial condition solution $y_{IC}(n)$.

b. Assume $y(-1) = 0$ and $y(-2) = 1$ and find an analytical expression for the initial condition response.

c. Use a MATLAB function to generate the IC response of this filter and compare the analytical and simulation results.

P7.4 Classification of systems. Indicate whether each system whose difference equation is given is recursive or nonrecursive and, if appropriate, find the system order.

a. $y(n) - 0.7y(n-1) + y(n-2) - y(n-3) = 0.5x(n) - x(n-2)$

b. $y(n) - y(n-4) = x(n)$

c. $y(n) - 0.3y(n-2) = x(n) + 0.3x(n-1) + 0.5x(n-4)$

d. $y(n) - 0.2x(n-1) + 0.3x(n-2) = x(n)$

P7.5 Characteristic equation and characteristic roots. Find the characteristic equation and the characteristic roots for each of the LTI causal systems whose DEs are given.

a. $y(n) - y(n-2) = x(n) + 0.1x(n-2)$

b. $y(n) + y(n-2) = x(n) + 0.1x(n-2)$

c. $y(n) - 0.64y(n-1) + 0.70y(n-2) = x(n) - x(n-3)$

d. $y(n) - 0.64y(n-1) - 0.70y(n-2) = x(n) - x(n-3)$

e. $y(n) - 0.5x(n-2) + x(n-3) = -0.9x(n)$

f. $y(n) + 0.65y(n-1) - 0.35y(n-2) - 0.13y(n-3) = x(n)$

P7.6 System stability from the CE. For each of the systems described in Problem P7.5, indicate whether or not the system is stable and explain.

P7.7 Frequency response. For each of the systems described in Problem P7.5, determine the expression for the system frequency response $H(e^{j\theta})$.

P7.8 Unit sample (impulse) response. Find an analytical expression for the unit sample response for each of the following systems.

a. $y(n) - 0.2y(n-1) = x(n)$
b. $y(n) - 0.2y(n-1) = 10x(n)$
c. $y(n) + 0.2y(n-1) = x(n)$
d. $y(n) + 0.2y(n-1) = 100x(n)$
e. $y(n) - 1.0y(n-1) = x(n)$
f. $y(n) - 1.1y(n-1) = x(n)$
g. $y(n) - 1.1y(n-1) = x(n) + x(n-1)$
h. $y(n) + 1.1y(n-1) = 17x(n)$
i. $y(n) = 3x(n) - 2x(n-4) + x(n-5)$
j. $y(n) = x(n) - 10x(n-1) + 8x(n-2) - 6x(n-4) + 4x(n-6)$

P7.9 System stability from the unit sample response. For each of the systems described in Problem P7.8, indicate whether the system is stable and explain your reasoning.

P7.10 Sinusoidal steady-state response: The unit sample response of a certain first-order system is $h(n) = (0.707)^n u(n)$. Compute the steady-state response of this system to the input $x(n) = 1 + 2\cos(\pi n/4) + 3\cos(3\pi n/4)$.

P7.11 Convolutions. Find analytic expressions for and sketch with labels the results of the following convolutions: (a) $u(n) * u(n)$; (b) $u(-n-1) * u(-n-1)$; (c) $u(-n-1) * a^n u(n)$, $0 < a < 1$; (d) $u(n) * a^n u(n)$, $0 < a < 1$.

P7.12 Computer or state models. For each of the systems characterized by the given difference equations, find a state-variable (first-order) representation. Include an output equation.

a. $y(n) - y(n-2) = x(n)$
b. $y(n) + y(n-2) = x(n)$
c. $y(n) - 0.64y(n-1) - 0.7y(n-2) = x(n)$
d. $y(n) + 0.65y(n-1) - 0.35y(n-2) - 0.11y(n-3) = x(n)$

P7.13 Simulation of linearity test. Consider the system characterized by the DE $y(n) = y(n-1) - 0.5y(n-2) + x(n)$, where we assume $y(-1) = y(-2) = 0$.

a. Let $x_a(n) = 2\delta(n) - 3\delta(n-1) + 5\delta(n-2) + 1\delta(n-3)$ and find $y_a(n)$ for $n = 0, 1, \ldots, 6$.
b. Let $x_b(n) = Kx_a(n)$, where K is a constant of your choosing and find $y_b(n)$. How are $y_a(n)$ and $y_b(n)$ related? What principle does this relationship demonstrate?
c. Let $x_c(n) = -\delta(n) - 2\delta(n-1) - 4\delta(n-2) + 8\delta(n-3)$ and find $y_c(n)$ for $n = 0, 1, \ldots, 6$.
d. Let $x_d(n) = x_a(n) + x_c(n)$ and find $y_d(n)$ for $n = 0, 1, \ldots, 6$. If possible, determine a relationship for $y_d(n)$ in terms of $y_a(n)$ and $y_c(n)$.
e. Let $x_e(n) = Ax_a(n) + Cx_c(n)$, where A and C are arbitrary constants of your choosing. Select values for A and C and find $y_e(n)$ for $n = 0, 1, \ldots, 6$. Is there a relationship involving $y_e(n)$, $y_a(n)$, and $y_c(n)$? If so, what is it?

P7.14 Simulation of linearity test using MATLAB. Consider the same system as in Problem P7.13 and repeat all parts of the problem using (sensibly) a MATLAB program to perform the calculations. Since it costs nothing more, use $n = 0, 1, \ldots, 20$ rather than $n = 0, 1, \ldots, 6$.

P7.15 Stability tests. An LTI system is modeled by the DE $y(n) + 0.5y(n-2) = x(n)$.

a. Let $x(n) = 0$, $y(-2) = 1$, $y(-1) = 1$ and find $y(n)$ for $n = 0, 1, \ldots, 6$.
b. Repeat part (a) for $x(n) = 0$, $y(-2) = -1$, $y(-1) = 2$.
c. From the results of parts (a) and (b), do you think the system is stable? If so, are the results sufficient to ensure stability? Explain.
d. Now let $y(-2) = 0$, $y(-1) = 0$, and $x(n) = \delta(n)$. Find $y(n)$ for $n = 0, 1, \ldots, 6$.
e. Repeat part (d) for $x(n) = u(n)$.
f. Repeat part (d) for $x(n) = nu(n)$.
g. From the results of parts (d)–(f), do you think that the system is stable or unstable?
h. Find the roots of the system's characteristic equation. Is the system stable? If so, how do you explain the behavior of the system in part (f)?

PROBLEMS

P7.16 Stability tests using MATLAB. The system of Problem P7.15 is quite simple, and stability was easily determined by hand calculations. Let's consider some more complex systems and use three different avenues to find stability. We'll find $y_{IC}(n)$, $h(n)$, and the characteristic roots using MATLAB functions and then show that all three approaches give consistent results. The systems are modeled by the following DEs:

a. $y(n) + 0.8y(n-1) - 0.25y(n-2) - 0.20y(n-3) = x(n)$
b. $y(n) - 0.1y(n-1) - 1.21y(n-2) - 0.121y(n-3) = 0.5x(n)$
c. $y(n) + 0.81y(n-4) = x(n)$
d. $y(n) - 0.71y(n-2) - 0.605y(n-4) = x(n)$

For each system plot the IC response $y_{IC}(n)$, the unit sample response $h(n)$, and find the characteristic roots. Do all three items give the same stability results?

Comment: Following are scripts that could be used to plot the IC and unit sample responses for the first model, part (a), where the values of the initial conditions are, of course, arbitrary.

MATLAB Script

```
%A7_16 Stability tests using MATLAB

%A7_16a System (a) from Problem P7.16
A=[0,1,0;0,0,1;0.2,0.25,-.8];           % A matrix
B=[0;0;1];                              % B matrix
C=[0.2,0.25,-.8];                       % C matrix
D=[1];                                  % D matrix
v0=[10;0;0];                            % initial state
```

Continues

```
n=0:1:20;
X=zeros(size(n))';                         % input column vector
[y,v]=ksdlsim(A,B,C,D,X,v0);               % call ksdlsim
%...plotting statements
pause;
n=0:1:20;
b=[0,0,0,1];                               % coefficients of input terms
a=[1,0.8,-.25,-.2];                        % coefficients of output terms
x=zeros(size(n));
x(1)=1;
y=filter(b,a,x);                           % call filter
%...plotting statements and pause
```

CHAPTER 7

P7.17 Time-invariance test. For the system of Problem P7.13 described by $y(n) = y(n-1) - 0.5y(n-2) + x(n)$ with the input sequence $x_a(n) = 2\delta(n) - 3\delta(n-1) + 5\delta(n-2) + 1\delta(n-3)$, the output sequence was found to be $y_a(n) = 2\delta(n) - 1\delta(n-1) + 3\delta(n-2) + 4.5\delta(n-3) + 3\delta(n-4) + 0.75\delta(n-5)$.

a. For a new input $x_b(n) = x_a(n-2)$, find the new output $y_b(n)$.
b. For a still different input $x_c(n) = x_a(n-5)$, find the new output $y_c(n)$.
c. Is the information in (a) and (b) sufficient for you to certify that the system is or is not time invariant? Explain.

EXPLORATION PROBLEMS

P7.18 Lowpass and highpass filters: Two different digital filters are described by

(i) $y(n) + 1.8y(n-1) + 0.81y(n-2) = x(n) - 2x(n-1) + x(n-2)$,
(ii) $y(n) - 1.8y(n-1) + 0.81y(n-2) = x(n) + 2x(n-1) + x(n-2)$.

a. Find the characteristic roots for each filter.
b. Use **filter** from MATLAB to plot $h(n)$ for each filter.
c. Can you tell from the unit sample responses of part (b) which filter describes a lowpass filter and which describes a highpass filter? Explain your reasoning.
d. Verify your answer to (c) by finding $y_{ss}(n)$ for $x(n) = A + B\cos(n\pi/2) + C\cos(n\pi)$, where A is the amplitude of the sequence component at zero frequency, B is the amplitude of the sequence component at $\theta = \pi/2$ (the midfrequency of the filter), and C is the amplitude of the sequence component at $\theta = \pi$ (the highest frequency applied to the filter).

Hint: Assume $A = B = C = 1$ and compare the output amplitudes at these frequencies with the amplitudes of the corresponding frequencies in $x(n)$.

P7.19 Bandpass and bandstop filters. Two different digital filters are described by

(i) $y(n) + 0.81y(n-2) = x(n) - x(n-2)$,

(ii) $y(n) - 0.81y(n-2) = x(n) + x(n-2)$.

a. Find the characteristic roots for each filter.

b. Use **filter** from MATLAB to plot $h(n)$ for each filter.

c. Can you tell from the unit sample responses of part (b) which filter describes a bandpass filter and which describes a bandstop filter? Explain your reasoning.

d. Verify your answer to (c) by finding $y_{ss}(n)$ for $x(n) = A + B\cos(n\pi/2) + C\cos(n\pi)$, where A is the amplitude of the sequence component at zero frequency, B is the amplitude of the sequence component at $\theta = \pi/2$ (the midfrequency of the filter), and C is the amplitude of the sequence component at $\theta = \pi$ (the highest frequency applied to the filter).

Hint: Assume $A = B = C = 1$ and compare the output amplitudes at these frequencies with the amplitudes of the corresponding frequencies in $x(n)$.

P7.20 Design a loan. Consider the difference equation

$$p(n) = p(n-1) + \frac{r}{100}p(n-1) - X$$

where $p(n)$ is the principal owed at time nT, T is one payment period, r is the interest rate in percent for the payment period, and X is the periodic loan payment. Typically, T is one month. It can be shown that at the end of the Lth period, the remaining principal is

$$p(L) = \beta^L P - X\left[\frac{1-\beta^L}{1-\beta}\right]$$

where P is the initial loan amount and $\beta = [1 + r/100]$. If the loan is to be completely repaid in N time periods, then

$$X = \frac{1-\beta}{\beta^{-N} - 1}P.$$

Suppose you want to design loans for a variety of situations. One possibility is to obtain plots that display trends.

a. Write a MATLAB script to obtain plots of X versus N for a \$1000 loan for $N = 60$ months, 72 months, . . . , 360 months with annual interest rates of 5%, 6%, . . . , 12%, i.e., $r = 0.05/12, 0.06/12, \ldots$

b. (i) Suppose you want to buy a house and decide that you can afford a monthly payment of \$1200. If the lowest interest rate available for a 30-year loan is 8% annually, what is the largest loan you could obtain?

(ii) What if you could obtain a 7% loan?

c. Suppose you decide to consider the possibility of getting a 15-year loan and that you still could afford a \$1200-per-month payment.

(i) What's the largest loan you could afford if the annual interest rate is 8%?

(ii) What if the interest rate is 7%?

d. Suppose you wanted to have a loan amount the same as found in (i) of part (b) but wish to pay the loan off in 15 years.
   (i) How much would your payments be with an annual interest rate of 8%?
   (ii) What are the total amounts you pay with the 15-year loan of (i) of part (d) and the 30-year loan of (i) of part (b)?

P7.21 Convolution. Develop and run a MATLAB script that demonstrates the identity

$$[x_1(n) * x_2(n)] * x_3(n) = x_1(n) * [x_2(n) * x_3(n)].$$

CHAPTER 7

DEFINITIONS, TECHNIQUES, AND CONNECTIONS

PROPERTIES

Linear system

$$x_1(n) \rightarrow y_1(n)$$
$$x_2(n) \rightarrow y_2(n)$$
$$x_3(n) = C_1 x_1(n) + C_2 x_2(n) \rightarrow y_3(n) = C_1 y_1(n) + C_2 y_2(n)$$

where C_1 and C_2 are arbitrary constants and $x_1(n)$ and $x_2(n)$ are arbitrary sequences.

Time-invariant system

$$x_1(n) \rightarrow y_1(n)$$
$$x_2(n) = x_1(n - n_0) \rightarrow y_2(n) = y_1(n - n_0)$$

where $x_1(n)$ is an arbitrary sequence and n_0 is an arbitrary integer.

Causal system
The system output does not precede the system input.

Stable system

1. Bounded inputs produce bounded outputs.
2. The IC response decays to zero as $n \rightarrow \infty$, or, in terms of the characteristic roots, the roots (the r_k's) must have magnitudes less than 1.
3. $\sum_{-\infty}^{\infty} |h(n)| < \infty$.

*N*th-ORDER DIFFERENCE EQUATION MODEL

Nth-order DE $\quad y(n) + \sum_{k=1}^{N} a_k y(n - k) = \sum_{k=0}^{L} b_k x(n - k)$

Characteristic equation, CE $1 + \sum_{k=1}^{N} a_k z^{-k} = 0$

Factored CE $(z - r_1)(z - r_2)\cdots(z - r_N) = 0$
or $(1 - r_1 z^{-1})(1 - r_2 z^{-1})\cdots(1 - r_N z^{-1}) = 0$

Characteristic roots $z_1 = r_1,\ z_2 = r_2, \ldots, z_N = r_N$

Initial condition (IC) solution $y_{\mathrm{IC}}(n) = \sum_{k=1}^{N} C_k (r_k)^n$

DEFINITIONS TECHNIQUES

UNIT IMPULSE RESPONSE MODEL

an input $x(n) = \delta(n) \xrightarrow{\text{produces}}$ the output $y(n) = h(n)$

CONVOLUTION

Convolution sum

$$c_3(n) = c_1(n) * c_2(n) = \sum_{m=-\infty}^{\infty} c_1(m) c_2(n - m)$$

$$= c_2(n) * c_1(n) = \sum_{m=-\infty}^{\infty} c_2(m) c_1(n - m)$$

Operational notation $c_3(n) = c_1(n) * c_2(n) = c_2(n) * c_1(n)$

Linear system output

$$y(n) = \sum_{m=-\infty}^{\infty} x(m) h(n - m) = \sum_{m=-\infty}^{\infty} h(m) x(n - m)$$

Methods Analytical: use the summation.
Numerical: use a computer.

Length For $c_1(n)$ of length N_1 samples
and $c_2(n)$ of length N_2 samples,
$c_3(n) = c_1(n) * c_2(n)$ is of length $N_1 + N_2 - 1$ samples.

CHAPTER 7

SINUSOIDAL STEADY-STATE RESPONSE

For $x(n) = A\cos(n\theta + \alpha)$, $-\infty \le n \le \infty$,

$$y_{ss}(n) = A|H(e^{j\theta})|\cos(n\theta + \alpha + \angle H(e^{j\theta})), \qquad -\infty \le n \le \infty$$

where $\sum_{m=-\infty}^{\infty} h(m)e^{-jm\theta} = H(e^{j\theta})$.

Alternative method for finding the frequency response $H(e^{j\theta})$

$$y(n) + \sum_{k=1}^{N} a_k y(n-k) = \sum_{k=0}^{L} b_k x(n-k)$$

$$H(e^{j\theta}) = \frac{\sum_{k=0}^{L} b_k e^{-j\theta k}}{1 + \sum_{k=1}^{N} a_k e^{-j\theta k}}$$

STATE-SPACE OR FIRST-ORDER MODEL FOR THE DE

General definition The state of a system is a minimum set of quantities $v_1(n), v_2(n), \ldots, v_N(n)$, which if known at $n = n_0$ are determined for $n > n_0$ by specifying the inputs to the system for $n \ge n_0$.

Matrix difference equation $\mathbf{v}(n+1) = \mathbf{A}\mathbf{v}(n) + \mathbf{B}x(n)$

Output equation $y(n) = \mathbf{C}\mathbf{v}(n) + \mathbf{D}x(n)$

SUMMARY OF RESULTS FOR A CAUSAL FIRST-ORDER SYSTEM

Difference equation $y(n) + a_1 y(n-1) = b_0 x(n)$

Unit sample response $h(n) = b_0(-a_1)^n u(n)$

State-space model $v(n+1) = -a_1 v(n) + b_0 x(n)$, $y(n) = v(n+1)$

Frequency response ($\|a_1\| < 1$)	$H(e^{j\theta}) = \dfrac{1}{1 + a_1 e^{-j\theta}}$

MATLAB FUNCTIONS USED

Function	*Purpose and Use*	*Toolbox*
conv	Given: unit sample response model of discrete system, input, **conv** returns system output.	MATLAB, Signal Processing
displot	Plots discrete functions.	California Functions
dlsim	Given: state or TF model of discrete system, input, ICs, **dlsim** returns system output.	Control System, Signals/Systems
filter	Given: DE or TF model of discrete system, input, **filter** returns system output.	Signals/Systems, Signal Processing
ksdlsim	Given: state or TF model of discrete system, input, ICs, **ksdlsim** returns system output.	California Functions
roots	Given: coefficients of polynomial p, **roots** returns roots of $p = 0$.	MATLAB

BIBLIOGRAPHY

ANNOTATED BIBLIOGRAPHY

1. Gabel, Robert A., and Richard A. Roberts, *Signals and Linear Systems, 3rd ed.,* John Wiley & Sons, New York, 1987. *Chapter Two treats convolution, impulse response, the solution of nonhomogeneous difference equations, frequency response, and the formulation of state-variable equations in the time domain for discrete-time systems.*

2. Kwakernaak, Huibert, and Raphael Sivan, *Modern Signals and Systems,* Prentice Hall, Inc., Englewood Cliffs, N.J., 1991. *In Chapter Three the authors present an introduction to systems that includes input-output mapping, system properties, convolution, and response to harmonic inputs. Chapter Four includes the solution of difference equations and frequency response of discrete systems. Both chapters have many interesting problems and computer exercises.*

3. Oppenheim, Alan V., and Alan S. Willsky with Ian T. Young, *Signals and Systems,* Prentice Hall, Inc., New York, 1983. *Chapter Three presents difference equations and the convolution sum along with the corresponding concepts for continuous-time systems.*

4. Reid, J. Gary, *Linear System Fundamentals,* McGraw-Hill Book Company, New York, 1983. *Chapters Five and Eight of this text, which emphasizes difference equations that describe approximately continuous-time systems, address important discrete-time issues, including difference equations, stability, solution of difference equations, sinusoidal steady-state analysis, and state difference equations and their solution.*

5. Strum, Robert D., and Donald E. Kirk, *First Principles of Discrete Systems and Digital Signal Processing,* Addison-Wesley Publishing Company, Reading, Mass., 1988. *Chapter Three of this introductory text presents material on difference equations, convolution, interconnected systems, solution of difference equations, and stability.*

6. [Historical Reference] Steiglitz, Kenneth, *An Introduction to Discrete Systems,* John Wiley & Sons, New York, 1974 (out of print). *This significant book was designed to introduce students to electrical engineering in terms of discrete signals and algorithms rather than by the conventional approach of continuous signals and differential equations. Thus it serves as an introduction to the use of the digital computer for solving problems associated with systems. Includes discussion of complex numbers, digital signals and phasors, z transforms, and moving average and recursive digital filters.*

ANSWERS

P7.1 a. $1 - 0.80z^{-1} + Kz^{-2} = 0$; b. $z_{1,2} = 0.40 \pm (0.16 - K)^{1/2}$; c. For negative values of K the roots will always be real. For positive values of K the roots will be real for $K \leq 0.16$; d. $-0.2 < K < 1$; e. For $K = 1$, $y_{IC}(n) = 1.092\cos(1.159n + 0.747)$; f. $v_1(n + 1) = v_2(n)$, $v_2(n + 1) = -Kv_1(n) + 0.8v_2(n)$, and $y(n) = -Kv_1(n) + 0.8v_2(n)$; g. For the script and plot, see A7_1 on the CLS disk. Notice that $y(n)$ is a sustained oscillation but it is *not* periodic.

CROSS-CHECK From part (e), $y(0) = 1.092\cos(0.747) = 0.80$, $y(1) = 1.092\cos(1.096) = -0.36$, $y(10) = 1.092\cos(12.337) = 1.06$. The values agree with the plot. ■

P7.2 a. $z_{1,2} = 0.900e^{\pm j1.047}$, $|z_{1,2}| < 1$, so the causal system is stable. b. $\theta = \pi/3$; $a = 0.90$; c. $y(-1) = 5.556$; $y(-2) = -6.173$; d. From the plot (A7_2), $\theta = \pi/3$ and $10(a)^6 \approx 0.55$, giving $a = 0.905$; from the DE, $a = 0.9$ and $\theta = \pi/3$.

P7.3 a. $y_{IC}(n) = A(0.8)^n + B(0.9)^n$; b. $y_{IC}(n) = 5.76(0.8)^n - 6.48(0.9)^n$; c. See A7_3 on the CLS disk.

P7.4 a. Recursive, third order; b. Recursive, fourth order; c. Recursive, second order; d. Nonrecursive

P7.5 a. $1 - z^{-2} = 0$, $z_{1,2} = 1, -1$; b. $1 + z^{-2} = 0$, $z_{1,2} = \pm j1 = 1e^{\pm j\pi/2}$; c. $1 - 0.64z^{-1} + 0.70z^{-2} = 0$, $z_{1,2} = 0.320 \pm j0.773 = 0.837e^{\pm j1.178}$;

d. $1 - 0.64z^{-1} - 0.70z^{-2} = 0$, $z_{1,2} = -0.576, 1.216$; e. Nonrecursive system, CE doesn't apply. f. $1 + 0.65z^{-1} - 0.35z^{-2} - 0.13z^{-3} = 0$, $z_1 = -0.880$, $z_2 = -0.286$, $z_3 = 0.516$

P7.6 a. Unstable, magnitude of roots = 1; b. Unstable, same as (a); c. Stable, both roots have a magnitude less than 1; d. Unstable, $|z_2| = 1.216$; e. Stable, a nonrecursive system; f. Stable, all roots have a magnitude less than 1.

P7.7 a. Unstable, not meaningful; b. Unstable, not meaningful; c. $H(e^{j\theta}) = (1 - e^{-j3\theta})/(1 - 0.64e^{-j\theta} + 0.70e^{-j2\theta})$; d. Unstable, not meaningful; e. $H(e^{j\theta}) = -0.9 + 0.5e^{-j2\theta} - e^{-j3\theta}$; f. $H(e^{j\theta}) = 1/(1 + 0.65e^{-j\theta} - 0.35e^{-j2\theta} - 0.13e^{-j3\theta})$

P7.8 a. $h(n) = 0.2^n u(n)$; b. $h(n) = 10 \cdot 0.2^n u(n)$; c. $h(n) = (-0.2)^n u(n)$; d. $h(n) = 100 \cdot (-0.2)^n u(n)$; e. $h(n) = 1^n u(n) = u(n)$; f. $h(n) = (1.1)^n u(n)$; g. $h(n) = (1.1)^n u(n) + (1.1)^{n-1} u(n-1)$; h. $h(n) = 17 \cdot (-1.1)^n u(n)$; i. $h(n) = 3\delta(n) - 2\delta(n-4) + \delta(n-5)$; j. $h(n) = \delta(n) - 10\delta(n-1) + 8\delta(n-2) - 6\delta(n-4) + 4\delta(n-6)$

P7.9 All answers are based on the stability criterion of $\sum |h(n)| < \infty$. Using this or any other criterion, (e), (f), (g), and (h) are unstable systems; the others are stable.

P7.10 $y_{ss}(n) = 3.413 + 2.828 \cos([n\pi/4] - 0.785) + 1.897 \cos([3\pi n/4] - 0.322)$

P7.11 a. $nu(n) + u(n)$; b. $-nu(-n-1) - u(-n-1)$; c. $[(a)/(1-a)]a^n u(n) + [1/(1-a)]u(-n-1)$; d. $[(1 - a^{n+1})/(1-a)]u(n)$

P7.12 For (a), (b), and (c) we define $v_1(n) = y(n-2)$, $v_2(n) = v_1(n+1)$, and $y(n) = v_2(n+1)$. a. $\mathbf{v}(n+1) = \begin{bmatrix} 0 & 1 \\ 1 & 0 \end{bmatrix}\mathbf{v}(n) + \begin{bmatrix} 0 \\ 1 \end{bmatrix}x(n)$, $y(n) = [1 \quad 0]\mathbf{v}(n) + x(n)$;

b. $\mathbf{v}(n+1) = \begin{bmatrix} 0 & 1 \\ -1 & 0 \end{bmatrix}\mathbf{v}(n) + \begin{bmatrix} 0 \\ 1 \end{bmatrix}x(n)$, $y(n) = [-1 \quad 0]\mathbf{v}(n) + x(n)$;

c. $\mathbf{v}(n+1) = \begin{bmatrix} 0 & 1 \\ 0.7 & 0.64 \end{bmatrix}\mathbf{v}(n) + \begin{bmatrix} 0 \\ 1 \end{bmatrix}x(n)$, $y(n) = [0.7 \quad 0.64]\mathbf{v}(n) + x(n)$;

d. $\mathbf{v}(n+1) = \begin{bmatrix} 0 & 1 & 0 \\ 0 & 0 & 1 \\ 0.11 & 0.35 & -0.65 \end{bmatrix}\mathbf{v}(n) + \begin{bmatrix} 0 \\ 0 \\ 1 \end{bmatrix}x(n)$, $y(n) = [0.11 \; 0.35 \; -0.65]\mathbf{v}(n)$

P7.13 a. $y_a(n) = \{2, -1, 3, 4.5, 3, 0.75, -0.75, \ldots\}$; b. for $K = 3$, $y_b(n) = \{6, -3, 9, 13.5, 9, 2.25, -2.25, \ldots\}$, $y_b(n) = 3y_a(n)$, homogeneity; c. $y_c(n) = \{-1, -3, -6.5, 3, 6.25, 4.75, \ldots\}$; d. $y_d(n) = y_a(n) + y_c(n)$; e. For $A = 2$ and $C = -3$, $y_e(n) = \{7, 7, 25.5, 0, -12.75, -12.75, -6.375, \ldots\}$, $y_e(n) = Ay_a(n) + Cy_c(n)$.

P7.14 See A7_14 on the CLS disk.

P7.15 a. $y(n) = \{-0.5000 \quad -0.5000 \quad 0.2500 \quad 0.2500 \quad -0.1250 \quad -0.1250 \quad 0.0625\}$; b. $y(n) = \{0.5000 \quad -1.0000 \quad -0.2500 \quad 0.5000 \quad 0.1250 \quad -0.2500 \quad -0.0625\}$; c. Seems to be stable; the IC responses $\rightarrow 0$ as $n \rightarrow \infty$. d. $y(n) = h(n) = \{1.0000 \quad 0 \quad -0.5000 \quad 0 \quad 0.2500 \quad 0 \quad -0.1250\}$; e. $y(n) = \{1.0000 \quad 1.0000 \quad 0.5000 \quad 0.5000 \quad 0.7500 \quad 0.7500 \quad 0.6250\}$; f. $y(n) = \{0 \quad 1.0000 \quad 2.0000 \quad 2.5000 \quad 3.0000 \quad 3.7500 \quad 4.5000\}$; g. Seems to be stable; $h(n) \rightarrow 0$ as $n \rightarrow \infty$. h. $z_{1,2} = \pm j0.707$, stable, response to an unbounded input. The file A7_15 on the CLS disk gives MATLAB solutions to the corresponding parts.

P7.16 a. See A7_16 on the CLS disk for the plots. Using **r=roots(p)** with **p** $= [1, .8, -.25, -.2]$ gives $z_{1,2} = \pm 0.5$ and $z_3 = -0.8$, stable; b. See A7_16, $z_1 = 1.20$, $z_2 = -0.99$, $z_3 = -0.10$, unstable; c. See A7_16; $z_{1,2} = -0.67 \pm j0.67$, $z_{3,4} = 0.67 \pm j0.67$, stable; d. See A7_16; $z_{1,2} = \pm 1.1$, $z_{3,4} = \pm j0.71$; unstable.

P7.17 a. $y_b(n) = 2\delta(n-2) - 1\delta(n-3) + 3\delta(n-4) + 4.5\delta(n-5) + 3\delta(n-6) + 0.75\delta(n-7)$; b. $y_c(n) = 2\delta(n-5) - 1\delta(n-6) + 3\delta(n-7) + 4.5\delta(n-8) + 3\delta(n-9) + 0.75\delta(n-10)$. c. The indications are that the system is time invariant, but we've tried only one input and two values of the delay n_0. In fact the system is time invariant because the difference equation has constant coefficients.

P7.18 a. (i) $z_{1,2} = -0.9$; (ii) $z_{1,2} = 0.9$; b. See A7_18 on the CLS disk. c. System (i) is a highpass filter. Notice that its unit sample response has alternating signs, indicating that rapid variations are present in the output. System (ii), on the other hand, has a unit sample response that varies slowly, indicating that any high frequencies present in the unit sample input are significantly attenuated. d. (i) $H(e^{j\theta}) = (1 - e^{-j\theta})^2/(1 + 0.9e^{-j\theta})^2$, $y_{ss}(n) = 0 + 1.10\cos(n\pi/2 + 3.04) + 400\cos(n\pi)$; (ii) $H(e^{j\theta}) = (1 + e^{-j\theta})^2/(1 - 0.9e^{-j\theta})^2$, $y_{ss}(n) = 400 + 1.10\cos(n\pi/2 - 3.04) + 0$. See A7_18 for the sinusoidal responses to $x(n) = 1 + \cos(n\pi/2) + \cos(n\pi)$.

P7.19 a. (i) $z_{1,2} = \pm j0.9$; (ii) $z_{1,2} = \pm 0.9$; b. See A7_19 on the CLS disk. c. You may have to use your imagination a bit, but looking at $h(n)$ for system (ii) you can see a low-frequency component with a superimposed high-frequency component. This is the band reject filter. System (i), on the other hand, indicates the presence of midrange frequencies without significant components of low and high frequencies. d. (i) $H(e^{j\theta}) = (1 - e^{-j2\theta})/(1 + 0.81e^{-j2\theta})$, $y_{ss}(n) = 0 + 10.53\cos(n\pi/2) + 0$; (ii) $H(e^{j\theta}) = (1 + e^{-j2\theta})/(1 - 0.81e^{-j2\theta})$, $y_{ss}(n) = 10.53 + 0 + 10.53\cos(n\pi)$. See A7_19 for the sinusoidal responses to $x(n) = 1 + \cos(n\pi/2) + \cos(n\pi)$.

P7.20 a. See A7_20 on the CLS disk; b. (i) \$163.5 K; (ii) \$180.5 K; c. (i) \$125.5 K; (ii) \$133.5 K; d. (i) \$1563.10/month; (ii) 15-yr, \$281.4 K, 30-yr, \$432.0 K.

P7.21 See A7_21 on the CLS disk.

z Transforms and Applications

PREVIEW

In Chapter 7 we established theory about and observed several applications of the analysis of linear time-invariant (LTI) systems. This was all accomplished in the discrete-time domain. It is also common practice to design and analyze LTI systems from the point of view of a transform or algebraic domain that enables us to determine response characteristics using familiar algebraic manipulations.

In this chapter we present the bilateral z transform[1] and its applications in the study of discrete systems. The development parallels that of Chapter 3,

FIGURE 8.1 *Karl Friedrich Gauss*

[1]The symbol z is used to denote this transformation, in contrast to the customary practice of associating a transformation with the name of a person, such as Laplace, Fourier, or Hilbert. This custom of using z, which is generally accepted, began in about 1950.

where the Laplace transform is presented, along with its applications to the study of continuous-time systems. A method for solving linear, constant-coefficient difference equations by *Laplace transforms* was introduced to graduate engineering students by Gardner and Barnes [7] in the early 1940s. They applied their procedure, which was based on jump functions, to ladder networks, transmission lines, and applications involving Bessel functions. This approach is quite complicated and in a separate attempt to simplify matters, a transform of a sampled signal or sequence was defined in 1947 by W. Hurewicz as

$$\mathcal{Z}[f(kT)] = \sum_{k=0}^{\infty} f(kT)z^{-k}$$

which was later denoted in 1952 as a "z transform" by a sampled-data control group at Columbia University led by Professor John R. Raggazini and including L. A. Zadeh, E. I. Jury, R. E. Kalman, J. E. Bertram, B. Friedland, and G. F. Franklin. Firmly established in sampled-data and digital control, z transforms became widely used in digital signal-processing applications and research that began flourishing in the late 1960s. Begun as a technique for solving difference equations, z-transform methods are important for other reasons because computers can solve DEs quite nicely and accurately where the order of the system is of little importance. By transforming DEs into algebraic equations, however, we gain insight into stability determination; the powerful transfer function concept is developed, including a pole-zero model of a system; we are able to use design procedures involving location of a system's characteristic roots that affect time response; and techniques are derived that allow us to "see" the system as a diagram or structure. Also, with the aid of the MATLAB function **residue** and **residuez,** an analytical solution for systems of relatively high order can be determined very easily.

This chapter begins by establishing the definition and deriving the transform pairs for a few commonly occurring sequences. Then some important properties of the z transform are developed, with a more complete listing of pairs and properties appearing in Tables 8.1, 8.2, and 8.3 at the end of the chapter. The role of the region of convergence (ROC) in the study of causal, anticausal, and noncausal sequences and systems is emphasized. Next, system characteristics and models, such as stability and transfer functions, are introduced from the point of view of the transform, or z domain. Inverse transform techniques are then developed and applied with the unilateral z transform to solve linear difference equations. Discrete convolution, first introduced in Chapter 7, is discussed from the point of view of the transform domain, and its ramifications and applications are coordinated with the time-domain counterpart. System structures or diagrams are introduced, and an algebraic method (Mason's Gain Rule) is presented for obtaining transfer functions from such diagrams.

BASIC CONCEPTS

The two-sided, or *bilateral*, z transform of the sequence $f(n)$ is defined as

bilateral transform

$$z[f(n)] = F(z) = \sum_{n=-\infty}^{\infty} f(n)z^{-n}$$

where z is a complex variable. The set of values of z for which the sum or power series[2] converges and $F(z)$ exists is called the region of convergence, known as the ROC. The notation

sequence ⇔ transform

$$f(n) \Leftrightarrow F(z)$$

is commonly used to denote a sequence and its corresponding transform (and vice versa).

BASIC CONCEPTS

TRANSFORM PAIRS

(a) Sample (Impulse) Sequence Using the bilateral transform definition, we have

$$z[A\delta(n)] = \sum_{n=-\infty}^{\infty} A\delta(n)z^{-n} = A$$

and in the common notation the pair is

sample (impulse) function

$$A\delta(n) \Leftrightarrow A$$

We notice that this transform is independent of the complex variable z and, consequently, we say that the transform converges for all z; that is, the ROC is everywhere.

[2]According to Gustav Doetsch, *Guide to the Applications of Laplace and Z-Transforms* (London: Van Nostrand Reinhold Company, 1961, 1971), it would be fitting to call this transformation the Laurent transformation, since the series is a Laurent series.

(b) Causal Exponential For $x(n) = Aa^n u(n)$, the transform is

lower limit 0 because of $u(n)$

$$X(z) = \sum_{n=-\infty}^{\infty} Aa^n u(n) z^{-n}$$
$$= A \sum_{n=0}^{\infty} (az^{-1})^n$$
$$= \frac{A}{1 - az^{-1}} = \frac{Az}{z - a}$$

where we have used the infinite geometric sum formula (see Appendix B) with the proviso that $|az^{-1}| < 1$. Thus the ROC is $|az^{-1}| < 1$ or $|z| > |a|$, where a may be real or complex, and the pair is

causal exponential

$$Aa^n u(n) \Leftrightarrow \frac{Az}{z - a}, \qquad |z| > |a|.$$

Using the Symbolic Math Toolbox, we enter `x=('A*a^n')` and **`X= ztrans(x)`** with the function to be transformed `x=A*a^n` and its transform `X=-A*z/(-z+a)` returned.

ILLUSTRATIVE PROBLEM 8.1 ***Derivation of Some Pairs***

Use the pair just derived for a causal exponential sequence to find the z transform of (a) a step sequence $Au(n)$ and (b) the cosine sequence $A \cos(n\theta)u(n)$.

Solution

a. For the step sequence we simply let $a = 1$, which yields the pair

$$Au(n) \Leftrightarrow \frac{Az}{z - 1}, \qquad |z| > 1.$$

b. For the cosine sequence we invoke Euler to write

$$x(n) = A \cos(n\theta)u(n) = 0.5A(e^{jn\theta} + e^{-jn\theta})u(n).$$

Using the exponential pair with $a = e^{\pm j\theta}$ gives us

$$X(z) = 0.5A\left(\frac{z}{z - e^{j\theta}} + \frac{z}{z - e^{-j\theta}}\right)$$
$$= A\frac{z(z - \cos\theta)}{z^2 - 2z(\cos\theta) + 1}, \qquad |e^{\pm j\theta}z^{-1}| < 1.$$

But $|e^{\pm j\theta}| = 1$, making the ROC $|z| > 1$; in the common notation

$$A \cos(n\theta)u(n) \Leftrightarrow A\frac{z(z - \cos\theta)}{z^2 - 2z(\cos\theta) + 1}, \quad |z| > 1.$$

(c) Anticausal Exponential An anticausal sequence can be described by $x(n) = Aa^n u(-n - 1)$, and its z transform is

$u(-n-1) = 0$, $n \geq 0$

$$X(z) = \sum_{n=-\infty}^{\infty} Aa^n u(-n - 1)z^{-n} = \sum_{n=-\infty}^{-1} Aa^n z^{-n}.$$

Adding and subtracting A, changing the sign on n inside the summation, using the infinite geometric sum formula, and doing some algebra gives the pair relationship

anticausal exponential

$$Aa^n u(-n-1) \Leftrightarrow \frac{-A}{1-az^{-1}} = \frac{-Az}{z-a}, \qquad |z| < |a|.$$

Notice that the pairs for the causal and anticausal exponentials are quite similar, the differences being a sign and the regions of convergence. Listings of some common pairs are given in Table 8.1 at the end of the chapter.

PROPERTIES AND RELATIONS

Some properties and relations for bilateral z transforms are derived in this section with a summary presented in Table 8.2 at the end of the chapter.

Linearity If $\mathcal{Z}[g_1(n)] = G_1(z)$ and if $\mathcal{Z}[g_2(n)] = G_2(z)$, then

linearity

$$\mathcal{Z}[ag_1(n) + bg_2(n)] = aG_1(z) + bG_2(z)$$

for all values of the constants a and b, where the region of convergence is the region where both $G_1(z)$ and $G_2(z)$ converge. Using the sequence $g_k(n)$ and its corresponding z transform $G_k(z)$, we can write the linearity property in a more general way as

generalized linearity

$$\mathcal{Z}\left[\sum_{k=1}^{p} C_k g_k(n)\right] = \sum_{k=1}^{p} C_k G_k(z).$$

Thus, the z transform of a sum is equal to the sum of the transforms.

ILLUSTRATIVE PROBLEM 8.2
Transform of an Exponential Pulse

Use the pairs derived for the causal and anticausal exponential sequences to find the z transform of the two-sided exponential sequence $x(n) = a^n u(n) + b^n u(-n-1)$.

Solution

Recalling the pairs for a causal and an anticausal exponential, we have

$$a^n u(n) \Leftrightarrow \frac{z}{z-a}, \quad |z| > |a|, \quad \text{and}$$

$$b^n u(-n-1) \Leftrightarrow -\frac{z}{z-b}, \quad |z| < |b|.$$

Thus, using linearity, we have

$$X(z) = \frac{z}{z-a} - \frac{z}{z-b}, \quad |z| > |a| \quad \text{and} \quad |z| < |b|.$$

The z transform of $x(n)$ will exist for all a and b such that $|a| < |b|$, in which case the ROC is $|a| < |z| < |b|$. Notice that if $|a| \geq |b|$, $X(z)$ does not exist because the ROCs are $|z| > |a|$ and $|z| < |b|$, and this means that the region of convergence is empty.

Shifting Property For the shifted sequence $g(n-m)$, where m is any integer, the z transform is

$$z[g(n-m)] = \sum_{n=-\infty}^{\infty} g(n-m)z^{-n}$$

and with $p = n - m$, we have

$$z[g(n-m)] = \sum_{p=-\infty}^{\infty} g(p)z^{-m}z^{-p}$$

$$= z^{-m} \sum_{p=-\infty}^{\infty} g(p)z^{-p} = z^{-m}G(z).$$

The shifting property can be described in the common notation as

shifting property

$$g(n-m) \Leftrightarrow z^{-m}G(z).$$

Multiplication by n and Derivatives in z This property is useful in finding transforms of certain sequences, such as a ramp $nu(n)$, and sequences of the form $na^n u(n)$. The property relates multiplication in the n, or time, domain to differentiation in the z, or transform, domain as

$$ng(n) \Leftrightarrow -z\frac{dG(z)}{dz} \quad \text{where } g(n) \Leftrightarrow G(z).$$

CHAPTER 8

TRANSFER FUNCTIONS

The transfer function of a linear time-invariant discrete system is defined in terms of the output and input sequences as

definition

$$\text{transfer function} = \frac{z \text{ transform of the output sequence } y(n)}{z \text{ transform of the input sequence } x(n)}$$

where the initial conditions are zero. In terms of symbols,

symbols

$$H(z) = \frac{Y(z)}{X(z)}.$$

Alternatively, starting with the system's difference equation,

difference equation (DE)

$$\sum_{k=0}^{N} a_k y(n-k) = \sum_{k=0}^{L} b_k x(n-k)$$

using the shifting property and linearity gives the algebraic or z domain equation

$$a_0Y(z) + a_1z^{-1}Y(z) + \cdots + a_Nz^{-N}Y(z)$$
$$= b_0X(z) + b_1z^{-1}X(z) + \cdots + b_Lz^{-L}X(z)$$

which can be written in terms of two summations as

transformed DE

$$Y(z)\left[\sum_{k=0}^{N} a_k z^{-k}\right] = X(z)\left[\sum_{k=0}^{L} b_k z^{-k}\right].$$

Solving for $H(z) = Y(z)/X(z)$, we find the transfer function to be

rational function

$$H(z) = \frac{Y(z)}{X(z)} = \frac{\sum_{k=0}^{L} b_k z^{-k}}{\sum_{k=0}^{N} a_k z^{-k}}$$

where we notice that the coefficients of the input terms in the difference equation determine the coefficients of the numerator polynomial of the transfer function and that the coefficients of the output terms in the difference equation determine the coefficients of the denominator polynomial of the transfer function. Recall that a delay of Q samples in a difference equation corresponds to z^{-Q} in the transformed difference equation.

ILLUSTRATIVE PROBLEM 8.3
Transfer Function from Difference Equation

Given the difference equation for a causal system

$$y(n) - 4y(n-1) + 6y(n-2) - 4y(n-3) = x(n) + x(n-5),$$

find the transfer function $H(z) = Y(z)/X(z)$.

Solution

Using linearity and the shift property we have

$$Y(z) - 4z^{-1}Y(z) + 6z^{-2}Y(z) - 4z^{-3}Y(z) = X(z) + z^{-5}X(z)$$

giving us the rational function

$$H(z) = \frac{1 + z^{-5}}{1 - 4z^{-1} + 6z^{-2} - 4z^{-3}} = \frac{z^5 + 1}{z^5 - 4z^4 + 6z^3 - 4z^2}.$$

What is the region of convergence for this transfer function? This question will be answered shortly.

If a unit sample sequence $\delta(n)$ is the input to an LTI system, the transform of the input is $X(z) = 1$ and $Y(z) = H(z) \cdot X(z) = H(z)$. Taking the inverse transform gives $y(n) = h(n)$. Thus the transfer function of an LTI system can be found by taking the z transform of the system's unit sample response $h(n)$ as

$h(n) \Leftrightarrow H(z)$

$$H(z) = z[h(n)].$$

Conversely, the unit sample response $h(n)$ can be found by evaluating the inverse z transform of the transfer function $H(z)$.

ILLUSTRATIVE PROBLEM 8.4
Transfer Function from Unit Sample Response

For the unit sample response $h(n) = [(0.5)^n + (-0.5)^n]u(n)$, find the transfer function model $H(z)$.

Solution

Using linearity and transform pair 2 from Table 8.1 for an exponential sequence gives

$$H(z) = Z[0.5^n u(n) + (-0.5)^n u(n)]$$
$$= \frac{z}{z - 0.5} + \frac{z}{z + 0.5} = \frac{2z^2}{z^2 - 0.25}$$

where the ROC for each component of $H(z)$ is $|z| > 0.5$. As we said earlier, we will discuss the region of convergence for transfer functions momentarily.

Poles and Zeros A transfer function can be factored into first-order terms as follows:

$$H(z) = \frac{b_0(1 - n_1 z^{-1})(1 - n_2 z^{-1})\cdots(1 - n_L z^{-1})}{a_0(1 - d_1 z^{-1})(1 - d_2 z^{-1})\cdots(1 - d_N z^{-1})}$$
$$= \frac{b_0 z^{N-L}(z - n_1)(z - n_2)\cdots(z - n_L)}{a_0(z - d_1)(z - d_2)\cdots(z - d_N)}.$$

The values of z that make $H(z)$ go to zero and infinity are called the *system zeros* and *system poles,* respectively. There are zeros at $n_1, n_2, \ldots, n_L$ and poles at $d_1, d_2, \ldots, d_N$. The term z^{N-L} governs the transfer function characteristics at $z = 0$. For $N - L > 0$, there are $N - L$ zeros at $z = 0$; for $N - L < 0$, there are $L - N$ poles at $z = 0$.

ILLUSTRATIVE PROBLEM 8.5
Determining Zeros and Poles

Find the zeros and poles of the system transfer function

$$H(z) = \frac{1 + z^{-5}}{1 - 4z^{-1} + 6z^{-2} - 4z^{-3}} = \frac{z^5 + 1}{z^5 - 4z^4 + 6z^3 - 4z^2}.$$

Solution

The zeros are the roots of the equation $z^5 + 1 = 0$, or $z^5 = -1$. But the complex number -1 can be written in exponential form as $-1 = 1e^{j(\pi + m2\pi)}$, $m = 0, 1, 2, 3, 4$. Thus, using De Moivre's[3] theorem to compute the five roots, we have $z_1 = 1e^{j\pi/5}$, $z_2 = 1e^{j3\pi/5}$, $z_3 = 1e^{j5\pi/5}$, $z_4 = 1e^{j7\pi/5}$, and $z_5 = 1e^{j9\pi/5}$. Notice

[3] Abraham De Moivre (1667–1754) was a French Protestant who was forced to seek asylum in London after the revocation of the Edict of Nantes and the expulsion of the Huguenots (1685). He had hoped to become a university professor, but he never succeeded in this, partly because of his non-British origin. A friend of Isaac Newton, he was appointed to the partisan commission by means of which the Royal Society of London sought to review the evidence in the Newton-Leibniz dispute over the origin of the calculus. De Moivre supported himself for many years by solving problems proposed to him by wealthy patrons who wanted to

that these zeros are located on the unit circle separated by $2\pi/5$ rad. The poles are the roots of the equation $z^2(z^3 - 4z^2 + 6z - 4) = 0$. The two poles at $z = 0$ are easy to find, but the other three require solving the cubic $z^3 - 4z^2 + 6z - 4 = 0$. Using MATLAB we enter the vector $\mathbf{p} = [1, -4, 6, -4]$ and invoke $\mathbf{r} = \mathbf{roots}(\mathbf{p})$, with the following results:

$$\begin{aligned} \mathbf{r} = {} & 2.0000 \\ & 1.0000 + 1.0000i \\ & 1.0000 - 1.0000i\,. \end{aligned}$$

Five zeros at $z = 1e^{j(\pi+m2\pi)/5}$, $m = 0, 1, 2, 3, 4$. Five poles: $z_{1,2} = 0$, $z_3 = 2$, and $z_{4,5} = 1 \pm j1 = 1.414e^{\pm j\pi/4}$.

Comment: The zeros and poles are portrayed graphically in Figure 8.2, where the zeros are marked with ○'s and the poles with ×'s. A plot like this can be made with the MATLAB function **zplane** from the Signals and Systems Toolbox. See F8_2 on the CLS disk.

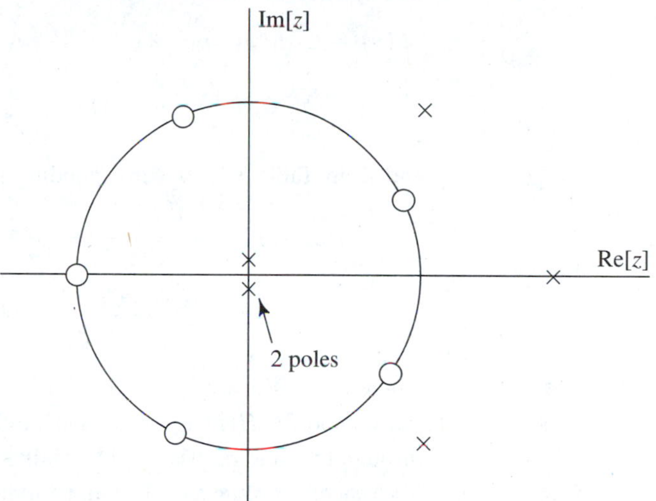

FIGURE 8.2 *Pole-zero plot for a fifth-order system*

BASIC CONCEPTS

Region of Convergence The ROC must be specified for any transform, and the transfer function is not excepted from this rule. Let's look at the ROCs of transfer functions for causal, anticausal, and noncausal systems.

---

know what stakes they should offer in games of chance. It is said that De Moivre predicted the day of his own death. Finding that each day he was sleeping a quarter of an hour longer than on the previous day, he calculated that he would die in this sleep on the very day in which he slept up to the limit of 24 h. He died at the age of 87; the cause of his death was recorded as "somnolence." From: *The History of Mathematics,* David M. Burton, Allyn and Bacon, Inc., Newton, Mass., 1985.

(a) Causal Systems The unit sample response of an Nth-order causal LTI system with distinct poles is

$$h(n) = C_1 d_1{}^n u(n) + C_2 d_2{}^n u(n) + \cdots + C_N d_N{}^n u(n) = \sum_{k=1}^{N} C_k d_k{}^n u(n)$$

and the corresponding transfer function, the z transform of $h(n)$, is

$$\begin{aligned} H(z) &= \frac{C_1}{1 - d_1 z^{-1}} + \frac{C_2}{1 - d_2 z^{-1}} + \cdots + \frac{C_N}{1 - d_N z^{-1}} \\ &= \sum_{k=1}^{N} \frac{C_k z}{z - d_k}, \qquad |z| > |d_k| \end{aligned}$$

where the ROCs are $|z| > |d_1|, |z| > |d_2|, \ldots, |z| > |d_N|$. Thus the ROC for the composite $H(z)$ is *outside* the circle centered at the origin and passing through the pole d_k having the largest magnitude, as shown in Figure 8.3a.

(b) Anticausal Systems The unit sample response of an Nth-order anticausal LTI system with distinct poles is

$$\begin{aligned} h(n) &= C_1 d_1{}^n u(-n-1) + C_2 d_2{}^n u(-n-1) + \cdots + C_N d_N{}^n u(-n-1) \\ &= \sum_{k=1}^{N} C_k d_k{}^n u(-n-1) \end{aligned}$$

and from Table 8.1, its corresponding transfer function is

$$\begin{aligned} H(z) &= \frac{-C_1}{1 - d_1 z^{-1}} + \frac{-C_2}{1 - d_2 z^{-1}} + \cdots + \frac{-C_N}{1 - d_N z^{-1}} \\ &= -\sum_{k=1}^{N} \frac{C_k z}{z - d_k}, \qquad |z| < |d_k| \end{aligned}$$

where the ROCs are $|z| < |d_1|, |z| < |d_2|, \ldots, |z| < |d_N|$. Thus the ROC for the composite $H(z)$ is *inside* the circle centered at the origin and passing through the pole d_k having the smallest magnitude, as shown in Figure 8.3b.

(c) Noncausal Systems The unit sample response of an Nth-order noncausal LTI system can be partitioned into its causal and anticausal parts,

$$h(n) = h_c(n) + h_{ac}(n) = \sum_{k=1}^{N_1} C_k d_k{}^n u(n) + \sum_{k=N_1+1}^{N} C_k d_k{}^n u(-n-1)$$

and the corresponding transfer function is

$$\begin{aligned} H(z) &= H_c(z) + H_{ac}(z) \\ &= \sum_{k=1}^{N_1} \frac{C_k z}{z - d_k} - \sum_{k=N_1+1}^{N} \frac{C_k z}{z - d_k}, \qquad |R_1| < |z| < |R_2| \end{aligned}$$

where $H_c(z)$ contains all poles of the composite transfer function $H(z)$ inside the radius R_1 and $H_{ac}(z)$ contains all the poles of $H(z)$ outside the radius R_2. See Figure 8.3c. Notice that if $|R_1| \geq |R_2|$, the z transform of $h(n)$ does not exist, because there is no common region of convergence.

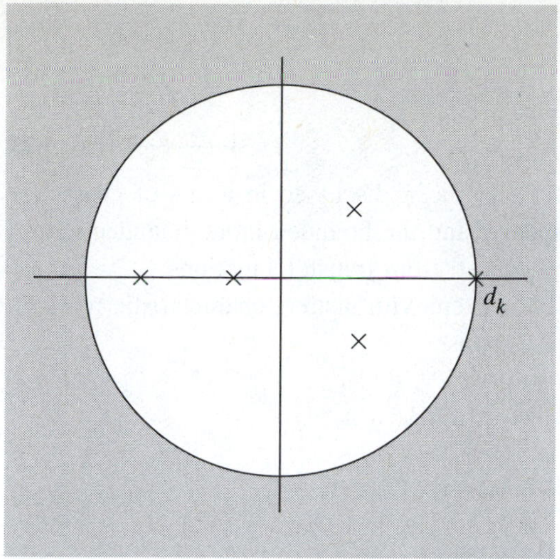

a. Causal system

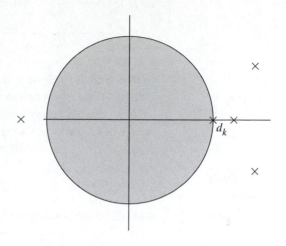

b. Anticausal system

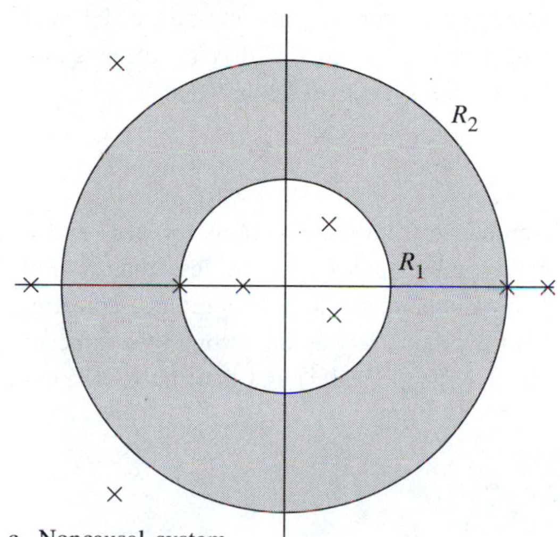

c. Noncausal system

FIGURE 8.3 *Regions of convergence*

ILLUSTRATIVE PROBLEM 8.6
Transfer Functions and ROC

Find the transfer functions that correspond to the following unit sample responses and state the ROC for each $H(z)$.

a. $h_1(n) = (-0.7)^n u(n)$

b. $h_2(n) = (-0.7)^n u(-n - 1)$

c. $h_3(n) = 3(0.5)^n u(n) + 3(2)^n u(-n - 1)$

Solution

From Table 8.1.

a. $H_1(z) = \dfrac{z}{z + 0.7}, \quad |z| > 0.7$

b. $H_2(z) = \dfrac{-z}{z + 0.7}, \quad |z| < 0.7$

c. $H_3(z) = \dfrac{3z}{z - 0.5} - \dfrac{3z}{z - 2}, \quad 0.5 < |z| < 2$

System Stability System stability has been discussed in terms of characteristic roots, the unit impulse response, and the bounded-input–bounded-output criterion. Stability may also be deduced from transfer functions.

(a) Causal Systems A causal LTI system with distinct characteristic roots d_k is described by the unit sample response

unit sample response

$$h(n) = \sum_{k=1}^{N} C_k d_k{}^n u(n)$$

with the corresponding transfer function

transfer function

$$H(z) = \sum_{k=1}^{N} \frac{C_k z}{z - d_k}, \quad |z| > |d_k|$$

where the region of convergence is outside the circle centered at the origin and passing through the pole d_k having the largest magnitude, as shown in Figure 8.4a. If the system is stable—that is, if

stability requirement

$$\sum_{n=-\infty}^{\infty} |h(n)| < \infty$$

—then we deduce from the equations for $h(n)$ and $H(z)$ that the magnitudes of all of the system poles must be less than 1. For this situation we have Figure 8.4b, where we see that the region of convergence is outside the circle centered at the origin and passing through the largest magnitude pole d_k of the system transfer function $H(z)$ and that the ROC includes the unit circle.[4]

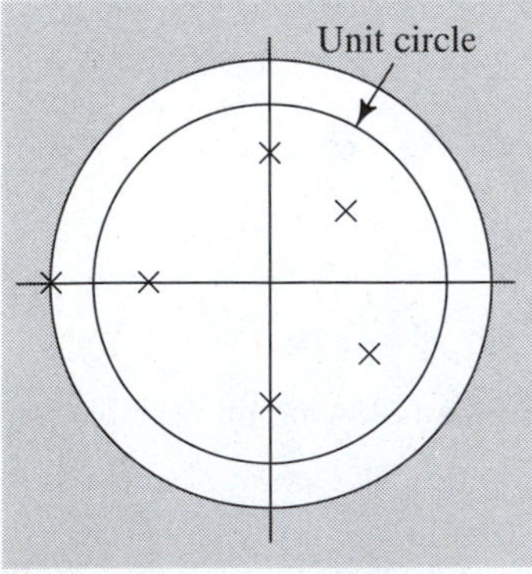

a. ROC for a causal system

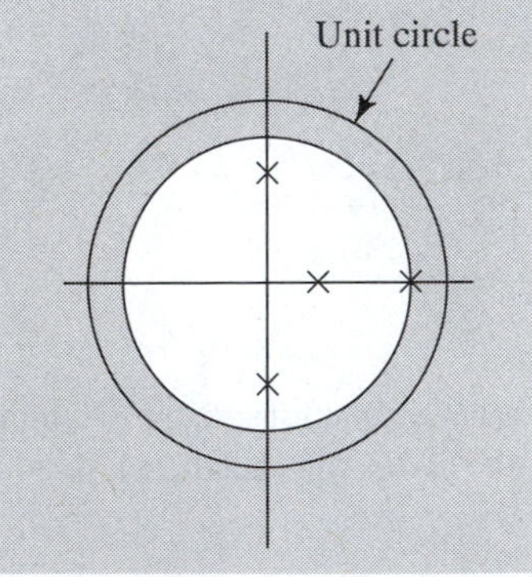

b. ROC for a stable causal system

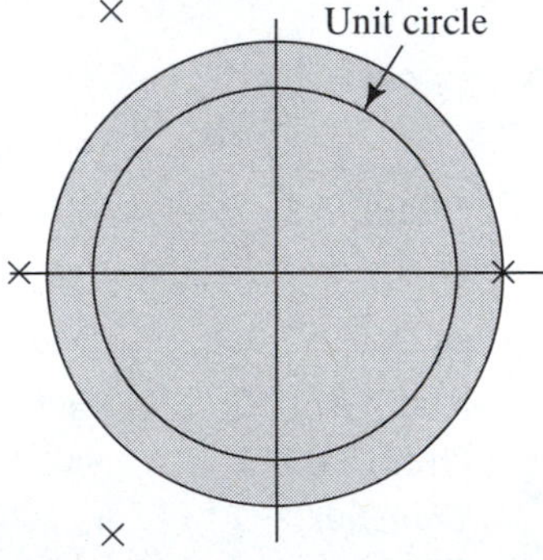

c. ROC for a stable anticausal system

FIGURE 8.4 *Stability and ROC*

[4]Although we've restricted consideration here to the predominant case of nonrepeated poles, the result that the ROC must contain the unit circle applies as well to the multiple-pole case.

CHAPTER 8

(b) Anticausal Systems An anticausal LTI system is described by the unit sample response

$$h(n) = \sum_{k=1}^{N} C_k d_k{}^n u(-n - 1)$$

and the corresponding transfer function

transfer function

$$H(z) = -\sum_{k=1}^{N} \frac{C_k z}{z - d_k}, \qquad |z| < |d_k|$$

where the region of convergence is inside of the circle centered at the origin and passing through the pole d_k having the smallest magnitude. If the system is stable—that is, if

summable sample response

$$\sum_{n=-\infty}^{\infty} |h(n)| < \infty$$

—then we see from the equations for $h(n)$ and $H(z)$ that the magnitudes of all of the system poles must be greater than 1. For this situation we have Figure 8.4c, where we see that the region of convergence is inside the circle centered at the origin and passing through the smallest pole d_k of the system transfer function $H(z)$ and that the ROC includes the unit circle.

BASIC CONCEPTS

Stable causal systems have all their poles inside the unit circle, whereas stable anticausal systems have all their poles outside the unit circle. For stable systems, consequently, the unit circle is included in the region of convergence.

(c) Noncausal Systems Noncausal systems are frequently used in an intermediate step in the design of signal-processing systems. The poles of the "causal" part of the system must lie inside the unit circle, while those of the "anticausal" part must lie outside for the overall noncausal system to be stable.

ILLUSTRATIVE PROBLEM 8.7
Stability

Given the system transfer function

$$H(z) = \frac{9z^2 - 12z}{z^2 - 2.5z - 1.5},$$

determine stability and the form of $h(n)$ for the following regions of convergence:

a. $|z| < 0.5$

b. $0.5 < |z| < 3$

c. $|z| > 3$

Solution The system poles are $z_1 = -0.5$ and $z_2 = +3$.

a. The unit circle is not included in the ROC. The system is unstable and anticausal with $h(n) = C_1(-0.50)^n u(-n-1) + C_2(3)^n u(-n-1)$.

b. The unit circle is included in the ROC. The system is stable and noncausal with $h(n) = C_3(-0.50)^n u(n) + C_4(3)^n u(-n-1)$.

c. The system is unstable and causal because the ROC does not include the unit circle. The unit sample response is $h(n) = C_5(-0.50)^n u(n) + C_6(3)^n u(n)$.

THE EVALUATION OF INVERSE TRANSFORMS

The inverse relationship that provides a way for finding a sequence from its z transform is

inverse transform

$$f(n) = \frac{1}{2\pi j} \oint_C F(z) z^{n-1} dz$$

where C is a closed contour that includes all poles of $F(z)$. If this route is taken, the integral is generally evaluated by means of the residue theorem of complex variable theory. An alternative, which we will emphasize, is to use a method based on the partial fraction expansion of $F(z)$.

CHAPTER 8

(a) Inverse Transforms from the Definition We recall the basic pair

shifted sample ⇔ its transform

$$C_k \delta(n-k) \Leftrightarrow C_k z^{-k}.$$

Consequently, if we have the transform

$$Y(z) = z^3 + 2z^2 + 3z^0 + 4z^{-1} + 5z^{-2}$$

the corresponding sequence is described by the sample (impulse) functions

$$y(n) = \delta(n+3) + 2\delta(n+2) + 3\delta(n) + 4\delta(n-1) + 5\delta(n-2).$$

(b) Inverse Transforms from Long Division Suppose we have the transform of a causal sequence written in the form

$$P(z) = \frac{N(z)}{D(z)}$$

where $N(z)$ and $D(z)$ are polynomials in z or z^{-1}. If we simply use long division (just as in grade school) to write $P(z)$ as a power series in *nonpositive* powers of z, we obtain

$$P(z) = \alpha_0 z^0 + \alpha_1 z^{-1} + \alpha_2 z^{-2} + \cdots + \alpha_{17} z^{-17} + \cdots,$$

and using the basic definition of the z transform yields the causal sequence

$$p(n) = \alpha_0 \delta(n) + \alpha_1 \delta(n-1) + \alpha_2 \delta(n-2) + \cdots + \alpha_{17}\delta(n-17) + \cdots.$$

ILLUSTRATIVE PROBLEM 8.8
Inverse Transforms by Long Division

For the transform $P(z) = (z + 1)/(z + 2)$, use long division to find the first few terms of the inverse transform when the ROC is $2 < |z|$.

Solution

By long division write $P(z)$ in nonpositive powers of z as

$$P(z) = \frac{z + 1}{z + 2} = 1 - z^{-1} + 2z^{-2} - 4z^{-3} + \cdots$$

which yields the time sequence

$$p(n) = \delta(n) - 1\delta(n - 1) + 2\delta(n - 2) - 4\delta(n - 3) + \cdots.$$

(c) Inverse Transform by Partial Fraction Expansion (PFE), Distinct Poles
Consider the z transform of a sequence $f(n)$ in the standard form

$$F(z) = K\frac{B(z)}{(z - d_1)(z - d_2)\cdots(z - d_N)}$$

where K is a scale factor, $B(z)$ is the numerator polynomial of order L, and we assume that all N poles are different. To ensure a proper fraction and to enable us to use Table 8.1, we divide by z to obtain

$$\frac{F(z)}{z} = K\frac{B(z)}{z(z - d_1)(z - d_2)\cdots(z - d_N)}.$$

Expanding this in partial fractions results in

$$\begin{aligned}\frac{F(z)}{z} &= K\frac{B(z)}{z(z - d_1)(z - d_2)\cdots(z - d_N)}\\ &= \frac{C_0}{z} + \frac{C_1}{z - d_1} + \frac{C_2}{z - d_2} + \cdots + \frac{C_N}{z - d_N}\end{aligned}$$

where for the pole at $z = d_j$ we have

$$\begin{aligned}C_j &= K\frac{B(z)\cdot(z - d_j)}{z(z - d_1)(z - d_2)\cdots(z - d_j)\cdots(z - d_N)}\Bigg|_{z=d_j}\\ &= K\frac{B(d_j)}{d_j(d_j - d_1)(d_j - d_2)(d_j - d_{j-1})(d_j - d_{j+1})\cdots(d_j - d_N)}.\end{aligned}$$

Then multiplying the expanded $F(z)/z$ by z yields

PF expansion of a transform

$$F(z) = C_0 + \sum_{k=1}^{N} \frac{C_k z}{z - d_k}.$$

BASIC CONCEPTS

We need to bring the region of convergence into the picture before it is possible to evaluate the inverse transform of each term in the preceding summation. If, for example, the ROC is outside the largest pole in $F(z)$, then the sequence is causal and we have

causal sequence

$$f(n) = C_0\delta(n) + \sum_{k=1}^{N} C_k d_k{}^n u(n).$$

If the ROC is inside the smallest pole in $F(z)$, then the sequence is anticausal and given by

anticausal sequence

$$f(n) = C_0\delta(n) - \sum_{r=1}^{N} C_r d_r{}^n u(-n-1).$$

For a mixture of these two situations, where the ROC is inside some poles and outside of the others, the inverse transform will be a sum of causal and anticausal sequences.

ILLUSTRATIVE PROBLEM 8.9
Inverse Transforms and ROC

Given the transform

$$Y(z) = \frac{6z^2 - 10z + 2}{z^2 - 3z + 2}.$$

a. Determine the partial fraction expansion of $Y(z)$.

b. Find the inverse transform $y(n)$ for the following regions of convergence: (i) $|z| > 2$, (ii) $|z| < 1$, and (iii) $1 < |z| < 2$.

Solution

a. We begin by first factoring the denominator of $Y(z)$ and then expanding $Y(z)/z$ as

$$\frac{Y(z)}{z} = \frac{6z^2 - 10z + 2}{z(z-1)(z-2)} = \frac{C_0}{z} + \frac{C_1}{z-1} + \frac{C_2}{z-2}.$$

The partial fraction constants are given by

$$C_0 = \left.\frac{6z^2 - 10z + 2}{(z-1)(z-2)}\right|_{z=0} = 1, \quad C_1 = \left.\frac{6z^2 - 10z + 2}{z(z-2)}\right|_{z=1} = 2, \quad \text{and}$$

$$C_2 = \left.\frac{6z^2 - 10z + 2}{z(z-1)}\right|_{z=2} = 3.$$

With the partial fraction constants determined, we multiply $Y(z)/z$ by z to obtain

$$Y(z) = 1 + \frac{2z}{z-1} + \frac{3z}{z-2}.$$

b. For the inverse transforms we make use of Table 8.1 and the stated regions of convergence:

(i) ROC $|z| > 2$: The region of convergence is outside both poles; consequently, the poles at $z = 1$ and $z = 2$ both yield causal sequences. The constant term 1 in $Y(z)$ will always give a unit sample function. Thus we have the causal sequence

$$y(n) = \delta(n) + [2(1)^n + 3(2)^n]u(n).$$

From the Symbolic Math Toolbox with `Y=('6*z^2-10*z+2)/(z^2-3*z+2)')` and **`Y=invztrans(y)`**, we obtain `y=Delta(n)+2+3*2^n`.

(ii) ROC $|z| < 1$: Now the ROC is inside both poles and we have anticausal sequences for the $+1$ and $+2$ poles and the overall noncausal sequence

$$y(n) = \delta(n) - [2(1)^n + 3(2)^n]u(-n-1).$$

(iii) ROC $1 < |z| < 2$: Now the ROC is outside the pole at $z = 1$, which yields a causal sequence, and inside the pole at $z = 2$, which produces an anticausal sequence, giving us the noncausal sequence

$$y(n) = \delta(n) + 2u(n) - 3(2)^n u(-n-1).$$

■

***(d) Inverse Transform Using the MATLAB functions* residue *and* residuez**

The partial fraction expansion is easily accomplished using **residue.** If $T(z)$ is the rational function to be expanded, the numerator and denominator coefficients of $T(z)/z$ are the input data in descending powers of z. **residue** returns the poles and the PF constants of $T(z)/z$. Alternatively, **residuez** permits the expansion of $T(z)$ directly. The numerator and denominator coefficients of $T(z)$ are input data, and **residuez** returns the poles and the PF constants of $T(z)$. For more details see **residuez** in the Signals and Systems and Signal Processing Toolboxes. The annotated script (PFE) below illustrates the use of **residue.**

BASIC CONCEPTS

MATLAB Script

```
%PFE...computes PFE constants, poles, and the direct term of a
%rational function.
b=[6,-10,2]              % numerator coefficients of Y(z)/z
a=[1,-3,2,0]             % denominator coefficients of Y(z)/z
[r,p,k]=residue(b,a)     % call residue, r=PFE constant, p=pole or
                         % characteristic root, k=direct term
```

Output

```
r=3                      the three partial fraction constants
  2
  1
p=2                      the three poles of Y(z)/z
  1
  0
k=[]                     the direct term, zero in this case
```

(e) Inverse Transform by Partial Fraction Expansion, Multiple Poles We have demonstrated the partial fraction method for the usual situation of distinct or simple poles of the transform that is to be inverted. If a transform has multiple poles, a different strategy must be employed to find the PFE constants. We show one procedure (there are several) and illustrate the method by an example. Consider a highpass digital filter described by the transfer function $H(z) = z/(z + 0.9)$, $|z| > 0.9$, that is subjected to the ramp input $x(n) = nu(n)$. With $X(z) = z/(z - 1)^2$, the transform of the output is

$$Y(z) = \frac{z}{z + 0.9} \cdot \frac{z}{(z - 1)^2}, \qquad |z| > 1$$

and the PFE for $Y(z)/z$ is

$$\frac{Y(z)}{z} = \frac{z}{z + 0.9} \cdot \frac{1}{(z - 1)^2} = \frac{A}{z + 0.9} + \frac{B}{(z - 1)^2} + \frac{C}{z - 1}$$

where $A = -0.2493$ and $B = 0.5263$. C cannot be evaluated in the normal manner (try it), but if we do some algebra *before* setting $z = 1$, we have

$$C = \left\{\frac{z}{z + 0.9} \cdot \frac{1}{z - 1} - \frac{0.5263}{z - 1}\right\}\Bigg|_{z=1} = \frac{0.4737(z - 1)}{(z + 0.9)(z - 1)}\Bigg|_{z=1} = 0.2493\,.$$

Using pairs 2, 3, and 9 in Table 8.1, the inverse transform is

$$y(n) = [-0.2493(-0.9)^n + 0.5263n + 0.2493]u(n).$$

To use this approach, there is a specific procedure that must be followed. Given

$$\frac{P(z)}{z} = \frac{N(z)}{(z + a)^3(z + b)(z + c)}$$
$$= \frac{C_1}{(z + a)^3} + \frac{C_2}{(z + a)^2} + \frac{C_3}{z + a} + \frac{C_4}{z + b} + \frac{C_5}{z + c}$$

C_1, C_2, and C_3 must be determined *in that order;* that is, we must find the numerator associated with the *highest power* of the repeated root *first,* and so on down the list. The partial fraction constants C_4 and C_5 for the simple roots may be found at any time. See Example 8.3 for a common situation where multiple roots (poles) occur.

Comment: Once again **residue** can be used on this example to find the PFE using the following script. Any confusion that might arise regarding the constants and the multiple poles can be resolved by making the easiest multiple-pole calculation by hand.

MATLAB Script

```
%PFE...computes PFE constants, poles, and the direct term of a rational
%function.
b=[1,0]                   % numerator coefficients of P(z)/z
a=[1,-1.1,-.8,.9]         % denominator coefficients of P(z)/z
[r,p,k]=residue(b,a)      % call residue, r=PFE constant, p=pole or
                          % characteristic root, k=direct term
```

Output

```
r= 0.2493                  the three partial fraction constants
    0.5263
   -0.2493
p= 1                       the three poles of P(z)/z
    1
   -0.9000
k=[]                       the direct term
```

SOLUTION OF LINEAR DIFFERENCE EQUATIONS BY *z* TRANSFORM

In this book we consider the solution of difference equations that represent causal systems only and are typically interested in the solutions for $n \geq 0$. Thus we need to consider the unilateral z transform

unilateral transform

$$G_u(z) = \sum_{n=0}^{\infty} g(n)z^{-n}$$

where the subscript u denotes the unilateral transform. The shifting property for bilateral transforms, property 2 in Table 8.2, needs to be modified for its application with the unilateral transform. Using $G_u(z)$ to denote the unilateral transform of $g(n)$, the corresponding unilateral transform of the sequence $g(n - m)$ for $m > 0$ is

$$\begin{aligned} Z[g(n - m)] &= \sum_{n=0}^{\infty} g(n - m)z^{-n} \\ &= \sum_{r=-m}^{\infty} g(r)z^{-(r+m)} \quad \text{where } r = n - m \\ &= \sum_{r=-m}^{-1} g(r)z^{-(r+m)} + \sum_{r=0}^{\infty} g(r)z^{-r}z^{-m}. \end{aligned}$$

The last term on the right is simply $z^{-m}G_u(z)$; hence the shifting theorem for *unilateral transforms* is

shifting theorem (unilateral transforms)

$$Z[g(n - m)] = \sum_{r=-m}^{-1} g(r)z^{-(r+m)} + z^{-m}G_u(z).$$

Comment: Other properties and relations for the unilateral z transform are found in Table 8.3 at the end of the chapter.

Turning now to the solution of the difference equation

$$\sum_{k=0}^{N} a_k y(n - k) = \sum_{k=0}^{L} b_k x(n - k)$$

we take the z transform term by term to obtain

$$\begin{aligned} &a_0Y(z) + a_1[z^{-1}Y(z) + z^{-0}y(-1)] + a_2[z^{-2}Y(z) + z^{-1}y(-1) \\ &\quad + z^{-0}y(-2)] + \cdots + a_N[z^{-N}Y(z) + z^{-(N-1)}y(-1) + \cdots + z^{-0}y(-N)] \\ &= b_0X(z) + b_1[z^{-1}X(z) + z^{-0}x(-1)] + b_2[z^{-2}X(z) + z^{-1}x(-1) \\ &\quad + z^{-0}x(-2)] + \cdots + b_L[z^{-L}X(z) + z^{-(L-1)}x(-1) + \cdots + z^{-0}x(-L)]. \end{aligned}$$

Next we collect terms and assume $L = N$:

$$\begin{aligned} &[a_0 + a_1z^{-1} + a_2z^{-2} + \cdots + a_N]Y(z) \\ &\qquad = [-a_1y(-1) - a_2y(-2) - \cdots - a_Ny(-N) \\ &\qquad\quad +b_1x(-1) + b_2x(-2) + \cdots + b_Lx(-L)] \\ &\qquad\quad +[-a_2y(-1) - a_3y(-2) - \cdots - a_Ny(-N+1) \\ &\qquad\quad +b_2x(-1) + b_3x(-2) + \cdots + b_Lx(-L+1)]z^{-1} \\ &\qquad\quad + \cdots + [-a_Ny(-1) + b_Lx(-1)]z^{-(N-1)} \\ &\qquad\quad +[b_0 + b_1z^{-1} + \cdots + b_Lz^{-L}]X(z). \end{aligned}$$

The right side of this equation consists of terms due to the ICs and terms due to the transform of the input $X(z)$, so we can write

$$Q(z)Y(z) = P_{\text{IC}}(z) + P(z) \cdot X(z)$$

which, solved for the z transform of the output $y(n)$, yields

transform of $y(n)$

$$Y(z) = \frac{P_{\text{IC}}(z)}{Q(z)} + \frac{P(z)}{Q(z)} \cdot X(z).$$

To complete the solution we need to know the initial conditions and $x(n)$, from which we obtain $X(z)$. At this point we can evaluate $Y(z)$ and expand it in partial fractions. Knowing the region of convergence, $y(n)$ can be determined using Table 8.1. A diagram of the procedure is given in Figure 8.5.

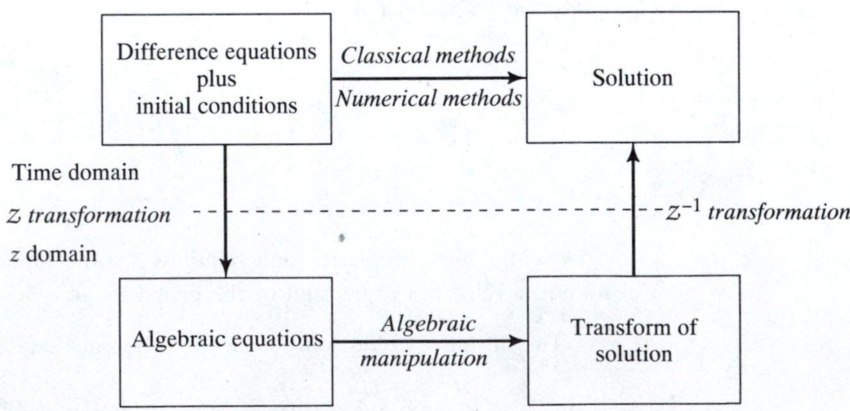

FIGURE 8.5 *A procedure for solving linear difference equations*

ILLUSTRATIVE PROBLEM 8.10 ***z-Transform Solution of a Difference Equation***

Find the complete solution of the difference equation

$$y(n) + 0.30y(n-1) - 0.40y(n-2) = 0.50[x(n) + x(n-1)]$$

for the initial conditions $y(-1) = y(-2) = 1$ and an input of $x(n) = u(n)$.

Solution

Taking the unilateral z transform gives

$$Y(z) + 0.30[z^{-1}Y(z) + z^{-0}y(-1)] - 0.40[z^{-2}Y(z) + z^{-1}y(-1) + z^{-0}y(-2)] = 0.50[X(z) + z^{-1}X(z) + z^{-0}x(-1)].$$

With $X(z) = 1/(1 - z^{-1})$ and $x(-1) = 0$, because $u(n) = 0$ for $n < 0$, we substitute the initial conditions for $y(-1)$ and $y(-2)$ and do some algebra to obtain the transform

$$Y(z) = \frac{0.40z^{-1} + 0.10}{1 + 0.30z^{-1} - 0.40z^{-2}} + \frac{0.50(1 + z^{-1})}{1 + 0.30z^{-1} - 0.40z^{-2}} \cdot \frac{1}{1 - z^{-1}}.$$

Combining the two preceding terms and putting the result in positive powers of z gives

$$Y(z) = \frac{0.60z^3 + 0.80z^2 - 0.40z}{(z^2 + 0.30z - 0.40)(z - 1)}, \qquad |z| > 1.$$

Using PFE on $Y(z)/z$ produces

$$\frac{Y(z)}{z} = \frac{0.60z^2 + 0.80z - 0.40}{(z + 0.80)(z - 0.50)(z - 1)} = \frac{C_1}{z + 0.80} + \frac{C_2}{z - 0.50} + \frac{C_3}{z - 1}$$

with the partial fraction constants given by

$$C_1 = \left.\frac{0.60z^2 + 0.80z - 0.40}{(z - 0.50)(z - 1)}\right|_{z=-0.80} = -0.280,$$

$$C_2 = \left.\frac{0.60z^2 + 0.80z - 0.40}{(z + 0.80)(z - 1)}\right|_{z=0.50} = -0.231, \quad \text{and}$$

$$C_3 = \left.\frac{0.60z^2 + 0.80z - 0.40}{(z + 0.80)(z - 0.50)}\right|_{z=1} = 1.111.$$

After multiplying through by z to be able to utilize the pairs of Table 8.1, we have

$$Y(z) = \frac{-0.280z}{z + 0.80} - \frac{0.231z}{z - 0.50} + \frac{1.111z}{z - 1}$$

and the solution

$$y(n) = [-0.280(-0.80)^n - 0.231(0.50)^n + 1.111]u(n).$$

CROSS-CHECK

1. We should check the partial fraction expansion by making use of the MATLAB function **residue.** The script for PFE (partial fraction expansion) follows along with the results that verify our longhand computations.

MATLAB Script

```
%PFE...computes PFE constants, poles, and the direct term of a rational
%function.
b=[0.6,0.8,-0.4]                  % numerator coefficients of Y(z)/z
a=[1,-0.7,-0.7,0.4]               % denominator coefficients of Y(z)/z
[r,p,k]=residue(b,a)              % call residue, r=PFE constant,
                                  % p=pole, k=direct term
```

Output

```
r=1.111
  -0.280                          the three partial fraction constants
  -0.231
p=1.0000
  -0.8000                         the three poles of Y(z)/z
  0.5000
```

2. A few values of the "time-domain" solution are also easily checked. First, from the iterative solution of the difference equation $y(n) = -0.3y(n-1) + 0.4y(n-2) + 0.5x(n) + 0.5x(n-1)$, we have $y(0) = -0.3(1) + 0.4(1) + 0.5(1) + 0.5(0) = 0.6$. From the solution found before, $y(0) = -0.280 - 0.231 + 1.111 = 0.600$, which agrees with the iterative solution. Second, from the DE, $y(1) = -0.3(0.6) + 0.4(1) + 0.5(1) + 0.5(1) = 1.22$. From the solution, $y(1) = -0.280(-0.80) - 0.231(0.50) + 1.111 = 1.22$, in agreement again. Since all poles of $(z-1) \cdot Y(z)$ of this causal system lie inside the unit circle, we can use the final value theorem to find $y(\infty) = (z-1) \cdot Y(z)|_{z=1} = 1.111$, which also agrees with the solution. We could also use **dlsim** or **ksdlsim,** but there is a good chance our result is correct. ■

Linear systems are often characterized by simultaneous difference equations. Illustrative Problem 8.11 demonstrates how to solve these difference equations using z transforms.

ILLUSTRATIVE PROBLEM 8.11
z-Transform Solution of Simultaneous DEs

A linear system is modeled by the simultaneous difference equations

$$3p(n-2) + 2p(n-1) + p(n) - 2q(n) = 5f(n) - 7g(n)$$
$$2q(n-2) - 3q(n-1) + 5p(n) = 3g(n)$$

where $f(n)$ and $g(n)$ are the system inputs and $p(n)$ and $q(n)$ its outputs. Taking the z transform of these equations gives us the two simultaneous algebraic equations

$$3\{z^{-2}P(z) + z^{-1}p(-1) + p(-2)\} + 2\{z^{-1}P(z) + p(-1)\} + P(z) - 2Q(z)$$
$$= 5F(z) - 7G(z)$$
$$2\{z^{-2}Q(z) + z^{-1}q(-1) + q(-2)\} - 3\{z^{-1}Q(z) + q(-1)\} + 5P(z) = 3G(z).$$

These equations can be written in matrix form as

$$\begin{bmatrix} 3z^{-2} + 2z^{-1} + 1 & -2 \\ 5 & 2z^{-2} - 3z^{-1} \end{bmatrix} \begin{bmatrix} P(z) \\ Q(z) \end{bmatrix} = \begin{bmatrix} -[3z^{-1} + 2]p(-1) - 3p(-2) + 5F(z) - 7G(z) \\ -[2z^{-1} - 3]q(-1) - 2q(-2) + 3G(z) \end{bmatrix}$$

and given the two input signals and the four initial conditions, we could use matrix methods or determinants to find the transforms $P(z)$ and $Q(z)$. Then, with the help of partial fractions (**residue**) and Table 8.1 (and lots of care!), the output sequences $p(n)$ and $q(n)$ are found. For $f(n) = u(n)$, $g(n) = \delta(n)$, $q(-1) = -1$, and $q(-2) = p(-1) = p(-2) = 0$, determine an analytical expression for the sequence $p(n)$.

Solution

Substituting the given IC data and the transforms of $f(n)$ and $g(n)$ gives

$$\begin{bmatrix} 3z^{-2} + 2z^{-1} + 1 & -2 \\ 5 & 2z^{-2} - 3z^{-1} \end{bmatrix} \begin{bmatrix} P(z) \\ Q(z) \end{bmatrix} = \begin{bmatrix} 5/(1 - z^{-1}) - 7 \\ 2z^{-1} - 3 + 3 \end{bmatrix}.$$

Solving this matrix equation we have

$$\begin{bmatrix} P(z) \\ Q(z) \end{bmatrix} = \begin{bmatrix} 3z^{-2} + 2z^{-1} + 1 & -2 \\ 5 & 2z^{-2} - 3z^{-1} \end{bmatrix}^{-1} \begin{bmatrix} 5/(1 - z^{-1}) - 7 \\ 2z^{-1} - 3 + 3 \end{bmatrix}$$
$$= \frac{1}{\Delta(z)} \begin{bmatrix} 2z^{-2} - 3z^{-1} & 2 \\ -5 & 3z^{-2} + 2z^{-1} + 1 \end{bmatrix}$$
$$\times \begin{bmatrix} 5/(1 - z^{-1}) - 7 \\ 2z^{-1} - 3 + 3 \end{bmatrix} \quad \text{where}$$
$$\Delta(z) = (2z^{-2} - 3z^{-1})(3z^{-2} + 2z^{-1} + 1) + 10$$
$$= 6z^{-4} - 5z^{-3} - 4z^{-2} - 3z^{-1} + 10.$$

Multiplying out the matrices and performing some algebra produces the vector of the z transforms of the output sequences $p(n)$ and $q(n)$:

$$\begin{bmatrix} P(z) \\ Q(z) \end{bmatrix} = \begin{bmatrix} \dfrac{z^4 - 2.9z^3 + 1.4z^2}{(z - 1)(z^4 - 0.3z^3 - 0.4z^2 - 0.5z + 0.6)} \\ \dfrac{z^5 - 3.3z^4 + 0.6z^3 + 2.2z^2 - 0.6z}{(z - 1)(z^4 - 0.3z^3 - 0.4z^2 - 0.5z + 0.6)} \end{bmatrix}.$$

Using **residue** to expand $P(z)/z$ produces

$$\frac{P(z)}{z} = \frac{-1.25}{z - 1} + \frac{-0.182 - j0.622}{z + 0.629 - j0.692} + \text{cplx conj.}$$
$$+ \frac{0.807 + j0.171}{z - 0.780 - j0.281} + \text{cplx conj.}$$

BASIC CONCEPTS

and using Table 8.1 gives the analytical expression $p(n) = -1.250 + 1.294(0.935)^n \cdot \cos(2.309n - 1.855) + 1.650(0.828)^n \cos(0.346n + 0.209)$, $n \geq 0$. Following a similar procedure we find that $q(n) = -6.250 + 1.286(0.935)^n \cos(2.309n + 0.111) + 6.232(0.828)^n \cos(0.346n - 0.289)$, $n \geq 0$.

UNIT SAMPLE RESPONSE OF A RECURSIVE SYSTEM

To find the unit sample response $h(n)$ for the causal Nth-order recursive system

$$\sum_{k=0}^{N} a_k y(n-k) = b_0 x(n),$$

we take the z transform, setting all ICs to zero with $X(z) = z[x(n)] = z[\delta(n)] = 1$. This results in the transform

$$Y(z) = H(z) = \frac{b_0}{a_0 + a_1 z^{-1} + \cdots + a_N z^{-N}} = \frac{b_0 z^N}{a_0 z^N + a_1 z^{N-1} + \cdots + a_N}$$

and assuming N distinct characteristic roots, the unit sample response is

$$h(n) = \sum_{k=1}^{N} C_k (r_k)^n u(n).$$

The situation becomes a bit more complex if there are delays of the input. To find a general expression for $h(n)$ that is similar to the preceding one, we use first-order systems to develop the concept. For the system $y(n) + a_1 y(n-1) = b_0 x(n) + b_1 x(n-1)$, ICs $= 0$, and $x(n) = \delta(n)$, we have

$$Y(z) = H(z) = \frac{b_0 + b_1 z^{-1}}{1 + a_1 z^{-1}} = \frac{b_0 z + b_1}{z + a_1}$$

and expanding $H(z)/z$ in partial fractions gives

$$\frac{H(z)}{z} = \frac{b_0 z + b_1}{z(z + a_1)} = \frac{A_0}{z} + \frac{C_1}{z + a_1}.$$

Returning to $H(z)$ and taking the inverse transform produces the results

$$H(z) = A_0 + \frac{C_1 z}{z + a_1} \quad \text{and} \quad h(n) = A_0 \delta(n) + C_1 (-a_1)^n u(n).$$

Now for $y(n) + a_1 y(n-1) = b_0 x(n) + b_1 x(n-1) + b_2 x(n-2)$, we have

$$H(z) = \frac{b_0 + b_1 z^{-1} + b_2 z^{-2}}{1 + a_1 z^{-1}} = \frac{b_0 z^2 + b_1 z + b_2}{z(z + a_1)}$$

and expanding $H(z)/z$ in partial fractions gives

$$\frac{H(z)}{z} = \frac{b_0 z^2 + b_1 z + b_2}{z^2 (z + a_1)} = \frac{A_1}{z^2} + \frac{A_0'}{z} + \frac{C_1'}{z + a_1}.$$

CHAPTER 8

Returning to $H(z)$ and taking the inverse transform produces the results

$$H(z) = \frac{A_1}{z} + A_0' + \frac{C_1'z}{z + a_1} \quad \text{and}$$
$$h(n) = A_1\delta(n-1) + A_0'\delta(n) + C_1'(-a_1)^n u(n).$$

For $y(n) + a_1 y(n-1) = b_0 x(n) + b_1 x(n-1) + b_2 x(n-2) + b_3 x(n-3)$, the pattern is apparent, and we can write the result as

$$h(n) = A_2\delta(n-2) + A_1'\delta(n-1) + A_0''\delta(n) + C_1''(-a_1)^n u(n).$$

Comments:

1. This result can be generalized for a first-order system with L delays of the input to

$$h(n) = C(-a_1)^n u(n) + \sum_{k=0}^{L-1} A_k\delta(n-k).$$

2. For an Nth-order system with the distinct characteristic roots $r_1, r_2, \ldots, r_N$ and with L delays of the input, the unit sample response is given by

$$h(n) = \sum_{k=1}^{N} C_k(r_k)^n u(n) + \sum_{k=0}^{L-N} A_k\delta(n-k).$$

This result was given in Chapter 7 in the section on unit sample response.

BASIC CONCEPTS

CONVOLUTION

In Chapter 7, "Discrete Systems," the forced response (ICs set to zero) of an LTI system was obtained from the convolution sum

$$y(n) = \sum_{m=-\infty}^{\infty} h(m)x(n-m) = \sum_{m=-\infty}^{\infty} h(n-m)x(m)$$

where $h(n)$ is the unit impulse response of the system and $x(n)$ represents the system input sequence. If the system is causal, $h(n) = 0$ for $n < 0$, and if the input is zero for $n < 0$, a causal sequence, the convolution $y(n) = h(n) * x(n) = x(n) * h(n)$ is given by

$$y(n) = \sum_{m=0}^{n} h(m)x(n-m) = \sum_{m=0}^{n} h(n-m)x(m).$$

Earlier in this chapter the system transfer function was defined as $H(z) = Y(z)/X(z)$, and we see that if the transform $X(z)$ of the input $x(n)$ is known, then the transform $Y(z)$ of the output sequence $y(n)$ is

$$Y(z) = H(z) \cdot X(z) = X(z) \cdot H(z)$$

and the output $y(n)$ can be found by the inverse transform

$$y(n) = z^{-1}[Y(z)] = z^{-1}[H(z) \cdot X(z)] = z^{-1}[X(z) \cdot H(z)].$$

Thus we have the convolution pair or property that allows the evaluation of a convolution in either the time domain directly or indirectly by the means of z transforms:

convolution property

$$y(n) = \sum_{m=-\infty}^{\infty} h(m)x(n - m) \Leftrightarrow H(z) \cdot X(z)$$

where the region of convergence for $H(z) \cdot X(z)$ will be determined as the intersection of the ROCs of $H(z)$ and $X(z)$.

CHAPTER 8

ILLUSTRATIVE PROBLEM 8.12
Convolution Using Transforms

A lowpass digital filter with the unit sample response $h(n) = [(0.9)^n - (0.8)^n]u(n)$ is subjected to a unit step input—that is, $x(n) = u(n)$.

a. Use z transforms to find the filter's output $y(n) = h(n) * x(n)$.

b. Verify your answer to part (a) by using the "time-domain" convolution sum.

Solution

a. We use the convolution pair or property $Y(z) = H(z) \cdot X(z)$. From Table 8.1,

$$H(z) = \frac{z}{z - 0.9} - \frac{z}{z - 0.8} = \frac{0.1z}{(z - 0.9)(z - 0.8)}, \qquad |z| > 0.9, \quad \text{and}$$

$$X(z) = \frac{z}{z - 1}, \qquad |z| > 1.$$

Thus the output transform is

$$\begin{aligned} Y(z) &= \frac{0.1z}{(z - 0.9)(z - 0.8)} \cdot \frac{z}{z - 1} \\ &= \frac{5z}{z - 1} + \frac{4z}{z - 0.8} - \frac{9z}{z - 0.9}, \qquad |z| > 1 \end{aligned}$$

and table lookup gives the output sequence

$$y(n) = [5 + 4(0.8)^n - 9(0.9)^n]u(n).$$

b. Using the convolution sum we have

$$\begin{aligned} y(n) &= \sum_{0}^{n} (0.9)^m - \sum_{0}^{n} (0.8)^m = \frac{1 - (0.9)^{n+1}}{1 - 0.9} - \frac{1 - (0.8)^{n+1}}{1 - 0.8} \\ &= 5 - 9(0.9)^n + 4(0.8)^n, \qquad n \geq 0. \end{aligned}$$

Comment: There are several MATLAB functions that can be used to obtain a graphical solution to this problem. Two of them are **filter** and **dstep** from the Signals and Systems Toolbox. Recall that the transfer function is

$$H(z) = \frac{0.1z}{z^2 - 1.7z + 0.72}.$$

The script using **filter** is next, and the plot is shown in Figure 8.6a.

MATLAB Script

```
%F8_6 System output using filter and dstep

%F8_6a Unit step response from filter
n=0:1:35;                    % evaluate for 36 samples
b=[0,.1,0];                  % coefficients of input terms, the x(n)'s
a=[1,-1.7,.72];              % coefficients of output terms, the y(n)'s
x=[1*ones(size(n))];         % constant input of 1
y=filter(b,a,x);             % call filter
%...plotting statements and pause
```

The script below for Figure 8.6b uses the transfer function input to **dstep** and the option that plots automatically.

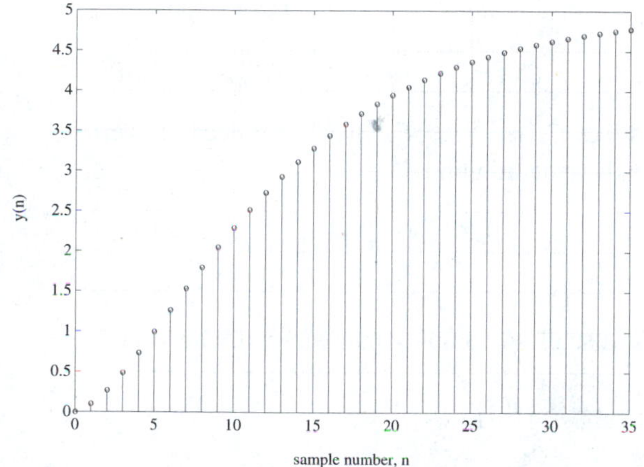

a. From **filter**

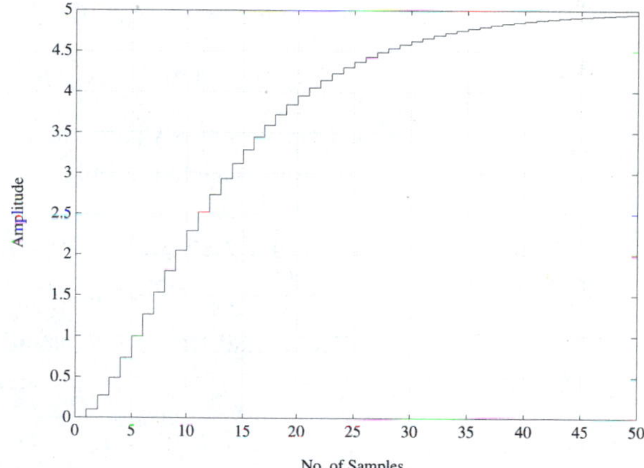

b. From **dstep**

FIGURE 8.6 *Unit step response*

MATLAB Script

```
%F8_6b Unit step response from dstep
num=[0,.1,0];                % numerator coefficients
den=[1,-1.7,0.72];           % denominator coefficients
dstep(num,den);              % call dstep
grid;
```

SINUSOIDAL STEADY-STATE RESPONSE

A very important and practical situation occurs when a stable causal system described by the transfer function $H(z) = Y(z)/X(z)$ is subjected to the sinusoidal input sequence $x(n) = A\cos(n\theta + \alpha)u(n)$. Let us show that the system's steady-state output response $y_{ss}(n)$ is given by the expression

$$y_{ss}(n) = A|H(e^{j\theta})|\cos(n\theta + \alpha + \angle H(e^{j\theta}))$$

where

$$H(e^{j\theta}) = H(z)|_{z=e^{j\theta}} = |H(e^{j\theta})|e^{j\angle H(e^{j\theta})}.$$

The z transform of the output caused by the input $x(n) = A\cos(n\theta + \alpha)u(n)$ is

$$Y(z) = H(z)\cdot X(z) = H(z)\cdot\left\{\frac{0.5Aze^{j\alpha}}{z - e^{j\theta}} + \frac{0.5Aze^{-j\alpha}}{z - e^{-j\theta}}\right\}.$$

Assuming nonrepeated poles and expanding in partial fractions produces

PFE

$$Y(z) = \frac{C_1 z}{z - p_1} + \frac{C_2 z}{z - p_2} + \cdots + \frac{C_N z}{z - p_N} + \frac{C_\theta z}{z - e^{j\theta}} + \frac{C_\theta^* z}{z - e^{-j\theta}}$$

where C_θ^* is the complex conjugate of C_θ. Taking the inverse transform yields the solution for $n \geq 0$:

$$y(n) = [C_1(p_1)^n + C_2(p_2)^n + \cdots + C_N(p_N)^n + C_\theta e^{jn\theta} + C_\theta^* e^{-jn\theta}]u(n).$$

If the system is stable, the terms like $C_1(p_1)^n, \ldots, C_N(p_N)^n$ will decay to zero as $n \to \infty$, leaving only the steady-state solution

using the Euler relation

$$\begin{aligned} y_{ss}(n) &= C_\theta e^{jn\theta} + C_\theta^* e^{-jn\theta} = |C_\theta|e^{j\angle C_\theta}e^{jn\theta} + |C_\theta|e^{-j\angle C_\theta}e^{-jn\theta} \\ &= 2|C_\theta|\cos(n\theta + \angle C_\theta). \end{aligned}$$

The partial fraction constant C_θ (usually a complex number) is found from

PF constant

$$\begin{aligned} C_\theta &= \frac{H(z)X(z)}{z}\cdot(z - e^{j\theta})]_{z=e^{j\theta}} \\ &= \frac{A}{2}H(e^{j\theta})e^{j\alpha} = \frac{A}{2}|H(e^{j\theta})|e^{j(\angle H(e^{j\theta})+\alpha)} \end{aligned}$$

Now we can write the steady-state response in the form

$$y_{ss}(n) = A|H(e^{j\theta})|\cos(n\theta + \alpha + \angle H(e^{j\theta}))$$

which is the result we set out to derive. This equation makes the *evaluation* of a discrete system's steady-state response to a sinusoidal input sequence very straightforward, and we refer to the result as the sinusoidal steady-state formula. In other words, it tells us that when a sinusoidal sequence $x(n) = A\cos(n\theta + \alpha)u(n)$ is applied to a *stable* linear time-invariant system, then the steady-state output $y_{ss}(n)$ has

1. as its amplitude the input amplitude A multiplied by the gain $|H(e^{j\theta})|$, and
2. as its phase the input phase α shifted by the angle $\angle H(e^{j\theta})$.

This is illustrated in Figure 8.7, where the input sequence and the steady-state output sequence of an LTI system subjected to a sinusoidal input are shown.

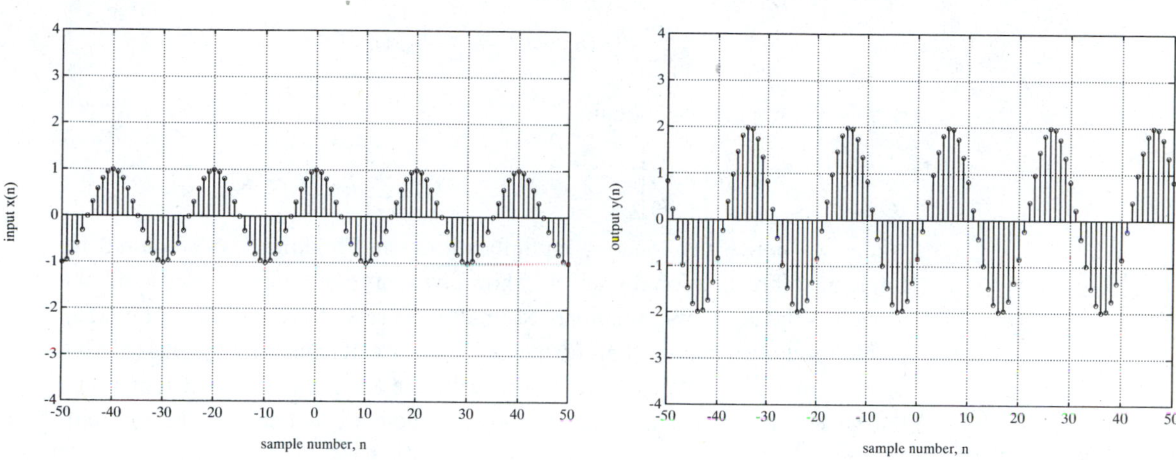

FIGURE 8.7 *Input sequence (left); output sequence (right)*

ILLUSTRATIVE PROBLEM 8.13 *Using the Sinusoidal Steady-State Formula*

A digital notch filter can be modeled by $y(n) + 0.90y(n-2) = x(n) + x(n-2)$, where $y(n)$ represents the filter output and $x(n)$ is its input. The input sequence $x(n) = A + B\cos(n\pi/2)$ is applied to this filter. Find the equation for the steady-state output voltage $y_{ss}(n)$.

Solution

Following the procedure just outlined, we first find the transfer function

$$H(z) = \frac{1 + z^{-2}}{1 + 0.90z^{-2}} = \frac{z^2 + 1}{z^2 + 0.90}.$$

Next $H(z)$ needs to be evaluated at the frequencies of the input sequence, which are $\theta = 0$ (a constant such as A can be considered a sinusoid of zero frequency) or $z = e^{j0} = 1$ and $\theta = \pi/2$ or $z = e^{j\pi/2} = j1$. Thus we have

$$H(1) = \frac{1^2 + 1}{1^2 + 0.90} = 1.053 \quad \text{and} \quad H(j1) = \frac{(j1)^2 + 1}{(j1)^2 + 0.90} = 0.$$

Finally, applying the sinusoidal steady-state formula gives $y_{ss}(n) = 1.053A$, and the filter does indeed notch out, or suppress, the frequency of $\theta = \pi/2$.

SYSTEM DIAGRAMS OR STRUCTURES

We have modeled linear discrete systems by constant-coefficient difference equations, by transfer functions, and by linear state-space equations. In this section we introduce an alternative model in the form of a system diagram, a graphical way of representing the same information contained in the difference equation, transfer function, or state-space model. Such diagrams can provide useful visualizations of system structure; in addition, there are graphical algorithms available to facilitate system analysis.

Recall that when the general form of a system's difference equation

$$\sum_{k=0}^{N} a_k y(n-k) = \sum_{k=0}^{L} b_k x(n-k)$$

is solved for $y(n)$, we obtain

$$y(n) = \frac{1}{a_0}\left[-\sum_{k=1}^{N} a_k y(n-k) + \sum_{k=0}^{L} b_k x(n-k)\right]$$

where we notice that delays, multiplications, and additions are required to implement this equation. Figure 8.8 shows two sets of commonly used symbols as well as the describing relations. Normally, the sets of symbols are not mixed, and it is strictly a matter of preference as to the use of signal flowgraphs (SFGs) or block diagrams. We will use both. In Figure 8.9 a diagram that represents the general difference equation is shown, where we notice that the top half

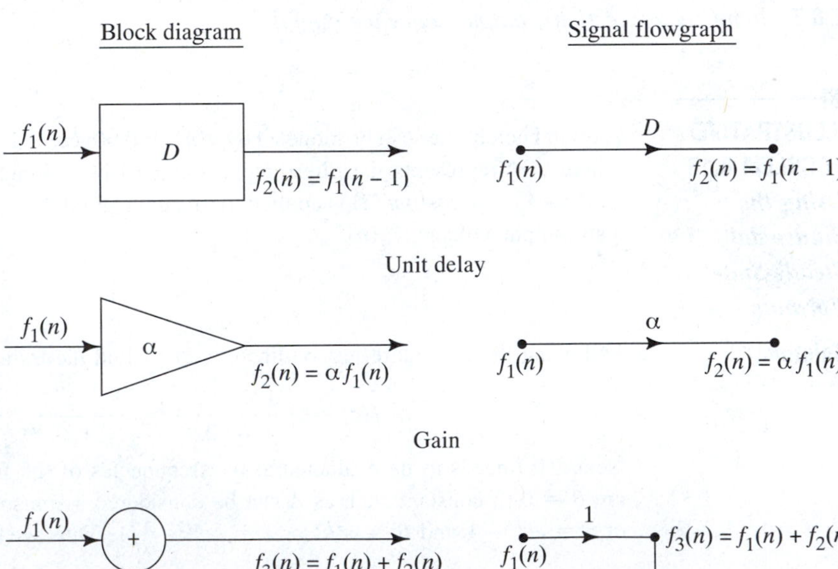

FIGURE 8.8 *System diagram symbols*

of the diagram represents (realizes) the input terms of the difference equation, and the bottom half shows the output terms.

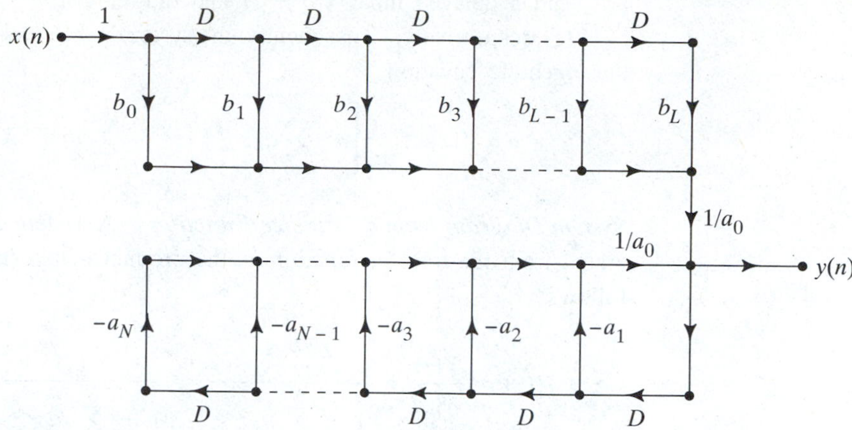

FIGURE 8.9 *System diagram for a DE*

ILLUSTRATIVE PROBLEM 8.14
Drawing a System Diagram

A bandpass digital filter is represented by the difference equation

$$y(n) = -0.81y(n-2) + 0.105x(n) - 0.085x(n-2).$$

Draw a signal flowgraph in the manner of Figure 8.9 that represents this filter.

Solution

Comparing the filter's difference equation with that of the general form of the difference equation, we see that $a_0 = 1$, $a_1 = 0$, $a_2 = 0.81$, $b_0 = 0.105$, $b_1 = 0$, and $b_2 = -0.085$. The system diagram marked with these gains is given in Figure 8.10. The multipliers showing zero gain could be eliminated, but this is simply a matter of choice.

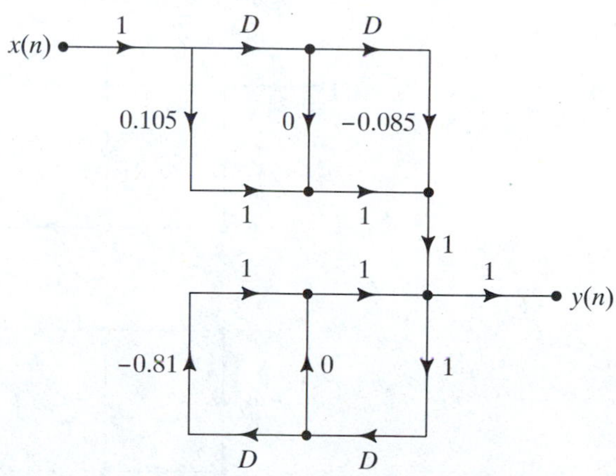

FIGURE 8.10 *System diagram for Illustrative Problem 8.14*

BASIC CONCEPTS

The general system diagram from the z-transform point of view is shown in Figure 8.11. There we have replaced $x(n)$ with its transform $X(z)$, $y(n)$ with its transform $Y(z)$, the unit delays of D with the transform equivalent z^{-1}, and a delayed input $x(n-i)$ and output $y(n-j)$ with $z^{-i}X(z)$ and $z^{-j}Y(z)$, respectively. Consequently, the system's difference equation becomes the algebraic equation

$$Y(z) = \frac{1}{a_0}\left[-\sum_{k=1}^{N} a_k z^{-k} Y(z) + \sum_{k=0}^{L} b_k z^{-k} X(z)\right].$$

System Diagram from a Transfer Function A system diagram can be generated from the transfer function as the product of two transfer functions as follows:

$$H(z) = \frac{Y(z)}{X(z)} = \frac{\sum_{k=0}^{L} b_k z^{-k}}{1 + \sum_{k=1}^{N} a_k z^{-k}} = \frac{1}{1 + \sum_{k=1}^{N} a_k z^{-k}} \cdot \frac{\sum_{k=0}^{L} b_k z^{-k}}{1},$$

where $a_0 = 1$ is assumed. Next we define a new variable $Q(z)$ such that

$$H(z) = \frac{Y(z)}{X(z)} = \frac{Q(z)}{X(z)} \cdot \frac{Y(z)}{Q(z)} = \frac{1}{1 + \sum_{k=1}^{N} a_k z^{-k}} \cdot \frac{\sum_{k=0}^{L} b_k z^{-k}}{1}$$

where

$$H_1(z) = \frac{Q(z)}{X(z)} = \frac{1}{1 + \sum_{k=1}^{N} a_k z^{-k}} \quad \text{and} \quad H_2(z) = \frac{Y(z)}{Q(z)} = \frac{\sum_{k=0}^{L} b_k z^{-k}}{1}.$$

$H_1(z)$ implements the poles of $H(z)$ and $H_2(z)$ implements the zeros of $H(z)$. Rearranging the equation for $H_1(z)$ and taking the inverse transform gives

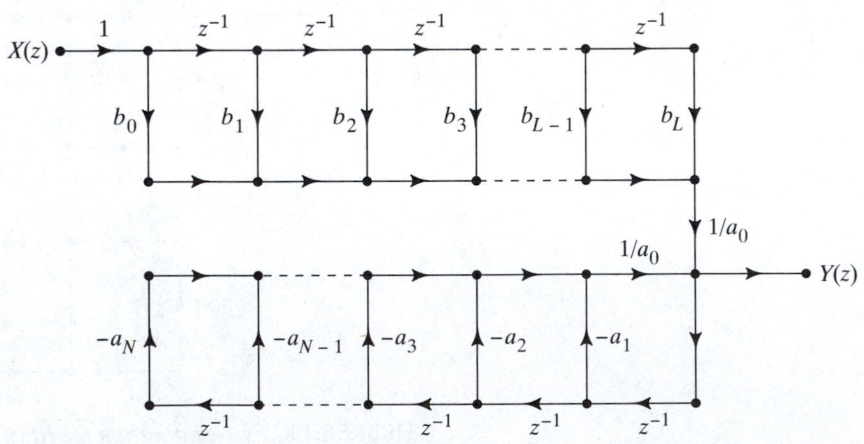

FIGURE 8.11 *System diagram for a transformed DE*

CHAPTER 8

$$Q(z) = X(z) - \sum_{k=1}^{N} a_k z^{-k} Q(z) \quad \text{and} \quad q(n) = x(n) - \sum_{k=1}^{N} a_k q(n-k).$$

See Figure 8.12a for the implementation of the equation for $Q(z)$ that realizes the poles of the system. Finally, rearranging the equation for $H_2(z)$ and taking the inverse transform gives the output relations

$$Y(z) = \sum_{k=0}^{L} b_k z^{-k} Q(z) \quad \text{and} \quad y(n) = \sum_{k=0}^{L} b_k q(n-k).$$

In Figure 8.12b we have shown the implementation of this output equation that realizes the system zeros, and Figure 8.12c (which assumes $N = L$) portrays

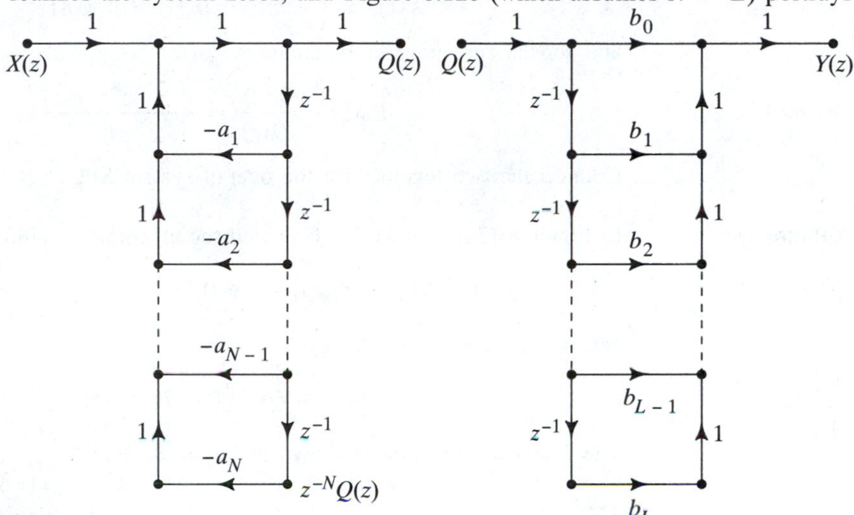

a. Poles of $H(z)$

b. Zeros of $H(z)$

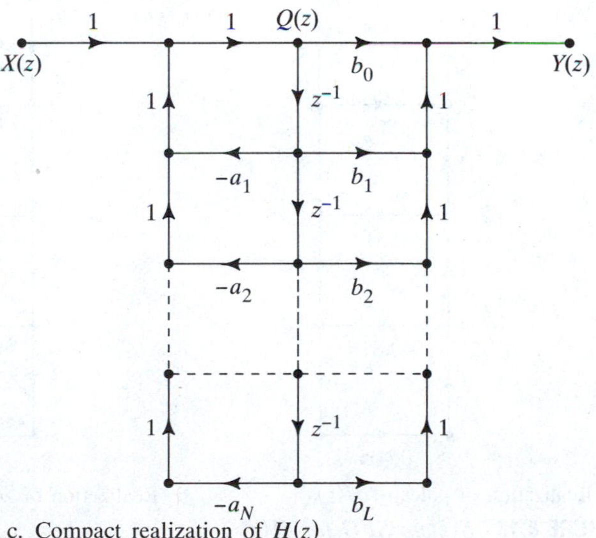

c. Compact realization of $H(z)$

FIGURE 8.12 *Realization of a transfer function*

the combination of the output and the input sections that share the delay elements to give a general diagram for a recursive system.

ILLUSTRATIVE PROBLEM 8.15
System Diagram from a Transfer Function

A digital filter is described by the fourth-order transfer function

$$H(z) = \frac{Y(z)}{X(z)} = \frac{1 - 2z^{-2} + z^{-4}}{1 - z^{-1} - 0.31z^{-2} + 0.81z^{-3} - 0.405z^{-4}}.$$

Draw a transform domain SFG for the pole transfer function

$$H_1(z) = \frac{Q(z)}{X(z)} = \frac{1}{1 - z^{-1} - 0.31z^{-2} + 0.81z^{-3} - 0.405z^{-4}}$$

and for the zero transfer function

$$H_2(z) = \frac{Y(z)}{Q(z)} = \frac{1 - 2z^{-2} + z^{-4}}{1}.$$

Connect the two together for the overall system SFG.

Solution

In Figure 8.13a we have implemented the algebraic equation

$$Q(z) = X(z) + Q(z)[z^{-1} + 0.31z^{-2} - 0.81z^{-3} + 0.405z^{-4}]$$

and in Figure 8.13b we have

$$Y(z) = Q(z)[1 - 2z^{-2} + z^{-4}].$$

The composite diagram is shown in Figure 8.13c.

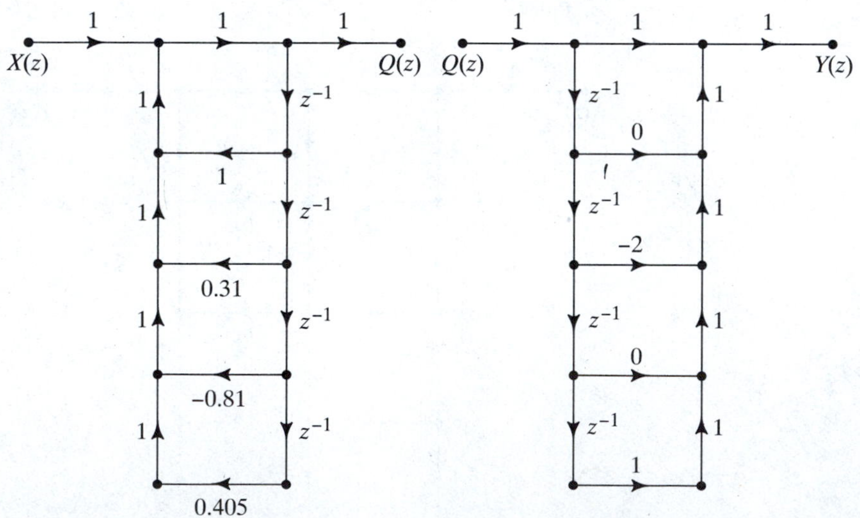

a. Realization of poles of $H(z)$ b. Realization of zeros of $H(z)$

FIGURE 8.13 *System SFG for $H(z)$*

Continues

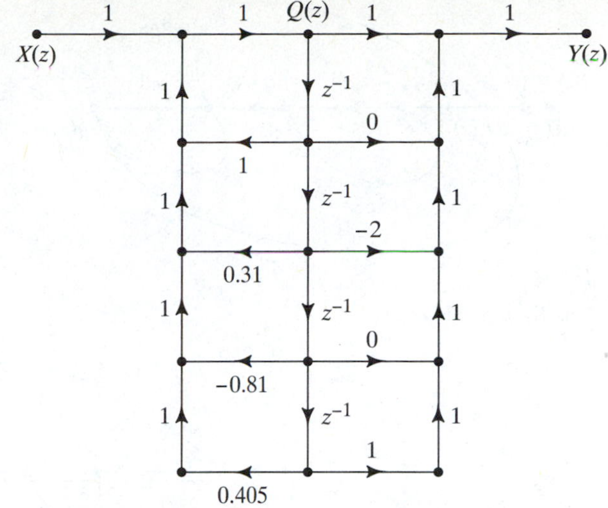

c. Composite diagram for $H(z)$

FIGURE 8.13 *Continued*

MASON'S GAIN RULE AND APPLICATIONS

In the early 1950s as part of his doctoral studies at the Massachusetts Institute of Technology, Professor Samuel J. Mason developed a rule (or algorithm) that was based on a graphical procedure for calculating transfer functions between input and output points of a signal flowgraph or an equivalent block diagram. The proof of this rule is based on Cramer's method for solving a set of simultaneous algebraic equations; consequently, we can apply the Mason Gain Rule (MGR) in the algebraic z domain. We explain the rule by using the SFG of Figure 8.14a, which models a sampled-data feedback control system.

Let us consider the following general definitions and their applicability to this signal flowgraph.

Nodes. Points on a graph where the signals appear. All the heavy dots are nodes, with the most important of these being $X(z)$, $E(z)$, $\Theta(z)$, and $\Omega(z)$. At any node, the quantities associated with the incoming branches are summed, whereas the outgoing branches have no effect on the signal at the node; that is, we don't apply Kirchhoff's law at the node.

Branch. A directed line segment, having an associated gain, that connects two nodes. An unmarked branch is assumed to have a gain of 1. Two easily identified branches are the delays with gains of z^{-1}.

Input node. Has no incoming branches. Obviously, $X(z)$ is the only input node.

Output node. Must have at least one incoming branch. All the rest of the nodes of the graph are output nodes. We use the node $\Theta(z) = Y(z)$ as the designated system output.

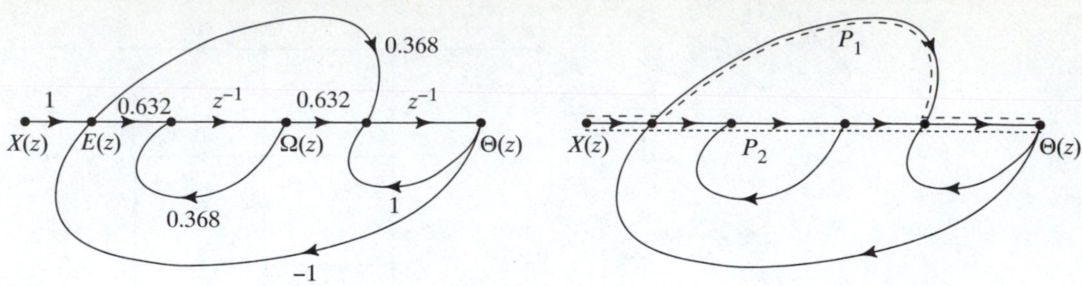

a. System SFG

b. SFG showing paths

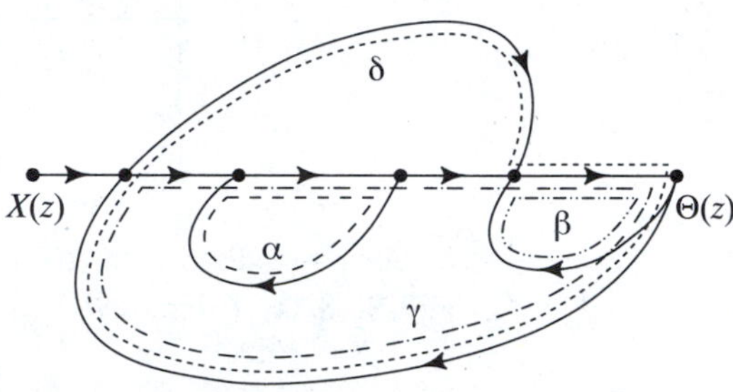

c. SFG showing loops

FIGURE 8.14 *SFG to illustrate Mason's Gain Rule*

CHAPTER 8

Path. A continuous sequence of branches, traversed in the indicated branch direction, along which no node is encountered more than once. From $X(z)$ to $Y(z) = \Theta(z)$ there are two paths (P_1 and P_2), as indicated in Figure 8.14b.

Loop. A continuous sequence of branches, traversed in the indicated branch directions from one node around a closed path back to the same node, along which no node is encountered more than once. This graph has four loops: α, β, γ, and δ, as indicated in Figure 8.14c.

The Mason Gain Rule is

MGR

$$H(z) = \frac{Y(z)}{X(z)} = \frac{\sum_{k=1}^{M} P_k(z) \cdot \Delta_k(z)}{\Delta(z)}$$

where

$H(z)$ = the transfer function relating an output node to an input node

$\Delta(z)$ = the graph determinant

$P_k(z)$ = the gain of the kth path from input to output

$\Delta_k(z)$ = cofactor of the kth path.

These terms will be defined as we find the transfer function $H(z) = Y(z)/X(z)$ for the SFG of Figure 8.14a, where there is a particular order required for the calculation of the three quantities $\Delta(z)$, $P_k(z)$, and $\Delta_k(z)$.

1. The first quantity to determine is the graph determinant $\Delta(z)$, where

$$\begin{aligned}\Delta(z) = 1 &- \sum \text{loop gains} \\ &+ \sum \text{products of the gains of nontouching loops taken two at a time} \\ &- \sum \text{products of the gains of nontouching loops taken three at a time} \\ &+ \cdots\end{aligned}$$

where the loop gain L_j is simply the product of the gains around the jth loop. For the example of Figure 8.14c, the loop gains are $L_1 = 0.368z^{-1}$ around the α loop, $L_2 = 1z^{-1}$ around the β loop, $L_3 = (0.632)(z^{-1})(0.632)(z^{-1})(-1) = -(0.632)^2 z^{-2}$ around the γ loop, and $L_4 = (0.368)(z^{-1})(-1) = -0.368z^{-1}$ around the δ loop. Loops α and β do not touch, nor do loops α and δ. In this graph, we cannot find three loops that do not touch, so the calculation of $\Delta(z)$ terminates at this point, giving

$$\Delta(z) = 1 - \underbrace{(L_1 + L_2 + L_3 + L_4)}_{\text{loop gains}} + \underbrace{(L_1L_2 + L_1L_4)}_{\text{gains of nontouching loops 2 at a time}} - 0 + 0 - \cdots$$

$$\begin{aligned}&= 1 - \{0.368z^{-1} + 1z^{-1} - (0.632)^2 z^{-2} - 0.368z^{-1}\} \\ &\quad + \{(0.368z^{-1})(1z^{-1}) + (0.368z^{-1})(-0.368z^{-1})\} \\ &= 1 - z^{-1} + 0.632z^{-2}.\end{aligned}$$

2. Next, we take up the path gains $P_k(z)$ and their cofactors $\Delta_k(z)$, where

$$\begin{aligned}P_k(z) &= \text{gain of } k\text{th path from input to output} \\ &= \text{product of branch gains in } k\text{th path}\end{aligned}$$

and

$$\Delta_k(z) = \text{cofactor of the } k\text{th path, formed by striking out from } \Delta(z) \text{ all terms associated with loops that are touched by the } k\text{th path.}$$

Consequently, for the SFG of Figure 8.14b where the paths are marked, we have

$$P_1(z) = (0.368)(z^{-1})(1) \quad \text{and} \quad \underbrace{\Delta_1(z) = 1 - 0.368z^{-1}}_{P_1 \text{ does not touch the } \alpha \text{ loop.}}$$

BASIC CONCEPTS

$$P_2(z) = (0.632)(z^{-1})(0.632)(z^{-1})(1) \quad \text{and} \quad \Delta_1(z) = 1$$

P_2 touches all the loops and all terms except the 1 are stricken from $\Delta(z)$.

3. Finally, the transfer function is

$$H(z) = \frac{Y(z)}{X(z)} = \frac{\sum_{k=1}^{2} P_k(z) \cdot \Delta_k(z)}{\Delta(z)}$$

$$= \frac{0.368z^{-1}(1 - 0.368z^{-1}) + (0.632)^2 z^{-2}}{1 - z^{-1} + 0.632z^{-2}}$$

$$= \frac{0.368z + 0.264}{z^2 - z + 0.632}.$$

ILLUSTRATIVE PROBLEM 8.16
Practice with Mason's Gain Rule

For the digital control system modeled by Figure 8.14a, use MGR to calculate the following transfer functions.

a. $H_a(z) = \dfrac{\Omega(z)}{X(z)}$

b. $H_b(z) = \dfrac{E(z)}{X(z)}$

Solution

a. Using the three steps outlined in the previous procedure gives the following:

1. $\Delta(z) = 1 - z^{-1} - 0.632z^{-2}$, as before, because the graph determinant $\Delta(z)$ is a function of the graph and is not affected by a designated input-output transfer function.
2. There is only one path from $X(z)$ to $\Omega(z)$, and it does not touch the β loop. Consequently, $P_1(z) = 0.632z^{-1}$ and $\Delta_1(z) = 1 - z^{-1}$.
3. Applying Mason's Gain Rule,

$$H_a(z) = \frac{\Omega(z)}{X(z)} = \frac{0.632z^{-1}(1 - z^{-1})}{1 - z^{-1} - 0.632z^{-2}} = \frac{0.632(z - 1)}{z^2 - z - 0.632}.$$

b. 1. There is no change.

2. The one path does not touch loops α and β, giving $P_1(z) = 1$ and $\Delta_1(z) = 1 - 0.368z^{-1} + 1z^{-1} + (0.368z^{-1})(1z^{-1})$.
3. Once again with MGR,

$$H_b(z) = \frac{E(z)}{X(z)} = \frac{1 - 0.368z^{-1} + 1z^{-1} + 0.368z^{-1}(z^{-1})}{1 - z^{-1} - 0.632z^{-2}}$$

$$= \frac{z^2 + 0.632z + 0.368}{z^2 - z - 0.632}.$$

Comment: A block diagram model that is equivalent to the SFG of Figure 8.14a is given in Figure 8.15. Mason's Gain Rule may be used to calculate transfer functions, but the signs at the summations must be included when calculating loop and path gains.

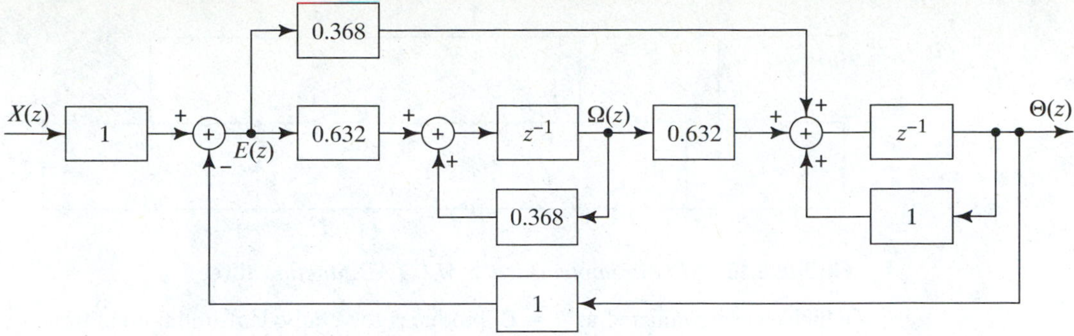

FIGURE 8.15 *Block diagram equivalent to Figure 8.14a*

System Simulation The MATLAB function **filter,** which appears in the Signals and Systems and Signal Processing Toolboxes, may also be used to simulate a system's response when there are nonzero initial conditions. (Previously, **filter** has been employed only for zero initial condition situations.) This function generates the response vector **y** to the input data in the vector **x** with the filter described by the coefficient vectors **a** and **b**. The data come from the difference equation

$$y(n) + a_1 y(n-1) + \cdots + a_N y(n-N) = b_0 x(n) + b_1 x(n-1) + \cdots + b_L x(n-L)$$

or from the z-transform equation,

$$Y(z) = \frac{b_0 + b_1 z^{-1} + \cdots + b_L z^{-L}}{1 + a_1 z^{-1} + \cdots + a_N z^{-N}} \cdot X(z)$$

with the coefficient vectors for **filter** becoming

$$\mathbf{a} = [1 \quad a_1 \quad \ldots \quad a_N] \text{ and } \mathbf{b} = [b_0 \quad b_1 \quad \ldots \quad b_L].$$

For problems with nonzero initial conditions, an initial condition vector, say **v0**, must be determined. The SFG of Figure 8.16 represents the algorithm used by the MATLAB function **filter**, where the output of each delay has been defined as a new variable $v_j(n)$. We need to determine the initial conditions, $v_j(0)$, for **filter** in terms of the given initial conditions of the system, $y(-1)$, $y(-2), \ldots, y(-N)$. Since we are interested in ICs, the input $x(n)$ is set to zero, and from Figure 8.16 we can write, in the time domain for $x(n) = 0$,

$$\begin{aligned}
v_1(n) &= -a_1 y(n-1) - a_2 y(n-2) - \cdots - a_N y(n-N) \\
v_2(n) &= -a_2 y(n-1) - a_3 y(n-2) - \cdots - a_N y(n-N+1) \\
&\vdots \\
v_N(n) &= -a_N y(n-1)
\end{aligned}$$

BASIC CONCEPTS

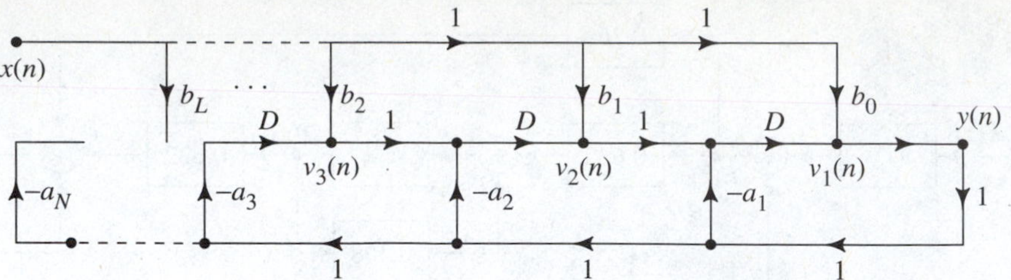

FIGURE 8.16 *SFG to implement the MATLAB function* **filter**

which, when evaluated at $n = 0$, produces the equivalent initial conditions

$$v_1(0) = -a_1y(-1) - a_2y(-2) - \cdots - a_Ny(-N)$$
$$v_2(0) = -a_2y(-1) - a_3y(-2) - \cdots - a_Ny(-N+1)$$
$$\vdots$$
$$v_N(0) = -a_Ny(-1)\,.$$

ILLUSTRATIVE PROBLEM 8.17
System Simulation with Filter

In Chapter 7 we used the functions **dlsim** and **ksdlsim,** which required state models to simulate the system described by the difference equation

$$y(n) - 0.25y(n-1) - 0.125y(n-2) + 0.5y(n-3) = 3x(n)\,.$$

The system input was $x(n) = u(n)$, a step function, and the ICs were $y(-1) = -1$, $y(-2) = -2$, and $y(-3) = -3$. Use **filter** to find the system output $y(n)$.

Solution

We need to find the components of the **v0** vector using the preceding relationships (or the SFG of Figure 8.16) with $a_1 = -0.25$, $a_2 = -0.125$, and $a_3 = 0.5$. Thus we have

$$v_1(0) = 0.25y(-1) + 0.125y(-2) - 0.5y(-3)$$
$$v_2(0) = 0.125y(-1) - 0.5y(-2)$$
$$v_3(0) = -0.5y(-1)\,.$$

Substituting the given values of $y(-1)$, $y(-2)$, and $y(-3)$ yields the desired IC vector

$$\mathbf{v0} = [1, 0.875, 0.5]^T$$

The following script produces the graph of the output $y(n)$ in Figure 8.17.

MATLAB Script

```
%F8_17 System simulation using filter with ICs
n=0:1:40;
b=[3];
a=[1,-0.25,-0.125,0.5];
v0=[1;0.875;0.5];
x=[ones(size(n))];
y=filter(b,a,x,v0);
%...plotting statements
```

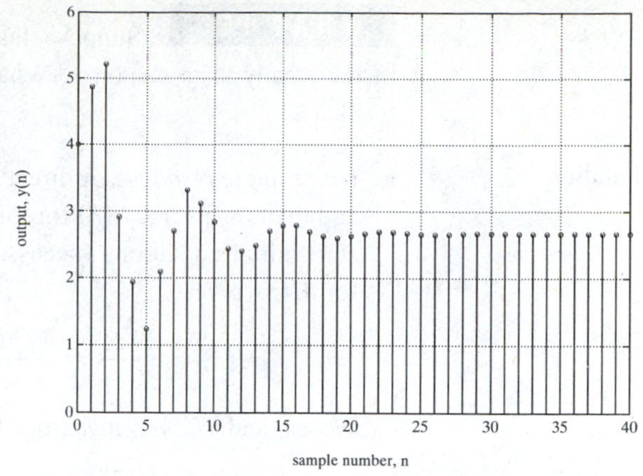

FIGURE 8.17 *Output sequence, Illustrative Problem 8.17*

CROSS-CHECK Are the results of the simulation reasonable? First, $y(0) = 0.25y(-1) + 0.125y(-2) - 0.5y(-3) + 3x(0) = 0.25(-1) + 0.125(-2) - 0.5(-3) + 3(1) = 4$, which agrees with the plot. At a large value of n (say, $n = 40$), since $y(n) = y(n-1) = y(n-2) = y(n-3) = y_\infty$, we have $y_\infty - 0.25y_\infty - 0.125y_\infty + 0.5y_\infty = 3(1)$ when this value is substituted in the given difference equation. Thus $y_\infty = 3/1.125 = 2.67$, which also matches Figure 8.17. Intermediate points could be checked by iteration or from a z-transform solution, but for the moment we'll assume that all is well. ■

EXAMPLES

SOLVED EXAMPLES AND MATLAB APPLICATIONS

EXAMPLE 8.1 *System Response to Different Inputs*

A causal recursive LTI system is described by the difference equation

$$y(n) = 0.5y(n-1) + x(n) + 0.5x(n-1).$$

a. Find its unit sample (impulse) response $h(n)$.

b. Find $y(n)$ for $x(n) = u(n)$, a unit step function, and $y(-1) = 0$.

c. Find the response to the sinusoidal input $x(n) = \cos(n\pi/12)u(n)$ with $y(-1) = 0$.

d. Verify the results of parts (a), (b), and (c) using MATLAB functions such as **filter, dimpulse, dstep, ksdlsim,** and **dlsim.**

WHAT IF? Suppose that in part (c) we had been interested only in the steady-state response. What approach should be used and what is the result? ■

Solution

a. *Unit sample response.* A direct approach is to find the transfer function $H(z)$ and then its inverse $h(n)$ for this causal system. Taking the z transform with zero initial conditions (necessary to find the forced response) and solving for $H(z)$ gives

$$H(z) = \frac{Y(z)}{X(z)} = \frac{1 + 0.5z^{-1}}{1 - 0.5z^{-1}} = \frac{z + 0.5}{z - 0.5}, \qquad |z| > 0.5\,.$$

We expand $H(z)/z$ in partial fractions:

$$\frac{H(z)}{z} = \frac{z + 0.5}{z(z - 0.5)} = \frac{-1}{z} + \frac{2}{z - 0.5}\,.$$

After multiplying this equation for $H(z)/z$ by z, we have

$$H(z) = -1 + \frac{2z}{z - 0.5}, \qquad |z| > 0.5 \quad \text{and}$$
$$h(n) = -\delta(n) + 2(0.5)^n u(n)\,.$$

b. *Unit step response.* With the IC $y(-1) = 0$, we can use the transfer function found in part (a) to find the output transform $Y(z)$. Thus, for $x(n) = u(n)$, $X(z) = z/(z - 1)$, giving

$$Y(z) = H(z) \cdot X(z) = \frac{z + 0.5}{z - 0.5} \cdot \frac{z}{z - 1}, \qquad |z| > 1\,.$$

Following the usual procedure for partial fraction expansion in the z domain yields

$$\frac{Y(z)}{z} = \frac{z + 0.5}{z - 0.5} \cdot \frac{1}{z - 1} = \frac{-2}{z - 0.5} + \frac{3}{z - 1}, \quad \text{and}$$
$$y(n) = [-2(0.5)^n + 3]u(n)\,.$$

c. *Sinusoidal response.* Following the approach of parts (a) and (b),

$$Y(z) = H(z) \cdot X(z) = \frac{z + 0.5}{z - 0.5} \cdot \frac{z(z - \cos(\pi/12))}{z^2 - 2z\cos(\pi/12) + 1}$$
$$= \frac{z + 0.5}{z - 0.5} \cdot \frac{z(z - 0.966)}{z^2 - 1.932z + 1}, \qquad |z| > 1\,.$$

In this situation $Y(z)/z$ produces

$$\frac{Y(z)}{z} = \frac{z + 0.5}{z - 0.5} \cdot \frac{z - 0.966}{z^2 - 1.932z + 1}$$
$$= \frac{-1.641}{z - 0.5} + \frac{1.397e^{-j0.332}}{z - 1e^{j0.262}} + \frac{1.397e^{j0.332}}{z - 1e^{-j0.262}},$$

and after multiplying this equation by z to obtain $Y(z)$ and using Table 8.1, we have

$$y(n) = [-1.641(0.5)^n + 2.794\cos(n\pi/12 - 0.332)]u(n)\,.$$

d. *Computer verification.* Unit sample response: We can use the function **dimpulse** or **filter** from the Signals and Systems Toolbox and put the input data in the form of a transfer function. The scripts are given next, and the responses are plotted in Figure E8.1a. The output from **filter** uses **displot,** while the unit sample found using **impulse** uses the automatic plotting option.

MATLAB Script

```
%E8_1 System response to different inputs

%E8_1a Unit sample response using filter
n=0:1:10;
b=[1,0.5];
a=[1,-0.5];
x=zeros(size(n));
x(1)=1;
y=filter(b,a,x);                    % call filter
%...plotting statements and pause

% Using dimpulse
num=[1,0.5];
den=[1,-0.5];
dimpulse(num,den);                  % call dimpulse
grid;
```

CROSS-CHECK In part (a) we found $h(n) = -\delta(n) + 2(0.5)^n u(n)$, which gives $h(0) = h(1) = 1$ and $h(2) = 2(0.50)^2 = 0.5$. These values agree with the plots in Figure E8.1a. ■

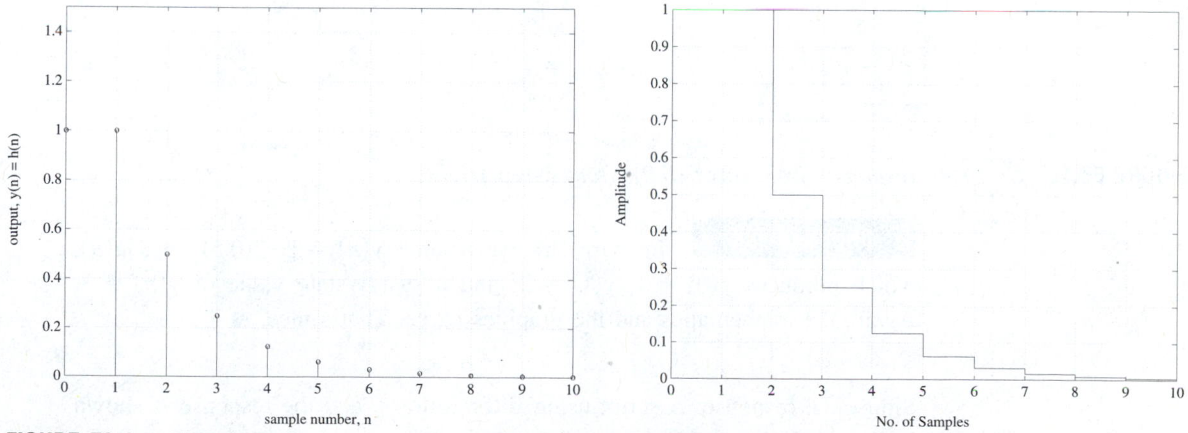

FIGURE E8.1a *Unit sample response: from* **filter** *(left); from* **dimpulse** *(right)*

Unit step response: Using **filter** the input becomes x = ones(n) rather than $x(1) = 1$. In the Control System Toolbox we can use the same script as for the unit sample response, changing the call to **dstep** rather than **dimpulse.** In either case, see Figure E8.1b for the plots.

MATLAB Script

```
%E8_1b Unit step response using filter
n=0:1:10;
b=[1,0.5];
a=[1,-0.5];
x=ones(size(n));
y=filter(b,a,x);                % call filter
%...plotting statements and pause

% Using dstep
num=[1,0.5];
den=[1,-0.5];
dstep(num,den);                 % call dstep
grid;
pause
```

C H A P T E R 8

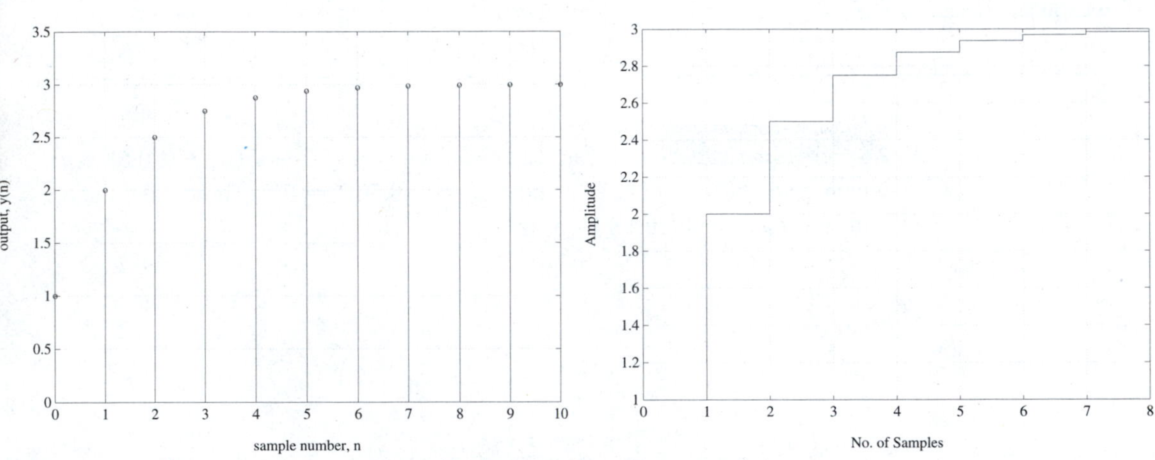

FIGURE E8.1b *Unit step response: from* **filter** *(left); from* **dstep** *(right)*

CROSS-CHECK In part (b) we found $y(n) = [-2(0.5)^n + 3]u(n)$, which produces $y(0) = 1$, $y(1) = 2$, and a steady-state value of $y(n) = 3$. Again, the mathematics and the graphics agree. That's nice. ■

Sinusoidal response: A script using **filter** follows, and the response is shown in Figure E8.1c.

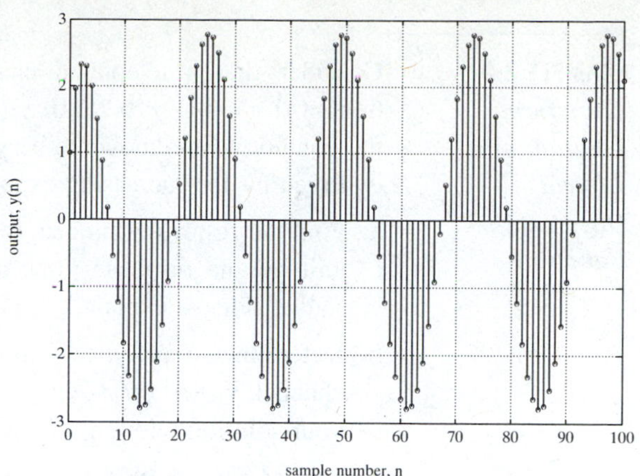

FIGURE E8.1c *Sinusoidal response*

MATLAB Script

```
%E8_1c Sinusoidal response
n=0:1:100;
b=[1,0.5];
a=[1,-0.5];
x=[1*cos(n*pi/12)];
y=filter(b,a,x);               % call filter
%...plotting statements
```

EXAMPLES

CROSS-CHECK In part (c) it was determined that $y(n) = [-1.641(0.5)^n + 2.794\cos(\pi n/12 - 0.332)]u(n)$, from which $y(0) = -1.641 + 2.794\cos(-0.332) = 1.000$ and the steady-state value is $2.794 \cdot \cos(\pi n/12 - 0.332)$. To the scale of the plot, these values appear to be reasonable. ■

WHAT IF? We could use the same approach as in part (c) and simply throw away terms that become vanishingly small as $n \longrightarrow \infty$. Alternatively, we can use the sinusoidal steady-state formula—i.e.,

$$H(e^{j\pi/12}) = \frac{0.5 + e^{j\pi/12}}{-0.5 + e^{j\pi/12}} = \frac{0.5 + 0.966 + j0.259}{-0.5 + 0.966 + j0.259}$$
$$= \frac{1.466 + j0.259}{0.466 + j0.259} = \frac{1.489e^{j0.175}}{0.533e^{j0.507}} = 2.794e^{-j0.332}.$$

Then,

$$y_{ss}(n) = |H(e^{j\pi/12})| \cos(n\pi/12 + \angle H(e^{j\pi/12}))$$
$$= 2.794\cos(n\pi/12 + -0.332).$$

Notice that this is the same as found in part (c) by letting $n \longrightarrow \infty$. ■

EXAMPLE 8.2 *Transform Solution of a Typical Difference Equation*

Consider that a telephone channel is described by the difference equation $y(n) = x(n) + 0.25x(n-1) + 1.2y(n-1) - 0.35y(n-2)$. The initial conditions are known to be $y(-1) = 0$ and $y(-2) = -2$, and the channel is driven by the increasing exponential sequence $x(n) = (-2)^n u(n)$.

a. Find the following models for this system: The transfer function $H(z)$, the unit sample response $h(n)$, and a signal flowgraph (SFG). Is the system stable? Please explain.

b. Determine an analytical expression for the initial condition response of the channel, $y_{IC}(n)$ for $n \geq 0$. Check your result for $n = 0$ and 1 by comparing your solution with an iterative solution of the given difference equation.

c. Find an analytical expression for the complete solution for $n \geq 0$. Verify your partial fraction expansion for $Y(z)$ by using the MATLAB function **residue.** Check a few values computed from the transform solution with those found by an iterative solution of the given difference equation.

d. Use MATLAB functions to find and plot the complete solution $y(n)$.

CHAPTER 8

WHAT IF? Suppose that a colleague or supervisor (spouse?) suggests that a better model of the system is obtained by adding two more delays of the input, so that the appropriate difference equation is now

$$y(n) = 1.2y(n-1) - 0.35y(n-2) + x(n) + 0.25x(n-1) + b_2 x(n-2) + b_3 x(n-3).$$

How will this affect the transfer function, the unit sample response, the system diagram, the system poles, and system stability? (For specific answers, use $b_2 = b_3 = 1$.) The original system zeros were $z = 0$ and $z = -0.25$. What are they with the added delays of the input? ■

Solution

a. *Stability.* Setting the ICs to zero and taking the z transform yields the system transfer function

$$H(z) = \frac{Y(z)}{X(z)} = \frac{1 + 0.25z^{-1}}{1 - 1.2z^{-1} + 0.35z^{-2}} = \frac{z(z+0.25)}{z^2 - 1.2z + 0.35}, \quad |z| > 0.7.$$

We expand $H(z)/z$ in partial fractions as

$$\frac{H(z)}{z} = \frac{z + 0.25}{z^2 - 1.2z + 0.35} = \frac{4.75}{z - 0.7} - \frac{3.75}{z - 0.5}$$

which leads to

$$H(z) = \frac{4.75z}{z - 0.7} - \frac{3.75z}{z - 0.5}, \qquad |z| > 0.7, \qquad \text{and}$$

$$h(n) = [4.75(0.7)^n - 3.75(0.5)^n]u(n).$$

A signal flowgraph for this system is shown in Figure E8.2a. The poles of $H(z)$ are $z_1 = 0.50$ and $z_2 = 0.70$, both inside the unit circle, and since this is a causal system, the system is stable. Also, the ROC is $|z| > 0.70$, which includes the unit circle.

b. *IC solution.* With $x(n) = 0$ the transformed DE is

$$Y_{IC}(z) = 1.2[z^{-1}Y_{IC}(z) + z^{-0}y(-1)] - 0.35[z^{-2}Y_{IC}(z) + z^{-1}y(-1) + z^{-0}y(-2)].$$

With $y(-1) = 0$ and $y(-2) = -2$, we solve for $Y_{IC}(z)$, getting

$$Y_{IC}(z) = \frac{-0.35(-2)}{1 - 1.2z^{-1} + 0.35z^{-2}} = \frac{0.7z^2}{z^2 - 1.2z + 0.35}, \qquad |z| > 0.7$$

and expanding $Y_{IC}(z)/z$ produces

$$\frac{Y_{IC}(z)}{z} = \frac{0.7z}{z^2 - 1.2z + 0.35z} = \frac{-1.75}{z - 0.5} + \frac{2.45}{z - 0.7}, \qquad |z| > 0.7.$$

Multiplying $Y_{IC}(z)/z$ by z and using Table 8.1 gives the initial condition solution

$$y_{IC}(n) = [-1.75(0.5)^n + 2.45(0.7)^n]u(n).$$

Comparison of results From the transform solution

$$y_{IC}(0) = -1.75 + 2.45 = 0.70 \quad \text{and}$$
$$y_{IC}(1) = -1.75(0.5) + 2.45(0.7) = 0.84,$$

and from the iterative solution of the difference equation with $x(n) = 0$,

$$y_{IC}(0) = -0.35y(-2) = 0.70 \quad \text{and}$$
$$y_{IC}(1) = 1.2y(0) - 0.35y(-1) = 1.2(0.70) = 0.84.$$

c. *Complete solution.* Adding in the term due to the input produces

$$\begin{aligned} Y(z) &= \frac{0.7}{1 - 1.2z^{-1} + 0.35z^{-2}} + \frac{1 + 0.25z^{-1}}{1 - 1.2z^{-1} + 0.35z^{-2}} \cdot X(z) \\ &= \frac{0.7}{1 - 1.2z^{-1} + 0.35z^{-2}} + \frac{1 + 0.25z^{-1}}{1 - 1.2z^{-1} + 0.35z^{-2}} \\ &\quad \cdot \frac{1}{1 + 2z^{-1}}, \qquad |z| > 2 \\ &= \frac{1.7z^3 + 1.65z^2}{(z - 0.5)(z - 0.7)(z + 2)}, \qquad |z| > 2. \end{aligned}$$

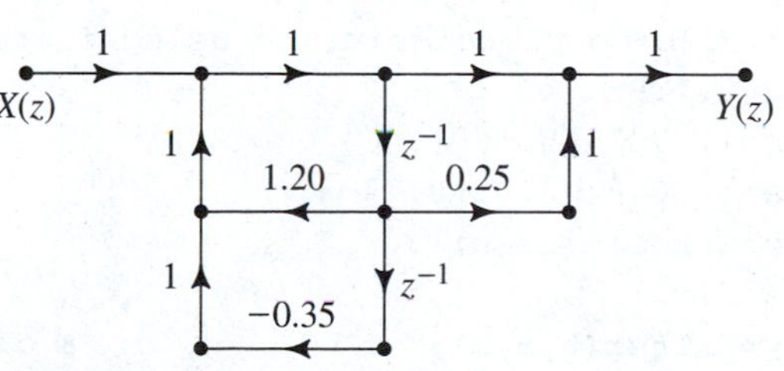

FIGURE E8.2a *System SFG*

EXAMPLES

Following the usual procedure, we expand $Y(z)/z$ as

$$\frac{Y(z)}{z} = \frac{1.7z^2 + 1.65z}{(z - 0.5)(z - 0.7)(z + 2)}$$
$$= \frac{-2.5}{z - 0.5} + \frac{3.6815}{z - 0.7} + \frac{0.5185}{z + 2}, \qquad |z| > 2.$$

After multiplying through this equation by z to obtain $Y(z)$, we use Table 8.1 to obtain

$$y(n) = [-2.5(0.5)^n + 3.6815(0.7)^n + 0.5185(-2)^n]u(n).$$

CROSS-CHECK Using the solution just derived, we have

$$y(0) = -2.5 + 3.6815 + 0.5185 = 1.7$$
$$y(1) = -2.5(0.5) + 3.6815(0.7) + 0.5185(-2) = 0.2901.$$

From the given difference equation,

$$y(0) = 1.2y(-1) - 0.35y(-2) + x(0) + 0.25x(-1)$$
$$= 0 + 0.7 + 1 + 0 = 1.7$$
$$y(1) = 1.2y(0) - 0.35y(-1) + x(1) + 0.25x(0)$$
$$= 2.0400 - 0 - 2 + 0.25 = 0.2900. \quad \blacksquare$$

CHAPTER 8

d. Usually a function such as **dlsim** or **ksdlsim** is used in the situation where the complete solution of a system's DE or state equation is required. As an alternative, we can think of the output transform

$$Y(z) = \frac{1.7z^3 + 1.65z^2}{(z - 0.5)(z - 0.7)(z + 2)}$$

as a transfer function $T(z)$, and use it to find the system output for an input $x(n) = \delta(n)$. In this vein, we reason that $Y(z) = T(z) \cdot X(z)$ with $X(z) = 1$. Thus we can use either **dimpulse** or **filter** as in the script below. In the plot of Figure E8.2b we notice that even though the system is stable, the output diverges rapidly due to the input sequence $x(n) = (-2)^n$. You may want to try some other inputs, such as $x(n) = (-0.5)^n u(n)$ and $x(n) = 0$, to observe the effect of various inputs on the complete response of a stable system.

MATLAB Script

```
%Computer solution of a DE

%E8_2b Complete solution using filter
n=0:1:8;
b=[1.7,1.65,0,0];
a=[1,0.8,-2.05,0.7];
x=zeros(size(n));
x(1)=1;
y=filter(b,a,x);                % call filter
%...plotting statements and pause
```

To show that this system is indeed stable, plot its unit sample response. About how many samples are needed for $h(n)$ to decay to essentially zero? See Figure E8.2c.

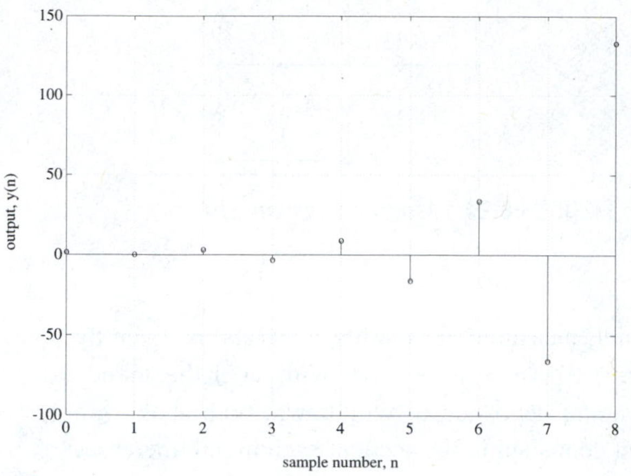

FIGURE E8.2b *Complete response*

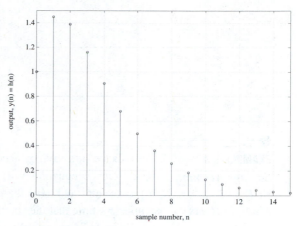

FIGURE E8.2c *Unit sample response*

EXAMPLES

MATLAB Script

```
%E8_2c Unit sample response
n=0:1:15;
b=[1,0.25];
a=[1,-1.2,0.35];
x=zeros(size(n));
x(1)=1;
y=filter(b,a,x);                   % call filter
%...plotting statements
```

WHAT IF? The new transfer function is $H(z) = (z^3 + 0.25z^2 + z + 1)/(z^3 - 1.2z^2 + 0.35z)$. Using **residue*** for the PFE of $H(z)/z$ helps us to find $h(n) = 12.65\delta(n) - 2.85\delta(n-1) + [22.10(0.70)^n - 33.75(0.5)^n]u(n)$. See Figure E8.2d for the new SFG. The system has the same original poles at $z = 0.5$ and $z = 0.7$ plus a pole at $z = 0$. The ROC is still described by $|z| > 0.7$, which includes the unit circle, so the system is still stable. Using **roots** the new zeros are $z_1 = -0.74$ and $z_{2,3} = 0.24 \pm j1.14$, where we see that there are complex zeros outside the unit circle. For stability, however—which is determined by pole locations—these zero locations are unimportant. ■

*We could also have used **residuez** for the PFE of $H(z)$.

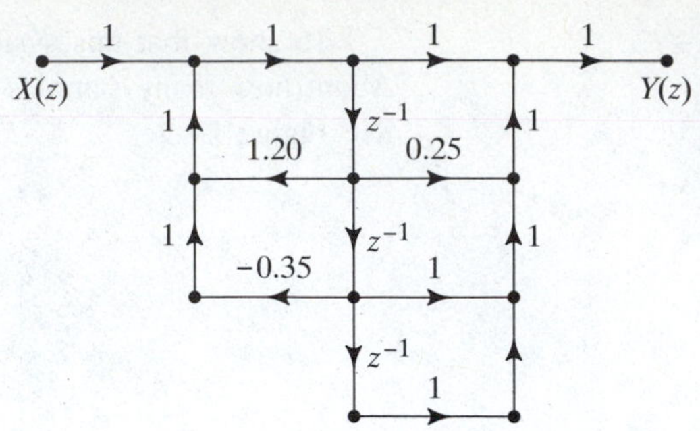

FIGURE E8.2d *Modified system SFG*

EXAMPLE 8.3
Savings Account, PFE with Multiple Roots

The balance in a savings account, measured at monthly intervals, is given by the difference equation $y(n) = 1.01y(n-1) + x(n)$, with $y(n)$ the money in the account and $x(n)$ the amount deposited or withdrawn, both at the nth sample. Assume that the amount deposited in the account each month increases in a linear fashion according to the rule $x(n) = 50 + 5n$ for $n \geq 0$.

a. For a beginning balance of zero—that is, $y(-1) = 0$—find a general expression for the balance $y(n)$ for $n \geq 0$.

b. Use the MATLAB function **filter** or **dimpulse** to plot the amount in savings for a two-year period.

WHAT IF? Investigate the effects of several different deposit strategies—for example, a constant amount each month, a gradually decreasing monthly deposit, a one-time deposit, etc. ■

CHAPTER 8

Solution

a. *Analytical solution.* Using the shifting theorem for unilateral transforms, we have

$$Y(z) - 1.01[z^{-1}Y(z) + y(-1)] = X(z)$$

and after setting $y(-1) = 0$, the transfer function relating $Y(z)$ and $X(z)$ is

$$\frac{Y(z)}{X(z)} = H(z) = \frac{1}{1 - 1.01z^{-1}} = \frac{z}{z - 1.01}, \qquad |z| > 1.01 .$$

With the z transform of the input given by

$$X(z) = \frac{50z}{z-1} + \frac{5z}{(z-1)^2}, \qquad |z| > 1$$

the transform of the output $y(n)$ is

$$Y(z) = H(z)X(z) = \frac{z}{z - 1.01} \cdot \left\{\frac{50z}{z-1} + \frac{5z}{(z-1)^2}\right\}$$

$$= \frac{50z^3 - 45z^2}{(z-1.01)(z-1)^2}, \qquad |z| > 1.01 .$$

Expanding $Y(z)/z$ in partial fractions yields

$$\frac{Y(z)}{z} = \frac{50z^2 - 45z}{(z - 1.01)(z - 1)^2} = \frac{A}{z - 1.01} + \frac{B}{(z - 1)^2} + \frac{C}{z - 1}$$

where, in the usual way, $A = 55,550$ and $B = -500$. Following the procedure outlined earlier for multiple poles,

$$C = \left\{ \frac{50z^2 - 45z}{(z - 1.01)(z - 1)} - \frac{-500}{z - 1} \right\}\Bigg|_{z=1}$$

$$= \left\{ \frac{50(z + 10.1)(z - 1)}{(z - 1.01)(z - 1)} \right\}\Bigg|_{z=1} = -55,500\,.$$

Summarizing these rather involved computations and multiplying $Y(z)/z$ by z gives

$$Y(z) = \frac{55,550z}{z - 1.01} - \frac{500z}{(z - 1)^2} - \frac{55,500z}{z - 1}, \ |z| > 1.01$$

and from Table 8.1

$$y(n) = [55,550(1.01)^n - 500n - 55,500]u(n)\,.$$

b. *Computer simulation.* Following is a script that uses **filter;** the plot is shown in Figure E8.3a on page 470.

MATLAB Script

```
%E8_3 Savings account analysis using filter and ksdlsim

%E8_3a Savings account history
n=0:1:24;
b=[1];                        % coefficients of input terms
a=[1,-1.01];                  % coefficients of output terms
x=[50*ones(size(n))+n*5];
y=filter(b,a,x);              % call filter
%...plotting statements and pause
```

WHAT IF? The MATLAB function **filter** can, of course, be used with different inputs and initial conditions. We can also describe this savings account as a state model, where $v_1(n) = y(n - 1)$, $v_1(n + 1) = 1.01v_1(n) + x(n)$, and $y(n) = v_1(n + 1) = 1.01v_1(n) + x(n)$. The script on the next page, which uses **ksdlsim,** allows us to handle different inputs very easily. Here we consider different investment strategies leading to the same total investment of \$2750 over 24 months (i.e., 25 deposits). We use (1) $x = \$50 + \$5n$; (2) $x = \$2750\delta(n)$; (3) $x = \$2750/25 = \$110$ per month; (4) $x = \$154.78(0.97)^n$.

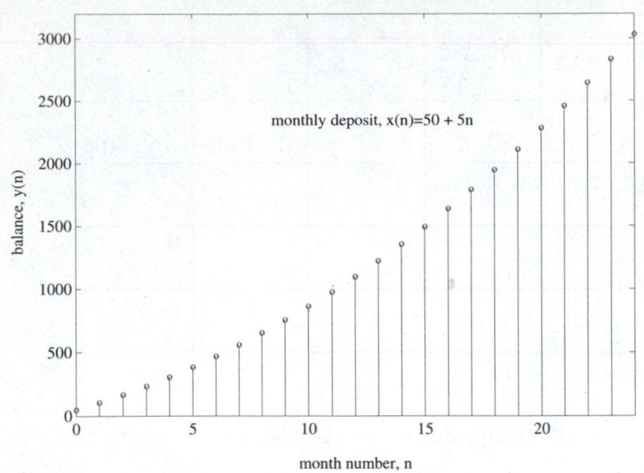

FIGURE E8.3a *Savings account history*

FIGURE E8.3b *Results of different investment strategies*

MATLAB Script

```
%E8_3b Results of different investment strategies
A=[1.01];
B=[1];
C=[1.01];
D=[1];
v0=[0];
n=0:1:24;
X=[50*ones(size(n))+5*n]';
[y,v]=ksdlsim(A,B,C,D,X,v0);
%...plotting statements
```

■

Comment: As your intuition would indicate, the earlier the money is invested, the better. So the best strategy is to invest it all at $n = 0$ and then let the interest accumulate. Notice in Figure E8.3b, above, that the other three strategies end up at $n = 24$ with similar total values.

EXAMPLE 8.4
Sinusoidal Steady-State Response

A causal bandstop digital filter is described by the transfer function

$$H(z) = \frac{1 + z^{-2}}{1 - 0.5z^{-2}} = \frac{z^2 + 1}{z^2 - 0.5}, \qquad |z| > 0.707.$$

a. Draw an SFG model, determine the zeros and poles, and find an analytical expression for the unit sample response $h(n)$ of this filter. Use **residue** or

residuez to verify your partial fraction expansion of $H(z)$. Use **filter** or **dimpulse** to plot $h(n)$, and from the plot show that the settling time of the filter is about 10 samples.

b. The input to this filter is the sinusoidal sequence $x(n) = 2.5\cos(\pi n/4)u(n)$. Determine the expression for the steady-state output $y_{ss}(n)$.

c. Plot the forced response using a MATLAB function and compare your results with the analytical expression found in part (b).

WHAT IF? Try some sinusoidal sequences of different frequencies as inputs to this filter in an attempt to verify if the bandstop label is appropriate. As a suggestion, the filter gains for the digital frequencies of $\theta = 0$, $\pi/2$, and π rad are very easy to calculate. ■

Solution

a. *SFG, zeros and poles, and $h(n)$.* See Figure E8.4a for a signal flowgraph representation of the filter. The zeros are $z_{1,2} = \pm j1 = 1e^{\pm j\pi/2}$, the poles are $z_{1,2} = \pm 0.707$, and the PFE for $H(z)/z$ is

$$\frac{H(z)}{z} = \frac{-2}{z} + \frac{1.5}{z - 0.707} + \frac{1.5}{z + 0.707}, \qquad |z| > 0.707$$

and from Table 8.1,

$$h(n) = -2\delta(n) + 1.5[(0.707)^n + (-0.707)^n]u(n).$$

From the plot in Figure E8.4b we see it takes about 10 samples for the $h(n)$ to decay to close to zero. Evaluating the analytical expression for $h(n)$ for $n = 10$ gives $h(n) = 1.5[(0.707)^{10} + (-0.707)^{10}] = 0.094$.

b. *Response to $x(n) = 2.5\cos(\pi n/4)u(n)$.* The digital frequency of the input is $\theta = \pi/4$, which means that we need to evaluate the transfer function $H(z)$ at $z = 1e^{j\pi/4}$, yielding

$$H(e^{j\pi/4}) = \left.\frac{1 + z^{-2}}{1 - 0.5z^{-2}}\right|_{z=e^{j\pi/4}} = \frac{1 + e^{-j\pi/2}}{1 - 0.5e^{-j\pi/2}}$$
$$= \frac{1.414e^{-j\pi/4}}{1.118e^{j0.464}} = 1.259e^{-j1.249}.$$

Thus, the magnitude of the transfer function at $z = 1e^{j\pi/4}$ is 1.259 and the phase (argument) is -1.249 rad. From the sinusoidal steady-state formula

$$y_{ss}(n) = (2.5)(1.259)\cos(n\pi/4 - 1.249) = 3.148\cos(n\pi/4 - 1.249).$$

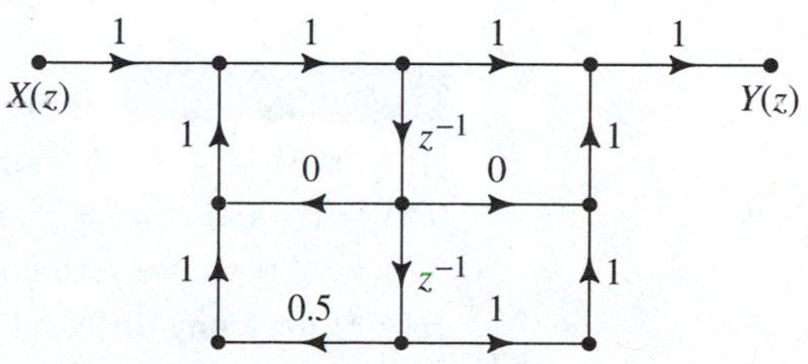

FIGURE E8.4a *SFG for a bandstop filter*

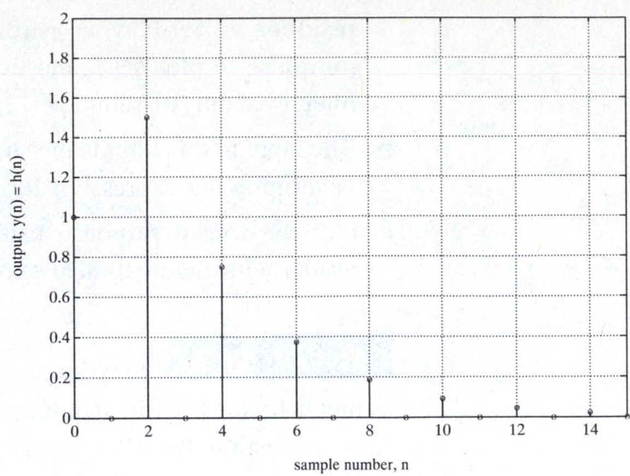

FIGURE E8.4b *Unit sample response for a bandstop filter*

c. *Forced response using* **filter** *or* **dlsim.** The input and output sequences are shown in Figure E8.4c, where it is seen that the first 10 samples or so include the transient part and that steady-state is not reached until about the tenth sample.

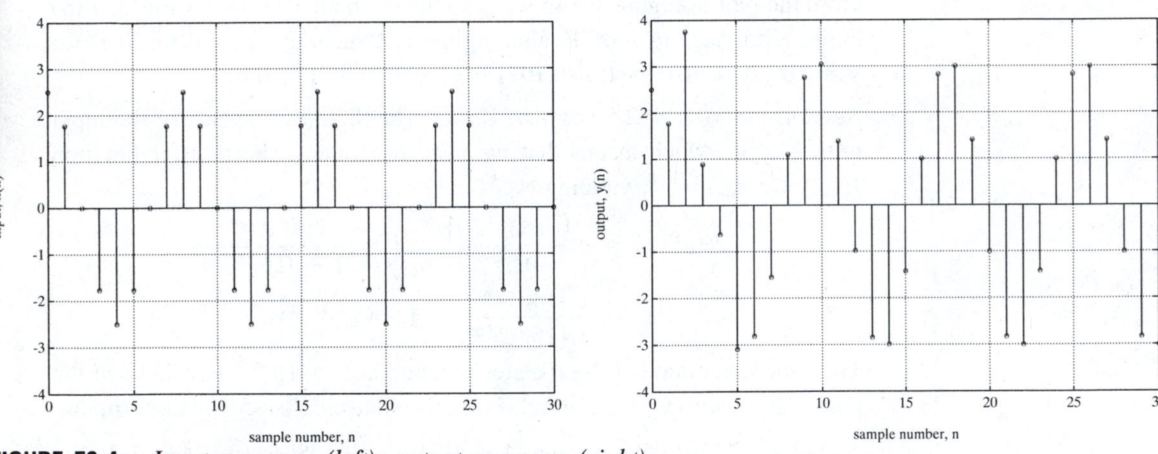

FIGURE E8.4c *Input sequence (left); output sequence (right)*

WHAT IF? For $\theta = 0$, we have $H(e^{j0}) = H(z)|_{z=1} = (1^2 + 1)/(1^2 - 0.5) = 4$, for $\theta = \pi/2$, $H(e^{j\pi/2}) = H(z)|_{z=j1} = ([j1]^2 + 1)/([j1]^2 - 0.5) = 0$, and for $\theta = \pi$, $H(e^{j\pi}) = H(z)|_{z=-1} = ([-1]^2 + 1)/([-1]^2 - 0.5) = 4$. If we were to continue (heaven forbid) to compute the magnitude of $H(e^{j\theta})$ for many different frequencies θ and then plot the results, a curve like that in Figure E8.4d would result. To obtain this plot, we used the MATLAB function **freqz,** which will be introduced in Chapter 9, "Frequency Response of Discrete Systems." ■

We will see in Chapter 9 that the design could be improved by moving the poles to $z_{1,2} = \pm j0.95$, for instance, with the resulting magnitude plot in Figure E8.4e.

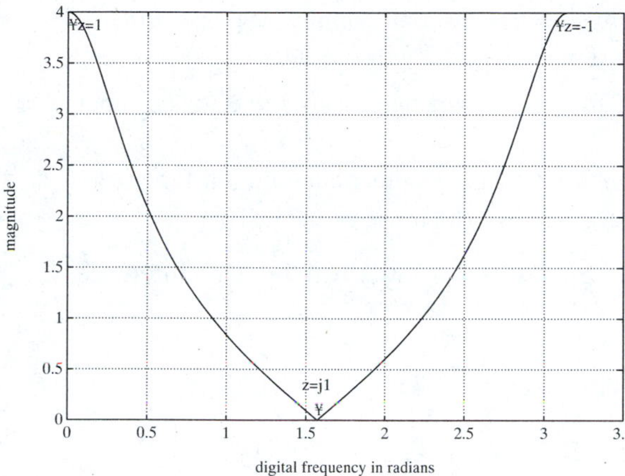

FIGURE E8.4d *Bandstop filter frequency response magnitude*

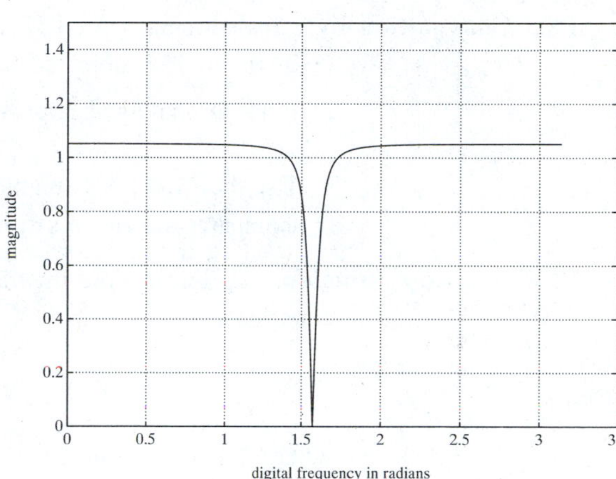

FIGURE E8.4e *A different bandstop filter's frequency response magnitude*

REINFORCEMENT PROBLEMS

P8.1 Stability and unit sample response. Consider the system transfer function $H(z) = z^{-1}/(1 - 2z^{-1})$, $|z| < 2$.

a. Is the system stable?
b. Find $h(n)$.

P8.2 Transform and computer solution of some DEs.

a. An LTI system is described by the difference equation $y(n) = -1.20y(n-1) + x(n)$ with the IC $y(-1) = -1$. The input is given as $x(n) = \delta(n) + (0.6)^n u(n)$. Obtain the total solution using the z-transform method.
b. Use z transforms to find the unit sample (impulse) response of the LTI system modeled by the second-order DE, $y(n) - 0.25y(n-2) = 4x(n)$.

c. Find the transfer function $H(z)$ and the unit sample response $h(n)$ for the first-order system $y(n) - y(n-1) = x(n) + x(n-2)$.

d. For the LTI system $y(n) + y(n-2) = x(n) + x(n-1)$ with the ICs $y(-1) = 0$, and $y(-2) = -10$, use z transforms to show that the response to the step input $x(n) = 10u(n)$ is given by $[10 + 10\sqrt{2}\cos(\pi n/2 - \pi/4)]u(n)$.

e. Use **filter** to obtain computer solutions to the linear systems of parts (a)–(d) above. Compare a few values of the analytical and computer solutions for each system.

P8.3 Convolution by z transforms. An LTI system that has the unit sample response $h(n) = (-0.80)^n u(n)$ is subjected to the input $x(n) = 2^n[u(n) - u(n-3)]$.

a. Find an analytical expression for the system output $y(n)$ due to the input $x(n)$.

b. Use the MATLAB function **conv** to obtain a numerical solution for $y(n)$.

c. Compare your results from parts (a) and (b) for $n = 0, 5, 10$.

P8.4 Transfer function. A linear time-invariant digital filter is described by the difference equation

$$y(n) - \sum_{k=1}^{3} \frac{1}{k} y(n-k) = \sum_{k=0}^{2} kx(n-k)$$

where k indicates the time delays of the filter.

a. Find the filter transfer function $H(z) = Y(z)/X(z)$.

b. What are the poles and zeros?

c. Is the system stable?

P8.5 Causal and anticausal systems. Suppose that a causal system has the unit sample response $h_c(n) = a^n u(n)$ and that an anticausal system is described by $h_{ac}(n) = b^n u(-n-1)$, with $|a| < |b|$.

a. If these two systems are connected together to form a new system, as shown in Figure P8.5, determine the transfer function $H(z)$ of this new system and give its region of convergence.

b. What is the unit sample response of the new system?

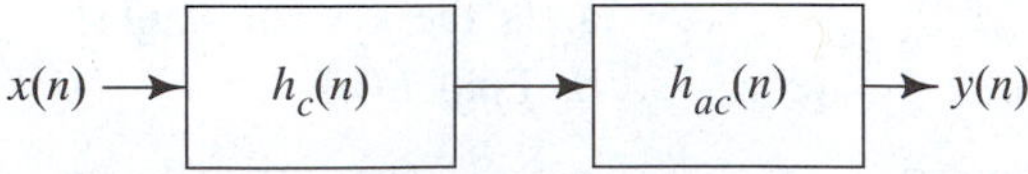

FIGURE P8.5 *Cascade connection of a causal and an anticausal system*

P8.6 Impulse and step response using MATLAB. A causal filter is described by

$$H(z) = \frac{Y(z)}{X(z)} = \frac{5z^2 - 0.9167z - 0.5417}{z^3 - 0.5833z^2 - 0.0417z + 0.0417}, \qquad |z| > 0.50.$$

CHAPTER 8

a. Find the zeros and poles. Is it stable?
b. Find an analytical expression for the unit sample response of this system. Use **residue** or **residuez** for the PFE. Use this result to show whether or not this filter is stable.
c. Determine an equation for the unit step response of this filter. Verify the steady-state output with the final value theorem.
d. Use **filter** or an equivalent function to plot the unit sample and step responses.

P8.7 Steady-state and forced responses A linear time-invariant system has the unit sample response $h(n) = a^n u(n)$ with $0 < a < 1$.

a. The system input is $x(n) = (-1)^n$, $-\infty \leq n \leq \infty$. Find the output $y(n)$ for $-\infty \leq n \leq \infty$.

Hint: Think of the input as the complex exponential $x(n) = e^{jn\theta}$, $-\infty \leq n \leq \infty$.

b. Now the input sequence is changed to $x(n) = (-1)^n u(n)$. Find the output $y(n)$.

P8.8 Step response of some filters. Find an analytical expression for the unit step response $y(n)$ of the following filters:

a. a lowpass filter described by $h(n) = (0.90)^n u(n)$;
b. a highpass filter described by $h(n) = (-0.90)^n u(n)$; and
c. a bandpass filter with $h(n) = 2(0.90)^n \cos(n\pi/2)u(n)$. This problem was solved by convolution in Example 7.4. You may want to use **filter** or **dstep** to graph the results.

P8.9 Cascade connection of two systems. Shown in Figure P8.9 is a cascade connection of two causal linear systems. The input to the first system is $x(n)$ and its output is $w(n)$; the input to the second system is $w(n)$ and its output is $y(n)$. System 1 is modeled by $h_1(n) = a^n u(n)$ and $h_2(n) = b^n u(n)$. Assume that a and b satisfy the condition $0 < a < b < 1$.

a. Find the transfer function $H(z) = Y(z)/X(z)$ and determine its poles and zeros. Is the overall system stable?
b. For $x(n) = A\cos(n\pi/2)u(n)$, find $y_{ss}(n)$.
c. Determine an analytical expression for the unit sample response $h(n)$ of the overall system. Use a MATLAB function to plot $h(n)$ for $a = 0.80$ and $b = 0.90$.
d. Find an analytical expression for the unit step response $y(n)$ of the overall system. Plot $y(n)$ using the values of a and b in part (c).

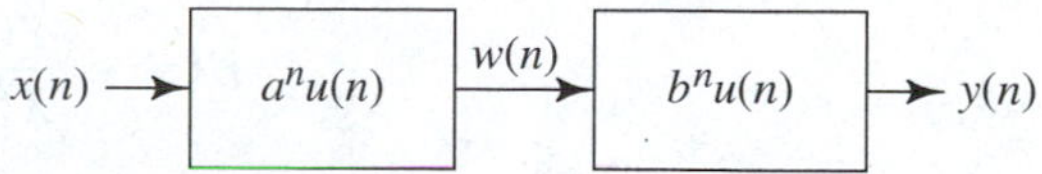

FIGURE P8.9 *Two causal systems in a cascade connection*

P8.10 Convolution and region of convergence. Use z transforms to evaluate the following convolutions. Check your results with Problem 7.11.

a. $c_1(n) = u(n) * u(n)$
b. $c_2(n) = u(-n - 1) * u(-n - 1)$
c. $c_3(n) = u(-n - 1) * a^n u(n),\ 0 < a < 1$
d. $c_4(n) = u(n) * a^n u(n),\ 0 < a < 1$

P8.11 Some extra transform pairs. Derive the z transforms and give the regions of convergence for the following sequences related to $a^n u(n)$ and $a^n u(-n - 1)$.

a. $f_1(n) = a^n u(n - 1)$
b. $f_2(n) = a^{n-1} u(n - 1)$
c. $f_3(n) = a^n u(-n)$

P8.12 Diagrams for moving average filters. In Chapter 7 we used weighted moving average filters described by

weighted MA filter

$$y(n) = \frac{G}{L} \sum_{k=0}^{L-1} w(k)x(n - k)$$

with different weighting sequences, $w(k)$. Draw a nonrecursive diagram (SFG) to realize this filter for the following weighting sequences with $0 \le k \le L - 1$. Use $L = 20$ and assign the gain factor G/L to an input or an output branch. See Example 7.3 to review the calculation of G/L: (a) $w(k) = 1$; (b) $w(k) = 1 - 0.05k$; (c) $w(k) = (0.86)^k$.

P8.13 Transfer functions for moving average filters. For each of the three systems in Problem P8.12, find the transfer function $H(z) = Y(z)/X(z)$.

P8.14 Response of a second-order system. A causal LTI system is described by $y(n) + a_1 y(n - 1) + a_2 y(n - 2) = b_0 x(n) + b_1 x(n - 1)$, where a_1, a_2, b_0, and b_1 are real constants.

a. Find the values of a_1, a_2, b_0, and b_1 such that the initial condition response will be of the form $y_{\text{IC}}(n) = A \cos(n\pi/3 + \alpha)$.
b. Repeat part (a) for $y_{\text{IC}}(n) = A \cos(2n\pi/3 + \alpha)$.
c. Repeat part (a) for $y_{\text{IC}}(n) = A(0.5)^n + B(2)^n$.
d. For $a_1 = -1$, $a_2 = 0.5$, and $b_0 = b_1 = 0.5$, find an analytical expression for $y(n)$ if $y(-1) = y(-2) = 0$ and $x(n) = 2\cos(n\pi/3)u(n)$.

P8.15 Sinusoidal steady state. For the system of Problem P8.14, (a) find $H(z)$ in terms of a_1, a_2, b_0, and b_1, and (b) use the sinusoidal steady-state formula to verify the steady-state part of the solution to Problem P8.14(d).

P8.16 Generic second-order filter. Second-order digital filters can be described by

$$H(z) = \frac{Y(z)}{X(z)} = \frac{1}{1 - [2r\cos\alpha]z^{-1} + r^2 z^{-2}}.$$

a. Draw an SFG that describes this filter.
b. What is the DE for this filter?

c. Show that the filter poles are $z_{1,2} = re^{\pm j\alpha}$.
d. Discuss stability from the point of view of r and α.
e. Find an analytical expression for the unit sample response $h(n)$.
f. Suppose that the numerator of the transfer function is multiplied by z^{-2}. Show the change required in the system diagram and determine the new unit sample response, denoted as $h'(n)$.
g. Use a MATLAB function to plot $h(n)$ and $h'(n)$ and comment on their similarities. (For the answer we used $r = 0.9$ and $\alpha = \pi/3$).

P8.17 System diagram and impulse response. A digital lowpass filter is modeled by the transfer function

$$H(z) = \frac{Y(z)}{X(z)} = \frac{0.067(z + 1)^2}{z^2 - 1.145z + 0.414}.$$

a. Draw a system diagram in the manner of Figure 8.12c.
b. Use Mason's Gain Rule to find $H(z) = Y(z)/X(z)$, which should yield the given transfer function.
c. Find an analytical expression for the unit sample response $h(n)$ of this filter.
d. Verify a few values of your calculation by comparing them with $h(n)$, determined by MATLAB's **filter** or equivalent.

P8.18 Transfer function, DE, zeros and poles, impulse response, and MATLAB. An SFG model for a Chebyshev digital filter is given in Figure P8.18 on page 478.

a. Use MGR to find the transfer function $H(z) = Y(z)/X(z)$.
b. What is the system's difference equation?
c. Use **roots** to determine the filter's poles and zeros.
d. Use a MATLAB function to plot the unit sample response $h(n)$. Would you estimate this to be a bandstop or a bandpass filter?

EXPLORATION PROBLEMS

P8.19 Poles and time response. The output of an LTI system is affected by the input, the initial conditions, and the system characteristics. In terms of transforms we can describe the situation as $Y(z) = Y_{\text{IC}}(z) + Y_{\text{F}}(z)$ or in the time domain, as $y(n) = y_{\text{IC}}(n) + y_{\text{F}}(n)$. In this problem let's look at the effect the poles of a transform have on the time-domain response. In any case, we have, for a typical term in the PF expansion,

$$W(z) = \sum_{k=1}^{R} \frac{C_k z}{z - p_k} \quad \text{or} \quad w(n) = \sum_{k=1}^{R} C_k (p_k)^n$$

where $w(n)$ is the time response of interest (IC, forced, or combined). We are assuming distinct poles and an ROC that yields causal terms. First, consider a term in $W(z)$ such as $Cz/(z - p)$: (1) give a description of and

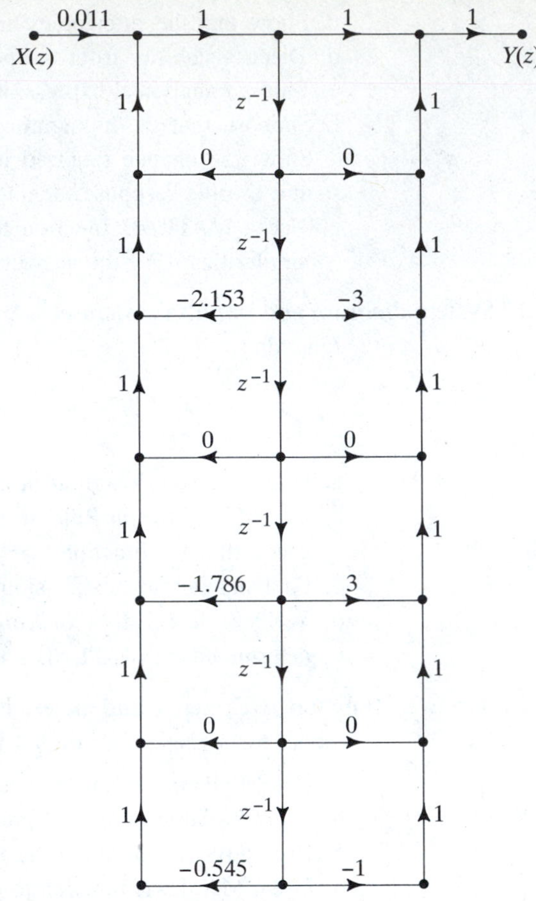

FIGURE P8.18

CHAPTER 8

(2) sketch the corresponding time-domain term for the following real values of p: (a) $0 < p < 1$; (b) $p > 1$; (c) $-1 < p < 0$; (d) $p < -1$; (e) $p = 1$; (f) $p = -1$. Next, consider two complex poles such as $p = re^{\pm j\theta}$ and repeat (1) and (2) for (g) $r = 1$, (h) $r < 1$, and (i) $r > 1$. Include at least two different values of θ in part (g).

P8.20 Poles and time response with MATLAB. We can use a MATLAB function such as **filter** to revisit Problem P8.19 and explore such items as the magnitude of the pole p, the magnitude of the radius r, and the size of the angle θ. We simply consider typical partial fraction terms such as

$$W(z) = \frac{Cz}{z - p} \quad \text{or} \quad W(z) = \frac{Cz}{z - re^{j\theta}} + \frac{C^*z}{z - re^{-j\theta}}$$

as transfer functions and use an impulse input to observe the corresponding time response. For instance, for complex poles we can assume C to be real, which gives

$$W(z) = \frac{Cz}{z - re^{j\theta}} + \frac{Cz}{z - re^{-j\theta}} = \frac{2Cz[z - r\cos(\theta)]}{z^2 - 2r\cos(\theta)z + r^2}$$

and the numerator and denominator polynomials are in the required form for MATLAB. Investigate parts (a)–(i) of Problem P8.19 by assuming $W(z) = H(z)$ and plotting $h(n)$ for different magnitudes of p, r, and θ.

P8.21 Locus of the roots of an equation with MATLAB. In Problems P8.19 and P8.20 we noticed the result of pole location on the time response. In system design we often want to see the effect of varying a system's parameter, called G, on the values of the system's roots. Starting with the characteristic equation

$$\sum_{k=0}^{N} a_k z^k = 0$$

we collect all terms that involve the varying parameter G and form the polynomial $G \cdot \text{num}(z)$ and put the rest of the terms in the polynomial $\text{den}(z)$. Now the characteristic equation is in the form

$$G \cdot \text{num}(z) + \text{den}(z) = 0 \quad \text{or} \quad 1 + \frac{G \cdot \text{num}(z)}{\text{den}(z)} = 0.$$

For instance, the CE $\alpha_3 z^3 + \alpha_2 z^2 + Gz^2 + \alpha_1 z + \alpha_0 + G = 0$ would be put in the form

$$G \cdot \{z^2 + 1\} + \alpha_3 z^3 + \alpha_2 z^2 + \alpha_1 z + \alpha_0 = 0 \quad \text{or}$$
$$1 + \frac{G \cdot \{z^2 + 1\}}{\alpha_3 z^3 + \alpha_2 z^2 + \alpha_1 z + \alpha_0} = 0$$

where, of course, some of the α's could be zero. A detailed discussion of this method is normally part of an undergraduate control systems course, and we will not proceed further in the theory and ramifications of the root locus procedure except to point out that the MATLAB function **rlocus** is found in the Signals and Systems and Control System Toolboxes. Let's use this function to plot the locus of roots for some equations of interest to you. Some relatively simple ones for starters are: (a) $z^2 + z + G = 0$, (b) $z^2 + Gz + 10 = 0$, and (c) $z^3 + 3z^2 + 2z + G = 0$. In each case you should check some easy-to-find roots to see if the computer-generated plot is reasonable.

P8.22 Multiple roots (poles). In Example 8.4 we determined the sinusoidal response of a causal bandstop filter described by the transfer function $H(z) = (1 + z^{-2})/(1 - 0.5z^{-2})$, $|z| > 0.707$.

a. Suppose that the filter input is a linearly increasing sequence such as $x(n) = 5nu(n)$. Find an analytical expression for the filter output $y(n)$.
b. A trick to avoid the entire problem of multiple roots is to make them simple or distinct by moving them a bit. In this case we can take the

PROBLEMS

two poles at $z = 1$ and leave one at $z = 1$ and put the second at, say, $z = 1 + 0.001 = 1.001$. We write $Y(z) = H(z) \cdot X(z)$ as

$$Q(z) = \frac{5z^3 + 5z}{(z + 0.707)(z - 0.707)(z - 1)(z - 1.001)}$$

and use the straightforward procedure for simple poles to determine $q(n)$. The plots for $y(n)$ and for $q(n)$ will be almost indistinguishable over a certain range of n.

P8.23 Another approach to multiple poles. In Problem P8.22a we had

$$\begin{aligned}\frac{Y(z)}{z} &= \frac{5z^2 + 5}{(z^2 - 0.5)(z - 1)^2} \\ &= \frac{A}{z - 0.707} + \frac{B}{z + 0.707} + \frac{C}{(z - 1)^2} + \frac{D}{z - 1}\end{aligned}$$

where by the usual approach $A = 61.82$, $B = -1.82$, and $C = 20$. By setting $z = z_k$, where z_k is any nonpole value of $Y(z)/z$, the partial fraction constant D is easily evaluated. Try this with a z_k of your choice.

P8.24 Polynomial multiplication using convolution (conv). The convolution property of z transforms provides a relatively easy procedure for finding the product of two polynomials such as $A(z)$ and $B(z)$, where $C(z) = A(z) \cdot B(z) = \mathcal{Z}[c(n) = a(n) * b(n)]$.

a. For $A(z) = B(z) = 1 + z^{-1} + z^{-2} + z^{-3}$, use **conv** to find the polynomial $C(z)$.

b. Keep $A(z)$ as in part (a), make $B(z) = 1 + z^1 + z^2 + z^3$, and find $C(z) = A(z) \cdot B(z)$ using **conv.**

c. For $A(x) = 1 + 2x + 3x^2 + 7x^3 + 12x^4 + 19x^5 + 21x^6 + 22x^7 + 18x^8 + 13x^9 + 7x^{10} + 5x^{11} + 3x^{12} + 2x^{13} + x^{14}$ and $B(x) = 1 - x$, use **conv** to find $C(x) = A(x) \cdot B(x)$.

P8.25 Inverse transform by long division. In Illustrative Problem 8.8 we had the transform $P(z) = (z + 1)/(z + 2)$, $|z| > 2$ and using long division wrote $P(z)$ in nonpositive powers of z as $P(z) = 1 - z^{-1} + 2z^{-2} - 4z^{-3} + \cdots$. Then the definition of the z transform was used to find the causal sequence $p(n)$. Now let's assume $P(z) = (z + 1)/(z + 2)$, $|z| < 2$.

a. Use long division to write $P(z)$ as a power series in nonnegative powers of z and then find the first three terms of the anticausal sequence $p(n)$.

Hint: Use long division on $P(z) = (1 + z)/(2 + z)$.

b. By PFE and table lookup, show that an analytical form for the sequence is $p(n) = 0.5\delta(n) - 0.5(-2)^n u(-n - 1)$.

P8.26 The Fibonacci sequence. Fibonacci[5] posed the sequence $\{1, 1, 2, 3, 5, 8, 13, 21, 34, 55, 89, 144, 233, \ldots\}$ when dealing with the number of offspring of a pair of rabbits. Notice that each number is found by summing the two previous numbers, which can be written as the Fibonacci equation $y(n+2) = y(n+1) + y(n)$ with $y(0) = y(1) = 1$. Use z transforms to find a general expression for $y(n)$.

Hint: One approach is to reference the given initial conditions to $n = -1$ and $n = -2$ and then shift the solution appropriately.

DEFINITIONS, TECHNIQUES, AND CONNECTIONS

Bilateral transform	$z[f(n)] = F(z) = \sum_{n=-\infty}^{\infty} f(n)z^{-n}$
Inverse transform	$z^{-1}[F(z)] = f(n) = \frac{1}{2\pi j}\oint_C F(z)z^{n-1}\,dz$
Region of convergence	ROC; region of z for which $F(z)$ exists
Notation	$f(n) \Leftrightarrow F(z)$
Pairs	Table 8.1
Unilateral transform	$z[f(n)u(n)] = F(z) = \sum_{n=0}^{\infty} f(n)z^{-n}$
Properties of bilateral transforms	Table 8.2

DEFINITIONS TECHNIQUES

[5]The greatest mathematician of the Middle Ages was Leonardo of Pisa, better known by the name Fibonacci. Born in Pisa about 1175 and educated in North Africa, he traveled widely in the countries of the Mediterranean observing and analyzing the arithmetical systems used. Fibonacci quickly recognized the enormous advantages of the Hindu-Arabic decimal system with its positional notation and zero symbol over the clumsy Roman system. In 1202 he wrote his famous book *Liber Abaci,* where he stated that with the nine figures of the Indians 9 8 7 6 5 4 3 2 1 and the sign 0, any number may be written. The second edition of this work (1228) acquainted Christian Europe with Arabic numerals. From *The History of Mathematics,* David M. Burton, Allyn and Bacon, Inc., Newton, Mass., 1985.

Transfer function $H(z) = \dfrac{Y(z)}{X(z)}, \quad \text{ICs} = 0$

1. $H(z) = \dfrac{z \text{ transform of the output sequence } y(n)}{z \text{ transform of the input sequence } x(n)}$

2. $H(z) = \dfrac{Y(z)}{X(z)} = \dfrac{\sum_{k=0}^{L} b_k z^{-k}}{\sum_{k=0}^{N} a_k z^{-k}}$

given the DE

$$\sum_{k=0}^{N} a_k y(n-k) = \sum_{k=0}^{L} b_k x(n-k)$$

3. $H(z) = z[h(n)]$

Zeros and poles $H(z) = \dfrac{b_0 z^{N-L}(z-n_1)(z-n_2)\cdots(z-n_L)}{(z-d_1)(z-d_2)\cdots(z-d_N)}$

CHAPTER 8

Stability from $H(z)$

1. Causal system: magnitudes of all poles are less than 1.
2. Anticausal system: magnitudes of all poles are greater than 1.
3. General: ROC includes the unit circle.

Inverse transforms

1. From the definition: $C_k \delta(n-k) \Leftrightarrow C_k z^{-k}$
2. From long division: $P(z) = N(z)/D(z)$, where

causal sequence $P(z) = \alpha_0 z^0 + \alpha_1 z^{-1} + \alpha_2 z^{-2} + \cdots + \alpha_{17} z^{-17} + \cdots$

anticausal sequence $P(z) = \beta_0 z^0 + \beta_1 z^1 + \beta_2 z^2 + \cdots + \beta_{17} z^{17} + \cdots$

3. Partial fraction expansion, distinct poles

$$F(z) = K \frac{B(z)}{(z-d_1)(z-d_2)\cdots(z-d_R)} = C_0 + \sum_{k=1}^{R} \frac{C_k z}{z-d_k}$$

where the partial fraction constants are determined from the expansion

$$\frac{F(z)}{z} = K \frac{B(z)}{z(z-d_1)(z-d_2)\cdots(z-d_R)} = \frac{C_0}{z} + \frac{C_1}{z-d_1} + \frac{C_2}{z-d_2} + \cdots + \frac{C_N}{z-d_R}$$

and knowing the ROC, the sequence $f(n)$ may be found using Table 8.1.

Convolution $$y(n) = \sum_{m=-\infty}^{\infty} h(m)x(n-m) \Leftrightarrow H(z) \cdot X(z)$$

Sinusoidal steady-state response For $x(n) = A\cos(n\theta + \alpha)u(n)$, $y_{ss}(n) = A|H(e^{j\theta})|\cos(n\theta + \alpha + \angle H(e^{j\theta}))$ where $H(e^{j\theta}) = |H(e^{j\theta})|e^{j\angle H(e^{j\theta})}$

Structure with delays
$$y(n) = \frac{1}{a_0}\left[-\sum_{k=1}^{N} a_k y(n-k) + \sum_{k=0}^{L} b_k x(n-k)\right]$$
$$Y(z) = \frac{1}{a_0}\left[-\sum_{k=1}^{N} a_k z^{-k} Y(z) + \sum_{k=0}^{L} b_k z^{-k} X(z)\right]$$

Mason's Gain Rule $$H(z) = \frac{Y(z)}{X(z)} = \frac{\sum_{k=1}^{M} P_k(z)\Delta_k(z)}{\Delta(z)}$$

MATLAB FUNCTIONS USED

Function	*Purpose and Use*	*Toolbox*
dimpulse	Given: state or TF model of discrete system, **dimpulse** returns impulse response.	Control System, Signals/Systems
dlsim	Given: state or TF model of discrete system, input, ICs, **dlsim** returns system output.	Control System, Signals/Systems
dstep	Given: state or TF model of discrete system, **dstep** returns step response.	Control System, Signals/Systems
filter	Given: DE or TF model of discrete system, input, ICs, **filter** returns system output.	Signals/Systems, Signal Processing
freqz	Given: TF of discrete system, **freqz** returns frequency response.	Signals/Systems, Signal Processing
invztrans	Given: z transform function F, **invztrans** returns the inverse z transform f.	Symbolic Math
ksdlsim	Given: state or TF model of discrete system, input, ICs, **ksdlsim** returns system output.	California Functions
residue	Given: rational function $T(\sigma) = N(\sigma)/D(\sigma)$, **residue** returns roots of $D(\sigma) = 0$ and PF constants of $T(\sigma)$.	MATLAB
residuez	Given: rational function $T(z) = N(z)/D(z)$, **residuez** returns roots of $D(z) = 0$ and PF constants of $T(z)$.	Signal Processing, Signals/Systems
rlocus	Given: an equation in the form $1 + K[\text{num}(\sigma)]/[\text{den}(\sigma)] = 0$, **rlocus** returns plot of locus of roots for K varying.	Signals/Systems, Control System
roots	Given: coefficients of polynomial p, **roots** returns roots of $p = 0$.	MATLAB
zplane	Given: TF or zeros and poles of discrete system, **zplane** returns display of zeros and poles.	Signal Processing, Signals/Systems
ztrans	Given: The symbolic expression f, **ztrans** returns the unilateral transform F.	Symbolic Math

ANNOTATED BIBLIOGRAPHY

1. Doetsch, Gustav, *Guide To The Applications of The Laplace and z-Transforms,* Van Nostrand Reinhold Company, London, 1961, 1971. *This is a lean but thorough treatise that was written early in the game (1961) to guide engineers through the pitfalls often encountered in applying Laplace transform methods to the solution of practical engineering problems. The second English edition of 1971 includes a treatment of the z-transform method (Chapter Eight) for handling discrete-time and computer-oriented systems. Professor Doetsch, University of Freiburg im Breisgau, introduces the z transformation by using the traditional (mid-20th century) approach of defining* $z = e^{sT}$*. This book is full of historical footnotes and the author's "Rules" that pertain to the use of transform methods, Laplace and z.*

2. Gabel, Robert A., and Richard A. Roberts, *Signals and Linear Systems, 3rd ed.,* John Wiley & Sons, 1987. *Chapter Four, "The z Transform," covers all the usual theorems and properties and also gives a concise treatment of the inversion of z transforms that includes a cleverly concealed use of the inversion integral. Applications to the steady-state analysis of linear systems and to the state-variable model are considered, and the chapter concludes with a short example that illustrates the use of the z transform in the numerical solution of partial differential equations.*

3. Kwakernaak, Huibert, and Raphael Sivan, *Modern Signals and Systems,* Prentice Hall, Inc., Englewood Cliffs, N.J., 1991. *A modern and comprehensive treatment of the basic concepts used in both the time- and frequency-domain analysis of systems is presented. In Chapter Eight, "The z and Laplace Transforms," the authors present a parallel treatment of z and Laplace transform topics, including a careful exposition of existence (ROC) regions and the solution of difference and differential equations.*

4. O'Flynn, Michael, and Eugene Moriarty, *Linear Systems, Time Domain and Transform Analysis,* John Wiley & Sons, New York, 1987. *This interesting book contains material suitable for a first course in linear systems as well as for a more advanced course. Methods of modeling and analysis are developed for both continuous- and discrete-time systems. Chapter Seven includes a thorough treatment of bilateral z transforms, including applications in linear systems with random and signal plus noise inputs, where correlation and spectral functions are considered.*

5. Oppenheim, Alan V., and Ronald W. Schafer, *Discrete-Time Signal Processing,* Prentice Hall, Inc., Englewood Cliffs, N.J., 1989. *In Chapter Four of this popular digital signal processing text, the z transform is developed as a generalization of the discrete-time Fourier transform. Evaluating the inverse transform by using contour integration is emphasized including applications to complex convolution and Parseval's theorem.*

6. Strum, Robert D., and Donald E. Kirk, *First Principles of Discrete Systems and Digital Signal Processing,* Addison-Wesley Publishing Company,

CHAPTER 8

Reading, Mass., 1988. *Chapter Six develops the theory and applications of z transforms. Included are detailed solutions of many examples.*

7. [Historical Reference] Gardner, Murray F., and John L. Barnes, *Transients in Linear Systems, Studied by the Laplace Transformation,* John Wiley & Sons, New York, 1942. *See Chapter Nine for a historical note about a method for solving difference equations that utilizes Laplace transforms rather than z transforms. The Annotated Bibliography in Chapter 3 contains more information about this important text.*

ANSWERS

P8.1 a. Yes, unit circle is included in the ROC.
b. $h(n) = -0.50\delta(n) - 0.50(2)^n u(-n-1)$.

P8.2 a. $y(n) = 0.33(0.6)^n + 2.87(-1.2)^n,\ n \geq 0$;
b. $h(n) = [2(-0.5)^n + 2(0.5)^n]u(n)$; c. $h(n) = -\delta(n) - \delta(n-1) + 2u(n)$;
e. See A8_2 on the CLS disk.

P8.3 a. $y(n) = -3.75\delta(n) + 5\delta(n-1) + 4.75(-0.8)^n u(n)$; b. See A8_3 on the CLS disk. c. From (a) $y(0) = 1$, $y(5) = -1.557$, $y(10) = 0.5100$; the results from A8_3 are identical.

P8.4 a. $H(z) = (z^2 + 2z)/(z^3 - z^2 - 0.500z - 0.333)$; b. Poles: $z_1 = 1.49$, $z_{2,3} = 0.473e^{\pm j2.111}$, zeros $z_1 = 0$, $z_2 = -2$; c. Causal system, unstable

P8.5 a. $H(z) = -z^2/[(z-a)(z-b)]$, $|b| < |z| < |a|$;
b. $h(n) = [1/(b-a)][a^{n+1}u(n) + b^{n+1}u(-n-1)]$.

P8.6 a. Poles: $z_1 = 0.500$, $z_2 = 0.334$, $z_3 = -0.250$; zeros: $z_1 = -0.250$, $z_2 = 0.433$; the system is stable because the unit circle is included in ROC. b. $h(n) = [4(0.50)^n + 9(0.33)^n]u(n) - 13\delta(n)$;
c. $y(n) = [8.5 - 4.5(0.334)^n - 4.0(0.5)^n]u(n)$, $y(\infty) = (z-1)Y(z)|_{z=1} = 8.50$ as in time domain; d. See A8_6 for both plots.

P8.7 a. $y(n) = [1/(1+a)](-1)^n,\ -\infty \leq n \leq \infty$;
b. $y(n) = [1/(1+a)][(a)^{n+1} + (-1)^n]u(n)$.

P8.8 a. $y(n) = [10 - 9(0.90)^n]u(n)$; b. $y(n) = [0.526 + 0.474(-0.90)^n]u(n)$;
c. $y(n) = [1.105 + 1.338(0.90)^n \cos(n\pi/2 - 0.838)]u(n)$

P8.9 a. $H(z) = H_1(z)H_2(z) = Y(z)/X(z) = z^2/[(z-a)(z-b)]$, 2 zeros at $z = 0$ and poles at $z_1 = a$ and $z_2 = b$; stable because both poles of this causal system are inside the unit circle. b. $y_{ss}(n) = AM\cos(n\pi/2 + P)$, where $M = 1/\sqrt{(1-ab)^2 + (a+b)^2}$ and $P = \tan^{-1}[(a+b)/(ab-1)]$;
c. $h(n) = [1/(b-a)]/(b^{n+1} - a^{n+1})u(n)$; see A8_9 on the CLS disk. d. $y(n) = 1/[(1-a)(1-b)] + [a^2/[(a-b)(a-1)]] \cdot a^n + [b^2/[(b-a)(b-1)]] \cdot b^n,\ n \geq 0$; see A8_9.

P8.10 a. $C_1(z) = z^2/(z - 1)^2$, $|z| > 1$, $c_1(n) = nu(n) + u(n)$;
b. $C_2(z) = z^2/(z - 1)^2$, $|z| < 1$, $c_2(n) = -nu(-n - 1) - u(-n - 1)$;
c. $C_3(z) = -z^2/[(z - 1)(z - a)]$, $|a| < |z| < 1$,
$c_3(n) = [a/(1 - a)]a^n u(n) + [1/(1 - a)]u(-n - 1)$; d. $C_4(z) = z^2/[(z - 1)(z - a)]$, $|z| > 1$, $c_4(n) = -[a/(1 - a)]a^n u(n) + [1/(1 - a)]u(n)$.

P8.11 a. $F_1(z) = a/(z - a)$, $|z| > |a|$; b. $F_2(z) = 1/(z - a)$, $|z| > |a|$;
c. $F_3(z) = -a/(z - a)$, $|z| < |a|$

P8.12 See Figure A8.12.

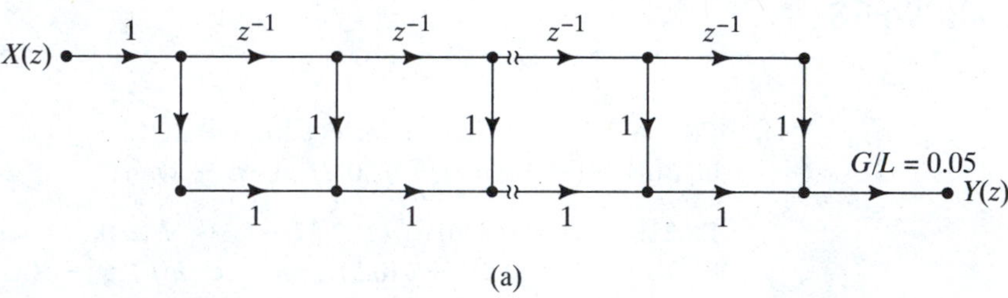

(a)

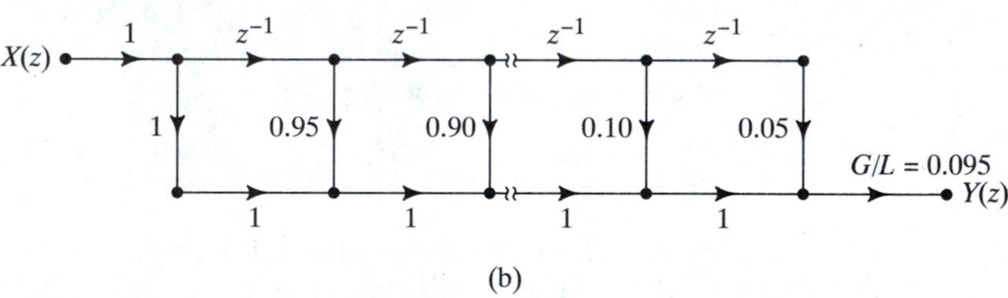

(b)

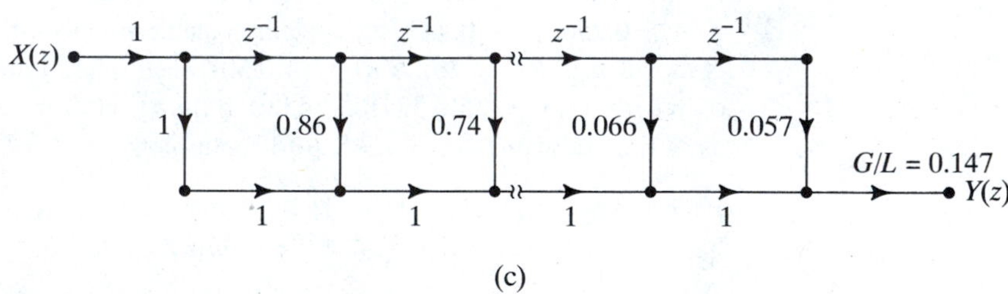

(c)

FIGURE A8.12

P8.13 a. $H(z) = 0.05[1 + z^{-1} + z^{-2} + \cdots + z^{-19}]$;
b. $H(z) = 0.095[1 + 0.95z^{-1} + 0.90z^{-2} + \cdots + 0.05z^{-19}]$;
c. $H(z) = 0.147[1 + 0.86z^{-1} + 0.74z^{-2} + \cdots + 0.057z^{-19}]$

P8.14 a. b_0, b_1 any values, $a_1 = -1$, $a_2 = 1$; b. b_0, b_1 any values, $a_1 = 1$, $a_2 = 1$; c. b_0, b_1 any values, $a_1 = -2.5$, $a_2 = 1$;
d. $2.24(0.707)^n \cos(n\pi/4 + \pi/3) + 3.46 \cos(n\pi/3 - \pi/2)$, $n \geq 0$

CHAPTER 8

P8.15 a. $H(z) = Y(z)/X(z) = (b_0 + b_1 z^{-1})/(1 + a_1 z^{-1} + a_2 z^{-2})$;
b. $H(z) = (0.5 + 0.5z^{-1})/(1 - z^{-1} + 0.5z^{-2})$, $H(e^{j\pi/3}) = 1.73e^{-j1.57}$, $y_{ss}(n) = 3.46\cos(n\pi/3 - 1.57)$

P8.16 a. See Figure A8.16a. b. $y(n) - [2r\cos\alpha]y(n-1) + r^2 y(n-2) = x(n)$; c. $z_{1,2} = r\cos\alpha \pm \sqrt{[r\cos\alpha]^2 - r^2} = r(\cos\alpha \pm j\sin\alpha) = re^{\pm j\alpha}$; d. α may take on any value with $|r| < 1$. e. $h(n) = [1/\sin\alpha]r^n \sin(\alpha[n+1])u(n)$; f. See Figure A8.16b. $h'(n) = [1/\sin\alpha]r^{n-2}\sin(\alpha[n-1])u(n-2)$; g. See A8_16 on the CLS disk. $h'(n)$ is the same as $h(n)$, only delayed by two samples.

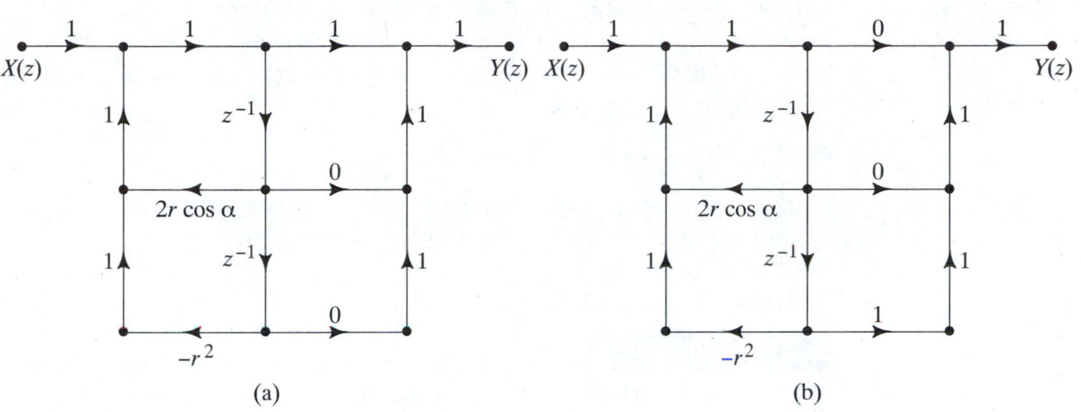

FIGURE A8.16

P8.17 a. See Figure A8.17a. b. MGR validates the SFG.
c. $h(n) = 0.162\delta(n) + 0.908(0.643)^n\cos(0.474n - 1.675), n \geq 0$ with $h(0) = 0.0675, h(1) = 0.211, \ldots$; see A8_17 on the CLS disk for the plot.

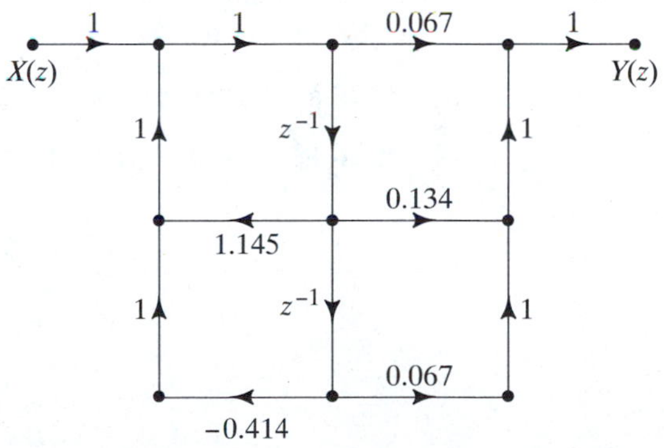

FIGURE A8.17a

P8.18 a. $H(z) = 0.011(z^6 - 3z^4 + 3z^2 - 1)/(z^6 + 2.153z^4 + 1.786z^2 + 0.545)$; b. $y(n) + 2.153y(n-2) + 1.786y(n-4) + 0.545y(n-6) = 0.011[x(n) - 3x(n-2) + 3x(n-4) - x(n-6)]$; c. Zeros: $z_{1,2,3} = +1$, $z_{4,5,6} = -1$; poles: $z_{1,2} = 0.276 \pm j0.888$, $z_{3,4} = \pm j0.853$, $z_{5,6} = -0.276 \pm j0.888$; d. see A8_18 on the CLS disk; probably bandpass.

P8.19 (1) a. $C(p)^n u(n)$, monotonically decreasing in amplitude; b. $C(p)^n u(n)$, monotonically increasing in amplitude; c. $C(p)^n u(n)$, alternating but decreasing in amplitude; d. $C(p)^n u(n)$, alternating but increasing is amplitude; e. $Cu(n)$, a step sequence; f. $C(-1)^n$, an alternating step sequence; g. $A\cos(\theta n + \alpha)u(n)$, undamped oscillation; h. $Ar^n \cos(\theta n + \alpha)u(n)$, decreasing oscillation; i. $Ar^n \cos(\theta n + \alpha)u(n)$, increasing oscillation (2) See the plots in the answers to P8.20.

P8.20 We used **filter** to plot the unit sample response. a. $h(n) = (0.9)^n u(n)$ and $h(n) = (0.5)^n u(n)$; b. $h(n) = (1.1)^n u(n)$ and $h(n) = (1.2)^n u(n)$; c. $h(n) = (-0.9)^n u(n)$ and $h(n) = (-0.5)^n u(n)$; d. $h(n) = (-1.1)^n u(n)$ and $h(n) = (-1.2)^n u(n)$; e. $h(n) = 1^n u(n)$; f. $h(n) = (-1)^n u(n)$; g. $h(n) = 2\cos(n\pi/6 - 1.047)$ and $h(n) = 2(0.577)\cos(n\pi/3 - 0.525)$; h. $h(n) = 2(0.9)^n \cos(n\pi/6 - 1.047)$; i. $h(n) = 2(1.1)^n \cos(n\pi/6 - 1.047)$; see A8_20 on the CLS disk.

P8.21 See A8_21 on the CLS disk.

P8.22 a. $y(n) = [20n - 60 + 61.82(0.707)^n - 1.82(-0.707)^n]u(n)$; b. $y'(n) = [19{,}934(1.001)^n - 19{,}993.9(1)^n + 61.6(0.707)^n - 1.82(-0.707)^n]u(n)$

CROSS-CHECK $y(5) = 67.5000$ and $y'(5) = 67.3537$; $y(10) = 161.4063$ and $y'(10) = 155.9346$. See A8_22 on the CLS disk. ■

P8.23 Using $z_k = 0$, $[(5z^2 + 5)/(z^2 - 0.500)(z - 1)^2]|_{z=0} = -10 = [61.82/(z - 0.707) - 1.82/(z + 0.707) + 20/(z - 1)^2 + D/(z - 1)]|_{z=0}$ gives $D = -60.01$ as compared with -60 in Problem P8.22.

P8.24 a. From A8_24, $C(z) = 1 + 2z^{-1} + 3z^{-2} + 4z^{-3} + 3z^{-4} + 2z^{-5} + z^{-6}$. b. From A8_24, $C(z) = z^3 + 2z^2 + 3z + 4 + 3z^{-1} + 2z^{-2} + z^{-3}$. c. See A8_24 for the coefficients of $C(x)$.

P8.25 a. $p(n) = 0.5\delta(n) + 0.25\delta(n + 1) - 0.125\delta(n + 2) + \cdots$

P8.26 For $n \geq 2$, $y(n) = [1.894(1.618)^{n-2} + 0.106(-0.618)^{n-2}]$

TABLE 8.1 *Bilateral z-transform pairs*

$$z[f(n)] = F(z) = \sum_{n=-\infty}^{\infty} f(n)z^{-n}$$

Sequence	*Transform*	*ROC*
1. $A\delta(n)$	A	all z
2. $Aa^n u(n)$	$\dfrac{Az}{z - a}$	$\|z\| > \|a\|$
3. $Au(n)$	$\dfrac{Az}{z - 1}$	$\|z\| > 1$
4. $Aa^n e^{jn\theta} u(n)$	$\dfrac{Az}{z - ae^{j\theta}}$	$\|z\| > \|a\|$
5. $Aa^n \cos(n\theta)u(n)$	$\dfrac{Az(z - a\cos\theta)}{z^2 - (2a\cos\theta)z + a^2}$	$\|z\| > \|a\|$
6. $A\cos(n\theta)u(n)$	$\dfrac{Az(z - \cos\theta)}{z^2 - (2\cos\theta)z + 1}$	$\|z\| > 1$
7. $A\sin(n\theta)u(n)$	$\dfrac{Az(\sin\theta)}{z^2 - (2\cos\theta)z + 1}$	$\|z\| > 1$
8. $Aa^n \cos(n\theta + \alpha)u(n)$	$\dfrac{Az[z\cos\alpha - a\cos(\alpha - \theta)]}{z^2 - (2a\cos\theta)z + a^2} = \dfrac{0.5Aze^{j\alpha}}{z - ae^{j\theta}} + \dfrac{0.5Aze^{-j\alpha}}{z - ae^{-j\theta}}$	$\|z\| > \|a\|$
9. $Anu(n)$	$\dfrac{Az}{(z - 1)^2}$	$\|z\| > 1$
10. $An^2u(n)$	$\dfrac{Az(z + 1)}{(z - 1)^3}$	$\|z\| > 1$
11. $Ana^n u(n)$	$\dfrac{Aaz}{(z - a)^2}$	$\|z\| > \|a\|$
12. $Aa^n u(-n - 1)$	$-\dfrac{Az}{z - a}$	$\|z\| < \|a\|$
13. $Au(-n - 1)$	$-\dfrac{Az}{z - 1}$	$\|z\| < 1$

TABLES

TABLE 8.2 *Bilateral z-transform properties and relations*

$$z[f(n)] = F(z) = \sum_{n=-\infty}^{\infty} f(n)z^{-n}$$

Property	$f(n)$	$F(z)$
1. Linearity	$af(n) + bg(n)$	$aF(z) + bG(z)$
2. Shifting theorem	$f(n - m)$	$z^{-m}F(z)$
3. Transfer function	$H(z) = \dfrac{Y(z)}{X(z)} = \dfrac{\sum_{k=0}^{L} b_0 z^{-k}}{\sum_{k=0}^{N} a_k z^{-k}}$ assuming	$\sum_{k=0}^{N} a_k y(n-k) = \sum_{k=0}^{L} b_k x(n-k)$
4. Convolution	$\sum_{m=-\infty}^{\infty} f_1(m) f_2(n-m)$	$F_1(z) \cdot F_2(z)$
5. Multiplication by n and derivatives in z	$nf(n)$	$-z\,\dfrac{dF(z)}{dz}$

CHAPTER 8

TABLE 8.3 *Unilateral z-transform properties and relations*

$$z[f(n)] = F(z) = \sum_{n=0}^{\infty} f(n)z^{-n}$$

Property	$f(n)$	$F(z)$
1. Linearity	$af(n) + bg(n)$	$aF(z) + bG(z)$
2. Shifting theorem	$f(n - m)$	$\sum_{r=-m}^{-1} f(r)z^{-(r+m)} + z^{-m}F_u(z)$
3. Final value theorem	$\lim_{n\to\infty} f(n)$	$\lim_{z\to 1} (z - 1)F(z)$
	provided that all poles of $(z - 1)F(z)$ lie inside the unit circle	
4. Initial value theorem	$f(0)$	$\lim_{z\to\infty} F(z)$

Frequency Response of Discrete Systems

CHAPTER 9

PREVIEW

The concept of frequency response is used extensively in many areas of engineering and applied technology. We have all encountered the idea of filtering in audio systems, where adjusting the bass and treble settings controls the frequency response of the CD or stereo system to better suit the fancy of the listener. Another example is the design of feedback control systems using frequency response methods. Mechanical engineers use frequency response methods in vibration and spectral analysis applications, and oceanographers model wave motion with frequency response techniques. Frequency methods are popular for several reasons, including their ability to reveal effects not discernible in the time domain; the availability of sinusoidal sources; ease of understanding the steady-state nature of the technique; ease of system testing; and the vast array of readily available design rules and algorithms.

If we make the input to a discrete system a sinusoidal sequence and observe the steady-state output magnitude and phase as the frequency of the input varies, the result is called the *frequency response.* Thus to perform a frequency response test on an LTI system, only a variable-frequency source and measuring instrumentation are required. A test setup to measure the frequency response of a digital filter is shown in Figure 9.1a, with the resulting plots of the magnitude $M = Y/X$ and phase $P = \beta - \alpha$ of the frequency response $H(e^{j\theta}) = Me^{jP}$ given in Figure 9.1b.

In this chapter, we will make use of several concepts, including transforms, transfer functions, and stability, that were developed earlier. These results and ideas will be restated here, but if you have doubts about them you should go back to the appropriate section and have another look. As in Chapter 4, we

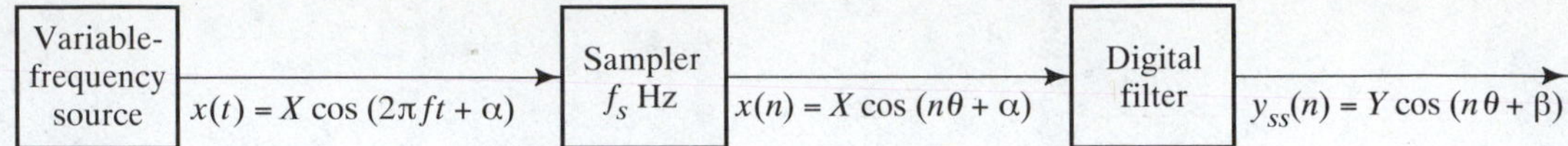

FIGURE 9.1a *Test setup for measuring frequency response of a filter*

include a description of the different ways of displaying frequency response data (rectangular, Bode, polar, and Nyquist plots). The concept of an ideal discrete-time filter is introduced, followed by a discussion of linear phase nonrecursive filters.

Frequency response methods were widely used in the analysis and design of analog communication systems in organizations such as the Bell Telephone Laboratories by engineers, mathematicians, and scientists, including Harold Black, Harry Nyquist, Hendrik Bode, and Claude Shannon. These frequency-oriented techniques were also employed by control system engineers during and following World War II in the design of continuous-time systems. But in the late 1940s and early 1950s, there were reportedly lunchtime conversations among the Bell Lab scientists on the possibility of using digital rather than analog elements to construct a filter. The IEEE-sponsored Arden House workshops in the late 1960s and early 1970s always seemed to include discussions comparing analog and digital filters. Also, the publication of several textbooks in the 1970s and 1980s ensured a place for courses on digital filters and digital signal processing in most engineering curricula.

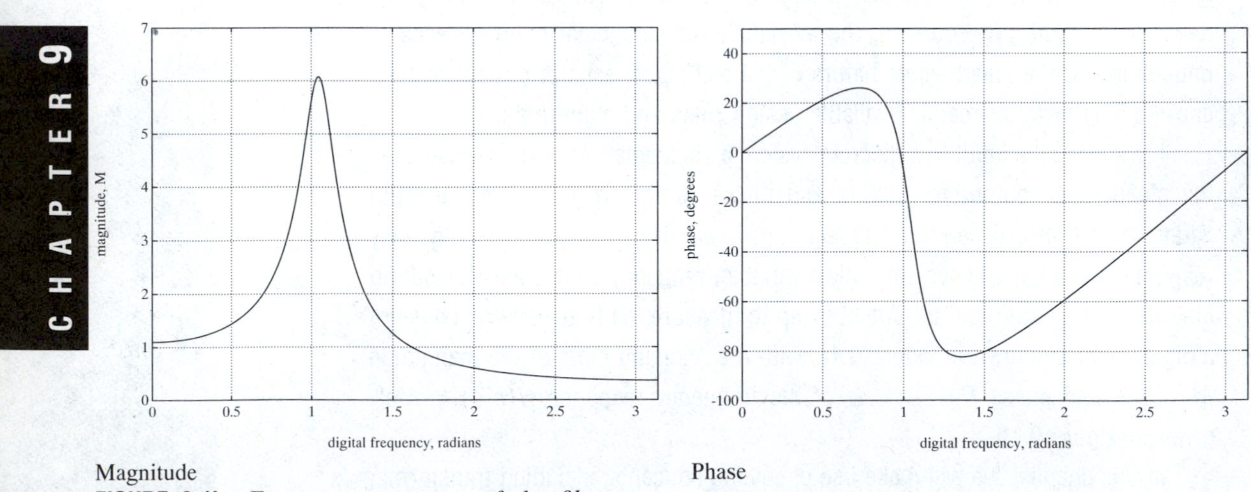

FIGURE 9.1b *Frequency response of the filter*

BASIC CONCEPTS

REVIEW

Transfer Function See Chapter 8 for a detailed discussion.

1. $H(z) = \dfrac{z \text{ transform of the output sequence } y(n)}{z \text{ transform of the input sequence } x(n)} = \dfrac{Y(z)}{X(z)}, \quad \text{ICs} = 0.$
2. $H(z) = \dfrac{Y(z)}{X(z)} = \dfrac{\sum_{k=0}^{L} b_k z^{-k}}{\sum_{k=0}^{N} a_k z^{-k}}$ given the DE $\sum_{k=0}^{N} a_k y(n-k) = \sum_{k=0}^{L} b_k x(n-k)$.
3. $H(z) = z[h(n)]$.

Stability See the appropriate sections in Chapters 7 and 8.
From $h(n)$:

1. All bounded inputs produce bounded outputs.
2. The initial condition (IC) response of a causal system decays to zero as $n \to \infty$, or, in terms of the characteristic roots, the roots (the r_k's) must have magnitudes less than 1.
3. $\sum_{n=-\infty}^{\infty} |h(n)| < \infty$.

From $H(z)$:

1. Causal system: The magnitudes of all poles are less than 1.
2. Anticausal system: The magnitudes of all poles are greater than 1.
3. General: The ROC includes the unit circle.

BASIC CONCEPTS

Sinusoidal Steady-State Response See the appropriate sections in Chapters 7 and 8.
From $h(n)$: For $x(n) = A\cos(n\theta + \alpha)$, $-\infty \le n \le \infty$,

$$y_{ss}(n) = A|H(e^{j\theta})|\cos(n\theta + \alpha + \angle H(e^{j\theta})), \qquad -\infty \le n \le \infty$$

where $H(e^{j\theta}) = \sum_{m=-\infty}^{\infty} h(m)e^{-jm\theta}$.
From $H(z)$: For $x(n) = A\cos(n\theta + \alpha)u(n)$,

$$y_{ss}(n) = A|H(e^{j\theta})|\cos(n\theta + \alpha + \angle H(e^{j\theta}))$$

where $H(e^{j\theta}) = H(z)|_{z=e^{j\theta}}$.

Sampling Theorem If an analog signal has no frequency components at frequencies greater than $f_{\max}$, the signal can be uniquely represented by equally spaced samples if the sampling frequency f_s is greater than two times $f_{\max}$.

Furthermore, the original analog signal can be recovered from the samples by passing them through an ideal lowpass filter having an appropriate bandwidth.

Frequency Response The function $H(e^{j\theta})$ is known as the frequency response function of the system, and it can be determined in several ways. The three following methods all result in an analytical expression.

1. From the transfer function $H(z)$ of a stable system, the frequency response is

$$H(e^{j\theta}) = H(z)\big|_{z=e^{j\theta}} = \left.\frac{\sum_{k=0}^{L} b_k z^{-k}}{\sum_{k=0}^{N} a_k z^{-k}}\right|_{z=e^{j\theta}} = \frac{\sum_{k=0}^{L} b_k e^{-jk\theta}}{\sum_{k=0}^{N} a_k e^{-jk\theta}}.$$

2. Given the unit sample response $h(n)$ of a stable LTI system, the frequency response is

$$H(e^{j\theta}) = \sum_{m=-\infty}^{\infty} h(m)e^{-jm\theta}.$$

3. Starting with the DE for a stable system,

$$\sum_{k=0}^{N} a_k y(n-k) = \sum_{k=0}^{L} b_k x(n-k), \quad \text{we have} \quad H(e^{j\theta}) = \frac{\sum_{k=0}^{L} b_k (e^{-j\theta})^k}{\sum_{k=0}^{N} a_k (e^{-j\theta})^k}.$$

CHAPTER 9

ILLUSTRATIVE PROBLEM 9.1
Evaluation of Frequency Response of Two Systems

Find the frequency response for each system described.

a. A nonrecursive filter with the unit impulse response $h(n) = 2\delta(n) - 3\delta(n-1) + 4\delta(n-2)$.

b. A bandpass filter modeled by the DE $y(n) + 0.25y(n-4) = x(n) - x(n-2)$.

Solution

a. This system is FIR; consequently it is stable and the frequency response exists. Thus

$$\begin{aligned} H(e^{j\theta}) &= \sum_{m=-\infty}^{\infty} h(m)e^{-jm\theta} \\ &= \sum_{m=-\infty}^{\infty} [2\delta(m) - 3\delta(m-1) + 4\delta(m-2)]e^{-jm\theta} \\ &= 2 - 3e^{-j\theta} + 4e^{-j2\theta}. \end{aligned}$$

b. In this case, $H(z) = z^2(z^2 - 1)/(z^4 + 0.25)$. Is this a stable system? The filter poles are the roots of $z^4 + 0.25 = 0$ or $z^4 = -0.25$. This fourth-degree equation is easy to solve using DeMoivre's theorem, where we write $-0.25 = 0.25e^{j(\pi+m2\pi)}$, $m = 0, 1, 2, 3$, giving $z^4 = 0.25e^{j(\pi+m2\pi)}$. Taking the fourth root of both sides, we have

$$(z^4)^{1/4} = (0.25)^{1/4} e^{j(\pi+m2\pi)/4}$$

and the four roots are

$$z_1 = 0.707e^{j\pi/4}, \quad z_2 = 0.707e^{j3\pi/4}, \quad z_3 = 0.707e^{j5\pi/4}, \quad \text{and}$$
$$z_4 = 0.707e^{j7\pi/4}.$$

Since these roots (which are also the poles of $H(z)$) have magnitudes that are less than 1, the causal system is stable and its frequency response can be found. Thus, by substituting $z = e^{j\theta}$ in $H(z)$, we have

$$H(e^{j\theta}) = \frac{e^{j2\theta}(e^{j2\theta} - 1)}{e^{j4\theta} + 0.25}.$$

Characteristics of the Frequency Response Function We have seen that the system frequency response function $H(e^{j\theta})$ totally determines the steady-state system output sequence for a known sinusoidal input sequence. The basis of many discrete-system design procedures, therefore, is to determine an acceptable frequency response function $H(e^{j\theta})$ that will satisfy both the magnitude and phase criteria of any particular design situation. The function $H(e^{j\theta})$ is very important in discrete system analysis and design, so let's look at its nature in greater detail.

Periodicity $H(e^{j\theta})$ is a periodic function with a period of 2π rad. With

$$H(e^{j\theta}) = \sum_{m=-\infty}^{\infty} h(m)e^{-jm\theta}$$

we substitute $\theta + 2\pi$ for θ and obtain

$$H(e^{j[\theta+2\pi]}) = \sum_{m=-\infty}^{\infty} h(m)e^{-j[\theta+2\pi]m} = e^{-jm2\pi}H(e^{j\theta}) = H(e^{j\theta})$$

because $e^{-jm2\pi} = 1$ for all integer values of m. This periodicity property is an alternative way of looking at the information provided by the sampling theorem, because if a continuous time signal is undersampled (i.e., creating digital frequencies greater than π), some frequencies will be treated by the system as if they were in the range $-\pi < \theta < \pi$.

Symmetry The magnitude M of $H(e^{j\theta})$ is an even function and the phase P is an odd function. To show these characteristics we start with

$$H(e^{j\theta}) = \sum_{m=-\infty}^{\infty} h(m)e^{-jm\theta} = \sum_{m=-\infty}^{\infty} h(m)[\cos(m\theta) - j\sin(m\theta)] = a + jb$$

and similarly

$$H(e^{-j\theta}) = \sum_{m=-\infty}^{\infty} h(m)e^{+jm\theta} = \sum_{m=-\infty}^{\infty} h(m)[\cos(m\theta) + j\sin(m\theta)] = a - jb.$$

Thus we see that

$$|H(e^{j\theta})| = |H(e^{-j\theta})| = \sqrt{a^2 + b^2} = M$$

which satisfies the definition of an even function. The phase of $H(e^{j\theta})$ is $\tan^{-1}(b/a)$, whereas that of $H(e^{-j\theta})$ is $\tan^{-1}(-b/a)$, and we find that

$$\angle H(e^{j\theta}) = -\angle H(e^{-j\theta}) = P$$

which satisfies the definition of an odd function.

ILLUSTRATIVE PROBLEM 9.2
Causal and Noncausal Filters

A noncausal filter has the unit sample response

$$h_{nc}(n) = 0.25\delta(n + 2) + 0.50\delta(n + 1) + 1.00\delta(n) + 0.50\delta(n - 1) + 0.25\delta(n - 2).$$

a. Find the transfer function $H_{nc}(z)$.

b. Determine $H_{nc}(e^{j\theta})$.

c. Shift the unit sample response $h_{nc}(n)$ two samples to the right to make it causal and find the causal frequency response $H_c(e^{j\theta})$ in terms of $H_{nc}(e^{j\theta}) = Me^{jP}$.

Solution

a. From the shifting property,

$$H_{nc}(z) = 0.25z^2 + 0.5z^1 + 1 + 0.5z^{-1} + 0.25z^{-2}.$$

b. Substituting $z = e^{j\theta}$ produces

$$H_{nc}(e^{j\theta}) = 0.25e^{j2\theta} + 0.5e^{j\theta} + 1 + 0.5e^{-j\theta} + 0.25e^{-j2\theta}.$$

With the help of the Euler relation, we can write

$$H_{nc}(e^{j\theta}) = 1 + \cos(\theta) + 0.50\cos(2\theta)$$

where we notice that the frequency response for the filter is always a real number.

c. The magnitude is the same but the phase is different, as seen by

$$H_c(z) = z^{-2}H_{nc}(z) \quad \text{and} \quad H_c(e^{j\theta}) = e^{-j2\theta}H_{nc}(e^{j\theta}) = Me^{j(P-2\theta)}.$$

CHAPTER 9

Time and frequency Notice that even though $h(n)$ is discrete (a sequence), the frequency response function $H(e^{j\theta})$ is a function of the continuous digital frequency θ.

Decibels (dB) The values of the gain M may vary over a very wide range, and it is often useful to define M in decibels as $M_{\text{dB}} = 20\log_{10} M$. We will use both versions of M in this chapter—that is, as a real number and in decibels.

Bandwidth If $M_{\max}$ is the maximum value of M, the range of frequencies $\theta_1 \le \theta \le \theta_2$ for which $0.707M_{\max} \le M \le M_{\max}$ is called the bandwidth of a passband. Filters may have one or several passbands, and in terms of decibels, a passband is a region where the frequency response magnitude is within -3 dB $= 20 \log_{10}(0.707)$ of the maximum value in decibels.

Discrete Time Fourier Transform It is common practice to call $H(e^{j\theta})$ the Discrete Time Fourier Transform (DTFT) of the unit sample response $h(n)$. In a similar way, for the sequences $x(n)$ and $y(n)$ we write

DTFT

$$X(e^{j\theta}) = \sum_{m=-\infty}^{\infty} x(m)e^{-jm\theta} \quad \text{and} \quad Y(e^{j\theta}) = \sum_{m=-\infty}^{\infty} y(m)e^{-jm\theta}$$

where $X(e^{j\theta})$ and $Y(e^{j\theta})$ are the DTFTs of the sequences $x(n)$ and $y(n)$, respectively. This transform, along with the Discrete Fourier Transform (DFT), is considered in detail in the next chapter.

Frequency Response Plots There are many different varieties of frequency response plots, and their designations are sometimes confusing. We now proceed to discuss several different types of plots, with their most common names.

Rectangular Plots Recalling that the frequency response can be expressed in terms of magnitude and phase as

$$H(e^{j\theta}) = |H(e^{j\theta})|e^{j\angle H(e^{j\theta})} = Me^{jP}$$

the most straightforward representation is to plot M versus θ and P versus θ on rectangular coordinates for the digital frequency range of $\theta_1 \leq \theta \leq \theta_2$. The usual range of frequencies used is $0 \leq \theta \leq \pi$, which, because of symmetry, can be extrapolated to the period $0 \leq \theta \leq 2\pi$ or $-\pi \leq \theta \leq \pi$. See Figure 9.2a on page 498 for the rectangular plots for a bandpass digital filter based on Chebyshev design that is described by the transfer function

$$H(z) = \frac{0.011(z+1)^3(z-1)^3}{z^6 + 2.153z^4 + 1.786z^2 + 0.545}.$$

To plot this with MATLAB we use (invoke) a function from the Signal Processing Toolbox or the Signals and Systems Toolbox named **freqz**, which stands for "frequency response in the z domain," or discrete-time domain. To use **freqz** we need to put the transfer function in the form

$$H(z) = \frac{B(z)}{A(z)} = \frac{b(1) + b(2)z^{-1} + \cdots + b(nb)z^{-(nb-1)}}{a(1) + a(2)z^{-1} + \cdots + a(na)z^{-(na-1)}}.$$

That is, the z^0 term in both the numerator and denominator polynomials must be the first term, with the other terms following in increasingly negative powers of z. Thus, after multiplying out the numerator factors (which could be done using the MATLAB function **conv**), we divide the numerator and denominator of the given transfer function by z^6 and enter the numerator and denominator coefficients in the vectors **b** and **a**, respectively. Running the program returns the frequency response vector **H** along with a frequency range of 0 to π rad in the frequency vector **q** (used to represent the vector $\boldsymbol{\theta}$), where we have specified $N = 512$ computation points, as seen in the call to **freqz** in the

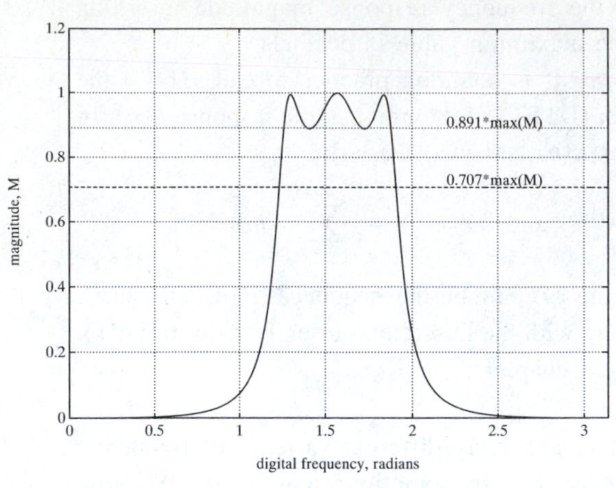

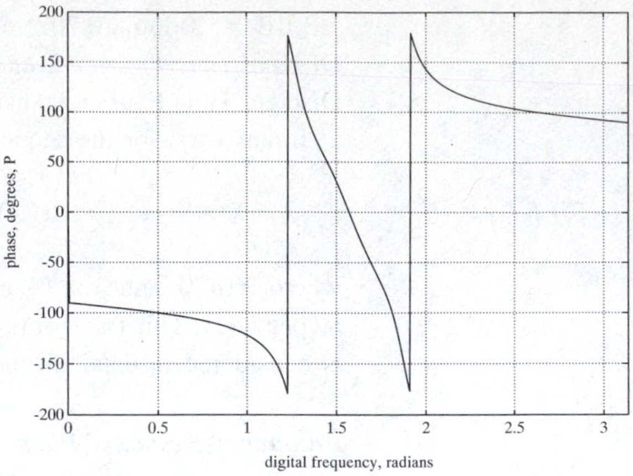

Magnitude Phase

FIGURE 9.2a *Frequency response for bandpass filter*

following script. This filter has a bandpass characteristic with the passband in the approximate range $1.25 \leq \theta \leq 1.88$ rad, that is, a bandwidth of 0.63 rad. It also exhibits a ripple band, where the magnitude varies between 1.000 and 0.891 or -1 dB. You will learn about designing filters such as this one in a later course.

MATLAB Script

```
%F9_2 Rectangular plots

%F9_2a...Frequency response for bandpass filter
b=[0.011,0,-.033,0,0.033,0,-.011];      % numerator coefficients in ascending
                                         % powers of z^-1
a=[1,0,2.153,0,1.786,0,0.545];          % denominator coefficients in
                                         % ascending powers of z^-1
[H,q]=freqz(b,a,512);                   % call freqz 512 frequency points
M=abs(H);                               % defines magnitude of H
P=angle(H);                             % defines angle of H
%...conversion of radians to degrees, plotting statements, and pause
```


CROSS-CHECK Are these plots reasonable? First, for $\theta = 0$ or $z = 1$, $H(1) = 0$, which agrees with the plot. Next, for $\theta = \pi$ or $z = -1$, $H(-1) = 0$, which is again acceptable. Checking at the midfrequency of $\theta = \pi/2$ or $z = +j1$ is not as easy, but it is not difficult. $H(j1) = 0.011(-1 - 3 - 3 - 1)/(-1 + 2.153 - 1.786 + 0.545) = 1e^{j0}$, which also matches the gain and phase of the plots. ■

It is often useful to plot the M and P curves for the entire period $0 \leq \theta \leq 2\pi$. To do this with **freqz,** we add 'whole' to the calling statement, giving

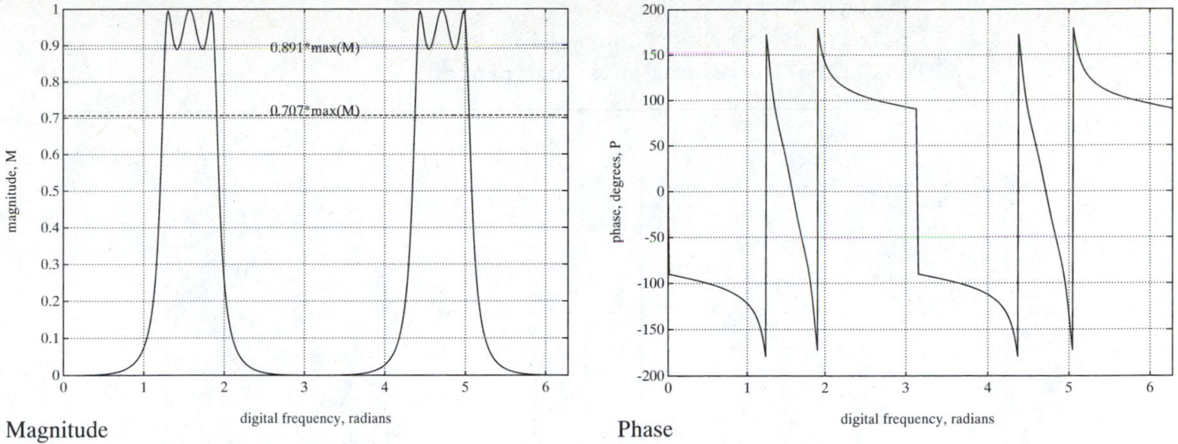

FIGURE 9.2b *Frequency response for bandpass filter for one period*

[H, q] = freqz(b, a, 512, 'whole'), with the plots shown in Figure 9.2b, where the magnitude and phase symmetry properties are easily verified.

Polar and Nyquist Plots If we plot $\text{Im}[H(e^{j\theta})]$ versus $\text{Re}[H(e^{j\theta})]$ with frequency as a parameter, a polar plot results. As an alternative to plotting the imaginary part versus the real part on rectangular scales, we could plot the magnitude versus the angle on polar graph paper. In either case the same result is obtained. The single polar plot represents the same data as the two curves of the rectangular plot, and we can extract the same information (e.g., bandwidth) *provided* the frequency values $\theta_a, \theta_b, \theta_c, \ldots$ are shown for each plotted point on the polar diagram. To illustrate this kind of diagram, we use the transfer function of a digital filter that was derived from an analog filter by the popular bilinear transformation method, which is usually treated in a subsequent course. See Figure 9.3a for a polar plot for the lowpass filter

$$H(z) = \frac{0.068(z + 1)^2}{z^2 - 1.142z + 0.413}$$

with the frequencies $\theta = 0$ $(z = 1)$, $\theta = \pi/6$ $(z = 0.866 + j0.500)$, and $\theta = \pi$ $(z = -1)$ marked on the plot. Notice that the locus proceeds in a clockwise sense for the frequency range $0 \leq \theta \leq \pi$.

MATLAB Script

```
%F9_3 Polar and Nyquist plots

%F9_3a...Polar plot for a lowpass filter
num=[0.068,0.136,0.068];
den=[1,-1.142,0.413];
[H,q]=freqz(num,den,512);
M=abs(H);
```

Continues

```
P=angle(H);
polar(P,M),title('Figure 9.3a Polar plot');
%...labeling statements and pause
```

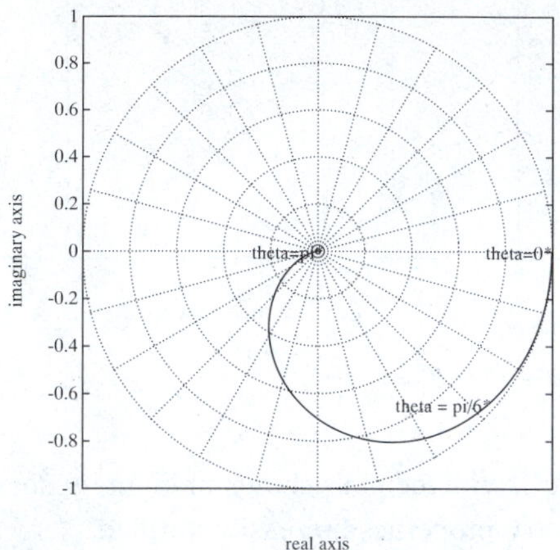

FIGURE 9.3a *Polar plot for a lowpass filter*

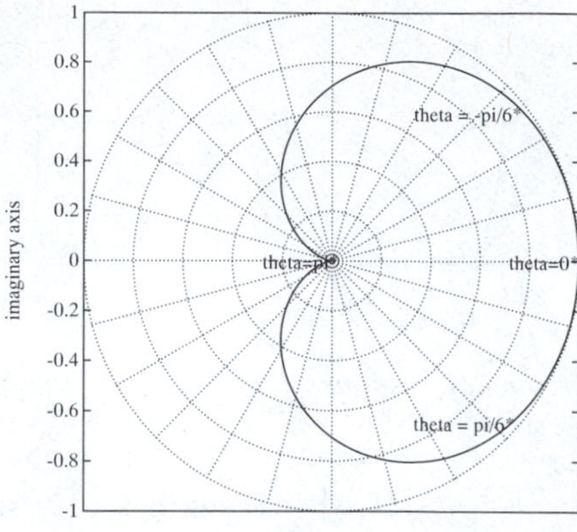

FIGURE 9.3b *Nyquist plot for a lowpass filter*

The frequency response vector **H** was obtained using **freqz**; the "polar" plotting routine then generated the plot.

If the polar diagram for $H(e^{j\theta})$ is plotted for $0 \leq \theta \leq 2\pi$, the resulting curve is called a *Nyquist diagram.*[1] See Figure 9.3b for the Nyquist diagram for the same $H(z)$ where 'whole' was added to the calling statement, giving **[H, q] = freqz(num, den, 512, 'whole')**. From symmetry we know that the plot in the frequency range from $-\pi$ to 0 is the same as in the range from π to 2π, which means that $\theta = -\pi/6$ corresponds to $\theta = 11\pi/6$. This also accounts for the fact that the top half $(-\pi \leq \theta \leq 0)$ is the mirror image of the bottom half $(0 \leq \theta \leq \pi)$; that is, $H(e^{j\theta}) = H^*(e^{-j\theta})$.

Logarithmic, or Bode, Plots Logarithmic, or Bode,[2] plots are used with well-known design rules to analyze certain system properties such as stability. To use these rules it is common to plot $M_{\text{dB}} = 20 \log_{10} M$ and P versus θ with M_{dB}

[1] Harry Nyquist (1889–1976), a native of Sweden, was an electrical engineer with the Bell Telephone Laboratories who made a very important contribution concerning the stability of feedback amplifiers. His result is contained in the seminal paper "Regeneration Theory," *Bell System Technical Journal,* 11: 126–147 (1932).

[2] Logarithmic scales for both magnitude and frequency were used extensively in the studies of Hendrik Wade Bode (1905–1984). Bode was a member of the technical staff at the Bell Telephone Laboratories and later joined the faculty at Harvard University. For further reading see *Network Analysis and Feedback Amplifier Design,* D. Van Nostrand Co., Princeton, N.J. (1945), pages 316 ff. or H. W. Bode, "Relations between attenuation and phase in feedback amplifier design," *Bell Systems Technical Journal,* 19: 421–454 (1940).

and P on linear scales and θ on a log scale. The quantity $20 \log_{10} M$ is called the *decibel*, or dB, gain of the frequency response $H(e^{j\theta})$. Figure 9.4 shows Bode plots for the sixth-order bandpass Chebyshev digital filter considered previously:

$$H(z) = \frac{0.011(z + 1)^3(z - 1)^3}{z^6 + 2.135z^4 + 1.768z^2 + 0.540}.$$

Notice that the use of decibels and a logarithmic frequency scale allows us to observe the response over a greater range of magnitudes and frequencies than in Figure 9.2. This is at the cost, however, of losing some detail in the ripple passband. Of course, we could always use MATLAB to "zoom in" on the frequency range if we wish. How would we do this?

MATLAB Script

```
%F9_4 Bode plot
num=[1,0,-3,0,3,0,-1];
den=[1,0,2.135,0,1.768,0,0.540];
[H,q]=freqz(num,den,512);
mag=0.011*abs(H);
mag=20*log10(mag);
g=ones(size(q))*max(mag)-3;
semilogx(q,mag,q,g),title('Magnitude');
%...labeling statements
pause;
phase=angle(H);
phase=phase*180/pi;
semilogx(q,phase),title('Phase');
%...labeling statements
```

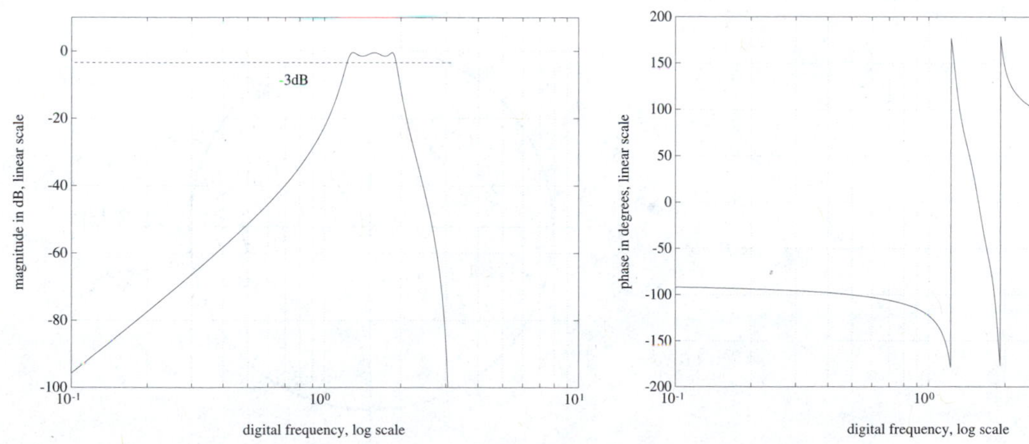

Magnitude Phase

FIGURE 9.4 *Bode plot for a bandpass Chebyshev filter*

CROSS-CHECK The given transfer function has three zeros at $z = 1$ ($\theta = 0$) and three zeros at $z = -1$ ($\theta = \pi = 3.14$) and so the magnitude should be zero at these frequencies. The Bode plot of Figure 9.4 shows a large negative decibel gain at $\theta = 0.1$ rad and $\theta = 3$ rad, which agrees with the transfer function results. Using the midfrequency of $\theta = \pi/2$ or $z = j1$, we have

$$H(j1) = \frac{0.011(j1 + 1)^3(j1 - 1)^3}{(j1)^6 + 2.135(j1)^4 + 1.768(j1)^2 + 0.540} = 0.946$$

or -0.48 dB, which is in close agreement with the magnitude part of the Bode plot. ■

Graphical Estimation of Frequency Response The transfer function $H(z)$ of a discrete-time system is described in terms of zeros $n_1, n_2, \ldots, n_L$ and poles $d_1, d_2, \ldots, d_N$ as

$$H(z) = \frac{Kz^{N-L}(z - n_1)(z - n_2)\cdots(z - n_L)}{(z - d_1)(z - d_2)\cdots(z - d_N)}$$

where K is a real constant. The frequency response $H(e^{j\theta})$ is then

$$H(e^{j\theta}) = \frac{Ke^{j\theta(N-L)}(e^{j\theta} - n_1)(e^{j\theta} - n_2)\cdots(e^{j\theta} - n_L)}{(e^{j\theta} - d_1)(e^{j\theta} - d_2)\cdots(e^{j\theta} - d_N)}$$
$$= \frac{Ke^{j\theta(N-L)}(N_1e^{j\alpha_1})(N_2e^{j\alpha_2})\cdots(N_Le^{j\alpha_L})}{(D_1e^{j\beta_1})(D_2e^{j\beta_2})\cdots(D_Ne^{j\beta_N})}.$$

The N's, D's, α's and β's can all be identified in the pole-zero plot of $H(z)$, as indicated in Figure 9.5 where $N_re^{j\alpha_r}$ is a vector from the zero n_r to the tip of the $e^{j\theta}$ vector, $r = 1, 2, \ldots, L$, and $D_pe^{j\beta_p}$ is the vector from the pole d_p to the tip of the $e^{j\theta}$ vector, $p = 1, 2, \ldots, N$. Remember that $z = e^{j\theta}$, where

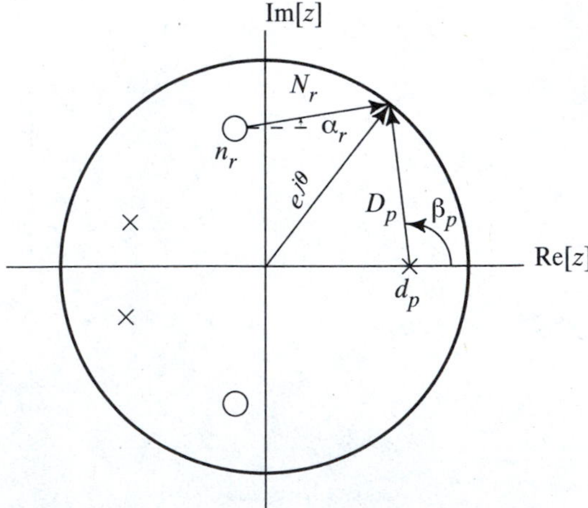

FIGURE 9.5 *Graphical estimation of frequency response magnitude*

CHAPTER 9

θ is the digital frequency of the input. Collecting the magnitude and phase components of the frequency response separately gives

$$M = \frac{KN_1N_2\cdots N_L}{D_1D_2\cdots D_N} \quad \text{and} \quad P = (N - L)\theta + \sum_{k=1}^{L} \alpha_k - \sum_{k=1}^{N} \beta_k .$$

magnitude phase

This graphical approach for determining the frequency response is most useful as an estimation procedure, since a computer should always be used to obtain the final plot. Although it is difficult, in general, to determine phase information for all frequencies, it is often possible to use the graphical approach to estimate the phase for a few selected frequencies.

ILLUSTRATIVE PROBLEM 9.3 ***Graphical Estimation of Frequency Response***

Use the graphical method to verify the magnitude characteristic of a simple filter with $H(z) = z^2/(z^2 - 0.40z + 0.60)$.

Solution

There are two zeros at $z = 0$ and poles at $z_{1,2} = 0.20 \pm j0.75 = 0.78e^{\pm j1.31}$, as plotted in Figure 9.6a, where the zero vectors $N_1e^{j\alpha} = N_2e^{j\alpha}$ and pole vectors

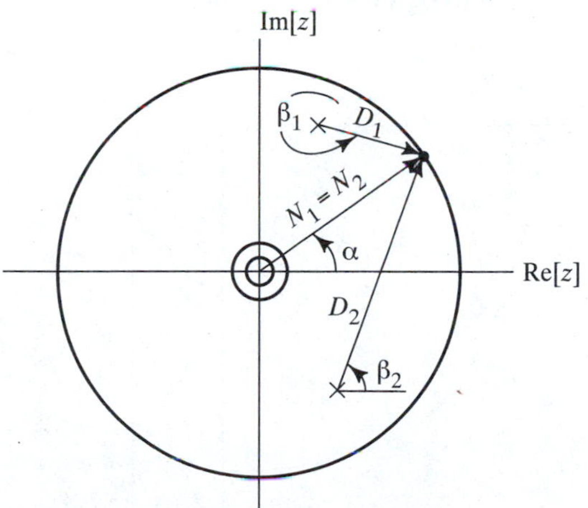

a. Pole and zero vectors for Illustrative Problem 9.3

FIGURE 9.6 *Graphical estimation of frequency response for Illustrative Problem 9.3*

Continues

BASIC CONCEPTS

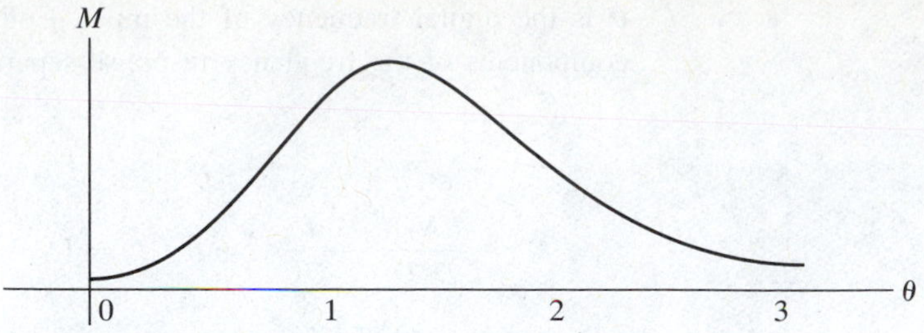

b. Estimate of magnitude plot for Illustrative Problem 9.3

FIGURE 9.6 *Continued*

$D_1e^{j\beta_1}$ and $D_2e^{j\beta_2}$ are also seen. As the digital frequency increases from 0 to π, we see that the length D_1 becomes short near the pole in the vicinity of the angle $\theta = 1.3$ rad, which tells us that the magnitude $M = N_1N_2/D_1D_2$ will reach its largest value in this region. Estimating the vector lengths at $\theta = 1.3$ gives $M = 1/((0.2)(1.8)) = 2.78$, and the estimated magnitude plot will be as in Figure 9.6b.

Comment: Although it wasn't emphasized in the preceding discussion, some phase properties also can be found from the graphical approach. The system phase is $P = \alpha_1 + \alpha_2 - (\beta_1 + \beta_2)$, and in this example we see that $\alpha_1 = \alpha_2 = 0$ and $\beta_2 + \beta_1 = 0$ at $\theta = 0$. Hence, the phase P of the frequency response is also zero. For $\theta = \pi$ the phase is $P = 2\pi$. These results may be verified in Figure 9.7, where the results from **freqz** are given. Notice that the phase from the MATLAB function is limited to the range $-\pi \leq \theta \leq \pi$ for the purpose of saving paper.

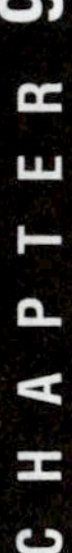

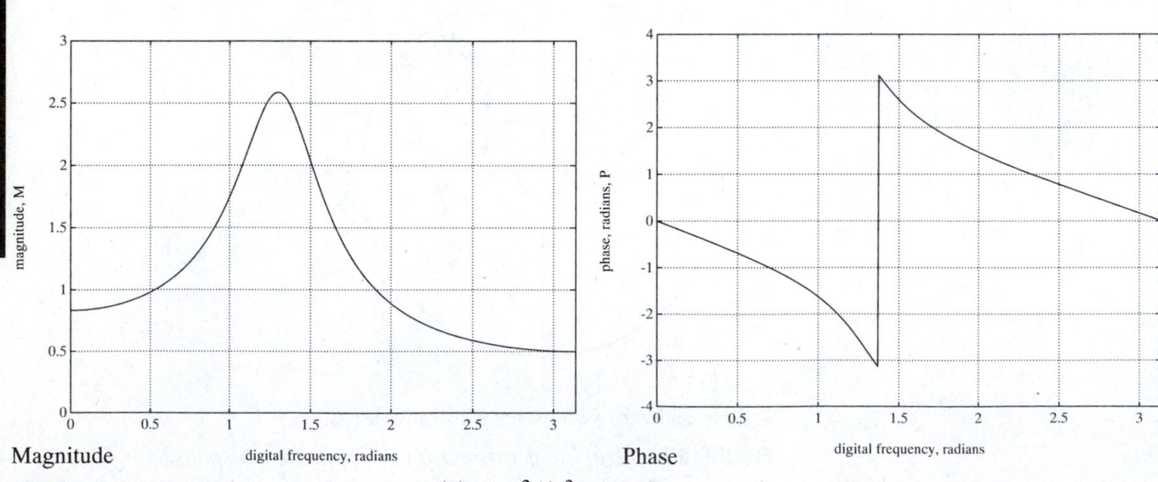

FIGURE 9.7 *Frequency response for* $H(z) = z^2/(z^2 - 0.42z + 0.60)$ *using* **freqz**

DISCRETE-TIME FILTER CHARACTERISTICS

Now let's consider a few important characteristics of digital filters. Detailed consideration of the design and application of these filters is normally treated in an introductory course in digital signal processing.

Ideal Filters The frequency response of an ideal filter is given by

ideal filter

$$H(e^{j\theta}) = \begin{cases} Ke^{-jl\theta}, & \text{in the passband} \\ 0, & \text{otherwise} \end{cases}$$

where the magnitude is a positive constant in the passband and zero elsewhere. The phase of an ideal filter is linear in the passband with a slope of $-l$ radians per radian (or degrees per radian). The magnitude and phase characteristics for lowpass, highpass, bandpass, and bandstop ideal filters are shown in Figure 9.8, where we recall that the discrete-time frequency response $H(e^{j\theta})$ has a period of 2π rad.

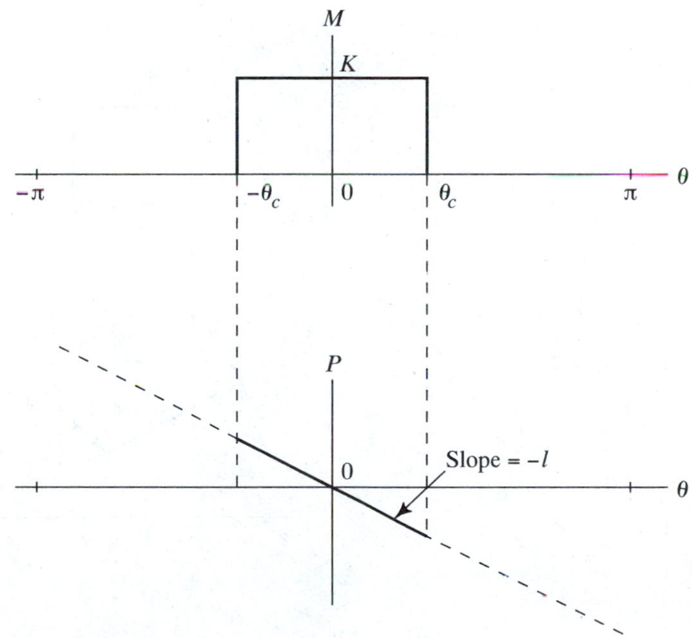

a. Lowpass filter

FIGURE 9.8 *Ideal filter frequency response characteristics*

Continues

BASIC CONCEPTS

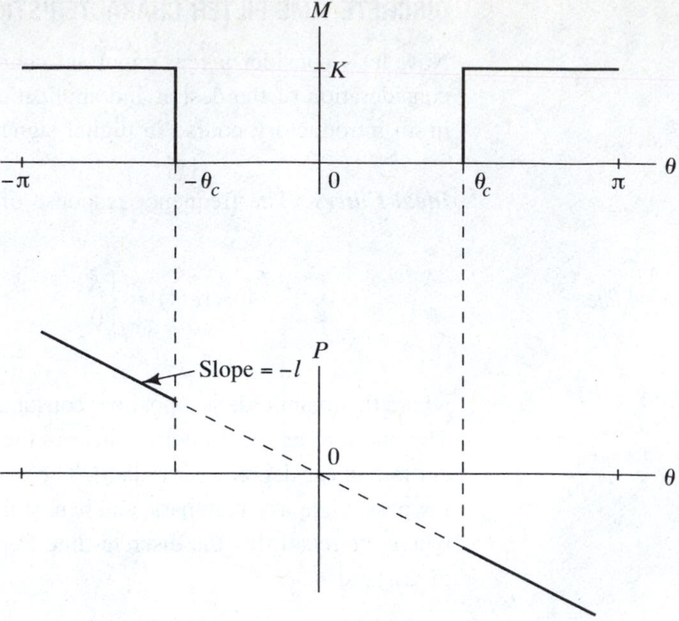

b. Highpass filter

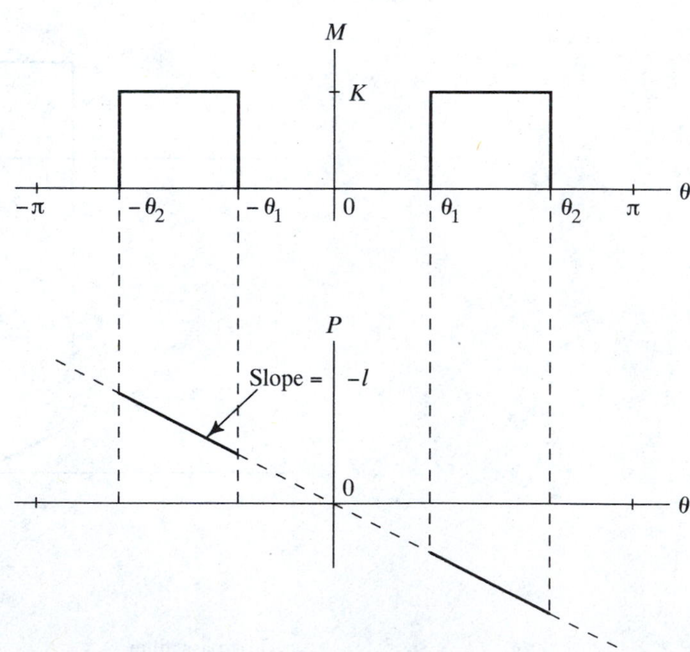

c. Bandpass filter

FIGURE 9.8 *Continued*

Continues

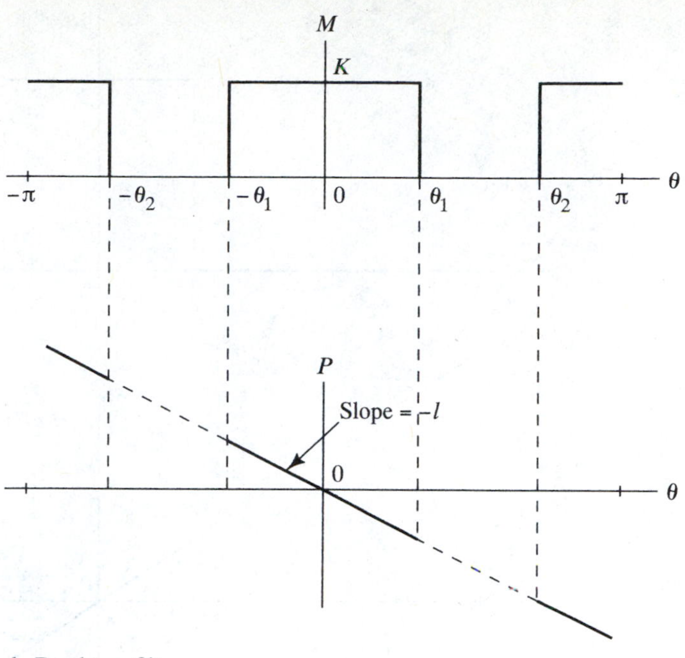

d. Bandstop filter

FIGURE 9.8 *Continued*

ILLUSTRATIVE PROBLEM 9.4
Response of an Ideal Filter

Consider the frequency response characteristics of the ideal filter of Figure 9.9 on page 508.

a. Find the steady-state response of this system to the input $x(n) = 10\cos(n\pi/8 - \pi/6) + 15\cos(n\pi/2 + \pi/3) + 20\cos(6n\pi/10) + 25\cos(8\pi n/10 + 0.3\pi)$.

b. Use MATLAB to plot the input and output sequences to illustrate the effects of ideal filtering.

Solution

a. The input frequencies are $\theta_1 = \pi/8$, $\theta_2 = \pi/2$, $\theta_3 = 6\pi/10$, and $\theta_4 = 8\pi/10$. From Figure 9.9 the ideal filter characteristics are

$$H(e^{j\theta}) = \begin{cases} 2e^{-j4\theta}, & |\theta| \le \pi/4 \\ 0, & \pi/4 < |\theta| \le \pi. \end{cases}$$

Thus, only the sequence with the frequency $\theta = \pi/8$ is passed by the filter and

$$y_{ss}(n) = (10)(2)\cos(n\pi/8 - \pi/6 - (4)(\pi/8)) = 20\cos(n\pi/8 - 2\pi/3).$$

b. From F9_10 on the CLS disk, we obtain the plots in Figure 9.10, which show the lowpass nature of the filter, resulting, in this case, in a pure sinusoid as the output sequence. The lowest frequency component of the input $x_1(n) = 10\cos(n\pi/8 - \pi/6)$ is changed in amplitude by the gain 2 and is delayed or shifted by an amount proportional to the frequency—that is, $-4 \cdot (\pi/8) = -\pi/2$ rad.

BASIC CONCEPTS

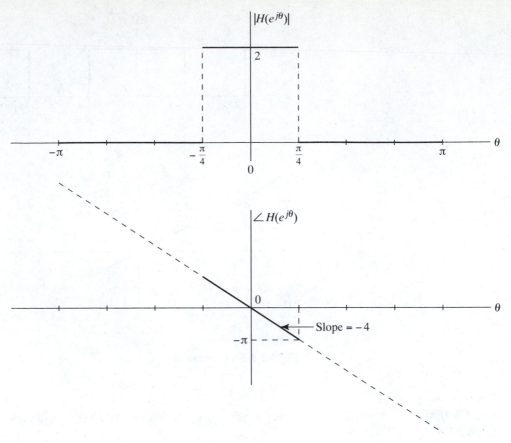

FIGURE 9.9 *Ideal filter characteristics for Illustrative Problem 9.4*

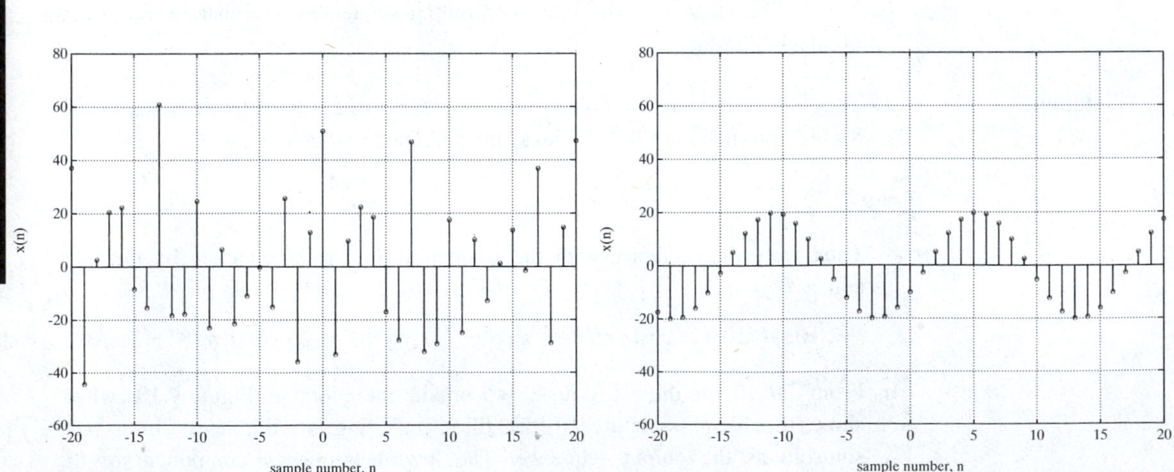

FIGURE 9.10 *Input sequence (left) and steady-state output sequence (right) for an ideal lowpass filter*

What is the time-domain consequence of the frequency-domain characteristic of linear phase? For a filter input sequence of $x(n) = s(n) + g(n)$, we would like to pass all $s(n)$ while preserving its original shape and completely reject or stop $g(n)$. The desirable part of the input is $s(n)$, the signal component of $x(n)$, and the unwanted part of $x(n)$ is $g(n)$, the garbage, or noise, component. Let us assume that $s(n)$ consists of sinusoidal components in one frequency band and that $g(n)$ is made up of sinusoidal components in another, nonoverlapping, band. In this case, we can write the input sequence as

$$x(n) = s(n) + g(n) = \sum_{m=1}^{M_1} S_m \cos(n\theta_m) + \sum_{m=M_1+1}^{M_2} G_m \cos(n\theta_m)$$

where the frequencies θ_m, $m = 1, 2, \ldots, M_1$ lie in the filter's passband and the frequencies θ_m, $m = M_1 + 1, \ldots, M_2$, lie in the filter's stop, or rejection, band. Assuming that the filter is ideal with frequency response

$$H(e^{j\theta}) = \begin{cases} Ke^{-jl\theta}, & \text{in the passband} \\ 0, & \text{otherwise} \end{cases}$$

application of the sinusoidal steady-state formula gives us the filter's steady-state output

$$y_{ss}(n) = \sum_{m=1}^{M_1} KS_m \cos(n\theta_m - l\theta_m) + 0 .$$

From this result, we see that each sinusoidal sequence in the passband is multiplied by the same gain K and is shifted or delayed by an amount proportional to its frequency θ_m, thus preserving the shape of the signal component $s(n)$ of the input.

Nonrecursive filters Let's consider a causal nonrecursive filter described by the DE

$$y(n) = \sum_{k=0}^{L} b_k x(n - k) .$$

The steady-state solution for an input $x(n) = e^{jn\theta}$ is $y_{ss}(n) = e^{jn\theta} H(e^{j\theta})$. Substituting this solution into the given difference equation gives

$$e^{jn\theta} H(e^{j\theta}) = b_0 e^{jn\theta} + b_1 e^{jn\theta} e^{-j\theta} + b_2 e^{jn\theta} e^{-j2\theta} + \cdots + b_L e^{jn\theta} e^{-jL\theta}$$

and the frequency response in terms of the coefficients of the filter's DE is

$$H(e^{j\theta}) = b_0 + b_1 e^{-j\theta} + b_2 e^{-j2\theta} + \cdots + b_L e^{-jL\theta} = \sum_{m=0}^{L} b_m e^{-jm\theta} .$$

But the frequency response for a causal filter with L delays is given by

$$H(e^{j\theta}) = \sum_{m=0}^{L} h(m) e^{-jm\theta}$$

and we have the important relationship for nonrecursive filters,

$$H(e^{j\theta}) = \sum_{m=0}^{L} b_m e^{-jm\theta} = \sum_{m=0}^{L} h(m) e^{-jm\theta} \quad \text{or} \quad b_m = h(m) .$$

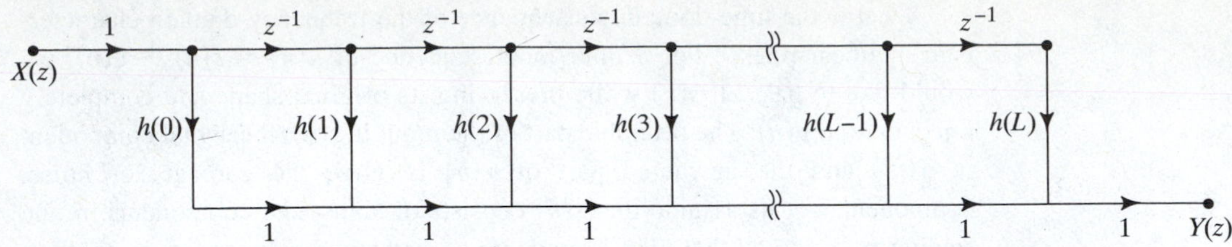

FIGURE 9.11 *Nonrecursive filter with L delays*

Thus, having used one of many design procedures available for finding the values of $h(n)$ needed to produce a given or wanted frequency response, the filter can be realized as in Figure 9.11 and implemented by a general-purpose computer and an appropriate program or by a DSP chip set.

Linear phase nonrecursive filters Now let's consider the characteristics of the unit sample response of a nonrecursive filter that will lead to linear phase in the frequency domain. Consider a unit sample response of a noncausal filter having an odd number of terms that exhibit the property of even symmetry. That is, centered about $h_{nc}(0)$ at $n = 0$, we have

$$h_{nc}(-1) = h_{nc}(1), h_{nc}(-2) = h_{nc}(2), \ldots, h_{nc}(-I) = h_{nc}(I)$$

as in Figure 9.12a, where we see that the number of terms in this filter is $2I + 1$. Taking advantage of the proposed symmetry in $h_{nc}(n)$, where $h_{nc}(r) = h_{nc}(-r)$, the frequency response of the noncausal filter is given by

$$H_{nc}(e^{j\theta}) = h_{nc}(0) + h_{nc}(1)\{e^{j\theta} + e^{-j\theta}\} + h_{nc}(2)\{e^{j2\theta} + e^{-2j\theta}\} + \cdots$$
$$+ h_{nc}(I)\{e^{jI\theta} + e^{-jI\theta}\}.$$

Using the Euler relation of $2M_k \cos(k\theta) = M_k(e^{jk\theta} + e^{-jk\theta})$, we can write

$$H_{nc}(e^{j\theta}) = h_{nc}(0) + 2h_{nc}(1)\cos(\theta) + 2h_{nc}(2)\cos(2\theta) + \cdots$$
$$+ 2h_{nc}(I)\cos(I\theta)$$
$$= h_{nc}(0) + \sum_{m=1}^{I} 2h_{nc}(m)\cos(m\theta)$$

which is a real function of frequency that can be either positive or negative. Thus we can recast the frequency response in the form

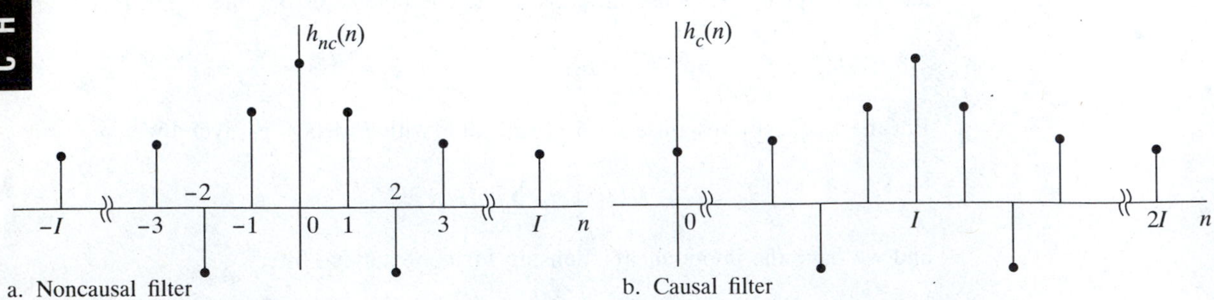

FIGURE 9.12 *Linear phase filter: even symmetry, odd number of terms*

CHAPTER 9

$$H_{nc}(e^{j\theta}) = \pm\left|h_{nc}(0) + \sum_{m=1}^{I} 2h_{nc}(m)\cos(m\theta)\right| = \pm M.$$

If all the elements in the unit sample response of the noncausal filter are shifted to the right by I samples, as in Figure 9.12b, the filter becomes causal. Recalling that a shift of I samples in the time domain produces a phase term of $e^{-jI\theta}$ in the frequency domain, the frequency response of the causal filter that now has linear phase is

$$H_c(e^{j\theta}) = \pm Me^{-jI\theta}.$$

In the filter's passband, the sign affixed to M does not change because $H_{nc}(e^{j\theta})$ is a continuous function of θ and a sign change implies that M passes through a zero in the passband—an impossibility.

ILLUSTRATIVE PROBLEM 9.5
One Kind of Linear Phase Filter

A noncausal filter has been designed with the unit sample response shown in Figure 9.13a on page 512.

a. Use **freqz** to plot the frequency response $H_{nc}(e^{j\theta})$.

b. Plot the frequency response of the causal filter derived from the filter of Figure 9.13a by shifting $h_{nc}(n)$ four samples to the right.

Solution

a. Taking the z transform of $h_{nc}(n)$ gives us

$$H_{nc}(z) = z^4 + 2z^3 + 3z^2 + 4z + 5 + 4z^{-1} + 3z^{-2} + 2z^{-3} + z^{-4}$$

which for the function **freqz** must be put into the form

$$H_{nc}(z) = \frac{1 + 2z^{-1} + 3z^{-2} + 4z^{-3} + 5z^{-4} + 4z^{-5} + 3z^{-6} + 2z^{-7} + z^{-8}}{z^{-4}}.$$

The partial script for the frequency response $H_{nc}(e^{j\theta})$ is given next and the plots are shown in Figure 9.13b, where we see that the magnitude is an even function and that the phase is zero.

b. Shifting the samples in the z domain produces the causal transfer function

$$H_c(z) = z^{-4} \cdot H_{nc}(z) = 1 + 2z^{-1} + 3z^{-2} + 4z^{-3} + 5z^{-4} + 4z^{-5} + 3z^{-6} + 2z^{-7} + z^{-8}$$

which is already in the required format for **freqz.** The script from F9_13b is altered by changing the **a** vector to **a** = [1], with the plots in Figure 9.13c,

MATLAB Script

```
%F9_13 Linear phase filter

%F9_13b Frequency response for noncausal filter
b=[1,2,3,4,5,4,3,2,1];
a=[0,0,0,0,1];
[H,q]=freqz(b,a,512);
%...plotting statements and pause
```

where we notice that the magnitude plot is unchanged from Figure 9.13b but the phase is now linear with a slope of -4 rad/rad.

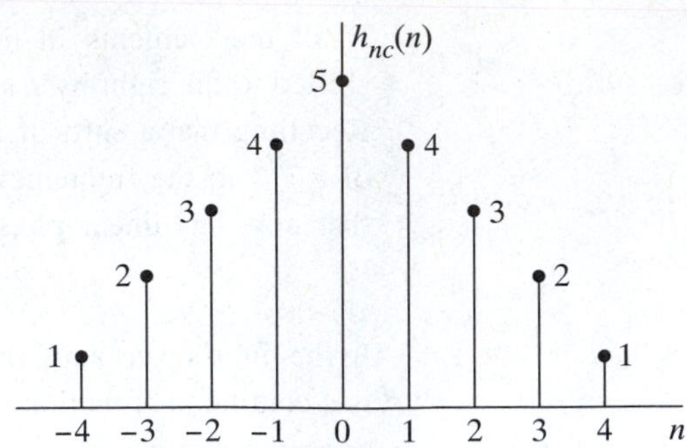

FIGURE 9.13a *Unit sample response for Illustrative Problem 9.5*

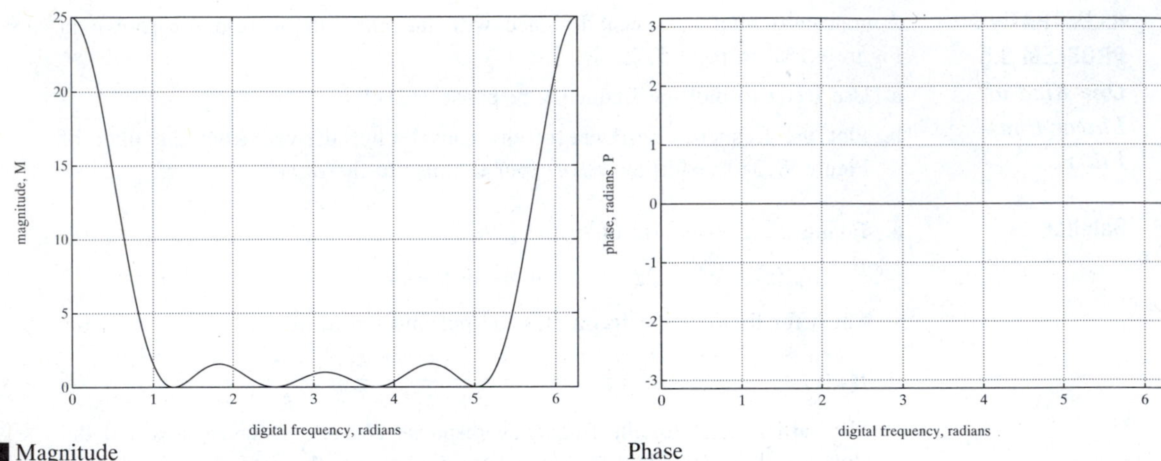

FIGURE 9.13b *Frequency response for noncausal filter: even symmetry for coefficients*

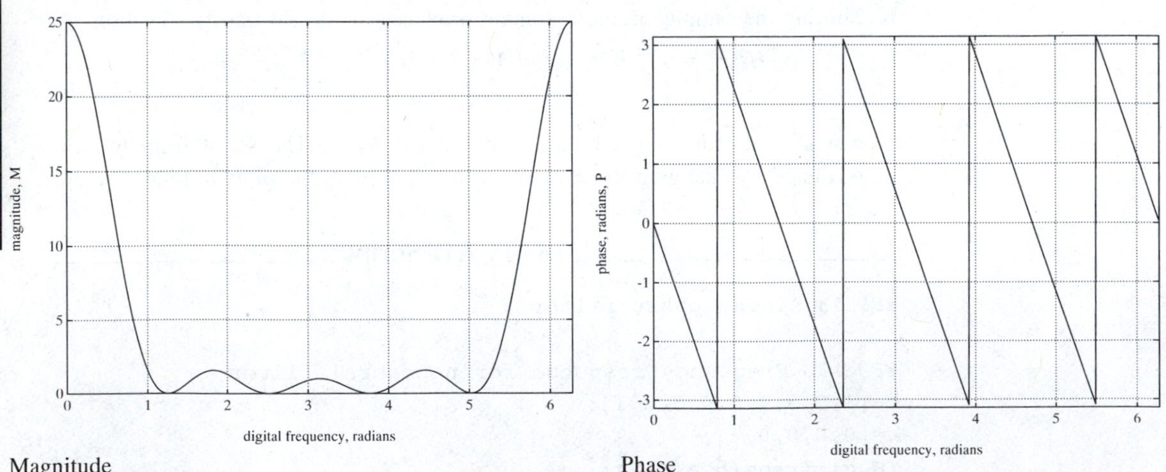

FIGURE 9.13c *Frequency response for causal filter: even symmetry for coefficients*

Next consider a unit sample response of a noncausal filter having an odd number of terms that exhibit the property of odd symmetry. That is, centered about $h_{nc}(0)$ at $n = 0$ we have

$$h_{nc}(-1) = -h_{nc}(1), h_{nc}(-2) = -h_{nc}(2), \ldots, h_{nc}(-I) = -h_{nc}(I)$$

as in Figure 9.14a, where we see that the number of terms in this filter is $2I + 1$. Taking advantage of the symmetry in $h_{nc}(n)$, where $h_{nc}(r) = -h_{nc}(-r)$, the frequency response of the noncausal filter is given by

$$H_{nc}(e^{j\theta}) = h_{nc}(1)\{e^{-j\theta} - e^{j\theta}\} + h_{nc}(2)\{e^{-j2\theta} - e^{2j\theta}\} + \cdots + h_{nc}(I)\{e^{-jI\theta} - e^{jI\theta}\}.$$

From the Euler relation we can write

$$\begin{aligned} H_{nc}(e^{j\theta}) &= -2jh_{nc}(1)\sin(\theta) - 2jh_{nc}(2)\sin(2\theta) - \cdots \\ &\quad - 2jh_{nc}(I)\sin(I\theta) \\ &= \sum_{m=1}^{I} -2jh_{nc}(m)\sin(m\theta) \end{aligned}$$

which is seen to be an imaginary function of frequency that can be either positive or negative. Thus we can recast the frequency response in the form

$$H_{nc}(e^{j\theta}) = \pm\left|\sum_{m=1}^{I} 2h_{nc}(m)\sin(m\theta)\right| e^{-j\pi/2} = \pm Me^{-j\pi/2}.$$

Again, there will be no changes of sign in the passband of the filter. If all the elements in the unit sample response of the noncausal filter are shifted to the right by I samples, as in Figure 9.14b, the filter becomes causal with the frequency response

$$H_c(e^{j\theta}) = \pm Me^{j(-\pi/2 - I\theta)}.$$

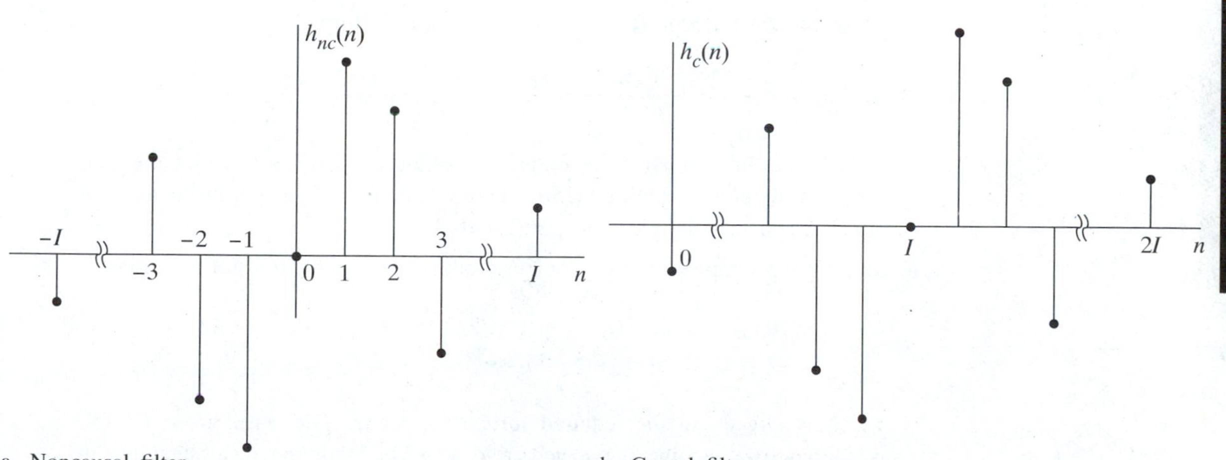

FIGURE 9.14 *Linear phase filter: odd symmetry, odd number of terms*

BASIC CONCEPTS

ILLUSTRATIVE PROBLEM 9.6
Another Kind of Linear Phase Filter

A noncausal filter has been designed with the unit sample response shown in Figure 9.15a.

a. Use **freqz** to plot the frequency response $H_{nc}(e^{j\theta})$.

b. Plot the frequency response of the causal filter that is derived from the filter of Figure 9.15a by shifting $h_{nc}(n)$ four samples to the right.

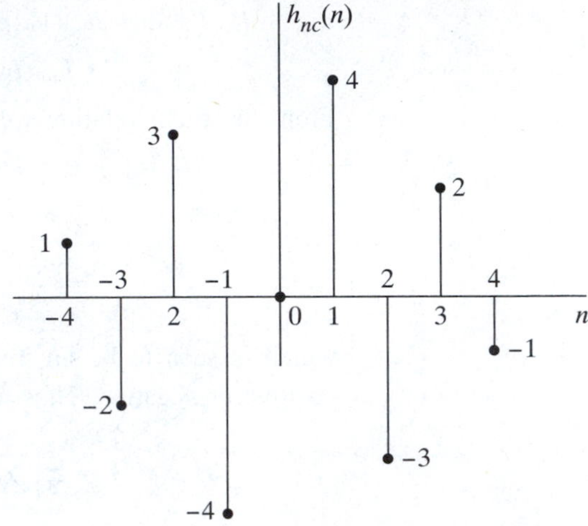

FIGURE 9.15a *Unit sample response for Illustrative Problem 9.6*

Solution

a. Taking the z transform of $h_{nc}(n)$ gives

$$H_{nc}(z) = z^4 - 2z^3 + 3z^2 - 4z^1 + 0z^0 + 4z^{-1} - 3z^{-2} + 2z^{-3} - z^{-4}$$

which for the function **freqz** must be put into the form

$$H_{nc}(z) = \frac{1 - 2z^{-1} + 3z^{-2} - 4z^{-3} + 0z^{-4} + 4z^{-5} - 3z^{-6} + 2z^{-7} - z^{-8}}{z^{-4}}$$

with the partial script for the frequency response $H_{nc}(e^{j\theta})$ given on the next page and the plots in Figure 9.15b, where we see that the magnitude is an even function and that the phase is $\pm\pi/2$ rad.

b. Shifting the samples in the z domain produces the causal transfer function

$$H_c(z) = z^{-4} \cdot H_{nc}(z) = 1 - 2z^{-1} + 3z^{-2} - 4z^{-3} + 4z^{-5} - 3z^{-6} + 2z^{-7} - z^{-8}$$

which is already in the required format for **freqz.** The script from F9_15b is altered by changing the **a** vector to **a** = [1], with the resulting plots in Figure 9.15c, where we notice that the magnitude plot is unchanged from Figure 9.15b but the phase has a constant slope of −4 rad/rad and an offset of $-\pi/2$ rad.

CHAPTER 9

MATLAB Script

```
%F9_15 Another linear phase filter

%F9_15b Frequency response for noncausal filter
b=[1,-2,3,-4,0,4,-3,2,-1];
a=[0,0,0,0,1];
[H,q]=freqz(b,a,512);
%...plotting statements and pause
```

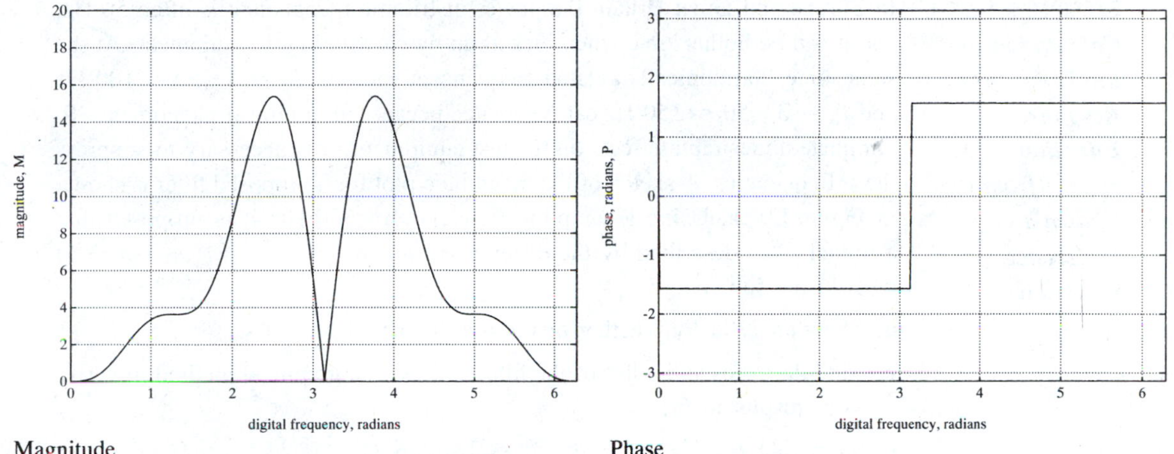

FIGURE 9.15b *Frequency response for noncausal filter: odd symmetry for coefficients*

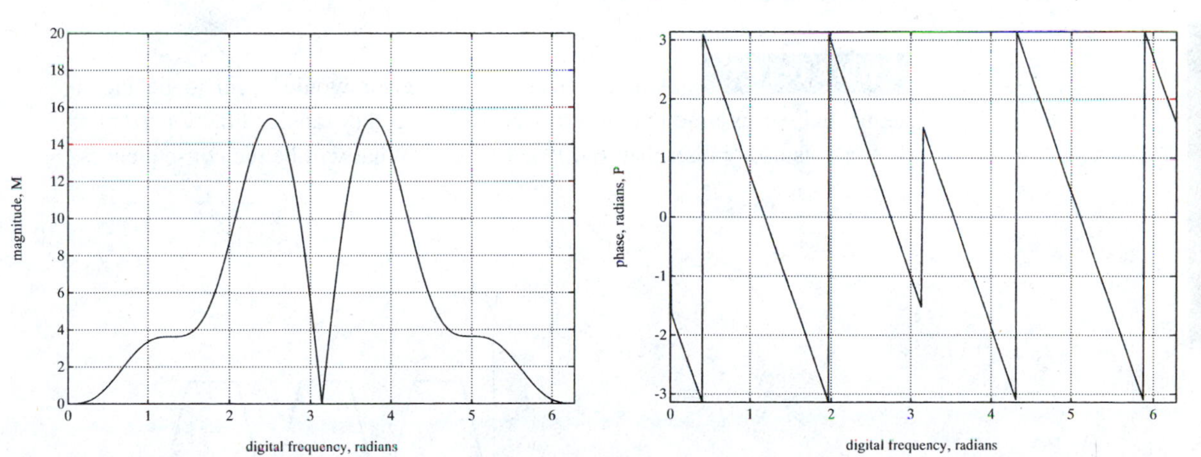

FIGURE 9.15c *Frequency response for causal filter: odd symmetry for coefficients*

BASIC CONCEPTS

There are two other configurations of the unit sample response of a nonrecursive filter that will produce linear phase: filters having a unit sample response with an even number of terms and with even or odd symmetry. See Exploration Problem P9.16.

SOLVED EXAMPLES AND MATLAB APPLICATIONS

EXAMPLE 9.1 ***Calculation of the Frequency Response Function $H(e^{j\theta})$ from the System's Difference Equation***

In Europe and Great Britain the power distribution frequency is often 50 Hz, and it can be bothersome when picked up by electronic test equipment. At the same time, harmonically related frequencies such as $f_2 = 2 \cdot 50 = 100$ Hz and $f_3 = 3 \cdot 50 = 150$ Hz can be created because of nonlinear phenomena. To eliminate undesirable effects on the test equipment, it is necessary to suppress these frequencies. A sketch of the magnitude plot for a proposed filter is shown in Figure E9.1a. Using a sampling frequency of 300 Hz, it is proposed that a digital filter described by the difference equation $y(n) = x(n) - x(n-6)$ will do the job.

a. Show analytically whether or not this filter will be satisfactory.

b. Find the zeros and poles of the filter and use the graphical method to verify your results in (a).

c. Show that $H(e^{j\theta}) = 1 - e^{-j6\theta} = 2\sin(3\theta)e^{j(\pi/2-3\theta)}$.

d. Use the MATLAB function **freqz** to calculate and plot the frequency response to verify your analysis.

WHAT IF? Among friends, a sign error would seem to be inconsequential, so assume that an error is made and the transfer function $H(z) = 1 + z^{-6}$ is used rather than $H(z) = 1 - z^{-6}$. What will be the consequences? ■

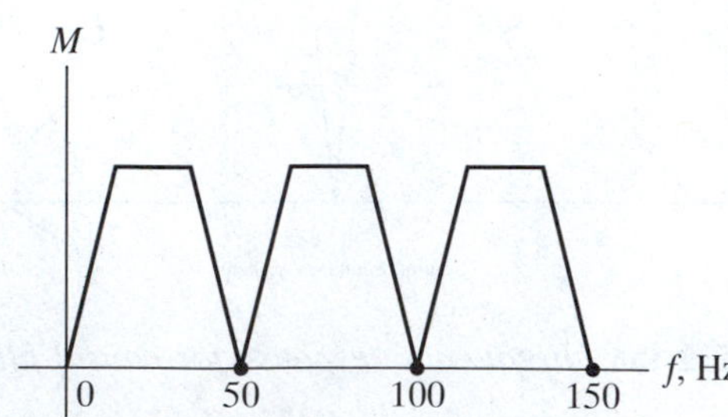

FIGURE E9.1a *Proposed frequency response magnitude*

Solution **a.** *Analytical approach.* The z transform of the given difference equation is $Y(z) = [1 - z^{-6}]X(z)$; consequently, the transfer function is

$$Y(z)/X(z) = H(z) = 1 - z^{-6}$$

where the frequency response function is found by substituting $e^{j\theta}$ for z in $H(z)$, giving

$$H(e^{j\theta}) = 1 - e^{-j6\theta}.$$

From Chapter 1 we recall that the digital frequency θ and the analog frequencies f and ω are related through $\theta = \omega T = 2\pi f T = 2\pi f/f_s$, where T is the sampling interval in time and f_s is the rate of sampling or sampling frequency. With $\theta = 2\pi f/f_s$ we have as the digital frequencies of interest $\theta_0 = 2\pi(0)/360 = 0$, $\theta_1 = 2\pi(50)/300 = \pi/3$, $\theta_2 = 2\pi(100)/300 = 2\pi/3$, and $\theta_3 = \pi$. For each of these frequencies we find

$$H(e^{j0}) = 1 - e^{j0} = 0, \qquad H(e^{j\pi/3}) = 1 - e^{-j6\pi/3} = 0,$$
$$H(e^{j2\pi/3}) = 1 - e^{-j12\pi/3} = 0, \qquad H(e^{j\pi}) = 1 - e^{-j6\pi} = 0,$$

and the desired frequencies have been "notched" out by this filter, as was desired.

b. *Graphical approach.* We have $H(z) = 1 - z^{-6} = (z^6 - 1)/z^6$. There are six poles at $z = 0$, and the zeros are the roots of the equation $z^6 - 1 = 0$, or $z^6 = 1$. The six roots, zeros of $H(z)$, are

$$z_{1,2,\ldots,6} = (z^6)^{1/6} = (1e^{jm2\pi})^{1/6}, \; m = 0, 1, 2, 3, 4, 5\,.$$

Thus the six zeros are all of magnitude 1 and are equally spaced on the unit circle at intervals of $\pi/3$ rad, starting at $\theta = 0$, as seen in Figure E9.1b. Using the graphical approach we see that a vector from each of the zeros to the unit circle is of zero length at the frequencies $\theta = 0, \pi/3, 2\pi/3, \ldots,$

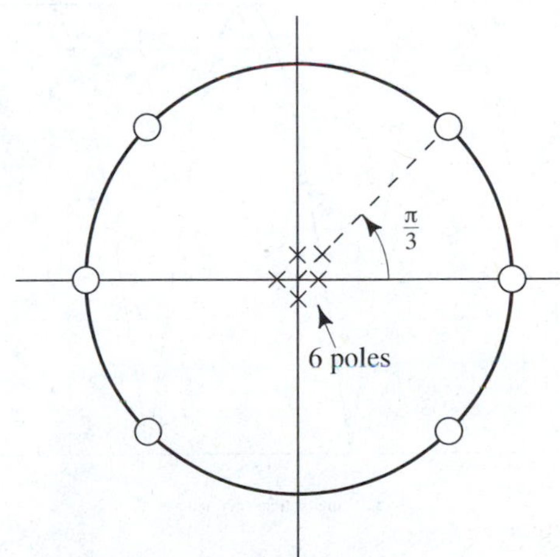

FIGURE E9.1b *Zeros and poles of $H(z)$*

EXAMPLES

which verifies the analytical results. This can also be seen from the factored transfer function

$$H(z) = \frac{(z - 1e^{j0})(z - 1e^{j\pi/3})(z - 1e^{j2\pi/3})(z - 1e^{j\pi})(z - 1e^{j4\pi/3})(z - 1e^{j5\pi/3})}{z^6}.$$

c. $H(e^{j\theta}) = 1 - e^{-j6\theta} = e^{-j3\theta}(e^{j3\theta} - e^{-j3\theta}) = 2je^{-j3\theta}\sin(3\theta) = 2\sin(3\theta)e^{j(\pi/2-3\theta)}$. Again, this clearly shows the magnitude $2\sin(3\theta)$ to be zero for $\theta = 0$, $\pi/3$, $2\pi/3$, and π.

d. *Computer approach.* To plot the frequency response, we again use the function **freqz**, with the plot of the notch filter shown in Figure E9.1c. (Notice that this is also a linear phase filter because of the symmetry in the filter coefficients.)

MATLAB Script

```
%E9_1 Frequency response from the DE

%E9_1c Notch filter
n=[1,0,0,0,0,0,-1];          % numerator coefficients
d=[1];                       % denominator coefficient
[H,q]=freqz(n,d,512);        % invokes freqz with 512 frequency points
mag=abs(H);
phase=angle(H);
%...changing radians to degrees and plotting statements
```

WHAT IF? With the sign error $H(e^{j\theta}) = 1 + e^{-j6\theta} = e^{-j3\theta}(e^{j3\theta} + e^{-j3\theta}) = 2\cos(3\theta)e^{-j3\theta}$. The magnitude is now zero at $\theta = \pi/6$, $\pi/2$, etc., clearly not what is wanted. ■

CHAPTER 9

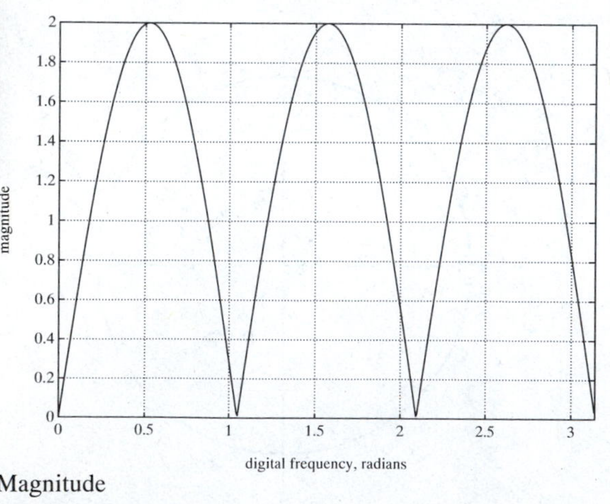

Magnitude

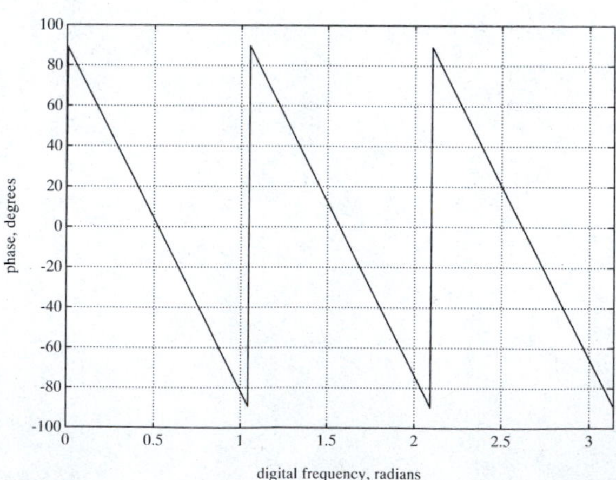

Phase

FIGURE E9.1c *Frequency response of notch filter*

EXAMPLE 9.2 *The Graphical Method for Estimating the Magnitude of the Frequency Response*

The transfer function models of some LTI systems follow:

a. For each system make a pole-zero plot and use the graphical method to estimate the frequency response (magnitude only). Use $0 \le \theta \le \pi$.

$$(1)\ H(z) = \frac{z}{z-0.9}; \qquad (2)\ H(z) = \frac{z}{z+0.9};$$

$$(3)\ H(z) = \frac{z^2}{z^2-0.9z+0.81}; \qquad (4)\ H(z) = \frac{z^2+1}{z^2+0.81}.$$

b. Verify your predictions in part (a) using MATLAB.

WHAT IF? Now let's see if we can use the estimation of the frequency response in an ad hoc design approach. Suppose that you have been asked to design a double bandpass filter that will pass easily the frequencies near $\theta = \pi/3$ and $2\pi/3$ rad and will block DC and $\theta = \pi$ rad. In the time domain, the unit sample response is to have a value of unity at $n = 0$. Can you find a design that will meet these requirements? ■

Solution

a. *Pole-zero plots and graphical analysis.*

(1) $H(z) = z/(z - 0.9)$ has a zero at $z = 0$ and a pole at $z = 0.9$. The length D of the pole vector in Figure E9.2a.1 will increase in length as the angle θ increases; consequently, the filter is lowpass. The magnitude at $\theta = 0$ is $M = 1/(0.10) = 10$ and at $\theta = \pi$, it is $M = 1/(1.9) = 0.526$.

(2) $H(z) = z/(z + 0.9)$ has a zero at $z = 0$ and a pole at $z = -0.9$. The pole vector in Figure E9.2a.2 (page 520) will decrease in length as the angle θ increases; consequently, the filter is highpass.

(3) $H(z) = z^2/(z^2 - 0.9z + 0.81)$ has two zeros at $z = 0$ and complex conjugate poles at $z_{1,2} = 0.9e^{\pm j1}$. The length of the pole vector D_1 in Figure E9.2a.3 will be shortest around $\theta = 1$ rad; consequently,

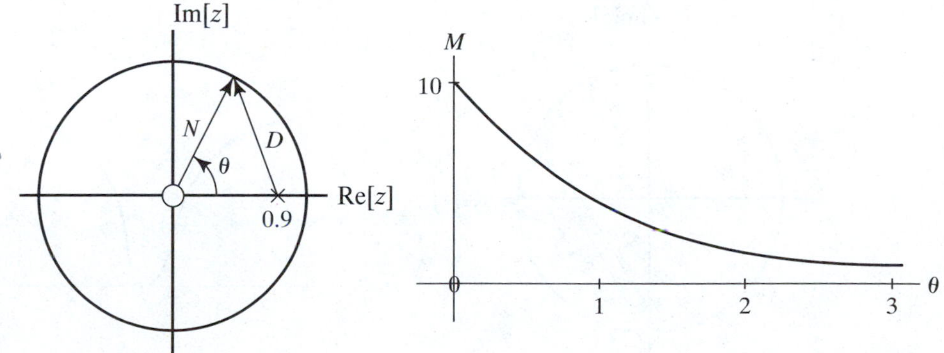

FIGURE E9.2a.1

EXAMPLES

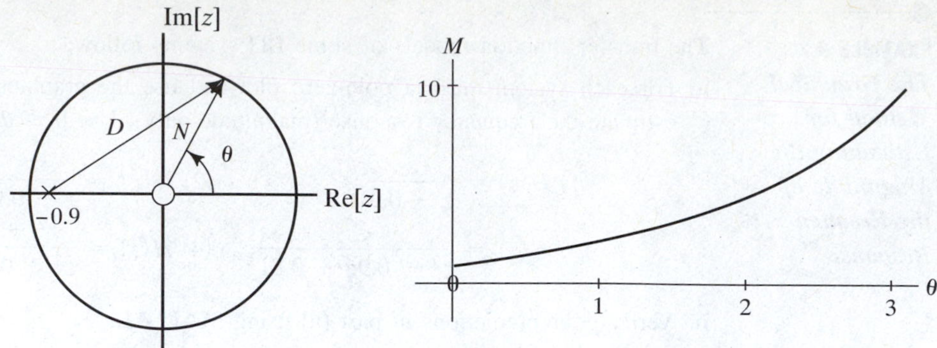

FIGURE E9.2a.2

the frequency response magnitude is maximum near there. From Figure E9.2a.3 we see that this is a bandpass filter.

(4) $H(z) = (z^2 + 1)/(z^2 + 0.81)$ has zeros at $z_{1,2} = 1e^{\pm j\pi/2}$ and poles at $z_{1,2} = 0.9e^{\pm j\pi/2}$. The filter gain will be zero at $\theta = \pi/2$ and because of the close proximity of the poles and zeros in Figure E9.2a.4, the gain will be relatively constant at all other frequencies.

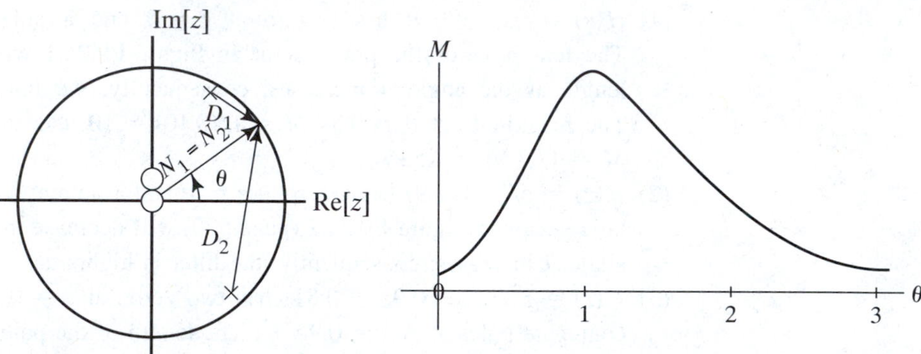

FIGURE E9.2a.3

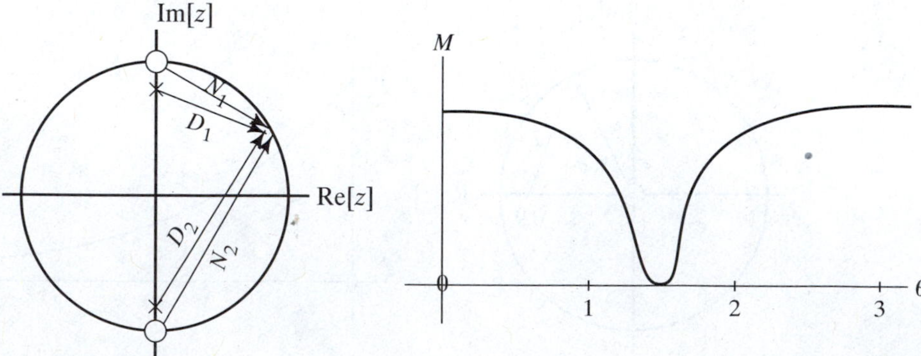

FIGURE E9.2a.4

b. *MATLAB plots.* The magnitude plots using **freqz** are given in Figure E9.2b.

MATLAB Script

```
%E9_2 Filter frequency response plots

%E9_2b_1 Lowpass filter
n=[1,0];
d=[1,-0.9];
[H,q]=freqz(n,d,512);
mag=abs(H);
%...plotting statements and pause

%E9_2b_2 Highpass filter
n=[1,0];
d=[1,0.9];
%...call to freqz, plotting statements, and pause

%E9_2b_3 Bandpass filter
n=[1,0,0];
d=[1,-0.9,0.81];
%...call to freqz, plotting statements, and pause

%E9_2b_4 Bandstop filter
n=[1,0,1];
d=[1,0,0.81];
%...call to freqz, plotting statements, and pause
```

EXAMPLES

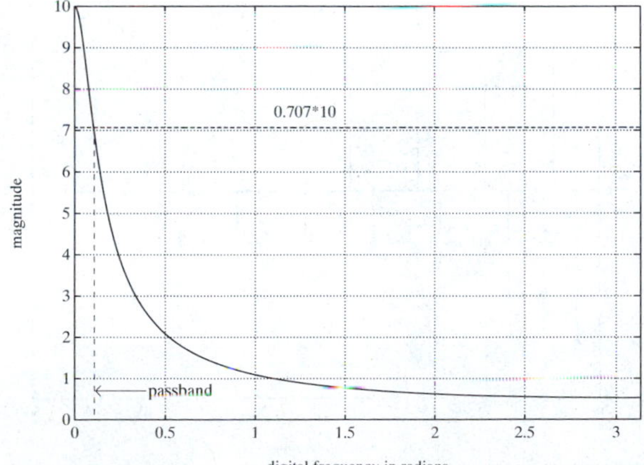

FIGURE E9.2b.1 *Lowpass filter frequency response magnitude*

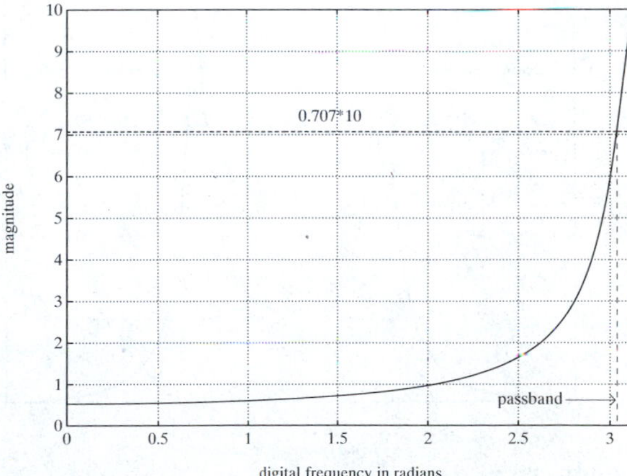

FIGURE E9.2b.2 *Highpass filter frequency response magnitude*

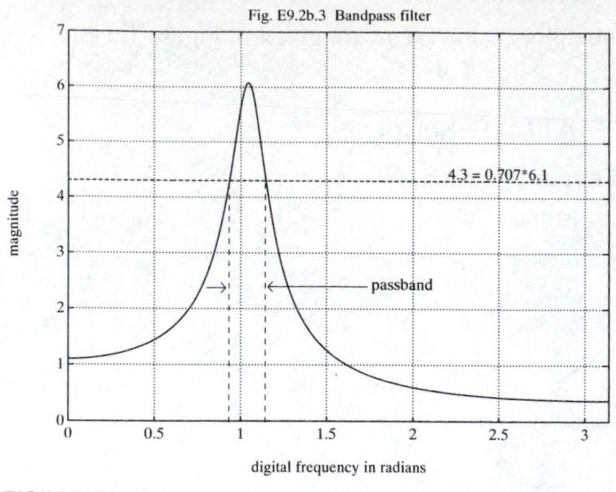

FIGURE E9.2b.3 *Bandpass filter frequency response magnitude*

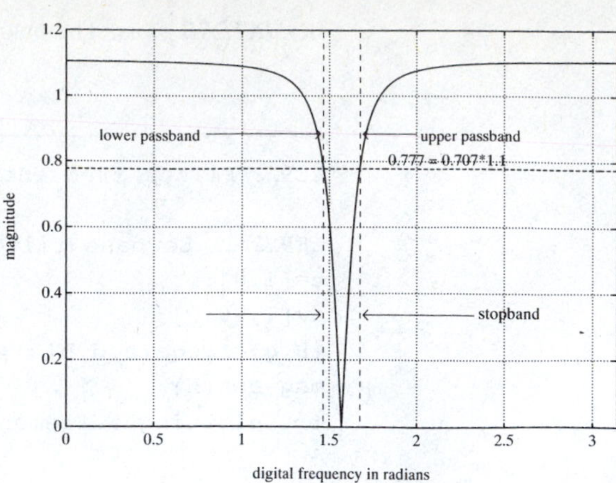

FIGURE E9.2b.4 *Bandstop filter frequency response magnitude*

WHAT IF? *Double bandpass.* Put double zeros at $z = \pm 1$ to block DC and the high frequency $\theta = \pi$ and poles at $\theta = \pm\pi/3$ and $\pm 2\pi/3$ rad, with the radius from the origin the designer's choice. For a selective filter we place the poles close to the unit circle, say at a radius of 0.95, giving

$$H(z) = \frac{(z^2 - 1)^2}{(z - 0.95e^{\pm j\pi/3})(z - 0.95e^{\pm j2\pi/3})} = \frac{z^4 - 2z^2 + 1}{z^4 + 0.9025z^2 + 0.815}.$$

The final test is to plot the frequency response, as in Figure E9.2c, where the double bandpass characteristic is seen; from Figure E9.2d, obtained using **filter**, we see that $h(0) = 1$ as requested. ■

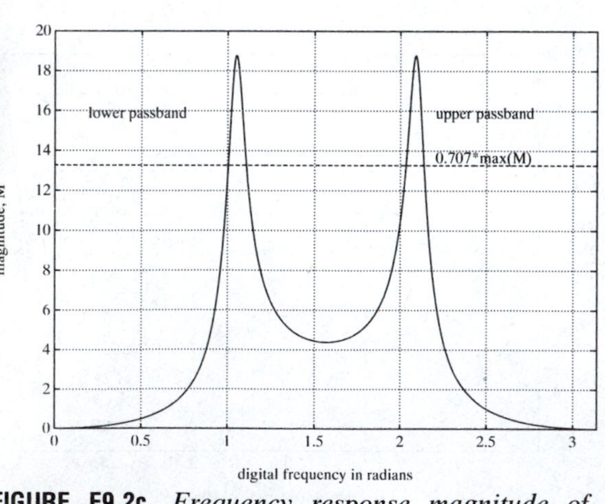

FIGURE E9.2c *Frequency response magnitude of double bandpass filter*

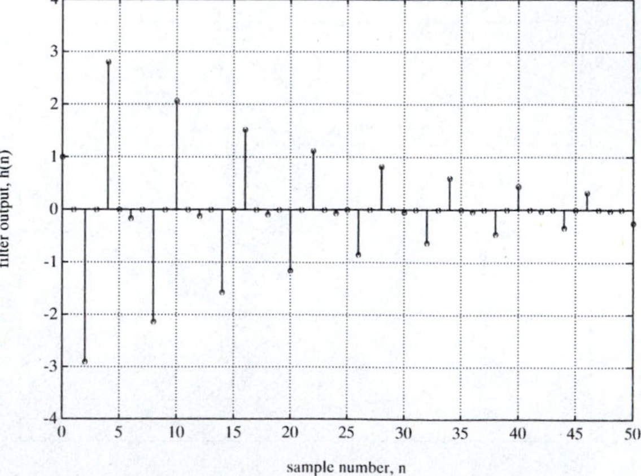

FIGURE E9.2d *Unit sample response of a double bandpass filter*

EXAMPLE 9.3 *Frequency Response of a Digital Differentiator*

In digital signal processing systems it is often necessary to differentiate electronically an analog signal. One way to do this is to sample the signal at intervals of T to obtain a digital sequence and then process this sequence with a digital filter often known as a differentiator. The ideal frequency response of a discrete-time differentiator can be shown to be $H_d(e^{j\theta}) = j\theta/T$, $|\theta| < \pi$. Using a Fourier series approach (see Hamming [2] or Strum and Kirk [5]), the unit sample response or impulse response of a discrete-time differentiator is given by

$$h(n) = \begin{cases} \dfrac{(-1)^n}{nT}, & n \neq 0 \\ 0, & n = 0 \end{cases}.$$

a. Write out the sequence for the unit sample response for this filter for $-5 \leq n \leq +5$, letting $T = 1$ s.

b. Find the transfer function $H_{nc}(z)$ of the noncausal filter based on the unit sample response $h(n)$ of part (a). Then find the transfer function $H_c(z)$ of a causal filter.

c. Plot the frequency response of the causal filter using **freqz**. Although perhaps impractical we will use a sampling interval of $T = 1$ yielding the desired frequency response $H_d(e^{j\theta}) = j\theta$.

WHAT IF? How will the frequency response of the noncausal filter differ from that of the causal filter? Show a plot (or plots) to complement your answer. ■

Solution

a. *Unit sample response.* Using $-5 \leq n < +5$ gives

$$h(n) = \{0.20, -0.25, 0.33, -0.50, 1.00, 0.00, -1.00, 0.50, -0.33, 0.25, -0.20\}.$$

Notice that this is a noncausal filter, since we have nonzero values for $h(n)$ for $n < 0$.

b. *Transfer function.* Taking the z transform of the unit sample sequence $h(n)$ gives

$$H_{nc}(z) = 0.20z^5 - 0.25z^4 + 0.33z^3 - 0.50z^2 + z^1 + 0 - z^{-1} + 0.50z^{-2} - 0.33z^{-3} + 0.25z^{-4} - 0.20z^{-5}.$$

The filter is made causal by simply shifting all the $h(n)$ terms to the right by five samples, which means that the causal transfer function can be obtained by multiplying $H_{nc}(z)$ by z^{-5}, giving

$$H_c(z) = 0.20 - 0.25z^{-1} + 0.33z^{-2} - 0.50z^{-3} + z^{-4} + 0 - z^{-6} + 0.50z^{-7} - 0.33z^{-8} + 0.25z^{-9} - 0.20z^{-10}.$$

EXAMPLES

c. *Frequency response plot.* The partial script is given next, and the plots are shown in Figure E9.3a, where the desired magnitude of θ and the desired phase of 90° are indicated.

MATLAB Script

```
%E9_3 Digital differentiators

%E9_3a Causal differentiator
n=[.20,-.25,.33,-.50,1.,0,-1.,.50,-.33,.25,-.20];
d=[1];
[H,q]=freqz(n,d,512);
%...plotting statements and pause
```

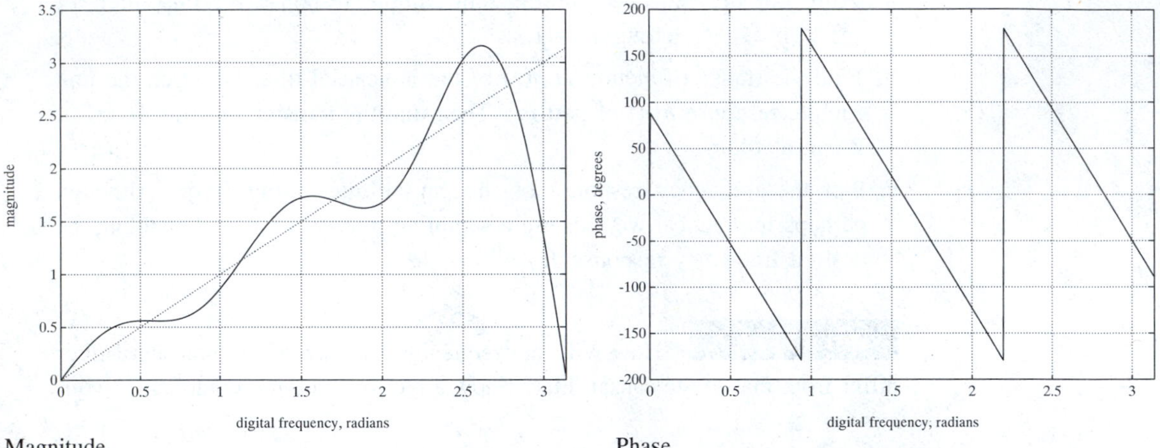

Magnitude Phase

FIGURE E9.3a *Frequency response of a causal differentiator*

Notice that the frequency response magnitude is "wiggly" and apparently contains some oscillatory terms that alter its smoothness. Using a "window" function to operate on the unit sample response $h(n)$ can help to alleviate this situation. We now briefly discuss this important topic. One well-known window sequence is due to Richard W. Hamming[3] and is described by the noncausal sequence

Hamming window

$$w(n) = \begin{cases} 0.54 + 0.46 \cos\left(\dfrac{n\pi}{I}\right), & |n| \leq I \\ 0, & |n| > I \end{cases}$$

[3]Richard W. Hamming (b. 1913) was a mathematician with the Bell Telephone Laboratories from 1946 until 1976, at which time he became a visiting professor at the Naval Postgraduate School. Hamming is the author of several well-known textbooks, including *Digital Filters, 3rd ed.,* Prentice Hall Inc. (1989) and *Coding and Information Theory, 2nd ed.,* Prentice Hall Inc. (1986).

where I denotes the number of terms on either side of $n = 0$. (For more on windows see Problem P9.10.) To find the values of the windowed coefficients, we multiply the original unit sample sequence, $h(n)$, by the Hamming window sequence, $w(n)$, giving the windowed coefficients $h_w(n)$ as

filter coefficients

$$h_w(n) = h(n) \cdot w(n) .$$

The result of using the Hamming window is tabulated next.

n	$h(n)$	$w(n)$	$h_w(n)$
0	0.200	0.080	0.016
1	−0.250	0.168	−0.042
2	0.333	0.398	0.133
3	−0.500	0.682	−0.341
4	1.000	0.912	0.912
5	0.000	1.000	0.000
6	−1.000	0.912	−0.912
7	0.500	0.682	0.341
8	−0.333	0.398	−0.133
9	0.250	0.168	0.042
10	−0.200	0.080	−0.016

Using these data the transfer function of the causal filter is

$$H_c(z) = 0.016 - 0.042z^{-1} + 0.133z^{-2} - 0.341z^{-3} + 0.912z^{-4} + 0 - 0.912z^{-6} + 0.341z^{-7} - 0.133z^{-8} + 0.042z^{-9} - 0.016z^{-10}.$$

The ultimate test is to plot the frequency response, and changing the elements of the vector **b** in **freqz** to reflect the effect of the Hamming window gives the plot of Figure E9.3b (page 526), where it can be seen that for at least two-thirds of the frequency range, the magnitude plot closely approaches the desired $|H_d(e^{j\theta})| = \theta$. The phase is unchanged and is not shown again. Notice that the departure from the desired straight-line characteristic begins at approximately $\theta = 2.3$ rad with the Hamming window, whereas the original differentiator begins its roll-off at about $\theta = 2.6$ rad.

EXAMPLES

WHAT IF? First, the transfer function of the noncausal filter with the Hamming window can be found by shifting the terms five samples to the left to orient the unit sample response about $n = 0$. This gives the transfer function

$$H_{nc}(z) = 0.016z^5 - 0.042z^4 + 0.133z^3 - 0.341z^2 + 0.912z^1 + 0 - 0.912z^{-1} + 0.341z^{-2} - 0.133z^{-3} + 0.042z^{-4} - 0.016z^{-5}.$$

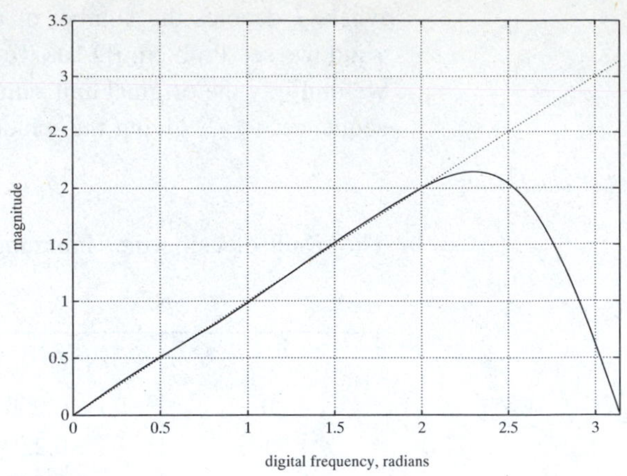

FIGURE E9.3b *Frequency response magnitude of causal differentiator: Hamming window*

We know that to use **freqz,** the first term in both the numerator and denominator polynomials must be the z^0 term, with the other terms following in increasingly negative powers of z. Thus we need to divide numerator and denominator by z^{-5}, which yields

$$\frac{0.016 - 0.042z^{-1} + 0.133z^{-2} - 0.341z^{-3} + \cdots + 0.341z^{-7} - 0.133z^{-8} + 0.042z^{-9} - 0.016z^{-10}}{0 + 0z^{-1} + 0z^{-2} + 0z^{-3} + 0z^{-4} + 1z^{-5}}.$$

The **n** vector, consequently, remains the same, but the **d** vector must be changed to $\mathbf{d} = [0, 0, 0, 0, 0, 1]$; the magnitude is unaffected, but the phase is now constant at $P = +90°$, as in Figure E9.3c. ■

CHAPTER 9

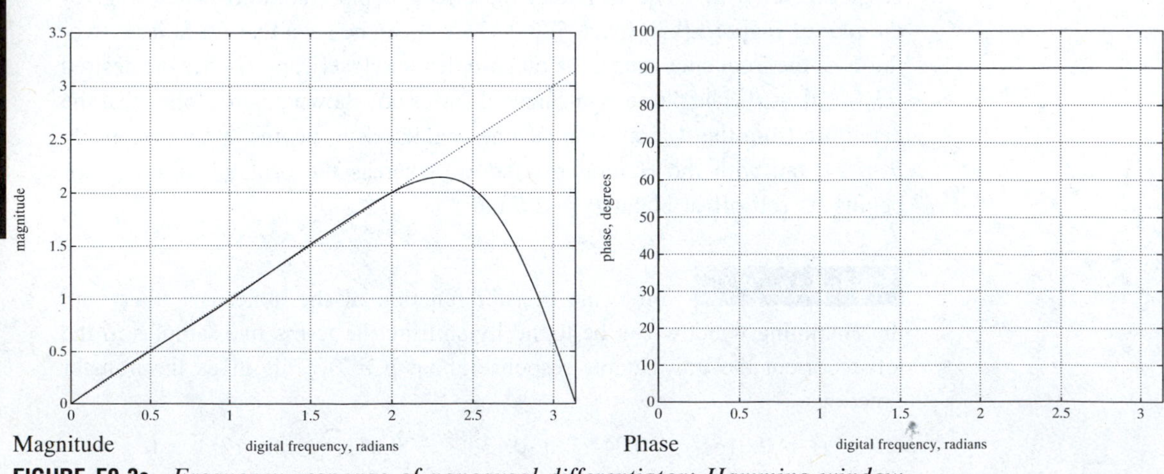

Magnitude Phase

FIGURE E9.3c *Frequency response of noncausal differentiator: Hamming window*

P9.1. Moving average filter. A moving average filter can be used to study time-series data generated by physical, biological, or economic processes. A running average is obtained by accumulating the average over the previous M samples and then, as each successive group of samples is taken, adding the new samples into the sum and subtracting the oldest samples from the sum. This filter is described by

$$y(n) = \frac{1}{M}\sum_{k=0}^{M-1} x(n-k)$$

and in a real application M will be a very large number to smooth the data and get a much better measure of the input.

a. Find the transfer function $H(z) = Y(z)/X(z)$ for this filter.

b. Show that the frequency response is given by

$$H(e^{j\theta}) = \frac{1}{M}\sum_{k=0}^{M-1} e^{-jk\theta}.$$

c. For $M = 5$, show that

$$H(e^{j\theta}) = \pm\left|\frac{1 + 2\cos\theta + 2\cos 2\theta}{5}\right| e^{-j2\theta}.$$

Plot the frequency response for this 5-term filter using **freqz** and verify that the magnitude goes to zero when it should and that the phase is indeed $P = -2\theta$.

d. Now use more terms, say $M = 19$, and plot the frequency response (magnitude only). Notice that the filter "gain" is much lower at higher frequencies and as such will filter out rapid fluctuations in the input time data.

P9.2 Recursive filter. Consider the system described by the difference equation $y(n) = 0.5y(n-1) + x(n) + 0.5x(n-1)$.

a. Determine its unit sample response $h(n)$.

b. What is the form of the initial condition response?

c. Find the response to a unit step sequence (zero ICs).

d. Find the frequency response $H(e^{j\theta})$.

e. Determine the steady-state response to the input, $x(n) = \cos(n\pi/2 + \pi/4)u(n)$.

P9.3 Filtering an analog signal. An analog signal is to be filtered to remove frequency components in the range 25 kHz $\leq f \leq$ 50 kHz with the maximum frequency component present in the input signal equal to 100 kHz. The filtering is to be accomplished by sampling the continuous-time signal and using a digital filter.

a. What is the minimum sampling rate that can be used?
b. For the minimum rate found in (a), what range of digital frequencies should be eliminated by the filter?
c. Find the magnitude characteristic of an ideal digital filter that will accomplish the desired filtering.

P9.4 Graphical estimate of frequency response. A linear filter is described by the transfer function $H(z) = (z^2 - 1.41z + 1)/z$.

a. Plot the poles and zeros.
b. Is this a causal filter?
c. Using the graphical method, estimate the magnitude of the frequency response $H(e^{j\theta})$. Four useful frequencies to calculate are $\theta = 0$, $\theta = \pi/4$, $\theta = \pi/2$, and $\theta = \pi$.
d. Verify your estimate with **freqz.**

P9.5 Linear phase filter. The transfer function of a causal filter is $H(z) = 0.25 + 0.5z^{-1} + z^{-2} + 0.5z^{-3} + 0.25z^{-4}$.

a. Find the unit sample response $h(n)$.
b. Find the frequency response $H(e^{j\theta})$.
c. Find an analytical expression for the phase of this filter.
d. Evaluate the magnitude and phase of the frequency response at $\theta = 0$, $\pi/2$, and π rad. From these results guess as to the kind of filter this is: bandpass, bandstop, lowpass, or highpass.
e. Verify your answer to (d) with **freqz.**

P9.6 An FIR filter. A finite impulse response filter is described by $h(n) = \delta(n) + 2\delta(n-2) + \delta(n-4)$.

a. Find the transfer function $H(z)$ and the zeros and poles.
b. Show that the frequency response is given by $H(e^{j\theta}) = [2 + 2\cos(2\theta)]e^{-j2\theta}$.
c. This filter is said to have linear phase. What does this imply in the time domain?
d. Find the steady-state filter output for an input $x(n) = A\cos(n\pi/2)u(n)$.

P9.7 A recursive filter. A recursive causal filter is described by the DE $y(n) + 0.656y(n-4) = x(n) - x(n-2)$.

a. Find the filter transfer function and its zeros and poles.
b. Sketch the magnitude of the frequency response.
c. Verify with **freqz.**

WHAT IF? We can modify a filter's frequency response by moving the poles and zeros around. Suppose that we want the filter's passbands to be broadened. Show how to change the filter's difference equation to achieve this. Use MATLAB to determine the revised frequency response. ■

P9.8 Lowpass and highpass filters. Two different digital filters are described by

$$\text{(i)}\ y(n) + 1.8y(n-1) + 0.81y(n-2) = x(n) - 2x(n-1) + x(n-2)$$

$$\text{(ii)}\ y(n) - 1.8y(n-1) + 0.81y(n-2) = x(n) + 2x(n-1) + x(n-2).$$

a. Find the transfer function and the poles and zeros for each filter.
b. From the pole-zero determinations of part (a), which filter describes a lowpass filter and which describes a highpass filter?
c. Plot the frequency response for each filter with a program such as **freqz.**
d. Use the graphical results of (c) to estimate $y_{ss}(n)$ for each filter with $x(n) = A + B\cos(n\pi/2) + C\cos(n\pi)$.

WHAT IF? Let's consider the possibility of changing the filter's frequency response by adjusting the zero and pole locations. Suppose that it is desirable to increase the width of the filter passbands without increasing the filter order. Use MATLAB to see how this can be done. What, if any, trade-offs are there? ■

P9.9 Bandpass and bandstop filters. Two different digital filters are described by

$$\text{(i)}\ y(n) + 0.81y(n-2) = x(n) - x(n-2)$$

$$\text{(ii)}\ y(n) - 0.81y(n-2) = x(n) + x(n-2).$$

a. Find the transfer function and the poles and zeros for each filter.
b. From the pole-zero determinations of part (a) which filter describes a bandpass filter and which describes a bandstop filter?
c. Plot the frequency response for each filter with a program such as **freqz.**
d. Use the graphical results of (c) to estimate $y_{ss}(n)$ for $x(n) = A + B\cos(n\pi/2) + C\cos(n\pi)$.
e. Repeat (d) for $x(n) = D\cos(0.5n) + E\cos(n) + F\cos(2n) + G\cos(2.5n)$.

WHAT IF? We can move zeros and poles around to modify a filter's frequency response. Let's consider this possibility to make the bandpass filter more selective by narrowing its bandwidth. Use MATLAB to investigate this. Do not change the number of poles or zeros. ■

P9.10 Frequency response of window functions.

a. A causal rectangular window $w(n)$ and its frequency response $W(e^{j\theta})$ are described by

$$w(n) = \begin{cases} 1, & 0 \le n \le N-1 \\ 0, & \text{elsewhere} \end{cases} \quad \text{and}$$

$$W(e^{j\theta}) = \frac{1 - e^{-jN\theta}}{1 - e^{-j\theta}} = \frac{\sin(N\theta/2)}{\sin(\theta/2)} \cdot e^{-j(N\theta/2 + \theta/2)}.$$

Use **freqz** to plot the frequency response for several different window lengths N. Show that the plots and the analytical expression for the frequency response agree.

PROBLEMS

b. A window sequence due to Julius von Hahn[4] is commonly known as the Hanning window, and for a noncausal sequence we have

$$w(n) = \begin{cases} 0.50 + 0.50\cos\left(\dfrac{n\pi}{I}\right), & |n| \le I \\ 0, & |n| > I. \end{cases}$$

Recast this formula into that of a causal window and write scripts to plot the sequence $w(n)$ and its frequency response $W(e^{j\theta})$. The formula for a Hanning window that appears in the Signals and Systems and Signal Processing Toolboxes is slightly different than the one given here, which is very similar to the Hamming formula of Example 9.3.

P9.11 Transmission through an ideal filter. An ideal digital filter is described by

$$H(e^{j\theta}) = \begin{cases} 2e^{-j2\theta}, & \pi/3 \le |\theta| \le 2\pi/3 \\ 0, & \text{for other values of } \theta \text{ in the interval } [-\pi, \pi]. \end{cases}$$

a. What kind of filter is this?

b. If the input to this filter is $x(n) = 2 + 4.5\cos(\pi n/4) + 9\cos(3\pi n/8) + 13\cos(n\pi/2) + 3\cos(3\pi n/4)$, find an analytical expression for the steady-state output $y_{ss}(n)$.

P9.12 Another ideal filter. Repeat Problem 9.11 for the filter characteristic

$$H(e^{j\theta}) = \begin{cases} 2e^{-j2\theta}, & \pi/3 \le |\theta| \le \pi \\ 0, & \text{for other values of } \theta \text{ in the interval } [-\pi, \pi]. \end{cases}$$

EXPLORATION PROBLEMS

P9.13 Filter design. An analog signal whose highest frequency component is at 2000 Hz is to be passed through a digital highpass filter after having been sampled at 10,000 samples per second. The highpass filter is to have the following characteristics:

$$\begin{array}{lr} |H| \le 0.3 & 0\text{ Hz} \le f \le 1000\text{ Hz} \\ |H| \le 0.707 & 1000\text{ Hz} \le f \le 1500\text{ Hz} \\ 0.75 \le |H| \le 1.0 & 1750\text{ Hz} \le f \le 2000\text{ Hz} \end{array}$$

Find the transfer function $H(z)$ and the corresponding difference equation for a digital filter that meets these specifications. Use **freqz** to verify your design.

P9.14 Design of two filters.

a. A lowpass filter is described by $H(z) = C_1/(1 - 0.81z^{-1})$. Find its zeros and poles and sketch $|H_1(e^{j\theta})|$ versus θ for $0 \le \theta \le \pi$ with C_1 as an unassigned scale factor.

[4] According to reliable sources, John Tukey of Bell Laboratories coined the name Hanning for the window function developed by Julius von Hahn, an Australian meteorologist.

b. A higher-order bandpass filter has

$$H_2(z) = \frac{C_2}{(1 - 0.31z^{-1} + 0.81z^{-2})\,(1 + 0.31z^{-1} + 0.81z^{-2})}.$$

Find its zeros and poles and sketch $|H_2(e^{j\theta})|$ versus θ for $0 \leq \theta \leq \pi$ with C_2 as an unassigned scale factor.

c. Adjust C_1 and C_2 so that the two filters have unity gain at $\theta = \pi/2$.

d. Use **freqz** to compare the frequency responses of the two filters with C_1 and C_2 as determined in part (c). Do the results fit with graphical estimates from the zero-pole plots?

P9.15 An unusual filter. The transfer function of an LTI digital filter is $H(z) = (1 - 2z^{-1} + 2z^{-2})/(1 - z^{-1} + 0.5z^{-2})$.

a. Find its zeros and poles. What do you notice about them?

b. Sketch the magnitude of the frequency response for $0 \leq \theta \leq \pi$. Can you think of any application for a filter such as this one?

c. Design another filter with these same characteristics.

P9.16 More linear phase filters.

a. Consider a filter described by the transfer function $H(z) = 1 + 2z^{-1} + 3z^{-2} + 4z^{-3} + 4z^{-4} + 3z^{-5} + 2z^{-6} + z^{-7}$. Plot the frequency response using **freqz** to verify the stated linear phase characteristic.

b. Use **freqz** to show that the filter whose transfer function is $H(z) = 1 - 2z^{-1} + 3z^{-2} - 4z^{-3} + 4z^{-4} - 3z^{-5} + 2z^{-6} - z^{-7}$ is also a linear phase filter.

P9.17 Time and frequency domain relationships for an LTI system. The California function **dzpresp** makes it easy to explore the characteristics of a discrete system in the time, frequency, and z domains. Running a script that specifies the numerator and denominator coefficients of a transfer function and invokes the California function **dzpresp** results in a display of the zero-pole diagram, the step and impulse responses, and the frequency response magnitude plot. The following script illustrates the use of **dzpresp** for the second-order bandpass digital filter $H(z) = (z^2 - 0.81)/(z^2 + 0.81)$ with the resulting plot given in Figure P9.17.

MATLAB Script

```
%P9_17 Time and frequency domain relationships
num=[1,0,-0.81];            % numerator coefficients
den=[1,0,0.81]              % denominator coefficients
dzpresp(num,den);           % call dzpresp
```

Use the function **dzpresp** with some transfer functions of your choice. You should verify your theoretical knowledge with the resulting graphical plots on items such as pole-zero locations with filter type, the effect of pole location on the settling time of the step and impulse responses, and the relationship between step and impulse responses.

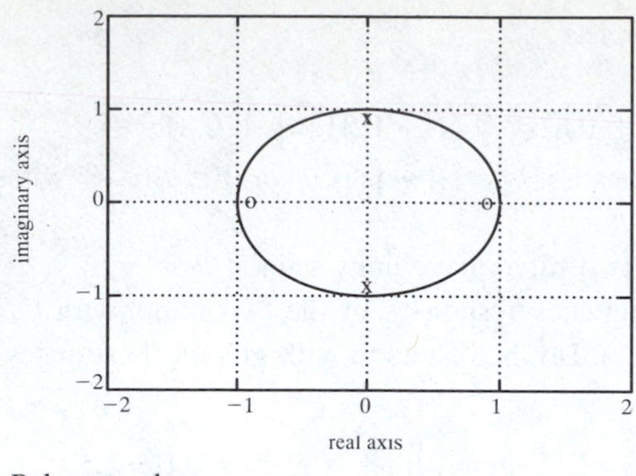

Pole-zero plot

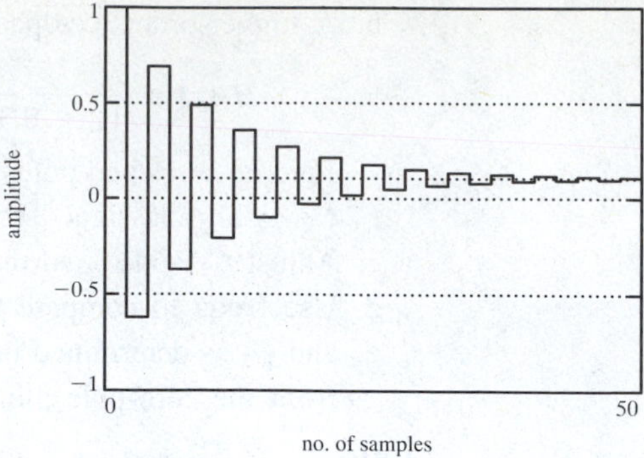

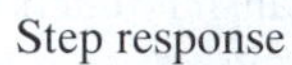

Step response

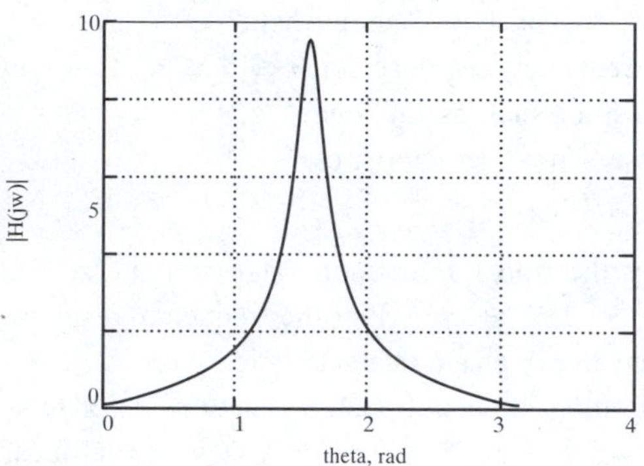

Frequency response magnitude

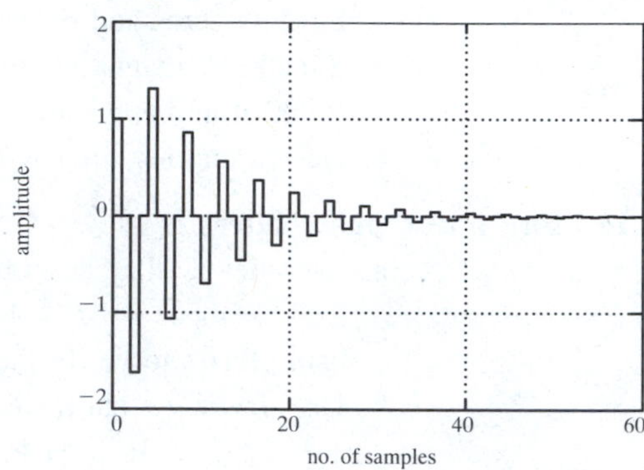

Unit sample (impulse) response

FIGURE P9.17 *Time and frequency domain relationships for* $H(z) = (z^2 - 0.81)/(z^2 + 0.81)$

DEFINITIONS, TECHNIQUES, AND CONNECTIONS

FINDING THE FREQUENCY RESPONSE FOR A STABLE SYSTEM

1. From the transfer function,

$$H(e^{j\theta}) = H(z)|_{z=e^{j\theta}} = \left.\frac{\sum_{k=0}^{L} b_k z^{-k}}{\sum_{k=0}^{N} a_k z^{-k}}\right|_{z=e^{j\theta}} = \frac{\sum_{k=0}^{L} b_k e^{-jk\theta}}{\sum_{k=0}^{N} a_k e^{-jk\theta}}.$$

2. Given the unit sample response $h(n)$ of a stable LTI system, the frequency response is

$$H(e^{j\theta}) = \sum_{m=-\infty}^{\infty} h(m)e^{-jm\theta}.$$

3. Starting with the DE for a stable system,

$$\sum_{k=0}^{N} a_k y(n-k) = \sum_{k=0}^{L} b_k x(n-k) \quad \text{gives } H(e^{j\theta}) = \frac{\sum_{k=0}^{L} b_k (e^{-j\theta})^k}{\sum_{k=0}^{N} a_k (e^{-j\theta})^k}.$$

4. Use a program such as MATLAB's **freqz,** where the data are entered as

$$H(z) = \frac{B(z)}{A(z)} = \frac{b(1) + b(2)z^{-1} + \cdots + b(nb)z^{-(nb-1)}}{a(1) + a(2)z^{-1} + \cdots + a(na)z^{-(na-1)}}.$$

CHARACTERISTICS OF THE FREQUENCY RESPONSE FUNCTION

1. *Periodicity.* $H(e^{j\theta})$ is a periodic function with a period of 2π rad.
2. *Symmetry.* The magnitude M of $H(e^{j\theta})$ is an even function and the phase P is an odd function.
3. *Time and frequency.* The unit sample response $h(n)$ is discrete (a sequence), but $H(e^{j\theta})$ is a function of the continuous digital frequency variable θ.
4. *Decibels.* The values of the gain M may vary over a very wide range and it is often useful to define M in decibels as $M_{\text{dB}} = 20 \log_{10} M$.
5. *Discrete Time Fourier Transform.* $H(e^{j\theta})$ is the Discrete Time Fourier Transform (DTFT) of the unit sample sequence $h(n)$. In a similar way, for the sequences $x(n)$ and $y(n)$ we write

$$X(e^{j\theta}) = \sum_{m=-\infty}^{\infty} x(m)e^{-jm\theta} \quad \text{and} \quad Y(e^{j\theta}) = \sum_{m=-\infty}^{\infty} y(m)e^{-jm\theta}$$

where $X(e^{j\theta})$ and $Y(e^{j\theta})$ are the DTFTs of the sequences $x(n)$ and $y(n)$, respectively.

FREQUENCY RESPONSE PLOTS

1. *Rectangular.* $|H(e^{j\theta})| = M$, linear scale, versus θ, linear scale; $\angle H(e^{j\theta}) = P$, linear scale, versus θ, linear scale.
2. *Polar.* $\text{Im}[H(e^{j\theta})]$ versus $\text{Re}[H(e^{j\theta})]$ for $0 \le \theta \le \pi$, with frequencies marked on locus.
3. *Nyquist.* $\text{Im}[H(e^{j\theta})]$ versus $\text{Re}[H(e^{j\theta})]$ for $0 \le \theta \le 2\pi$, with frequencies marked on locus.
4. *Bode.* $20 \log_{10}|H(e^{j\theta})|$ versus θ and $\angle H(e^{j\theta})$ versus θ with $20 \log_{10}|H(e^{j\theta})|$ and $\angle H(e^{j\theta})$ on linear scales and θ on a log scale, where $20 \log_{10}|H(e^{j\theta})|$ is called the decibel (dB) gain.

GRAPHICAL ESTIMATION

$M = \dfrac{KN_1N_2\cdots N_L}{D_1D_2\cdots D_N}$ and $P = (N - L)\theta + \sum_{k=1}^{L} \alpha_k - \sum_{k=1}^{N} \beta_k$, with the N's, D's, α's, and β's identified in Figure 9.5.

MATLAB FUNCTIONS USED

Function	*Purpose and Use*	*Toolbox*
dzpresp	Given: TF model of continuous system, **dzpresp** returns zero-pole plot, step and sample responses, and frequency response magnitude plot.	California Functions
filter	Given: DE or TF model of discrete system, input, ICs, **filter** returns system output.	Signals/Systems, Signal Processing
find	Given: a vector **v**, **find** returns a vector containing the indices of the nonzero elements of **v**.	MATLAB
freqz	Given: TF of discrete system, **freqz** returns frequency response.	Signals/Systems, Signal Processing
hamming	Given: length of window *N*, **hamming** returns window coefficients.	Signals/Systems, Signal Processing
hanning	Given: length of window *N*, **hanning** returns window coefficients.	Signals/Systems, Signal Processing

CHAPTER 9

ANNOTATED BIBLIOGRAPHY

1. Gabel, Robert A., and Richard A. Roberts, *Signals and Linear Systems, 3rd ed.,* John Wiley & Sons, New York, 1987. *Chapter Seven of this thorough text is a brief but complete treatment of an important application of frequency response, the design of digital filters.*

2. Hamming, Richard W., *Digital Filters, 3rd ed.,* Prentice Hall, Inc., Englewood Cliffs, N.J., 1989. *This book is a specialized text by an author who has made many significant contributions to mathematics and linear systems theory.* Digital Filters *presents the fundamental ideas of digital filter theory, including design methods to approximate given frequency response specifications. It can be understood by readers with a wide range of backgrounds.*

3. Oppenheim, Alan V., and Alan S. Willsky, with Ian T. Young, *Signals and Systems,* Prentice Hall, Inc., 1983. *Continuous-time and discrete-time systems are treated in a parallel fashion by this well-known text. Frequency response from differential and difference equation characterizations of systems is considered in Chapters Four and Five, and Chapter Six presents a brief but insightful view of basic filtering principles.*

4. Parks, T. W., and C. S. Burrus, *Digital Filter Design,* John Wiley & Sons, New York, 1987. *This text is entirely dedicated to a discussion of the design of linear, constant-coefficient digital filters, with the main focus on the use of frequency-domain techniques.*

5. Strum, Robert D., and Donald E. Kirk, *First Principles of Discrete Systems and Digital Signal Processing,* Addison-Wesley Publishing Company, Reading, Mass., 1988. *In Chapter Four there are many examples of the calculation of the frequency response of discrete-time systems, including computer evaluation. Ideal filters are defined, and a typical filtering problem is examined. Chapter Five contains a detailed explanation and examples that illustrate the practicality of the graphical procedure for estimating frequency response.*

ANSWERS

P9.1 a. $H(z) = \frac{1}{M}\sum_{k=0}^{M-1} z^{-k}$; b. $H(e^{j\theta}) = H(z)|_{z=e^{j\theta}}$; c., d. See A9_1.

P9.2 a. $h(n) = \delta(n) + 2(0.5)^n u(n)$ or $h(n) = \delta(n) + (0.5)^{n-1}u(n-1)$; b. $y_{IC}(n) = C(0.5)^n u(n)$; c. $y(n) = [-2(0.5)^n + 3(1)^n]u(n)$; d. $H(e^{j\theta}) = (e^{j\theta} + 0.5)/(e^{j\theta} - 0.5)$; e. $y_{ss}(n) = 1.0\cos(n\pi/2 - 0.142)$.

P9.3 a. $f_s > 200$ kHz; b. Using $\theta = 2\pi f/f_s$, the range is $0.25\pi \leq \theta \leq 0.50\pi$. c. $M = 0$, $0.25\pi \leq |\theta| \leq 0.50\pi$, $M = 1$ elsewhere in the range $-\pi \leq \theta \leq \pi$.

P9.4 a. Zeros at $z_{1,2} = 1e^{\pm j\pi/4}$, pole at $z = 0$; b. No, $h(-1) = 1$; c. $H(e^{j0}) = 0.59$, $H(e^{j\pi/4}) = 0$, $H(e^{j\pi/2}) = -1.41$, $H(e^{j\pi}) = -3.41$; d. See A9_4 on the CLS disk.

P9.5 a. $h(n) = 0.25\delta(n) + 0.5\delta(n-1) + \delta(n-2) + 0.5\delta(n-3) + 0.25\delta(n-4)$; b. $H(e^{j\theta}) = 0.25 + 0.5e^{-j\theta} + e^{-j2\theta} + 0.5e^{-j3\theta} + 0.25e^{-j4\theta}$; c. $P = -2\theta$; d. $H(e^{j0}) = 2.5$, $H(e^{j\pi/2}) = -0.5$, $H(e^{j\pi}) = 0.5$, guess lowpass. e. See A9_5 on the CLS disk.

P9.6 a. $H(z) = 1 + 2z^{-2} + z^{-4}$, two zeros at $z = +j1$ and two at $z = -j1$, four poles at $z = 0$; b. $H(e^{j\theta}) = 1 + 2e^{-j2\theta} + 1e^{-j4\theta} = e^{-j2\theta}[2 + 2\cos(2\theta)]$; c. The output frequencies are all delayed by the same amount in time, that is, $Ke^{-jm\theta} \rightarrow Kx(n-m)$; d. $y_{ss}(n) = 0$.

P9.7 a. $H(z) = (1 - z^{-2})/(1 + 0.656z^{-4})$; zeros at $z_{1,2} = \pm 1$, $z_{3,4} = 0$; poles $z_{1,2,3,4} = 0.900e^{j(\pi + m2\pi)/4}$, $m = 0, 1, 2, 3$; b. From zero-pole plot, M will be 0 at $\theta = 0$ and π with peaks at $\theta = \pm\pi/4 \pm 3\pi/4$. c. See A9_7.

WHAT IF? By making the coefficient 0.656 smaller, the poles move farther away from the unit circle; this broadens the peaks, as shown in A9_7b. The coefficient of the $y(n-4)$ term is changed, in this case to 0.5 from 0.656. ■

ANSWERS

P9.8 a. (i) $H(z) = (z - 1)^2/(z + 0.9)^2$; two zeros at $z = 1$ and two poles at $z = -0.9$. (ii) $H(z) = (z + 1)^2/(z - 0.9)^2$, two zeros at $z = -1$ and two poles at $z = 0.9$. b. (i) is highpass and (ii) is lowpass. c. See A9_8 for verification. d. From the plots of (c): (i) Highpass filter $H(e^{j0}) = H(e^{j\pi/2}) = 0$ and $H(e^{j\pi}) = 400e^{j0}$, $y_{ss}(n) = 400C\cos(n\pi)$. (ii) Lowpass filter $H(e^{j\pi}) = H(e^{j\pi/2}) = 0$ and $H(e^{j0}) = 400e^{j0}$, $y_{ss}(n) = 400A$.

WHAT IF? See A9_8, where the poles have been moved to -0.7 for the highpass filter and $+0.7$ for the lowpass. Additional gain has been provided to aid comparisons. ■

P9.9 a. (i) $H(z) = (z^2 - 1)/(z^2 + 0.81)$, zeros at $z = \pm 1$ and poles at $z = \pm j0.9$ (ii) $H(z) = (z^2 + 1)/(z^2 - 0.81)$, zeros at $z = \pm j1$ and poles at $z = \pm 0.9$; b. (i) is bandpass and (ii) is bandstop. c. See A9_9 for verification. d. From the plots of (c): (i) bandpass filter $H(e^{j0}) = H(e^{j\pi}) = 0$ and $H(e^{j\pi/2}) = 10.5e^{j0}$, $y_{ss}(n) = 10.5B\cos(n\pi/2)$; (ii) bandstop filter $H(e^{j\pi/2}) = 0$ and $H(e^{j0}) = H(e^{j\pi}) = 10.5e^{j0}$, $y_{ss}(n) = 10.5A + 10.5C\cos(n\pi)$; e. (i) bandpass $H(e^{j0.5}) = 0.6e^{j87°}$, $H(e^{j1}) = 1.6e^{j80°}$, $H(e^{j2}) = 2.4e^{-j75°}$, $H(e^{j2.5}) = 0.8e^{-j90°}$, $y_{ss}(n) = 0.6D\cos(0.5n + 87°) + 1.6E\cos(n + 80°) + 2.4F\cos(2n - 75°) + 0.8G\cos(2.5n - 90°)$; bandstop $H(e^{j0.5}) = 2e^{-j75°}$, $H(e^{j1}) = 0.7e^{-j85°}$, $H(e^{j2}) = 0.5e^{j90°}$, $H(e^{j2.5}) = 1.5e^{j80°}$, $y_{ss}(n) = 2D\cos(0.5n - 75°) + 0.7E\cos(n - 85°) + 0.5F\cos(2n + 90°) + 1.5G\cos(2.5n + 80°)$.

WHAT IF? The gain of the original plot was adjusted so that the maximum value was 1.0 and the measured bandwidth was 0.21 rad from the data file. Moving the poles to $0.85e^{\pm j\pi/2}$ increased the bandwidth to 0.32 rad. See A9_9 on the CLS disk. ■

Comment: Determining the bandwidth can be done easily by using the MATLAB **find** command. First, use **find**(mag > 0.707). This returns the indices of the mag array for which mag > 0.707. Then select the min array index, imin, and max array index, imax, and q(imin) and q(imax). The bandwidth is approximately q(imin) − q(imax).

P9.10 See A9_10 on the CLS disk.

Comment: An ideal window would have a frequency response of 1.0 at $\theta = 0$ and 0 for all other values of θ.

P9.11 a. Bandpass; b. $y_{ss}(n) = 18\cos(3\pi n/8 - 3\pi/4) - 26\cos(\pi n/2)$.

P9.12 a. Highpass; b. $y_{ss}(n) = 18\cos(3\pi n/8 - 3\pi/4) - 26\cos(\pi n/2) + 6\cos(3\pi n/4 - 3\pi/2)$.

P9.13 Use cut and try with poles placed near passband frequencies and zeros near stopband frequencies. One possibility is $H(z) = 0.159z(z - 1)/$

$(z^2 - 0.714z + 0.81)$ with $y(n) - 0.714y(n-1) + 0.81y(n-2) = 0.159x(n) - 0.159x(n-1)$; see A9_13 on the CLS disk. Notice that for frequencies greater than $\theta = 0.4\pi$, the response drops off significantly, and since this is above the useful range of frequencies anticipated, it is OK.

P9.14 a. Zero at $z = 0$, pole at $z = 0.81$; b. Four zeros at $z = 0$, poles at $0.155 \pm j0.887$ and $-0.155 \pm j0.887$; c. $C_1 = 1.287$ and $C_2 = 0.132$ with the plots in A9_14 on the CLS disk.

P9.15 a. Zeros at $1 \pm j1$ and poles at $0.5 \pm j0.5$. The poles and zeros are reciprocals. b. It is an allpass filter; that is, the gain is constant for all frequencies. Such filters are used with other filters to adjust (equalize) the phase without affecting the overall magnitude characteristic. c. $H(z) = (z + 2)/(z + 0.5)$.

P9.16 See A9_16.

A N S W E R S

RETROSPECTIVE

Chapters 7, 8, and 9

This is the third of four Retrospectives, which have been designed to insert a break in the action to allow us to gather our thoughts and to summarize what has gone before. Many different concepts and techniques have been presented so far in our study of LTI discrete systems, and we recognize that one of the important issues is when to apply each of these approaches. To a great extent this depends on what the input sequence is, or how it is specified (sinusoidal, tabular data, analytical expression, etc.); the system model (frequency response, difference equations, transfer function, unit sample (impulse) response, etc.); and, the desired form of the output (forced and/or initial condition response, steady-state response, analytical versus tabular numerical form, etc.). Although it would be naive to think that the selection of an appropriate choice can be made in a mechanical fashion, the alternatives can be summarized in the first two sections that follow. In the third section, we summarize our knowledge concerning how to find a particular form of a system's model, starting with another form. References to text material are given where appropriate.

SUMMARY OF DISCRETE SYSTEM SOLUTION METHODS

INPUT	SYSTEM MODEL	SOLUTION METHOD	RESULT
General $x(n)$ analytical or tabular	Difference equation	Iterative numerical solution	Total solution in tabular form
$x(n)$ analytical	Difference equation	Classical solution	Analytical solution (not covered in text)
$x(n)$ analytical	Difference equation	z transforms	Total solution in analytical form $y(n) = y_{\mathrm{IC}}(n) + y_{\mathrm{F}}(n)$
$x(n)$ analytical	Transfer function	z transforms	Forced response in analytical or tabular form, $y(n) = y_{\mathrm{F}}(n)$

INPUT	SYSTEM MODEL	SOLUTION METHOD	RESULT
$x(n)$ tabular	Unit sample response $h(n)$, tabular form	Convolution, tabular	Forced response in tabular form
$x(n)$ analytical	Unit sample response $h(n)$, analytical form	Convolution, analytical	Forced response in analytical form
$x(n) = X \cos(n\theta + \alpha)$	Frequency response $H(e^{j\theta})$, stable system	Sinusoidal steady-state formula	Steady-state forced response $y(n) = y_{ss}(n)$

SOLUTION CHOICES FOR GIVEN INPUT-OUTPUT CONDITIONS

INPUT	DESIRED OUTPUT	SOLUTION METHOD	SYSTEM MODEL
General $x(n)$; analytical or tabular	Total solution	Iterative numerical, tabular	DE
General $x(n)$; analytical	Total solution	Classical, analytical; z transforms, analytical	DE DE
General $x(n)$; analytical	Forced response	z transforms, analytical	$H(z)$
General $x(n)$; tabular	Forced response	Convolution, tabular	$h(n)$
General $x(n)$; analytical	Forced response	Convolution, analytical	$h(n)$
$x(n) = X \cos(n\theta + \alpha)$	Forced response (steady-state)	Sinusoidal steady-state formula	$H(e^{j\theta})$

HOW TO FIND ONE MODEL FROM ANOTHER

DESIRED MODEL	STARTING POINT	LIKELY APPROACH
Unit sample response $h(n)$	Difference equation	1. By inspection, nonrecursive 2. Classical, recursive 3. $\mathcal{Z}^{-1}[H(z)]$, recursive
	System diagram	1. By inspection, nonrecursive 2. Obtain $H(z)$, find $\mathcal{Z}^{-1}[H(z)]$.
	Transfer function	$\mathcal{Z}^{-1}[H(z)]$
	Frequency response	Depends on form given. If in rational form, find $H(z)$. Then $\mathcal{Z}^{-1}[H(z)] = h(n)$.

DESIRED MODEL	STARTING POINT	LIKELY APPROACH
Transfer function $H(z)$	Difference equation	z transforms
	System diagram	Via difference equation or MGR
	$h(n)$	$H(z) = Z[h(n)]$
	$H(e^{j\theta})$	Depends on form. If in rational form, find $H(z)$ directly via $e^{j\theta} = z$.
Difference equation	System diagram	By inspection
	$h(n)$	z transforms; that is, $Z^{-1}[H(z) = Y(z)/X(z)]$
	$H(z)$	By inspection
	$H(e^{j\theta})$	By inspection
Frequency response	Difference equation	By inspection
	System diagram	Via difference equation
	$h(n)$	Difference equation or $H(z)$, $z = e^{j\theta}$
	$H(z)$	$H(e^{j\theta}) = H(z)$, $z = e^{j\theta}$
System diagram	Difference equation	Solve for $y(n)$ and implement with unit delays and gains.
	$h(n)$	Find $H(z)$ and implement as $H(z) = H_a(z) \cdot H_b(z)$.
	$H(z)$	See $h(n)$.
	$H(e^{j\theta})$	Depends on form. If in rational form, find $H(z)$ directly and implement as before.

SUMMARY OF MATLAB FUNCTIONS USED

Function	*Purpose and Use*	*Toolbox*
conv	Given: unit sample response model of discrete system, input, **conv** returns system output.	Signals/Systems, Signal Processing
dimpulse	Given: state or TF model of discrete system, **dimpulse** returns impulse response.	Control System, Signals/Systems
displot	Plots discrete functions.	California Functions
dlsim	Given: state or TF model of discrete system, input, ICs, **dlsim** returns system output.	Control System, Signals/Systems
dstep	Given: state or TF model of discrete system, **dstep** returns step response.	Control System, Signals/Systems
filter	Given: DE or TF model of discrete system, input, ICs, **filter** returns system output.	Signals/Systems, Signal Processing

Function	*Purpose and Use*	*Toolbox*
freqz	Given: TF of discrete system, **freqz** returns frequency response.	Signals/Systems, Signal Processing
hamming	Given: length of window N, **hamming** returns window coefficients.	Signals/Systems, Signal Processing
hanning	Given: length of window N, **hanning** returns window coefficients.	Signals/Systems, Signal Processing
invztrans	Given: z transform function F, **invztrans** returns the inverse z transform f.	Symbolic Math
ksdlsim	Given: TF model of discrete system, input, ICs, **ksdlsim** returns system output.	California Functions
residue	Given: rational function $T(\sigma) = N(\sigma)/D(\sigma)$, **residue** returns roots of $D(\sigma) = 0$ and PF constants of $T(\sigma)$.	MATLAB
residuez	Given: rational function $T(z) = N(z)/D(z)$, **residuez** returns roots of $D(z) = 0$ and PF constants of $T(z)$.	Signal Processing, Signals/Systems
rlocus	Given: an equation in the form $1 + K[\text{num}(\sigma)]/[\text{den}(\sigma)] = 0$, **rlocus** returns plot of locus of roots for K varying.	Signals/Systems, Signal Processing
roots	Given: coefficients of polynomial p, **roots** returns roots of $p = 0$.	MATLAB
zplane	Given: TF or zeros and poles of discrete system, **zplane** returns display of zeros and poles.	Signal Processing, Signals/Systems
ztrans	Given: The symbolic expression f, **ztrans** returns the unilateral transform F.	Symbolic Math

Toolbox

MATLAB—Version 4.1, October 1, 1993
Signals/Systems—, January 1, 1992
Control System—Version 3.06, October 13, 1993
Signal Processing—Version 3.06, February 7, 1994
California Functions, June 20, 1992

Discrete Fourier Transforms

PREVIEW

A central topic in Chapters 7, 8, and 9 was the analysis of systems in the frequency domain, where we found the important result that when a sinusoidal sequence is the input to an LTI discrete system, the resulting steady-state output sequence is also sinusoidal with the same frequency as the input, but with its amplitude and phase modified by the characteristics of the system at the particular frequency. In the present chapter, we consider input and output sequences that are not single sinusoids. We will see that it is possible to represent an arbitrary sequence as a summation of sinusoids (or complex exponentials). One application of such an approach is to determine those

FIGURE 10.1 *James W. Cooley and John W. Tukey*

frequency components that are contained in a sequence. This is known as *spectrum analysis* and may be very useful in discerning sequence characteristics not observable in the time domain.

We begin by developing the Discrete Time Fourier Transform (DTFT) and then restrict the situation to one where the sequences are of finite duration. This leads to the Discrete Fourier Transform (DFT). Several properties and characteristics of DFTs are developed, including the relationships among sampling period, record length, and frequency resolution when a sequence is generated by sampling a continuous-time signal. Thus the DFT provides a computationally tractable way of approximating the Continuous-Time Fourier Transform (CTFT) of an analog signal. We also show that convolution can be performed using DFTs and that, with the aid of the Fast Fourier Transform (FFT), this may be far better than doing convolution using the definition.

One of the first major applications of the DFT was in the area of digital filter design and synthesis by J. F. Kaiser at Bell Laboratories. Applications of DFTs expanded rapidly with the impetus provided by the Cooley-Tukey (1965) paper on a fast method of computing the Discrete Fourier Transform. This method was popularized and extended via many papers in the IEEE *Transactions of the Group on Audio and Electroacoustics* and other journals, with the set of techniques becoming known as the FFT. In courses on digital signal processing, the theory of the DFT is connected with applications from seismic, speech, image, radar, sonar, music, and medical signal processing. For a historical perspective see M. T. Heideman, D. H. Johnson, and C. S. Burrus, "Gauss and the History of the Fast Fourier Transform," *The ASSP Magazine,* (4): 14–21, 1984.

BASIC CONCEPTS

THE DISCRETE TIME FOURIER TRANSFORM

In Chapters 7 and 8 we found that the frequency response of a discrete system is given by

frequency response

$$H(e^{j\theta}) = \sum_{m=-\infty}^{\infty} h(m)e^{-jm\theta},$$

where $h(n)$ is the unit sample response of a stable system and θ is frequency. Generalizing this definition to an arbitrary sequence $x(n)$ gives

DTFT

$$\mathcal{D}[x(n)] = X(e^{j\theta}) = \sum_{m=-\infty}^{\infty} x(m)e^{-jm\theta}$$

where $X(e^{j\theta})$ is called the Discrete Time Fourier Transform (DTFT) of the sequence $x(n)$. We recall that $X(e^{j\theta})$ is periodic in θ with a period of 2π. The inverse relationship is given by

inverse DTFT

$$\mathcal{D}^{-1}[X(e^{j\theta})] = x(n) = \frac{1}{2\pi}\int_{\theta_0}^{\theta_0+2\pi} X(e^{j\theta})e^{jn\theta}\,d\theta$$

with the DTFT pair given in the familiar notation

DTFT pair

$$x(n) \Leftrightarrow X(e^{j\theta}).$$

The expression for $X(e^{j\theta})$ is referred to as the *analysis* relationship because it describes how the sequence $x(n)$ can be analyzed into its frequency components. The equation for $x(n)$ is called the *synthesis* relationship because it indicates how the sequence can be synthesized from its frequency components.

ILLUSTRATIVE PROBLEM 10.1
DTFT of a Pulse Sequence

Find the DTFT of the pulse sequence shown in Figure 10.2a, page 546.

Solution

From the definition of the DTFT we have

$$X(e^{j\theta}) = \sum_{m=-4}^{4} x(m)e^{-jm\theta} = \sum_{m=-4}^{4} 1e^{-jm\theta}.$$

Defining $r = m + 4$ gives

$$\sum_{r=0}^{8} e^{-j(r-4)\theta} = e^{j4\theta}\sum_{r=0}^{8} e^{-jr\theta}$$

which, after using the finite geometric series formula, yields

$$X(e^{j\theta}) = \sin(4.5\theta)/\sin(0.5\theta).$$

This DTFT is shown in Figure 10.2b. Notice that by using l'Hôpital's rule

$$X(e^{j0}) = \lim_{\theta\to 0}\frac{\sin(4.5\theta)}{\sin(0.5\theta)} = 9.0.$$

Also, $\sin(0.5\theta)$ is periodic with period θ_a, found by setting $0.5\theta_a = 2\pi$, or $\theta_a = 4\pi$, and $\sin(4.5\theta)$ is periodic with period θ_b, found from $4.5\theta_b = 2\pi$, or $\theta_b = 2\pi/4.5 = 0.44\pi = 1.396$. The values of θ for which the DTFT are zero correspond to the zeros of $\sin(4.5\theta)$ which occur for $\theta = 0.70, 1.40, 2.09, 2.79, 3.49, \ldots$.

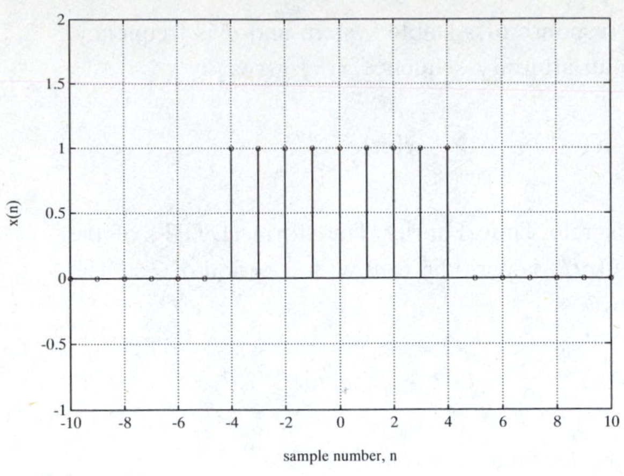

a. Pulse sequence

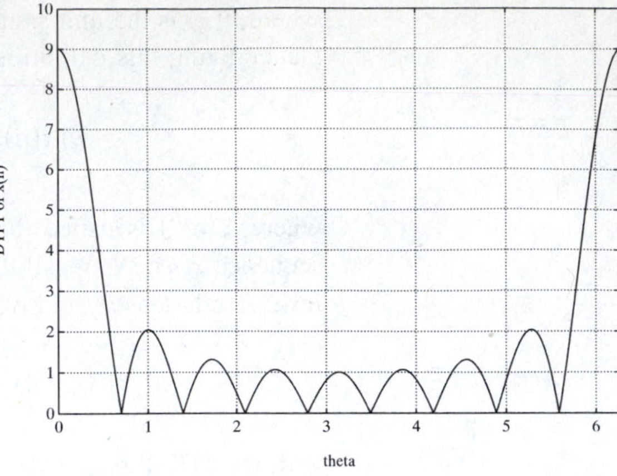

b. Discrete Time Fourier Transform

FIGURE 10.2 *A pulse sequence and its Discrete Time Fourier Transform*

THE DISCRETE FOURIER TRANSFORM

If we have a sequence $x(n)$ that is nonzero only for a finite number of samples in the interval $0 \leq n \leq N - 1$ and we choose to represent the sequence only over this interval, then the analysis relationship is given by

DFT

$$X(k) = \sum_{n=0}^{N-1} x(n)e^{-j(2\pi/N)nk} \qquad k = 0, 1, \ldots, N - 1$$

where $X(k)$ is referred to as the Discrete Fourier Transform (DFT) of the sequence $x(n)$. The inverse transform (IDFT), or synthesis relationship, is given by

IDFT

$$x(n) = \frac{1}{N} \sum_{k=0}^{N-1} X(k)e^{j(2\pi/N)nk} \qquad n = 0, 1, \ldots, N - 1 .$$

The index k corresponds to frequency, and we will see later how to associate values of k with analog frequencies in the case of a sampled analog time signal. Another version of notation used is

$$X(k) = \text{DFT}[x(n)] \quad \text{and} \quad x(n) = \text{IDFT}[X(k)]$$

or, in the familiar pair notation

DFT pair

$$x(n) \Leftrightarrow X(k) .$$

The terms

$$e^{-j(2\pi/N)nk}, \qquad n = 0, 1, \ldots, N - 1 ,$$

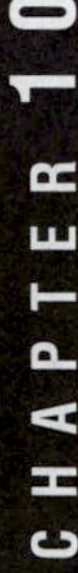

represent complex exponentials that rotate clockwise in the complex plane with an angular velocity of $2\pi k/N$ rad/sample. Notice that the larger the value of k, the greater the angular velocity. A similar interpretation may be made of the exponentials that appear in the expression for $x(n)$. Comparing the DFT and its inverse with the corresponding DTFT relationships, we note the following.

1. The DFT is generated by a finite sum and consists only of N frequency values.
2. The Inverse DFT (IDFT) is a finite sum.
3. $X(k)$ is periodic in k with a period equal to N.
4. The DFT representation of $x(n)$ is periodic in n with a period equal to N.
5. The DFT values, the $X(k)$'s, for a finite-duration sequence are the values that would be obtained by sampling the DTFT $X(e^{j\theta})$ of the sequence with values of θ equal to $2\pi k/N$.

The periodicity of $X(k)$ and $x(n)$ imply that if we were to allow the indices k or n to assume values outside the range $0 \le n \le N-1$, then the resulting values for $X(k)$ and $x(n)$ would simply repeat the values within the interval $0 \le n \le N-1$. For this reason, we can consider $x(n)$ to be one period of a periodic sequence $x_p(n)$; that is,

$$x(n) = x_p(n), \qquad n = 0, 1, \ldots, N-1 .$$

Conversely, we can view a periodic sequence $x_p(n)$ as the *periodic extension* of a finite-duration sequence $x(n)$; i.e.,

$$x_p(n) = x(n + mN), \qquad n = 0, 1, \ldots, N-1, \qquad m = 0, \pm 1, \pm 2, \ldots .$$

ILLUSTRATIVE PROBLEM 10.2
Finding a DFT

Find the DFT of the sequence $x(0) = 1$, $x(1) = 2$, $x(2) = 3$, $x(3) = 4$.

Solution

From the definition

$$X(k) = \sum_{n=0}^{3} x(n) e^{-j(2\pi/4)nk}, \quad k = 0, 1, 2, 3$$

and with $k = 0$

$$X(0) = \sum_{n=0}^{3} x(n) = 10 .$$

Letting $k = 1$ gives

$$\begin{aligned} X(1) &= x(0)e^{-j(0)} + x(1)e^{-j(\pi/2)} + x(2)e^{-j(\pi)} + x(3)e^{-j(3\pi/2)} \\ &= 1 - j2 - 3 + j4 = -2 + j2 . \end{aligned}$$

Similar calculations for $k = 2$ and $k = 3$ give $X(2) = -2$ and $X(3) = -2 - j2$. In exponential form we have

$$X(0) = 10e^{j0}, \qquad X(1) = 2.83e^{j2.36}, \qquad x(2) = 2e^{j\pi} \quad \text{and} \quad X(3) = 2.83e^{-j2.36}.$$

Comment: A MATLAB script for accomplishing these calculations and its output follow.

MATLAB Script and Output

```
%IP10_2 DFT via function fft            Output
m=[0 1 2 3];    % vector of time samples   Xk=
xn=[1 2 3 4];   % x(n) vector
Xk=fft(xn)      % call fft to get X(k)     10.0000   -2.0000+2.0000i-2.0000
                                           -2.0000-2.0000i

%convert X(k) to magnitude and angle
Xmag=abs(Xk)
Xphase=angle(Xk)                           Xmag=

                                           10.0000  2.8284  2.0000  2.8284


                                           Xphase=


                                             0  2.3562  3.1416  -2.3562
```

SAMPLING PERIOD, RECORD LENGTH, AND FREQUENCY RESOLUTION

A sequence often results from sampling a continuous-time (analog) signal. In this case it is reasonable to ask which frequencies are present in the sequence's DFT. Assume that the analog signal is given by $x(t) = 2\cos(20\pi t)$ and that samples are taken at intervals of $T = 0.0125$ s during one period $T_0 = 0.1$ s of the sinusoid, as shown in Figure 10.3a. Evaluating the DFT of these sample values, we find the plot of $X(k)$ shown in Figure 10.3b. (The DFT is real here, so $X(k)$ can be shown using only one plot.) Notice that there are two frequencies indicated by the plot, but we know that only $f = 10$ Hz is present in the original signal. Actually, the line at $k = 1$ represents the frequency $f = 10$ Hz; the line at $k = 7$ represents the negative frequency image of 10 Hz. This is in the same vein as noting that $x(t)$ can be written as (Euler)

$$x(t) = e^{j20\pi t} + e^{-j20\pi t}$$

where we see the frequencies 10 Hz and -10 Hz.

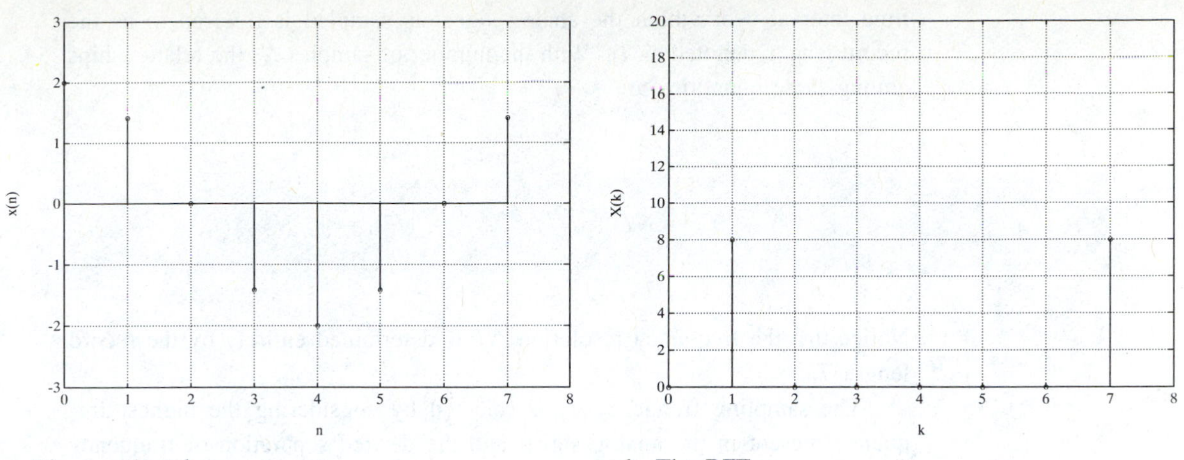

a. The sample values

b. The DFT

FIGURE 10.3 *A sampled sinuosid and its DFT*

We know that $X(k)$ is periodic in k with period N. Thus the plot of $X(k)$ can be represented as either alternative shown in Figure 10.4a and b, which accounts for the frequency labels shown in Figure 10.4a.

Let us now explore the more general relationships among sampling period, record length, and frequency resolution.

The sampling interval, denoted by T, is related to the sampling frequency f_s by $T = 1/f_s$. From the Sampling Theorem (see Chapter 5 or 1) we know that f_s should be greater than twice the highest frequency present in the waveform to be sampled in order to avoid aliasing. Frequency resolution is concerned with the ability of the DFT to detect the presence of adjacent frequencies present in the analog waveform to be sampled; the smaller the frequency separation Δf in the DFT, the better (higher) the resolution. The

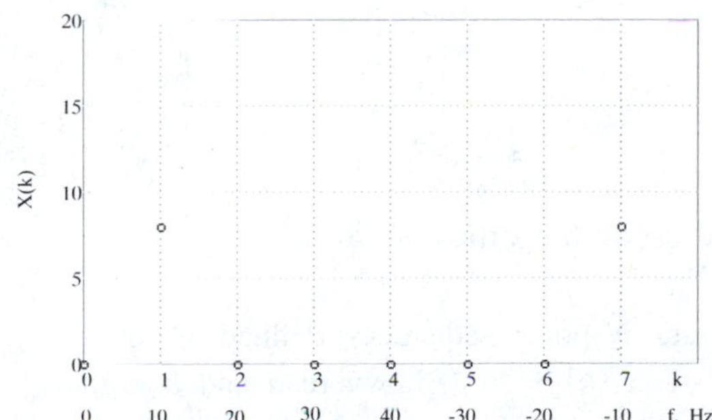

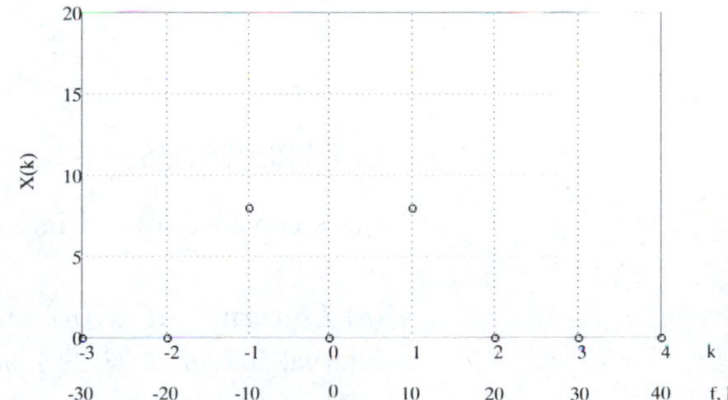

a. The DFT, first alternative

b. The DFT, second alternative

FIGURE 10.4 *DFT with analog frequencies shown*

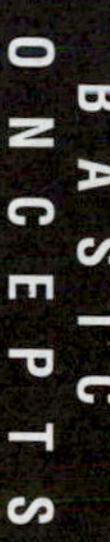

time interval over which the analog signal is sampled is referred to as the record length, denoted as T_0. With the number of samples N, the relationships among these quantities are

$$T_0 = NT = N/f_s \quad \Delta f = 1/T_0 = f_s/N\,.$$

Notice that the frequency resolution Δf is determined entirely by the record length T_0.

The sampling frequency f_s is selected by considering the highest frequency present in the analog signal and the desired separation of frequency replicas introduced by the sampling process. Next, the record length T_0 is selected to achieve a desired resolution. The required number of samples is then determined from the relationship $N = T_0 f_s$.

ILLUSTRATIVE PROBLEM 10.3 ***Finding Record Length, Sampling Frequency, and Frequency Resolution***

An analog signal having frequencies of up to 600 Hz is to be sampled at the minimum possible frequency, and the frequency resolution is to be 0.5 Hz. Find the sampling frequency, record length, and number of samples.

Solution

The minimum sampling frequency is $f_s > 2 \cdot 600 = 1200$ Hz. To achieve a frequency resolution of 0.5 Hz requires a record length of $T_0 = 1/0.5 = 2$ s. The resulting number of samples is then $N = T_0 \cdot f_s = 2 \cdot 1200 = 2400$ samples.

DFT PROPERTIES

Let's consider several interesting and useful properties of DFTs.

(a) Linearity If $x_1(n)$ and $x_2(n)$ are N-point sequences defined in the interval $0 \leq n \leq N - 1$ and $x_3(n) = ax_1(n) + bx_2(n)$, where a and b are arbitrary constants, then $X_3(k) = aX_1(k) + bX_2(k)$. In words, the transform of a weighted sum of sequences is the weighted sum of the transforms. Of course, this result also includes the special cases

$$a = b = 1\text{: } X_3(k) = X_1(k) + X_2(k)\,,$$

the transform of a sum is the sum of the transforms;

$$b = 0\colon X_3(k) = aX_1(k),$$

DFT[constant · sequence] = constant · DFT[sequence].

(b) Symmetry For $x(n)$ a *real* sequence of length N,

symmetry, magnitude/angle

$$|X(k)| = |X(N-k)| \quad \text{and} \quad \angle X(k) = -\angle X(N-k)$$

where these relationships hold for $k = 1, 2, \ldots, N-1$. An alternative, but equivalent, form of these relationships is

symmetry—real and imaginary

$$\text{Re}[X(k)] = \text{Re}[X(N-k)] \quad \text{and} \quad \text{Im}[X(k)] = -\text{Im}[X(N-k)]$$

again for the same range of values of k.

We have already seen two examples where this symmetry property is illustrated. For example, in Illustrative Problem 10.2, we notice that the symmetry is about the point $k = 2 = N/2$, and that the point $k = 0$ is excluded from the symmetry relationships because it represents the zero-frequency component, which has no negative frequency image. The second example is the cosine function's DFT shown in Figure 10.3b.

A consequence of these symmetry conditions is that if $x(n)$ is a real and even function of n—that is, $x(n) = x(-n)$—then the DFT $X(k)$ is a real and even sequence of k.[1] In a similar way, if $x(n)$ is a real and odd function of n—that is, $x(n) = -x(-n)$ and $x(0) = 0$—then the DFT $X(k)$ is an odd and imaginary sequence of k.

(c) Circular shift of a sequence Consider a sequence defined in the interval $0 \le n \le N-1$. If this sequence is circularly shifted by m samples, we denote the result as $x(n \oplus m)$. By circular shift we mean that as values "fall off" from one end of the sequence, they are appended to the opposite end. If m is positive, the shift is toward the left, whereas a negative value of m indicates a rightward shift. An example is shown in Figure 10.5, on page 552.

The DFT property of a circularly shifted sequence is

$$x_2(n) = x_1(n \oplus m) \Leftrightarrow X_2(k) = X_1(k)e^{j(2\pi/N)(km)}.$$

This indicates that a circular shift in time results in a DFT having the same magnitude as the unshifted time sequence but with a phase shift that depends linearly on the index k.

[1]The condition $x(n) = x(-n)$ is equivalent to $x(n) = x(N-n)$, with $n = 1, 2, \ldots, N-1$.

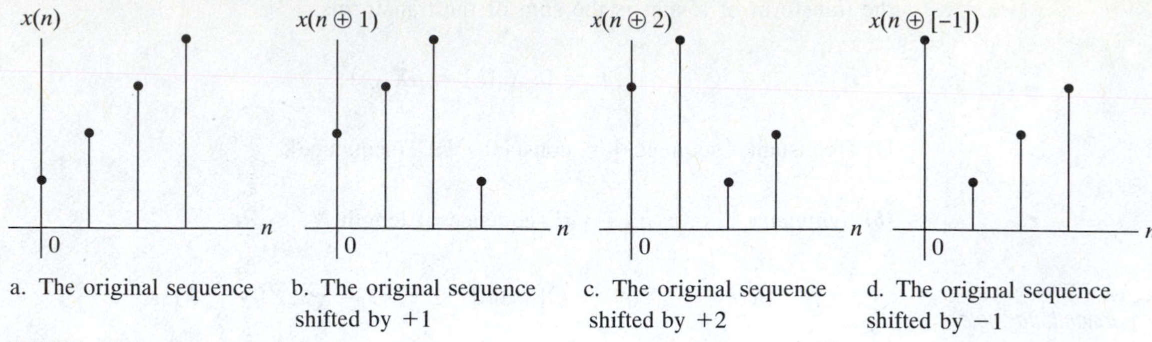

a. The original sequence

b. The original sequence shifted by +1

c. The original sequence shifted by +2

d. The original sequence shifted by −1

FIGURE 10.5 *Circular shift of a sequence*

duality

(d) Duality For each DFT determined, a second DFT comes for free. This "two-for-one" property is called *duality.* In the usual pair notation we have

$$x(n) \Leftrightarrow X(k)$$

$$\frac{1}{N} X(n) \Leftrightarrow x(-k)\,.$$

The starting point is a sequence $x(n)$ and its DFT $X(k)$. The property says that the DFT of the sequence $(1/N)X(n)$—formed by treating $X(k)$ as a time sequence—is found by replacing n in $x(n)$ with $-k$. An illustrative group of four plots is shown in Figure 10.6. Notice that the original sequence $x(n)$ is real and even, a consequence of which is that $X(k)$ is real and even.

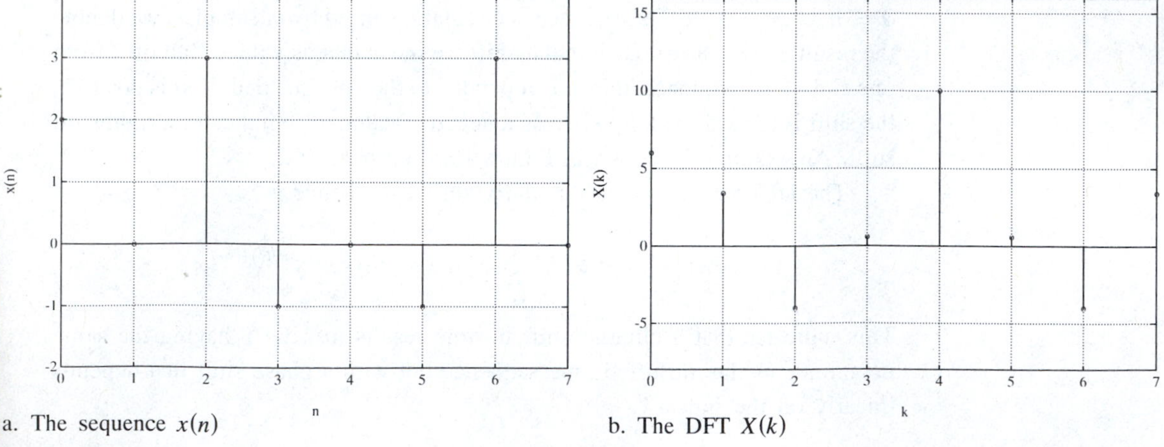

a. The sequence $x(n)$

b. The DFT $X(k)$

FIGURE 10.6 *Illustration of the duality property*

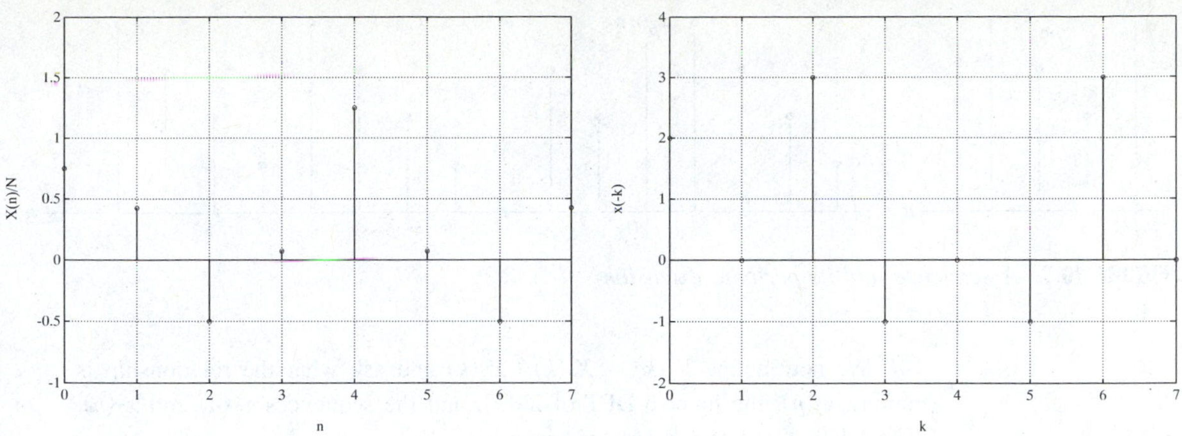

c. The sequence $X(n)/N$

d. The DFT $x(-k)$

FIGURE 10.6 *Continued*

(e) Relationship to z Transform Comparing the definition of the DFT with that of the z transform for a sequence that is nonzero only in the interval $0 \leq n \leq N-1$, we see that

$$X(k) = X(z)|_{z=e^{j(2\pi k)/N}} \qquad k = 0, 1, \ldots, N-1 .$$

In other words, the DFT is the z transform evaluated at points on the unit circle separated in angle by $2\pi/N$ radians.

CONVOLUTION AND THE DFT

There is a relationship, which we will now develop, between convolution and DFTs. The starting point is to consider the product of two DFTs $X_1(k)$ and $X_2(k)$, which are the DFTs of the N-point sequences $x_1(n)$ and $x_2(n)$, respectively, defined in the interval $0 \leq n \leq N-1$. We need to keep in mind that the inverse DFT

$$x(n) = \frac{1}{N} \sum_{k=0}^{N-1} X(k) e^{j(2\pi/N)nk}$$

not only represents the sequence $x(n)$ in the interval $0 \leq n \leq N-1$, but, because of the periodicity of the exponentials, also yields replicas of $x(n)$ outside the interval $0 \leq n \leq N-1$. The resulting sequence is called the *periodic extension* of $x(n)$ and is denoted by $x_p(n)$. Figure 10.7 shows a sequence and its periodic extension.

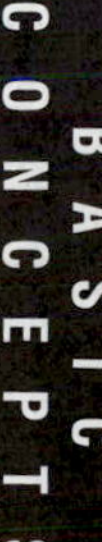

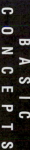

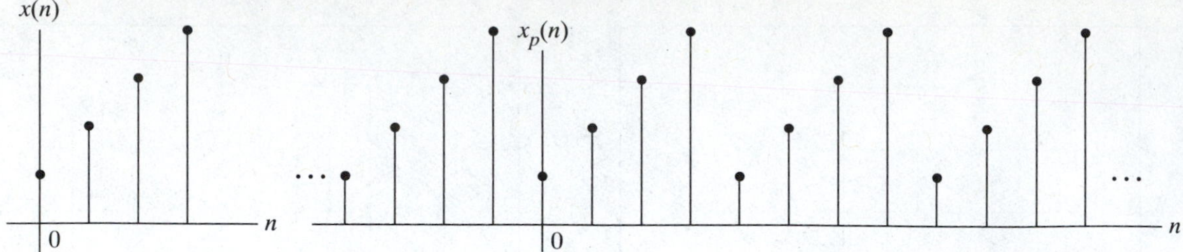

FIGURE 10.7 *A sequence and its periodic extension*

We now define $X_3(k) = X_1(k) \cdot X_2(k)$ and ask what the relationship is among $x_3(n)$, the inverse DFT of $X_3(k)$, and the sequences $x_1(n)$ and $x_2(n)$. The DFTs $X_1(k)$ and $X_2(k)$ are given by

$$X_1(k) = \sum_{m=0}^{N-1} x_1(m) W_N{}^{mk} \quad \text{and} \quad X_2(k) = \sum_{l=0}^{N-1} x_2(l) W_N{}^{lk}$$

where m and l are dummy variables of summation and the notation $W_N = e^{-j2\pi/N}$ has been introduced to make the expressions more compact. Notice that with this definition,

$$W_N{}^{mk} = e^{-j(2\pi/N)mk} \quad \text{and} \quad W_N{}^{lk} = e^{-j(2\pi/N)lk}.$$

Using these DFT definitions we can write

$$X_3(k) = X_1(k) \cdot X_2(k) = \sum_{m=0}^{N-1} \sum_{l=0}^{N-1} x_1(m) x_2(l) W_N{}^{mk} W_N{}^{lk}$$

and the IDFT is

$$\begin{aligned} x_3(n) &= \text{IDFT}[X_3(k)] = \text{IDFT}[X_1(k)X_2(k)] \\ &= \frac{1}{N} \sum_{k=0}^{N-1} \left[\sum_{m=0}^{N-1} \sum_{l=0}^{N-1} x_1(m) x_2(l) W_N{}^{mk} W_N{}^{lk} \right] W_N{}^{-nk}, \\ n &= 0, 1, \ldots, N-1. \end{aligned}$$

This looks pretty complicated, but if we keep our wits about us it will simplify considerably. First, we interchange the order of the summations and write the W factors all together by using the properties of exponentials:

$$x_3(n) = \frac{1}{N} \sum_{m=0}^{N-1} \sum_{l=0}^{N-1} \sum_{k=0}^{N-1} x_1(m) x_2(l) W_N{}^{(m+l-n)k} \qquad n = 0, 1, \ldots, N-1.$$

Noticing that the innermost summation is over the index k and that $x_1(m)$ and $x_2(l)$ do not depend on k, we have

$$x_3(n) = \frac{1}{N} \sum_{m=0}^{N-1} \sum_{l=0}^{N-1} x_1(m) x_2(l) \left[\sum_{k=0}^{N-1} W_N{}^{(m+l-n)k} \right] \qquad n = 0, 1, \ldots, N-1.$$

Although it may not be obvious at first glance, the innermost summation is a finite geometric sum. If we rewrite the summation as

$$\left[\sum_{k=0}^{N-1} W_N{}^{(m+l-n)k}\right] = \sum_{k=0}^{N-1} V^k$$

and use

$$\sum_{k=0}^{N-1} V^k = \frac{1 - V^N}{1 - V}, \qquad V \neq 1,$$

we obtain

$$\left[\sum_{k=0}^{N-1} W_N{}^{(m+l-n)k}\right] = \begin{Bmatrix} 0, & \text{if } m + l - n \neq 0 \\ N, & \text{if } m + l - n = 0 \end{Bmatrix}$$
$$= N\delta(m + l - n).$$

At this point we have

$$x_3(n) = \sum_{m=0}^{N-1} \sum_{l=0}^{N-1} x_1(m)x_2(l)\delta(m + l - n), \qquad n = 0, 1, \ldots, N - 1$$

which, using the definition of the unit sample function, becomes

$$x_3(n) = \sum_{m=0}^{N-1} x_1(m)x_2(n - m)$$
$$= \sum_{l=0}^{N-1} x_1(n - l)x_2(l), \qquad n = 0, 1, \ldots, N - 1.$$

Although this looks just like convolution, it's not quite the same. First, notice that $x_3(n)$ is defined only over the interval $0 \leq n \leq N - 1$; consequently, according to Chapter 7, convolution would result in a sequence of length $2N - 1$ samples. In addition, there is a question about what to use for $x_2(n - m)$ in the first version when $n - m < 0$. The temptation is to say that the values are zero, but, keeping in mind the underlying periodicity of the DFT, we must use the values of the periodic extension of the sequence $x_3(n)$. As a result, the shift that appears as $x_2(n - m)$ or $x_1(n - l)$ is a circular shift. Extending our earlier notation $\oplus$ for a circular shift, we write

circular convolution

$$x_3(n) = \sum_{m=0}^{N-1} x_1(m)x_2(n \ominus m) = \sum_{l=0}^{N-1} x_1(n \ominus l)x_2(l),$$
$$n = 0, 1, \ldots, N - 1,$$

where $x_2(n \ominus m)$ corresponds to a circular shift to the right for positive m. The operation indicated by these equations is known as *circular convolution* to distinguish it from the convolution operation of Chapter 7, which from here on we'll refer to as *linear convolution.* A way to think of the circular shift found in circular convolution is to represent the sequences on circles, one fixed and the other rotating. This approach is illustrated in Illustrative Problem 10.4.

ILLUSTRATIVE PROBLEM 10.4 ***Evaluating Circular Convolution in the Time and DFT Domains***

Use a time-domain formula to find the circular convolution of the four-point sequences

$$x_1(n) = 1, 2, 3, 4, \quad \text{for } n = 0, 1, 2, 3, \quad \text{and}$$
$$x_2(n) = -5, 2, -1, 4, \quad \text{for } n = 0, 1, 2, 3.$$

Verify the answer by using the DFT relationship derived previously.

Solution

To begin, we display the sample values of the sequences on concentric circles, as shown in Figure 10.8a. We choose to write the sequence values for $x_1(m)$ at equally spaced intervals moving counterclockwise around the outer circle, which will be stationary. The sequence values for $x_2(n-m)$, on the other hand, are written clockwise around the inner circle, which will be allowed to rotate by one-quarter of a revolution in the counterclockwise direction for each increase by one in the value of n. Figure 10.8b shows the four relative positions of the concentric circles. To evaluate the circular convolution for $n = 0$, we simply multiply the values lying adjacent to one another on the two circles and sum these products. We then repeat this procedure for $n = 1, 2$, and 3. Thus we have

$$n = 0\text{: } x_3(n) = (1)\cdot(-5) + (2)\cdot(4) + (3)\cdot(-1) + (4)\cdot(2) = 8$$
$$n = 1\text{: } x_3(n) = (1)\cdot(2) + (2)\cdot(-5) + (3)\cdot(4) + (4)\cdot(-1) = 0$$
$$n = 2\text{: } x_3(n) = (1)\cdot(-1) + (2)\cdot(2) + (3)\cdot(-5) + (4)\cdot(4) = 4$$
$$n = 3\text{: } x_3(n) = (1)\cdot(4) + (2)\cdot(-1) + (3)\cdot(2) + (4)\cdot(-5) = -12$$

Shown on the next page are the MATLAB script and its output, which yield the same results.

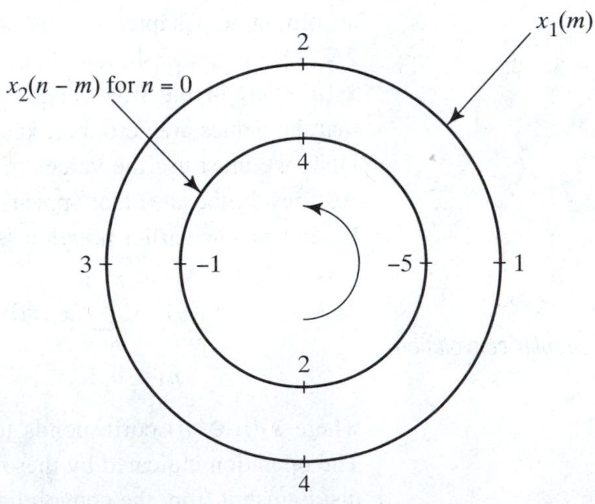

a. Initial representation of the sequences

FIGURE 10.8 *Evaluating circular convolution using concentric circles*

CHAPTER 10

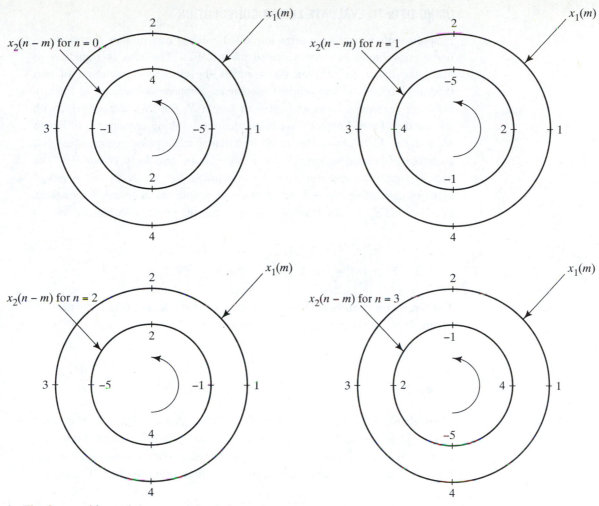

b. The four positions of the concentric circles

FIGURE 10.8 *Continued*

MATLAB Script and Output

```
%IP10_4 Circular convolution via DFT
N=4;              % number of points
x1n=[1 2 3 4];    % x1(n) vector
x2n=[-5 2 -1 4];% x2(n) vector
m=[0:1:N-1];      % vector of time samples
                  % used in plotting
X1k=fft(x1n);     % call fft to get X1(k)
X2k=fft(x2n);     % call fft to get X2(k)
X3k=X1k.*X2k      % multiply DFTs to get X3(k)
x3n=ifft(X3k)     % find inverse DFT to get x3(n)
%...plotting statements
```

Output

```
x3n=

 8.0000  0+0.0000i  4.0000
-12.0000 -0.0000i
```

USING DFTs TO EVALUATE LINEAR CONVOLUTION

In applications we are most often interested in linear convolution. Surprisingly, linear convolution can be evaluated using DFTs. The idea is somehow to make the linear convolution the same as the circular convolution of two modified versions of the original sequences. Suppose we wanted to linearly convolve sequences $x_1(n)$ and $x_2(n)$ of length N_1 and N_2, respectively, with one another. From Chapter 7 we know that the resulting sequence is of length $N_1 + N_2 - 1$. To obtain this result by circular convolution requires that two sequences, both having length $N_1 + N_2 - 1$, be circularly convolved. The trick is simply to augment $x_1(n)$ by appending $N_2 - 1$ zeros and augment $x_2(n)$ by appending $N_1 - 1$ zeros. We will denote the augmented sequences by $x_1'(n)$ and $x_2'(n)$. For example, suppose that the original sequences are

$$x_1(n) = 1, 2, 3, 4, \qquad \text{for } n = 0, 1, 2, 3, \qquad \text{and}$$
$$x_2(n) = -3, 5, -4, \qquad \text{for } n = 0, 1, 2.$$

The linear convolution of these sequences will have length 6 and is shown in Figure 10.9. The augmented sequences are given by

$$x_1'(n) = 1, 2, 3, 4, 0, 0 \quad \text{for } n = 0, 1, 2, 3, 4, 5, \qquad \text{and}$$
$$x_2'(n) = -3, 5, -4, 0, 0, 0, \qquad \text{for } n = 0, 1, 2, 3, 4, 5.$$

Evaluating the DFTs of $x_1'(n)$ and $x_2'(n)$ yields $X_1'(k)$ and $X_2'(k)$. The result of multiplying these six-point DFTs together is denoted by $X_3'(k)$. Taking the inverse DFT then gives $x_3'(n)$, which is the linear convolution of $x_1(n)$ and $x_2(n)$. The details of doing this are given in Illustrative Problem 10.5.

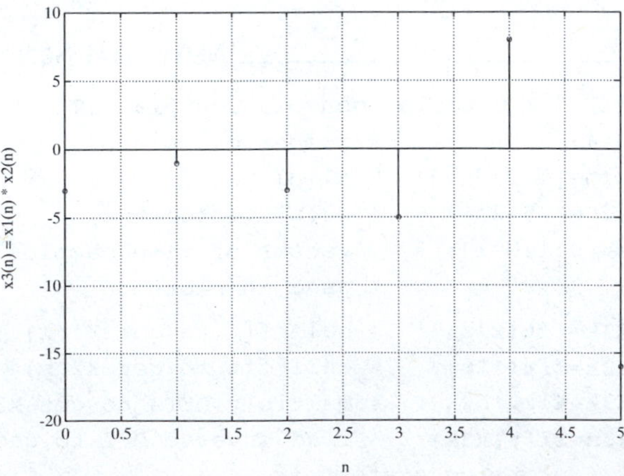

FIGURE 10.9 *Linear convolution of $x_1(n)$ and $x_2(n)$*

CHAPTER 10

ILLUSTRATIVE PROBLEM 10.5
Evaluating Linear Convolution Using DFTs

Use DFTs to find the linear convolution of the sequences

$$x_1(n) = 1, 2, 3, 4, \quad \text{for } n = 0, 1, 2, 3, \quad \text{and}$$
$$x_2(n) = -3, 5, -4, \quad \text{for } n = 0, 1, 2.$$

Solution

The length of the linear convolution of $x_1(n)$ *and* $x_2(n)$ will be $4 + 3 - 1 = 6$, so we need to augment $x_1(n)$ and $x_2(n)$ by appending enough zeros to make $x'_1(n)$ and $x'_2(n)$ both of length 6. The resulting sequences, which we refer to as being *zero padded,* are

$$x'_1(n) = 1, 2, 3, 4, 0, 0 \quad \text{for } n = 0, 1, 2, 3, 4, 5, \quad \text{and}$$
$$x'_2(n) = -3, 5, -4, 0, 0, 0, \quad \text{for } n = 0, 1, 2, 3, 4, 5.$$

The MATLAB script and output data are shown next.

MATLAB Script and Output

```
%IP10_5 Linear convolution via DFT
n=6;                            % number of points
x1prime=[1 2 3 4 0 0];
x2prime=[-3 5 -4 0 0 0];
m=[0 1 2 3 4 5];                % vector of time
                                % samples
format compact                  % compact display
X1pk=fft(x1prime)               % find X1prime(k)
X2pk=fft(x2prime)               % find X2prime(k)
X1pkmag=abs(X1pk)               % mag of X1prime(k)
X1pkangle=angle(X1pk)           % angle of X1prime(k)
X2pkmag=abs(X2pk)               % mag of X2prime(k)
X2pkangle=angle(X2pk)           % angle of X2prime(k)
X3pk=X1pk.*X2pk                 % find X3prime(k)
X3pkmag=abs(X3pk)               % mag of X3prime(k)
X3pkangle=angle(X3pk)           % angle of X3prime(k)
```

Output

```
X1pk=
 Columns 1 through 4
 10.0000  -3.5000-4.3301i
  2.5000+0.8660i
-2.0000+0.0000i
 Columns 5 through 6
 2.5000-0.8660i-3.5000
  +4.3301i
X2pk=
 Columns 1 through 4
 -2.0000  1.5000-0.8660i
  -3.5000-7.7942i
-12.0000-0.0000i
 Columns 5 through 6
 -3.5000+7.7942i
  1.5000+0.8660i
X1pkmag=
  10.0000 5.5678 2.6458
  2.0000 2.6458
5.5678
X1pkangle=
   0 -2.2506 0.3335
   3.1416 -0.3335
2.2506
```

Continues

BASIC CONCEPTS

```
x3prime=ifft(X3pk)          % find x3prime(n)
```

```
X2pkmag=
  2.0000 1.7321 8.5440
   12.0000 8.5440
1.7321
X2pkangle=
  3.1416 -0.5236 -1.9929
   -3.1416 1.9929
0.5236
X3pk=
 Columns 1 through 4
-20.0000   -9.0000-3.4641i
  -2.0000-22.5167i
24.0000-0.0000i
 Columns 5 through 6
 -2.0000+22.5167i -9.0000
   +3.4641i
X3pkmag=
 20.0000 9.6437 22.6053
   24.0000 22.6053 9.6437
X3pkangle=
 3.1416 -2.7742 -1.6594
   -0.0000 1.6594 2.7742
x3prime=
 Columns 1 through 4
 -3.0000+0.0000i -1.0000
   +0.0000i -3.0000+0.0000i
   -5.0000+0.0000i
 Columns 5 through 6
 8.0000-0.0000i-16.0000
 -0.0000i
```

■

To help you check these results, the DFTs have been converted from rectangular to exponential form. Thus it can easily be seen, for example, that

$$X_3'(2) = X_1'(2) \cdot X_2'(2) = 2.6458e^{j0.3335} \cdot 8.5440e^{-j1.9929}$$
$$= 22.6053e^{-j1.6594}.$$

To understand why the zero padding works, it is helpful to appeal again to the concentric circles artifact shown in Figure 10.8b, suitably modified for the case at hand. Figure 10.10 shows four positions of the sequences from Illustrative Problem 10.5 arranged on concentric circles. By multiplying the numbers opposite one another on the circles and adding, it is easily verified that

$$x_3'(0) = -3, \qquad x_3'(1) = -1, \qquad x_3'(3) = -5, \qquad x_3'(5) = -16.$$

The other positions and calculations are left to you. It should be clear that

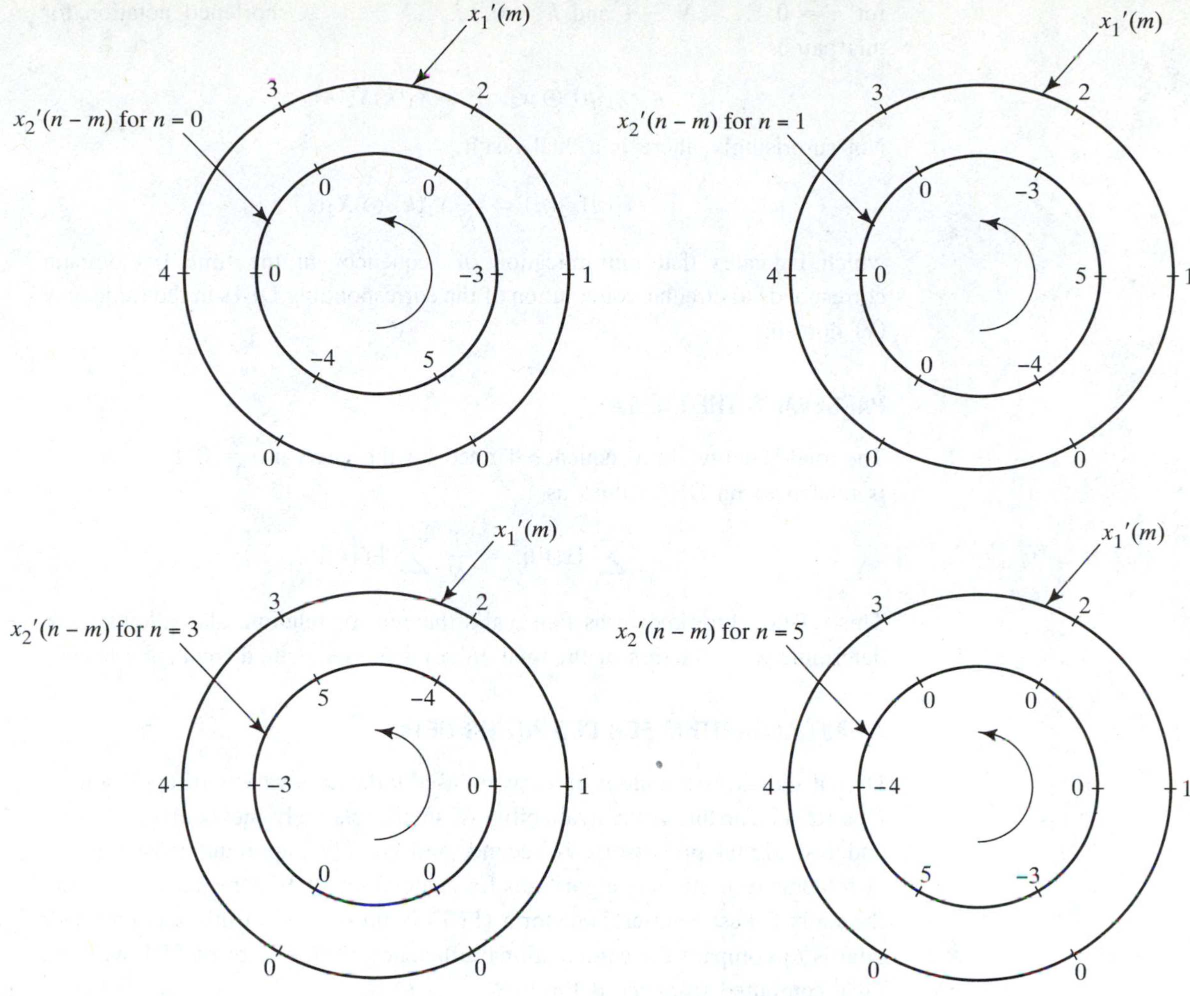

FIGURE 10.10 *Four positions of the concentric circles*

the effect of the zero padding is to prevent the "wraparound" effect that is a consequence of the underlying periodicity of DFTs and the sequences they represent.

FREQUENCY DOMAIN CONVOLUTION

We have seen that multiplication of DFTs corresponds to circular convolution of the corresponding time-domain sequences; that is,

$$x_3(n) = \sum_{m=0}^{N-1} x_1(m)x_2(n \ominus m)$$

$$= \sum_{l=0}^{N-1} x_1(n \ominus l)x_2(l) \Leftrightarrow X_3(k) = X_1(k)X_2(k),$$

for $n = 0, 1, \ldots, N - 1$ and $k = 0, 1, \ldots, N - 1$. A shortened notation for this pair is

$$x_1(n) \circledast x_2(n) \Leftrightarrow X_1(k)X_2(k).$$

Not surprisingly, there is a dual result,

$$x_1(n)x_2(n) \Leftrightarrow \frac{1}{N} X_1(k) \circledast X_2(k)$$

which indicates that multiplication of sequences in the time (n) domain corresponds to circular convolution of the corresponding DFTs in the frequency (k) domain.

PARSEVAL'S THEOREM

The total "energy" in a sequence defined for the interval $n = 0, 1, \ldots, N - 1$ is related to its DFT values as

$$\sum_{n=0}^{N-1} |x(n)|^2 = \frac{1}{N} \sum_{k=0}^{N-1} |X(k)|^2 .$$

This relationship, known as Parseval's theorem or relation, also allows us to determine what fraction of the total energy lies in specified frequency bands.

A FAST ALGORITHM FOR COMPUTING DFTs

Digital signal processing is extensively used today in a variety of applications. One reason for this is the availability of small, relatively inexpensive, reliable and fast digital processors. A second, and equally important, reason is the development of efficient algorithms for calculating DFTs. One such algorithm, the radix-2 Fast Fourier Transform (FFT) is introduced in this section.[2] Our goal is to compare the computational efficiency of this form of FFT with the DFT computed from the definition

$$X(k) = \sum_{n=0}^{N-1} x(n)e^{-j(2\pi/N)nk} \qquad k = 0, 1, \ldots, N - 1 .$$

As a starting point, let's consider the number of multiplications and additions required to compute an N-point DFT from the definition. For convenience we'll assume that the elements of the sequence $x(n)$ may be complex numbers. This assumption is made to enable us to compute Inverse DFTs (IDFTs) where the data samples are, in fact, complex numbers and also to allow additional generality that may be helpful in certain situations.

---

[2]The radix-2 FFT was introduced by J.W. Cooley and J.W. Tukey in their paper "An Algorithm for the Machine Calculation of Complex Fourier Series," *Math. Comput.* 19, no. 2, April 1965: 297–301. However, as described by M.T. Heideman, D.H. Johnson, and C.S. Burrus in "Gauss and the History of the Fast Fourier Transform," *The ASSP Magazine,* 1, no. 4 (1984): 14–21, the roots of the algorithm go back to the great German mathematician Carl Freidrich Gauss in the early 1800s.

From the preceding expression we see that for each value of k, evaluating $X(k)$ requires N complex multiplications and $N - 1$ complex additions. Each complex multiplication, as in

$$(a + jb) \cdot (c + jd) = ac - bd + j(bc + ad)$$

requires four real multiplications and two real additions, and each complex addition involves two real additions. Thus for each k the evaluation of $X(k)$ requires $4N$ real multiplications and $2[N + (N - 1)] = 2(2N - 1)$ real additions. For a complete DFT, N values of $X(k)$ are needed, so the total requirement is $4N^2$ real multiplications and $2N(2N - 1)$ real additions. For large N, $2N(2N - 1) \approx 4N^2$.

Now let's try another way that leads to an FFT algorithm. Consider the two-point DFT

$$X(k) = \sum_{m=0}^{1} x(m){W_2}^{mk} = x(0){W_2}^{0k} + x(1){W_2}^{1k} \qquad k = 0, 1$$

$$\text{where } {W_2}^{0k} = 1 \quad \text{and} \quad {W_2}^{k} = e^{-j(2\pi/2)k} = e^{-j\pi k} = (-1)^k.$$

So, we can write

$$X(0) = x(0) + x(1) \qquad X(1) = x(0) - x(1).$$

These calculations can be represented by the signal flowgraph in Figure 10.11, which is known (using a bit of imagination) as a *butterfly*. This process can be extended to DFTs of length 4, 8, 16, 32, . . . ($N = 2^r$, where r is a positive integer). The general approach starting with an $N = 2^r$-point DFT is to decompose the N-point DFT into two $(N/2)$-point DFTs. The process continues by decomposing each $(N/2)$-point DFT into two $(N/4)$-point DFTs, and so on, until eventually reaching the situation of having $(N/2)$ two-point DFTs. The algorithm then consists of computing each of the $(N/2)$ two-point DFTs, recombining them into $(N/4)$ four-point DFTs, which are then recombined into $(N/8)$ eight-point DFTs, and so forth, until the complete DFT is evaluated. The result of this process for $N = 8$ is shown in Figure 10.12, where we notice the replication of the butterfly structure of Figure 10.11 in both the calculation of the two-point DFTs and the subsequent recombinations (or recompositions) into four- and eight-point DFTs.

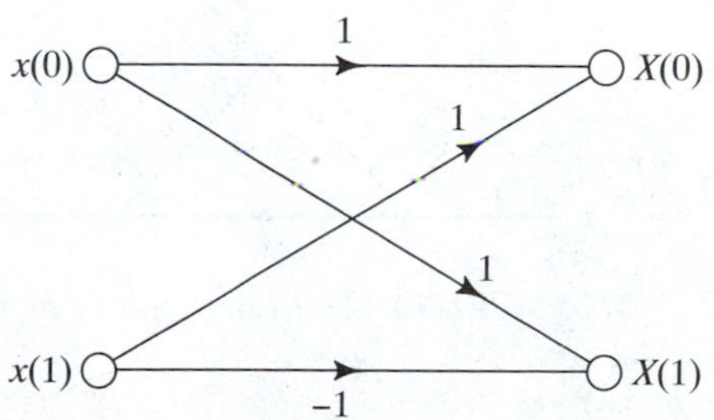

FIGURE 10.11 *Butterfly for a 2-point DFT*

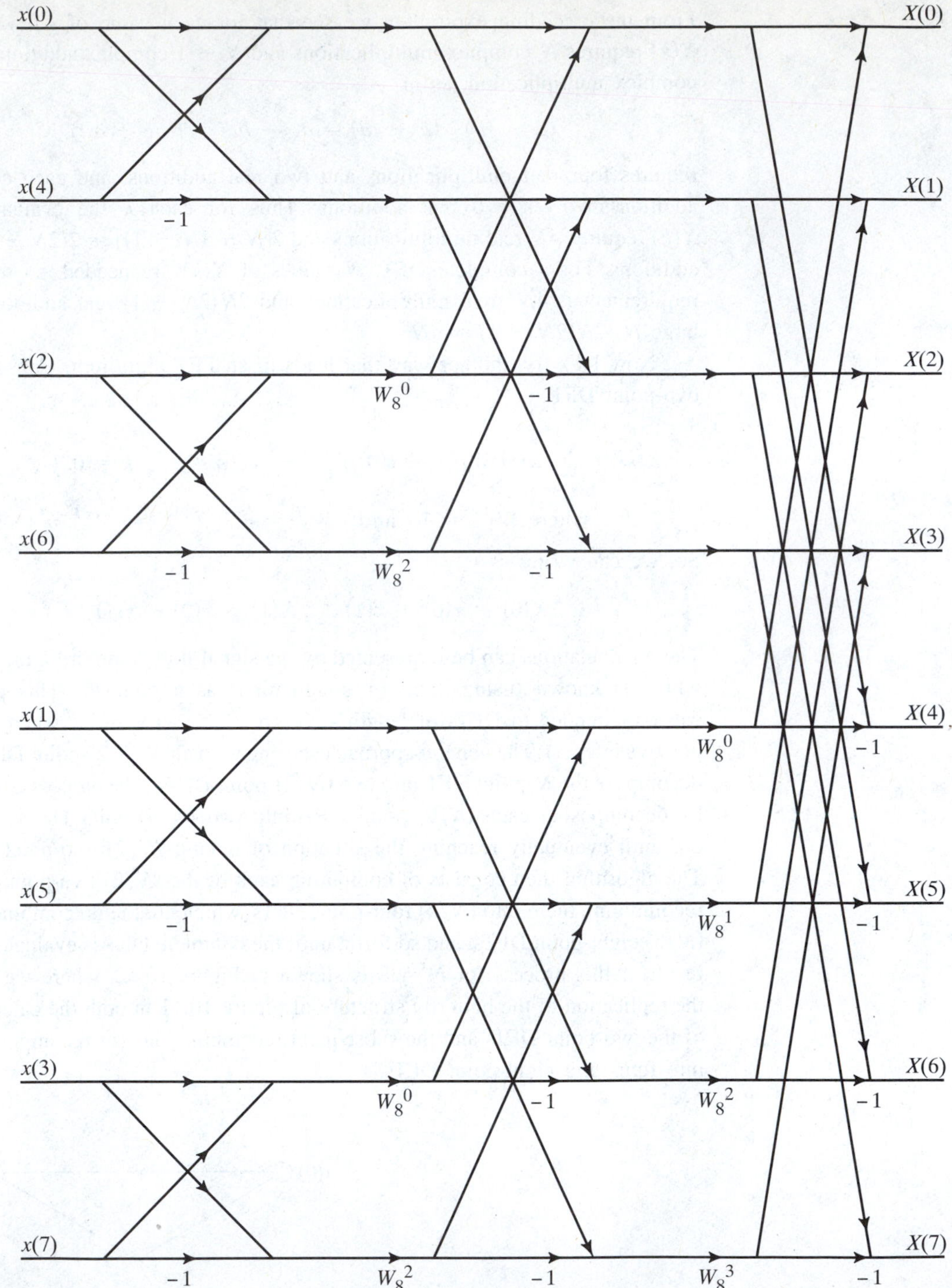

FIGURE 10.12 *Flowgraph of 8-point DFT by an FFT algorithm*

This version of the FFT algorithm, which is known as the radix-2, decimation-in-time FFT, is but one of many fast algorithms that have been developed for computing DFTs. Packaged programs are available for actually carrying out the computations, and MATLAB has one available as the function **fft.** For the algorithm just described, N must be an integer power of 2. If this is not the case, there are other fast algorithms that may be used. For example, in MATLAB a radix-2 FFT is used if N is an integer power of 2, whereas a mixed-radix algorithm is used if this is not the case.

The question remains as to what we've gained by using the FFT to compute the DFT. For an N-point FFT, where N is an integer power of 2, there are $\log_2 N$ stages of butterfly computations—one stage of two-point DFTs and $\log_2 N - 1$ stages of recompositions. Figure 10.12 illustrates this for $N = 8$, where there are three stages. Each stage has $N/2$ butterflies to evaluate, and each butterfly requires one complex multiplication and two complex additions. Table 10.1 illustrates the computational advantage provided by the FFT.

TABLE 10.1 *Computations for an N-point DFT*

N	*Real Multiplications*		*Real Additions*	
	DFT $4N^2$	*FFT* $2N \log_2 N$	*DFT* $2N(2N-1)$	*FFT* $3N \log_2 N$
2	16	4	12	6
4	64	16	56	24
8	256	48	240	72
16	1,024	128	992	192
32	4,096	320	4,032	480
64	16,384	768	16,256	1,152
128	65,536	1,792	65,280	2,638
256	262,144	4,096	261,632	6,144
512	1,048,576	9,216	1,047,552	13,824
1,024	4,194,304	20,480	4,192,256	30,720

The reason the FFT has such an advantage is that the ratio of DFT operations to FFT operations is proportional to $N/\log_2 N$, and N grows much faster as N increases than does $\log_2 N$. Of course, multiplications and additions are not the only operations needed to compute DFTs and FFTs, but they provide a reasonable indication of the relative computational speed of the two algorithms.

One application of FFTs that may be surprising is in evaluating the linear convolution of two finite-length sequences. This process is known as *fast convolution.* Suppose, for example, we have two N-point sequences $x_1(n)$ and $x_2(n)$, to be linearly convolved, and assume that N is a power of 2.

The resulting convolution will have $2N - 1$ points. Thus, to accomplish the convolution using FFTs (or DFTs), we need to zero pad the sequences to lengths of at least $2N - 1$. To keep the arithmetic simple and to enable use of a radix-2 FFT, we'll zero pad both sequences with N zeros to obtain sequences of length $2N$. If the zero-padded sequences are denoted by $x_1'(n)$ and $x_2'(n)$, the convolution evaluation proceeds as shown in Figure 10.13. The inverse FFT shown requires the same number of operations as an FFT, so focusing on multiplications only, we require $3 \cdot (2N/2) \cdot \log_2 2N$ complex multiplications to evaluate the FFTs and $2N$ more complex multiplications to evaluate the product of FFTs $X_1'(k)$ and $X_2'(k)$. Thus, a total of $3N \cdot \log_2 N + 2N$ complex multiplications, or $12N \cdot \log_2 N + 8N$ real multiplications, is needed for the linear convolution of two N-point sequences using FFTs. If the convolution is done directly by using the definition

$$x_3(n) = \sum_{m=-\infty}^{\infty} x_1(m)x_2(n-m)$$

then, assuming that the sequences $x_1(n)$ and $x_2(n)$ have real elements, evaluating the convolution requires that each of the N points of $x_1(n)$ be multiplied by the N points of $x_2(n)$, a total of N^2 real multiplications. If $N \geq 128$, the FFT has a computational advantage; the larger N becomes, the greater the FFT's advantage.[3]

In summary, the radix-2 FFT and other fast transform algorithms offer significant computational advantages over DFTs calculated from the definition. Nevertheless, the properties developed for DFTs all still apply; FFTs simply offer a fast way to compute DFTs.

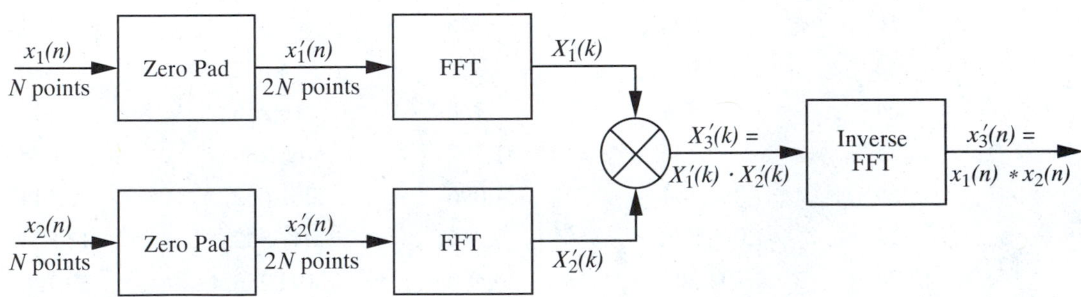

FIGURE 10.13 *Fast convolution using FFTs*

SOLVED EXAMPLES AND MATLAB APPLICATIONS

EXAMPLE 10.1
Finding a DFT from the DTFT

a. Find and plot the Discrete Time Fourier Transform (DTFT) of the four-term sequence $x(0) = x(1) = x(2) = x(3) = 1$.

[3] If the data are real, a slick trick allows one to find a $2N$-point FFT by using an N-point FFT (see Brigham [1], Chapter 9), further increasing the FFT advantage.

b. Sample the DTFT with $\theta = 2\pi k/N$ to find the DFT of the sequence $x(n)$. To keep things simple, use $N = 8$.

c. Use the DFT definition to find $X(0)$ and $X(4)$, and check the results against the values found in part (b).

d. Use MATLAB to find the DFT $X(k)$ and compare results with those found in part (b).

WHAT IF? Investigate the effects of changing N on the DFT of the sequence $x(n)$, where

$$x(n) = \begin{cases} 1, & 0 \le n \le 3 \\ 0, & 4 \le n \le N - 1. \end{cases} \blacksquare$$

Solution

a. From the definition of the DTFT,

$$X(e^{j\theta}) = \sum_{m=-\infty}^{\infty} x(m)e^{-jm\theta} = \sum_{m=0}^{3} 1 \cdot e^{-jm\theta} = \sum_{m=0}^{3} (e^{-j\theta})^m.$$

Using the finite geometric series formula, this result can be written as

$$\begin{aligned} X(e^{j\theta}) &= \frac{1 - e^{-j4\theta}}{1 - e^{-j\theta}} = \frac{e^{-j2\theta}[e^{j2\theta} - e^{-j2\theta}]}{e^{-j\theta/2}[e^{j\theta/2} - e^{-j\theta/2}]} \\ &= [\sin(2\theta)/\sin(\theta/2)]e^{-j3\theta/2}. \end{aligned}$$

Notice that this is also the DTFT of

$$x(n) = \begin{cases} 1, & 0 \le n \le 3 \\ 0, & 4 \le n \le N - 1 \end{cases}$$

for arbitrary N. We could write a MATLAB script to evaluate this function. An alternative, however, is to use the MATLAB function **freqz** with $X(z) = (1 + z^{-1} + z^{-2} + z^{-3})/1$. The result is shown in Figure E10.1a.

Comment: The jumps in the DTFT phase angle we observe are caused by the changing sign of the $\sin(2\theta)$ term at $\theta = \pi/2$, π, and $3\pi/2$.

b. For $N = 8$ we sample $X(e^{j\theta})$ at $\theta = 0, \pi/4, \pi/2, \ldots, 3\pi/2, 7\pi/4$, which gives

$$\begin{aligned} X(k) &= X(e^{j2\pi k/8}) = [\sin(4\pi k/8)/\sin(\pi k/8)]e^{-j(3\pi k/8)}, \\ k &= 0, 1, \ldots, 7. \end{aligned}$$

The resulting values are tabulated next.

EXAMPLES

k	$\|X(k)\|$	*Angle of* $X(k)$	$\text{Re}[X(k)]$	$\text{Im}[X(k)]$
0	4.0	0.0	4.0	0.0
1	2.613	−1.178	1.0	−2.414
2	0.0	0.0	0.0	0.0
3	1.082	−0.393	1.0	−0.414
4	0.0	0.0	0.0	0.0
5	1.082	0.393	1.0	0.414
6	0.0	0.0	0.0	0.0
7	2.613	1.178	1.0	2.414

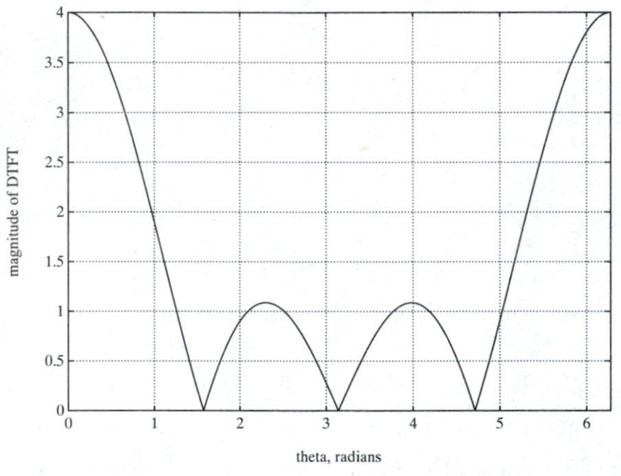

The DTFT magnitude

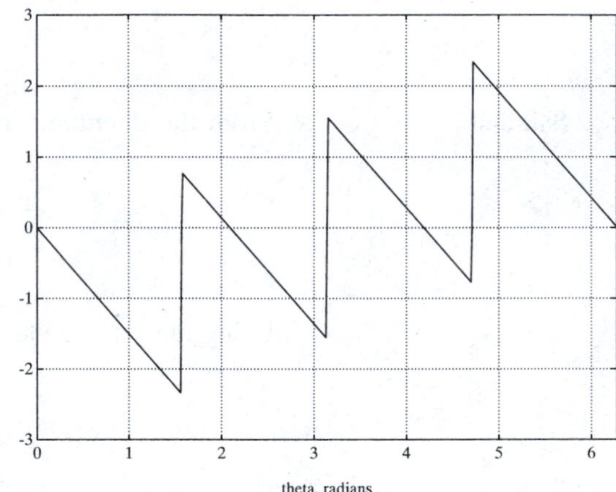

The DTFT phase

FIGURE E10.1a *The Discrete Time Fourier Transform of the sequence* $x(n) = \{1, 1, 1, 1\}$

c. The DFT definition is

$$X(k) = \sum_{n=0}^{N-1} x(n)e^{-j(2\pi/N)nk}, \qquad k = 0, 1, \ldots, N-1 .$$

With $N = 8$ and $x(n)$ as specified,

$$X(0) = \sum_{n=0}^{3} 1 = 4 \quad \text{and} \quad X(4) = \sum_{n=0}^{3} 1e^{-j(n\pi)} = 1 - 1 + 1 - 1 = 0 .$$

d. The MATLAB script follows and the resulting plot is in Figure E10.1b.

MATLAB Script

```
% E10_1 Using the function fft to calculate DFTs

%E10_1b An 8-point DFT
n=[0 1 2 3 4 5 6 7];
```

```
xn=[1 1 1 1 0 0 0 0];                    % x(n)
Xk=fft(xn);                              % call fft to get X(k)
MagXk=abs(Xk);
PhaseXk=angle(Xk);
%...plotting statements and pause
```

The values in these plots agree with those obtained previously in parts (b) and (c).

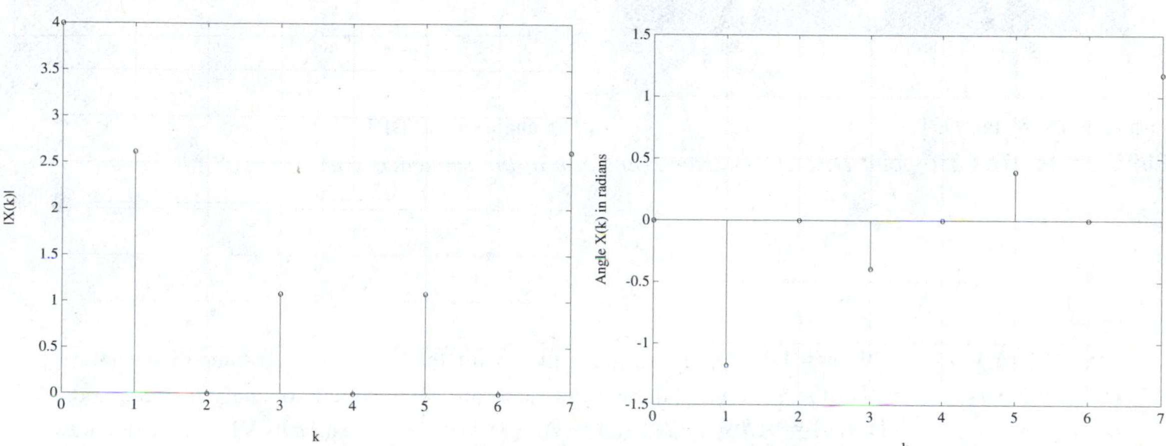

The magnitude of the DFT The angle of the DFT

FIGURE E10.1b *The 8-point Discrete Fourier Transform of the sequence $x(n)$*

WHAT IF? A possible script is given next, with the results for $N = 256$ given in Figure E10.1c page 570, where we notice that the envelopes of the plots $|X(k)|$ and $\angle X(k)$ versus k remain the same as before. The envelope of the DFT—that is, the DTFT—remains the same, but the two DFTs are quite different. ■

MATLAB Script

```
%E10_1c A 256-point DFT
n=[0:1:255];
k=n;
x=zeros(size(n));                        % zero the x vector
x(1:4)=ones(1,4);                        % set nonzero values to 1
Xk=fft(x);                               % call fft to get X(k)
MagXk=abs(Xk);
PhaseXk=angle(Xk);
%...plotting statements
```

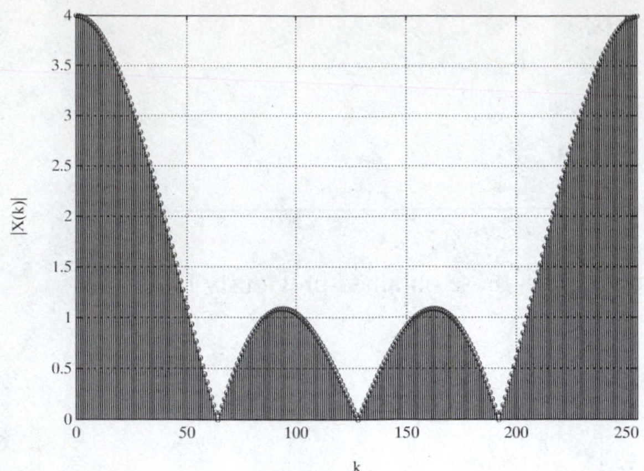

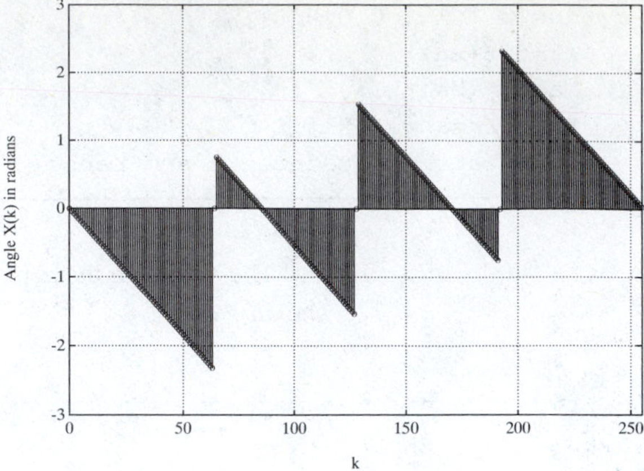

The magnitude of the DFT The angle of the DFT

FIGURE E10.1c *The 256-point Discrete Fourier Transform of the sequence $x(n)$*

EXAMPLE 10.2
DFT of Three Important Sequences

Although DFT pairs are generally evaluated numerically rather than analytically, the following three sequences are well suited to analytical methods: (1) $x(n) = A\delta(n)$; (2) $x(n) = A$; (3) $x(n) = A\cos(2\pi n/N)$. All sequences are defined for $0 \le n \le N-1$.

a. Find analytical expressions for the DFTs.

b. Use the MATLAB function **fft** to find the DFTs for values of A and N of your choice. Use at least two different values for N.

WHAT IF? a. Make use of the results of (a), part (3), to find the DFT of the cosine sequence $x(n) = A\cos(2 \cdot 2\pi n/N)$. Also find the DFT of $x(n) = A\cos(3 \cdot 2\pi n/N)$. b. Suppose that the cosine sequence in part (3) is now changed to $x(n) = A\cos(2\pi n/M)$, where M, an integer equal to the period of the sequence, is not a submultiple of the implicit period, N, of the DFT; that is, $M \neq N/L$, where L is an integer. Use MATLAB to investigate what happens. ■

Solution

a.

(1) Using the DFT definition we have

$$X(k) = \sum_{n=0}^{N-1} A\delta(n)e^{-j(2\pi/N)nk} = A, \quad 0 \le k \le N-1.$$

In other words, the DFT of a unit sample sequence contains equal amounts of all frequencies. Also, notice that the DFT does not depend on the value selected for N.

(2) Proceeding as before,

$$X(k) = \sum_{n=0}^{N-1} Ae^{-j(2\pi/N)nk} = A \sum_{n=0}^{N-1} (e^{-j(2\pi/N)k})^n$$

$$= A\left[\frac{1 - e^{-j2\pi k}}{1 - e^{-j2\pi k/N}}\right], \qquad 0 \le k \le N-1,$$

where we have used the finite geometric sum formula. For $k = 0$, we have 0/0, which is an indeterminate form. By l'Hôpital's formula, or by returning to the summation and letting $k = 0$, we obtain $X(0) = AN$. For $k \neq 0$, $X(k) = 0$, because the numerator is zero and the denominator is not. We can, therefore, write that $X(k) = AN\delta(k)$. In this case we observe that the DFT of a constant contains only a zero-frequency (DC) component and that the amplitude of this DC component depends on the value selected for N. The duality property can be used to illustrate the results of (1) and (2), as follows.

DFT of sample sequence

DFT of a constant

$$A\delta(n) \Leftrightarrow A$$

$$A \Leftrightarrow NA\delta(k)$$

(3) Writing the cosine sequence as $A\cos(2\pi n/N) = 0.5A[e^{j2\pi n/N} + e^{-j2\pi n/N}]$, we have

$$X(k) = A\sum_{n=0}^{N-1}\left[\frac{e^{j(2\pi/N)n} + e^{-j(2\pi/N)n}}{2}\right]e^{-j(2\pi/N)nk}$$

$$= \frac{A}{2}\sum_{n=0}^{N-1} e^{j(2\pi/N)n[1-k]} + \frac{A}{2}\sum_{n=0}^{N-1} e^{-j(2\pi/N)n[1+k]}$$

$$= \frac{A}{2}\sum_{n=0}^{N-1} (e^{j(2\pi/N)[1-k]})^n + \frac{A}{2}\sum_{n=0}^{N-1} (e^{-j(2\pi/N)[1+k]})^n,$$

$k = 0, 1, \ldots, N-1$.

Again using the finite geometric series formula, we obtain

$$X(k) = \frac{A}{2}\left[\frac{1 - e^{j2\pi[1-k]}}{1 - e^{j2\pi[1-k]/N}}\right] + \frac{A}{2}\left[\frac{1 - e^{j2\pi[1+k]}}{1 - e^{j2\pi[1+k]/N}}\right].$$

The first term is 0 when $k \neq 1$ and equal to $NA/2$ when $k = 1$, whereas the second term equals 0 when $k \neq N-1$ and equals $NA/2$ when $k = N-1$. Thus the final result is

DFT of cosine

$$X(k) = \frac{NA}{2}[\delta(k-1) + \delta(k-[N-1])].$$

The unit sample at $k = 1$ is the positive frequency component, and the unit sample at $k = N-1$ represents the negative frequency image.

b.

(1) Using **fft** with $N = 64$ and $A = 1.5$ we have from file E10_2 on the CLS disk, $X(k) = 1.5$, $k = 0, 1, \ldots, 63$.

EXAMPLES

(2) For a 128-point constant sequence of amplitude 2 we see from file E10_2 that the amplitude of the DFT at $k = 0$ is $NA = 128 \cdot 2 = 256$ and is zero for all other values of k as predicted by the analytical results.

(3) For the cosine sequence $x(n) = 5\cos(2\pi n/32)$ the results from file E10_2 show that the nonzero DFT samples have the values $NA/2 = (32)(5)/2 = 80$ at $k = 1$ and $k = 31$ as predicted by the analytical result

$$X(k) = \frac{NA}{2}[\delta(k-1) + \delta(k - [N-1])].$$

Comment: Notice that the DFTs in all three cases are real sequences and, hence, the plots of these DFTs that also may be found in file E10_2 include magnitudes only.

WHAT IF? a. Writing the cosine function as $A\cos(2 \cdot 2\pi n/N) = 0.5A[e^{j2\cdot 2\pi n/N} + e^{-j2\cdot 2\pi n/N}]$ we have

$$X(k) = A\sum_{n=0}^{N-1}\left[\frac{e^{j(2\cdot 2\pi/N)n} + e^{-j(2\cdot\pi/N)n}}{2}\right]e^{-j(2\pi/N)nk}$$

$$= \frac{A}{2}\sum_{n=0}^{N-1} e^{j(2\pi/N)n[2-k]} + \frac{A}{2}\sum_{n=0}^{N-1} e^{-j(2\pi/N)n[2+k]}$$

$$= \frac{A}{2}\sum_{n=0}^{N-1} (e^{j(2\pi/N)[2-k]})^n + \frac{A}{2}\sum_{n=0}^{N-1} (e^{-j(2\pi/N)[2+k]})^n.$$

Using the finite geometric series formula once again, we obtain

$$X(k) = \frac{A}{2}\left[\frac{1 - e^{j2\pi[2-k]}}{1 - e^{j2\pi[2-k]/N}}\right] + \frac{A}{2}\left[\frac{1 - e^{j2\pi[2+k]}}{1 - e^{j2\pi[2+k]/N}}\right].$$

The first term is 0 when $k \neq 2$ and equal to $NA/2$ when $k = 2$, whereas the second term equals 0 when $k \neq N-2$ and equals $NA/2$ when $k = N-2$. Thus the final result is

$$X(k) = \frac{NA}{2}[\delta(k-2) + \delta(k - [N-2])].$$

The unit sample at $k = 2$ is the positive frequency component, and the unit sample at $k = N-2$ represents the negative frequency image. For $x(n) = A\cos(3 \cdot 2\pi n/N)$,

$$X(k) = \frac{NA}{2}[\delta(k-3) + \delta(k - [N-3])].$$

b. Again we build on the results of (3) of part (b) by letting $M = 32$ and finding the DFTs for $N = 48$ and 57. The results are shown in Figures E10.2a and b.

Closer inspection of the DFTs in Figures E10.2a and b reveals some interesting characteristics. We note that even though the sequence contained only the single frequency $\pi/16$, the 48- and 57-point DFTs show nonzero components for several frequencies. This is a phenomenon known as *leakage*—the energy at the single frequency in the sequence spills over, or leaks, into the other frequencies. This is actually a consequence of the fact that the numbers of points in the DFTs are not integer multiples of the period ($M = 32$) of the sinusoidal sequence; hence the frequency $\pi/16$ is not among the frequencies represented by the DFT. We also observe that the phases of the 48- and 57-point DFTs are not zero. ■

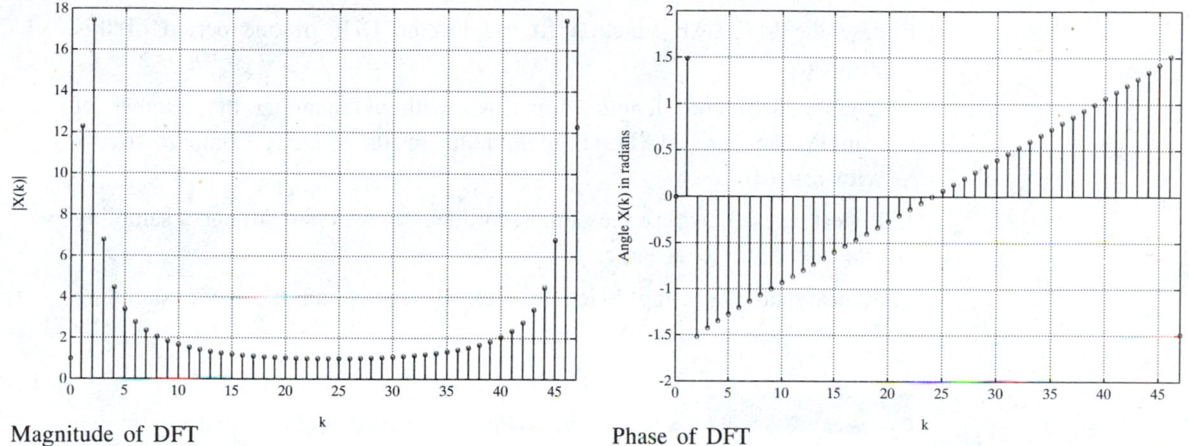

Magnitude of DFT Phase of DFT

FIGURE E10.2a *The 48-point DFT of* $x(n) = \cos(2\pi n/32)$

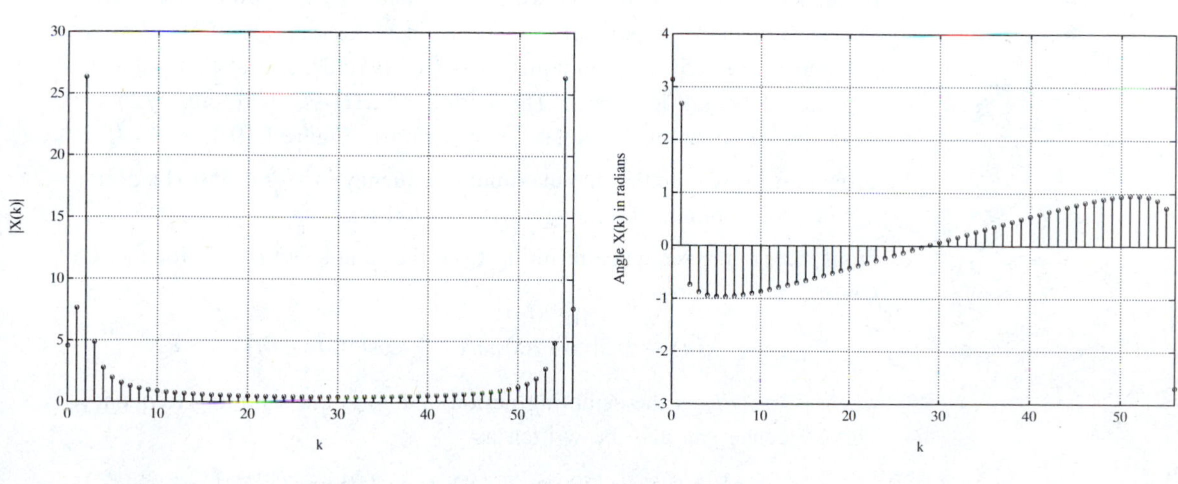

Magnitude of DFT Phase of DFT

FIGURE E10.2b *The 57-point DFT of* $x(n) = \cos(2\pi n/32)$

EXAMPLES

EXAMPLE 10.3 *Record Length, Resolution, and Sampling Frequency*

A periodic analog sinusoidal signal at a frequency of 300 Hz is modulated by another sinusoid having a frequency of 50 Hz and the resulting waveform is $x(t) = \cos(600\pi t) \cdot \cos(100\pi t)$.

a. Determine the period of $x(t)$ and plot one period of $x(t)$.

b. Determine the minimum acceptable sampling frequency for $x(t)$.

c. Using a sampling frequency of four times the minimum sampling frequency from part (b) and a record length of one period from part (a), find an expression for the sampled sequence $x(n)$. Plot this sequence.

d. Use the MATLAB function **fft** to plot the DFT of one period of this sequence.

e. Double the record length of part (c) without changing the number of samples, and use MATLAB to find the resulting DFT. Compare results with part (d).

f. Repeat part (e), but use the same record length as in part (d) and a sampling rate that is twice as large.

g. Repeat part (e) using twice the record length and the same sampling frequency as in part (d).

WHAT IF? Suppose we change the record length so that it is not an integral number of periods of the analog signal. You could start by making $T_0 = 1.5$ times the period of the analog signal and find and plot the DFT. Explain the results. ■

Solution

a. Using the trigonometric identity $\cos(\alpha) \cdot \cos(\beta) = 0.5[\cos(\alpha + \beta) + \cos(\alpha - \beta)]$, we have $x(t) = 0.5[\cos(350 \cdot 2\pi t) + \cos(250 \cdot 2\pi t)]$. The frequencies contained in $x(t)$ are 350 Hz and 250 Hz, which are related to one another by a ratio of 7:5. Thus, in seven periods of the 350-Hz component, the 250-Hz component has five periods, and at this point the signal $x(t)$ begins to repeat. The period of $x(t)$ is, therefore, $5/250 = 7/350 = 20$ ms, with the plot for one period in Figure E10.3a.

b. From part (a) we see that the maximum frequency in $x(t)$ is 350 Hz, giving a minimum sampling frequency of $f_s > 700$ Hz.

c. From part (a) the sequence resulting from the samples of $x(t)$ is found with $t = nT$ as

$$x(n) = 0.5[\cos(700\pi nT) + \cos(500\pi nT)],$$

where $T = 1/f_s$ is the sampling period and f_s is the sampling frequency. This sequence can also be written as

$$\begin{aligned} x(n) &= 0.5[\cos(700\pi nT_0/N) + \cos(500\pi nT_0/N)] \\ &= 0.5[\cos(700\pi n/N\Delta f) + \cos(500\pi n/N\Delta f)]. \end{aligned}$$

CHAPTER 10

The sampling frequency is to be $f_s = 4 \cdot 700$ Hz $= 2800$ Hz. The record length is $T_0 = 20$ ms, which corresponds to a frequency resolution of $\Delta f = 1/(20 \cdot 10^{-3}) = 50$ Hz, and $N = f_s \cdot T_0 = 2800 \cdot 20 \cdot 10^{-3} = 56$ samples. This leads to the sequence

$$x(n) = 0.5[\cos(n\pi/4) + \cos(n5\pi/28)]$$

with the plot for one period given in Figure E10.3b.

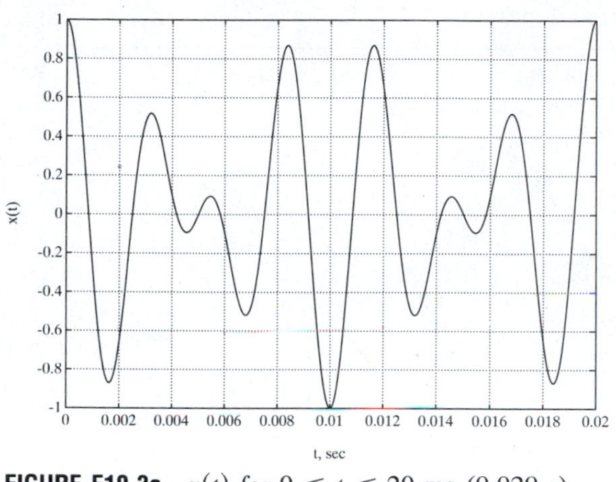

FIGURE E10.3a $x(t)$ *for* $0 \leq t \leq 20$ ms (0.020 s)

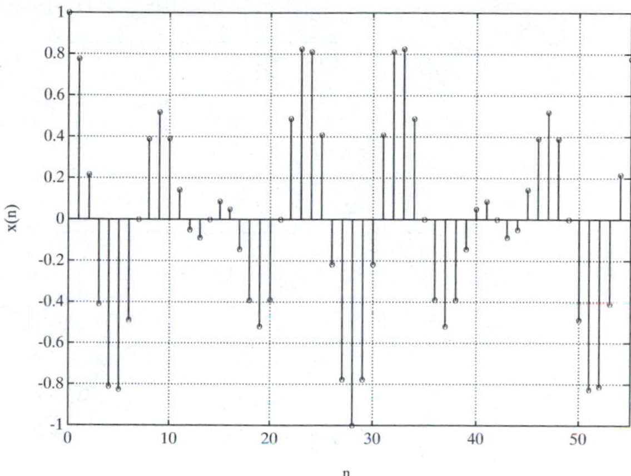

FIGURE E10.3b *Sample values of* $x(t)$ *for* $N = 56$

Comment: It is useful to display an analog signal $x(t)$ and its samples $x(n)$ on the same axes. We can do this by scaling the analog signal with a new variable that we call *tbar*. The original analog signal is given by $x(t) = \cos(600\pi t) \cdot \cos(100\pi t)$, and when $t = T_0$ the scaled time variable *tbar* is to be equal to the number of samples N—that is, $tbar = N$. Thus the time-scaled representation is

$$x(t) = \cos(600\alpha\pi\bar{t}) \cdot \cos(100\alpha\pi\bar{t})$$

and in the MATLAB script this becomes

$$xt = \cos(600\alpha\pi tbar) \cdot \cos(100\alpha\pi tbar)\,.$$

To find α we simply form the equalities

$$600\pi T_0 = 600\alpha\pi N \quad \text{and} \quad 100\pi T_0 = 100\alpha\pi N$$

both of which lead to the relationship $\alpha = T_0/N$. This gives $\alpha = 1/2800$, as shown in the MATLAB equation for *xt* in the script on the next page. The signal and its samples are plotted in Figure E10.3c.

EXAMPLES

```
%E10_3 DFT of sampled analog signals

%E10_3c x(t) and its sampled values
tbar=0.0:0.5:56;
n=[0:1:55];
xn=0.5*cos(pi*n/4)+0.5*cos(pi*n*5/28);
% Scale x(t) to fit on the same time axis as x(n)
xt=cos((600/2800)*pi*tbar).*cos((100/2800)*pi*tbar);
%...plotting statements and pause
```

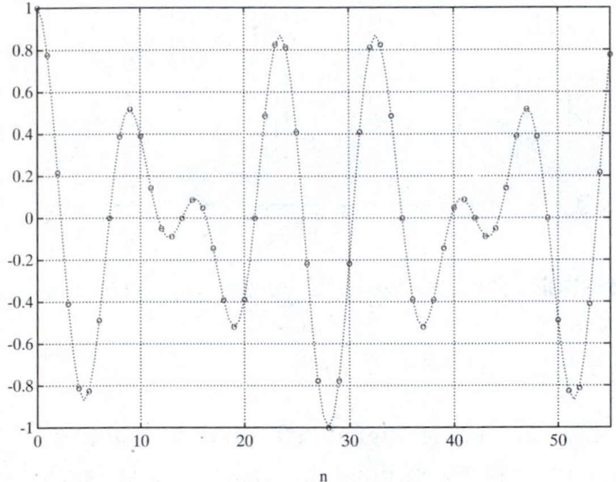

FIGURE E10.3c *$x(t)$ and its sample values*

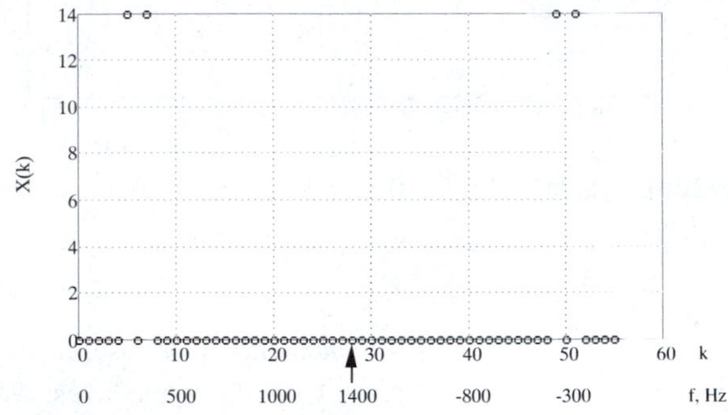

FIGURE E10.3d *The DFT $X(k)$ for a 20-ms record length and $N = 56$*

d. The DFT is shown in Figure E10.3d, where we notice that

$$X(k) = 14[\delta(k-5) + \delta(k-7) + \delta(k-49) + \delta(k-51)].$$

This result appears to be reasonable, because in Example 10.2 we found the following pairs for the N-point DFTs of sequences, each with a period that is an integer multiple of N:

$$A\cos(1 \cdot 2\pi n/N) \Leftrightarrow [NA/2]\{\delta(k-1) + \delta(k-[N-1])\}$$
$$A\cos(2 \cdot 2\pi n/N) \Leftrightarrow [NA/2]\{\delta(k-2) + \delta(k-[N-2])\}$$
$$A\cos(3 \cdot 2\pi n/N) \Leftrightarrow [NA/2]\{\delta(k-3) + \delta(k-[N-3])\}.$$

Putting the given sequence into a comparable form gives

$$\begin{aligned} x(n) &= 0.5[\cos(n\pi/4) + \cos(n5\pi/28)] \\ &= 0.5[\cos(7 \cdot 2\pi n/56) + \cos(5 \cdot 2\pi n/56)]. \end{aligned}$$

We see that $NA/2 = 56(0.5)/2 = 14$ and the DFT samples of this amplitude should appear at $k = 5$, $k = 56 - 5 = 51$, $k = 7$, and $k = 56 - 7 = 49$, as indeed they do. We also observe, from the analog frequency scale that's been added to the MATLAB plot, that the frequencies present in the DFT are 250 and 350 Hz, and their images are -250 and -350 Hz.

The derivation of the N-point DFT for the general sequence $v(n) = A\cos(Ln\pi/M)$, where L, n, and M are integers and N is equal to an integer multiple of one period of $v(n)$, is the focus of Problem P10.23.

e. By doubling the record length to $T_0 = 40$ ms and retaining the same number of samples ($N = 56$) we have

$$\begin{aligned} x(n) &= 0.5[\cos(14 \cdot 2\pi n/56) + \cos(10 \cdot 2\pi n/56)] \\ &= 0.5[\cos(n\pi/2) + \cos(5n\pi/14)]. \end{aligned}$$

The MATLAB plots of $x(n)$ and the corresponding DFT $X(k)$ are found in file E10_3 on the CLS disk. Notice that doubling the record length and retaining the same number of samples as in part (d) means that the frequency resolution becomes halved to $\Delta f = 1/T_0 = 25$ Hz, which means that $k = 10$ represents $10 \cdot 25 = 250$ Hz, $k = 14$ represents $14 \cdot 25 = 350$ Hz, and that the maximum frequency represented in the DFT is now 700 Hz, rather than the 1400 Hz of part (d).

f. With $T_0 = 20$ ms and $N = 112$, the sequence of samples is

$$\begin{aligned} x(n) &= 0.5[\cos(7 \cdot 2\pi n/112) + \cos(5 \cdot 2\pi n/112)] \\ &= 0.5[\cos(n\pi/8) + \cos(5n\pi/56)]. \end{aligned}$$

In the file E10_3 are the MATLAB plots of the signal $x(t)$, the sequence of samples $x(n)$, and the 112-point DFT $X(k)$. Again, there are lines at ± 250 Hz ($k = 5$ and 106) and ± 350 Hz ($k = 7$ and 104) and a frequency resolution of 50 Hz, but a maximum frequency represented of 2800 Hz corresponding to a sampling frequency of 5600 Hz.

g. Now, with a record length of $T_0 = 40$ ms and a sampling frequency of $f_s = 2800$ Hz, which together imply $N = 112$ samples, the sequence $x(n)$ is given by

$$\begin{aligned} x(n) &= 0.5[\cos(14 \cdot 2\pi n/112) + \cos(10 \cdot 2\pi n/112)] \\ &= 0.5[\cos(n\pi/4) + \cos(n5\pi/28)]. \end{aligned}$$

Here we have a frequency resolution of $\Delta f = 25$ Hz and a maximum frequency of 1400 Hz. As found in file E10_3, there are frequency lines at ± 250 Hz and ± 350 Hz.

Comment: The following table summarizes our observations in parts (d) through (g).

Part	N	f_s	T_0	Δf	f_{max}
d	56	2800 Hz	20 ms	50 Hz	1400 Hz
e	56	1400 Hz	40 ms	25 Hz	700 Hz
f	112	5600 Hz	20 ms	50 Hz	2800 Hz
g	112	2800 Hz	40 ms	25 Hz	1400 Hz

WHAT IF? Here we have a record length of 30 ms and, again, 56 samples. The analog signal and its samples are shown, along with the resulting DFT magnitude, in Figure E10.3e. We observe that the 30-ms record contains $1\frac{1}{2}$ periods of $x(t)$. Notice that with this record length, the frequency resolution is $\Delta f = 1/T_0 = 33.3$ Hz. With this frequency resolution the frequencies 250 Hz and 350 Hz are not contained in the DFT. The DFT frequencies nearest 250 Hz are 233 Hz and 267 Hz, and those closest to 350 Hz are 333 Hz and 367 Hz. It is at these frequencies that the largest components are present, but notice that there are also smaller amplitudes present at the other frequencies. What we are seeing here is the leakage effect. We have shown only the magnitude of the DFT in Figure E10.3e, but the phase is not zero in this case. ■

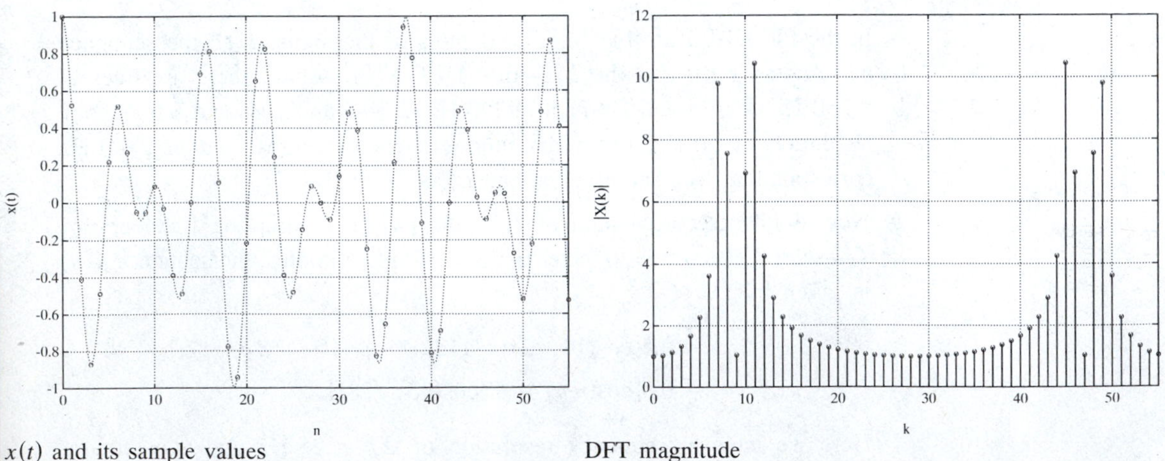

FIGURE E10.3e *$x(t)$, $x(n)$, and $|X(k)|$ for a 30-ms record length and $N = 56$*

EXAMPLE 10.4
Convolution of Two Sequences

Two sequences are described by

$$x(n) = \begin{cases} \cos(21\pi n/56) + \cos(15\pi n/56), & 0 \le n \le 25 \\ 0, & \text{elsewhere} \end{cases}$$

$$y(n) = \begin{cases} 0.1n, & 0 \le n \le 15 \\ 0, & \text{elsewhere} \end{cases}$$

a. Determine the maximum number of nonzero sample values for the (linear) convolution $z(n) = x(n) * y(n)$.

b. Using the MATLAB function **fft** with the number of points equal to the number found in part (a), find and plot the sequence $z(n)$.

c. Repeat part (b) with the number of points equal to the smallest power of 2 that can be used to evaluate the convolution. Compare the results to those of part (b).

d. Use **conv** to determine $z(n)$. Compare the results with parts (b) and (c).

e. A partial table of values of $x(n)$ and $y(n)$ follows. Check a few values of $z(n)$ by hand.

$$x(n) = \{\ldots \quad 0 \quad 2.0 \quad 1.049 \quad -0.819$$
$$\ldots \quad 0.493 \quad -0.778 \quad -0.961 \quad 0 \quad 0 \quad \ldots\}$$
$$y(n) = \{\ldots \quad 0 \quad 0 \quad 0.1 \quad 0.2 \quad 0.3$$
$$\ldots \quad 1.3 \quad 1.4 \quad 1.5 \quad 0 \quad 0 \quad \ldots\}$$

Solution

a. $x(n)$ and $y(n)$ have intervals of 26 and 16 points containing all the nonzero values. (Note that even though $y(0) = 0$, we will not attempt to make use of this fact.) Thus, the maximum number of nonzero points in the linear convolution of $x(n)$ and $y(n)$ is $26 + 16 - 1 = 41$.

b. The MATLAB script for plotting the sequences (Figure E10.4a, on page 580) and using **fft** to find $z(n)$ (Figure E10.4b) is shown here.

MATLAB Script

```
%E10_4 Linear convolution using the function fft

%E10_4a and b The original sequences x(n) and y(n) and the
%convolution z(n)
n=[0:1:25];
xn=cos(21*pi*n/56)+cos(15*pi*n/56);
r=[0:1:15];
yn=0.1*r;
% append zeros to yn for plotting purposes
lendif=length(xn)-length(yn);
ynprime=[yn zeros(1,lendif)];
lenzn=length(xn)+length(yn)-1; % length of convolution
```

Continues

```
nr=[0:1:lenzn-1]; % time sample vector for plotting
Xk=fft(xn,lenzn);
Yk=fft(yn,lenzn);
Zk=Xk.*Yk;
zn=ifft(Zk,lenzn);
%...plotting statements and pause
```

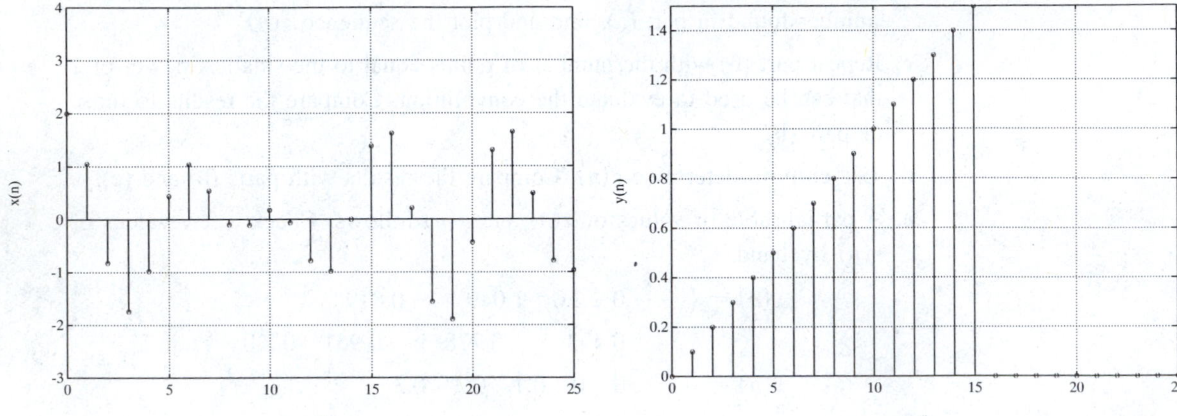

The sequence $x(n)$ The sequence $y(n)$

FIGURE E10.4a *The sequences $x(n)$ and $y(n)$ for Example 10.4*

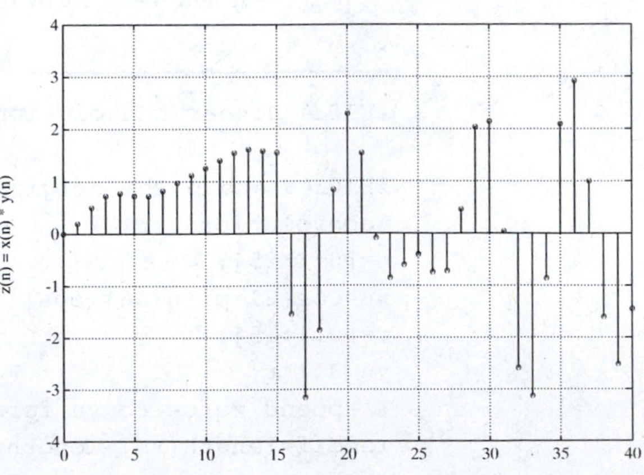

FIGURE E10.4b *The convolution $z(n)$ obtained using 41-point DFTs*

CHAPTER 10

c. Here we use 64 points for the DFT, which enables MATLAB to use a radix-2 FFT algorithm, though this is transparent to us as users. Figure E10.4c shows the resulting convolution, and we observe the same values as shown on the plot of Figure E10.4b. We also notice the trailing zeros for $41 \leq n$. This is a consequence of the zero padding done when using the **fft** function to fill out the 64 points specified. Notice that we can always use a DFT/FFT with more points than are needed to evaluate a linear convolution. It doesn't work the other way around, however. If we use fewer than the minimum number of points necessary, we generate erroneous results because of the circular convolution effects of the algorithm.

d. The MATLAB script is shown below, and the resulting plot is in Figure E10.4d.

e. Given the table of values, it is relatively easy to calculate a few points, especially at the beginning and end of the interval, where the convolution is nonzero. This is accomplished by folding and shifting one of the

MATLAB Script

```
%E10_4d Convolution using conv and 41 points
n=[0:1:25];
xn=cos(21*pi*n/56)+cos(15*pi*n/56);
r=[0:1:15];
yn=0.1*r;
zn=conv(xn, yn);
%...plotting statements and pause
```

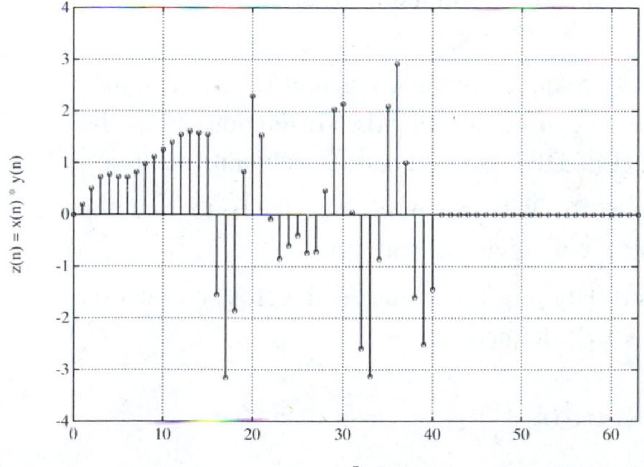

FIGURE E10.4c *The convolution $z(n)$ obtained using 64-point DFTs*

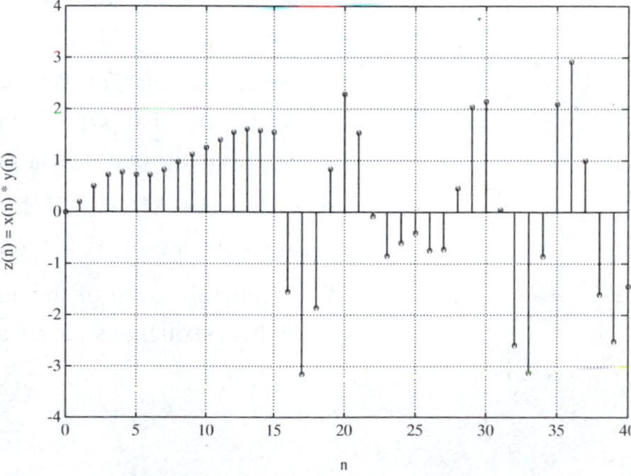

FIGURE E10.4d *The convolution $z(n)$ obtained using the MATLAB function* **conv**

EXAMPLES

sequences, forming the product sequence and summing. For example, at $n = 0$ we have (folding $x(n)$ and shifting it 0 units) $z(n) = 0 \times 0 = 0$. At $n = 1$, $z(n) = 0.1 \times 2.0 = 0.2$, and for $n = 2$, $z(n) = 0.1 \times 1.049 + 0.2 \times 2.0 = 0.5049, \ldots$. For $n = 39$, $z(n) = 1.5 \times (-0.778) + 1.4 \times (-0.961) = -2.512$, and for $n = 40$, $z(n) = 1.5 \times (-0.961) = -1.442$. Inspection of the corresponding values generated by MATLAB reveals that they are the same.

EXAMPLE 10.5
Correlation

Correlation functions are very important in communications, radar detection, and control systems. In this problem we introduce correlation and illustrate its use. The sequence

$$x(n) = \sin(\pi n/16)[u(n) - u(n - 64)]$$

is plotted in Figure E10.5a. The autocorrelation of $x(n)$ is defined as

$$R_{xx}(p) = \sum_{n=-\infty}^{\infty} x(n)x(n + p), \quad -\infty \le p \le \infty$$

where p is the relative shift and the autocorrelation is defined in terms of this shift. Notice the similarity of this definition to (linear) convolution.

a. Use the California function **ksxcorr** to determine and plot the autocorrelation of the sequence $x(n)$.

b. Find an expression for the autocorrelation of $x(n)$ in terms of the convolution of $x(n)$ with itself.

c. Using the relationship found in part (b) and the MATLAB function **conv,** determine the autocorrelation of $x(n)$. Compare results with those of part (a).

d. Since we already know how to do convolution using DFTs, it stands to reason that DFTs can also be used to perform correlation. Find the relationship between a product of DFTs and the autocorrelation function.

e. Using the MATLAB functions **fft, ifft,** and **conj,** use DFTs to find the autocorrelation of the sequence $x(n)$ given in part (a).

f. A generalization of the autocorrelation of a sequence is the cross correlation of two sequences $x_1(n)$ and $x_2(n)$, defined as

$$R_{x_1x_2}(p) = \sum_{n=-\infty}^{\infty} x_1(n)x_2(n + p), \quad -\infty \le p \le \infty$$

where, as before, p is the relative shift. Use **xcorr** or the California function **ksxcorr** to determine and plot the cross correlation of the sequences $x_1(n) = \{2\ 2\ 2\ 2\ 0\}$ and $x_2(n) = \{0\ 1\ 3\ 2\ 0\}$.

Solution **a.** The script is shown next, and the MATLAB output is in Figure E10.5b. Notice in the script that the vectors **m** and **p** have been defined to facilitate the plotting. It is not difficult to show that the autocorrelation has its largest value at $p = 0$ and that it is an even function of p. (See Problem P10.22.)

MATLAB Script

```
%E10_5 Correlation

%E10_5ab A sequence and its autocorrelation
n=[0:1:63];
xn=sin(pi*n/16);
% append zeros to xn for plotting purposes
xnprime=[zeros(1,20) xn zeros(1,20)];
m=[-20:1:83]; % time sample vector for plotting xn
p=[-63:1:63]; % time sample vector for plotting
autocorrelation
znn=ksxcorr(xn);  % call California function ksxcorr
%...plotting statements and pause
```

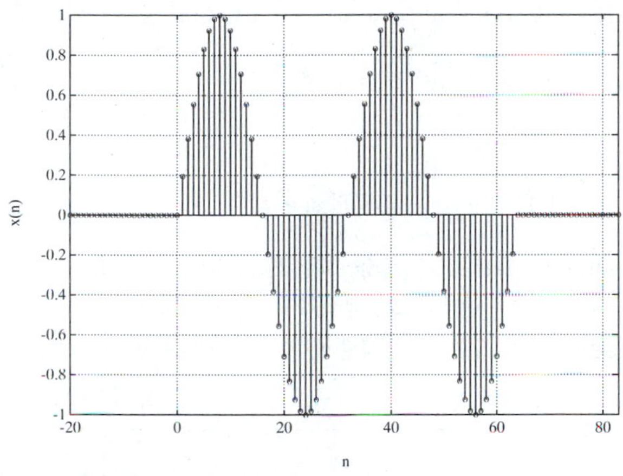

FIGURE E10.5a *The sequence x(n)*

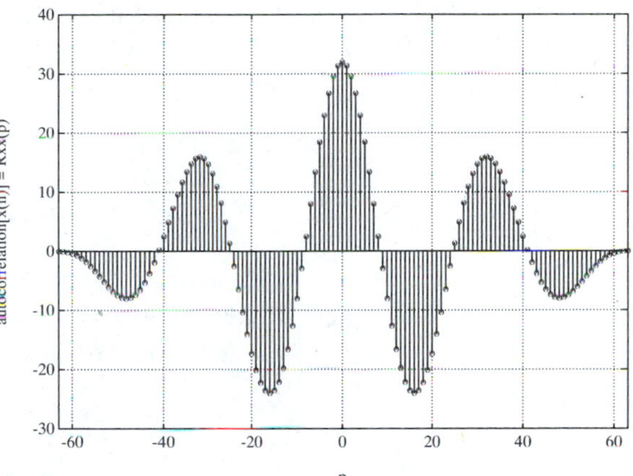

FIGURE E10.5b *Autocorrelation sequence of x(n)*

b. Starting with the convolution definition,

$$x(p) * x(p) = \sum_{m=-\infty}^{\infty} x(m)x(p - m), \qquad -\infty \leq p \leq \infty,$$

this allows us to write

$$x(-p) * x(p) = \sum_{m=-\infty}^{\infty} x(-m)x(p - m), \qquad -\infty \leq p \leq \infty.$$

Substituting the change of variable $n = -m$ into this expression gives

$$x(-p) * x(p) = \sum_{n=-\infty}^{\infty} x(n)x(p + n) = R_{xx}(p), \qquad -\infty \leq p \leq \infty.$$

EXAMPLES

Thus, another way to evaluate the autocorrelation is to reverse the time argument of one of the $x(n)$'s and perform convolution.

c. The script is shown next and the MATLAB plot is in Figure E10.5c. Again notice in the script that the vector **p** is defined to locate the autocorrelation appropriately on the abscissa of the plot. As expected, we observe that the results of parts (b) and (c) are the same.

MATLAB Script

```
%E10_5c Autocorrelation via convolution
n=[0:1:63];
xn=sin(pi*n/16);
xnrev=sin(-pi*n/16);
p=[-64:1:62];           % time sample vector for plotting autocorrelation
zn=conv(xnrev, xn);     % call conv
%...plotting statements and pause
```

d. We can make use of the relationship found in part (b). In particular, since

$$R_{xx}(p) = x(-p) * x(p), \qquad -\infty \leq p \leq \infty,$$

we see that

$$\mathrm{DFT}[\tilde{R}_{xx}(p)] = \mathrm{DFT}[x(-p)] \cdot \mathrm{DFT}[x(p)],$$

where $\tilde{R}_{xx}(p)$ is the *circular* autocorrelation sequence evaluated in a similar way to circular convolution. In other words, to use DFTs to evaluate the (linear) autocorrelation, we again have to do zero padding, as was the case

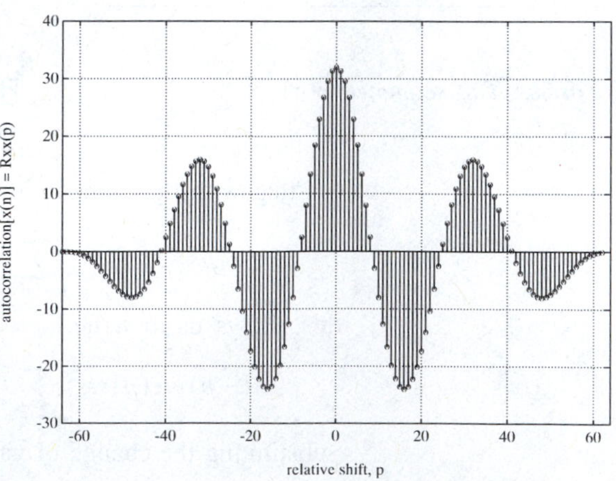

FIGURE E10.5c *Autocorrelation sequence of $x(n)$ found by convolution*

CHAPTER 10

with convolution. There is one additional detail to look after. What is the DFT of $x(-p)$? Returning to the DFT definition, we have

$$\mathrm{DFT}[x(p)] = \sum_{p=0}^{N-1} x(p)e^{-j(2\pi pk/N)},$$

and changing the time argument for the sequence $x(p)$ gives

$$\mathrm{DFT}[x(-p)] = \sum_{p=0}^{N-1} x(-p)e^{-j(2\pi pk/N)}.$$

Making the change of variable $m = -p$ results in

$$\mathrm{DFT}[x(-p)] = \sum_{m=0}^{N-1} x(m)e^{+j(2\pi mk/N)} = X^*(k),$$

where $X^*(k)$ is the complex conjugate of the DFT of $x(p)$. To summarize, we have

$$\mathrm{DFT}[\tilde{R}_{xx}(p)] = X^*(k) \cdot X(k).$$

e. The MATLAB script follows and the autocorrelation is shown in Figure E10.5d, page 586. To obtain the effect of zero padding, we have simply specified an FFT of length 128, which automatically fills in the zeros. Notice that the autocorrelation result looks similar to but also different from our earlier results. This is caused by the DFT index, which runs from 0 to $N - 1$. The first 64 points in Figure E10.5d are the values for $0 \leq p \leq 63$. The values for $64 \leq p \leq 127$ correspond, because of the periodicity of the DFT, to the time points $-64 \leq p \leq -1$. The vector **Rxxp** formed in the script simply places the values in the sequence as we are used to seeing them. This vector is plotted against the vector **pn**, also defined in the script, with the result shown in Figure E10.5e.

MATLAB Script

```
%E10_5de Correlation via FFTs
n=[0:1:63];
xn=sin(pi*n/16);
p=[0:1:127]; % time sample vector for plotting
%autocorrelation
pn=[-64:1:63]; % time sample vector for plotting shifted
%autocorrelation
yn=[yn zeros(1,lendif)]; % lendif defined earlier

Xk = fft(xn,128);
Yk = fft(xn,128);
Zk = (conj(Xk)).*Yk;
zn = ifft(Zk,128);
Rxxp = [zn(65:128) zn(1:64)]; % shift output for plotting
%...plotting statements and pause
```

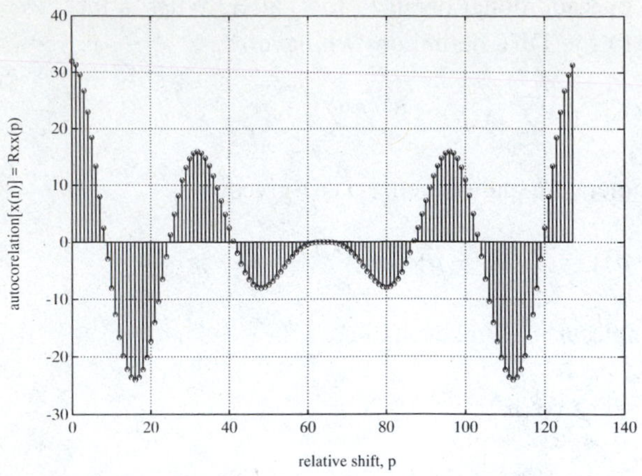

FIGURE E10.5d *Autocorrelation sequence of $x(n)$ found by using DFTs*

FIGURE E10.5e $R_{xx}(p)$*, using DFTs and shifting output*

f. The following MATLAB script includes the solution for the cross correlation sequence $R_{x_1x_2}(p)$ in three ways: (1) using the California function **ksxcorr**; (2) using DFTs (notice the shift caused by using the interval 0 to $N - 1$ for the DFT); (3) using the convolution relation, which can be shown in a way similar to the one in part (d). The data are in Figure E10.5f with the results given in Figures E10.5g, h, and i.

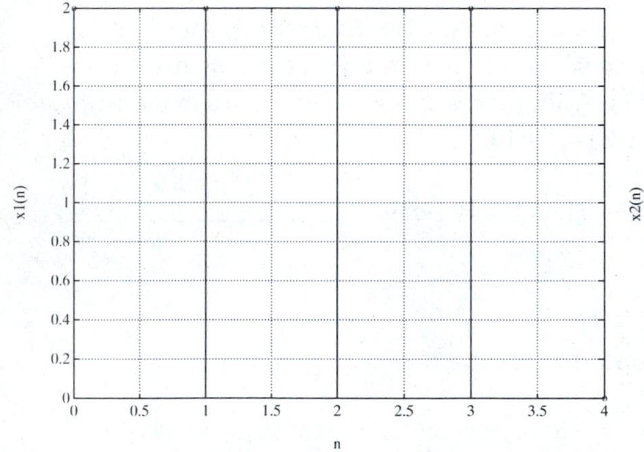

The sequence $x1(n)$

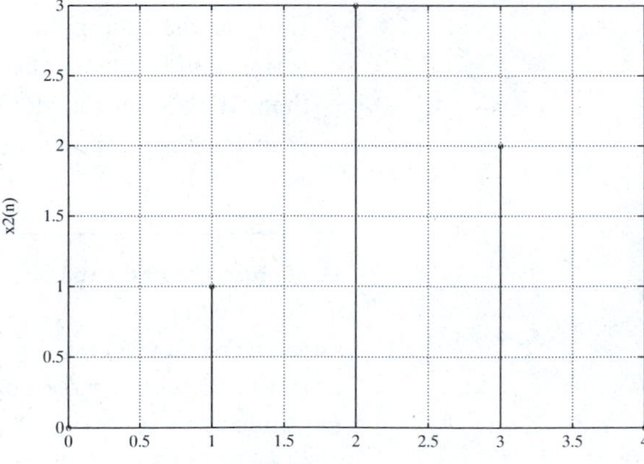

The sequence $x2(n)$

FIGURE E10.5f

MATLAB Script

```
%E10_5f_i Cross correlation
n=[0:1:4];
p=[-3:1:5];
x1n=[2 2 2 2 0];
x2n=[0 1 3 2 0];
```

Continues

```
Rx1x2p = ksxcorr (x1n,x2n);
len=length(x1n)+length(x2n)-1; % length of correlation
X1k = fft(x1n,len);
X2k = fft(x2n,len);
Rx1x2k = conj(X1k) .* X2k;
crossdft = ifft(Rx1x2k,len);
y1n=fliplr(x1n); % reverse the sequence x1(n)
crossconv = conv(y1n,x2n);
%...plotting statements
```

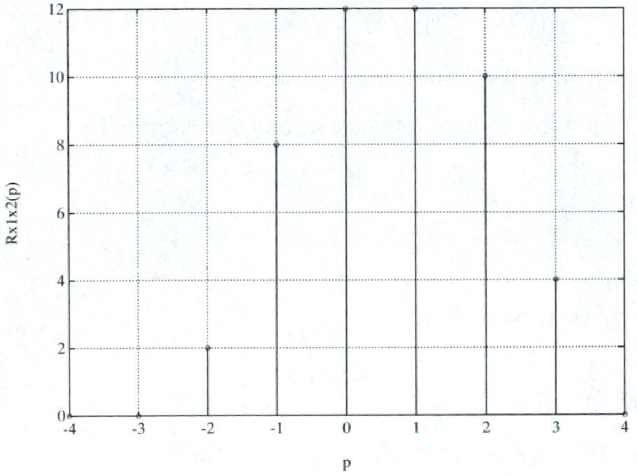

FIGURE E10.5g *The cross correlation* $R_{x_1x_2}(p)$

FIGURE E10.5h *The cross correlation* $R_{x_1x_2}(p)$ *obtained by DFTs*

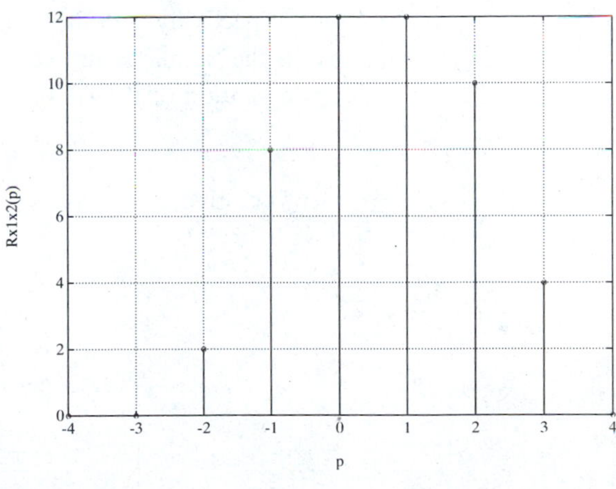

FIGURE E10.5i *The cross correlation* $R_{x_1x_2}(p)$ *obtained by convolution*

EXAMPLE 10.6
DFT Properties

In this example we make use of several MATLAB functions to illustrate and illuminate some important DFT properties.

a. Use the property of linearity to determine the 16-point DFT for the sequence

$$x(n) = 2 - 7 \cdot 2^{-n} + 3\cos(n\pi/4),$$

plot the result, and compare by using the MATLAB function **fft** to determine and plot the DFT.

b. Find an analytical expression for the 16-point DFT of the sequence

$$x(n) = 3\cos([n - 2]\pi/4),$$

and also use MATLAB to determine $X(k)$ and compare results.

c. Show that the DFTs of the following real sequences satisfy the symmetry condition:
(i) $1,\ 0 \le n \le 15$
(ii) $2^{-n},\ 0 \le n \le 15$
(iii) $\cos(n\pi/4),\ 0 \le n \le 15$
(iv) $2 - 7 \cdot 2^{-n} + 3\cos(n\pi/4),\ 0 \le n \le 15$
(v) $2^{-n},\ 0 \le n \le 14$

d. Find the 8-point DFT of the sequence

$$g(n) = A[\delta(n) + \delta(n-1) + \delta(n-2) + \delta(n-6) + \delta(n-7)],$$

and use the duality theorem to find the dual DFT pair.

Solution

a. Linearity enables us to find the DFTs of the sequences $x_1(n) = 1$, $x_2(n) = 2^{-n}$, and $x_3(n) = \cos(n\pi/4)$ separately, multiply them by 2, -7, and 3, respectively, and then add them to find the DFT of $x(n)$. The DFT of $x_1(n)$ is the sample sequence $N\delta(k)$. To find the DFT of $x_2(n)$ we go back to the definition of the DFT, and derive the general result for the sequence

$$B^{-n}, \qquad 0 \le n \le N - 1.$$

This gives

$$\begin{aligned} X_2(k) &= \sum_{n=0}^{N-1} x(n)e^{-j(2\pi nk/N)} = \sum_{n=0}^{N-1} B^{-n}e^{-j(2\pi nk/N)} \\ &= \sum_{n=0}^{N-1} [B^{-1}e^{-j(2\pi k/N)}]^n \\ &= \frac{1 - B^{-N}e^{-j2\pi k}}{1 - B^{-1}e^{-j(2\pi k/N)}} = \frac{1 - B^{-N}}{1 - B^{-1}e^{-j(2\pi k/N)}}, \end{aligned}$$

where $B \ne 1$ is a (possibly complex) constant, and we have again made use of the finite geometric series formula. To find the DFT of $x_3(n)$, we use the results of Examples 10.2 and 10.3 and write $x_3(n)$ in the form

$$x_3(n) = \cos(n\pi/4) = \cos(2 \cdot 2\pi n/16)$$

and with $NA/2 = 16/2 = 8$; the DFT follows as

$$X_3(k) = 8\delta(k-2) + 8\delta(k-14).$$

Thus we have, letting $B = 2$ in $X_2(k)$,

$$\begin{aligned} X(k) &= 2X_1(k) - 7X_2(k) + 3X_3(k) \\ &= 32\delta(k) - 7\frac{1-2^{-16}}{1-2^{-1}e^{-j(2\pi k/16)}} + 24\delta(k-2) + 24\delta(k-14). \end{aligned}$$

The MATLAB script for finding this DFT is shown on page 590; the plots of the sequence $x(n)$ and the magnitude and phase of the DFT $X(k)$ are given in Figure E10.6a.

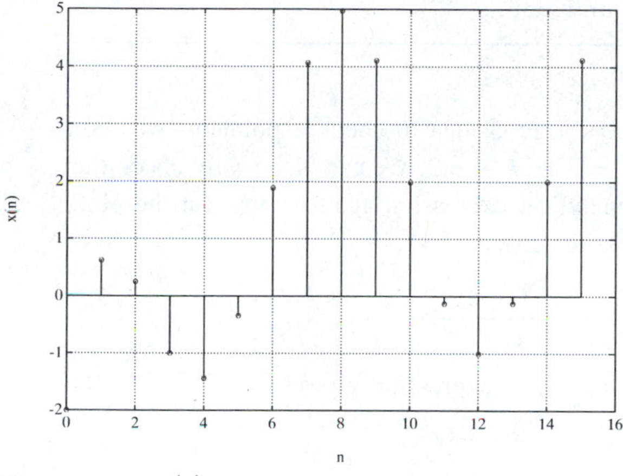

The sequence $x(n)$

Magnitude of $X(k)$

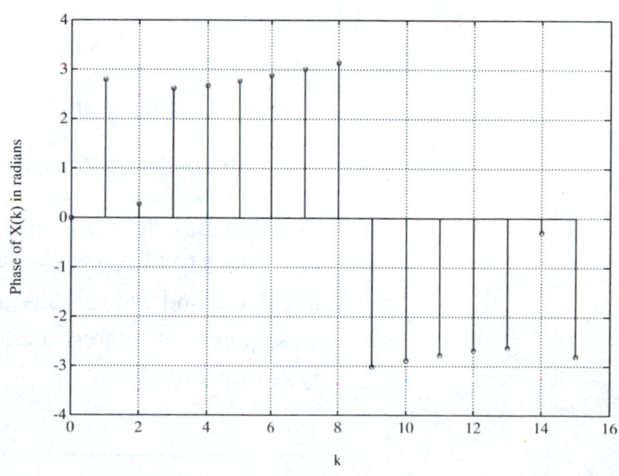

Phase of $X(k)$

FIGURE E10.6a *A sequence and its DFT*

MATLAB Script

```
%E10_6 DFT properties

%E10_6a Linearity property
N=16;
n=[0:1:N-1];
k=n;
vones=[ones(1,N)];
xn=2*vones-7*2.^(-n)+3*cos(n*pi/4);
Xk = fft(xn,N); % call fft
magXk=abs(Xk);
phaseXk=angle(Xk);
%...plotting statements and pause
```

CROSS-CHECK It's always a good idea to check a point or two. For example, the value of $x(0)$ is $2 - 7 + 3 = -2$. We can also easily check the values for $X(k)$ given by the analytical expression against those on the plot. For $k = 0$ the analytical expression gives

$$X(0) = 32 - 7\frac{1 - 2^{-16}}{1 - 2^{-1}} \approx 18\,.$$

Similarly, letting $k = 8$ in the analytical expression gives

$$X(8) = -7\frac{1 - 2^{-16}e^{-j16\pi}}{1 - 2^{-1}e^{-j\pi}} \approx -4.67\,.$$

These values agree with the values given in the plots of Figure E10.6a. ■

b. From part (a) we know that

$$3\cos(n\pi/4) \Leftrightarrow 24\delta(k - 2) + 24\delta(k - 14)\,.$$

From the circular shifting theorem, then,

$$3\cos([n \ominus 2]\pi/4) \Leftrightarrow [24\delta(k - 2) + 24\delta(k - 14)]e^{-j\pi k/4}\,.$$

This DFT has the same magnitude but has a phase shift that depends linearly on k. The MATLAB script is shown next, and the sequence $x(n)$ and the DFT magnitude and phase plots are shown in Figure E10.6b. Notice that the only phase points of interest are at $k = 2$ and 14, where the phases are $-\pi/2$ and $\pi/2$, respectively.

MATLAB Script

```
%E10_6b DFT of a shifted sequence
N=16;
n=[0:1:N-1];
```

```
k=n;
xn=3*cos((n-2)*pi/4);
Xk=fft(xn,N); % call fft
magXk=abs(Xk);
phaseXk=angle(Xk);
%...plotting statements and pause
```

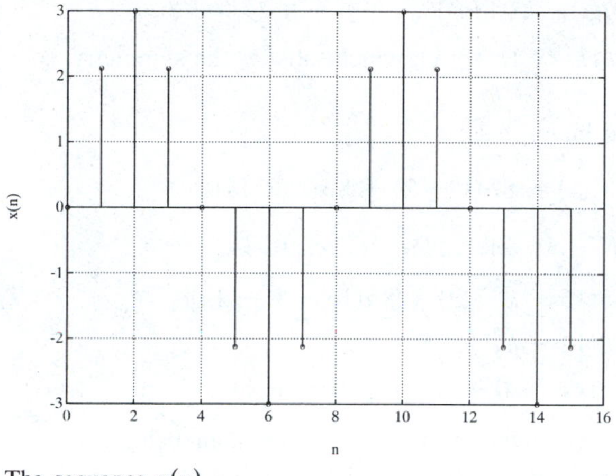

The sequence $x(n)$

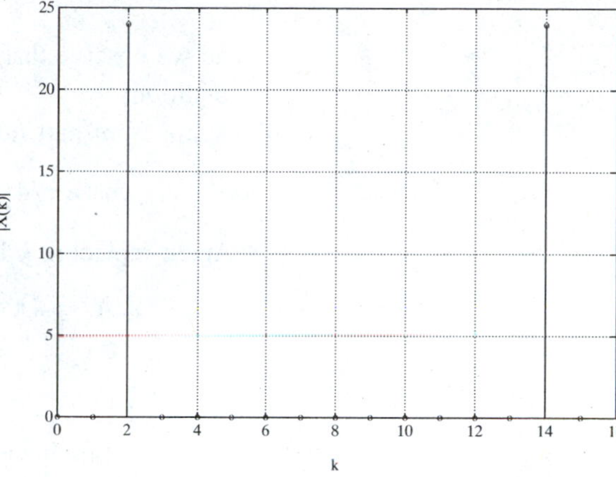

Magnitude of $X(k)$

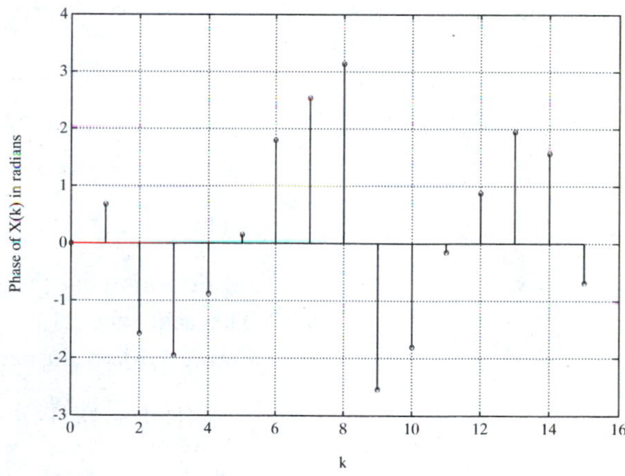

Phase of $X(k)$

FIGURE E10.6b *A shifted sinusoidal sequence and its DFT*

c. The symmetry condition we want to verify is $X(k) = X^*(N - k)$, or

$$|X(k)| = |X(N - k)| \quad \text{and} \quad \angle X(k) = -\angle X(N - k)$$

for $k = 1, 2, \ldots, N - 1$.

(i) The DFT of a constant sequence is simply a sample sequence with its nonzero value at $k = 0$. Observing that the DFT value at $k = 0$ is

EXAMPLES

excluded from the symmetry condition, it is clear that since the other DFT values are zero, the symmetry condition is satisfied.

(ii) From part (a) we know that

$$B^{-n} \Leftrightarrow X(k) = \frac{1 - B^{-N}}{1 - B^{-1}e^{-j(2\pi k/N)}},$$

where in this case $B = 2$. The expression for $X(N - k)$ is then

$$X(N - k) = \frac{1 - B^{-N}}{1 - B^{-1}e^{-j(2\pi[N-k]/N)}} = \frac{1 - B^{-N}}{1 - B^{-1}e^{j(2\pi k/N)}}$$

and we observe that $X(k) = X^*(N - k)$, which satisfies the symmetry condition.

(iii) Again from part (a) we know that

$$\cos(n\pi/4) \Leftrightarrow V(k) = 8\delta(k - 2) + 8\delta(k - 14).$$

Again replacing k by $N - k$, where $N = 16$, results in

$$\begin{aligned} V(N - k) &= 8\delta(16 - k - 2) + 8\delta(16 - k - 14) \\ &= 8\delta(14 - k) + 8\delta(2 - k) \\ &= 8\delta(k - 14) + 8\delta(k - 2) = V(k). \end{aligned}$$

(iv) Here we can interpret symmetry graphically, rather than using the analytical approach of parts (ii) and (iii). The DFT of interest is shown in Figure E10.6a. Comparing the DFT magnitude values with one another, we observe that

$$|X(1)| = |X(15)|, \qquad |X(2)| = |X(14)|, \ldots, |X(7)| = |X(9)|.$$

Notice that the point $X(8)$ is not included because N is even. Now consider the phase portion of the DFT. Here we find that

$$\angle X(1) = -\angle X(15), \qquad \angle X(2) = -\angle X(14), \ldots, \angle X(7) = -\angle X(9),$$

again satisfying the symmetry conditions.

(v) The sequence $x(n) = 2^{-n}$, $0 \le n \le 14$, and its DFT are shown in Figure E10.6c. Inspecting the DFT plots reveals that

$$|X(1)| = |X(14)|, \qquad |X(2)| = |X(13)|, \ldots, |X(7)| = |X(8)|$$

and

$$\angle X(1) = -\angle X(14), \qquad \angle X(2) = -\angle X(13), \ldots, \angle X(7) = -\angle X(8),$$

which again indicate that the symmetry conditions are met. Notice that in this case N is odd and the point of symmetry is actually at $N/2$, halfway between the points $k = 7$ and $k = 8$.

Comment: Symmetry allows us to plot only approximately half of the points in a DFT if we wish. Notice that this result applies only when the sequence $x(n)$ is real.

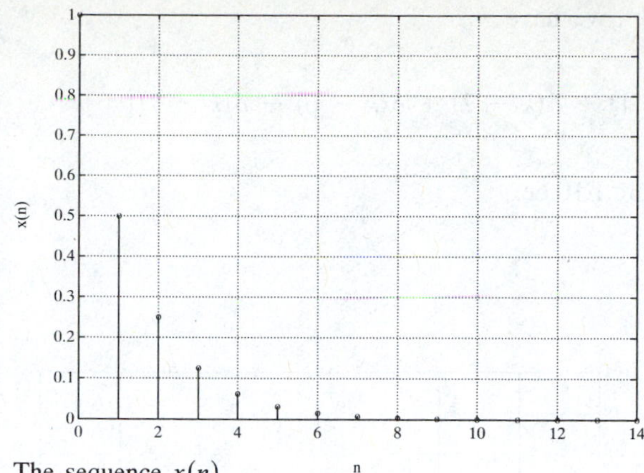

The sequence $x(n)$

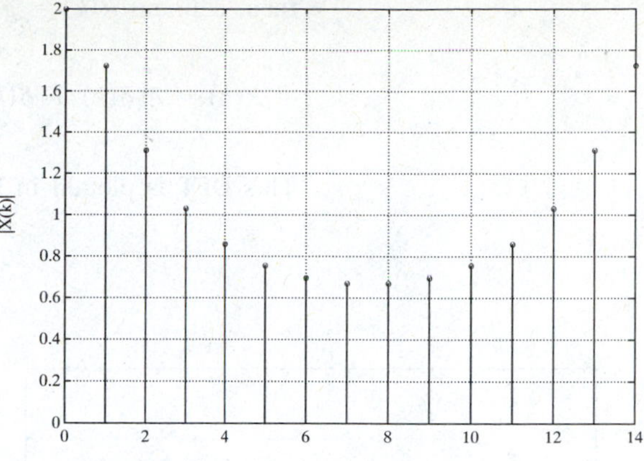

Magnitude of $X(k)$

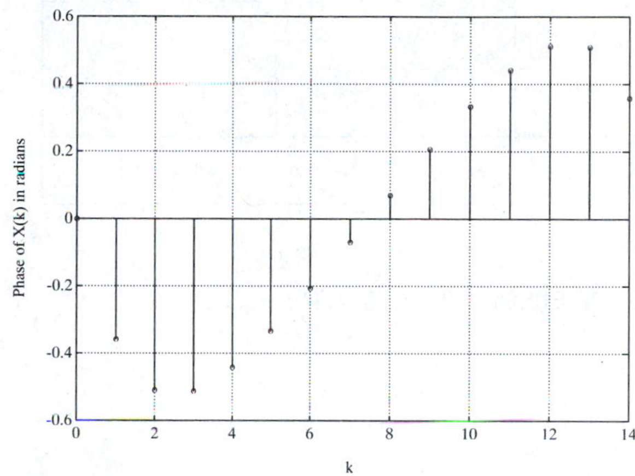

Phase of $X(k)$

FIGURE E10.6c *The sequence $x(n) = 2^{-n}$ and its 15-point DFT*

d. From the DFT definition,

$$G(k) = A\left[1 + e^{-j(\pi k/4)} + e^{-j(\pi k/2)} + e^{-j(6\pi k/4)} + e^{-j(7\pi k/4)}\right].$$

Using the facts that

$$e^{-j(7\pi k/4)} = e^{-j(8\pi k/4)}e^{j(\pi k/4)} = e^{j(\pi k/4)} \quad \text{and}$$
$$e^{-j(6\pi k/4)} = e^{-j(8\pi k/4)}e^{j(\pi k/2)} = e^{j(\pi k/2)}$$

and doing a modest bit of algebra yields

$$G(k) = A[1 + 2\cos(\pi k/4) + 2\cos(\pi k/2)], \qquad k = 0, 1, \ldots, 7.$$

The resulting DFT, which is real, is shown for $A = 1$ in Figure E10.6d, page 594. Using the duality property with $r(n) = G(n)$,

$$r(n) = \frac{A}{8}[1 + 2\cos(\pi n/4) + 2\cos(\pi n/2)], \qquad k = 0, 1, \ldots, 7,$$

then, defining $R(k) = g(-k)$, we have

$$R(k) = A[\delta(k) + \delta(k-1) + \delta(k-2) + \delta(k-6) + \delta(k-7)].$$

This DFT is plotted in Figure E10.6e.

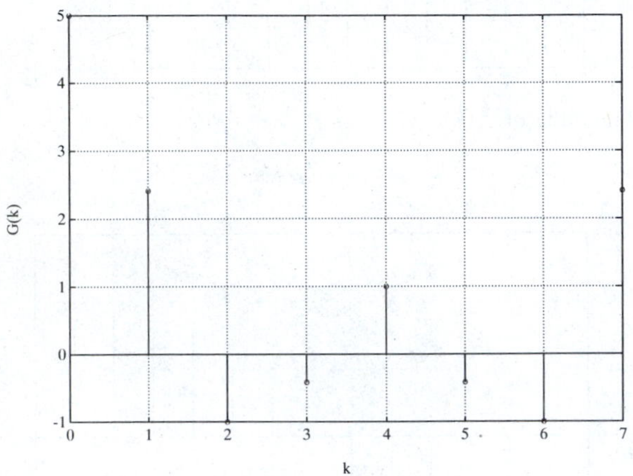

FIGURE E10.6d *The DFT $G(k)$*

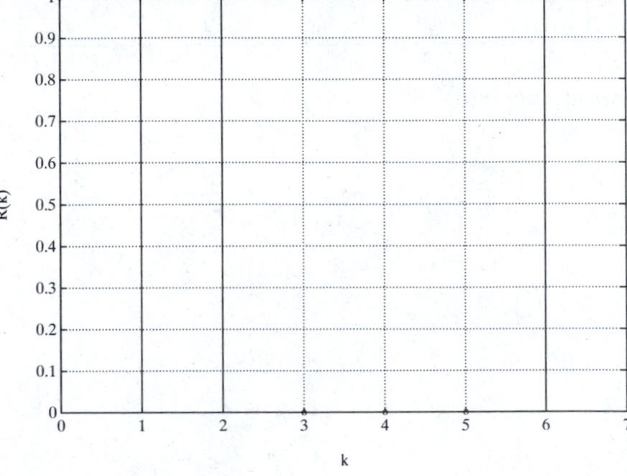

FIGURE E10.6e *The DFT $R(k)$*

EXAMPLE 10.7
Using the DFT to Approximate a Continuous-Time Fourier Transform

In this example we will derive the approximation of a CTFT $X(\omega)$ by a DFT $X(k)$ and with an example illustrate the quality of the approximation.

a. With the Continuous-Time Fourier Transform of a causal signal $x(t)$ given by

$$X(\omega) = \int_0^{\infty} x(t)e^{-j\omega t}\,dt\,,$$

break this integral up into N pieces and indicate how the evaluation can be carried out over the finite time interval $0 \le t \le N\Delta t$, a record length of T_0.

b. Substitute $t = 0,\ \Delta t,\ 2\Delta t, \ldots, (N-1)\Delta t$, and determine a piecewise approximation for the integrals of part (a).

c. Use the results of (b) to show that

$$\mathcal{F}[x(t)] \approx \Delta t \sum_{n=0}^{N-1} x(n)e^{-j2\pi nk/N} = \Delta t \cdot \text{DFT}[x(n)].$$

d. Summarizing our results, we have for the causal analog signal $x(t)$

approximation of Fourier transform by DFT

$$x(n) = x(t)|_{t=nT_0/N} \quad \text{and}$$

$$\text{DFT}[x(n)] = X(k) \approx \frac{1}{\Delta t} \cdot X(\omega)|_{\omega=2\pi k/T_0}, \qquad k = 0, 1, \ldots, N-1.$$

Write a script to find the DFT of the samples of the analog signal $x(t) = 5e^{-5t}u(t)$ with $T_0 = 2$ s and $N = 256$. Compare results with $X(\omega)$ at $\omega = 0$ rad/s and $\omega = \pi$ rad/s.

WHAT IF? Approximating a Fourier transform by a DFT works well, but we must be careful to keep in mind the conditions of the sampling theorem. In part (d) the DFT's frequency range is $0 \leq \omega \leq 128\pi$, and at $\omega = 128\pi$, $|X(\omega)| = 0.0124$, which is quite small compared to $|X(0)| = 130.51$. Here let's assume that $x(t)$ is the pulse $x(t) = u(t + 0.1) - u(t - 0.1)$. Use the approach developed in part (c) to approximate the Fourier transform $X(\omega) = 0.2 \sin(0.2\omega)/(0.2\omega)$ by a DFT for two cases: (i) A record length of 2 s with $N = 100$ samples and (ii) a record length of 2 s with $N = 400$ samples. Use MATLAB to plot each DFT, with $X(\omega)$ shown on the same plot. Compare the results. ■

Solution

a. *Breaking up the integral.* Integrating over N segments with the time interval for each of Δt gives

$$X(\omega) = \int_0^{\Delta t} x(t)e^{-j\omega t}\,dt + \int_{\Delta t}^{2\Delta t} x(t)e^{-j\omega t}\,dt + \cdots + \int_{(N-1)\Delta t}^{N\Delta t} x(t)e^{-j\omega t}\,dt.$$

b. *Piecewise approximation.* For $t = 0, \Delta t, 2\Delta t, \ldots, (N-1)\Delta t$, we have, with the help of Figure E10.7a, page 596,

$$X(\omega) \approx x(0)\Delta t e^{-j\omega 0} + x(1)\Delta t e^{-j\omega\Delta t} + x(2)\Delta t e^{-j\omega 2\Delta t} + \cdots + x(N-1)\Delta t e^{-j\omega[N-1]\Delta t}$$
$$= \Delta t \sum_{n=0}^{N-1} x(n)e^{-jn\omega\Delta t}.$$

c. *Making the transform discrete.* Knowing that $\Delta t = T_0/N$ and considering only the frequency values $\omega = k\Delta\omega = k2\pi/T_0$, the term $\omega\Delta t$ in the preceding exponential can be written as $k2\pi/N$, giving

$$X(k\Delta\omega) \approx \Delta t \sum_{n=0}^{N-1} x(n)e^{-j2\pi nk/N} = \Delta t \cdot \text{DFT}[x(n)].$$

EXAMPLES

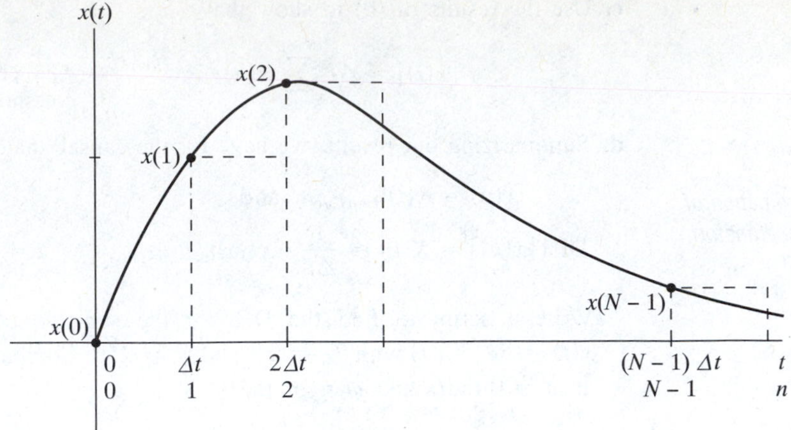

FIGURE E10.7a *Piecewise approximation of* $x(t)$

d. *DFT and theoretical results.* See Figure E10.7b for the sequence and its DFT. The theoretical transform is $X(\omega) = 5/(5 + j\omega)$, which gives $|X(0)| = 1$ and $|X(\pi)| = 0.85$. From the DFT we have $\Delta t = 2/256 = 0.0078125$, $\Delta\omega = \pi$, and—from the file—$X(0) = 130.51$, which compares with $|X(0)|/\Delta t = 1/0.0078125 = 128$. This is not bad, and using a larger value of N would give even better results. To check at $\omega = \pi$ for the DFT, we set $k = 1$ or $|X(1)| = 110.50$, which compares with $|X(\pi)|/\Delta t = 0.85/0.0078125 = 108.8$, also reasonably close.

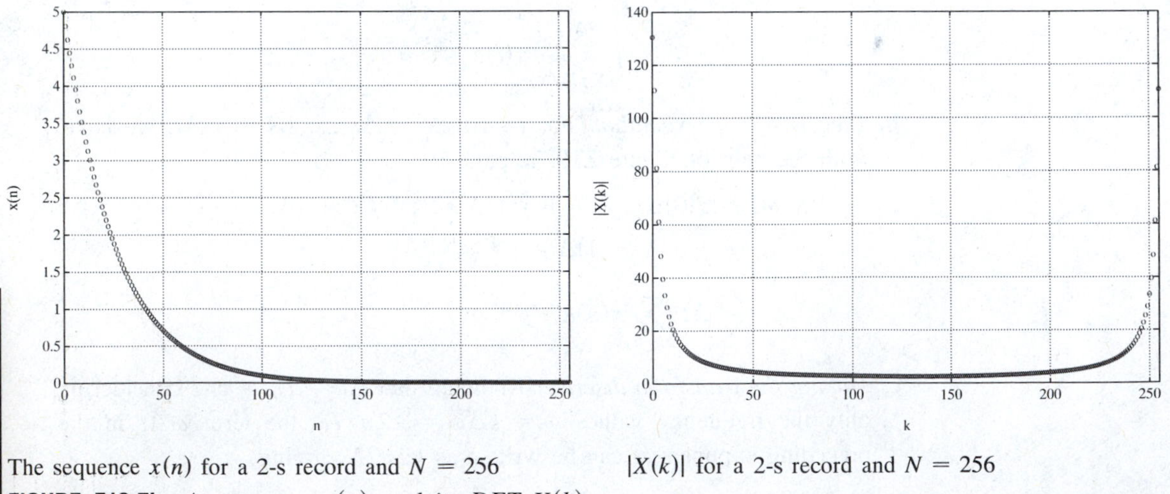

The sequence $x(n)$ for a 2-s record and $N = 256$ $\qquad$ $|X(k)|$ for a 2-s record and $N = 256$

FIGURE E10.7b *A sequence* $x(n)$ *and its DFT* $X(k)$

WHAT IF? With 100 points and a record length of 2 s, we have a frequency resolution of $\Delta f = 0.5$ Hz and a maximum frequency of $50 \cdot 0.5 = 25$ Hz represented. A plot of the sampled signal is not shown here, but can be seen by running the script in file E10_7. In Figure E10.7c we observe that the DFT is a good approximation for relatively low frequencies, but it is poor at the higher frequencies. The reason is that the DFT with this sampling frequency implicitly assumes that the highest frequency present in $X(\omega)$ is 25 Hz, and this is not true. Note that the values of $X(k)$ shown for $k = 51, \ldots, 99$ are the negative frequencies represented by the DFT.

By retaining the 2-s record length and increasing the number of samples to $N = 400$, the highest frequency represented becomes $200 \cdot 0.5 = 100$ Hz. The DFT is now as shown in Figure E10.7d (see E10_7 for the sampled signal) and the approximation is quite good. One final word: The difficulty with the smaller sampling frequency is that the sampling process introduces *aliasing*—the frequency spectrum of the sampled signal has portions where frequencies have been distorted due to "foldover" effects. ■

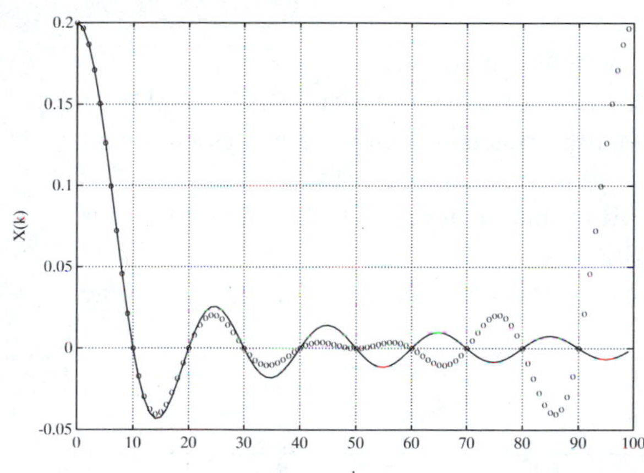

FIGURE E10.7c *$X(k)$ for a 2-s record and $N = 100$*

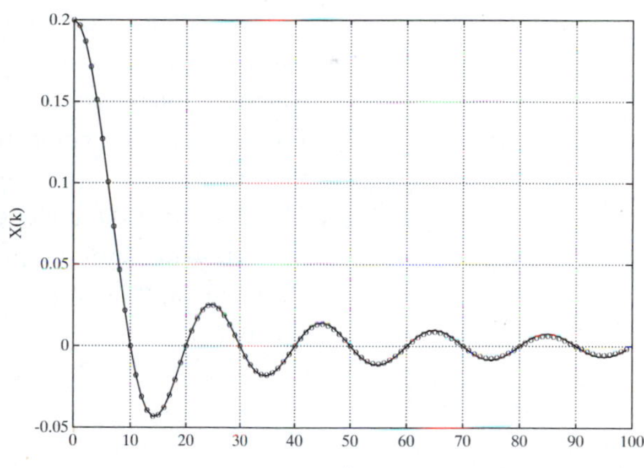

FIGURE E10.7d *$X(k)$ (partial) for a 2-s record and $N = 400$*

PROBLEMS

REINFORCEMENT PROBLEMS

P10.1. Sampling frequency, resolution, and record length. It is desired to determine the DFT of the analog signal $x(t) = 25\sin(10\pi t)\cos(400\pi t)$.

a. Find the minimum sampling frequency in hertz.

b. What is the record length if the desired frequency resolution is 5 Hz?

c. If the minimum sampling frequency from part (a) is used and the record length found in part (b) is used, find the number of points that will be contained in the DFT.

P10.2 Sampling frequency, resolution, and record length. The samples in the 15-point sequence listed here are spaced 0.5 ms apart.

$$x(n) = \{3, -2, 1, -1, 4, 3, 1, -4, -2, -1, 2, 1, -1, 1, -3\}.$$

a. What is the frequency resolution of the DFT?
b. What is the maximum frequency that this sequence's DFT would contain?
c. Determine the value of $X(0)$.
d. Use the MATLAB function **fft** to find and plot the DFT $X(k)$. Add the frequency scale (in hertz) to the diagram, labeling all points corresponding to $0 \leq k \leq 14$.

P10.3 Frequency resolution and inverse DFTs. A periodic continuous-time signal is sampled at a frequency of $f_s = 10^3$ Hz, and the resulting DFT coefficients for one period of the signal are shown in Figure P10.3. Notice that the DFT coefficients are real numbers.

a. Add the frequency scale (in hertz) to the diagram, labeling all points corresponding to $0 \leq k \leq 5$.
b. Write an expression for $x(n)$ in terms of cosines.
c. Calculate one easy-to-check value of $x(n)$ from the DFT samples in Figure P10.3 and compare with the results given by the expression found in part (b).
d. Use the MATLAB function **ifft** to find the inverse DFT and compare your answers with those of part (b).
e. Write an expression for the continuous-time signal $x(t)$ in terms of cosine functions.

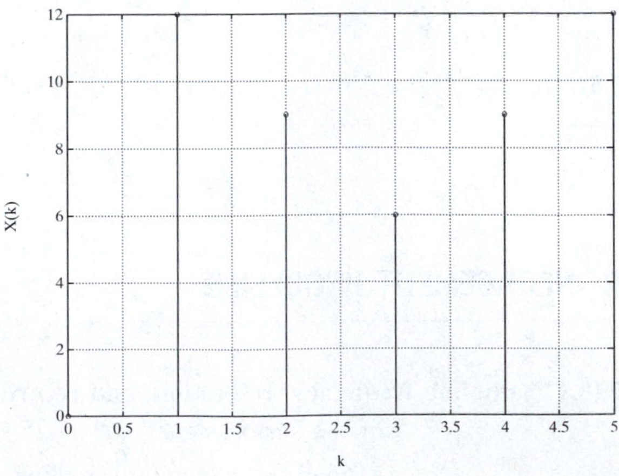

FIGURE P10.3 *The DFT $X(k)$*

P10.4 Frequency resolution and inverse DFTs. A periodic analog signal is sampled at a frequency of $f_s = 200$ Hz. The record length is one period of the analog signal. The resulting DFT coefficients, which are real, are shown in Figure P10.4.

a. What is the period of the analog signal?
b. Add the frequency scale (in hertz) to the diagram, labeling all points corresponding to $0 \le k \le 9$.
c. Write an expression for $x(n)$ in terms of cosines.
d. Calculate one easy-to-check value of $x(n)$ from the DFT samples in Figure P10.4 and compare with the results given by the expression found in part (c).
e. Use MATLAB to find the Inverse DFT and compare your answers with those of part (c).
f. Determine the amplitudes and frequencies that the DFT indicates are present in the original analog signal.

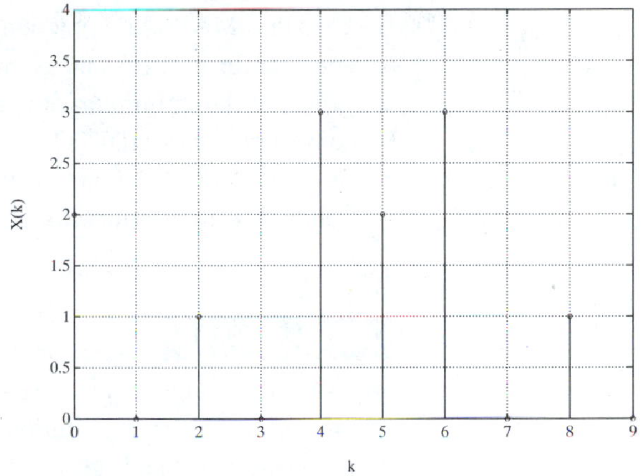

FIGURE P10.4 *The DFT X(k)*

P10.5 Convolution—linear and circular. Given the following sequences:

$$x_1(n) = \{2, -1, 0, 4\} \quad \text{and} \quad x_2(n) = \{3, -2, -1, 1\}.$$
$$\uparrow n = 0 \qquad\qquad\qquad \uparrow n = 0$$

a. Calculate the circular convolution of $x_1(n)$ and $x_2(n)$ using a paper-and-pencil approach.
b. Repeat part (a) using the MATLAB function **fft** and compare your answers.
c. Determine the linear convolution of $x_1(n)$ and $x_2(n)$ using a paper-and-pencil approach.
d. What is the minimum number of points needed to accomplish linear convolution using DFTs?
e. Use MATLAB functions **conv** and/or **fft** to find the linear convolution $x_1(n)*x_2(n)$ and compare your results with those of part (c).

PROBLEMS

WHAT IF? Suppose that you attempted to do linear convolution using the DFT approach and MATLAB but used too few points in your DFTs. Try doing the linear convolution of part (e) using 6-point DFTs. Check a few of the values by hand. ■

P10.6 Convolution—linear and circular. Given the following sequences:

$$x_1(n) = \{5, 2, -3, -1, 4\} \quad \text{and} \quad x_2(n) = \{2, 6, 0, 4, 1\}.$$
$$\uparrow n = 0 \qquad\qquad\qquad\qquad \uparrow n = 0$$

a. Calculate the circular convolution of $x_1(n)$ and $x_2(n)$ using a paper-and-pencil approach.
b. Repeat part (a) using the MATLAB function **fft** and compare your answers.
c. Calculate the linear convolution of $x_1(n)$ and $x_2(n)$ using a paper-and-pencil approach.
d. Use the MATLAB function **conv** to calculate the linear convolution specified in part (c) and compare your results.
e. What is the minimum number of points needed to accomplish linear convolution using DFTs?
f. Use the MATLAB function **fft** to find the linear convolution $x_1(n)*x_2(n)$. Compare your results with those of parts (c) and (d).

WHAT IF? Suppose that you attempted to do linear convolution using the DFT approach and MATLAB but used too few points in your DFTs. Try doing the linear convolution of part (e) using 7-point DFTs. Check a few of the values by hand. ■

P10.7 Linear convolution. Two sequences $x_1(n)$ and $x_2(n)$ have the following values:

$$x_1(n) = \{1, -2, -5, 3, -2, 3, 5, 10, -5, -8, -1\} \quad \text{and}$$
$$\uparrow n = 0$$
$$x_2(n) = \{3, -1, 0, -4, 3, 6\}.$$
$$\uparrow n = 0$$

a. Find the linear convolution of $x_1(n)$ and $x_2(n)$ using the MATLAB function **conv**.
b. Repeat part (a) using the **fft** function of MATLAB.
c. Verify the first and last points of the convolution by hand calculation.
d. Can you evaluate the circular convolution of $x_1(n)$ and $x_2(n)$? Explain.

P10.8 Use of the symmetry property.

a. The 8-point DFT of a real sequence $x(n)$ was evaluated, but some of the results were misplaced. The values remaining are

$$X(k) = \begin{cases} 0, & k = 3, 5 \\ 2 + j0, & k = 6 \\ 0.5 + j0.5, & k = 7. \end{cases}$$

Use the symmetry property to determine as many of the missing values as possible. If any values cannot be found, indicate why this is so.

b. A 7-point DFT of a real sequence has the values

$$X(k) = \{3, 1e^{-j\pi/3}, 2e^{-j\pi/4}, 5e^{-j\pi/8}, \ldots\}.$$
$$\uparrow k = 0$$

Sketch the magnitude and phase of the entire DFT.

c. You are tutoring a student and before turning her loose with MATLAB, you provide a practice exercise to see if she can calculate a 4-point DFT. The sequence $x(n)$ is real and she says the DFT should be

$$X(k) = \{2e^{j\pi/6}, 3e^{j\pi/3}, 3e^{-j\pi/3}, 2e^{-j\pi/6}\}.$$

Is she ready to use MATLAB?

P10.9 DFT of sinc sequence. An analog signal described by

$$x(t) = \frac{2f_0 \sin(2\pi f_0 t)}{2\pi f_0 t}$$

is sampled at a frequency of $f_s = 32$ Hz, and only the 64 samples in the interval $-1 \leq t < 1$ are retained. Use MATLAB to find and plot the magnitudes of the DFT coefficients for f_0 equal to (a) 0.5 Hz, (b) 2.0 Hz, (c) 4.0 Hz, and (d) 8.0 Hz.

WHAT IF? Discuss the effect of changing the number of samples to 256 without changing the sampling frequency f_s. ■

P10.10 DFTs of some sequences.

a. Find analytically the 8-point DFT of the sequence $x(n) = 3\delta(n) - 5\delta(n-6)$. Do this two ways, once using the DFT definition and a second time using the known DFT of the sequence $\delta(n)$ and the shifting property and linearity.

b. Use MATLAB to find the DFT of the sequence in part (a) and compare your results.

c. Find the 8-point DFT of the sequence $y(n) = 5\delta(n) + 5\delta(n-1) + 5\delta(n-7)$ using analytical methods. Is the resulting DFT (1) real? (2) An even sequence in k? (3) An odd sequence in k? Explain.

d. Use MATLAB to find the DFT of part (c) and compare results.

e. Repeat part (c) for the sequence $r(n) = 5\delta(n-1) - 5\delta(n-7)$.

P10.11 DFT of a modulated sequence. An analog signal to be sampled is given by $x(t) = 7\cos(20\pi t)\cdot\cos(200\pi t)$.

a. What is the minimum sampling frequency that should be used?
b. What record length should be used so that it is an integer multiple of the period of $x(t)$?
c. Using a sampling frequency of twice the minimum value found in part (a), the record length found in part (b), and MATLAB, find and plot the DFT.
d. Verify the DFT of part (c) analytically.
e. Repeat the MATLAB portion of part (c) but with a record length 50% longer. Comment on your observations.

P10.12 DFT of another modulated sequence. An analog signal to be sampled given by $x(t) = [1 + 0.2\cos(20\pi t)]\cos(200\pi t)$.

a. What is the minimum sampling frequency that should be used?
b. What record length should be used so that it is an integer multiple of the period of $x(t)$?
c. Using a sampling frequency of twice the minimum value found in part (a), the record length found in part (b), and MATLAB, find and plot the DFT.
d. Verify the DFT of part (c) analytically.
e. Repeat the MATLAB portion of part (c) but with a record length 50% longer. Comment on your observations.

P10.13 Evaluation of correlations.

a. Given the following sequences $x(n)$ and $y(n)$:

$$x(n) = \{2, -1, 0, 4\} \quad \text{and} \quad y(n) = \{3, -2, -1, 1\}.$$
$$\uparrow n = 0 \qquad\qquad\qquad \uparrow n = 0$$

Evaluate and sketch: (1) the autocorrelation $R_{xx}(p)$; (2) the autocorrelation $R_{yy}(p)$; (3) the cross correlation $R_{xy}(p)$; (4) the cross correlation $R_{yx}(p)$.

b. Use the MATLAB functions **fft** and **ifft** to evaluate the correlations of part (a) and compare answers.
c. Use **xcorr** or **ksxcorr** to evaluate the correlations of part (a) and compare your answers with those of parts (a) and (b).

P10.14 Evaluation of circular correlations. Repeat parts (a) and (b) of Problem P10.13, but find the circular correlations in each case.

P10.15 Approximation of a Continuous-Time Fourier Transform using a DFT.

a. Using the definition of the Fourier transform given in Chapter 5, find an analytical expression for $\mathcal{F}[x(t) = te^{-t}u(t)]$.
b. Use the technique of Example 10.7 to find the DFT approximation to the transform $X(\omega)$ of part (a). Let the record length be 5 s and use $N = 256$.
c. Now use $T_0 = 10$ s and $N = 1024$. Compare values from the expression of part (a) with the DFT values of parts (b) and (c).

EXPLORATION PROBLEMS

P10.16 Additional DFT properties. Show the following DFT properties.

a. *Even sequences.* If $x(n)$ is a real and even function of n—that is, $x(n) = x(-n)$—the DFT $X(k)$ is real and is an even function of k.

b. *Odd sequences.* If $x(n)$ is a real and odd function of n—that is, $x(n) = -x(-n)$—the DFT $X(k)$ is imaginary and is an odd function of k.

c. *Frequency shifting.* If $X(k \ominus r)$ is the circularly shifted version of $X(k)$, the inverse DFT of $X(k \ominus r)$ is $x(n)e^{j2\pi rn/N}$.

d. *Cross correlation.* The cross correlation of two sequences is defined as

$$R_{x_1x_2}(p) = \sum_{n=-\infty}^{\infty} x_1(n)x_2(p+n), \quad -\infty \le p \le \infty,$$

and the circular cross correlation of two sequences is defined as

$$\tilde{R}_{x_1x_2}(p) = \sum_{n=0}^{N-1} x_1(n)x_2(p \oplus n), \quad p = 0, 1, \ldots, N-1;$$

the DFT of $\tilde{R}_{x_1x_2}(p)$ is given by $X_1{}^*(k)X_2(k)$, where $X_1{}^*(k)$ is the complex conjugate of $X_1(k)$.

P10.17 Demonstration of DFT properties.

a. Consider the 128-point sequences

$$x_1(n) = \begin{cases} 1, & \text{for } n = 0, 1, 127 \\ 0, & \text{otherwise} \end{cases} \qquad x_2(n) = \begin{cases} 1, & \text{for } n = 0, 1, 2 \\ 0, & \text{otherwise.} \end{cases}$$

(i) Use MATLAB to find and plot the DFTs of these two sequences.

(ii) Using the implicit periodicity of a sequence and its DFT, express $x_2(n)$ in terms of $x_1(n)$. Use a DFT property to express $X_2(k)$ in terms of $X_1(k)$.

b. Consider the 16-point sequences

$$r_1(n) = \begin{cases} 1, & \text{for } n = 0, 1, 2 \\ 0, & \text{otherwise} \end{cases} \qquad r_2(n) = \begin{cases} 1, & \text{for } n = 14, 15 \\ 0, & \text{otherwise.} \end{cases}$$

(i) Use MATLAB to find and plot the DFTs of these two sequences.

(ii) Write a MATLAB script to add the two DFTs found in part (i).

(iii) Use MATLAB to compute the DFT of the sequence

$$r_3(n) = \begin{cases} 1, & \text{for } n = 0, 1, 2, 14, 15 \\ 0, & \text{otherwise} \end{cases}$$

and compare results to those of part (ii).

(iv) What DFT property have you demonstrated?

(v) How could you extend what has been done here to demonstrate the appropriate property in more generality?

c. Use the duality property to find the DFTs that are the duals of the DFT pairs

(i) $x(n) = A\cos(L2n\pi/N) \Leftrightarrow \dfrac{NA}{2}\delta(k-L) + \dfrac{NA}{2}\delta(k-[N-L])$,

where N is an integer multiple of L, and

(ii) $r(n) = A \Leftrightarrow NA\delta(k)$.

d. Verify that the following symmetry relations are satisfied for the sequences given in parts (a), (b), and (c): For $x(n)$ a *real* sequence of length N

$$|X(k)| = |X(N-k)| \quad \text{and} \quad \angle X(k) = -\angle X(N-k)$$

for $k = 1, 2, \ldots, N-1$.

P10.18 Proof of symmetry relations. Show that for $x(n)$ a *real* sequence of length N,

$$|X(k)| = |X(N-k)| \quad \text{and} \quad \angle X(k) = -\angle X(N-k)$$

for $k = 1, 2, \ldots, N-1$.

P10.19 Some DFT characteristics.

a. Assume that $x(n)$ is an 8-point sequence whose DFT is given in Figure P10.19 and a second sequence $r(n)$ is formed by separating each element of $x(n)$ by a zero value. Thus $r(n)$ has 16 points and can be written as

$$r(n) = \begin{cases} x(n/2), & \text{for } n \text{ even} \\ 0, & \text{for } n \text{ odd.} \end{cases}$$

Find an expression for the DFT $R(k)$ in terms of $X(k)$ and sketch $R(k)$ as a function of k.

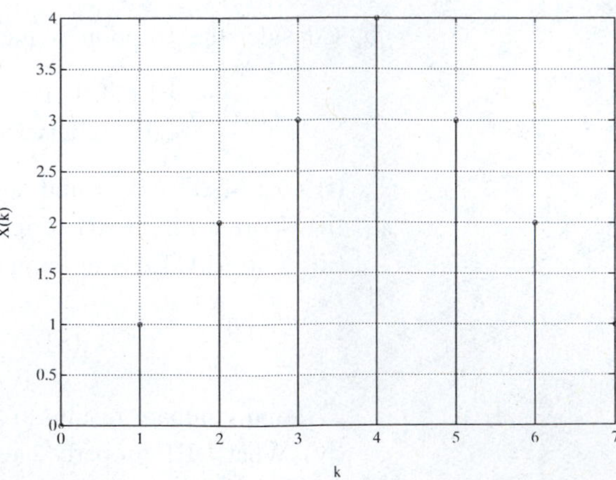

FIGURE P10.19 *The DFT $X(k)$*

CHAPTER 10

b. Assume that a sequence $v(n)$ of length N has all its nonzero values in the interval $n = 0, 1, \ldots, N-1$. Another sequence $s(n)$ is formed by appending N zeros to $v(n)$ to obtain

$$s(n) = \begin{cases} v(n), & 0 \le n \le N-1 \\ 0, & N \le n \le 2N-1. \end{cases}$$

Find an expression relating the even-indexed components of the DFT $S(k)$ to the DFT $V(k)$; that is, express $S(2m)$ in terms of $V(m)$ for $m = 0, 1, \ldots, N-1$.

P10.20 Checking DFTs. It is always helpful to be able to check DFTs, and the symmetry conditions and properties of Problems P10.16 and P10.18 provide ways of doing this. There are some other ways also, and one of these is developed in part (a).

a. Show the following DFT properties:

(i) $X(0) = \sum_{n=0}^{N-1} x(n)$

(ii) $X(N/2) = \sum_{n=0}^{N-1} (-1)^n x(n)$ for N even

(iii) $X(N/4) = \sum_{n=0}^{N-1} (-j)^n x(n)$ for N divisible by 4 with zero remainder

(iv) $X(3N/4) = \sum_{n=0}^{N-1} (+j)^n x(n)$ for N divisible by 4 with zero remainder

b. Use MATLAB to find the DFT of the 128-point sequence

$$x(n) = \begin{cases} 1, & \text{for } n = 0, 1, 127 \\ 0, & \text{otherwise} \end{cases}$$

and check the values using the conditions shown in part (a).

P10.21 A simplified application of correlation. One application of correlation is to detect the presence of signals in noise. Since we're not concerning ourselves with noise yet, we ignore this aspect of the application. We have a discrete system consisting of two correlators, each driven by the same input sequence $r(n)$, as shown in Figure P10.21 on page 606. The purpose of the system is to determine whether $r(n)$ is equal to either $w(n)$ or $z(n)$. (For simplicity, we'll assume that $r(n)$ is either a sequence of zeros or is equal to $w(n)$ or $z(n)$.) The sequences $w(n)$ and $z(n)$ are defined as

$$w(n) = \begin{cases} 1, & n = 0, 2 \\ -1, & n = 1, 3 \\ 0, & \text{otherwise} \end{cases} \qquad z(n) = \begin{cases} 1, & n = 0, 1 \\ -1, & n = 2, 3 \\ 0, & \text{otherwise.} \end{cases}$$

a. If $r(n) = w(n)$, use MATLAB to find and plot the correlations $R_{wr}(p)$ and $R_{zr}(p)$.

PROBLEMS

b. Repeat part (a) for $r(n) = z(n)$.

c. Using your plots of parts (a) and (b), describe how the system allows you to observe the outputs $R_{wr}(p)$ and $R_{zr}(p)$ and to determine whether $r(n) = w(n)$, or $r(n) = z(n)$.

d. Could the system still be used to determine if $w(n)$ or $z(n)$ has been received if these sequences were delayed—that is, if $r(n) = w(n - L)$ or $r(n) = z(n - L)$? Explain.

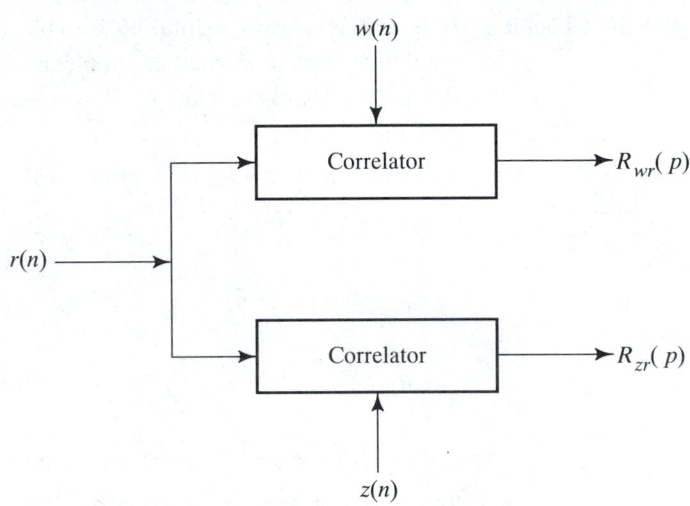

FIGURE P10.21 *Correlation detection*

P10.22 Correlation properties. Show that correlation sequences have the following properties for real sequences.

a. The autocorrelation is an even sequence—i.e., $R_{xx}(p) = R_{xx}(-p)$.

b. The maximum value of the autocorrelation occurs at $p = 0$—that is, $\max R_{xx}(p) = R_{xx}(0)$ for $-\infty \leq p \leq \infty$.

c. The cross correlation sequence depends on the order of the sequences, because $R_{xy}(p) = R_{yx}(-p)$.

d. Correlation and convolution are related by $R_{xy}(p) = x(-p)*y(p)$ or $R_{yx}(p) = x(p)*y(-p)$.

P10.23 DFT of the general sinusoidal sequence. In this problem we want to derive the N-point DFT for the general sequence $v(n) = A\cos(Ln\pi/M)$, where L, n, and M are integers and N is equal to an integer multiple of one period of $v(n)$.

a. Show that the period of this sequence is $P = 2MR/L$, where R is selected to make P equal to the smallest possible integer.

b. In Example 10.3, part (c), we had the sequence $x(n) = 0.5[\cos(n\pi/4) + \cos(5n\pi/28)]$. Find the periods P_1 and P_2 for each term in the sequence.

c. As stated, N is assumed to be an integer multiple of the period P, or $N = qP$. Find the integer q for each term in the sequence $x(n)$ of part (b).

d. The N-point DFT of $v(n) = A\cos(Ln\pi/M)$ is

$$V(k) = \sum_{n=0}^{N-1} A\cos(Ln\pi/M)e^{-j2\pi nk/N}, \qquad k = 0, 1, \ldots, N-1$$
$$= \frac{A}{2}\sum_{n=0}^{N-1} e^{j(L\pi/M - 2\pi k/N)n} + \frac{A}{2}\sum_{n=0}^{N-1} e^{-j(L\pi/M + 2\pi k/N)n},$$
$$k = 0, 1, \ldots, N-1.$$

Knowing that $P = 2MR/L$ and $N = qP$, which gives $N = q \cdot 2MR/L$, use the finite geometric series formula to show that

$$V(k) = \frac{A}{2}\frac{1 - e^{j2\pi[qR-k]}}{1 - e^{jL\pi/(M[1-k/(qR)])}} + \frac{A}{2}\frac{1 - e^{-j2\pi[qR+k]}}{1 - e^{-jL\pi/(M[1+k/(qR)])}},$$
$$k = 0, 1, \ldots, N-1.$$

e. Find the values of k for which the first term in the preceding equation is zero. Find the value of the first term and the value of k for which it is nonzero.

f. Repeat part (e) for the second term.

g. Show that the DFT pair for a sinusoidal sequence is

$$v(n) = A\cos(Ln\pi/M) \Leftrightarrow V(k)$$
$$= \frac{NA}{2}\delta(k - qR) + \frac{NA}{2}\delta(k - N + qR),$$
$$k = 0, 1, \ldots, N-1.$$

DEFINITIONS, TECHNIQUES, AND CONNECTIONS

DISCRETE TIME FOURIER TRANSFORM (DTFT) PAIR

$$X(e^{j\theta}) = \sum_{m=-\infty}^{\infty} x(m)e^{-jm\theta}$$

analysis expression

$$x(n) = \frac{1}{2\pi}\int_{\theta_0}^{\theta_0+2\pi} X(e^{j\theta})e^{jn\theta}\,d\theta$$

synthesis expression

DISCRETE FOURIER TRANSFORM (DFT) PAIR

$$X(k) = \sum_{n=0}^{N-1} x(n)e^{-j(2\pi/N)nk} \qquad k = 0, 1, \ldots, N-1$$

analysis expression

$$x(n) = \frac{1}{N}\sum_{k=0}^{N-1} X(k)e^{j(2\pi/N)nk} \qquad n = 0, 1, \ldots, N-1$$

synthesis expression

SAMPLING PERIOD, RECORD LENGTH, AND FREQUENCY RESOLUTION

$$T_0 = NT = N/f_s \quad \Delta f = 1/T_0 = f_s/N\,.$$

T_0 = record length; N = number of samples; $f_s = 1/T$ = sampling frequency; Δf = frequency resolution.

DFT PROPERTIES

Linearity $ax_1(n) + bx_2(n) \Leftrightarrow aX_1(k) + bX_2(k)$

Symmetry For $x(n)$ a sequence with real values:

symmetry—magnitude and angle $|X(k)| = |X(N-k)|$ and $\angle X(k) = -\angle X(N-k)$ for $k = 1, 2, \ldots, N-1$.

symmetry—real and imaginary $\text{Re}[X(k)] = \text{Re}[X(N-k)]$ and $\text{Im}[X(k)] = -\text{Im}[X(N-k)]$ for $k = 1, 2, \ldots, N-1$.

Circular shift $x_2(n) = x_1(n \oplus m) \Leftrightarrow X_2(k) = X_1(k)e^{j(2\pi/N)(km)}$.

Duality

$$x(n) \Leftrightarrow X(k)$$

$$\frac{1}{N}X(n) \Leftrightarrow x(-k)$$

Relationship to z transform $X(k) = X(z)|_{z=e^{j(2\pi k)/N}} \qquad k = 0, 1, \ldots, N-1$

Real and even or odd sequences $x(n)$ real and even; i.e., $x_1(n) = x(-n) \Rightarrow X(k)$ real and even function of k. $x(n)$ real and odd; i.e., $x_1(n) = -x(-n) \Rightarrow X(k)$ imaginary and odd function of k.

Frequency shifting $x(n)e^{j2\pi rk/N} \Leftrightarrow X(k \ominus r)$

Convolution and DFTs

$$\sum_{m=0}^{N-1} x_1(m)x_2(n \ominus m) = \sum_{l=0}^{N-1} x_1(n \ominus l)x_2(l) \Leftrightarrow X_1(k)X_2(k).$$

Circular convolution in the time domain corresponds to multiplying DFTs. By zero padding the sequences $x_1(n)$ and $x_2(n)$ to appropriate length to obtain $x_1'(n)$ and $x_2'(n)$, linear convolution is given by

$$\sum_{m=0}^{2N-1} x_1'(m)x_2'(n-m) = \sum_{l=0}^{2N-1} x_1'(n-l)x_2'(l) \Leftrightarrow X_1'(k)X_2'(k)$$

FAST FOURIER TRANSFORMS (FFTs)

Computationally efficient ways of computing DFTs: $2N \log_2 N$ real multiplications and $3N \log_2 N$ real additions for N-point DFT by radix-2 algorithm, compared with $4N^2$ real multiplications and $2N(2N-1)$ real additions for using the definition to find the DFT.

FREQUENCY DOMAIN CONVOLUTION

$$x_1(n)x_2(n) \Leftrightarrow \frac{1}{N} X_1(k)¿X_2(k)$$

PARSEVAL'S THEOREM

$$\sum_{n=0}^{N-1} |x(n)|^2 = \frac{1}{N} \sum_{k=0}^{N-1} |X(k)|^2$$

CORRELATION

Autocorrelation $R_{xx}(p) = \sum_{n=-\infty}^{\infty} x(n)x(n+p), \qquad -\infty \leq p \leq \infty$

Cross correlation $R_{x_1x_2}(p) = \sum_{n=-\infty}^{\infty} x_1(n)x_2(n+p), \qquad -\infty \leq p \leq \infty$

Relationship between correlation and convolution	$R_{xy}(p) = x(-p) * y(p), \qquad -\infty \le p \le \infty$
Correlation via DFTs	$\mathrm{DFT}[\tilde{R}_{xy}(p)] = X^*(k) \cdot Y(k)$ where $\tilde{R}_{xy}(p)$ is the circular correlation and $X^*(k)$ is the complex conjugate of $X(k)$.

TABLE 10.2 *The Forms of Fourier*

Discrete Time Fourier Transform (DTFT)

$$x(n) = \frac{1}{2\pi}\int_{\theta_0}^{\theta_0+2\pi} X(e^{j\theta})e^{jn\theta}\,d\theta \Leftrightarrow X(e^{j\theta}) = \sum_{n=-\infty}^{\infty} x(n)e^{-jn\theta}$$

Nonperiodic in the discrete variable n — Periodic in the continuous variable θ

Discrete Fourier Transform (DFT)

$$x(n) = \frac{1}{N}\sum_{k=0}^{N-1} X(k){W_N}^{-nk} \Leftrightarrow X(k) = \sum_{n=0}^{N-1} x(n){W_N}^{nk}, \qquad W_N = e^{-j2\pi/N}$$

$n = 0, 1, \ldots, N-1$ — $k = 0, 1, \ldots, N-1$

Periodic in the discrete variable n — Periodic in the discrete variable k

Two additional forms of Fourier discussed in Chapter 5, "Continuous-Time Fourier Series and Transforms," follow.

Fourier Series

$$x(t) = \sum_{k=-\infty}^{\infty} \mathbf{X_k} e^{jk\omega_0 t} \Leftrightarrow \mathbf{X_k} = \frac{1}{T_0}\int_{t_0}^{t_0+T_0} x(t)e^{-jk\omega_0 t}\,dt$$

Periodic in the continuous variable t — Nonperiodic in the discrete variable k

Continuous-Time Fourier Transform (CTFT)

$$x(t) = \frac{1}{2\pi}\int_{-\infty}^{\infty} X(\omega)e^{j\omega t}\,d\omega \Leftrightarrow X(\omega) = \int_{-\infty}^{\infty} x(t)e^{-j\omega t}\,dt$$

Nonperiodic in the continuous variable t — Nonperiodic in the continuous variable ω

MATLAB FUNCTIONS USED

Function	*Purpose and Use*	*Toolbox*
conj, imag, real	Given: a matrix (or vector), **conj, imag,** and **real** return the conjugate, imaginary part, and real part, respectively.	MATLAB

Function	*Purpose and Use*	*Toolbox*
freqz	Given: TF model of discrete system, **freqz** returns frequency response; *also* Given: the sequence $x(n)$, **freqz** returns the DTFT of $x(n)$.	Signals/Systems, Signal Processing
fft	Given: the sequence $x(n)$, **fft** returns the DFT of $x(n)$.	MATLAB, Signal Processing
ifft	Given: the sequence $X(k)$, **ifft** returns the IDFT of $X(k)$.	MATLAB, Signal Processing
ksxcorr	Given: the sequences $a(n)$ and $b(n)$, **ksxcorr** returns the cross correlation.	California Functions
xcorr	Given: the sequences $a(n)$ and $b(n)$, **xcorr** returns the cross correlation.	Signal Processing

ANNOTATED BIBLIOGRAPHY

1. Brigham, E. Oran, *The Fast Fourier Transform and Its Applications,* Prentice Hall, Inc. Englewood Cliffs, N.J., 1988. *This compact treatment of Fourier methods covers not only DFTs and FFTs, but Fourier series and Fourier transforms also. A large number of illustrations provide pictures of connections between signals in the time and frequency domains. See the Annotated Bibliography of Chapter 5 for more details about this book.*

2. Oppenheim, Alan V., and Ronald W. Schafer, *Discrete-Time Signal Processing,* Prentice Hall, Inc., Englewood Cliffs, N.J., 1989. *Discrete Fourier Transforms are developed in Chapter Eight and the computation of the DFT using various FFT algorithms follows in Chapter Nine. Applications to the Fourier analysis of signals using DFTs makes up Chapter Eleven. This book is a new version of the important text* Digital Signal Processing *written by the same authors and published in 1975.*

3. Proakis, John G., and Dimitris G. Manolakis, *Introduction to Digital Signal Processing,* Macmillan Publishing Company, 1988. *This is an extensive text, which introduces DFTs in Chapter Four and then provides additional detail on their properties, calculation by fast algorithms, and applications in Chapter Nine.*

4. Soliman, Samir S., and Mandyam D. Srinath, *Continuous and Discrete Signals and Systems,* Prentice Hall, Inc., Englewood Cliffs, N.J., 1990. *Chapter Nine includes a concise treatment of DFTs and FFTs, with some applications to digital filter design included in Chapter Ten.*

5. Strum, Robert D., and Donald E. Kirk, *First Principles of Discrete Systems and Digital Signal Processing,* Addison-Wesley Publishing Company, Reading, Mass., 1988. *Chapter Seven contains an extensive introduction to Discrete Fourier Transforms. Included is an expanded treatment of many of the topics considered in the present chapter, along with some more advanced topics, such as spectrum estimation. Chapter Eight develops two forms of radix-2 FFT algorithms and describes FFT applications such as fast convolution.*

ANSWERS

P10.1 a. 410 Hz; b. 0.2 s; c. 82 samples.

P10.2 a. 133.33 Hz; b. 933.33 Hz; c. $X(0) = 2$; d. See A10_2 on the CLS disk; corresponding to $k = \{0, 1, 2, \ldots, 13, 14\}$ are $\{0, 133.33, 266.67, 400, \ldots, -266.67, -133.33\}$ in Hz.

P10.3 a. Corresponding to $k = \{0, 1, 2, 3, 4, 5\}$ are the frequency values in Hz: $\{0, 167, 334, 500, -334, -167\}$. b. $x(n) = 4\cos(n\pi/3) + 3\cos(n2\pi/3) + \cos(n\pi)$; c. $x(0) = 8$; d. $x(n) = \{8, -0.5, -2.5, -2.0, -2.5, -0.5\}$—the same as found by letting $n = 0, 1, 2, 3, 4, 5$ in the expression of (b). See A10_3 on the CLS disk. e. $x(t) = 4\cos(2\pi \cdot 167t) + 3\cos(2\pi \cdot 333t) + \cos(2\pi \cdot 500t)$.

P10.4 a. 0.05 s; b. Corresponding to $k = \{0, 1, 2, 3, 4, 5, 6, 7, 8, 9\}$ are the frequency values in Hz $\{0, 20, 40, 60, 80, 100, -80, -60, -40, -20\}$. c. $x(n) = 0.2 + 0.2\cos(2n\pi/5) + 0.6\cos(4n\pi/5) + 0.2\cos(n\pi)$; d. $x(0) = 1.2$; e. $x(n) = \{1.2, -0.424, 0.424, 0.024, -0.024, 0.8, -0.024, 0.024, 0.424, -0.424\}$; See A10_4. f. Frequencies are 0, 40, 80, and 100 Hz, with amplitudes 0.2, 0.2, 0.6, and 0.2, respectively.

P10.5 a. $\{-3, -11, 4, 15\}$; b. See A10_5. c. $\{6, -7, 0, 15, -9, -4, 4\}$; d. 7 points. e. See A10_5.

WHAT IF? $\{10, -7, 0, 15, -9, -4\}$ ■

P10.6 a. $\{24, 27, 21, 4, 15\}$; b. See A10_6. c. $\{10, 34, 6, 0, 15, 14, -7, 15, 4\}$; d. See A10_6. e. 9 points; f. Same as (c).

WHAT IF? $\{25, 38, 6, 0, 15, 14, -7\}$ ■

P10.7 a. and b. See A10_7; $\{3, -7, -13, 10, 2, 31, -27, 12, -25, -42, -2, 81, 77, -50, -51, -6\}$. c. At $n = 0$, $1 \cdot 3 = 3$; at $n = 15$, $-1 \cdot 6 = -6$. d. The circular convolution cannot be evaluated because the sequences don't have the same number of points. This would be equivalent to multiplying together two DFTs with different numbers of points—a meaningless operation.

P10.8 a. $X(0)$ cannot be determined because it is not included in any symmetry relations, and $X(4)$ is not included in the symmetry relations for $N = 8$. Symmetry indicates that $X(1) = 0.707e^{-j\pi/4}$ and $X(2) = 2$. b. $X(k) = \{3, 1e^{-j\pi/3}, 2e^{-j\pi/4}, 5e^{-j\pi/8}, 5e^{j\pi/8}, 2e^{j\pi/4}, 1e^{j\pi/3}\}$. c. $X(k)$ can't be correct: $X(0)$ must be real; $X(1)$ should equal $X^*(3)$; $X(2)$ should be real.

P10.9 A MATLAB script for determining the sample values, finding the DFT, and plotting is shown next. The plots for $f_0 = 4.0$ Hz are given in A10_9. You

CHAPTER 10

should find as the value of f_0 increases from 0.5 Hz to 8 Hz, the width of the $\sin(x)/x$ pulse decreases, and the DFT approaches a rectangular pulse in the k domain. The frequency domain pulse becomes wider as f_0 increases. The reason that the DFT is not perfectly rectangular is that the $\sin(x)/x$ time-domain pulse has been truncated and values outside the $-1 \le t < 1$ interval have been discarded.

MATLAB Script

```
%A10_9 DFT of sinc sequence
N=64;
f0=4.0;
delt=1/32.0; % sample interval
n=[0:1:N-1];
k=n;
nneg=[-N/2:1:-1]; % negative time indices
npos=[1:1:N/2-1]; % positive time indices
% determine sample values for negative time
xneg=(2*f0*sin(2*pi*f0*nneg*delt))./(2*pi*f0*nneg*delt);
% determine value at t=0
xzro=2*f0;
% determine sample values for positive time
xpos=(2*f0*sin(2*pi*f0*npos*delt))./(2*pi*f0*npos*delt);
% stack together values to obtain x(n)
% note: the implicit periodicity of the DFT is being used
%here
x=[xzro xpos xneg];
Xk=fft(x,N);
magXk=abs(Xk);
%...plotting statements
```

WHAT IF? By increasing the number of samples while keeping a constant sampling frequency, we lengthen the record length T_0 by a factor of 4. As a result, we discard only values outside the range $-4 \le t < 4$, and the frequency domain pulse becomes closer to a true rectangular shape. MATLAB plots for $f_0 = 4.0$ Hz are shown in A10_9. ■

P10.10 a. $X(k) = 3 - 5e^{-j3\pi k/2}$, $k = 0, 1, \ldots, 7$; b. See A10_10.
c. $Y(k) = 5 + 5e^{-j2\pi k/8} + 5e^{-j14\pi k/8} = 5 + 10\cos(2\pi k/8)$, $k = 0, 1, \ldots, 7$. Real because imaginary part is zero; even because $Y(k) = Y(-k) = Y(N - k)$. d. See A10_10;
e. $R(k) = 5e^{-j2\pi k/8} - 5e^{-j14\pi k/8} = 5e^{-j2\pi k/8} - 5e^{j2\pi k/8} = -j10\sin(2\pi k/8)$, $k = 0, 1, \ldots, 7$. Imaginary because real part is zero; odd because $R(k) = -R(-k) = -R(N - k)$. See A10_10.

ANSWERS

P10.11 a. >220 Hz; b. 0.1 s; c. $f_s = 440$ Hz, $T_0 = 0.1$ s, plot is in A10_11. d. DFT$[3.5\cos(9\pi n/22) + 3.5\cos(\pi n/2)] = 77[\delta(k-9) + \delta(k-35) + \delta(k-11) + \delta(k-33)]$; e. Plot is in A10_11. Notice the leakage and that $\Delta f = 6.67$ Hz.

P10.12 a. >220 Hz; b. 0.1 s; c. $f_s = 440$ Hz, $T_0 = 0.1$ s, plot is in A10_12. d. DFT$[\cos(5\pi n/11) + 0.1\cos(9\pi n/22) + 0.1\cos(\pi n/2)] = 22[\delta(k-10) + \delta(k-34)] + 2.2[\delta(k-9) + \delta(k-35)] + 2.2[\delta(k-11) + \delta(k-33)]$, as shown in A10_12. e. The plot is in A10_12. Notice the leakage and that $\Delta f = 6.67$ Hz.

P10.13 a. For $p = \{-3, -2, -1, 0, 1, 2, 3\}$, (1) $R_{xx}(p) = \{8, -4, -2, 21, -2, -4, 8\}$; (2) $R_{yy}(p) = \{3, -5, -5, 15, -5, -5, 3\}$; (3) $R_{xy}(p) = \{12, -8, -7, 12, -3, -3, 2\}$; (4) $R_{yx}(p) = \{2, -3, -3, 12, -7, -8, 12\}$; b. and c. See A10_13 on the CLS disk. Notice that the results appear with $0 \le p \le 6$, or $p = 4, 5, 6$, corresponding to $p = -3, -2, -1$ of part (a).

P10.14 a. (1) $\tilde{R}_{xx}(p) = \{21, 6, -8, 6\}$; (2) $\tilde{R}_{yy}(p) = \{15, -2, -10, -2\}$; (3) $\tilde{R}_{xy}(p) = \{12, 9, -11, -5\}$; (4) $\tilde{R}_{yx}(p) = \{12, -5, -11, 9\}$; b. See A10_14 on the CLS disk.

P10.15 See A10_15.

P10.17 a. (i) See A10_17, (ii) From the plots $|X_1(k)| = |X_2(k)|$; (ii) $x_2(n) = x_1(n \oplus [-1])$ (circular shift) $\Rightarrow X_2(k) = X_1(k)\, e^{-j2\pi k/128}$. (*Note:* At $k = 64$, $X_2(k) = -X_1(k)$, and MATLAB verifies this.) b. (i–iii) See A10_17b; (iv) DFTs are the same, which illustrates linearity; (v) Form the sequences $ax_1(n) + bx_2(n)$ and repeat the operations of part (b) (i–iii). c. $(A/2)\,\delta(n-L) + (A/2)\,\delta(n-[N-L]) \Leftrightarrow A\cos(L2k\pi/N)$, $A\delta(n) \Leftrightarrow A$; d. symmetry conditions are satisfied.

P10.18 *Hint:* Use the DFT definition and the Euler relation to write as sine and cosine, evaluate Re$[X(k)]$, Re$[X(N-k)]$ and compare. Repeat for Im$[X(k)]$.

P10.19 a. $R(k + mN) = X(k)$, $k = 0, 1, \ldots, N-1$, $m = 0, 1$. See A10_19. b. $S(2m) = V(m)$, $m = 0, 1, \ldots, N-1$.

P10.20 b. $X(0) = 3$, $X(64) = 1$, $X(32) = 1$, $X(96) = 1$. See A10_20.

P10.21 a. and b. See A10_21. c. The autocorrelations $R_{ww}(p)$ and $R_{zz}(p)$ have values of 4.0 at $p = 0$ and relatively larger values for other values of p than the cross correlations $R_{zw}(p)$ and $R_{wz}(p)$, which are both 0 at $p = 0$ and range between -1 and 1 for other values of p. d. Yes; in fact, the delay would show up as a shift, relative to $p = 0$, of the point where the autocorrelations have their maximum values. Thus, the autocorrelation peaks could be used to measure or estimate the delay L.

P10.23 b. $P_1 = 2 \cdot 4 \cdot R_1/1 = 8$ and $P_2 = 2 \cdot 28 \cdot R_2/5 = 56$; c. $q_1 = 7$, $q_2 = 1$; e. Zero for $k \ne qR$, at $k = qR$ its value is $NA/2$. f. Zero for $k \ne N - qR$; at $k = N - qR$ its value is $NA/2$.

CHAPTER 11

State-Space Topics for Discrete Systems

PREVIEW

In earlier chapters we used state-space (state-variable) functions such as **dlsim** and **ksdlsim** to simulate linear time-invariant discrete systems with MATLAB. Indeed, a strong motivation for the state form of difference equations is the ease with which such systems can be simulated. There are, however, other good reasons for state equations. We can effectively describe Multiple-Input–Multiple Output (MIMO) systems using state variables, and the well-developed concepts of linear algebra can be a valuable aid in analyzing and designing LTI systems. Finally, state equations also allow a unified treatment of nonlinear and

FIGURE 11.1 *Lotfi A. Zadeh*

time-varying systems. These observations now lead us in this chapter to a more complete development of state-variable theory for discrete systems.

We begin by evaluating the time-domain response of LTI systems described in state-space form to find that the result is an initial condition solution and a forced response in the form of a convolution sum. Closed-form analytical solutions of state equations can be obtained using z transforms, though this is feasible by hand only for systems of relatively low order. Finally, we consider the manipulation of various forms of system models: transfer functions, state equations, and system diagrams. Throughout we will observe that although state equations are capable of describing high-order MIMO systems, they work just as well in characterizing the Single-Input–Single-Output systems that we often encounter.

The state-space approach to analysis and design of discrete linear systems has its primary applications in control theory and in the description of digital filters. Its leading advocate was Rudolf E. Kalman, who, as a graduate student at Columbia University under the leadership of Professor John R. Ragazzini in the late 1950s, demonstrated the need for an alternative to traditional frequency-domain methods that had their roots in communication systems using the results of Bode, Nyquist, and others. The application of state-variable methods broadened with the appearance of the landmark work by Zadeh and Desoer in 1963 [5]. Today, state-variable methods provide an important tool for system design.

BASIC CONCEPTS

COMPLETE SOLUTION OF THE STATE EQUATION IN THE TIME DOMAIN

Starting with $\mathbf{v}(n+1) = \mathbf{A}\mathbf{v}(n) + \mathbf{B}\mathbf{x}(n)$, we will assume that the $\mathbf{A}$ and $\mathbf{B}$ matrices are known. Also known are the initial condition (IC) vector $\mathbf{v}(0)$ with N components and the input vector $\mathbf{x}(n)$, where we allow the possibility of M system inputs. The state vector $\mathbf{v}(n)$ has the same dimensions as the IC vector $\mathbf{v}(0)$, $N \times 1$, while $\mathbf{A}$ is a matrix with N rows and N columns, an $N \times N$ matrix, and $\mathbf{B}$ is $N \times M$. First, for $n = 0$

$$\mathbf{v}(1) = \mathbf{A}\mathbf{v}(0) + \mathbf{B}\mathbf{x}(0)$$

and for $n = 1$

$$\begin{aligned}\mathbf{v}(2) &= \mathbf{A}\mathbf{v}(1) + \mathbf{B}\mathbf{x}(1) = \mathbf{A}\{\mathbf{A}\mathbf{v}(0) + \mathbf{B}\mathbf{x}(0)\} + \mathbf{B}\mathbf{x}(1) \\ &= \mathbf{A}^2\mathbf{v}(0) + \mathbf{A}\mathbf{B}\mathbf{x}(0) + \mathbf{B}\mathbf{x}(1).\end{aligned}$$

Continuing this iterative process leads to

$$\begin{aligned}\mathbf{v}(3) &= \mathbf{A}^3\mathbf{v}(0) + \mathbf{A}^2\mathbf{B}\mathbf{x}(0) + \mathbf{A}\mathbf{B}\mathbf{x}(1) + \mathbf{B}\mathbf{x}(2) \\ &\;\vdots \\ \mathbf{v}(n) &= \mathbf{A}^n\mathbf{v}(0) + \sum_{m=0}^{n-1} \mathbf{A}^{n-m-1}\mathbf{B}\mathbf{x}(m) \\ &= \mathbf{v}_{\mathbf{IC}}(n) + \mathbf{v}_{\mathbf{F}}(n)\end{aligned}$$

where we see the tidy separation of initial condition solution $\mathbf{v}_{\mathbf{IC}}(n)$ and the forced solution $\mathbf{v}_{\mathbf{F}}(n)$—that part of the solution due to the input vector $\mathbf{x}(n)$.[1]

ILLUSTRATIVE PROBLEM 11.1
Iterative Solution of the State Equation

For the system described by the state equation

$$\mathbf{v}(n+1) = \begin{bmatrix} 0 & 1 \\ -0.25 & 0 \end{bmatrix}\mathbf{v}(n) + \begin{bmatrix} 0 \\ 1 \end{bmatrix}x(n)$$

with the initial condition $\mathbf{v}(0) = [2 \quad 3]^T$ and the scalar exponential input sequence $x(n) = (0.5)^n u(n)$, use the iterative process to compute the state vector $\mathbf{v}(4)$.

Solution

For $n \geq 0$, the input sequence is $x(n) = \{1, 0.5, 0.25, 0.125, \ldots\}$.

For $n = 0$, $\mathbf{v}(1) = \begin{bmatrix} 0 & 1 \\ -0.25 & 0 \end{bmatrix}\begin{bmatrix} 2 \\ 3 \end{bmatrix} + \begin{bmatrix} 0 \\ 1 \end{bmatrix}(1) = \begin{bmatrix} 3 \\ 0.5 \end{bmatrix}.$

For $n = 1$, $\mathbf{v}(2) = \begin{bmatrix} 0 & 1 \\ -0.25 & 0 \end{bmatrix}\begin{bmatrix} 3 \\ 0.5 \end{bmatrix} + \begin{bmatrix} 0 \\ 1 \end{bmatrix}(0.5) = \begin{bmatrix} 0.5 \\ -0.25 \end{bmatrix}.$

For $n = 2$, $\mathbf{v}(3) = \begin{bmatrix} 0 & 1 \\ -0.25 & 0 \end{bmatrix}\begin{bmatrix} 0.5 \\ -0.25 \end{bmatrix} + \begin{bmatrix} 0 \\ 1 \end{bmatrix}(0.25) = \begin{bmatrix} -0.250 \\ 0.125 \end{bmatrix}.$

For $n = 3$, $\mathbf{v}(4) = \begin{bmatrix} 0 & 1 \\ -0.25 & 0 \end{bmatrix}\begin{bmatrix} -0.250 \\ 0.125 \end{bmatrix} + \begin{bmatrix} 0 \\ 1 \end{bmatrix}(0.125) = \begin{bmatrix} 0.1250 \\ 0.1875 \end{bmatrix}.$

[1] $\mathbf{v}_{\mathbf{IC}}(n)$ is also known as the zero-input response, and $\mathbf{v}_{\mathbf{F}}(n)$ is called the zero-state response.

We can compute the IC solution and the forced solution separately by using

$$\mathbf{v}(n) = \mathbf{v}_{\mathbf{IC}}(n) + \mathbf{v}_{\mathbf{F}}(n) = \mathbf{A}^n\mathbf{v}(0) + \sum_{m=0}^{n-1} \mathbf{A}^{n-m-1}\mathbf{B}x(m).$$

$$\mathbf{v}_{\mathbf{IC}}(4) = \mathbf{A}^4\mathbf{v}(0) = \begin{bmatrix} 0 & 1 \\ -0.25 & 0 \end{bmatrix}^4 \begin{bmatrix} 2 \\ 3 \end{bmatrix} = \begin{bmatrix} 0.0625 & 0 \\ 0 & 0.0625 \end{bmatrix}\begin{bmatrix} 2 \\ 3 \end{bmatrix} = \begin{bmatrix} 0.1250 \\ 0.1875 \end{bmatrix}.$$

$$\mathbf{v}_{\mathbf{F}}(4) = \sum_{m=0}^{n-1} \mathbf{A}^{n-m-1}\mathbf{B}x(m) = \mathbf{A}^3\mathbf{B}x(0) + \mathbf{A}^2\mathbf{B}x(1) + \mathbf{A}^1\mathbf{B}x(2) + \mathbf{A}^0\mathbf{B}x(3)$$

$$= \begin{bmatrix} 0 & -0.25 \\ 0.0625 & 0 \end{bmatrix}\begin{bmatrix} 0 \\ 1 \end{bmatrix}(1) + \begin{bmatrix} -0.25 & 0 \\ 0 & -0.25 \end{bmatrix}\begin{bmatrix} 0 \\ 1 \end{bmatrix}(0.5)$$

$$+ \begin{bmatrix} 0 & 1 \\ -0.25 & 0 \end{bmatrix}\begin{bmatrix} 0 \\ 1 \end{bmatrix}(0.25) + \begin{bmatrix} 1 & 0 \\ 0 & 1 \end{bmatrix}\begin{bmatrix} 0 \\ 1 \end{bmatrix}(0.125) = \begin{bmatrix} 0 \\ 0 \end{bmatrix}.$$

STABILITY—REVISITED

As discussed in Chapter 7, one definition of stability for a causal LTI system is the following:

The initial condition response of a stable causal system approaches zero as $n \to \infty$ for all initial conditions.

We now investigate this from the point of view of the state equation. First, the initial condition response is given by

$$\mathbf{v}_{\mathbf{IC}}(n) = \mathbf{A}^n\mathbf{v}(0)$$

and in order that

$$\mathbf{v}_{\mathbf{IC}}(n) \to \mathbf{0} \quad \text{as } n \to \infty$$

for arbitrary $\mathbf{v}(0)$, we see that $\mathbf{A}^n$ must approach the null matrix $\mathbf{0}$ as $n \to \infty$; that is, all the elements of $\mathbf{A}^n$ must approach zero as n approaches infinity.

ILLUSTRATIVE PROBLEM 11.2 *Evaluation of Stability*

Given the state equation $\mathbf{v}(n+1) = \mathbf{A}\mathbf{v}(n) + \mathbf{B}x(n)$ with different coefficient ($\mathbf{A}$) matrices, as shown here. Use the concept of raising $\mathbf{A}$ to the nth power to estimate the stability of the following:

$$\text{(a) } \mathbf{A} = \begin{bmatrix} 0 & 1 \\ 0 & 0 \end{bmatrix}; \quad \text{(b) } \mathbf{A} = \begin{bmatrix} 1 & 1 \\ 0 & 1 \end{bmatrix}; \quad \text{(c) } \mathbf{A} = \begin{bmatrix} 1 & 0.632 \\ 0 & 0.368 \end{bmatrix};$$

$$\text{(d) } \mathbf{A} = \begin{bmatrix} 0.632 & 0.632 \\ 0.632 & 0.368 \end{bmatrix}.$$

Solution Using $n = 10$ (any fairly large integer will do), we find the following by using the matrix power **A**^*n* from MATLAB.

a. $\mathbf{A}^{10} = \begin{bmatrix} 0 & 0 \\ 0 & 0 \end{bmatrix}$, stable. In fact $\mathbf{A}^n = \mathbf{0}$ for $n \geq 2$.

b. $\mathbf{A}^{10} = \begin{bmatrix} 1 & 10 \\ 0 & 1 \end{bmatrix}$, unstable.

c. $\mathbf{A}^9 = \begin{bmatrix} 1 & 0.9999 \\ 0 & 0.0001 \end{bmatrix}$, $\mathbf{A}^{10} = \mathbf{A}^{11} = \begin{bmatrix} 1 & 1 \\ 0 & 0 \end{bmatrix}$, unstable.

d. $\mathbf{A}^{10} = \begin{bmatrix} 2.3455 & 1.9062 \\ 1.9062 & 1.5492 \end{bmatrix}$, unstable.

There are better methods for checking stability forthcoming.

SOLUTION OF THE STATE EQUATION BY *z* TRANSFORMS

Given the state equation for a linear time-invariant system

state equation

$$\mathbf{v}(n + 1) = \mathbf{A}\mathbf{v}(n) + \mathbf{B}\mathbf{x}(n)$$

with the initial condition vector $\mathbf{v}(0)$. Taking the unilateral z transform of this matrix equation and making use of the time-shift property that was derived in Chapter 8 for unilateral transforms gives

transformed state equation

$$z\mathbf{V}(z) - z\mathbf{v}(0) = \mathbf{A}\mathbf{V}(z) + \mathbf{B}\mathbf{X}(z).$$

The vector $z\mathbf{V}(z)$ $(N \times 1)$ is multiplied by the identity matrix $\mathbf{I}$ $(N \times N)$ to form the square matrix $z\mathbf{I}\mathbf{V}(z)$, which is combined with the $\mathbf{A}$ matrix to yield

$$(z\mathbf{I} - \mathbf{A})\mathbf{V}(z) = z\mathbf{v}(0) + \mathbf{B}\mathbf{X}(z)$$

and premultiplying both sides of the equation by$(z\mathbf{I} - \mathbf{A})^{-1}$ gives

z transform of v(n)

$$\mathbf{V}(z) = z(z\mathbf{I} - \mathbf{A})^{-1}\mathbf{v}(0) + (z\mathbf{I} - \mathbf{A})^{-1}\mathbf{B}\mathbf{X}(z).$$

Defining $\mathbf{\Phi}(z) = z(z\mathbf{I} - \mathbf{A})^{-1}$ gives the notationally neater form

$$\begin{aligned} \mathbf{V}(z) &= \mathbf{\Phi}(z)\mathbf{v}(0) + \frac{\mathbf{\Phi}(z)}{z}\mathbf{B}\mathbf{X}(z) \\ &= \mathbf{V}_{\mathbf{IC}}(z) + \mathbf{V}_{\mathbf{F}}(z). \end{aligned}$$

Taking the inverse z transform of the transformed state vector $\mathbf{V}(\mathbf{z})$ gives

$$\mathcal{Z}^{-1}[\mathbf{V}(z)] = \mathcal{Z}^{-1}[\mathbf{\Phi}(z)\mathbf{v}(0)] + \mathcal{Z}^{-1}\left[\frac{\mathbf{\Phi}(z)\mathbf{B}\mathbf{X}(z)}{z}\right]$$

or, since $\mathbf{v}(0)$ is a matrix of constants,

$$\mathbf{v}(n) = \boldsymbol{\phi}(n)\mathbf{v}(0) + \sum_{m=1}^{n} \boldsymbol{\phi}(n-m)\mathbf{B}\mathbf{x}(m-1)$$

$$= \boldsymbol{\phi}(n)\mathbf{v}(0) + \sum_{m=0}^{n-1} \boldsymbol{\phi}(n-m-1)\mathbf{B}\mathbf{x}(m)$$

where $\boldsymbol{\phi}(n)$ is the inverse z transform of the matrix $\boldsymbol{\Phi}(z)$; i.e., the ijth element of $\boldsymbol{\phi}(n)$ is the inverse z transform of the ijth element of $\boldsymbol{\Phi}(z)$. The second term on the right side of this equation is the *matrix convolution sum.*

THE TRANSITION MATRIX $\boldsymbol{\phi}(n)$

The matrix $\boldsymbol{\phi}(n)$ is called the state transition matrix and its z transform is

$$z[\boldsymbol{\phi}(n)] = \boldsymbol{\Phi}(z) = z(z\mathbf{I} - \mathbf{A})^{-1}.$$

The steps in determining $\boldsymbol{\phi}(n)$ using z transforms are as follows:

1. Calculate $(z\mathbf{I} - \mathbf{A})$.
2. Obtain the matrix $(z\mathbf{I} - \mathbf{A})^{-1} = z^{-1}\boldsymbol{\Phi}(z)$ by forming the matrix inverse of $(z\mathbf{I} - \mathbf{A})$.
3. Find the state transition matrix $\boldsymbol{\phi}(n)$ by taking the *inverse z transform* of each element of $\boldsymbol{\Phi}(z) = z(z\mathbf{I} - \mathbf{A})^{-1}$.

ILLUSTRATIVE PROBLEM 11.3 *Calculating the State Transition Matrix $\boldsymbol{\phi}(n)$*

An LTI discrete system is described by the state equation

$$\mathbf{v}(n+1) = \begin{bmatrix} 0 & 1 \\ -0.72 & 1.7 \end{bmatrix}\mathbf{v}(n) + \begin{bmatrix} 0 \\ 1 \end{bmatrix}x(n).$$

Find the state transition matrix $\boldsymbol{\phi}(n)$.

Solution

Following the steps just given:

1. $z\mathbf{I} - \mathbf{A} = \begin{bmatrix} z & 0 \\ 0 & z \end{bmatrix} - \begin{bmatrix} 0 & 1 \\ -0.72 & 1.7 \end{bmatrix} = \begin{bmatrix} z & -1 \\ 0.72 & z-1.7 \end{bmatrix}.$
2. The inverse of any matrix $\mathbf{M}$ can be written as the adjoint matrix, **adj M**, divided by the determinant of $\mathbf{M}$ (see Appendix B), or

$$\mathbf{M}^{-1} = \frac{\mathbf{adj\,M}}{|\mathbf{M}|}.$$

Thus for the inverse of $(z\mathbf{I} - \mathbf{A})$, we have

$$(z\mathbf{I} - \mathbf{A})^{-1} = \frac{\mathbf{adj}(z\mathbf{I} - \mathbf{A})}{|z\mathbf{I} - \mathbf{A}|} = \frac{1}{z(z - 1.7) + 0.72} \cdot \begin{bmatrix} z - 1.7 & 1 \\ -0.72 & z \end{bmatrix}.$$

3. $$\mathbf{\Phi}(z) = z(z\mathbf{I} - \mathbf{A})^{-1} = \begin{bmatrix} \dfrac{z(z - 1.7)}{z^2 - 1.7z + 0.72} & \dfrac{z}{z^2 - 1.7z + 0.72} \\ \dfrac{-0.72z}{z^2 - 1.7z + 0.72} & \dfrac{z^2}{z^2 - 1.7z + 0.72} \end{bmatrix}.$$

Using **residue** from MATLAB to expand each element in the $\mathbf{\Phi}(z)/z$ matrix in partial fractions, and then multiplying the resulting fractions by z gives

$$\mathbf{\Phi}(z) = \begin{bmatrix} \dfrac{-8z}{z - 0.9} + \dfrac{9z}{z - 0.8} & \dfrac{10z}{z - 0.9} - \dfrac{10z}{z - 0.8} \\ \dfrac{-7.2z}{z - 0.9} + \dfrac{7.2z}{z - 0.8} & \dfrac{9z}{z - 0.9} - \dfrac{8z}{z - 0.8} \end{bmatrix}.$$

Taking the inverse transform with the use of Table 8.1 produces the transition matrix

$$\boldsymbol{\phi}(n) = \begin{bmatrix} -8.0(0.9)^n + 9.0(0.8)^n & 10(0.9)^n - 10(0.8)^n \\ -7.2(0.9)^n + 7.2(0.8)^n & 9(0.9)^n - 8(0.8)^n \end{bmatrix}.$$

■

CROSS-CHECK A good check here is that $\boldsymbol{\phi}(0) = \mathbf{I}$, which is true for all $\boldsymbol{\phi}(n)$ matrices. ■

CHARACTERISTIC EQUATION AND CHARACTERISTIC ROOTS (EIGENVALUES)

For a general Nth-order system, the matrix $z\mathbf{I} - \mathbf{A}$ has the following appearance:

$$z\mathbf{I} - \mathbf{A} = \begin{bmatrix} z - a_{11} & -a_{12} & \cdots & -a_{1N} \\ -a_{21} & z - a_{22} & \cdots & -a_{2N} \\ \vdots & \vdots & \cdots & \vdots \\ -a_{N1} & -a_{N2} & \cdots & z - a_{NN} \end{bmatrix}$$

and for the inverse of $(z\mathbf{I} - \mathbf{A})$, we have

$$(z\mathbf{I} - \mathbf{A})^{-1} = \frac{\mathbf{adj}(z\mathbf{I} - \mathbf{A})}{|z\mathbf{I} - \mathbf{A}|}.$$

In calculating the determinant $|z\mathbf{I} - \mathbf{A}|$, we see that one of the terms will be the product of the diagonal elements of $z\mathbf{I} - \mathbf{A}$:

$$(z - a_{11})(z - a_{22})\cdots(z - a_{NN}) = z^N + c_{N-1}z^{N-1} + \cdots + c_0$$

a polynomial of degree N with the leading coefficient of unity. There will also be other terms coming from the off-diagonal elements of $z\mathbf{I} - \mathbf{A}$, but none will have a degree as high as N. Thus we conclude that the determinant of $z\mathbf{I} - \mathbf{A}$ is of the form

characteristic polynomial

$$|z\mathbf{I} - \mathbf{A}| = z^N + \alpha_{N-1}z^{N-1} + \cdots + \alpha_1 z + \alpha_0 .$$

This is known as the *characteristic polynomial* of the matrix **A**, and it plays a vital role in the dynamic behavior of the system. When equated to zero, this characteristic polynomial becomes the characteristic equation (CE)

characteristic equation

$$z^N + \alpha_{N-1}z^{N-1} + \cdots + \alpha_1 z + \alpha_0 = 0 .$$

This CE can be written in factored form as

factored CE

$$(z - r_1)(z - r_2)\cdots(z - r_N) = 0$$

where $z_1 = r_1, z_2 = r_2, \ldots, z_N = r_N$ are the *characteristic roots,* or *eigenvalues,* of the system. These roots determine the essential features of the unforced (initial condition) behavior of the system, and for a stable causal system all these roots must have magnitudes less than 1. These characteristic roots also create the terms in the partial fraction expansion of the matrix $\mathbf{\Phi}(z)$ whose inverse z transform is the *transition matrix* $\boldsymbol{\phi}(n)$. The eigenvalues, usually denoted as λ_k, $k = 1, 2, \ldots, N$, and the characteristic roots are the roots of the characteristic equation $|\lambda\mathbf{I} - \mathbf{A}| = 0$.

ILLUSTRATIVE PROBLEM 11.4 *System Characteristics from the State Equation*

A causal system is described by the state and output equations

$$\mathbf{v}(n+1) = \begin{bmatrix} 0 & 1 \\ -1 & 1 \end{bmatrix}\mathbf{v}(n) + \begin{bmatrix} 0 \\ 1 \end{bmatrix}x(n) \quad \text{and} \quad y(n) = v_2(n) .$$

a. What is the characteristic equation and what are the characteristic roots?

b. Is the system stable?

Solution

a. The characteristic equation is found from $|z\mathbf{I} - \mathbf{A}| = 0$ as $z^2 - z + 1 = 0$, and the characteristic roots are $z_{1,2} = 0.50 \pm j0.866 = 1e^{\pm j\pi/3}$.

b. The system is unstable because for a stable causal system the magnitude of all of the system roots must be less than 1. These roots lie on the unit circle in the z plane, and the IC response of this system is an undamped oscillation with a digital frequency of $\theta = \pi/3$ rad or a period of 6 samples. This is shown in Figure 11.2, where $y(n)$ is plotted for an initial condition vector of $\mathbf{v}(0) = [1 \quad 0]^T$ and $x(n) = 0$.

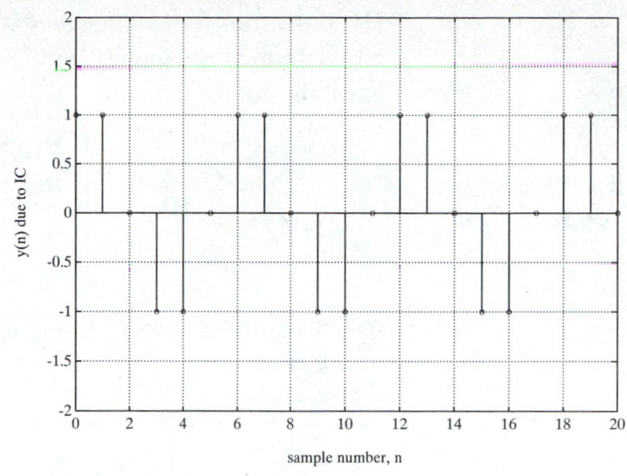

FIGURE 11.2 *IC response for Illustrative Problem 11.4*

MATLAB Script

```
%F11_2 IC response for Illustrative Problem 11.4
A=[0,1;-1,1];                              % A matrix
B=[0;1];                                   % B matrix
C=[0,1];                                   % C matrix
D=[0];                                     % D, a scalar
v0=[1,0]';                                 % IC vector
n=0:1:20;                                  % start:increment:stop
X=[0*n]';                                  % input vector—zero
[y,v]=ksdlsim(A,B,C,D,X,v0);               % call ksdlsim
%...plotting statements
```

TRANSFER FUNCTION MATRIX

For an initial state $\mathbf{v}(0)$ assumed to be zero, the z transform of the state equation is

$$\mathbf{V}(z) = (z\mathbf{I} - \mathbf{A})^{-1}\mathbf{B}\mathbf{X}(z) = \frac{\mathbf{\Phi}(z)}{z}\mathbf{B}\mathbf{X}(z).$$

The multiple system outputs (P in number) defined in the vector $\mathbf{y}(n)$ are given by

$$\mathbf{y}(n) = \mathbf{C}\mathbf{v}(n) + \mathbf{D}\mathbf{x}(n)$$

and the z transform is

$$\mathbf{Y}(z) = \mathbf{C}\mathbf{V}(z) + \mathbf{D}\mathbf{X}(z) = \mathbf{C}\frac{\mathbf{\Phi}(z)}{z}\mathbf{B}\mathbf{X}(z) + \mathbf{D}\mathbf{X}(z)$$
$$= \left[\mathbf{C}\frac{\mathbf{\Phi}(z)}{z}\mathbf{B} + \mathbf{D}\right]\mathbf{X}(z) = \mathbf{H}(z)\mathbf{X}(z).$$

$\mathbf{H}(z)$, the *transfer function matrix,* relates the z transform of the outputs to the z transform of the inputs. With $\mathbf{Y}(z) = \mathbf{H}(z)\mathbf{X}(z)$, the transfer function matrix is of the form

transfer function matrix

$$\mathbf{H}(z) = \begin{bmatrix} H_{11}(z) & H_{12}(z) & \cdots & H_{1M}(z) \\ H_{21}(z) & H_{22}(z) & \cdots & H_{2M}(z) \\ \vdots & \vdots & \cdots & \vdots \\ H_{P1}(z) & H_{P2}(z) & \cdots & H_{PM}(z) \end{bmatrix}$$

where the transforms of the $M \times 1$ input vector $\mathbf{x}(n)$ and the $P \times 1$ output vector $\mathbf{y}(n)$ are

$$\mathbf{X}(z) = [X_1(z) \quad X_2(z) \quad \ldots \quad X_M(z)]^T \quad \text{and}$$
$$\mathbf{Y}(z) = [Y_1(z) \quad Y_2(z) \quad \ldots \quad Y_P(z)]^T.$$

Thus the particular transfer function that relates the ith output to the jth input is $H_{ij}(z)$, and for M possible inputs and P possible outputs, we see that a system has $M \cdot P$ different transfer functions. In reality, therefore, the complete transfer function model for an LTI system is the transfer function matrix $\mathbf{H}(z)$ rather than a scalar transfer function $H(z)$, as discussed in Chapter 8.

The inverse z transform of the transfer function matrix is, of course, also a matrix, known as the *impulse response matrix,*

impulse response matrix

$$\mathbf{h}(n) = \mathcal{Z}^{-1}[\mathbf{H}(z)].$$

FREQUENCY RESPONSE MATRIX

In Chapter 9, "Frequency Response of Discrete Systems," one method used to find a system's frequency response was to substitute $e^{j\theta}$ for z in the transfer function $H(z)$. The same idea applies to the state-space model; the frequency response matrix is

frequency response matrix

$$\mathbf{H}(e^{j\theta}) = \mathbf{C}e^{-j\theta}\boldsymbol{\phi}(e^{j\theta})\mathbf{B} + \mathbf{D}.$$

ILLUSTRATIVE PROBLEM 11.5 ***Transfer Functions and Stability***

A causal system is described by

$$\mathbf{v}(n+1) = \begin{bmatrix} 0 & 1 \\ -1 & 1 \end{bmatrix}\mathbf{v}(n) + \begin{bmatrix} 0 \\ 1 \end{bmatrix}x(n) \quad \text{and} \quad y(n) = v_2(n).$$

a. Find the scalar transfer function $H(z) = Y(z)/X(z)$ and the unit sample response $h(n)$.

b. What is the characteristic equation and what are the characteristic roots?

c. Is the system stable?

Solution

a. With $y(n) = v_2(n)$ we have $\mathbf{C} = [0 \quad 1]$, $\mathbf{D} = \mathbf{0}$, and

$$\mathbf{H}(z) = \mathbf{C}\frac{\mathbf{\Phi}(z)}{z}\mathbf{B} = \mathbf{C}(z\mathbf{I} - \mathbf{A})^{-1}\mathbf{B}$$

$$= [0 \quad 1](z\mathbf{I} - \mathbf{A})^{-1}\begin{bmatrix} 0 \\ 1 \end{bmatrix} \quad \text{where}$$

$$(z\mathbf{I} - \mathbf{A})^{-1} = \frac{\mathbf{adj}(z\mathbf{I} - \mathbf{A})}{|z\mathbf{I} - \mathbf{A}|} = \frac{1}{z(z-1)+1} \cdot \begin{bmatrix} z-1 & 1 \\ -1 & z \end{bmatrix}.$$

Multiplying out the three matrices $\mathbf{C}$, $(z\mathbf{I} - \mathbf{A})^{-1}$, and $\mathbf{B}$ produces the scalar transfer function

$$H(z) = Y(z)/X(z) = z/(z^2 - z + 1)$$

and after expanding $H(z)/z$ and using Table 8.1, we have

$$h(n) = 1.15\cos(n\pi/3 - \pi/2)u(n) = 1.15\sin(n\pi/3)u(n).$$

b. The characteristic equation can be found from the denominator of $H(z)$ as $z^2 - z + 1 = 0$, and the characteristic roots are $z_{1,2} = 0.50 \pm j0.866 = 1e^{\pm j\pi/3}$.

c. The system is unstable, because for a stable causal system the magnitude of all of the system roots must be less than 1. These roots lie on the unit circle in the z plane, and the unit sample response is an undamped oscillation with a digital frequency of $\theta = \pi/3$ rad or a period of 6 samples.

CROSS-CHECK The period of oscillation of 6 samples is shown in Figure 11.3, where the unit sample response $h(n)$ is plotted. ■

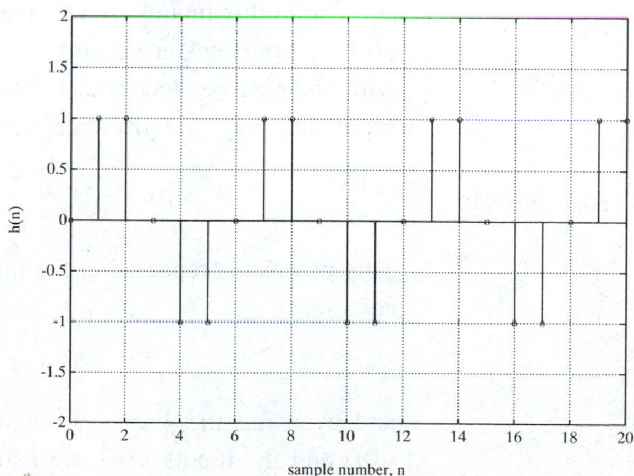

FIGURE 11.3 *Unit sample response*

MATLAB Script

```
%F11_3 Unit sample response of Illustrative Problem 11.5
a=[0,1;-1,1];
b=[0;1];
n=0:1:20;
x=zeros(size(n));
x(1)=1;
y=filter(b,a,x);
%...plotting statements
```

MANIPULATION OF SYSTEM MODELS

In Chapter 8 we started with an SISO system's difference equation (DE) and determined the corresponding scalar transfer function $H(z)$ and/or a system diagram (SFG) that related the output transform $Y(z)$ to the input transform $X(z)$. In this same chapter on z-transform applications, we moved from the unit sample response model $h(n)$ to the system transfer function model $H(z)$ and vice versa. Now, with a background in state-space concepts, let's look at a few other possibilities for moving about or manipulating these system models that can now include the state model. Frequently, we may know or be supplied with one particular system description and want to convert it to a different one for easier design or analysis computations. In contrast with Chapter 8, we will include Multiple-Input–Multiple-Output (MIMO) systems.

STATE AND OUTPUT EQUATIONS FROM THE SYSTEM DIAGRAM

Often we have a system diagram for an LTI system and we need to write a set of state equations as well as the system output equations. The state model is useful for simulating MIMO systems with nonzero initial conditions. We can derive this model directly from the signal flowgraph or block diagram. To do this, the states are designated as the outputs of the delays and the signals with their associated branch gains are summed at the inputs of the delays producing, say, the jth state equation

jth state equation

$$v_j(n+1) = \sum_{k=1}^{N} a_{jk} v_k(n) + \sum_{k=1}^{M} b_{jk} x_k(n) .$$

This process is repeated at the input of all N delays, yielding the matrix state equation

$$\mathbf{v}(n+1) = \mathbf{A}\mathbf{v}(n) + \mathbf{B}\mathbf{x}(n)$$

and by using the graph to obtain the indicated linear combination of states $\mathbf{v}(n)$ and the inputs $\mathbf{x}(n)$, we can write the output equation:

$$\mathbf{y}(n) = \mathbf{C}\mathbf{v}(n) + \mathbf{D}\mathbf{x}(n) .$$

ILLUSTRATIVE PROBLEM 11.6
State Model from an SFG

A discrete system is modeled by the SFG of Figure 11.4, where the states have been arbitrarily assigned as indicated. Find the **A**, **B**, **C**, and **D** matrices of the state-space model.

Solution

Following the previous line of reasoning at each of the delays gives the state equations: $v_1(n+1) = v_1(n) + 4x_1(n)$; $v_2(n+1) = 2v_2(n) + 5x_1(n) + 6x_2(n)$; $v_3(n+1) = 3v_3(n) + 7x_2(n)$. Summing the signals incoming to the output nodes gives the output equations: $y_1(n) = v_1(n) - v_2(n)$; $y_2(n) = v_1(n) + v_3(n)$. Thus the desired matrices are

$$\mathbf{A} = \begin{bmatrix} 1 & 0 & 0 \\ 0 & 2 & 0 \\ 0 & 0 & 3 \end{bmatrix}, \quad \mathbf{B} = \begin{bmatrix} 4 & 0 \\ 5 & 6 \\ 0 & 7 \end{bmatrix}, \quad \mathbf{C} = \begin{bmatrix} 1 & -1 & 0 \\ 1 & 0 & 1 \end{bmatrix}, \quad \mathbf{D} = \mathbf{0}.$$

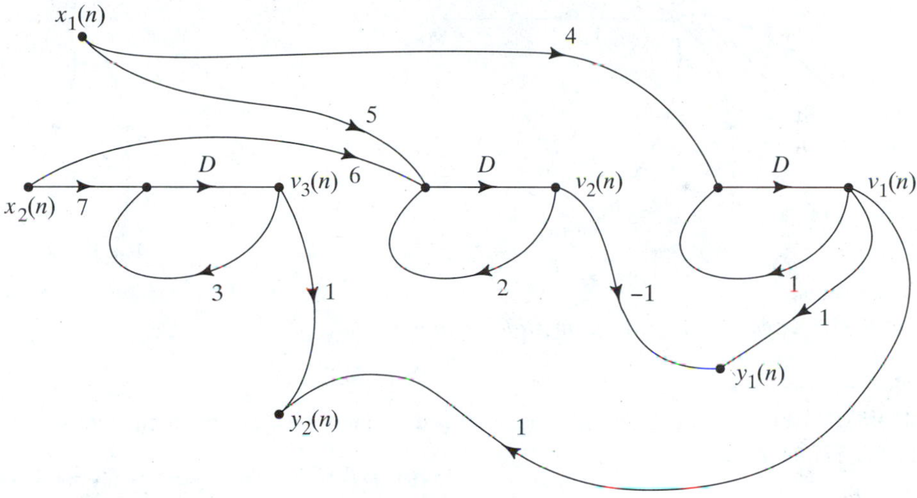

FIGURE 11.4 *SFG for Illustrative Problem 11.6*

SYSTEM DIAGRAM FROM STATE EQUATION

We may want to reverse the preceding process and obtain a pictorial representation of a system from a set of state and output equations. To develop this SFG or block diagram we start with the matrix state equation

$$\mathbf{v}(n+1) = \mathbf{A}\mathbf{v}(n) + \mathbf{B}\mathbf{x}(n)$$

and write out the jth state equation, giving

$$v_j(n+1) = \sum_{k=1}^{N} a_{jk} v_k(n) + \sum_{k=1}^{M} b_{jk} x_k(n).$$

To solve for $v_j(n)$ we make use of the idea that $v_j(n)$ can be found from $v_j(n+1)$ by sending the signal $v_j(n+1)$ through a unit delay, as pictured in Figure 11.5a in the time domain and in Figure 11.5b for its transform-domain counterpart. Thus we need $N + M$ (N for the states and M for the inputs) branches (some of which have zero gains) in the SFG for the inputs to each delay unit. This process needs to be repeated for each of the N states. The final step is to add the graphical representation of the output equation $\mathbf{y}(n) = \mathbf{Cv}(n) + \mathbf{Dx}(n)$ to the signal flowgraph.

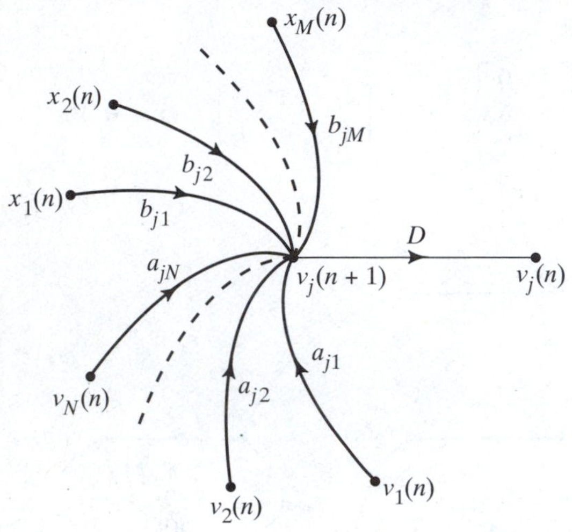

a. Time domain

b. Transform domain

FIGURE 11.5 *System diagram from state equations*

ILLUSTRATIVE PROBLEM 11.7
System Diagram from State and Output Equations

A digital filter is described by the state and output equations

$$\mathbf{v}(n+1) = \mathbf{Av}(n) + \mathbf{Bx}(n) \quad \text{and} \quad \mathbf{y}(n) = \mathbf{Cv}(n) + \mathbf{Dx}(n)$$

where

$$\mathbf{A} = \begin{bmatrix} 1 & 0.500 & 0.333 \\ 1 & 0 & 0 \\ 0 & 1 & 0 \end{bmatrix}, \quad \mathbf{B} = \begin{bmatrix} 1 \\ 0 \\ 0 \end{bmatrix}, \quad \mathbf{C} = [1 \quad 2 \quad 0], \quad \text{and} \quad \mathbf{D} = \mathbf{0}.$$

a. Draw a signal flowgraph that models this digital filter.

b. Draw a z-domain block diagram model.

c. Use either diagram from part (a) or (b) and MGR to find the transfer function $H(z) = Y(z)/X(z)$.

Solution

a. The primary nodes are $v_1(n+1)$, $v_1(n)$, $v_2(n+1)$, $v_2(n)$, $v_3(n+1)$, and $v_3(n)$ for the states; $x(n)$ for the input; $y(n)$ for the output. See Figure 11.6a for this realization.

b. Replacing $v_j(n)$ with $V_j(z)$, $v_j(n+1)$ with $zV_j(z)$, and using the block diagram symbols from Chapter 8 gives the diagram of Figure 11.6b.

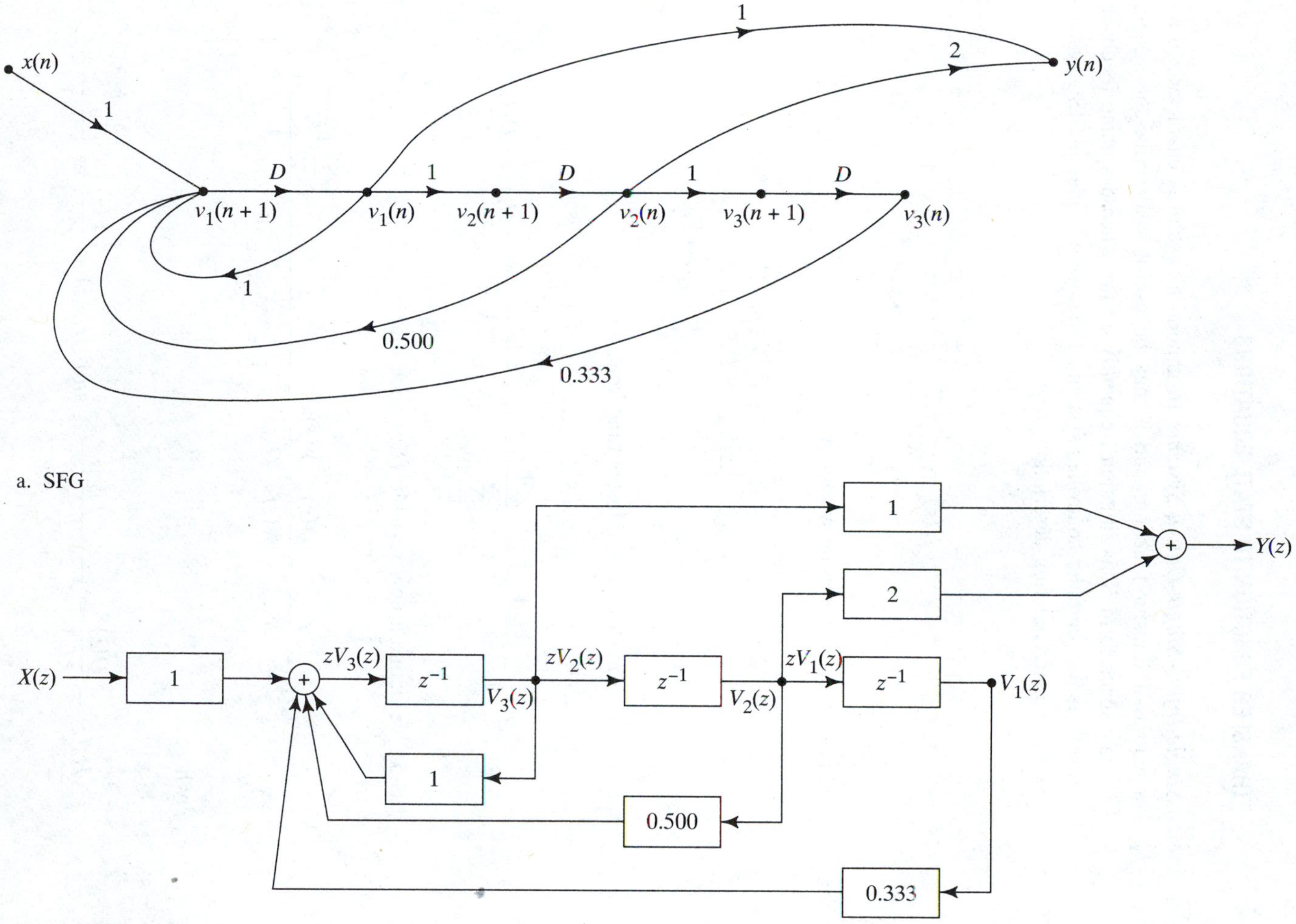

FIGURE 11.6 *SFG and block diagram realizations from state and output equations*

BASIC CONCEPTS

c. There are three touching loops and the two paths from $X(z)$ to $Y(z)$ touch all these loops, so we have $H(z) = (1z^{-1} + 2z^{-2})/(1 - z^{-1} - 0.500z^{-2} - 0.333z^{-3}) = (z^2 + 2z)/(z^3 - z^2 - 0.500z - 0.333)$.

TRANSFER FUNCTION TO STATE EQUATIONS

Single-Input–Single-Output System In general a system is characterized by a matrix of transfer functions and it may be useful to describe the system in state-space form. This situation frequently occurs when designing feedback controllers for computer-controlled systems. First, we consider an SISO system whose transfer function description is

$$H(z) = \frac{Y(z)}{X(z)} = \frac{\sum_{k=0}^{L} b_k z^{-k}}{\sum_{k=0}^{N} a_k z^{-k}}$$

which for $a_0 \equiv 1$ can be written as

$$H(z) = \frac{Y(z)}{X(z)} = \frac{\sum_{k=0}^{L} b_k z^{-k}}{1 + \sum_{k=1}^{N} a_k z^{-k}}$$

or as the product of two transfer functions, namely,

$$H(z) = \frac{Y(z)}{X(z)} = \frac{1}{1 + \sum_{k=1}^{N} a_k z^{-k}} \cdot \frac{\sum_{k=0}^{L} b_k z^{-k}}{1} = H_a(z) \cdot H_b(z).$$

Next we define a new variable $Q(z)$ such that

$$H(z) = \frac{Y(z)}{X(z)} = \frac{Q(z)}{X(z)} \cdot \frac{Y(z)}{Q(z)} = \frac{1}{1 + \sum_{k=1}^{N} a_k z^{-k}} \cdot \frac{\sum_{k=0}^{L} b_k z^{-k}}{1}$$

where

$$H_a(z) = \frac{Q(z)}{X(z)} = \frac{1}{1 + \sum_{k=1}^{N} a_k z^{-k}} \quad \text{and} \quad H_b(z) = \frac{Y(z)}{Q(z)} = \frac{\sum_{k=0}^{L} b_k z^{-k}}{1}.$$

We notice that $H_a(z)$ implements the poles of $H(z)$ and $H_b(z)$ implements the zeros of $H(z)$. Rearranging the equation for $H_a(z)$ and taking the inverse z transform gives

$$Q(z) = X(z) - \sum_{k=1}^{N} a_k z^{-k} Q(z) \quad \text{or} \quad q(n) = x(n) - \sum_{k=1}^{N} a_k q(n-k).$$

Referring to the section "The State-Space Model for Nth-Order Systems" in Chapter 7, we define the state variables as $v_1(n), v_2(n), \ldots, v_N(n)$ and write

$$v_1(n) = q(n-N), \qquad v_2(n) = q(n-N+1),$$
$$v_3(n) = q(n-N+2), \ldots, v_N(n) = q(n-1).$$

Then, we can establish the following relationships

$$\begin{aligned}
v_1(n+1) &= q(n-N+1) = v_2(n) \\
v_2(n+1) &= q(n-N+1+1) = v_3(n) \\
&\vdots \\
v_N(n+1) &= q(n) = x(n) - a_1 v_N(n) - a_2 v_{N-1}(n) - \cdots - a_N v_1(n).
\end{aligned}$$

Putting all this into matrix form, the state vector is defined as

state vector

$$\mathbf{v}(n) = [v_1(n) \quad v_2(n) \quad \cdots \quad v_{N-1}(n) \quad v_N(n)]^T$$

and the matrix state equation follows as

state equation

$$\begin{aligned}
\mathbf{v}(n+1) &= \begin{bmatrix} 0 & 1 & 0 & 0 & \cdots & 0 \\ 0 & 0 & 1 & 0 & \cdots & 0 \\ \vdots & \vdots & \vdots & \vdots & \vdots & \vdots \\ 0 & 0 & 0 & 0 & \cdots & 1 \\ -a_N & -a_{N-1} & -a_{N-2} & -a_{N-3} & \cdots & -a_1 \end{bmatrix} \begin{bmatrix} v_1(n) \\ v_2(n) \\ \vdots \\ v_{N-1}(n) \\ v_N(n) \end{bmatrix} \\
&\quad + \begin{bmatrix} 0 \\ 0 \\ \vdots \\ 0 \\ 1 \end{bmatrix} x(n) \\
&= \mathbf{A}\mathbf{v}(n) + \mathbf{B}x(n).
\end{aligned}$$

Now we need to concern ourselves with obtaining an equation for the system output $y(n)$. From the transfer function

$$H_b(z) = \frac{Y(z)}{Q(z)} = \frac{\sum_{k=0}^{L} b_k z^{-k}}{1}$$

we can solve for $Y(z)$ and take the inverse z transform to obtain

$$Y(z) = \sum_{k=0}^{L} b_k z^{-k} Q(z) \quad \text{or} \quad y(n) = \sum_{k=0}^{L} b_k q(n-k).$$

But, $v_1(n) = q(n-N)$, $v_2(n) = q(n-N+1)$, $v_3(n) = q(n-N+2), \ldots,$ $v_N(n) = q(n-1)$, giving

$$y(n) = b_0 v_N(n+1) + b_1 v_N(n) + b_2 v_{N-1}(n) + \cdots + b_L v_{N-L+1}(n).$$

We need to eliminate the $v_N(n+1)$ term by using the last row of the matrix state equation. Thus, for $L = N$,

$$\begin{aligned} y(n) = {} & b_0[-a_N v_1(n) - a_{N-1} v_2(n) - a_{N-2} v_3(n) - \cdots - a_1 v_N(n) + x(n)] \\ & + b_1 v_N(n) + b_2 v_{N-1}(n) + \cdots + b_L v_{N-L+1}(n) \\ = {} & [-b_0 a_1 + b_1] v_N(n) + [-b_0 a_2 + b_2] v_{N-1}(n) \\ & + [-b_0 a_3 + b_3] v_{N-2}(n) + \cdots + [-b_0 a_N + b_L] v_1(n) + b_0 x(n). \end{aligned}$$

In matrix form the output equation is

output equation

$$\begin{aligned} y(n) &= [-b_0 a_N + b_L, -b_0 a_{N-1} + b_{L-1}, \ldots, -b_0 a_2 + b_2, \\ &\qquad - b_0 a_1 + b_1] \mathbf{v}(n) + b_0 x(n) \\ &= \mathbf{C}\mathbf{v}(n) + \mathbf{D}x(n) \end{aligned}$$

where **C** is the row vector shown and **D** is the scalar b_0 in this case.

ILLUSTRATIVE PROBLEM 11.8 ***Transfer Function to State and Output Equations***

Find a state-space model for the transfer function $H(z) = Y(z)/X(z) = 2z^2/(z^2 - 0.25) = 2/(1 - 0.25z^{-2})$.

Solution

With $H(z) = H_a(z) \cdot H_b(z)$, we have $H_a(z) = Q(z)/X(z) = 1/(1 - 0.25z^{-2})$. Following the procedure outlined previously, $q(n-2) = v_1(n)$, $q(n-1) = v_1(n+1) = v_2(n)$, and $q(n) = v_2(n+1) = 0.25v_1(n) + x(n)$. From $H_b(z) = Y(z)/Q(z) = 2$ we have $Y(z) = 2Q(z)$, or $y(n) = 2q(n) = 2v_2(n+1) = 2[0.25v_1(n) + x(n)]$. In matrix form the results are

$$\mathbf{v}(n+1) = \begin{bmatrix} 0 & 1 \\ 0.25 & 0 \end{bmatrix} \mathbf{v}(n) + \begin{bmatrix} 0 \\ 1 \end{bmatrix} x(n) \quad \text{and} \quad y(n) = [0.50 \quad 0]\mathbf{v}(n) + 2x(n).$$

Single-Input–Multiple-Output System When a system has one input and several outputs, it is called a Single-Input–Multiple-Output (SIMO) system. The system output vector (P by 1) is given by

$$\mathbf{Y}(z) = \mathbf{H}(z)X(z)$$

and the transfer function matrix is a column vector of the form

$$\mathbf{H}(z) = \begin{bmatrix} H_1(z) \\ H_2(z) \\ \vdots \\ H_P(z) \end{bmatrix} = \begin{bmatrix} N_1(z)/D(z) \\ N_2(z)/D(z) \\ \vdots \\ N_P(z)/D(z) \end{bmatrix}$$

where the denominator polynomial of each transfer function is denoted as $D(z)$ and the numerator polynomials are $N_1(z), N_2(z), \ldots, N_P(z)$, respectively. Consider the first transfer function $H_1(z) = N_1(z)/D(z) = H_a(z) \cdot H_b(z)$, where we let $H_a(z) = 1/D(z)$ and $H_b(z) = N_1(z)$ and then follow the procedures of the previous section to create the state and output equations for $H_1(z)$. We repeat this procedure for each of the P transfer functions, and the result is one state equation, $\mathbf{v}(n+1) = \mathbf{A}\mathbf{v}(n) + \mathbf{B}x(n)$, and P scalar output equations, each of the form $y_i(n) = \mathbf{C}_i\mathbf{v}(n) + d_i x(n)$. The P output equations are then simply grouped into the matrix equation

$$\mathbf{y}(n) = \mathbf{C}\mathbf{v}(n) + \mathbf{D}x(n)$$

where $\mathbf{y}(n)$ is a P by 1 output vector, $\mathbf{C}$ is a P by N matrix, $\mathbf{v}(n)$ is the N by 1 state vector, $\mathbf{D}$ is a P by 1 column vector, and $x(n)$ is the scalar or single input.

Comment: For a general discussion of determining state-space realizations from a transfer function matrix see Thomas Kailath, *Linear Systems,* (Englewood Cliffs, N.J.: Prentice Hall, Inc., 1980), Chapter 6.

SOLVED EXAMPLES AND MATLAB APPLICATIONS

EXAMPLE 11.1 *Initial Condition Solution of a State Equation*

A discrete system is described by the state and output equations

$$\mathbf{v}(n+1) = \begin{bmatrix} k & -1 \\ 1 & 0 \end{bmatrix}\mathbf{v}(n) + \begin{bmatrix} 3 & 0 \\ 0 & -4 \end{bmatrix}x(n) \quad \text{and} \quad y(n) = [0 \quad 1]\mathbf{v}(n).$$

a. Estimate the algebraic form of the initial condition response from simply knowing the characteristic roots. Determine the expression for $y_{\text{IC}}(n)$ for the following values of k: $0, 1, -1, 2, -2$. Use the MATLAB function **eig** (eigenvalues) to find (or check) the system eigenvalues (roots).

b. Use **dlsim** or **ksdlsim** to plot $y_{\text{IC}}(n)$ for the different situations of part (a).

WHAT IF? A quadratic equation can be written as $(z - r_1)(z - r_2) = z^2 + z(-r_1 - r_2) + r_1 r_2 = 0$, where the coefficient of the z^1 term equals the negative of the sum of the roots, and the constant term is the product of the roots. Use this information to locate the regions of the z plane where roots may be found for positive and negative values of k. Can this system ever have a decaying oscillatory response? ■

Solution

a. *Algebraic form for* $y_{IC}(n)$. The CE is $|z\mathbf{I} - \mathbf{A}| = 0$, or $z^2 - kz + 1 = 0$.

(i) $k = 0$: $z^2 + 1 = 0$, $z_{1,2} = \pm j1 = 1e^{\pm j\pi/2}$, and $y_{IC}(n) = C_1(1e^{j\pi/2})^n + C_1{}^*(1e^{-j\pi/2})^n = M\cos(\pi n/2 + P)$. An undamped oscillatory response with period $N = 4$.

(ii) $k = 1$: $z^2 - z + 1 = 0$, $z_{1,2} = 0.5 \pm j0.866 = 1e^{\pm j\pi/3}$, and $y_{IC}(n) = C_1(1e^{j\pi/3})^n + C_1{}^*(1e^{-j\pi/3})^n = M\cos(\pi n/3 + P)$. An undamped oscillatory response with a period of 6.

(iii) $k = -1$: $z^2 + z + 1 = 0$, $z_{1,2} = -0.5 \pm j0.866 = 1e^{\pm j2.09} = 1e^{\pm j2\pi/3}$, $y_{IC}(n) = C_1(1e^{j2\pi/3})^n + C_1{}^*(1e^{-j2\pi/3})^n = M\cos(2\pi n/3 + P)$. An undamped oscillatory response with a period of 3.

(iv) $k = 2$: $z^2 - 2z + 1 = 0$, $z_{1,2} = 1, 1$, and $y_{IC}(n) = C_1(1)^n + C_2 n(1)^n$. A linearly increasing sequence.

(v) $k = -2$: $z^2 + 2z + 1 = 0$, $z_{1,2} = -1, -1$, and $y_{IC}(n) = C_1(-1)^n + C_2 n(-1)^n$. An alternating sequence whose amplitudes are increasing in a linear manner.

b. *Computer solution.* We used $\mathbf{v0} = [0; -1]$ for the plots in Figure E11.1a.

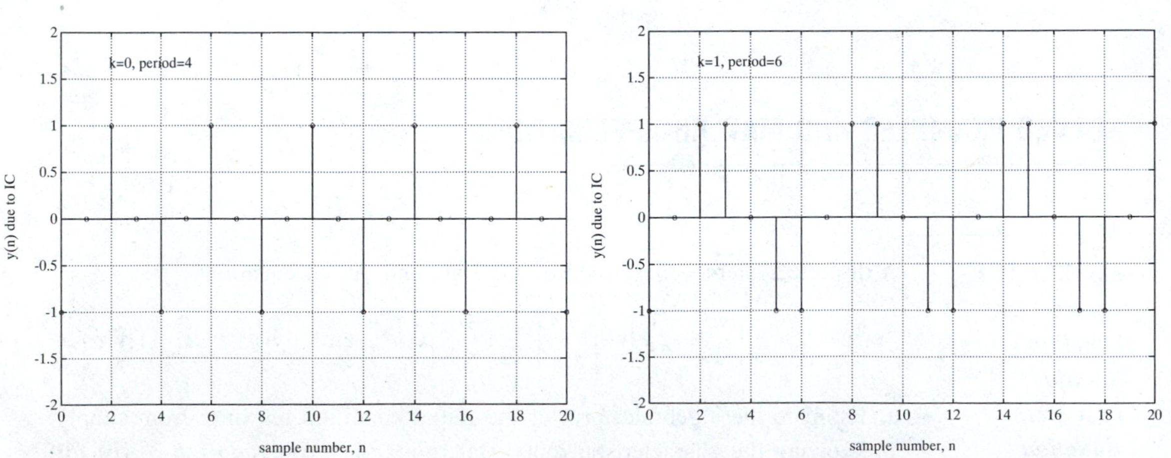

Undamped oscillatory response: $z_{1,2} = 1e^{\pm j\pi/2}$ Undamped oscillatory response: $z_{1,2} = 1e^{\pm j\pi/3}$

FIGURE E11.1a *Initial condition response for different characteristic roots (eigenvalues)*

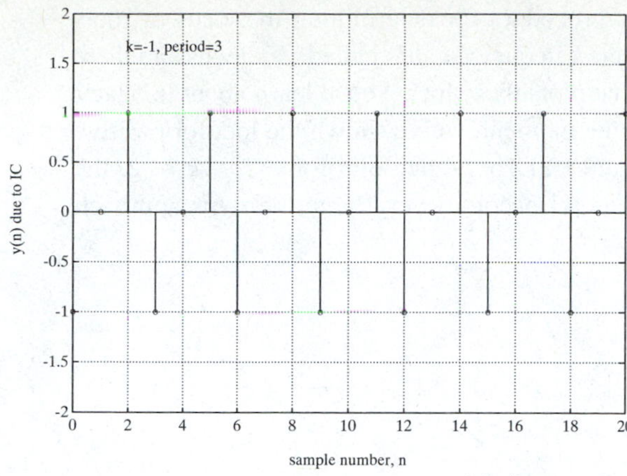

Undamped oscillatory response: $z_{1,2} = 1e^{\pm j2\pi/3}$

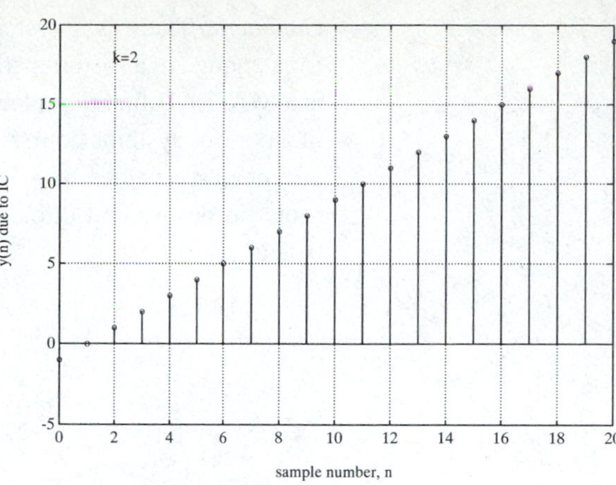

Linearly increasing sequence: $z_{1,2} = 1, 1$

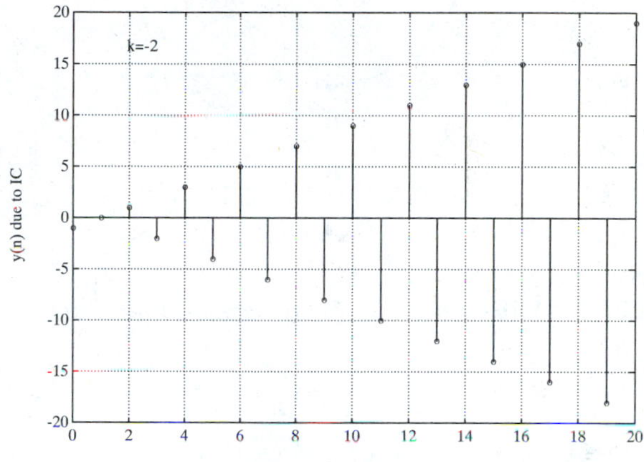

Alternating linearly increasing sequence: $z_{1,2} = 1, -1$

FIGURE E11.1a *Continued*

WHAT IF? In terms of the variable k, the system roots are

$$z_{1,2} = 0.5k \pm \sqrt{(0.5k)^2 - 1}$$

and for $-2 < k < 2$, the roots are complex and equal to

$$z_{1,2} = 0.5k \pm j\sqrt{1 - (0.5k)^2}\,.$$

The product of these complex conjugate roots is given by

$$z_1 \cdot z_1^{*} = \left(0.5k + j\sqrt{1 - (0.5k)^2}\right) \cdot \left(0.5k - j\sqrt{1 - (0.5k)^2}\right)$$
$$= 0.25k^2 + 1 - 0.25k^2 = 1\,.$$

With the product of the roots equal to 1, and since the roots are complex conjugates, we have $|z_1| = |z_1^{*}| = \sqrt{1} = 1$, which means that the roots lie on the unit circle; hence, the system cannot have a decaying oscillatory response. ■

Comment: There is a systematic procedure for determining the locus of roots for a change in a system parameter, k in our case. It's called root locus and there is a MATLAB function **rlocus** that plots these loci. You'll learn about this later in a control systems course. For the moment, we've shown the loci for positive and negative values of k in Figure E11.1b. Notice that for $-2 < k < 2$, the roots lie on the unit circle, and as $|k|$ becomes very large the roots approach 0 and $\pm\infty$.

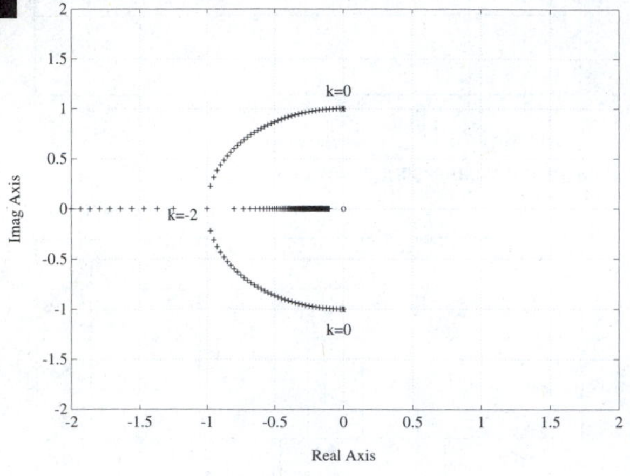

Locus of roots, $k \leq 0$

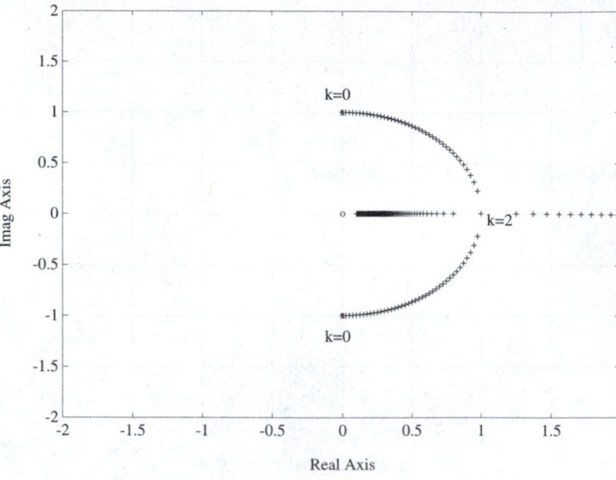

Locus of roots, $k \geq 0$

FIGURE E11.1b *Root loci for a second-order system*

EXAMPLE 11.2
State-Space Representation and Solution by z Transforms

Consider the causal recursive system of Example 8.2 that was described by the DE

$$y(n) - 1.2y(n-1) + 0.35y(n-2) = x(n) + 0.25x(n-1).$$

a. Find the transfer function $H(z) = Y(z)/X(z)$ and draw a signal flowgraph that represents this system.

b. Label a set of states on the SFG; then write a matrix state equation and an output equation for this second-order system.

c. Find the state transition matrix $\boldsymbol{\phi}(n)$ by first determining the $\boldsymbol{\Phi}(z)$ matrix.

d. Next determine the initial condition response $y_{\text{IC}}(n)$ for $y(-1) = 0$ and $y(-2) = -2$. You will have to find the state initial condition vector $\mathbf{v}(0)$ from the given values of $y(-1)$ and $y(-2)$ and the state and output equations.

e. Finally determine the complete solution for an input of $x(n) = (-2)^n u(n)$ and the initial conditions of part (d). You may want to use **residue** or **residuez** to check the PFE in both parts (c) and (e).

f. Use a MATLAB function to verify any or all of the analytical results above.

Solution

a. *Transfer function and SFG.* Taking the z transform gives

$$\frac{Y(z)}{X(z)} = H(z) = \frac{1 + 0.25z^{-1}}{1 - 1.2z^{-1} + 0.35z^{-2}}, \qquad |z| > 0.70\,.$$

Using the procedure outlined earlier, one possible SFG is given in Figure E11.2a

b. *State-space description.* Assigning states as the outputs of the delays in Figure E11.2a (the labeling is arbitrary), we have

$$v_1(n+1) = v_2(n), \qquad v_2(n+1) = -0.35v_1(n) + 1.2v_2(n) + x(n),$$

and $\; y(n) = 0.25v_2(n) + v_2(n+1) = -0.35v_1(n) + 1.45v_2(n) + x(n)\,.$

In matrix form the state and output equations are

$$\mathbf{v}(n+1) = \begin{bmatrix} 0 & 1 \\ -0.35 & 1.2 \end{bmatrix}\mathbf{v}(n) + \begin{bmatrix} 0 \\ 1 \end{bmatrix}x(n) \quad \text{and}$$

$$y(n) = [-0.35 \quad 1.45]\mathbf{v}(n) + x(n).$$

c. *State transition matrix.* First we form $(z\mathbf{I} - \mathbf{A})$ as

$$(z\mathbf{I} - \mathbf{A}) = \begin{bmatrix} z & -1 \\ 0.35 & z - 1.2 \end{bmatrix}.$$

The inverse of this matrix is

$$(z\mathbf{I} - \mathbf{A})^{-1} = \frac{1}{(z-0.5)(z-0.7)} \cdot \begin{bmatrix} z - 1.2 & 1 \\ -0.35 & z \end{bmatrix}.$$

Recalling that $\mathbf{\Phi}(z) = z(z\mathbf{I} - \mathbf{A})^{-1}$, we have

$$\mathbf{\Phi}(z) = \begin{bmatrix} \dfrac{z(z-1.2)}{(z-0.5)(z-0.7)} & \dfrac{z}{(z-0.5)(z-0.7)} \\ \dfrac{-0.35z}{(z-0.5)(z-0.7)} & \dfrac{z^2}{(z-0.5)(z-0.7)} \end{bmatrix}.$$

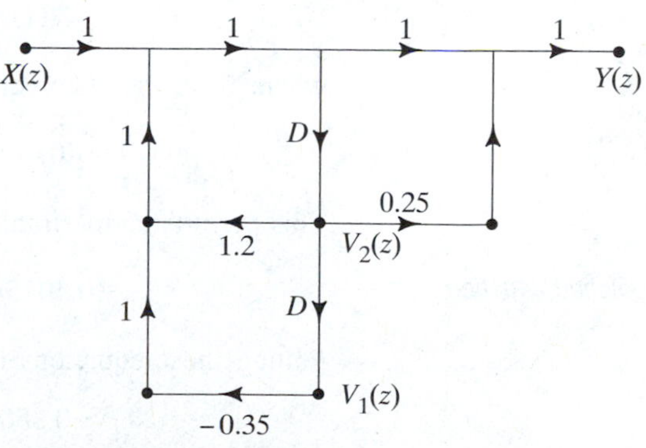

FIGURE E11.2a *SFG model*

Now we must do a PFE for each of the four elements in the $\mathbf{\Phi}(z)$ matrix. Recall that the PF expansion is performed on $\phi_{11}(z)/z = (z-1.2)/[(z-0.5)(z-0.7)]$, for instance, rather than on $\phi_{11}(z) = z(z-1.2)/[(z-0.5)(z-0.7)]$. Being very careful with hand calculations and checking with **residue** or **residuez** gives

$$\mathbf{\Phi}(z) = \begin{bmatrix} \frac{3.5z}{z-0.5} - \frac{2.5z}{z-0.7} & \frac{-5z}{z-0.5} + \frac{5z}{z-0.7} \\ \frac{1.75z}{z-0.5} - \frac{1.75z}{z-0.7} & \frac{-2.5z}{z-0.5} + \frac{3.5z}{z-0.7} \end{bmatrix}.$$

The transition matrix $\boldsymbol{\phi}(n)$ is the inverse z transform of $\mathbf{\Phi}(z)$; from Table 8.1

$$\boldsymbol{\phi}(n) = \begin{bmatrix} 3.5(0.5)^n - 2.5(0.7)^n & -5(0.5)^n + 5(0.7)^n \\ 1.75(0.5)^n - 1.75(0.7)^n & -2.5(0.5)^n + 3.5(0.7)^n \end{bmatrix}.$$

d. *The initial condition solution.* The IC solution is $y_{\text{IC}}(n) = \mathbf{C}\boldsymbol{\phi}(n)\mathbf{v}(0)$, so $\mathbf{C}\boldsymbol{\phi}(n)$ is

$$\begin{aligned} \mathbf{C}\boldsymbol{\phi}(n) &= [-0.35 \quad 1.45]\begin{bmatrix} 3.5(0.5)^n - 2.5(0.7)^n & -5(0.5)^n + 5(0.7)^n \\ 1.75(0.5)^n - 1.75(0.7)^n & -2.5(0.5)^n + 3.5(0.7)^n \end{bmatrix}. \\ &= [1.3125(0.5)^n - 1.6625(0.7)^n \quad -1.8750(0.5)^n + 3.3250(0.7)^n]. \end{aligned}$$

We need to compute the initial condition vector $\mathbf{v}(0)$ using the state equation and the output equation. The given initial conditions are $y(-1) = 0$ and $y(-2) = -2$, and from these values and the system equation with the input set to zero, $y(n) = 1.2y(n-1) - 0.35y(n-2)$, we find that

$$\begin{aligned} y(0) &= 1.2y(-1) - 0.35y(-2) = 0 + 0.7 = 0.70 \quad \text{and} \\ y(1) &= 1.2y(0) - 0.35y(-1) = 0.84 + 0 = 0.84\,. \end{aligned}$$

Using the output equation we have $y(0) = 0.70 = -0.35v_1(0) + 1.45v_2(0)$, and we obtain one equation in two unknowns, namely,

first equation

$$-0.3500v_1(0) + 1.4500v_2(0) = 0.70\,.$$

For the second equation we use the equation for $y(1)$ in the same manner:

$$y(1) = -0.35v_1(1) + 1.45v_2(1)$$

where from the state equation we know that

$$v_1(1) = v_2(0) \quad \text{and} \quad v_2(1) = -0.35v_1(0) + 1.2v_2(0)\,.$$

This permits us to eliminate $v_1(1)$ and $v_2(1)$, and with $y(1) = 0.84$,

second equation

$$-0.5075v_1(0) + 1.3900v_2(0) = 0.8400\,.$$

Putting these equations in matrix form yields

$$\begin{bmatrix} -0.3500 & 1.4500 \\ -0.5075 & 1.3900 \end{bmatrix}\mathbf{v}(0) = \begin{bmatrix} 0.7000 \\ 0.8400 \end{bmatrix}.$$

We will show the matrix inverse but will use MATLAB to perform the arithmetic:

initial condition vector

$$\mathbf{v}(0) = \frac{1}{0.2494}\begin{bmatrix} 1.3900 & -1.4500 \\ 0.5075 & -0.3500 \end{bmatrix}\begin{bmatrix} 0.7000 \\ 0.8400 \end{bmatrix} = \begin{bmatrix} -0.9824 \\ 0.2456 \end{bmatrix}.$$

Comment: A general approach for finding $\mathbf{v}(0)$ from the output values $y(0), y(1), \ldots, y(N-1)$ is developed in Problem P11.13. Now we are finally able to find $y_{\mathrm{IC}}(n)$ as

$$\begin{aligned} y_{\mathrm{IC}}(n) &= \mathbf{C}\boldsymbol{\phi}(n)\mathbf{v}(0) \\ &= [1.3125(0.5)^n - 1.6625(0.7)^n \quad -1.8750(0.5)^n + 3.3250(0.7)^n] \\ &\quad \times \begin{bmatrix} -0.9824 \\ 0.2456 \end{bmatrix} \\ &= -1.7498(0.5)^n + 2.4498(0.7)^n, \qquad n \geq 0. \end{aligned}$$

This compares favorably with the result in Example 8.2, where it was found that

$$y_{\mathrm{IC}}(n) = [-1.75(0.5)^n + 2.45(0.7)^n]u(n).$$

e. *Complete solution.* The z transform of the part of the solution due to the input $x(n) = (-2)^n u(n)$ is given by

$$\begin{aligned} \mathbf{v}_\mathbf{F}(z) &= \frac{\boldsymbol{\Phi}(z)}{z}\mathbf{B}X(z) \\ &= \begin{bmatrix} \dfrac{z - 1.2}{(z-0.5)(z-0.7)} & \dfrac{1}{(z-0.5)(z-0.7)} \\ \dfrac{-0.35}{(z-0.5)(z-0.7)} & \dfrac{z}{(z-0.5)(z-0.7)} \end{bmatrix}\begin{bmatrix} 0 \\ 1 \end{bmatrix}\frac{z}{z+2} \\ &= \begin{bmatrix} \dfrac{z}{(z-0.5)(z-0.7)(z+2)} \\ \dfrac{z^2}{(z-0.5)(z-0.7)(z+2)} \end{bmatrix} \\ &= \begin{bmatrix} \dfrac{-2z}{z-0.5} + \dfrac{1.8519z}{z-0.7} + \dfrac{0.1481z}{z+2} \\ \dfrac{-z}{z-0.5} + \dfrac{1.2963z}{z-0.7} - \dfrac{0.2963z}{z+2} \end{bmatrix}. \end{aligned}$$

Taking the inverse transform yields the state vector $\mathbf{v}_\mathbf{F}(n)$—i.e., that part of $\mathbf{v}(n)$ due to the input $x(n)$, or

$$\mathbf{v}_\mathbf{F}(n) = \begin{bmatrix} -2(0.5)^n + 1.8159(0.7)^n + 0.1481(-2)^n \\ -(0.5)^n + 1.2963(0.7)^n - 0.2963(-2)^n \end{bmatrix}.$$

The output due to the input $x(n)$ is given by

$$\begin{aligned} y_F(n) &= \mathbf{C}\mathbf{v}_F(n) + x(n) \\ &= [-0.35 \quad 1.45]\begin{bmatrix} -2(0.5)^n + 1.8159(0.7)^n + 0.1481(-2)^n \\ -(0.5)^n + 1.2963(0.7)^n - 0.2963(-2)^n \end{bmatrix} + (-2)^n \\ &= -0.75(0.5)^n + 1.232(0.7)^n + 0.519(-2)^n, \qquad n \geq 0. \end{aligned}$$

The complete solution is

$$\begin{aligned} y(n) &= y_{IC}(n) + y_F(n) \\ &= -1.75(0.5)^n + 2.45(0.7)^n - 0.75(0.5)^n \\ &\quad + 1.23(0.7)^n + 0.52(-2)^n \\ &= -2.50(0.5)^n + 3.68(0.7)^n + 0.52(-2)^n, \qquad n \geq 0. \end{aligned}$$

f. Let's look at a computer simulation to verify the results for $y(n)$. A plot of $y(n)$ is given in Figure E11.2b, and the MATLAB script with a listing of the output follows.

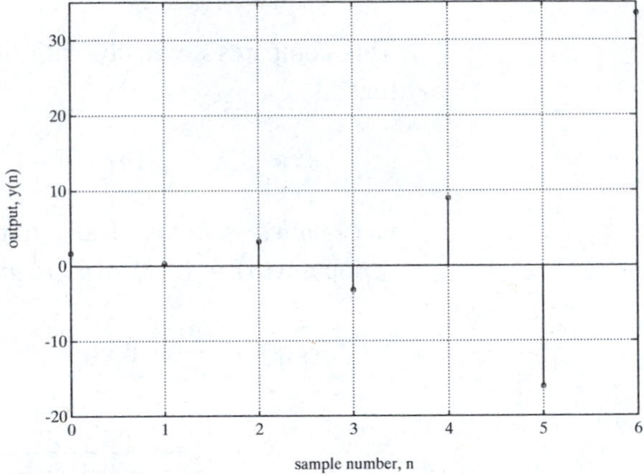

FIGURE E11.2b *Total response*

MATLAB Script and Output

```
%E11_2 Solution of the state equation

%E11_2b Total response                    Output
n=0:1:6;                                  y=1.7000
A=[0,1;-0.35,1.2];                          0.2900
B=[0;1];                                    3.2530
C=[-0.35,1.45];                            -3.1979
D=[1];                                      9.0239
v0=[-0.9824;0.2456];                      -16.0520
X=[(-2).^n]';                              33.5792
[y,v]=ksdlsim(A,B,C,D,X,v0);
%...plotting statements
```

EXAMPLE 11.3 *Computer Control of an Analog System*

A continuous-time linear system is supplied with a piecewise-constant input, $x(t) = x(nT)$, $nT \le t < nT + T$, that comes from a digital controller. The system is described at the sampling instants by the matrix difference equation and by the output equation

$$\mathbf{v}(n+1) = \begin{bmatrix} 1 & 1 & 0.5 \\ 0 & 1 & 1 \\ 0 & 0 & 1 \end{bmatrix} \mathbf{v}(n) + \begin{bmatrix} 0.1667 \\ 0.5000 \\ 1 \end{bmatrix} x(n) \quad \text{and}$$

$$y(n) = [1 \quad 0 \quad 0]\mathbf{v}(n).$$

a. Find the system's characteristic equation. What are the eigenvalues? Is the system stable? Reinforce your answer to the stability issue by using **dlsim/ksdlsim** to plot a few initial condition responses.

b. To improve this situation we modify the system according to the control law

$$x(n) = -[1 \quad 2 \quad 1.833]\mathbf{v}(n) + r(n)$$

where $r(n)$ is a new reference input. Without worrying too much about it at this time, we are simply sensing or measuring the states and modifying the input sequence $x(n)$ to change the dynamic response of the system. The procedure used to calculate the feedback gains $\mathbf{F} = -[1 \quad 2 \quad 1.833]$ is usually an important part of a course in the design of control systems. At any rate, this change yields the *modified* state equation with the same output equation:

$$\mathbf{v}(n+1) = \begin{bmatrix} 0.8333 & 0.6667 & 0.1944 \\ -0.5 & 0 & 0.0833 \\ -1 & -2 & -0.8333 \end{bmatrix} \mathbf{v}(n) + \begin{bmatrix} 0.1667 \\ 0.5000 \\ 1 \end{bmatrix} r(n) \quad \text{and}$$

$$y(n) = [1 \quad 0 \quad 0]\mathbf{v}(n).$$

What are the eigenvalues of the modified system? Is it stable? Plot the IC response for a few different initial conditions.

WHAT IF? System designers claim that characteristic roots at the origin of the z plane ($z = 0$) produce a "nice" step response. Can you verify this? ■

Solution

a. *Analysis of original system.* The characteristic polynomial is formed from $|z\mathbf{I} - \mathbf{A}|$ or $|\lambda\mathbf{I} - \mathbf{A}|$ as

$$\begin{vmatrix} z-1 & -1 & -0.5 \\ 0 & z-1 & -1 \\ 0 & 0 & z-1 \end{vmatrix} = (z-1)(z-1)(z-1) + 1(0) - 0.5(0)$$

which gives the characteristic equation

$$(z-1)(z-1)(z-1) = 0 \quad \text{or} \quad (z-1)^3 = 0 \quad \text{or}$$
$$z^3 - 3z^2 + 3z - 1 = 0.$$

We find the characteristic roots or eigenvalues to be $z_{1,2,3} = 1$. The system is, consequently, unstable, as can be seen in Figure E11.3a, where the IC response, apparently increasing without bound, is plotted for $\mathbf{v}(0) = [1, 1, 1]^T$.

b. *Analysis of modified system.* The characteristic polynomial for the modified system is formed from $|z\mathbf{I} - \mathbf{A}|$ as

$$\begin{vmatrix} z - 0.8333 & -0.6667 & -0.1944 \\ 0.5 & z & -0.0833 \\ 1 & 2 & z + 0.8333 \end{vmatrix} = (z - 0.8333)\left[z^2 + 0.8333z + 2(0.0833)\right] + 0.6667[0.5(z + 0.8333) + 0.0833] - 0.1944[1 - z].$$

Multiplying out the characteristic polynomial and equating the result to zero produces the unusual characteristic equation

$$z^3 + 0z^2 + 0z + 0 = 0 \quad \text{with the eigenvalues (roots) of } z_{1,2,3} = 0.$$

The three roots at the origin are about as far inside the unit circle as one can get, so the modified system is indeed stable, as can be seen in Figure E11.3b.

WHAT IF? The step response from **ksdlsim** is shown in Figure E11.3c. Notice that the step response is "very nice"; the system reaches the value of unity in three samples and stays there. This deadbeat response is popular when designing systems that are meant to follow a reference input, which in this case was a step sequence. ■

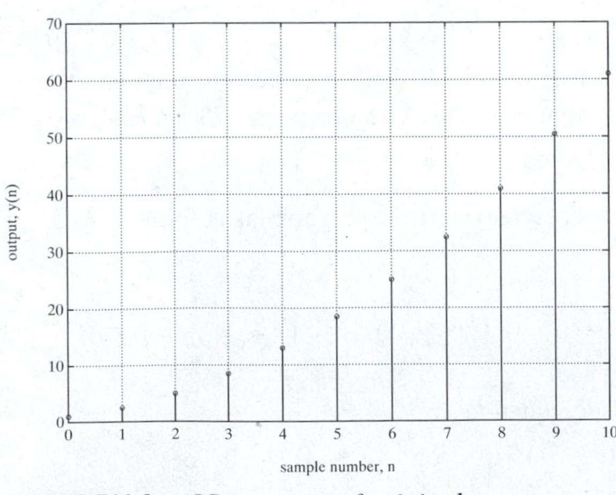

FIGURE E11.3a *IC response of original system*

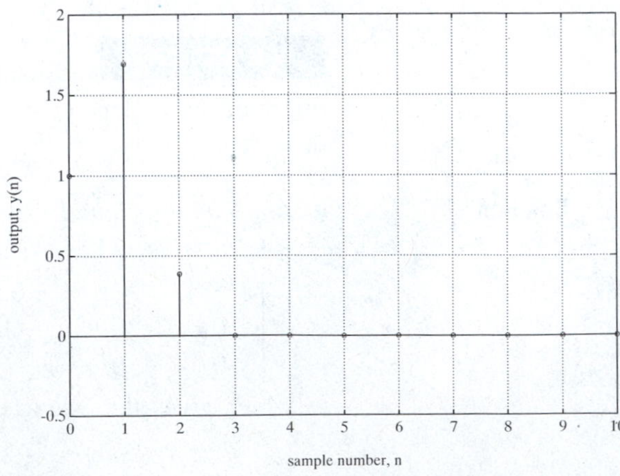

FIGURE E11.3b *IC response of modified system*

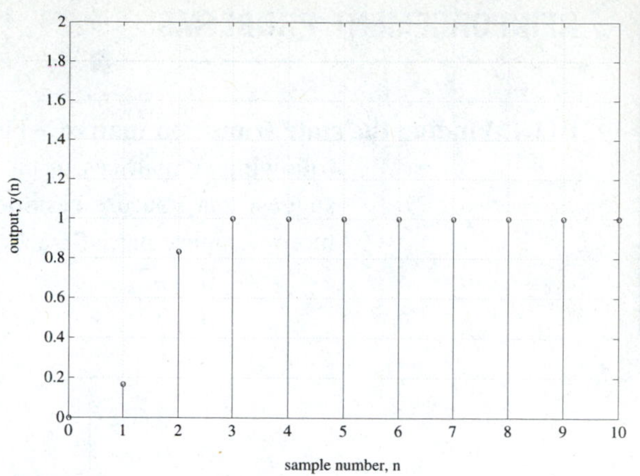

FIGURE E11.3c *Step response of modified system*

CROSS-CHECK To verify this response analytically, let's get some help from MATLAB and use the function **ss2tf** (state-space to transfer function) from the Signals and Systems Toolbox.[2] Entering the **A**, **B**, **C**, and **D** matrices returns

$$\text{num} = 0 \quad 0.1667 \quad 0.6666 \quad 0.1666$$
$$\text{den} = 1 \quad 0 \quad 0.0005 \quad 0$$

giving the transfer function

$$H(z) = \frac{Y(z)}{X(z)} = \frac{0.1667z^2 + 0.6666z + 0.1666}{z^3 + 0z^2 + 0.0005z + 0}.$$

Finally, for $x(n) = u(n)$, $X(z) = z/(z-1)$ and the z transform of the output is

$$\begin{aligned} Y(z) = H(z) \cdot X(z) &= \frac{0.1667z^2 + 0.6666z + 0.1666}{z^3 + 0z^2 + 0.0005z + 0} \cdot \frac{z}{z-1} \\ &\approx \frac{0.167z^3 + 0.667z^2 + 0.167z}{z^4 - z^3} \\ &\approx 0.167z^{-1} + 0.834z^{-2} + z^{-3} + z^{-4} + \cdots \quad \text{and} \end{aligned}$$

$$y(n) = 0.167\delta(n-1) + 0.834\delta(n-2) + 1\delta(n-3) + 1\delta(n-4) + \cdots$$

which agrees very well with Figure E11.3c. ■

[2]**ss2tf** works for both continuous and discrete systems, as does the function **tf2ss** (Signals and Systems Toolbox), which does the inverse of **ss2tf**.

REINFORCEMENT PROBLEMS

P11.1. Finding the state transition matrix. Find the state transition matrix $\boldsymbol{\phi}(n)$ for each of the following **A** matrices in the state equation $\mathbf{v}(n+1) = \mathbf{A}\mathbf{v}(n) + \mathbf{B}\mathbf{x}(n)$. We suggest that you use **residue** or **residuez** to at least check your work for the more complex partial fraction expansions.

a. $\mathbf{A} = \begin{bmatrix} 0 & 1 \\ 0 & 0 \end{bmatrix}$

b. $\mathbf{A} = \begin{bmatrix} 1 & 1 \\ 0 & 1 \end{bmatrix}$

c. $\mathbf{A} = \begin{bmatrix} 1 & 0.632 \\ 0 & 0.368 \end{bmatrix}$

d. $\mathbf{A} = \begin{bmatrix} 0.632 & 0.632 \\ -0.632 & 0.368 \end{bmatrix}$.

P11.2 System eigenvalues (characteristic roots) and stability. For each of the systems (parts (a)–(d)) of P11.1, find the eigenvalues (characteristic roots) and determine stability.

P11.3 Initial condition response. For each of the systems in P11.1, assume $\mathbf{B} = [0 \quad 1]^T$, $\mathbf{C} = [0 \quad 1]$, and $\mathbf{D} = d = 0$.

(i) Find the initial condition response $y_{\text{IC}}(n)$ at $n = 5$ for the initial conditions $\mathbf{v}(0) = [1 \quad 2]^T$. For $\mathbf{v}_{\text{IC}}(n)$ try both analytical methods, $\mathbf{v}_{\text{IC}}(n) = \mathbf{A}^n\mathbf{v}(0)$ and $\mathbf{v}_{\text{IC}}(n) = \boldsymbol{\phi}(n)\mathbf{v}(0)$. Use the $\boldsymbol{\phi}(n)$ matrices found in Problem P11.1.

(ii) Then use **dlsim/ksdlsim** to illustrate the approach of the 21st century.

P11.4 Unit sample response. For each of the systems in Problem P11.1, assume $\mathbf{B} = [0 \quad 1]^T$, $\mathbf{C} = [0 \quad 1]$, and $\mathbf{D} = d = 0$.

(i) Find an analytical expression for the unit sample response $h(n)$ for each of these systems.

(ii) Use a function such as **ksdlsim** to verify the results in (i).

P11.5 System transfer functions. For each of the systems in Problems P11.1 and P11.4, find the scalar transfer function $Y(z)/X(z) = H(z)$ by using $\mathbf{H}(z) = \mathbf{C}z^{-1}\boldsymbol{\Phi}(z)\mathbf{B} + \mathbf{D}$. Check your results with MATLAB's **ss2tf**.

P11.6 System diagrams. For each of the systems in Problems P11.1 and P11.4, draw an SFG and use Mason's Gain Rule to verify the transfer functions computed in Problem P11.5.

P11.7 Sinusoidal steady-state response. For the stable systems in Problem P11.4, find an analytical expression for the steady-state value of the output $y_{ss}(n)$ for the input $x(n) = A\cos(n\pi/8)u(n)$. Plot the responses for $n \geq 0$ and estimate the number of samples before steady-state is reached.

EXPLORATION PROBLEMS

P11.8 System design with feedback. A continuous-time system with a piecewise-constant input is described at the sampling instants by

$$\mathbf{v}(n+1) = \begin{bmatrix} -1 & 1 \\ -1 & -1 \end{bmatrix}\mathbf{v}(n) + \begin{bmatrix} 1 \\ 1 \end{bmatrix}x(n) \quad \text{and}$$

$$y(n) = [1 \quad 0]\mathbf{v}(n).$$

a. Find the eigenvalues (characteristic roots) of this system.
b. Determine the transfer function $H(z) = Y(z)/X(z)$.
c. The system input $x(n)$ is modified by the algorithm

$$x(n) = -f_1 v_1(n) - f_2 v_2(n) + r(n)$$

where $r(n)$ is a new reference signal. Find the required values of f_1 and f_2 to obtain eigenvalues of $\lambda_{1,2} = \pm 0.5$.
d. Although the eigenvalues of part (c) yield a nice step response, try some different feedback gains. What gains f_1 and f_2 are needed to have eigenvalues of $\lambda_{1,2} = \pm j0.50$?
e. For the gains that you found in part (d), find $y(n) = v_1(n)$ for $n = 0, 1, 2, 3$ if $r(n)$ is a unit step sequence and both initial conditions are zero.

P11.9 Second-order filter. A digital filter is described by the transfer function

$$H(z) = \frac{Y(z)}{X(z)} = \frac{z^2}{z^2 - [2r\cos\alpha]z + r^2},$$

$$-\pi \le \alpha \le \pi \quad \text{and} \quad r > 0.$$

a. Draw an SFG for this filter.
b. Use the SFG to write a set of state and output equations for the filter.
c. Find the filter's eigenvalues and show that they are always complex.
d. Find an analytical expression for the filter's unit sample response $h(n)$.
e. Find the values of α and r that will ensure a stable filter.
f. We want to design a bandpass filter with its passband centered at $\theta = \pi/3$, where the gain is to be about 10. Find suitable values of α and r and verify your selections with a computer plot using **freqz.**

P11.10 A more general second-order filter. Consider the filter described by

$$H(z) = \frac{Y(z)}{X(z)} = \frac{b_0 + b_1 z^{-1} + b_2 z^{-2}}{1 + a_1 z^{-1} + a_2 z^{-2}}, \qquad \text{all coefficients real.}$$

a. Draw an SFG for this filter.
b. Use the SFG to write a set of state and output equations for the filter.
c. Find the coefficients a_1 and a_2 for purely imaginary system eigenvalues of $\lambda_{1,2} = \pm j0.866$.

d. We want to design a bandpass filter with its passband centered at $\theta = \pi/2$. The filter gain values at $\theta = 0$ and $\theta = \pi$ are to be zero. Use a_1 and a_2 from part (c) and determine b_0, b_1, and b_2. Verify your design with **freqz** or equivalent.

WHAT IF? Having designed a bandpass filter, let's design a bandstop filter that eliminates the digital frequency of $\theta = \pi/2$ rad. Find a set of filter coefficients that will accomplish this and, again, verify your design with a computer plot. ■

P11.11 A parallel system and potpourri. A causal system is modeled by the signal flowgraph of Figure P11.11a.

a. What is the order of the system?
b. Write a set of state and output equations for this system.
c. Find the characteristic equation for the system.
d. What are the characteristic roots? Is the system stable?

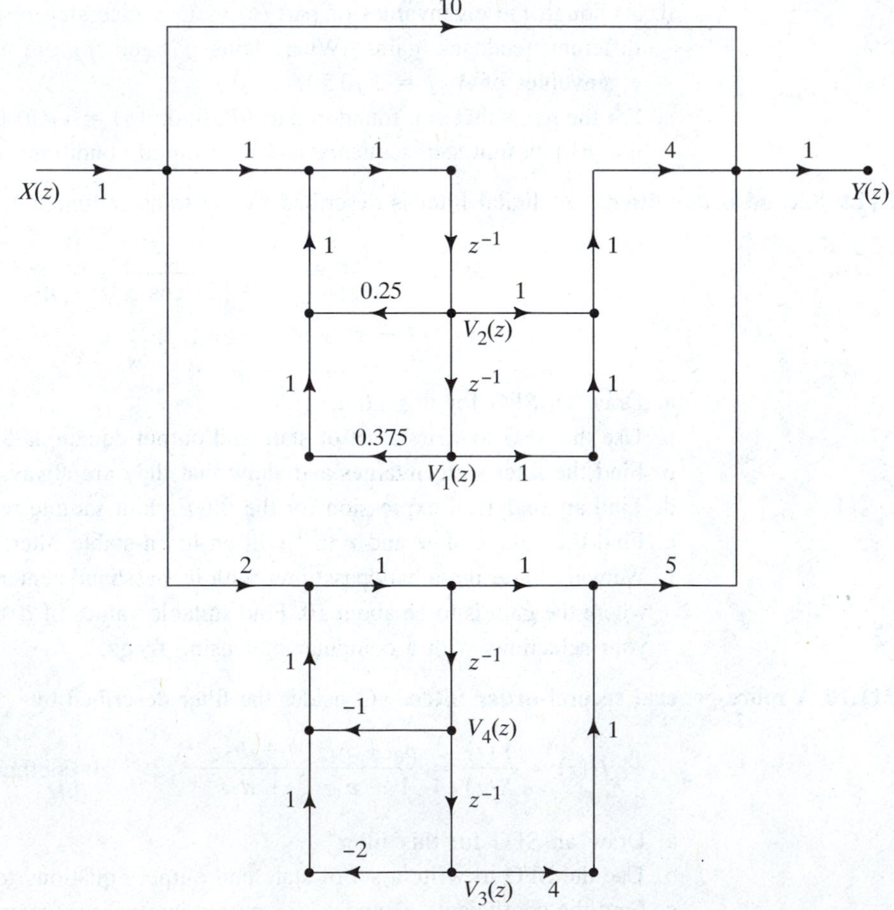

FIGURE P11.11a

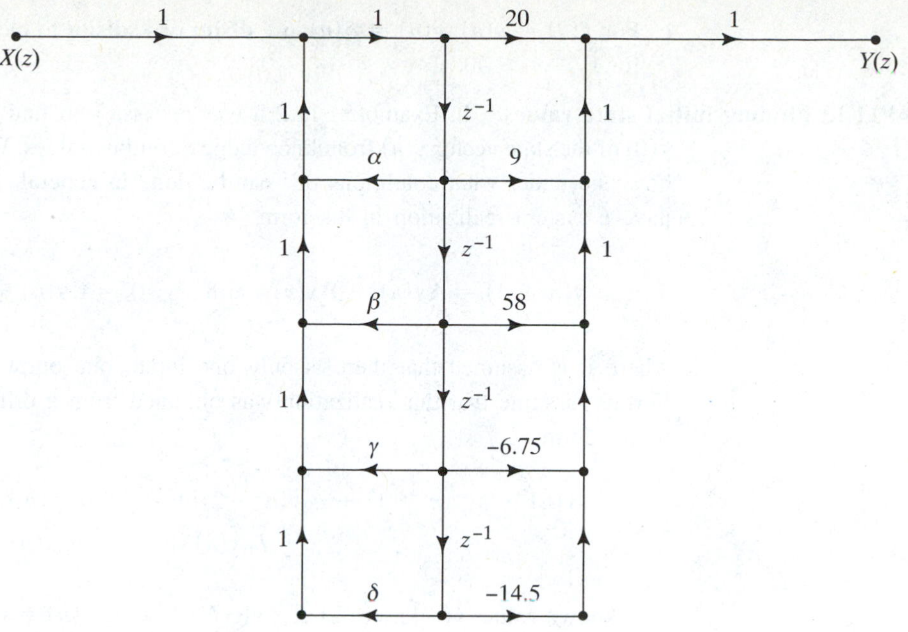

FIGURE P11.11b

e. The SFG for another implementation of this same system is shown in Figure P11.11b, with some unspecified gains. Determine the values of α, β, γ, and δ to make the two realizations have the same characteristic equation.

f. Use MGR to find the transfer function $H(z) = Y(z)/X(z)$ of the graph in (e).

g. From the SFG given in Figure P11.11a, find the first two values of the unit sample response $h(n)$ by tracing the paths with the corresponding gains of a unit sample through zero and one delay from input to output. Verify your computation by using long division on $H(z)$ of part (f).

P11.12 A Multiple-Input–Multiple-Output system. The LTI system of Illustrative Problem 8.11 was modeled by the simultaneous difference equations

$$3p(n-2) + 2p(n-1) + p(n) - 2q(n) = 5f(n) - 7g(n)$$
$$2q(n-2) - 3q(n-1) + 5p(n) = 3g(n)$$

where $f(n)$ and $g(n)$ are the system inputs and $p(n)$ and $q(n)$ are its outputs.

a. Put these equations in state-variable form with the states defined as $v_1(n) = p(n-2)$, $v_2(n) = p(n-1)$, $v_3(n) = q(n-2)$, and $v_4(n) = q(n-1)$. For the inputs let $x_1(n) = f(n)$ and $x_2(n) = g(n)$; that is, find the **A**, **B**, **C**, and **D** matrices of the state-space model.

b. For the initial conditions $q(-1) = -1$, $q(-2) = p(-1) = p(-2) = 0$, determine the components of the initial condition state vector $\mathbf{v}(0)$.

c. For $f(n) = u(n)$, $g(n) = \delta(n)$, use **dlsim** or **ksdlsim** to plot $p(n)$ and $q(n)$ for $0 \le n \le 40$.

P11.13 Finding initial state values. In Example E11.2 it was necessary to find the initial value $\mathbf{v}(0)$ of the state vector $\mathbf{v}(n)$ from knowledge of output values. We now consider how and under what conditions this can be done in general. Assume that we have a system realization in the form

$$\mathbf{v}(n+1) = \mathbf{A}\mathbf{v}(n) + \mathbf{B}\mathbf{x}(n) \quad \text{and} \quad y(n) = \mathbf{C}\mathbf{v}(n) + \mathbf{D}\mathbf{x}(n)$$

where it is assumed that there is only one input, one output, and N states. Further, assume that this realization was obtained from a difference equation of the form

$$y(n) + a_1 y(n-1) + a_2 y(n-2) + \cdots + a_N y(n-N) = b_0 x(n) + \cdots + b_L x(n-L)$$

with known values $y(-1), y(-2), \ldots, y(-N)$. Setting $x(n) = 0$ for all n allows us to compute the values $y(0), y(1), \ldots, y(N-1)$ from the difference equation. Also setting $x(n) = 0$ for all n in the Nth-order difference equation gives

$$\mathbf{v}(n+1) = \mathbf{A}\mathbf{v}(n) \quad \text{and} \quad y(n) = \mathbf{C}\mathbf{v}(n).$$

a. Find an equation of the form

$$\begin{bmatrix} y(0) \\ y(1) \\ \vdots \\ y(N-1) \end{bmatrix} = \mathbf{P} \begin{bmatrix} v_1(0) \\ v_2(0) \\ \vdots \\ v_N(0) \end{bmatrix} \quad \text{or} \quad \mathbf{y}^{(N)} = \mathbf{P}\mathbf{v}(0)$$

and express the $N \times N$ matrix $\mathbf{P}$ in terms of the matrices $\mathbf{A}$ and $\mathbf{C}$.

b. Under what conditions can we solve this equation for $\mathbf{v}(0)$?

c. Use the expression for $\mathbf{P}$ found in part (a) to verify the results obtained in Example E11.2.

DEFINITIONS, TECHNIQUES, AND CONNECTIONS

State equation $\quad \mathbf{v}(n+1) = \mathbf{A}\mathbf{v}(n) + \mathbf{B}\mathbf{x}(n)$

Output equation $\quad \mathbf{y}(n) = \mathbf{C}\mathbf{v}(n) + \mathbf{D}\mathbf{x}(n)$

IC solution $\mathbf{v}_{\mathbf{IC}}(n) = \mathbf{A}^n\mathbf{v}(0) = \boldsymbol{\phi}(n)\mathbf{v}(0)$

Forced solution
$$\mathbf{v}_{\mathbf{F}}(n) = \sum_{m=0}^{n-1} \mathbf{A}^{n-m-1}\mathbf{B}\mathbf{x}(m) = \sum_{m=1}^{n} \boldsymbol{\phi}(n-m)\mathbf{B}\mathbf{x}(m-1) = \sum_{m=0}^{n-1} \boldsymbol{\phi}(n-m-1)\mathbf{B}\mathbf{x}(m)$$

z transform solution
$$\mathbf{V}(z) = z(z\mathbf{I} - \mathbf{A})^{-1}\mathbf{v}(0) + (z\mathbf{I} - \mathbf{A})^{-1}\mathbf{B}\mathbf{X}(z) = \boldsymbol{\Phi}(z)\mathbf{v}(0) + \frac{\boldsymbol{\Phi}(z)}{z}\mathbf{B}\mathbf{X}(z) \quad \text{where} \quad \boldsymbol{\Phi}(\mathbf{z}) = z(z\mathbf{I} - \mathbf{A})^{-1}$$
$$\mathbf{v}(n) = \boldsymbol{\phi}(n)\mathbf{v}(0) + \sum_{m=1}^{n} \boldsymbol{\phi}(n-m)\mathbf{B}\mathbf{x}(m-1) = \boldsymbol{\phi}(n)\mathbf{v}(0) + \sum_{m=0}^{n-1} \boldsymbol{\phi}(n-m-1)\mathbf{B}\mathbf{x}(m)$$

$\boldsymbol{\Phi}(z)$ matrix $\boldsymbol{\Phi}(z) = z(z\mathbf{I} - \mathbf{A})^{-1} = z\,\dfrac{\mathbf{adj}(z\mathbf{I} - \mathbf{A})}{|z\mathbf{I} - \mathbf{A}|}$

Characteristic equation $|z\mathbf{I} - \mathbf{A}| = 0$

Characteristic roots zeros of $|z\mathbf{I} - \mathbf{A}| = 0$ or eigenvalues of the **A** matrix

Transition matrix $\boldsymbol{\phi}(n) = \mathcal{Z}^{-1}[\boldsymbol{\Phi}(z)]$

Transfer function matrix $\mathbf{H}(z) = \mathbf{C}z^{-1}\boldsymbol{\Phi}(z)\mathbf{B} + \mathbf{D}$

Frequency response matrix $\mathbf{H}(e^{j\theta}) = \mathbf{C}e^{-j\theta}\boldsymbol{\Phi}(e^{j\theta})\mathbf{B} + \mathbf{D}$

MATLAB FUNCTIONS USED

Function	*Purpose and Use*	*Toolbox*
dlsim	Given: state or TF model of discrete system, input, ICs, **dlsim** returns system output.	Control System, Signals/Systems
eig	Given: N by N matrix A, **eig** returns the eigenvalues of A.	MATLAB, Signals/Systems

CHAPTER 11

Function	*Purpose and Use*	*Toolbox*
freqz	Given: TF of discrete system, **freqz** returns frequency response.	Signals/Systems, Signal Processing
ksdlsim	Given: state or TF model of discrete system, input, ICs, **ksdlsim** returns system output.	California Functions
residue	Given: rational function $T(\sigma) = N(\sigma)/D(\sigma)$, **residue** returns roots of $D(\sigma) = 0$ and PF constants of $T(\sigma)$.	MATLAB
residuez	Given: rational function $T(z) = N(z)/D(z)$, **residuez** returns roots of $D(z) = 0$ and PF constants of $T(z)$.	Signal Processing, Signals/Systems
rlocus	Given: equation in the form $1 + K[\text{num}(\sigma)]/[\text{den}(\sigma)] = 0$, **rlocus** returns plot of locus of roots for K varying.	Signals/Systems, Control System
roots	Given: coefficients of polynomial p, **roots** returns roots of $p = 0$.	MATLAB
ss2tf	Given: State model of discrete system, **ss2tf** returns TF model.	Control System, Signals/Systems
tf2ss	Given: TF model of discrete system, **tf2ss** returns state model.	Signals/Systems, Control System

ANNOTATED BIBLIOGRAPHY

1. Astrom, Karl J., and Bjorn Wittenmark, *Computer Controlled Systems, Theory and Design,* Prentice Hall, Inc., Englewood Cliffs, N.J., 1984. *Chapter Nine discusses state-space design methods for systems under computer control that include deadbeat control and the use of Ackermann's formula (Control System Toolbox in MATLAB) for the general case of pole-placement design. Observer design, output feedback, and the servo problem are included in this chapter.*

2. Gabel, Robert A., and Richard A. Roberts, *Signals and Linear Systems, 3rd ed.,* John Wiley & Sons, New York, 1987. *The use of the state-space model to describe linear systems is introduced in Chapter Two and is followed directly with a detailed discussion of the solution of the state matrix equation using time-domain methods only. A practical example of the usefulness of the state model is given in the section describing limit cycles in digital filters. Solution of state equations by z transforms is completed in Chapter Four.*

3. Kwakernaak, Huibert, and Raphael Sivan, *Modern Signals and Systems,* Prentice Hall, Inc., Englewood Cliffs, N.J., 1991. *In Chapter Five, "State Descriptions of Systems," the authors present a parallel treatment of discrete and continuous state-space topics, including several comprehensive (and historical) computer projects. Twenty end-of-chapter problems illustrate the text material.*

4. Strum, Robert D., and Donald E. Kirk, *First Principles of Discrete Systems and Digital Signal Processing,* Addison-Wesley Publishing Company, Reading, Mass., 1988. *Chapter Eleven includes a thorough discussion of the discrete-time and z-transform solution of the discrete state model of a linear system. The important application of digital control of a continuous-time system is illustrated with several examples, including the use of state feedback to modify an unruly system's behavior.*

5. [*Historical Reference*]: Zadeh, Lotfi, A., and Charles A. Desoer, *Linear System Theory, The State Space Approach,* Krieger Publishing Company, Melbourne, Florida, 1979. Reprint of original, McGraw-Hill Book Company, New York, 1963. *See comments in the Annotated Bibliography of Chapter 6 of this text.*

ANSWERS

ANSWERS

P11.1 a. $\boldsymbol{\phi}(n) = \begin{bmatrix} \delta(n) & \delta(n-1) \\ 0 & \delta(n) \end{bmatrix}$; b. $\boldsymbol{\phi}(n) = \begin{bmatrix} 1^n & n \\ 0 & 1^n \end{bmatrix}$;

c. $\boldsymbol{\phi}(n) = \begin{bmatrix} 1^n & 1^n - (0.368)^n \\ 0 & (0.368)^n \end{bmatrix}$; d. $\boldsymbol{\phi}(n) =$

$$\begin{bmatrix} 1.022(0.795)^n \cos(0.891n - 0.210) & 1.022(0.795)^n \cos(0.891n - 1.571) \\ 1.022(0.795)^n \cos(0.891n + 1.571) & 1.022(0.795)^n \cos(0.891n + 0.210) \end{bmatrix}.$$

P11.2 a. $\lambda_{1,2} = 0$; system is stable. b. $\lambda_{1,2} = 1$; system is unstable. c. $\lambda_{1,2} = 1, 0.368$; system is unstable. d. $\lambda_{1,2} = 0.500 \pm j0.618 = 0.795e^{\pm j0.891}$; system is stable.

P11.3 (i) a. $\mathbf{v}_{\mathbf{IC}}(5) = \mathbf{A}^5\mathbf{v}(0) = \begin{bmatrix} 0 & 0 \\ 0 & 0 \end{bmatrix}\begin{bmatrix} 1 \\ 2 \end{bmatrix} = \mathbf{0}$;

$\mathbf{v}_{\mathbf{IC}}(5) = \boldsymbol{\phi}(5)\mathbf{v}(0) = \begin{bmatrix} 0 & 0 \\ 0 & 0 \end{bmatrix}\begin{bmatrix} 1 \\ 2 \end{bmatrix} = \mathbf{0}$; b. $\mathbf{v}_{\mathbf{IC}}(5) = \mathbf{A}^5\mathbf{v}(0) =$

$\begin{bmatrix} 1 & 5 \\ 0 & 1 \end{bmatrix}\begin{bmatrix} 1 \\ 2 \end{bmatrix} = \begin{bmatrix} 11 \\ 2 \end{bmatrix}$; $\mathbf{v}_{\mathbf{IC}}(5) = \boldsymbol{\phi}(5)\mathbf{v}(0) = \begin{bmatrix} 1^5 & 5 \\ 0 & 1^5 \end{bmatrix}\begin{bmatrix} 1 \\ 2 \end{bmatrix} = \begin{bmatrix} 11 \\ 2 \end{bmatrix}$;

c. $\mathbf{v}_{\mathbf{IC}}(5) = \mathbf{A}^5\mathbf{v}(0) = \begin{bmatrix} 1 & 0.9933 \\ 0 & 0.0067 \end{bmatrix}\begin{bmatrix} 1 \\ 2 \end{bmatrix} = \begin{bmatrix} 2.9865 \\ 0.0135 \end{bmatrix}$;

$\mathbf{v}_{\mathbf{IC}}(5) = \boldsymbol{\phi}(5)\mathbf{v}(0) = \begin{bmatrix} 1^5 & 1^5 - (0.368)^5 \\ 0 & (0.368)^5 \end{bmatrix}\begin{bmatrix} 1 \\ 2 \end{bmatrix} = \begin{bmatrix} 2.9865 \\ 0.0135 \end{bmatrix}$;

d. $\mathbf{v}_{\mathbf{IC}}(5) = \mathbf{A}^5\mathbf{v}(0) = \begin{bmatrix} 0.632 & 0.632 \\ -0.632 & 0.368 \end{bmatrix}^5\begin{bmatrix} 1 \\ 2 \end{bmatrix} =$

$\begin{bmatrix} -0.147 & -0.314 \\ 0.314 & -0.016 \end{bmatrix}\begin{bmatrix} 1 \\ 2 \end{bmatrix} = \begin{bmatrix} -0.775 \\ 0.282 \end{bmatrix}$; $\mathbf{v}_{\mathbf{IC}}(5) = \boldsymbol{\phi}(5)\mathbf{v}(0) =$

$$\begin{bmatrix} 1.022(0.795)^5 \cos(0.891(5) - 0.210) & 1.022(0.795)^5 \cos(0.891(5) - 1.571) \\ 1.022(0.795)^5 \cos(0.891(5) + 1.571) & 1.022(0.795)^5 \cos(0.891(5) + 0.210) \end{bmatrix}\begin{bmatrix} 1 \\ 2 \end{bmatrix} =$$

$\begin{bmatrix} -0.775 \\ 0.282 \end{bmatrix}$ (ii) See A11_3 on the CLS disk where plots are generated for $y_{\mathrm{IC}}(n) = v_2(n)$.

P11.4 (i) a. $h(n) = \delta(n-1)$; b. $h(n) = u(n-1)$; c. $h(n) = (0.368)^{n-1}u(n-1)$; d. $h(n) = -\delta(n) + 1.286(0.795)^n \cos(0.891n - 0.68)u(n)$; (ii) See A11_4 on the CLS disk.

P11.5 a. $H(z) = z^{-1}$; b. $H(z) = 1/(z-1)$; c. $H(z) = 1/(z-0.368)$; d. $H(z) = (z-0.632)/(z^2 - z + 0.632)$. Using **[num,den]=ss2tf(a,b,c,d,1)** yields, for (d):

```
num = 0         1.0000    -0.6320
den = 1.0000   -1.0000    0.6320
```

P11.6 See Figure A11.6 for the SFGs. The transfer functions are verified by MGR.

P11.7 a. $y_{ss}(n) = A\cos(n\pi/8 - \pi/8)$. This system simply delays the input by 1 sample and steady-state is achieved beginning at $n = 1$, as in A11_7 on the CLS disk. b. Unstable, c. Unstable but unstable root at $z = 1$ does not affect output. About 4 samples to steady-state of $y_{ss}(n) = 1.483A\cos(n\pi/8 - 0.6029)$ as in A11_7. Transient due to pole at $z = 0.368$. d. $y_{ss}(n) = 0.913A\cos(n\pi/8 - 0.256)$. About 10 samples occur before steady-state, as in A11_7. Transient due to the poles at $z_{1,2} = 0.795e^{\pm j0.891}$.

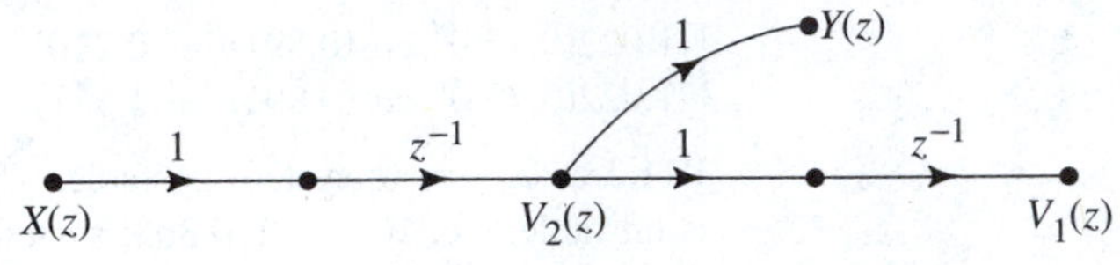

(a)

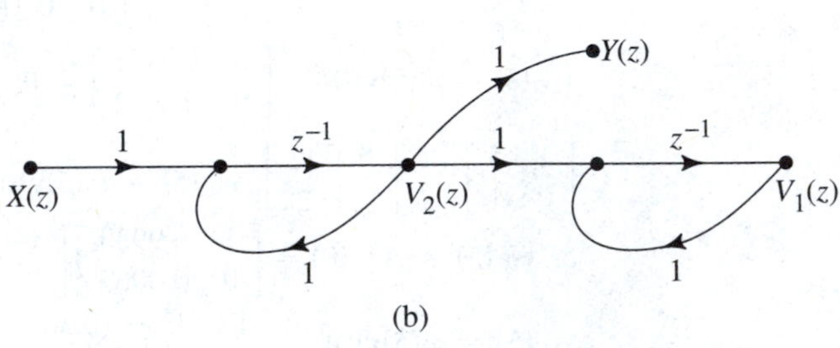

(b)

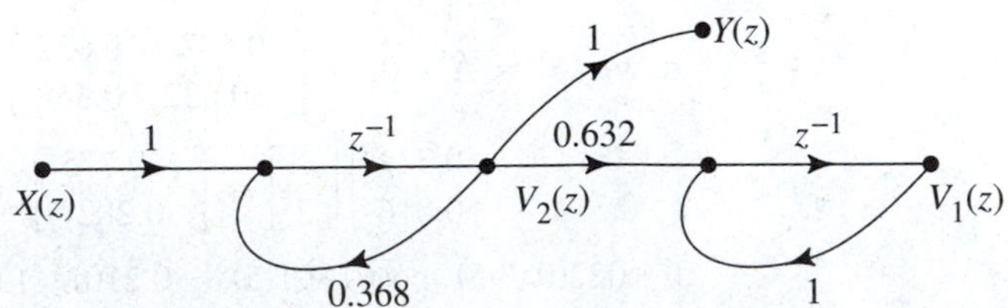

FIGURE A11.6 (c)

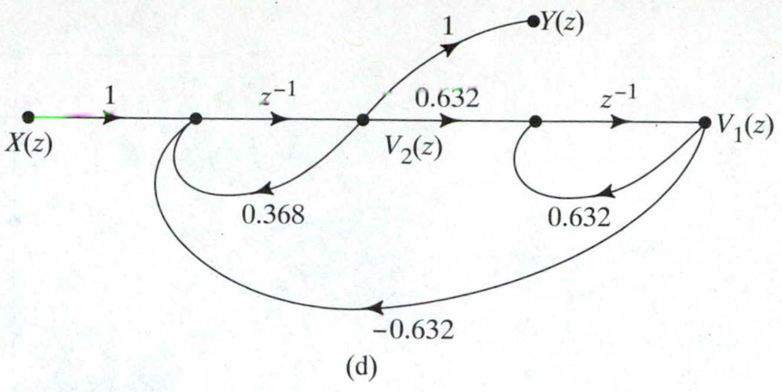

FIGURE A11.6 *Continued*

P11.8 a. $-1 \pm j1$; b. $Y(z)/X(z) = (z + 2)/(z^2 + 2z + 2)$;
c. $f_1 = -1.125$, $f_2 = -0.875$; d. $f_1 = -0.875$, $f_2 = -1.125$;
e. $y(n) = 0, 1, 3, 2.75$ for $n = 0, 1, 2, 3$

P11.9 a. See Figure A11.9. b. $v_1(n + 1) = v_2(n)$,
$v_2(n + 1) = -r^2 v_1(n) + [2r \cos \alpha]v_2(n) + x(n)$,
$y(n) = v_2(n + 1) = -r^2 v_1(n) + [2r \cos \alpha]v_2(n) + x(n)$;
c. $\lambda_{1,2} = r \cos \alpha \pm \sqrt{r^2 \cos^2 \alpha - r^2} = r(\cos \alpha \pm j \sin \alpha) = re^{\pm j\alpha}$;
d. $h(n) = [1/\sin \alpha]r^n(\sin[n + 1]\alpha)u(n)$; e. $|r| < 1$; f. $\alpha = \pi/3$, $r = 0.94$.
See A11_9 on the CLS disk.

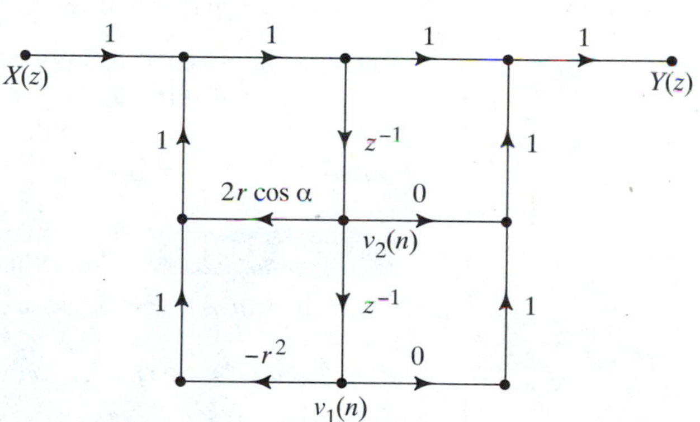

FIGURE A11.9

P11.10 a. See Figure A11.10. b. $v_1(n + 1) = v_2(n)$,
$v_2(n + 1) = -a_2 v_1(n) - a_1 v_2(n) + x(n)$, $y(n) =$
$[b_2 - b_0 a_2]v_1(n) + [b_1 - b_0 a_1]v_2(n) + b_0 x(n)$; c. $a_1 = 0$, $a_2 = 0.75$;
d. $b_0 = -b_2$, $b_1 = 0$; see A11_10 scaled for max response = 1.

WHAT IF? See A11_10 scaled for $|H(e^{j0})| = |H(e^{j\pi})| = 1$, $b_0 = b_2 = 0.095$, $b_1 = 0$, $a_1 = 0$, $a_2 = -0.81$. ■

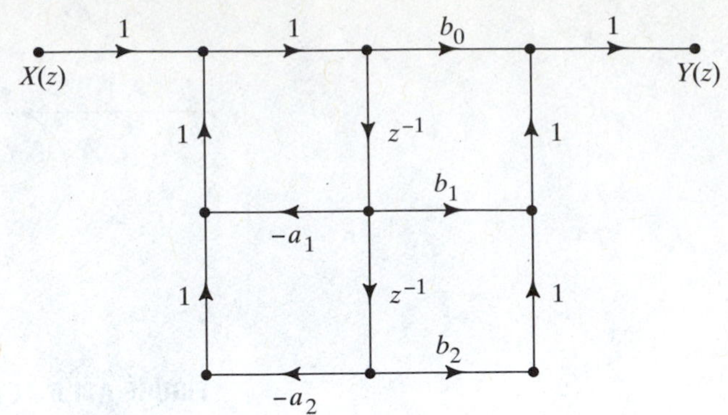

FIGURE A11.10

P11.11 a. Fourth order. b. The states were assigned as the outputs of the delays from top to bottom in the SFG as $v_2(n)$, $v_1(n)$, $v_4(n)$, $v_3(n)$, respectively. $v_1(n+1) = v_2(n)$, $v_2(n+1) = 0.375v_1(n) + 0.25v_2(n) + x(n)$, $v_3(n+1) = v_4(n)$, $v_4(n+1) = -2v_3(n) - v_4(n) + 2x(n)$, $y(n) = 4v_1(n) + 4v_2(n) + 10v_3(n) - 5v_4(n) + 20x(n)$. c. $z^4 + 0.75z^3 + 1.375z^2 - 0.875z - 0.75 = 0$; d. $z_{1,2} = -0.500 \pm j1.323$, $z_3 = 0.750$, $z_4 = -0.500$; unstable causal system; e. $A = -0.75$, $B = -1.375$, $C = 0.875$, $D = 0.75$. f. $H(z) = (20z^4 + 9z^3 + 58z^2 - 6.75z - 14.5)/(z^4 + 0.75z^3 + 1.375z^2 - 0.875z - 0.75)$. g. $h(0) = 20$, $h(1) = -6, \ldots$

P11.12 a. $\mathbf{A} = [0,1,0,0;0,0,-0.4,0.6;0,0,0,1;1.5,1,-0.2,0.3]$, $\mathbf{B} = [0,0;0,0.6;0,0;-2.5,3.8]$, $\mathbf{C} = [0,0,-0.4,0.6;1.5,1,-0.2,0.3]$, $\mathbf{D} = [0,0.6;-2.5,3.8]$; b. $\mathbf{v}(0) = [0;0;0;-1]$; c. See A11_12 on the CLS disk.

CROSS-CHECK From Illustrative Problem 8.11 on page 440, $p(n) \to -1.25$ and $q(n) \to -6.25$ as $n \to \infty$. ■

P11.13 a. $[\,y(0) \quad y(1) \quad \ldots \quad y(N-1)] =$ $\mathbf{v}^T(0)\,[\mathbf{C}^T \vdots \; \mathbf{A}^T\mathbf{C}^T \vdots \; (\mathbf{A}^T)^2\mathbf{C}^T \vdots \ldots \vdots \; (\mathbf{A}^T)^{N-1}\mathbf{C}^T] = \mathbf{v}^T(0)\mathbf{P}^T$; b. $\mathbf{P}$ must be nonsingular. c. $\mathbf{P} = [-0.35 \quad 1.45; \quad -0.5075 \quad 1.39]$, $\mathbf{v}(0) = [-0.9825 \quad 0.2456]^T$

RETROSPECTIVE

Discrete Systems

Having completed the chapters devoted to discrete-time LTI systems, it's a good time to take another look backward to see where we've been. The retrospective following Chapter 9 emphasized analytical methods and when they may be used. An important factor is deciding what approach to apply. In Chapter 10 we encountered the Discrete Fourier Transform and in Chapter 11, the solution methods for the state-space description for linear systems. Thus we have added to our repertoire of models, and at this point we can use as models a system's difference equation(s), transfer function, unit sample (impulse) response, frequency response, block diagram or signal flowgraph, and state and output equations. Furthermore, we now are able to convert from one model to another.

SYSTEM MODELS

Difference equation	$\sum_{k=0}^{N} a_k y(n-k) = \sum_{k=0}^{L} b_k x(n-k)$
Transfer function	$H(z) = \dfrac{Y(z)}{X(z)}$, ICs $= 0$
Unit sample response	$h(n)$
Frequency response	$H(e^{j\theta})$
State-space	$\mathbf{v}(n+1) = \mathbf{A}\mathbf{v}(n) + \mathbf{B}\mathbf{x}(n)$ $\mathbf{y}(n) = \mathbf{C}\mathbf{v}(n) + \mathbf{D}\mathbf{x}(n)$
Block diagram or SFG	

FINDING ONE MODEL FROM ANOTHER

See Figure R.2 for a pictorial view of the possibilities and procedures.

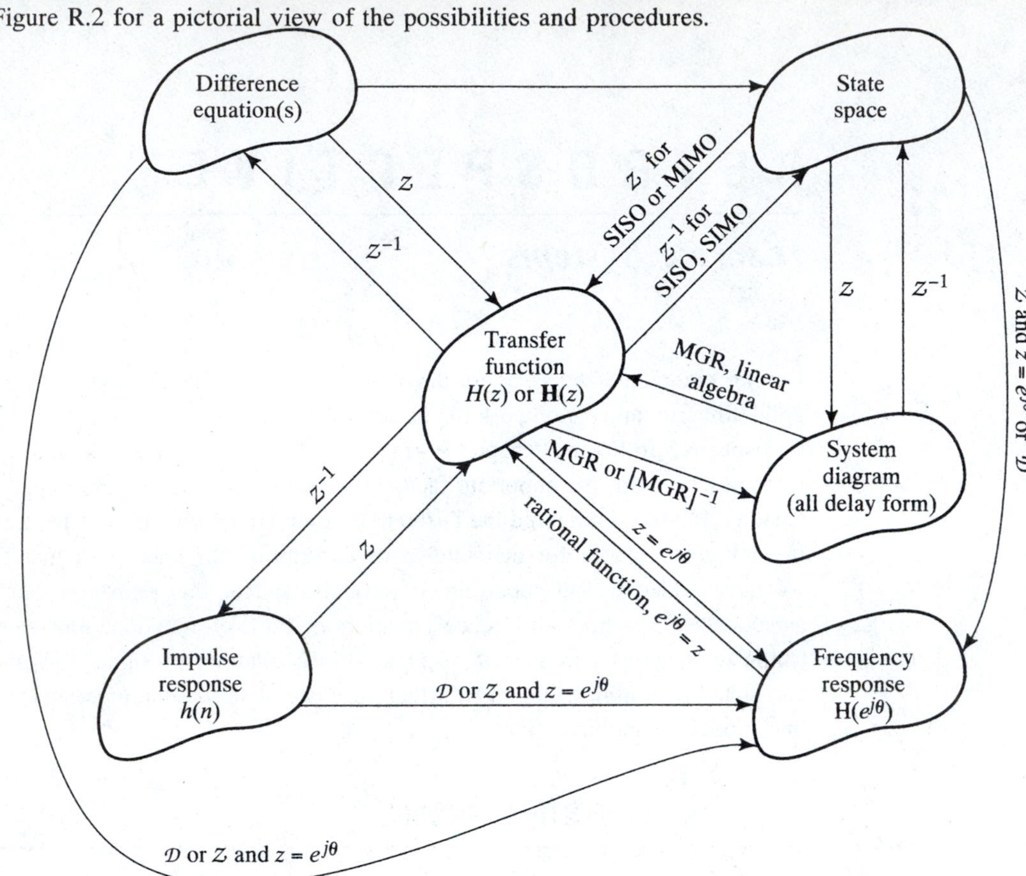

FIGURE R.2 *Finding one model from another*

PROPERTIES AND PERTINENT CHARACTERISTICS

Linearity	Homogeneity and additivity $\leftrightarrow$ superposition
Causality	Causal, anticausal, and noncausal
Time invariance	Fixed system parameters
Characteristic equation	$\sum_{k=0}^{N} a_k y(n-k) = 0 \rightarrow \sum_{k=0}^{N} a_k z^{-k} = 0$ by assuming $y(n) = Cz^n$
Characteristic roots (eigenvalues)	$(z - r_1)(z - r_2)\cdots(z - r_N) = 0$
Stability (causal system)	$y_{\text{IC}}(n) \rightarrow 0 \quad \text{as} \quad n \rightarrow \infty$ $\sum_{n=0}^{\infty} \lvert h(n)\rvert < \infty$ $\lvert v_i\rvert < 1, \qquad i = 1, 2, \ldots N$

TECHNIQUES FOR FINDING SYSTEM OUTPUT

Convolution	$y(n) = \sum_{m=-\infty}^{\infty} h(m)x(n-m)$, ICs = 0
z transform	$Y(z)$ found from DE(s) with ICs $\neq 0$ or $H(z)$, ICs = 0
(for distinct roots)	$Y(z) = \sum_{k=1}^{N} \frac{C_k z}{z - d_k}$, $y(n) = \sum_{k=1}^{N} C_k (d_k)^n$
Sinusoidal steady-state	$y_{ss}(n) = A\lvert H(e^{j\theta})\rvert \cos(n\theta + \alpha + \angle H(e^{j\theta}))$ for $x(n) = A\cos(n\theta + \alpha)$, $-\infty \leq n \leq \infty$
Discrete Fourier Transform	$Y(k) = H(k)X(k)$

MISCELLANEOUS

Alternative path to transfer function using Mason's Gain Rule

$$H(z) = \frac{\sum_{k=1}^{M} P_k(z)\Delta_k(z)}{\Delta(z)}$$

ADDITIONAL MATLAB FUNCTIONS USED

The Retrospective at the end of Chapter 9 (page 539) contains a list of the MATLAB functions used in Chapters 7, 8, and 9. Descriptions of additional functions that appear in Chapters 10 and 11 follow.

Function	*Purpose and Use*	*Toolbox*
conj, imag, real	Given: a matrix (or vector) **conj, imag,** and **real** return the conjugate, imaginary part, and real part, respectively.	MATLAB
eig	Given: an N by N matrix **A**, **eig** returns the eigenvalues of **A**.	MATLAB, Signals/System
fft	Given: the sequence $x(n)$, **fft** returns the DFT of $x(n)$.	MATLAB, Signal Processing
ifft	Given: the sequence $X(k)$, **ifft** returns the IDFT of $X(k)$.	MATLAB, Signal Processing

Function	*Purpose and Use*	*Toolbox*
ksxcorr	Given: the sequences $a(n)$ and $b(n)$, **ksxcorr** returns the cross correlation.	California Functions
ss2tf	Given: State model of system, **ss2tf** returns the TF model.	Control System, Signals/Systems
tf2ss	Given: TF model of system, **tf2ss**returns equivalent state model.	Signals/Systems, Control System
xcorr	Given: the sequences $a(n)$ and $b(n)$, **xcorr** returns the cross correlation.	Signal Processing

Appendix A

MATLAB®: *An Overview*[1]

PREVIEW

This appendix is designed to help you get started with MATLAB. We begin by suggesting an installation procedure for the files on the *Contemporary Linear Systems* (CLS) disk that accompanies the book. A brief tutorial overview of a few of MATLAB's features follows, including information on how to execute, modify, and develop scripts (programs). A few examples are given and many more are available on the CLS disk.

INSTALLATION SUGGESTIONS

MATLAB, or the student version of MATLAB, should be installed on your hard drive in a directory (for a DOS machine) or folder (for a Macintosh) named MATLAB. To create this directory or folder and complete the installation process, follow the instructions given in the MATLAB documentation. Take care to follow the guidelines given for setting the path.

Like MATLAB the CLS disk has two versions: DOS and Macintosh. To install the DOS version, copy (use of the command xcopy is recommended) the two files (CAFCNS and SCRPTFLS) to your MATLAB directory. Then set MATLABPATH in the MATLAB.BAT file of the subdirectory of MATLAB named BIN. MATLABPATH should include paths to MATLAB\SCRPTFLS and to MATLAB\CAFCNS. See the MATLAB documentation if you need help doing this. If you're using a Macintosh, copy the CLS files (CAFCNS and CHAPTER1 through CHAPTER11) into your MATLAB folder by clicking and dragging. With these files in the MATLAB folder they should be automatically included in the search path.

---

[1]An adaptation of "Getting Started with MATLAB" by Professor Murali Tummala, Naval Postgraduate School.

BASIC CONCEPTS

The best source of information on MATLAB is the MATLAB *User's Guide* or *The Student Edition* of MATLAB. Usually MATLAB is found in a directory on the C: drive on DOS machines and in a folder on the hard drive of a Macintosh. In either case, once you are in the directory (or folder) simply type (double click) MATLAB to start the program.

Operations and commands in MATLAB are intended to be used in much the same way as writing a formula on paper. MATLAB works with essentially one kind of object: a rectangular numerical **matrix** that may possibly contain complex terms.

Scalar A one-by-one matrix.

Vector A matrix with one row or one column.

Operators MATLAB usually deals with vectors or matrices. There is a method of identifying arithmetic operations that act on a "member-by-member" basis rather than on the vector or matrix as a whole. The operator that specifies this action is the period (.) used in conjunction with one of the other standard MATLAB operators, such as * (multiply) and / (divide). For example, to multiply each component of a vector **a** by the corresponding components of a vector **b** having the same dimension, the appropriate operation is **a .*b**. Writing **a*b** will generate an error message. Notice the "space" after **a** and before the period (.). Although not always required, it's a good idea to insert this space because if instead of **a** we had a number, the "." would be interpreted as a decimal point if there's no space. For example, try the commands **3./a** and **3 ./a** with a vector **a**.

Delimiter The semicolon (;) may be used to end a statement. If it is omitted, the result of executing the statement will be shown on the screen, and, consequently, the delimiter may also be used to suppress display of MATLAB calculations.

EXAMPLE M.1 ***Use of Delimiter and Division Operator***

Suppose we want to compute sin(x)/x for the vector **x** = [1 2 3 4 5]. We set the vector **x** by the statement "**x** = [1 2 3 4 5];" where the semicolon suppresses the screen display. Now after setting the vector **x** we write the MATLAB command "**y = sin(x) ./x**;" where the use of the ./ operator is needed because **sin(x)** is the vector **sin(x)** = [0.8415 0.9093 0.1411 −0.7568 −0.9589] and it is meaningless to divide the vector **sin(x)** by the vector **x**. It is meaningful, however, to divide the first component of the vector **sin(x)** by the first component of the vector **x**, and so forth.

Workspace Place where data are stored. If you type **A** = [1 2 3; 4 5 6; 7 8 9] and then [return] or [enter], the result displayed will be

```
A =
    1  2  3
    4  5  6
    7  8  9
```

The command **A** = [1 2 3; 4 5 6; 7 8 9] would give the same result. If you wish to access a particular component of a matrix, this can be done by specifying the appropriate row and column indices, as in A(2,3) which in this case yields the display `ans=6`. You can also access an entire row or column, for example, **A(3,:)** displays the third row of **A**. *Note that in* MATLAB *row and column indices always begin with the number* 1. The matrix **A** is *saved* for your later use *as long as MATLAB is running.* You could go on to add additional matrices and vectors, which would also be saved for later use.

who Used to see what is currently stored in the workspace with a listing of the names of the variables stored.

clear Used to clear the entire workspace

clear x Removes just the variable **x** from the workspace.

save <filename> Workspace is saved in a file (no file extension is needed for DOS machines—the extension **.mat** is automatically assigned). To recall the workspace, use the **load** command.

load <filename> Again no extension is needed as MATLAB will automatically look for your <filename> among the files with **.mat** extensions. For Macintosh machines, MATLAB can be restarted by simply double clicking on <filename>.

exit or ***quit*** Used to leave MATLAB; the workspace is cleared

FUNCTIONS

Much of MATLAB's power is derived from its extensive set of functions and the capability for the user to add new functions. Some functions are intrinsic to the main MATLAB program, and others are available from the library of external files called ***M-files.*** Whether a file is intrinsic or contained in an M-file is transparent to the user. In either case, the user can call, or invoke, the function directly from within MATLAB. Refer to the user's manual or utilize the help facility of MATLAB to get more information on specific functions.

help Produces a list of help topics. To get more information on a specific function, type **help** followed by the function name. For example, for help on eigenvalue functions, type: **help eig.**

Script file An M-file without arguments. Similar to a macro in computer programming where a list of statements is inserted into a program and then executed as a group. When a **script** is invoked, MATLAB simply executes the commands found in the file rather than waiting for input from the keyboard.

Function file An M-file with arguments. Allows arguments to be passed to the function, unlike scripts that have no calling arguments. Additionally, variables may be defined and manipulated inside the function file that are local to the function and do not operate globally on the workspace. Function files are useful for creating new MATLAB functions written in the MATLAB language.

MODES OF OPERATION

command or ***interpreter*** MATLAB commands are executed as they are entered into the keyboard. This is useful when trying new commands, executing short sequences of operations, troubleshooting, or in taking a more microscopic or macroscopic view of computed results.

script For reusing or modifying a sequence of MATLAB commands. Scripts can be written using any editor or word-processing software available with your computer. The idea is that you can create a script (or MATLAB program) off line, store it, and then execute it. If you need to change the input data, or modify the script, simply re-enter the editor, make the desired changes, save the result, and execute the script. (On DOS machines, save script files with a **.m** extension and place them in the MATLAB directory. On a Macintosh, when you open a new file while in MATLAB you will open the **edit window**; you can type the script into this window and then give a **save as . . .** command to attach a name to the script.) To execute a script on a DOS machine, type the filename without extension when in MATLAB. On a Macintosh, open the script from within MATLAB and it will appear in the edit window. Then select the **save and go** command from the file menu.

RESIZING OPERATIONS ON VECTORS AND MATRICES

Resizing operations including the operations of appending, extracting, and indexing can be performed on vectors and matrices.

PLOTTING

One of the principal attractions of MATLAB is its integrated plotting and graphics capability. There are various ways in which to plot data to the screen.

plot An intrinsic MATLAB function. For example, for the vector **y** = [0.0 0.48 0.84 1.0 0.91 0.6 0.14] with the indices [1 2 3 4 5 6 7], "**plot**(y);" produces a linear plot of the elements of **y** versus the indices of the elements of **y** as shown in Figure A.1. It is possible to label the axes, provide a title, annotate the plot with notes, and use other types of plots. It is also possible to vary the lines used for the plot, or to add a grid and show the plot as unconnected points, as in the command sequence: "**plot**(y,'o'); grid;" which gives the plot in Figure A.2.

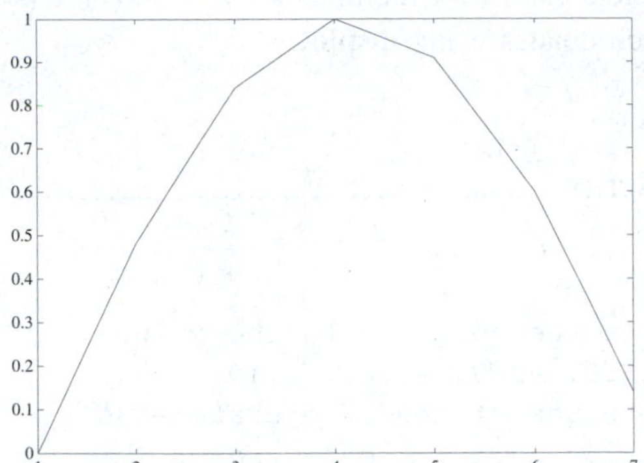

FIGURE A.1 *Basic plot*

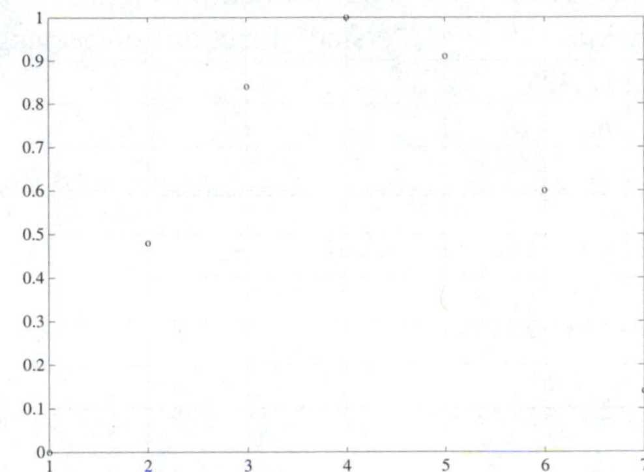

FIGURE A.2 *Basic plot with grid and unconnected points*

displot Found in the California Functions file, this function provides the capability to make a point plot easier to read. Example A.2 on page 664 illustrates this feature.

subplot Allows four different graphs on a single page. See the California functions **zpresp** or **dzpresp**.

clg Means clear graphics and can be typed at the keyboard, or placed in a script to clear a graphical display from the screen.

stem Plots sequence data as "stems" in a manner similar to **displot**.

FORMATTING OUTPUT AND AUTOMATICALLY SAVING THE MATLAB SESSION

Once you become more familiar with using MATLAB, you will most likely want to assure that your work is automatically saved. To save your workspace in a compact format in memory that omits the empty lines between calculations, type the following immediately after entering MATLAB:

```
format compact; diary <filename.txt>;
```

Notice that there are two separate commands, each separated with a semicolon (;). (On a Macintosh the **diary** command is simply followed by the filename.) The first command specifies the format for storage as compact to save space in memory. The second command establishes a diary (a record) for everything you type in MATLAB and automatically stores this information in a filename with the extension you specify. (The extension does not have to be **.txt**. You can give another extension if you wish.) To view this file on a DOS machine, use the DOS type command or an editor. On a Macintosh, simply double click on the file.

EXAMPLE A.2 *Typical MATLAB Script*

Shown below is a plot of the discrete values for the function $y = \sin(x)/x$ (a "sinc" function) on rectangular coordinates using **displot**.

MATLAB Script

```
%Sinc function plot

x=-10:0.2:10;                            % sets the vector x to the values
                                         % -10.0, -9.8, -9.6,...,10.
y=sin(x)./x;                             % note use of the ./ operator—can't
                                         % divide a vector by a vector
xlabel='Discrete time,n';                % label for abscissa
ylabel='y';                              % label for ordinate
title='sinc function';                   % title for graph
displot(x,y,xlabel,ylabel,title);        % invoke the displot function
%displot is a California function
```

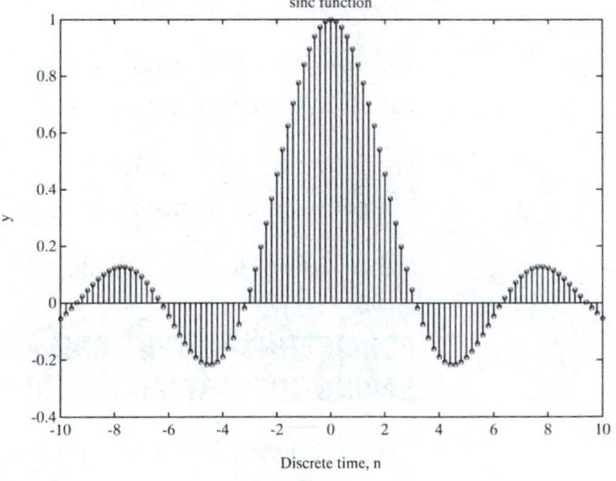

FINAL COMMENT

Experience has shown that MATLAB is easy to use, the biggest hurdle being to just get started. One way to do this is by viewing and executing script files on the CLS disk. You can learn a lot by exploring MATLAB's user manual and trying things out. Don't try to learn everything at once. Just get going and as you need new capabilities find them, or, if necessary, make them yourself.

Appendix B
Useful Formulae and Definitions

Complex numbers

Rectangular to exponential

$$z = x + jy = re^{j\theta},\ r = \sqrt{x^2 + y^2},$$
$$\theta = \tan^{-1}(y/x)$$

Sum and difference

For $z_1 = x_1 + jy_1$ and $z_2 = x_2 + jy_2$,
$$z_3 = z_1 + z_2 = x_1 + x_2 + j(y_1 + y_2)$$
$$z_4 = z_1 - z_2 = x_1 - x_2 + j(y_1 - y_2)$$

Product and quotient

For $z_1 = r_1 e^{j\theta_1}$ and $z_2 = r_2 e^{j\theta_2}$,
$$z_3 = z_1 z_2 = r_1 r_2 e^{j(\theta_1 + \theta_2)}$$
$$z_4 = z_1/z_2 = [r_1/r_2]e^{j(\theta_1 - \theta_2)}$$

Roots of a complex number

$$z^{1/N} = (r)^{1/N} e^{j(\theta + 2\pi k/N)},$$
$$k = 0, 1, \ldots, N - 1$$

DeMoivre's formula

$$(\cos\theta + j\sin\theta)^n = \cos(n\theta) + j\sin(n\theta)$$

Euler's relation

$$e^{\pm j\theta} = \cos\theta \pm j\sin\theta,$$
θ any real number

Finite geometric sum

$$S_N = \sum_{m=0}^{N-1} \alpha^m = \frac{1 - \alpha^N}{1 - \alpha}, \qquad \alpha \neq 1$$

Infinite geometric sum

$$S_N = \sum_{m=0}^{\infty} \alpha^m = \frac{1}{1 - \alpha}, \qquad |\alpha| < 1$$

l'Hôpital's rule

$$\lim_{\sigma \to a} \frac{f(\sigma)}{g(\sigma)} = \lim_{\sigma \to a} \frac{df(\sigma)/d\sigma}{df(\sigma)/d\sigma}$$

Matrix properties for **A**, m rows, n columns

Transpose	$\mathbf{A}^T$, the rows of **A** become the columns of $\mathbf{A}^T$
Adjoint of a square matrix	$\mathbf{adj\ A} = \mathbf{C}^T$, where **C** is the matrix of cofactors of **A**
Inverse of a square matrix	$\mathbf{A}^{-1} = \dfrac{\mathbf{adj\ A}}{\lvert\mathbf{A}\rvert}$ for $\lvert\mathbf{A}\rvert \neq 0$

Quadratic formula

For $ax^2 + bx + c = 0$,

$$x_{1,2} = \frac{1}{a}\left\{-(b/2) \pm \sqrt{(b/2)^2 - ac}\right\}$$

Index

IMPORTANT! READ CAREFULLY BEFORE BREAKING SEAL

PWS PUBLISHING COMPANY LICENSE AGREEMENT This is a legal agreement between you, the program user, and PWS Publishing Company (the Publisher). By opening the attached sealed disk package, you are agreeing to the terms of this agreement. If you do not agree to these terms, promptly return the sealed disk package and all accompanying materials.

Grant of License PWS Publishing Company grants you the right to use one copy of the enclosed software and data files ("Software") on one microcomputer at a time. You may not network the Software or otherwise use it on more than one computer at a time without obtaining a site license from the Publisher.

Copyright The Software is owned by the Publisher and is protected by United States copyright. You may not copy the Software in whole or in part except that you may: (a) make one copy of the Software solely for back-up purposes, or (b) transfer the Software to a single hard disk or other mass-storage device provided that you keep the original solely for backup purposes. You may not copy any written materials that accompany the Software.

Other Restrictions You may not rent, lease, lend, or otherwise distribute copies of the Software to others. The Software license may not be transferred to others except by written permission of the Publisher. You may not reverse engineer, decompile, or disassemble the Software.

Limited Warranty The warranty for the enclosed disk(s) is for ninety (90) days. If, during that time period, you find defects in workmanship or material, the Publisher will replace the defective item. THE PUBLISHER AND THE AUTHOR PROVIDE NO OTHER WARRANTIES, EXPRESSED OR IMPLIED, AND SHALL NOT BE LIABLE FOR ANY DAMAGES, SPECIAL, INDIRECT, INCIDENTAL, CONSEQUENTIAL, OR OTHERWISE ARISING OUT OF THE USE OR INABILITY TO USE THIS SOFTWARE.

For warranty service or additional information about PWS Publishing Company's licensing policies, contact:

Technology Services
511 Forest Lodge Road
Pacific Grove, CA 93950-5098
(408) 373-6515 Fax: (408) 373-0351

For damaged disk replacement *only*, contact:

ITP Educational Division Customer Service
(800) 354-9706